EPIGENETICS OF CANCER PREVENTION

Translational Epigenetics Series

Trygve O. Tollefsbol, PhD, Series Editor

Transgenerational Epigenetics
Edited by Trygve O. Tollefsbol, 2014

Personalized Epigenetics
Edited by Trygve O. Tollefsbol, 2015

Epigenetic Technological Applications
Edited by Y. George Zheng, 2015

Epigenetic Cancer Therapy
Edited by Steven G. Gray, 2015

DNA Methylation and Complex Human Disease
By Michel Neidhart, 2015

Epigenomics in Health and Disease
Edited by Mario F. Fraga and
Agustin F. F Fernández, 2015

Epigenetic Gene Expression and Regulation
Edited by Suming Huang, Michael Litt and
C. Ann Blakey, 2015

Epigenetic Biomarkers and Diagnostics
Edited by Jose Luis García-Giménez, 2015

Drug Discovery in Cancer Epigenetics
Edited by Gerda Egger and
Paola Barbara Arimondo, 2015

Medical Epigenetics
Edited by Trygve O. Tollefsbol, 2016

Chromatin Signaling and Diseases
Edited by Olivier Binda and
Martin Fernandez-Zapico, 2016

Genome Stability
Edited by Igor Kovalchuk and
Olga Kovalchuk, 2016

Chromatin Regulation and Dynamics
Edited by Anita Göndör, 2016

Neuropsychiatric Disorders and Epigenetics
Edited by Dag H. Yasui, Jacob Peedicayil and
Dennis R. Grayson, 2016

Polycomb Group Proteins
Edited by Vincenzo Pirrotta, 2016

Epigenetics and Systems Biology
Edited by Leonie Ringrose, 2017

Cancer and Noncoding RNAs
Edited by Jayprokas Chakrabarti and
Sanga Mitra, 2017

Nuclear Architecture and Dynamics
Edited by Christophe Lavelle and
Jean-Marc Victor, 2017

Epigenetic Mechanisms in Cancer
Edited by Sabita Saldanha, 2017

Epigenetics of Aging and Longevity
Edited by Alexey Moskalev and
Alexander M. Vaiserman, 2017

The Epigenetics of Autoimmunity
Edited by Rongxin Zhang, 2018

Epigenetics in Human Disease, Second Edition
Edited by Trygve O. Tollefsbol, 2018

TRANSLATIONAL EPIGENETICS

VOLUME 8

EPIGENETICS OF CANCER PREVENTION

Edited by

Anupam Bishayee, PhD

Lake Erie College of Osteopathic Medicine, Bradenton, FL, United States

Deepak Bhatia, PhD

Department of Pharmacogenomics, Bernard J Dunn School of Pharmacy
Shenandoah University, Fairfax, VA, United States

ELSEVIER

ACADEMIC PRESS

An imprint of Elsevier

Academic Press is an imprint of Elsevier
125 London Wall, London EC2Y 5AS, United Kingdom
525 B Street, Suite 1650, San Diego, CA 92101, United States
50 Hampshire Street, 5th Floor, Cambridge, MA 02139, United States
The Boulevard, Langford Lane, Kidlington, Oxford OX5 1GB, United Kingdom

Notices

Knowledge and best practice in this field are constantly changing. As new research and experience broaden our understanding, changes in research methods, professional practices, or medical treatment may become necessary.

Practitioners and researchers must always rely on their own experience and knowledge in evaluating and using any information, methods, compounds, or experiments described herein. In using such information or methods they should be mindful of their own safety and the safety of others, including parties for whom they have a professional responsibility.

To the fullest extent of the law, neither the Publisher nor the authors, contributors, or editors, assume any liability for any injury and/or damage to persons or property as a matter of products liability, negligence or otherwise, or from any use or operation of any methods, products, instructions, or ideas contained in the material herein.

Library of Congress Cataloging-in-Publication Data
A catalog record for this book is available from the Library of Congress

British Library Cataloguing-in-Publication Data
A catalogue record for this book is available from the British Library

ISBN: 978-0-12-812494-9

For information on all Academic Press Publications visit our website at https://www.elsevier.com/books-and-journals

Publisher: Stacy Masucci
Acquisition Editor: Rafael E. Teixeira
Editorial Project Manager: Timothy Bennett
Production Project Manager: Poulouse Joseph
Designer: Mark Rogers

Typeset by TNQ Technologies

Contents

5. Epigenetic Effects of Curcumin in Cancer Prevention

MUTHU K. SHANMUGAM, FRANK ARFUSO, JUDY C. SNG, ANUPAM BISHAYEE, ALAN PREM KUMAR, AND GAUTAM SETHI

6. The Epigenetic Targets of Berry Anthocyanins in Cancer Prevention

DINA MOUSTAFA ABO EL-ELLA AND ANUPAM BISHAYEE

7. Isothiocyanates, Epigenetics, and Cancer Prevention

SILVIA NOVÍO, MARÍA JESÚS NÚÑEZ-IGLESIAS, AND MANUEL FREIRE-GARABAL

8. Epigenetic Mechanisms of Resveratrol and Its Analogs in Cancer Prevention and Treatment

AVINASH KUMAR AND ANAIT S. LEVENSON

9. Epigenetic Mechanisms of Quercetin and Other Flavonoids in Cancer Therapy and Prevention

GIAN LUIGI RUSSO AND PAOLA UNGARO

II
ORGAN SPECIFIC EPIGENETIC TARGETS OF NATURAL PRODUCTS

10. Dietary Phytochemicals, Epigenetics, and Colon Cancer Chemoprevention

RITIKRAJ ARYA AND SABITA N. SALDANHA

11. Epigenetic Aspects of Hepatocellular Carcinoma Chemoprevention

RENATO HEIDOR, ERNESTO VARGAS-MENDEZ, AND FERNANDO SALVADOR MORENO

12. Epigenetic Control of Pancreatic Carcinogenesis and Its Regulation by Natural Products

HASEEB ZUBAIR, SHAFQUAT AZIM, MOHAMMAD ASLAM KHAN, GIRIJESH KUMAR PATEL, AAMIR AHMAD, SACHIN PAI, SEEMA SINGH, AND AJAY P. SINGH

13. Phytochemicals as Epigenetic Modifiers for Cancer Management With Special Reference to Lung Cancer

TAMILSELVAM RAJAVEL AND KASI PANDIMA DEVI

14. Epigenetics of Prostate Cancer and Novel Chemopreventive and Therapeutic Approaches

DIVYA BHAGIRATH, THAO L YANG, RAJVIR DAHIYA, SHAHANA MAJID, AND SHARANJOT SAINI

15. Prevention of Breast Cancer by Food Bioactives in Relation to Cancer Subtypes: Epigenetic Mechanisms

MICAH G. DONOVAN, ORNELLA I. SELMIN, AND DONATO F. ROMAGNOLO

16. Epigenomics of Ovarian Cancer and Its Chemoprevention

OURANIA KOUKOURA, STAVROS SIFAKIS, NIKOLAOS GOUTSIAS, IRONTIANTA GKOREZI-NTAVELA, AND JIANNIS HAJIIOANNOU

III
MISCELLANEOUS

Contributors

Dina Moustafa Abo El-Ella Department of Pharmacology and Toxicology, Faculty of Pharmacy, October 6 University, 6th October City, Egypt

Aamir Ahmad Department of Oncologic Sciences, Mitchell Cancer Institute, University of South Alabama, Mobile, AL, United States

Kristina Andrijauskaite Department of Molecular Medicine, UT Health San Antonio, San Antonio, TX, United States

Frank Arfuso Stem Cell and Cancer Biology Laboratory, School of Biomedical Sciences, Curtin Health Innovation Research Institute, Curtin University, Perth, WA, Australia

Ritikraj Arya Loveless Academic Magnet Program High School, Montgomery, AL, United States

Shafquat Azim Department of Oncologic Sciences, Mitchell Cancer Institute, University of South Alabama, Mobile, AL, United States

Barbara Banelli Laboratory of Tumor Epigenetics, IRCCS Ospedale Policlinico San Martino, Genova, Italy; Department of Health Sciences, University of Genova, Genova, Italy

Divya Bhagirath Department of Urology, Veterans Affairs Medical Center, San Francisco and University of California, San Francisco, CA, United States

Deepak Bhatia Department of Pharmacogenomics, Bernard J Dunn School of Pharmacy, Shenandoah University, Fairfax, VA, United States

Anupam Bishayee Lake Erie College of Osteopathic Medicine, Bradenton, FL, United States

Rajvir Dahiya Department of Urology, Veterans Affairs Medical Center, San Francisco and University of California, San Francisco, CA, United States

Laxmidhar Das Epigenetics and Cancer Research Laboratory, Biochemistry and Molecular Biology Group, Department of Life Science, National Institute of Technology, Rourkela, India

Moonmoon Deb Epigenetics and Cancer Research Laboratory, Biochemistry and Molecular Biology Group, Department of Life Science, National Institute of Technology, Rourkela, India

Kasi Pandima Devi Department of Biotechnology, Alagappa University (Science Campus), Karaikudi, India

Micah G. Donovan Department of Nutritional Sciences, The University of Arizona, Tucson, AZ, United States; The University of Arizona Cancer Center, The University of Arizona, Tucson, AZ, United States

Manuel Freire-Garabal SNL Laboratory, School of Medicine and Dentistry, University of Santiago de Compostela, Santiago de Compostela, Spain

Irontianta Gkorezi-Ntavela Department of Obstetrics and Gynecology, University Hospital of Larissa, Larissa, Greece

Nikolaos Goutsias Department of Obstetrics and Gynecology, University Hospital of Larissa, Larissa, Greece

Karishma Gupta Department of Urology, The James and Eilleen Dicke Research Laboratory, Case Western Reserve University, Cleveland, OH, United States; Department of Urology, The Urology Institute, University Hospitals Cleveland Medical Center, Cleveland, OH, United States

Sanjay Gupta Department of Urology, The James and Eilleen Dicke Research Laboratory, Case Western Reserve University, Cleveland, OH, United States; Department of Urology, The Urology Institute, University Hospitals Cleveland Medical Center, Cleveland, OH, United States; Department of Nutrition, Case Western Reserve University, Cleveland, OH, United States; Division of General Medical Sciences, Case Comprehensive Cancer Center, Cleveland, OH, United States; Department of Urology, Louis

Stokes Cleveland Veterans Affairs Medical Center, Cleveland, OH, United States

Jiannis Hajiioannou Department of ENT, Head and Neck Surgery, University Hospital of Larissa, Larissa, Greece

Renato Heidor Department of Food and Experimental Nutrition, Faculty of Pharmaceutical Sciences of University of São Paulo, São Paulo, Brazil

Swayamsiddha Kar Epigenetics and Cancer Research Laboratory, Biochemistry and Molecular Biology Group, Department of Life Science, National Institute of Technology, Rourkela, India

Mohammad Aslam Khan Department of Oncologic Sciences, Mitchell Cancer Institute, University of South Alabama, Mobile, AL, United States

Ourania Koukoura Department of Obstetrics and Gynecology, University Hospital of Larissa, Larissa, Greece

Avinash Kumar Arnold and Marie Schwartz College of Pharmacy and Health Sciences, Long Island University, Brooklyn, NY, United States

Alan Prem Kumar Department of Pharmacology, Yong Loo Lin School of Medicine, National University of Singapore, Singapore, Singapore; Cancer Science Institute of Singapore, National University of Singapore, Singapore, Singapore; Curtin Medical School, Faculty of Health Sciences, Curtin University, Perth, WA, Australia; National University Cancer Institute, National University Health System, Singapore, Singapore; Department of Biological Sciences, University of North Texas, Denton, TX, United States

Anait S. Levenson Arnold and Marie Schwartz College of Pharmacy and Health Sciences, Long Island University, Brooklyn, NY, United States; College of Veterinary Medicine, Long Island University, Brookville, NY, United States

Thao L Yang Department of Urology, Veterans Affairs Medical Center, San Francisco and University of California, San Francisco, CA, United States

Shahana Majid Department of Urology, Veterans Affairs Medical Center, San Francisco and University of California, San Francisco, CA, United States

Fernando Salvador Moreno Department of Food and Experimental Nutrition, Faculty of Pharmaceutical Sciences of University of São Paulo, São Paulo, Brazil

Jay Morris Department of Molecular Medicine, UT Health San Antonio, San Antonio, TX, United States

Shahrzad Movafagh Department of Pharmacogenomics, Bernard J Dunn School of Pharmacy, Shenandoah University, Fairfax, VA, United States

Amanda Munson Department of Pharmacogenomics, Bernard J Dunn School of Pharmacy, Shenandoah University, Fairfax, VA, United States

Silvia Novío SNL Laboratory, School of Medicine and Dentistry, University of Santiago de Compostela, Santiago de Compostela, Spain

María Jesús Núñez-Iglesias SNL Laboratory, School of Medicine and Dentistry, University of Santiago de Compostela, Santiago de Compostela, Spain

Sachin Pai Department of Interdisciplinary Clinical Oncology, Mitchell Cancer Institute, University of South Alabama, Mobile, AL, United States

Sabnam Parbin Epigenetics and Cancer Research Laboratory, Biochemistry and Molecular Biology Group, Department of Life Science, National Institute of Technology, Rourkela, India

Girijesh Kumar Patel Department of Oncologic Sciences, Mitchell Cancer Institute, University of South Alabama, Mobile, AL, United States

Samir Kumar Patra Epigenetics and Cancer Research Laboratory, Biochemistry and Molecular Biology Group, Department of Life Science, National Institute of Technology, Rourkela, India

Nibedita Pradhan Epigenetics and Cancer Research Laboratory, Biochemistry and Molecular Biology Group, Department of Life Science, National Institute of Technology, Rourkela, India

Tamilselvam Rajavel Department of Biotechnology, Alagappa University (Science Campus), Karaikudi, India

Donato F. Romagnolo Department of Nutritional Sciences, The University of Arizona, Tucson, AZ, United States; The University of Arizona Cancer Center, The University of Arizona, Tucson, AZ, United States

Massimo Romani Laboratory of Tumor Epigenetics, IRCCS Ospedale Policlinico San Martino, Genova, Italy

Gian Luigi Russo National Research Council, Institute of Food Sciences, Avellino, Italy

Sharanjot Saini Department of Urology, Veterans Affairs Medical Center, San Francisco and University of California, San Francisco, CA, United States

Sabita N. Saldanha Department of Biological Sciences, Alabama State University, Montgomery, AL, United States

Ornella I. Selmin Department of Nutritional Sciences, The University of Arizona, Tucson, AZ, United States; The University of Arizona Cancer Center, The University of Arizona, Tucson, AZ, United States

Dipta Sengupta Epigenetics and Cancer Research Laboratory, Biochemistry and Molecular Biology Group, Department of Life Science, National Institute of Technology, Rourkela, India

Gautam Sethi Department of Pharmacology, Yong Loo Lin School of Medicine, National University of Singapore, Singapore, Singapore; School of Pharmacy and Biomedical Sciences, Curtin Health Innovation Research Institute, Curtin University, Perth, WA, Australia

Eswar Shankar Department of Urology, The James and Eilleen Dicke Research Laboratory, Case Western Reserve University, Cleveland, OH, United States; Department of Urology, The Urology Institute, University Hospitals Cleveland Medical Center, Cleveland, OH, United States

Muthu K. Shanmugam Department of Pharmacology, Yong Loo Lin School of Medicine, National University of Singapore, Singapore, Singapore

Qiwen Shi Collaborative Innovation Center of Yangtze River Delta Region Green Pharmaceuticals, Hangzhou, China; Zhejiang University of Technology, Hangzhou, China

Stavros Sifakis Department of Obstetrics and Gynecology, "Mitera" Hospital, Herakleion, Greece

Seema Singh Department of Oncologic Sciences, Mitchell Cancer Institute, University of South Alabama, Mobile, AL, United States; Department of Biochemistry and Molecular Biology, College of Medicine, University of South Alabama, Mobile, AL, United States

Ajay P. Singh Department of Oncologic Sciences, Mitchell Cancer Institute, University of South Alabama, Mobile, AL, United States; Department of Biochemistry and Molecular Biology, College of Medicine, University of South Alabama, Mobile, AL, United States

Judy C. Sng Department of Pharmacology, Yong Loo Lin School of Medicine, National University of Singapore, Singapore, Singapore

Paola Ungaro National Research Council, Institute of Experimental Endocrinology and Oncology 'G. Salvatore', Naples, Italy

Ernesto Vargas-Mendez Department of Food and Experimental Nutrition, Faculty of Pharmaceutical Sciences of University of São Paulo, São Paulo, Brazil

Michael J. Wargovich Department of Molecular Medicine, UT Health San Antonio, San Antonio, TX, United States

Haseeb Zubair Department of Oncologic Sciences, Mitchell Cancer Institute, University of South Alabama, Mobile, AL, United States

Foreword

Environmental factors such as diet frequently influence gene expression and contribute to epigenetic processes that consist of changes in gene expression that are heritable but are not encoded in the DNA itself. The impact of dietary components such as bioactive phytochemicals on gene expression has garnered a tremendous amount of interest largely because the expression of genes encoding for tumor suppressors and oncogenes is often altered in cancers. One of the most notable and encouraging aspects of chemoprevention is that epigenetic aberrations are readily reversible and can be corrected by a myriad of mechanisms through which bioactive phytochemicals influence the epigenetic machinery in cancer cells. These basic tenets of the epigenetics of chemoprevention have opened a rapidly expanding field of investigation not only of the underlying mechanisms through which these epigenetic-modifying compounds confer their efficacy in preventing and treating cancer and other diseases, but also of the translational potential of this approach in controlling the incidence of disease. This book provides numerous reviews on a number of different bioactive compounds that act through the correction of epigenetic defects such as alterations in DNA methylation, histone modifications, and noncoding RNAs that are characteristic of a variety of different cancers.

The epigenetic machinery is often a driving force in many diseases, although efforts to neutralize epigenetic defects that contribute to disease through the use of bioactive phytochemicals have been the topic of intense investigation for only about the last 15 years. As a result of these concerted efforts, we have learned that many of these compounds can prevent the initiation (i.e., primary prevention) or slow the progression (i.e., secondary prevention) of cancer development. Another notable finding has been that unlike what we too often observe with standard cancer treatment using drugs, many phytochemicals are able to confer cancer chemoprevention with negligible adverse effects on normal cells and tissues. Therefore, in light of the high incidence of epigenetic aberrations in cancer combined with the reversibility of these defects and the relative lack of toxicity associated with the use of bioactive phytochemicals, the potential for these compounds to have a significant impact on cancer prevention seems particularly high.

Despite these advances, however, many challenges remain. For instance, although a wealth of information about the use of phytochemicals in preventing the epigenetic defects of cancer has been forthcoming, our understanding of the potential for combinatorial approaches that use multiple compounds to prevent cancer is relatively limited when compared with the field of cancer therapy, where combinations of anticancer drugs are frequently employed. Recent studies have found that certain combinations of epigenetic-aberration-neutralizing phytochemicals can act synergistically without added toxicity, which opens numerous possibilities for the increased efficacy of epigenetic-based cancer chemoprevention while maintaining the relatively high safety of these compounds. There are also challenges ahead in understanding the temporal aspects of cancer chemoprevention using epigenetic-modulating phytochemicals, in that many epigenetic processes change during key stages of the life span. Efforts to determine when during the life span these compounds are the most efficacious will therefore be especially useful, and clinical trials will likely address many of these challenges in the near future.

The chapters contained within this volume will lead us to further overcoming many of the challenges that we are facing in the epigenetics of chemoprevention and open new opportunities to efficaciously and safely prevent diseases such as cancer. The most effective approach to health management is obviously to prevent these diseases before they take their toll in human suffering and rising medical costs, and epigenetic-intervention approaches to chemoprevention as covered in this volume are primed to contribute greatly to this goal.

Professor Trygve Tollefsbol, PhD
Department of Biology
College of Arts and Sciences
University of Alabama at Birmingham
Birmingham, AL 35233
USA

PART I

MOLECULAR EPIGENETIC TARGETS OF NATURAL PRODUCTS

CHAPTER

1

Genetic and Epigenetic Targets of Natural Dietary Compounds as Anticancer Agents

Qiwen Shi[1,2], *Anupam Bishayee*[3], *Deepak Bhatia*[4]

[1]Collaborative Innovation Center of Yangtze River Delta Region Green Pharmaceuticals, Hangzhou, China; [2]Zhejiang University of Technology, Hangzhou, China; [3]Lake Erie College of Osteopathic Medicine, Bradenton, FL, United States; [4]Department of Pharmacogenomics, Bernard J Dunn School of Pharmacy, Shenandoah University, Fairfax, VA, United States

Abstract

Cancer is a heterogenous disease characterized by changes in the expression of multiple genes caused by genetic alternations and epigenetic modifications. Studies of the changes in genetic pathways and epigenetic regulation in tumor development and progression have provided a large pool of potential targets for anticancer therapy. In recent decades, numerous natural dietary products, such as curcumin, theanine, luteolin and epigallocatechin-3-gallate, have been identified as anticancer lead molecules with the ability to affect gene expression through both genetic and epigenetic mechanisms and have attracted considerable attentions in cancer treatment and prevention. This chapter focuses on the main genetic and epigenetic targets of anticancer dietary molecules, covering three of the most common genetic mutations (i.e. mutations in epidermal growth factor receptor, p53 and K-Ras) in cancer patients and epigenetic targets, such as DNA methyltransferases, histone deacetylases, histone acetyltransferases and changes in noncoding RNA expression. It summarizes miRNA dysregulation according to cancer type as well as the epigenetic targets of 11 common natural dietary compounds. The potential and current limitations of bioactive dietary molecules for the transition from laboratory to clinic in cancer management are also discussed.

Keywords

Cancer; Epigenetic signaling; Genetic targets; DNA methyltransferases; Histone acetyltransferases; Histone deacetylases; miRNA; Natural dietary compounds.

INTRODUCTION

Cancer, a leading cause of death worldwide with an increasing incidence and mortality, remains a major public health problem in our society. It is widely accepted that cancer is a heterogeneous group of disorders driven by both genetic lesions and epigenetic plasticity. Acquired genetic mutations are the primary cause of cancer initiation and progression, while microenvironment-mediated epigenetic perturbations play a vital role in neoplastic development.

Natural dietary compounds have received tremendous attention for their efficacy in cancer

Epigenetics of Cancer Prevention
https://doi.org/10.1016/B978-0-12-812494-9.00001-9

chemoprevention due to the demonstration, from epidemiological studies, of an association between consumption of fruits and vegetables and the reduced risk of various cancers. In past decades, accumulated evidence has shown that natural active compounds in the diet can exhibit anticancer properties through multiple mechanisms, including modifying epigenetic patterns of oncogenes and tumor suppressor genes, targeting molecules and signaling pathways associated with tumor development, optimizing immune responses and tumor microenvironments, and reversing of drug resistance (Huang et al., 2016).

As a consequence of evolution, natural origin molecules are often complex and may affect more than one target. This chapter discusses different mechanisms of genetics changes and epigenetic modifications in cancer and describes the main genetic and epigenetic modifications targeted by natural dietary compounds in cancer prevention and treatment.

GENETIC MUTATIONS IN CANCER TARGETED BY NATURAL DIETARY PRODUCTS

Aberrant gene expression due to specific activating mutations in oncogenes and prometastatic genes, or silencing mutations in tumor suppressor genes, is believed to be the driver of cancer initiation and progression. A certain proportion of mutations are inherited, while the majority of mutations are considered to be the result of accumulated changes in response to endogenous factors and environmental factors throughout one's life. The identification and verification of these so-called "driver mutations" have provided a pool of promising drug targets for which cancer treatment could achieve considerable progress. Table 1.1 lists the important genetic mutations in lung, breast, and colorectal cancers as well as acute myeloid leukemia, and the following subsections cover three common genetic mutations targeted by dietary compounds.

Epidermal Growth Factor Receptor

The epidermal growth factor receptor (EGFR) is a class of membrane receptor tyrosine kinase that is frequently amplified and overexpressed in a variety of tumors. Genetic aberration in EGFR has been validated as a key driver in the progression and malignancy of various cancers. Currently, the Food and Drug Administration has approved a dozen small-molecule tyrosine kinase inhibitors (TKIs). Although these inhibitors have exhibited excellent potency and strong efficacy, new resistance and a narrow therapeutic index have been realized (Shao et al., 2016). Herein,

TABLE 1.1 List of Important Genetic Mutations in Lung, Breast, and Colorectal Cancers and Acute Myeloid Leukemia

Types of Cancer	Mutations	References
Lung cancer	KRAS, EGFR, AKT1, ALK, BRAF, DDR2, MEK1, PTEN, PIK3CA, NRAS, ROS1, RET	Lovly et al. (2016)
Breast cancer	BRCA1, BRCA2, ESR1, HER2/neu, AKT1, PIK3CA, PTEN, RB1, TP53	Balko et al. (2016) and Mehrgou and Akouchekian (2016)
Acute myeloid leukemia	FLT3, KIT, CBFB-MYH11, CEBPA, NPM1, PML-RARA	Stahl et al. (2016) and Wheeler et al. (2017)
Colorectal cancer	KRAS, BRAF, PIK3CA, AKT1, SMAD4, PTEN, NARS, TGFBR2	Chan (2018)

new inhibitors with alternative strategy and phytochemicals are being developed.

Curcumin, a natural compound present in turmeric, has been used as an herbal medicine for centuries and extensively studied for its pharmacological activity against lung, cardiovascular, metabolic and autoimmune diseases, inflammation, neurodegeneration and cancer. One of its versatile mechanisms is to overcome EGFR overexpression. In several lung adenocarcinoma cell lines, curcumin downregulated EGFR expression and phosphorylation by inducing EGFR degradation, eventually leading to apoptosis and antiproliferation (Wada et al., 2015). When combined with gefitinib, a clinically approved TKI, curcumin showed the potential to reverse gefitinib-resistance and ameliorated gefitinib-induced gastrointestinal damage (Lee et al., 2011). A study in human epidermal growth factor receptor 2 (HER2)/neu-overexpressing breast cancer cells demonstrated that curcumin suppressed the tyrosine kinase activity of HER2/neu, which is a member of EGFR without a known natural ligand, and cancer cell growth (Chen et al., 2003). Moreover, Pculin02H, a curcumin derivative, performed high efficacy in HER2 depletion and subsequently enhanced the antitumor activity of clinical drugs in HER2-overexpressing cancer cells (Lien et al., 2015). To date, a number of curcumin derivatives have been synthesized and modified and their pharmacological effects on different diseases are being investigated.

Theanine, a characteristic amino acid from tea, and its derivatives have been shown to exhibit repressive effects on the phosphorylation and expression of EGFR, Met, Akt, and nuclear factor-κB (NF-κB) in cervical cancer cells, and thereby to inhibit the EGFR + HGF-activated EGFR/Met-Akt/NF-κB signaling pathway (Liu et al., 2016). The antitumor property of theaflavin, a polyphenol from black tea, has been exemplified by evidence that theaflavin has reduced matrix metalloproteinase-2 (MMP-2) by downregulating the expression of focal adhesion kinase, EGFR, and extracellular signal-regulated kinase (ERK) in human melanoma cells (Sil et al., 2010). It has been reported that in addition to (−)-epigallocatechin-3-gallate (EGCG), the major polyphenol from green tea, theaflavin-3,3′-digallate strongly inhibited EGFR autophosphorylation in A431 epidermoid carcinoma cells, leading to antiproliferation (Liang et al., 1999).

p53

p53 is the most frequently altered tumor suppressor gene in human cancer, and the inactivation of p53 has been observed in nearly all cancers. Overexpression of the wild-type p53 gene can retard cancer cell growth in vitro, but the mutant p53 protein serves as an oncogene that enhances cancer development and hinders cancer therapeutic response. Many natural agents exhibit their antitumor and chemopreventive properties through p53-dependent induction of cell cycle arrest and/or apoptosis.

Pectin is a complex polysaccharide commonly found in fruits and vegetables, especially apples. In female BALB/c mice implanted with breast cancer cells, pectin significantly upregulated p53 expression and subsequently induced apoptosis and inhibited tumor growth (Delphi and Sepehri, 2016). Luteolin is a multifunctional flavonoid existing in many fruits and medicinal herbs. The activation of p53 and its target p21, Bax, and PUMA by luteolin resulted in cell cycle arrest, apoptosis, and chemosensitization (Wu et al., 2008). In lung cancer, luteolin and the combination of luteolin and EGCG led to increased mitochondrial translocation of p53 (Amin et al., 2010). It is important to note that EGCG alone also can activate p53, p21, and Bax in wild-type p53 prostate cancer cells (Roy et al., 2005). Other extensively investigated dietary chemopreventive agents that are able to activate p53 pathway include resveratrol from red wine and grape skin, genistein from soy, and curcumin (Schmidt et al., 2008; Shankar et al., 2007; Tian et al., 2008).

Besides direct activation of p53, upstream regulators and downstream targets of the p53 pathway provide a number of targets to suppress carcinogenesis. α-mangostin derived from mangosteen inhibited the interaction between p53 and the mouse double minute 2 homolog that is critical for p53 degradation, contributing to the stabilization of p53 protein and further enhancement of p53 activity (Leão et al., 2013). p53 activation is regulated by several kinases that are critical in the DNA damage-response pathway, including ataxia telangiectasia-and-Rad3 related (ATR), checkpoint kinase 1 (Chk1), and ataxia telangiectasia mutated (ATM). Therefore, chemopreventive agents, such as selenium existing in seafood, nuts, and vegetables; diallyl trisulfide from garlic; vitamin D; and luteolin, which can activate ATR, Chk1, or ATM; also can exert their antitumor effects through p53 activation (Amin et al., 2015).

K-Ras

Ras belongs to a large family of small molecular weight GTP-binding proteins typically 20–25 kDa in size. The Ras kinase family consists of four members termed H-Ras, K-Ras, N-Ras and R-Ras, which play critical roles in differentiation and proliferation. Oncogenic K-Ras mutation is predominant in pancreatic, colon and lung carcinomas, and most mutations happen at codon 12 where a glycine is replaced by a valine or aspartate (Asati et al., 2017).

The regulatory regions of K-Ras contain G-quadruplexes, which are special secondary structures adopted in guanine (G)-rich DNA sequences; the stabilization of G-quadruplex structures within the promoter region can cause downregulation of K-Ras (Morgan et al., 2016). Pattanayak et al. (2016) reported that K-Ras G-quadruplex stabilizing activity was enhanced in the presence of natural polyphenols, such as curcumin and ellagic acid, indicating a new mechanism for their antiproliferative activity. In addition to dietary compounds, several alkaloids, such as berberine and sanguinarine, have shown a strong binding capacity with K-Ras G-quadruplex to suppress tumor growth (Wen and Xie, 2017).

The mitogen-activated protein kinase/ERK pathway and the phosphatidylinositol 3-kinase (PI3K)/Akt pathway are classic downstream targets of K-Ras activation. In K-Ras mutated colorectal cancer cells, quercetin and luteolin significantly decreased ERK phosphorylation, followed by the antiproliferative and apoptotic activity (Xavier et al., 2009). Oleanolic acid present in various vegetables and medicinal herbs induced autophagy to impair the proliferation and invasion of K-Ras-transformed normal cells (Liu et al., 2014). Lupeol, which is found in vegetables and fruits, such as cabbage, pepper, cucumber, strawberry and mango, displayed Ras activity in pancreatic cancer cells as well as inhibitory effects on PI3K/Akt and NF-κB signaling networks that are activated by Ras and may contribute to chemoresistance (Saleem, 2009). Interestingly, lupeol has a better antitumor efficacy on highly malignant and invasive pancreatic cancer cells harboring mutated K-Ras than less-aggressive Bx-PC-3 cells with wild-type K-Ras (Murtaza et al., 2009). Supplementation with resveratrol restrained the formation of colon tumors in mice with APC locus knockout and activated K-Ras by inducing the level of miR-96, which is a regulator of K-Ras translation, and subsequent K-Ras downregulation (Saud et al., 2014).

MECHANISMS OF EPIGENETIC MODIFICATIONS

Epigenetics is defined as the study of inheritable modifications in gene expression that do not involve changes in DNA nucleotide sequences (Morgan et al., 2005). Conrad Waddington first established the concept in

1942, which was derived from combination of the words "epigenesist" and "genetics" (Van Speybroeck, 2002). Epigenetic mechanisms are essential in the developmental stages as well as the regulation of normal cellular functions, and aberrations in epigenetic modifications can lead to diseases, including cancer (Füllgrabe et al., 2011). For instance, among all aberrants discovered in acute myeloid leukemia (AML), more than 30% are epigenetically based (Cancer Genome Atlas Research Network, 2013). Epigenetic modifications include DNA methylation, histone modification and changes in noncoding RNAs (ncRNAs), which can be influenced by environmental, dietary, and lifestyle factors and are potentially reversible. Thus they can serve as attractive and promising targets for cancer preventive and therapeutic strategies.

DNA Methylation

DNA methylation, the most extensively studied epigenetic changes in cancer, is termed the transfer of a methyl moiety from the donor S-adenosyl methionine (SAM) to the 5′-position of the cytosine residues within CpG dinucleotides, forming 5-methylcytosines (5mC). CpG dinucleotides are not distributed evenly throughout the human genome and are frequently enriched in the gene promoters as well as the regions of large repetitive sequences, such as LINE and ALU retrotransposon elements (Bird, 2002). Short CpG-rich regions are referred to as CpG islands and are present in more than 50% of human gene promoters. Interestingly, most CpG dinucleotides in the genome are methylated, while the majority of CpG islands in the promoter region often remain unmethylated, especially those in the promoter regions of transcriptionally active genes. The methylation of CpG islands can result in the silence of gene expression through a variety of mechanisms, including preventing or promoting the recruitment of regulatory elements to their binding sites. In cancer cells, CpG islands usually are hypermethylated, causing tumor suppressor gene silencing.

The reaction of DNA methylation is catalyzed by DNA methyltransferases (DNMTs). So far, there are three main DNMTs identified in mammals namely, DNMT1, DNMT3a and DNMT3b. DNMT1 regulates the maintenance of methylation, preserving existing methylation patterns in the daughter DNA strands by adding methyl gourds to the hemimethylated CpG sites. DNMT3a and DNMT3b are necessary for global de novo methylation and show an equal preference for both unmethylated and hemimethylated DNA. Other family members of DNMTs include DNMT3L, which lacks intrinsic DNMT activity, and DNMT2, which also lacks any biochemically detectable DNMT activity, though it is well conserved in eukaryotes (Müller et al., 2013). The potential functions of DNMT3L and DNMT2 are still underexplored.

DNA demethylation is associated with DNA repair and oxidative pathways that are controlled by ten-eleven translocation (TET) and activation-induced deaminase (AID)/apolipoprotein B mRNA-editing catalytic component (APOBEC). TET enzymes convert 5mC to 5-hydroxymethylcytosine (5hmC), 5-formylcytosine and 5-carboxylcytosine, and AID/APOBEC family members deaminate 5 mC and 5hmC to 5-methyluracil and 5-hydroxymethyluracil. All the above cytosine derivatives are further targeted by the uracil-DNA glycosylase family of base excision repair glycosylases, resulting in cytosine replacement and DNA demethylation (Schnekenburger et al., 2014). In fact, TET-modulated DNA demethylation recently has been found altered in several types of cancer, further indicating the complex network in cancer (Chen et al., 2014).

Histone Modification

In eukaryotic cells, DNA is packaged with both histone and nonhistone proteins into a higher-order structure of chromatin. Histone modifications typically occur at the N-terminal via posttranslational modifications, such as acetylation, methylation, phosphorylation, ubiquitylation, biotinylation, and sumoylation, frequently influencing chromatin structure and all DNA-templated processes, including transcription, repair and replication. The results of histone modifications can be either activation or repression depending on the type and position of the modification, the residue involved, and side chain modifications. For instance, lysine acetylation associates with transcriptional activation, whereas lysine methylation can either activate or suppress transcription determined by the specific lysine modified.

Histone modifications are dynamically governed by groups of enzymes and are potentially reversible. Based on their function, enzymes can be classified as "readers" that recognize histone modifications and regulate the recruitment, substrate specificity, and catalysis of chromatin-modifying enzymes, "writers" that introduce chromatin marks, and "erasers" that remove marks (Torres and Fujimori, 2015). The best-characterized modifications within histone tails are acetylation/deacetylation and methylation/demethylation processes that are catalyzed by histone acetyltransferases (HATs)/histone deacetylases (HDACs) and histone methyltransferase (HMTs)/histone demethylases (HDMs), respectively. So far, 25 HATs have been identified and classified into five families: Gcn5-related N-acetyl-transferase, MYST, p300/CBP, SRC and TAFII250; all HATs use acetyl-coenzyme A as an acetyl donor. In a broad range of cancers, HATs have exhibited abnormal expression and chromosomal translocations, and some of them undergo coding mutations (Avvakumov and Côté, 2007). HDACs have been classified into four classes: class I (HDAC 1–2 and HDAC 8), class II (HDAC 4–7 and HDAC 9–10), class III (sirtuins 1–7), and class IV (HDAC 11) (Wang et al., 2005). All members of HDAC classes I, II, and IV are zinc-dependent enzymes that do not need a cofactor when catalyzing their targets, while sirtuins require nicotinamide adenine dinucleotide as a cofactor (Wang et al., 2005). Similar to DNA methylation, HMTs use S-adenosyl methionine (SAM) as a cofactor and can be divided into SET-domain-containing proteins and DOT1-like proteins targeting lysines, and protein arginine N-methyltransferase acting on arginines (Greer and Shi, 2012). HDMs are a large group of enzymes that have two classes: the flavin adenine dinucleotide–dependent monoamine oxidase family and both iron- and α-ketoglutarate-dependent dioxygenase jumonji (jmj) C-domain-containing proteins (Greer and Shi, 2012). Although proteins termed "readers" are incapable of modifying posttranslational histone modifications, they play a crucial role in epigenetically controlled gene expression. Currently, several proteins have been identified as potential targets for treatment, such as bromodomain-containing proteins, plant homeodomain–containing proteins, and methyl-lysine- and/or methyl-arginine-binding domain-containing proteins (Arrowsmith et al., 2012).

Noncoding RNAs

Initially, ncRNAs were proposed to perform catalytic functions in facilitating RNA splicing, and recently, much evidence indicates that they are key mediators in epigenetic regulation of gene expression. Based on size, ncRNAs can be divided into two main groups, small ncRNAs generally less than 200 nucleotides in length and long ncRNAs (lncRNAs) containing more than 200 nucleotides. The major classes of small ncRNAs include microRNAs (miRNAs), small interfering RNAs, and piwi-interacting RNAs.

Typically 19–24 nucleotides in length, miRNAs are derived from longer precursors. miRNAs mostly act by binding to complementary mRNA and ultimately lead to cleavage or degradation of mRNA or prevention of mRNA translation. The activity of miRNAs has been suggested to play an essential role in the regulation of various biological processes, including development, differentiation, apoptosis and proliferation. Deregulation of miRNA expression patterns has been implicated in tumor initiation and development. Table 1.2 indicates the expression levels of important miRNAs in various cancers. Furthermore, miRNA profiles are now being studied as markers to classify human cancers (Dvinge et al., 2013). Of note, miRNAs can modulate DNA methylation by targeting DNMTs, and reciprocally, the expression of miRNA can be regulated by DNA methylation and other epigenetic mechanisms, further suggesting the complexity of epigenetic modifications (Godlewski et al., 2008; Saito and Jones, 2006).

Besides miRNAs, lncRNAs are aberrantly expressed during tumorigenesis, offering promising targets for anticancer therapy. IncRNAs can interact with proteins, RNA and DNA to regulate chromatin remodeling and gene expression. For example, the α-isoform of two antisense RNA transcripts encoded by the phosphatase and tension homolog (PTEN) pseudogen PTENP1 can recruit DNMT3A, the polycomb group protein EZH2, and the HMT G9A-to-PTEN promoter to block its transcription (Deng and Sui, 2013).

However, the information regarding ncRNAs is far from conclusive. The list of tumor-associated ncRNAs is growing, and knowledge about each ncRNA is being updated all the time.

TABLE 1.2 Principal miRNAs Deregulated in Cancers

Types of Cancer	Downregulated miRNAs	Upregulated miRNAs	References
Lung cancer	let-7 family, miR-34 family, miR-200 family, miR-152, miR-92-3p, miR-148/152 family	miR-21, miR-17-92 cluster, miR-221/222, miR-34b, miR-15b-15p, miR-3662, miR-7, miR-146a	Hou et al. (2016) and Inamura and Ishikawa (2016)
Breast cancer	let-7a, miR-9-1, miR-10b, miR-17-5p, miR-20b, miR-27b, miR-125a/b, miR-126, miR-130a, miR-140, miR-145, miR-146, miR-205, miR-206	miR-7, miR-128, miR-21, miR-27a, miR-29a/b-1/b-2/c, miR-34, miR-98, miR-122, miR-155, miR-181b-1, miR-182, miR206, miR-210, miR-213, miR-221, miR-365, miR-516-3p, miR-520	Chang et al. (2016)
Colorectal cancer	miR-29b, miR-34a, miR-124a, miR-130b, miR-139-3p	miR-21, miR-29a, miR-155, miR-224, miR-378	Yi et al. (2016)
Prostate cancer	miR-200a, miR-200b, miR-429, miR-100, miR-145, miR-143, miR-205, miR-23b, miR-27b, let-7b, miR-203, miR-34a, miR-101, miR-146a	miR-141, miR-21, miR-574, miR-200c, miR-221/222, miR-663, miR-182, miR-187	Filella and Foj (2016) and Wen et al. (2014)
Chronic myeloid leukemia	miR-328, miR-150, miR-10a	miR-17/92	Yeh et al. (2016)
Acute myeloid leukemia	let-7b, let-7c, miR-210, miR-23a, miR-181b, miR-495, miR-34b, miR-16	miR-382, miR-134, miR-376a, miR-127, miR-299-5p, miR-323, miR-126, miR-224, miR-368, miR-17–92 cluster, miR-196b, miR-22	Shahjahani et al. (2015) and Song and Pandolfi (2014)

EPIGENETIC TARGETS MODIFIED BY NATURAL DIETARY PRODUCTS AGAINST CANCER

Epigenetic therapy is a fascinating area for drug development in the field of cancer chemoprevention and treatment due to the reversible characteristic of epigenetic defects. The major natural dietary compounds acting as epigenetic anticancer agents are listed in Table 1.3. This section introduces four primary targets for epigenetic therapy: DNMTs, HDACs, HATs, and miRNAs.

DNMT

Reversal of gene hypermethylation through the inhibiting of DNMT is a plausible and promising strategy for developing anticancer drugs due to emerging evidence that hypermethylation-induced transcriptional silencing of tumor suppressor genes constitutes a frequent epigenetic defect in many human cancers.

Various catechol-containing dietary polyphenols are excellent substrates for the catechol-*O*-methyltransferase–mediated *O*-methylation, which is responsible for SAM-dependent methylation reactions. Therefore, tea polyphenols, such as catechin, epicatechin, and EGCG, have been extensively investigated in epigenetic approaches for cancer. EGCG reversed the expression of various tumor suppressor genes by inhibiting the expression and enzymatic activity of DNMT in cervical cancer cells (Khan et al., 2015). The inhibitory activity of EGCG on DNMT associated with demethylation of CpG islands in the promoters and the reactivation of silenced tumor suppressor genes, such as p16INK4a, retinoic acid receptor-β, and O^6-methylguanine methyltransferase, have been observed in multiple types of cancer cells, including human esophageal, colon, prostate, and mammary cancer cell lines (Fang et al., 2007). Since EGCG targets a variety of enzymes, a number of EGCG analogs have been developed and screened to enhance specificity and efficacy (Shilpi et al., 2015).

Lycopene, a member of the carotenoid family that is abundant in tomatoes, is a potential antioxidant and nutrient protector inversely related to the risk of prostate adenocarcinoma. It has been noted that lycopene has reduced DNMT3a protein levels in the androgen-independent PC-3 cell line, but it has not affected DNMT1 and DNMT3B levels. In contrast, in androgen-dependent LNCaP cells, no DNMT levels were affected by lycopene, but the methylation levels of LINE-1 and ALU were decreased (Fu et al., 2014).

Mahanine, a naturally occurring carbazole alkaloid in various Asian vegetables, is a potent anticancer and antimicrobial agent. In both LNCaP and PC3 prostate cancer cells, mahanine increased the expression of Ras-association domain family 1A (RASSF1A) by inhibiting DNMT activity (Jagadeesh et al., 2007). It also induced the degradation of DNMT1 and DNMT3b, but not DNMT3a, through the ubiquitin–proteasome pathway and a disruption in the interaction of Akt with DNMT1 and DNMT3b (Agarwal et al., 2013).

Genistein, an estrogen agonist, is the predominant and most biologically active isoflavone in soybeans. One proposed mechanism for its chemopreventive action is DNMT inhibition. Fang et al. (2005) reported that compared with other soy isoflavones, such as biochanin A and daidzein, genistein was more effective in suppressing DNMT activity and its inhibitory effect was dose-dependent (Fang et al., 2005). In breast cancer cells, genistein treatment reduced the activity of DNMT1, DNMT3a, and DNMT3b, and might inhibit the transcription of human telomerase reverse transcriptase (hTERT) through chromatin remodeling (Li et al., 2009a).

TABLE 1.3 Epigenetic Targets Regulated by Natural Dietary Compounds

				miRNA Regulation	
Compounds	**Dietary Source**	**DNA Methylation**	**Histone Modifications**	**Upregulated**	**Downregulated**
Apigenin	Parsley, celery	DNMT inhibitor (Fang et al., 2007)	HDAC inhibitor (Pandey et al., 2012)	miR-183 (Biersack, 2016a)	
Butyrate	Cheese, butter		HDAC inhibitor (Wei et al., 2016)	miR-203 (Han et al., 2016)	miR-17, miR-17-92, miR-18b-106a, miR-20a, miR-25-106b, miR-93, miR-106a/b (Karius et al., 2012)
Curcumin	Turmeric	DNMT inhibitor (Fang et al., 2007)	HAT and HDAC inhibitor (Chen et al., 2007)	miR-15a, miR-16, miR-103, miR-22, miR-27a, miR-181b/d, miR-81a, miR-200 (Ali et al., 2010; Karius et al., 2012)	miR-21(Ali et al., 2010)
Diallyl trisulfide	Garlic		HDAC inhibitor (Wallace et al., 2013)	miR-34a, miR-143, miR-145, miR-200b/c (Li et al., 2013)	
DIM	Broccoli		↓HDAC expression (Bhatnagar et al., 2009)	miR-200b/c, let-7b/c/d/e, miR-21, miR-34, miR-146a, miR-212/132 cluster (Biersack, 2016a)	miR-92, miR-221 (Biersack, 2016b)
EGCG	Green tea	DNMT inhibitor (Khan et al., 2015)	HAT and HMT inhibitor (Choi et al., 2009; Pandey et al., 2010)	let-7 family, miR-20a, miR-16, miR-221, miR-330, miR-1, miR-7-1, miR-18, miR-25, miR-34a/b, miR-92, miR-99a, miR-126, miR-144-3p, miR-210, miR-330 (Biersack, 2016a; Karius et al., 2012)	miR-21, miR-30, miR-92, miR-92, miR-98-5p, miR-106b, miR-374c-5p, miR-449c-5p, miR-450a-2-3p, miR-453, miR-520-e, miR-608, miR-629 (Biersack, 2016b)
Ellagic acids	Berries	DNMT inhibitor (Paluszczak et al., 2010)	HDAC inhibitor (Kowshik et al., 2014)	miR-122, miR-127, miR-206 (Munagala et al., 2013)	miR-182, miR-182, miR-375 (Munagala et al., 2013)
Genistein	Soy	DNMT inhibitor (Fang et al., 2005)	HAT activator, HDAC inhibitor (Majid et al., 2009; Wu et al., 2016)	miR-1296, let-7, miR-23b, miR-34a, miR-146a, miR-200, miR-574-3p (Biersack, 2016a)	miR-23b-3p, miR-155, miR-221, miR-222, miR-1260b (de la Parra et al., 2016; Phuah and Nagoor, 2014)
I3C	Cruciferous vegetables			miR-34a (Hargraves et al., 2016)	miR-21, miR-31, miR-130a, miR-146b, miR-377 (Melkamu et al., 2010)
Luteolin	Parsley, celery	DNMT inhibitor (Fang et al., 2007)	SIRT1 activator, HAT inhibitor (Ma et al., 2015; Selvi et al., 2015)	let-7c, miR-195/215, miR-34a, miR-181a, miR-139-5p, miR-224, miR-246 (Selvi et al., 2015; Sun et al., 2015; Wu et al., 2015)	miR-135a (Selvi et al., 2015)

Continued

TABLE 1.3 Epigenetic Targets Regulated by Natural Dietary Compounds—cont'd

				miRNA Regulation	
Compounds	**Dietary Source**	**DNA Methylation**	**Histone Modifications**	**Upregulated**	**Downregulated**
Lycopene	Tomatoes	DNMT inhibitor (Fu et al., 2014)		let-7f-1 (Biersack, 2016a)	
Oleanolic acid	Olive oil, garlic			miR-122 (Zhao et al., 2015)	miR-27a (Jutooru et al., 2010)
Quercetin	Citrus	DNMT inhibitor (Tan et al., 2009)	SIRT1 activator, HAT inhibitor (Ruiz et al., 2007; Wood et al., 2004)	let-7, miR-16, miR-26, miR-34a, miR-146a, miR-142-3p (Biersack, 2016a)	miR-27a, miR-125b-3p (Biersack, 2016b)
Resveratrol	Grapes, wines, eucalyptus	DNMT inhibitor (Paluszczak et al., 2010)	SIRT1 activator (Buhrmann et al., 2016)	miR-1, miR-30c-1, miR-146b-5p, miR-194-2, miR-206, miR-323, miR-340, miR-494, miR-497, miR-506, miR-560, miR-572, miR-574, miR-615, miR-622, miR-638, miR-639, miR-659, miR-801 (Karius et al., 2012)	miR-16-1, miR-17, miR-17-92, miR-21, miR-23b, miR-25, miR-26a, miR-29c, miR-30a-3p, miR-30d, miR-30e-5p, miR-92a-2, miR-100-1/2, miR-102, miR-103-2, miR-146b, miR-181a2, miR-196a1, miR-205, miR-424, miR-594, miR-629, miR-631, miR-657 (Karius et al., 2012)
Selenium	Meat, seafood, eggs	↓DNMT1 and 3a expression (Xiang et al., 2008)	HADC inhibitor (Xiang et al., 2008)		
Sulforaphane	Broccoli, cabbage	↓DNMTs expression (Fang et al., 2005)	HDAC inhibitor (Fang et al., 2005)	let-7a, miR-9, miR-23b, miR-27b, miR-140, miR-145, miR-146a, miR-200c, miR-342-3p, miR-372, miR-486-5p, miR-505, miR-629, miR-758 (Biersack, 2016a)	miR-155, miR-633, miR-106a (Biersack, 2016b)
Vitamin A	Animal livers, tuna fish, butter	↓DNMT expression (Fazi et al., 2005)		let-7 family, miR-10a, miR-15a/b, miR-16-1, miR-34a, miR-107, miR-125b, miR-223, miR-342 (Karius et al., 2012)	miR-17, miR-25, miR-93, miR-130a, miR-181b, miR-193 (Biersack, 2016b)
Vitamin D	Cod liver oil, salmon, herring	↑DNA demethylation (Fetahu et al., 2014)	↓HMT expression (Fetahu et al., 2014)	let-7a-2, miR-15a, miR-20b, miR-22, miR-100, miR-125b (Biersack, 2016a)	miR-17-92 cluster, miR-181a/b (Biersack, 2016b)

HDAC

Numerous studies have been conducted to illustrate the mechanisms of the preventive nature of dietary components as HDAC inhibitors that increase global and local histone acetylation of a number of gene inhibitors in cancer.

Sulforaphane (SFN) is an isothiocyanate present in cruciferous vegetables, such as broccoli and cabbage. It is produced by enzymatic conversion of metabolites upon damage to the plant as a protective mechanism. In cells, SFN is converted into SFN-cysteine and SFN-N-acetylcysteine, which are responsible for potential HDAC-inhibitory activity (Myzak et al., 2004). Long-term treatment with SFN in the diet (more than 10 weeks) showed a trend toward increased acetylated histones and p21 expression in the ileum, colon, prostate and peripheral blood mononuclear cells from *Apc*^{min} mice (Myzak et al., 2006). Meeran et al. (2010) revealed that SFN increased the level of active chromatin markers acetyl-H3, acetyl-H3K9, and acetyl-H4 and reduced the level of inactive chromatin markers trimethyl-H3K9, and trimethyl-H3K27 (Meeran et al., 2010). The hyperacetylation induced by SFN further facilitated the binding of hTERT repressors to the hTERT regulatory region, decreasing the expression of hTERT in human breast cancer cells.

Apigenin is a dietary plant flavonoid present in common fruits and vegetables, such as parsley, onions, and oranges. Prostate cancer cells treated with apigenin showed inhibited HDAC1 and HDAC3 enzyme activity followed by global histone H3 and H4 acetylation and localized hyperacetylation of histone H3 on the p21/waf1 promoter (Pandey et al., 2012). Consistently, athymic nude mice with PC-3 xenografts that were fed apigenin orally for 8 weeks exhibited tumor growth arrest, inhibition of HDAC1 and HDAC3 expression and proapoptotic features (Pandey et al., 2012). Similarly, in MDA-MB-231 breast cancer cells as well as a breast cancer xenograft model, apigenin significantly reduced HDAC activity and induced H3 acetylation on the p21/waf1 promoter, resulting in the increase of p21/waf1 transcription against cancer development (Tseng et al., 2017).

SIRTs, a special group of HDAC, modify a broad spectrum of protein substrates, including cytoskeletal proteins, signaling components, transcription factors, and histone, and have been implicated in tumorigenesis and the sensitivity of tumors to treatment (Olmos et al., 2011). Interventions that disrupt SIRT activity are regarded as effective methods against tumor growth and/or chemoresistance in cancer. In particular, SIRT1 activation is required for resveratrol-mediated chemopreventive effects in multiple types of cancers, such as colorectal and breast carcinomas (Buhrmann et al., 2016; Lin et al., 2010). Resveratrol activated SIRT1 to repress tumorigenesis in SIRT1 mutant mice with a p53 null background (Wang et al., 2008). In addition, SIRT2 activity was necessary for resveratrol-induced glioma stem cell proliferation arrest, because the blockade of SIRT2 activity or expression ameliorated the inhibitory effect of resveratrol on cell proliferation (Sayd et al., 2014). It is worth mentioning that SIRT3 was involved in resveratrol-attenuated oxidative stress in mitochondrial complex I deficiency; however, the interaction of SIRT3 and resveratrol remains to be understood in cancer (Mathieu et al., 2016).

Luteolin is an SIRT1 activator as well. Ma et al. (2015) reported that luteolin induced SIRT1-mediated apoptosis in H460 human non-small-cell lung cancer cells as well as inhibition of cell migration. In mice fed an ethanol diet, luteolin supplementation possibly reversed alcohol-reduced SIRT activity, as hyperacetylation was observed in SIRT1 targets, indicating that the SIRT1 signaling pathway is potentially involved in the preventive effect of luteolin on alcohol-promoted preneoplastic lesions (Rafacho et al., 2015).

HAT

In cancer, HATs may suppress or stimulate tumor growth and disease progression depending on the genes and proteins that are hyperacetylated. For example, P300/CBP-associated factor (PCAF), also known as K (lysine) acetyltransferase 2B (KAT2B), has been shown to acetylate c-MYC, leading to increased stability of the c-MYC oncoprotein responsible for cancer progression; in contrast, PCAF has acetylated the tumor-suppressor protein p53 and activated its transcriptional activity (Wapenaar and Dekker, 2016). HAT inhibition may contribute to cancer treatment. The inhibition of p300 has enhanced gemcitabine-induced apoptosis in pancreatic cancer and also has exhibited antitumor ability in GI stromal tumor cells (Gu et al., 2016; Ono et al., 2016).

Several natural products have been reported to be HAT inhibitors. Structural analysis revealed that the side chain of curcumin contains α, β-unsaturated carbonyl groups, which function as Michael reaction sites responsible for the inhibitory effect on HAT (Marcu et al., 2006). Indeed, curcumin has promoted proteasome-dependent degradation of p300/CBP without affecting HATs, such as PCAF or GCN5 and has inhibited the acetyltransferase activity of purified p300, indicating that curcumin is possibly a specific inhibitor of p300/CBP (Marcu et al., 2006). HAT rather than HDAC was involved in curcumin-induced histone hypoacetylation, because curcumin significantly inhibited HAT activity both in vivo and in vitro but did not affect HDAC activity in vitro, and interestingly, reactive oxygen species seemed to be required for curcumin-generated histone hypoacetylation (Kang et al., 2005). Anacardic acid (AA), an active compound of cashew nuts, is also a potent inhibitor of p300, PCAF and Tip60 HAT factors (Link et al., 2010). It inhibited both inducible and constitutive NF-κB activation as well as acetylation and nuclear translocation of p65 by suppressing p300 activity. Synthetic compounds that share a similar chemical formula with AA have been investigated for the development of new HAT inhibitors (Ghizzoni et al., 2010).

Because the exact role of HATs in cancer remains unclear, HAT activators may also mediate anticancer activity. For instance, genistein has been shown to enhance trichostatin A (TSA)-induced apoptosis via increasing p300 expression, p53 acetylation, and TSA-induced histone H3/H4 acetylation (Wu et al., 2016). Natural compounds have many targets, and thus it is always difficult to determine whether the underlying mechanism of observed effects is HAT inhibition, activation, or antioxidation.

miRNA

Many natural dietary compounds with known anticancer activity can modulate the expression of miRNAs to modify cancer-related processes, including cell death, cell proliferation, cell differentiation, angiogenesis, and metastasis.

All the above-mentioned natural agents from the diet are capable of influencing the expression of some miRNAs in addition, to the previously described epigenetic mechanisms. For instance, curcumin has been reported to induce the expression of miR-15a and miR-16, the first identified tumor-suppressing miRNAs, accompanied by antiapoptotic protein Bcl-2 suppression in breast cancer cells (Yang et al., 2010). The expression of the oncogenic miR-21 was reduced in colon cancer cells when treated with curcumin, resulting in inhibition of both tumor growth and metastasis formation (Mudduluru et al., 2011). The following section focuses on the regulation of miRNAs by natural dietary compounds covered in previous sections.

The condensation product of indole-3-carbinol (I3C), 3,3′-diindolylmethane (DIM), has been detected in various Brassica vegetables in the stomach after consumption due to the acidic environment. DIM has upregulated let-7b/c/d/e and miR-200b/c to reverse epithelial-to-mesenchymal transition in pancreatic cancer cells with gemcitabine resistance (Li et al., 2009b). Similarly, in HER-2-positive breast cancer cells, DIM has enhanced the effect of herceptin via upregulation of miR-200 and subsequent FoxM1 suppression (Ahmad et al., 2013).

Butyrate is a short-chain fatty acid mainly formed by the fermentation of starch, cheese, and butter (Gao et al., 2009). One explanation for the association between compromised colonic butyrate production and colon neoplasia is the repressive effect of butyrate on the expression of the oncogenic miR-17-92a cluster via reducing c-Myc level (Hu et al., 2015). MicroRNA-203 was also a target upregulated by butyrate against colorectal cancer cell proliferation, colony formation, and invasion (Han et al., 2016). Still, in colon cancer cells, butyrate has modified colon cancer-relevant miRNAs such as miR-106b and miR-135a to alter the expression of genes involved in cell cycle regulation and to inhibit cell proliferation (Schlörmann et al., 2015).

The anticancer effects of vitamins have been observed in various cancer models (Biersack et al., 2016). Treatment of AML cells with vitamin A, also known as retinol, induced the levels of several miRNAs, such as miR-15a, miR-16-1, and let-7c/d, and also reduced the expression of oncogenic miRNAs, including miR-17-92 and miR-181b (Garzon et al., 2007; Yan et al., 2016). Vitamin C, also called ascorbic acid, is a potent antioxidant present in many fruits and vegetables. Clinical trials administering a high-dose vitamin C infusion together with chemotherapeutic agents registered remarkable tumor remission and reduced side effects (Fritz et al., 2014). In rats, vitamin C has prevented 17β-estradiol (E2)-induced breast tumor development by reverting elevated miR-93 levels as well as the expression of downstream target nuclear factor erythroid 2-related factor 2 (NRF2) and NRF2-related genes (Singh et al., 2013). Dietary deficiencies in tocopherol (vitamin E) and folic acid (vitamin B9) have led to downregulation of miR-122, a liver-specific miRNA that is underexpressed in patients with hepatocellular carcinomas, suggesting that induction of miR-122 might be involved in the anticancer effects of these two vitamins (Gaedicke et al., 2008; Kutay et al., 2006).

CONCLUSION

Compounds derived from natural origins have been shown to possess pleiotropic effects with high efficacy and low toxicity. This chapter provides a comprehensive summary of natural dietary molecules with anticancer effects on various cellular targets, including both genetic and epigenetic pathways and enzymes. Currently, the development of dietary components as effective chemopreventive and/or chemotherapeutic agents has been limited by insufficient preclinical and clinical data. Most studies on natural dietary products were conducted in cancer cells and animal models with high concentrations that may not be achieved in humans. Therefore, the relevance and validity of previous in vitro and in vivo experiments must be assessed in preclinical and clinical studies along with pharmacokinetic aspects and safety issues.

With further understanding of the epigenetics and nutrigenomics of dietary components, future research may focus on the combination of dietary compounds and chemotherapeutic agents as a promising strategy to slow, prevent, and reverse cancer progression; personalized regimens with natural dietary agents for cancer prevention and treatment; development of nanomaterials and targeted drug delivery systems to improve the bioavailability of nutritional molecules; cancer-related immune pathways modified by dietary products; and pathways altered by natural dietary products against cancer stem cells.

In summary, dietary agents hold great potential for cancer management. Although mechanistic studies of the antitumor consequences of naturally occurring products are limited, the emergence of novel technologies and development of bioinformatics tools will facilitate the possibility for some dietary compounds and their derivatives to become indispensable parts of individualized anticancer approaches in the future.

References

Agarwal, S., Amin, K.S., Jagadeesh, S., Baishay, G., Rao, P.G., Barua, N.C., Bhattacharya, S., Banerjee, P.P., 2013. Mahanine restores RASSF1A expression by down-regulating DNMT1 and DNMT3B in prostate cancer cells. Mol. Cancer 12, 99.

Ahmad, A., Ali, S., Ahmed, A., Ali, A.S., Raz, A., Sakr, W.A., Rahman, K.M.W., 2013. 3, 3′-Diindolylmethane enhances the effectiveness of herceptin against HER-2/neu-expressing breast cancer cells. PLoS One 8, e54657.

Ali, S., Ahmad, A., Banerjee, S., Padhye, S., Dominiak, K., Schaffert, J.M., Wang, Z., Philip, P.A., Sarkar, F.H., 2010. Gemcitabine sensitivity can be induced in pancreatic cancer cells through modulation of miR-200 and miR-21 expression by curcumin or its analogue CDF. Cancer Res. 70, 3606–3617.

Amin, A.R.M.R., Wang, D., Zhang, H., Peng, S., Shin, H.J.C., Brandes, J.C., Tighiouart, M., Khuri, F.R., Chen, Z.G., Shin, D.M., 2010. Enhanced anti-tumor activity by the combination of the natural compounds (-)-epigallocatechin-3-gallate and luteolin: potential role of p53. J. Biol. Chem. 285, 34557–34565.

Amin, A.R.M.R., Karpowicz, P.A., Carey, T.E., Arbiser, J., Nahta, R., Chen, Z.G., Dong, J.-T., Kucuk, O., Khan, G.N., Huang, G.S., et al., 2015. Evasion of anti-growth signaling: a key step in tumorigenesis and potential target for treatment and prophylaxis by natural compounds. Semin. Cancer Biol. 35 (Suppl.), S55–S77.

Arrowsmith, C.H., Bountra, C., Fish, P.V., Lee, K., Schapira, M., 2012. Epigenetic protein families: a new frontier for drug discovery. Nat. Rev. Drug Discov. 11, 384–400.

Asati, V., Mahapatra, D.K., Bharti, S.K., 2017. K-Ras and its inhibitors towards personalized cancer treatment: pharmacological and structural perspectives. Eur. J. Med. Chem. 125, 299–314.

Avvakumov, N., Côté, J., 2007. The MYST family of histone acetyltransferases and their intimate links to cancer. Oncogene 26, 5395–5407.

Balko, J.M., Schwarz, L.J., Luo, N., Estrada, M.V., Giltnane, J.M., Dávila-González, D., Wang, K., Sánchez, V., Dean, P.T., Combs, S.E., et al., 2016. Triple-negative breast cancers with amplification of JAK2 at the 9p24 locus demonstrate JAK2-specific dependence. Sci. Transl. Med. 8, 334ra53.

Bhatnagar, N., Li, X., Chen, Y., Zhou, X., Garrett, S.H., Guo, B., 2009. 3,3′-diindolylmethane enhances the efficacy of butyrate in colon cancer prevention through down-regulation of survivin. Cancer Prev. Res. 2, 581–589.

Biersack, B., 2016a. Current state of phenolic and terpenoidal dietary factors and natural products as non-coding RNA/microRNA modulators for improved cancer therapy and prevention. Non Coding RNA Res. 1, 12–34.

Biersack, B., 2016b. Non-coding RNA/microRNA-modulatory dietary factors and natural products for improved cancer therapy and prevention: alkaloids, organosulfur compounds, aliphatic carboxylic acids and water-soluble vitamins. Non Coding RNA Res. 1, 51–63.

Biersack, M.G., Hajdukiewicz, M., Uebelhack, R., Franke, L., Piazena, H., Klaus, P., Höhne-Zimmer, V., Braun, T., Buttgereit, F., Burmester, G.-R., et al., 2016. Sustained increase of 25-hydroxyvitamin D levels in healthy young women during wintertime after three Suberythemal UV irradiations-the MUVY pilot study. PLoS One 11, e0159040.

Bird, A., 2002. DNA methylation patterns and epigenetic memory. Genes Dev. 16, 6–21.

Buhrmann, C., Shayan, P., Popper, B., Goel, A., Shakibaei, M., 2016. Sirt1 is required for resveratrol-mediated chemopreventive effects in colorectal cancer cells. Nutrients 8, 145.

Cancer Genome Atlas Research Network, 2013. Genomic and epigenomic landscapes of adult de novo acute myeloid leukemia. N. Engl. J. Med. 368, 2059–2074.

Chan, E. Molecular Profiling of Colorectal Cancer - My Cancer Genome, 2018.

Chang, J.T.-H., Wang, F., Chapin, W., Huang, R.S., 2016. Identification of MicroRNAs as breast cancer prognosis markers through the cancer genome Atlas. PLoS One 11, e0168284.

Chen, J.-S., Lan, K., Hung, M.-C., 2003. Strategies to target HER2/neu overexpression for cancer therapy. Drug Resist. Updates 6, 129–136.

Chen, Y., Shu, W., Chen, W., Wu, Q., Liu, H., Cui, G., 2007. Curcumin, both histone deacetylase and p300/CBP-specific inhibitor, represses the activity of nuclear factor kappa B and Notch 1 in Raji cells. Basic Clin. Pharmacol. Toxicol. 101, 427–433.

Chen, H., Kazemier, H.G., de Groote, M.L., Ruiters, M.H.J., Xu, G.-L., Rots, M.G., 2014. Induced DNA demethylation by targeting Ten-Eleven Translocation 2 to the human ICAM-1 promoter. Nucleic Acids Res. 42, 1563–1574.

Choi, K.-C., Jung, M.G., Lee, Y.-H., Yoon, J.C., Kwon, S.H., Kang, H.-B., Kim, M.-J., Cha, J.-H., Kim, Y.J., Jun, W.J., et al., 2009. Epigallocatechin-3-gallate, a histone acetyltransferase inhibitor, inhibits EBV-induced B lymphocyte transformation via suppression of RelA acetylation. Cancer Res. 69, 583–592.

de la Parra, C., Castillo-Pichardo, L., Cruz-Collazo, A., Cubano, L., Redis, R., Calin, G.A., Dharmawardhane, S., 2016. Soy isoflavone genistein-mediated downregulation of miR-155 contributes to the anticancer effects of genistein. Nutr. Cancer 68, 154–164.

Delphi, L., Sepehri, H., 2016. Apple pectin: a natural source for cancer suppression in 4T1 breast cancer cells in vitro and express p53 in mouse bearing 4T1 cancer tumors, in vivo. Biomed. Pharmacother. 84, 637–644.

Deng, G., Sui, G., 2013. Noncoding RNA in oncogenesis: a new era of identifying key players. Int. J. Mol. Sci 14, 18319–18349.

Dvinge, H., Git, A., Gräf, S., Salmon-Divon, M., Curtis, C., Sottoriva, A., Zhao, Y., Hirst, M., Armisen, J., Miska, E.A., et al., 2013. The shaping and functional consequences of the microRNA landscape in breast cancer. Nature 497, 378–382.

Fang, M.Z., Chen, D., Sun, Y., Jin, Z., Christman, J.K., Yang, C.S., 2005. Reversal of hypermethylation and reactivation of p16INK4a, RARbeta, and MGMT genes by genistein and other isoflavones from soy. Clin. Cancer Res. Res 11, 7033–7041.

Fang, M., Chen, D., Yang, C.S., 2007. Dietary polyphenols may affect DNA methylation. J. Nutr. 137, 223S–228S.

Fazi, F., Travaglini, L., Carotti, D., Palitti, F., Diverio, D., Alcalay, M., McNamara, S., Miller, W.H., Lo Coco, F., Pelicci, P.G., et al., 2005. Retinoic acid targets DNA-methyltransferases and histone deacetylases during APL blast differentiation in vitro and in vivo. Oncogene 24, 1820–1830.

Fetahu, I.S., Höbaus, J., Kállay, E., 2014. Vitamin D and the epigenome. Front. Physiol. 5, 164.

Filella, X., Foj, L., 2016. Prostate cancer detection and prognosis: from prostate specific antigen (PSA) to exosomal biomarkers. Int. J. Mol. Sci. 17, E1784.

Fritz, H., Flower, G., Weeks, L., Cooley, K., Callachan, M., McGowan, J., Skidmore, B., Kirchner, L., Seely, D., 2014. Intravenous vitamin C and cancer: a systematic review. Integr. Cancer Ther. 13, 280–300.

Fu, L.-J., Ding, Y.-B., Wu, L.-X., Wen, C.-J., Qu, Q., Zhang, X., Zhou, H.-H., 2014. The effects of lycopene on the methylation of the GSTP1 promoter and global methylation in prostatic cancer cell lines PC3 and LNCaP. Int. J. Endocrinol. 2014, 620165.

Füllgrabe, J., Kavanagh, E., Joseph, B., 2011. Histone oncomodifications. Oncogene 30, 3391–3403.

Gaedicke, S., Zhang, X., Schmelzer, C., Lou, Y., Doering, F., Frank, J., Rimbach, G., 2008. Vitamin E dependent microRNA regulation in rat liver. FEBS Lett. 582, 3542–3546.

Gao, Z., Yin, J., Zhang, J., Ward, R.E., Martin, R.J., Lefevre, M., Cefalu, W.T., Ye, J., 2009. Butyrate improves insulin sensitivity and increases energy expenditure in mice. Diabetes 58, 1509–1517.

Garzon, R., Pichiorri, F., Palumbo, T., Visentini, M., Aqeilan, R., Cimmino, A., Wang, H., Sun, H., Volinia, S., Alder, H., et al., 2007. MicroRNA gene expression during retinoic acid-induced differentiation of human acute promyelocytic leukemia. Oncogene 26, 4148–4157.

Ghizzoni, M., Boltjes, A., de Graaf, C., Haisma, H.J., Dekker, F.J., 2010. Improved inhibition of the histone acetyltransferase PCAF by an anacardic acid derivative. Bioorg. Med. Chem. 18, 5826–5834.

Godlewski, J., Nowicki, M.O., Bronisz, A., Williams, S., Otsuki, A., Nuovo, G., Raychaudhury, A., Newton, H.B., Chiocca, E.A., Lawler, S., 2008. Targeting of the Bmi-1 oncogene/stem cell renewal factor by microRNA-128 inhibits glioma proliferation and self-renewal. Cancer Res. 68, 9125–9130.

Greer, E.L., Shi, Y., 2012. Histone methylation: a dynamic mark in health, disease and inheritance. Nat. Rev. Genet. 13, 343–357.

Gu, M.-L., Wang, Y.-M., Zhou, X.-X., Yao, H.-P., Zheng, S., Xiang, Z., Ji, F., 2016. An inhibitor of the acetyltransferases CBP/p300 exerts antineoplastic effects on gastrointestinal stromal tumor cells. Oncol. Rep. 36, 2763–2770.

Han, R., Sun, Q., Wu, J., Zheng, P., Zhao, G., 2016. Sodium butyrate upregulates miR-203 expression to exert antiproliferation effect on colorectal cancer cells. Cell. Physiol. Biochem. 39, 1919–1929.

Hargraves, K.G., He, L., Firestone, G.L., 2016. Phytochemical regulation of the tumor suppressive microRNA, miR-34a, by p53-dependent and independent responses in human breast cancer cells. Mol. Carcinog. 55, 486–498.

Hou, J., Meng, F., Chan, L.W.C., Cho, W.C.S., Wong, S.C.C., 2016. Circulating plasma MicroRNAs as diagnostic markers for NSCLC. Front. Genet. 7, 193.

Hu, S., Liu, L., Chang, E.B., Wang, J.-Y., Raufman, J.-P., 2015. Butyrate inhibits pro-proliferative miR-92a by diminishing c-Myc-induced miR-17-92a cluster transcription in human colon cancer cells. Mol. Cancer 14, 180.

Huang, Z., Huang, Q., Ji, L., Wang, Y., Qi, X., Liu, L., Liu, Z., Lu, L., 2016. Epigenetic regulation of active Chinese herbal components for cancer prevention and treatment: a follow-up review. Pharmacol. Res. 114, 1–12.

Inamura, K., Ishikawa, Y., 2016. MicroRNA in lung cancer: novel biomarkers and potential tools for treatment. J. Clin. Med. 5, E36.

Jagadeesh, S., Sinha, S., Pal, B.C., Bhattacharya, S., Banerjee, P.P., 2007. Mahanine reverses an epigenetically silenced tumor suppressor gene RASSF1A in human prostate cancer cells. Biochem. Biophys. Res. Commun. 362, 212–217.

Jutooru, I., Chadalapaka, G., Abdelrahim, M., Basha, M.R., Samudio, I., Konopleva, M., Andreeff, M., Safe, S., 2010. Methyl 2-cyano-3,12-dioxooleana-1,9-dien-28-oate decreases specificity protein transcription factors and inhibits pancreatic tumor growth: role of microRNA-27a. Mol. Pharmacol. 78, 226–236.

Kang, J., Chen, J., Shi, Y., Jia, J., Zhang, Y., 2005. Curcumin-induced histone hypoacetylation: the role of reactive oxygen species. Biochem. Pharmacol. 69, 1205–1213.

Karius, T., Schnekenburger, M., Dicato, M., Diederich, M., 2012. MicroRNAs in cancer management and their modulation by dietary agents. Biochem. Pharmacol 83, 1591–1601.

Khan, M.A., Hussain, A., Sundaram, M.K., Alalami, U., Gunasekera, D., Ramesh, L., Hamza, A., Quraishi, U., 2015. (-)-Epigallocatechin-3-gallate reverses the expression of various tumor-suppressor genes by inhibiting DNA methyltransferases and histone deacetylases in human cervical cancer cells. Oncol. Rep. 33, 1976–1984.

Kowshik, J., Giri, H., Kishore, T.K.K., Kesavan, R., Vankudavath, R.N., Reddy, G.B., Dixit, M., Nagini, S., 2014. Ellagic acid inhibits VEGF/VEGFR2, PI3K/Akt and MAPK signaling cascades in the hamster cheek pouch carcinogenesis model. Anticancer Agents Med. Chem. 14, 1249–1260.

Kutay, H., Bai, S., Datta, J., Motiwala, T., Pogribny, I., Frankel, W., Jacob, S.T., Ghoshal, K., 2006. Downregulation of miR-122 in the rodent and human hepatocellular carcinomas. J. Cell. Biochem. 99, 671–678.

Leão, M., Gomes, S., Pedraza-Chaverri, J., Machado, N., Sousa, E., Pinto, M., Inga, A., Pereira, C., Saraiva, L., 2013. A-mangostin and gambogic acid as potential inhibitors of the p53-MDM2 interaction revealed by a yeast approach. J. Nat. Prod. 76, 774–778.

Lee, J.-Y., Lee, Y.-M., Chang, G.-C., Yu, S.-L., Hsieh, W.-Y., Chen, J.J.W., Chen, H.-W., Yang, P.-C., 2011. Curcumin induces EGFR degradation in lung adenocarcinoma and modulates p38 activation in intestine: the versatile adjuvant for gefitinib therapy. PLoS One 6, e23756.

Li, Y., Liu, L., Andrews, L.G., Tollefsbol, T.O., 2009a. Genistein depletes telomerase activity through cross-talk between genetic and epigenetic mechanisms. Int. J. Cancer 125, 286–296.

Li, Y., VandenBoom, T.G., Kong, D., Wang, Z., Ali, S., Philip, P.A., Sarkar, F.H., 2009b. Up-regulation of miR-200 and let-7 by natural agents leads to the reversal of epithelial-to-mesenchymal transition in gemcitabine-resistant pancreatic cancer cells. Cancer Res. 69, 6704–6712.

Li, Y., Zhang, J., Zhang, L., Si, M., Yin, H., Li, J., 2013. Diallyl trisulfide inhibits proliferation, invasion and angiogenesis of osteosarcoma cells by switching on suppressor micro-RNAs and inactivating of Notch-1 signaling. Carcinogenesis 34, 1601–1610.

Liang, Y.C., Chen, Y.C., Lin, Y.L., Lin-Shiau, S.Y., Ho, C.T., Lin, J.K., 1999. Suppression of extracellular signals and cell proliferation by the black tea polyphenol, theaflavin-3,3′-digallate. Carcinogenesis 20, 733–736.

Lien, J.-C., Hung, C.-M., Lin, Y.-J., Lin, H.-C., Ko, T.-C., Tseng, L.-C., Kuo, S.-C., Ho, C.-T., Lee, J.-C., Way, T.-D., 2015. Pculin02H, a curcumin derivative, inhibits proliferation and clinical drug resistance of HER2-overexpressing cancer cells. Chem. Biol. Interact. 235, 17–26.

Lin, J.-N., Lin, V.C.-H., Rau, K.-M., Shieh, P.-C., Kuo, D.-H., Shieh, J.-C., Chen, W.-J., Tsai, S.-C., Way, T.-D., 2010. Resveratrol modulates tumor cell proliferation and protein translation via SIRT1-dependent AMPK activation. J. Agric. Food Chem. 58, 1584–1592.

Link, A., Balaguer, F., Goel, A., 2010. Cancer chemoprevention by dietary polyphenols: promising role for epigenetics. Biochem. Pharmacol. 80, 1771–1792.

Liu, J., Zheng, L., Ma, L., Wang, B., Zhao, Y., Wu, N., Liu, G., Lin, X., 2014. Oleanolic acid inhibits proliferation and invasiveness of Kras-transformed cells via autophagy. J. Nutr. Biochem. 25, 1154–1160.

Liu, J., Sun, Y., Zhang, H., Ji, D., Wu, F., Tian, H., Liu, K., Zhang, Y., Wu, B., Zhang, G., 2016. Theanine from tea and its semi-synthetic derivative TBrC suppress human cervical cancer growth and migration by inhibiting EGFR/Met-Akt/NF-κB signaling. Eur. J. Pharmacol. 791, 297–307.

Lovly, C., Horn, L., and Pao, W. Molecular Profiling of Lung Cancer - My Cancer Genome, 2016.

Ma, L., Peng, H., Li, K., Zhao, R., Li, L., Yu, Y., Wang, X., Han, Z., 2015. Luteolin exerts an anticancer effect on NCI-H460 human non-small cell lung cancer cells through the induction of Sirt1-mediated apoptosis. Mol. Med. Rep. 12, 4196–4202.

Majid, S., Dar, A.A., Ahmad, A.E., Hirata, H., Kawakami, K., Shahryari, V., Saini, S., Tanaka, Y., Dahiya, A.V., Khatri, G., et al., 2009. BTG3 tumor suppressor gene promoter demethylation, histone modification and cell cycle arrest by genistein in renal cancer. Carcinogenesis 30, 662–670.

Marcu, M.G., Jung, Y.-J., Lee, S., Chung, E.-J., Lee, M.-J., Trepel, J., Neckers, L., 2006. Curcumin is an inhibitor of p300 histone acetylatransferase. Med. Chem. 2, 169–174.

Mathieu, L., Costa, A.L., Le Bachelier, C., Slama, A., Lebre, A.-S., Taylor, R.W., Bastin, J., Djouadi, F., 2016. Resveratrol attenuates oxidative stress in mitochondrial Complex I deficiency: involvement of SIRT3. Free Radic. Biol. Med. 96, 190–198.

Meeran, S.M., Patel, S.N., Tollefsbol, T.O., 2010. Sulforaphane causes epigenetic repression of hTERT expression in human breast cancer cell lines. PLoS One 5, e11457.

Mehrgou, A., Akouchekian, M., 2016. The importance of BRCA1 and BRCA2 genes mutations in breast cancer development. Med. J. Islam. Repub. Iran 30, 369.

Melkamu, T., Zhang, X., Tan, J., Zeng, Y., Kassie, F., 2010. Alteration of microRNA expression in vinyl carbamate-induced mouse lung tumors and modulation by the chemopreventive agent indole-3-carbinol. Carcinogenesis 31, 252–258.

Morgan, H.D., Santos, F., Green, K., Dean, W., Reik, W., 2005. Epigenetic reprogramming in mammals. Hum. Mol. Genet. 14 (Spec No 1), R47–R58.

Morgan, R.K., Batra, H., Gaerig, V.C., Hockings, J., Brooks, T.A., 2016. Identification and characterization of a new G-quadruplex forming region within the kRAS promoter as a transcriptional regulator. Biochim. Biophys. Acta 1859, 235–245.

Mudduluru, G., George-William, J.N., Muppala, S., Asangani, I.A., Kumarswamy, R., Nelson, L.D., Allgayer, H., 2011. Curcumin regulates miR-21 expression and inhibits invasion and metastasis in colorectal cancer. Biosci. Rep. 31, 185–197.

Müller, S., Windhof, I.M., Maximov, V., Jurkowski, T., Jeltsch, A., Förstner, K.U., Sharma, C.M., Gräf, R., Nellen, W., 2013. Target recognition, RNA methylation activity and transcriptional regulation of the *Dictyostelium discoideum* Dnmt2-homologue (DnmA). Nucleic Acids Res. 41, 8615–8627.

Munagala, R., Aqil, F., Vadhanam, M.V., Gupta, R.C., 2013. MicroRNA "signature" during estrogen-mediated mammary carcinogenesis and its reversal by ellagic acid intervention. Cancer Lett. 339, 175–184.

Murtaza, I., Saleem, M., Adhami, V.M., Hafeez, B.B., Mukhtar, H., 2009. Suppression of cFLIP by lupeol, a dietary triterpene, is sufficient to overcome resistance to TRAIL-mediated apoptosis in chemoresistant human pancreatic cancer cells. Cancer Res. 69, 1156–1165.

Myzak, M.C., Karplus, P.A., Chung, F.-L., Dashwood, R.H., 2004. A novel mechanism of chemoprotection by sulforaphane: inhibition of histone deacetylase. Cancer Res. 64, 5767–5774.

Myzak, M.C., Dashwood, W.M., Orner, G.A., Ho, E., Dashwood, R.H., 2006. Sulforaphane inhibits histone deacetylase in vivo and suppresses tumorigenesis in Apc-minus mice. FASEB J. 20, 506–508.

Olmos, Y., Brosens, J.J., Lam, E.W.-F., 2011. Interplay between SIRT proteins and tumour suppressor transcription factors in chemotherapeutic resistance of cancer. Drug Resist. Updates 14, 35–44.

Ono, H., Basson, M.D., Ito, H., 2016. P300 inhibition enhances gemcitabine-induced apoptosis of pancreatic cancer. Oncotarget 7, 51301–51310.

Paluszczak, J., Krajka-Kuźniak, V., Baer-Dubowska, W., 2010. The effect of dietary polyphenols on the epigenetic regulation of gene expression in MCF7 breast cancer cells. Toxicol. Lett. 192, 119–125.

Pandey, M., Shukla, S., Gupta, S., 2010. Promoter demethylation and chromatin remodeling by green tea polyphenols leads to re-expression of GSTP1 in human prostate cancer cells. Int. J. Cancer 126, 2520–2533.

Pandey, M., Kaur, P., Shukla, S., Abbas, A., Fu, P., Gupta, S., 2012. Plant flavone apigenin inhibits HDAC and remodels chromatin to induce growth arrest and apoptosis in human prostate cancer cells: in vitro and in vivo study. Mol. Carcinog. 51, 952–962.

Pattanayak, R., Basak, P., Sen, S., Bhattacharyya, M., 2016. Interaction of KRAS G-quadruplex with natural polyphenols: a spectroscopic analysis with molecular modeling. Int. J. Biol. Macromol. 89, 228–237.

Phuah, N.H., Nagoor, N.H., 2014. Regulation of microRNAs by natural agents: new strategies in cancer therapies. BioMed Res. Int. 2014, 804510.

Rafacho, B.P.M., Stice, C.P., Liu, C., Greenberg, A.S., Ausman, L.M., Wang, X.-D., 2015. Inhibition of diethylnitrosamine-initiated alcohol-promoted hepatic inflammation and precancerous lesions by flavonoid luteolin is associated with increased sirtuin 1 activity in mice. Hepatobiliary Surg. Nutr. 4, 124–134.

Roy, A.M., Baliga, M.S., Katiyar, S.K., 2005. Epigallocatechin-3-gallate induces apoptosis in estrogen receptor-negative human breast carcinoma cells via modulation in protein expression of p53 and Bax and caspase-3 activation. Mol. Cancer Ther. 4, 81–90.

Ruiz, P.A., Braune, A., Hölzlwimmer, G., Quintanilla-Fend, L., Haller, D., 2007. Quercetin inhibits TNF-induced NF-kappaB transcription factor recruitment to proinflammatory gene promoters in murine intestinal epithelial cells. J. Nutr. 137, 1208–1215.

Saito, Y., Jones, P.A., 2006. Epigenetic activation of tumor suppressor microRNAs in human cancer cells. Cell Cycle Georget. Tex. 5, 2220–2222.

Saleem, M., 2009. Lupeol, a novel anti-inflammatory and anti-cancer dietary triterpene. Cancer Lett. 285, 109–115.

Saud, S.M., Li, W., Morris, N.L., Matter, M.S., Colburn, N.H., Kim, Y.S., Young, M.R., 2014. Resveratrol prevents tumorigenesis in mouse model of Kras activated sporadic colorectal cancer by suppressing oncogenic Kras expression. Carcinogenesis 35, 2778–2786.

Sayd, S., Thirant, C., El-Habr, E.A., Lipecka, J., Dubois, L.G., Bogeas, A., Tahiri-Jouti, N., Chneiweiss, H., Junier, M.-P., 2014. Sirtuin-2 activity is required for glioma stem cell proliferation arrest but not necrosis induced by resveratrol. Stem Cell Rev. 10, 103–113.

Schlörmann, W., Naumann, S., Renner, C., Glei, M., 2015. Influence of miRNA-106b and miRNA-135a on butyrate-regulated expression of p21 and Cyclin D2 in human colon adenoma cells. Genes Nutr. 10, 50.

Schmidt, F., Knobbe, C.B., Frank, B., Wolburg, H., Weller, M., 2008. The topoisomerase II inhibitor, genistein, induces G2/M arrest and apoptosis in human malignant glioma cell lines. Oncol. Rep. 19, 1061–1066.

Schnekenburger, M., Dicato, M., Diederich, M., 2014. Epigenetic modulators from "The Big Blue": a treasure to fight against cancer. Cancer Lett. 351, 182–197.

Selvi, R.B., Swaminathan, A., Chatterjee, S., Shanmugam, M.K., Li, F., Ramakrishnan, G.B., Siveen, K.S., Chinnathambi, A., Zayed, M.E., Alharbi, S.A., et al., 2015. Inhibition of p300 lysine acetyltransferase activity by luteolin reduces tumor growth in head and neck squamous cell carcinoma (HNSCC) xenograft mouse model. Oncotarget 6, 43806–43818.

Shahjahani, M., Khodadi, E., Seghatoleslami, M., Asl, J.M., Golchin, N., Zaieri, Z.D., Saki, N., 2015. Rare cytogenetic abnormalities and alteration of microRNAs in acute myeloid leukemia and response to therapy. Oncol. Rev. 9, 261.

Shankar, S., Singh, G., Srivastava, R.K., 2007. Chemoprevention by resveratrol: molecular mechanisms and therapeutic potential. Front. Biosci. J. Virtual Libr. 12, 4839–4854.

Shao, J., Chen, E., Shu, K., Chen, W., Zhang, G., Yu, Y., 2016. 6-Oxooxazolidine-quinazolines as noncovalent inhibitors with the potential to target mutant forms of EGFR. Bioorg. Med. Chem. 24, 3359–3370.

Shilpi, A., Parbin, S., Sengupta, D., Kar, S., Deb, M., Rath, S.K., Pradhan, N., Rakshit, M., Patra, S.K., 2015. Mechanisms of DNA methyltransferase-inhibitor interactions: procyanidin B2 shows new promise for therapeutic intervention of cancer. Chem. Biol. Interact. 233, 122–138.

Sil, H., Sen, T., Moulik, S., Chatterjee, A., 2010. Black tea polyphenol (theaflavin) downregulates MMP-2 in human melanoma cell line A375 by involving multiple regulatory molecules. J. Environ. Pathol. Toxicol. Oncol. 29, 55–68.

Singh, B., Ronghe, A.M., Chatterjee, A., Bhat, N.K., Bhat, H.K., 2013. MicroRNA-93 regulates NRF2 expression and is associated with breast carcinogenesis. Carcinogenesis 34, 1165–1172.

Song, S.J., Pandolfi, P.P., 2014. MicroRNAs in the pathogenesis of myelodysplastic syndromes and myeloid leukaemia. Curr. Opin. Hematol. 21, 276–282.

Stahl, M., Kohrman, N., Gore, S.D., Kim, T.K., Zeidan, A.M., Prebet, T., 2016. Epigenetics in cancer: a hematological perspective. PLoS Genet. 12, e1006193.

Sun, D.-W., Zhang, H.-D., Mao, L., Mao, C.-F., Chen, W., Cui, M., Ma, R., Cao, H.-X., Jing, C.-W., Wang, Z., et al., 2015. Luteolin inhibits breast cancer development and progression in vitro and in vivo by suppressing notch signaling and regulating MiRNAs. Cell. Physiol. Biochem. 37, 1693–1711.

Tan, S., Wang, C., Lu, C., Zhao, B., Cui, Y., Shi, X., Ma, X., 2009. Quercetin is able to demethylate the p16INK4a gene promoter. Chemotherapy 55, 6–10.

Tian, B., Wang, Z., Zhao, Y., Wang, D., Li, Y., Ma, L., Li, X., Li, J., Xiao, N., Tian, J., et al., 2008. Effects of curcumin on bladder cancer cells and development of urothelial tumors in a rat bladder carcinogenesis model. Cancer Lett. 264, 299–308.

Torres, I.O., Fujimori, D.G., 2015. Functional coupling between writers, erasers and readers of histone and DNA methylation. Curr. Opin. Struct. Biol. 35, 68–75.

Tseng, T.-H., Chien, M.-H., Lin, W.-L., Wen, Y.-C., Chow, J.-M., Chen, C.-K., Kuo, T.-C., Lee, W.-J., 2017. Inhibition of MDA-MB-231 breast cancer cell proliferation and tumor growth by apigenin through induction of G2/M arrest and histone H3 acetylation-mediated p21WAF1/CIP1 expression. Environ. Toxicol. 32, 434–444.

Van Speybroeck, L., 2002. From epigenesis to epigenetics: the case of C. H. Waddington. Ann. N.Y. Acad. Sci. 981, 61–81.

Wada, K., Lee, J.-Y., Hung, H.-Y., Shi, Q., Lin, L., Zhao, Y., Goto, M., Yang, P.-C., Kuo, S.-C., Chen, H.-W., et al., 2015. Novel curcumin analogs to overcome EGFR-TKI lung adenocarcinoma drug resistance and reduce EGFR-TKI-induced GI adverse effects. Bioorg. Med. Chem. 23, 1507–1514.

Wallace, G.C., Haar, C.P., Vandergrift, W.A., Giglio, P., Dixon-Mah, Y.N., Varma, A.K., Ray, S.K., Patel, S.J., Banik, N.L., Das, A., 2013. Multi-targeted DATS prevents tumor progression and promotes apoptosis in ectopic glioblastoma xenografts in SCID mice via HDAC inhibition. J. Neurooncol. 114, 43–50.

Wang, J., Iwasaki, H., Krivtsov, A., Febbo, P.G., Thorner, A.R., Ernst, P., Anastasiadou, E., Kutok, J.L., Kogan, S.C., Zinkel, S.S., et al., 2005. Conditional MLL-CBP targets GMP and models therapy-related myeloproliferative disease. EMBO J. 24, 368–381.

Wang, R.-H., Sengupta, K., Li, C., Kim, H.-S., Cao, L., Xiao, C., Kim, S., Xu, X., Zheng, Y., Chilton, B., et al., 2008. Impaired DNA damage response, genome instability, and tumorigenesis in SIRT1 mutant mice. Cancer Cell 14, 312–323.

Wapenaar, H., Dekker, F.J., 2016. Histone acetyltransferases: challenges in targeting bi-substrate enzymes. Clin. Epigenet. 8, 59.

Wei, W., Sun, W., Yu, S., Yang, Y., Ai, L., 2016. Butyrate production from high-fiber diet protects against lymphoma tumor. Leuk. Lymphoma 57, 2401–2408.

Wen, L.-N., Xie, M.-X., 2017. Spectroscopic investigation of the interaction between G-quadruplex of KRAS promoter sequence and three isoquinoline alkaloids. Spectrochim. Acta. A. Mol. Biomol. Spectrosc. 171, 287–296.

Wen, X., Deng, F.-M., Wang, J., 2014. MicroRNAs as predictive biomarkers and therapeutic targets in prostate cancer. Am. J. Clin. Exp. Urol. 2, 219–230.

Wheeler, S., Seegmiller, A., Vnencak-Jones, C. FLT3 Internal Tandem Duplication in Acute Myeloid Leukemia - My Cancer Genome, 2017.

Wood, J.G., Rogina, B., Lavu, S., Howitz, K., Helfand, S.L., Tatar, M., Sinclair, D., 2004. Sirtuin activators mimic caloric restriction and delay ageing in metazoans. Nature 430, 686–689.

Wu, B., Zhang, Q., Shen, W., Zhu, J., 2008. Anti-proliferative and chemosensitizing effects of luteolin on human gastric cancer AGS cell line. Mol. Cell. Biochem. 313, 125–132.

Wu, H., Huang, M., Liu, Y., Shu, Y., Liu, P., 2015. Luteolin induces apoptosis by up-regulating miR-34a in human gastric cancer cells. Technol. Cancer Res. Treat. 14, 747–755.

Wu, T.-C., Lin, Y.-C., Chen, H.-L., Huang, P.-R., Liu, S.-Y., Yeh, S.-L., 2016. The enhancing effect of genistein on apoptosis induced by trichostatin A in lung cancer cells with wild type p53 genes is associated with upregulation of histone acetyltransferase. Toxicol. Appl. Pharmacol. 292, 94–102.

Xavier, C.P.R., Lima, C.F., Preto, A., Seruca, R., Fernandes-Ferreira, M., Pereira-Wilson, C., 2009. Luteolin, quercetin and ursolic acid are potent inhibitors of proliferation and inducers of apoptosis in both KRAS and BRAF mutated human colorectal cancer cells. Cancer Lett. 281, 162–170.

Xiang, N., Zhao, R., Song, G., Zhong, W., 2008. Selenite reactivates silenced genes by modifying DNA methylation and histones in prostate cancer cells. Carcinogenesis 29, 2175–2181.

Yan, Y., Hanse, E.A., Stedman, K., Benson, J.M., Lowman, X.H., Subramanian, S., Kelekar, A., 2016. Transcription factor C/EBP-β induces tumor-suppressor phosphatase PHLPP2 through repression of the miR-17-92 cluster in differentiating AML cells. Cell Death Differ. 23, 1232–1242.

Yang, J., Cao, Y., Sun, J., Zhang, Y., 2010. Curcumin reduces the expression of Bcl-2 by upregulating miR-15a and miR-16 in MCF-7 cells. Med. Oncol. Northwood Lond. Engl. 27, 1114–1118.

Yeh, C.-H., Moles, R., Nicot, C., 2016. Clinical significance of microRNAs in chronic and acute human leukemia. Mol. Cancer 15, 37.

Yi, R., Li, Y., Wang, F.-L., Miao, G., Qi, R.-M., Zhao, Y.-Y., 2016. MicroRNAs as diagnostic and prognostic biomarkers in colorectal cancer. World J. Gastrointest. Oncol. 8, 330–340.

Zhao, X., Liu, M., Li, D., 2015. Oleanolic acid suppresses the proliferation of lung carcinoma cells by miR-122/Cyclin G1/MEF2D axis. Mol. Cell. Biochem. 400, 1–7.

CHAPTER

2

Epigenetic Dietary Interventions for Prevention of Cancer

Nibedita Pradhan, Swayamsiddha Kar, Sabnam Parbin, Dipta Sengupta, Moonmoon Deb, Laxmidhar Das, Samir Kumar Patra

Epigenetics and Cancer Research Laboratory, Biochemistry and Molecular Biology Group, Department of Life Science, National Institute of Technology, Rourkela, India

Abstract

Epigenetic regulation is a dynamic and reversible process that plays an important role in the normal development and control of gene expression in vertebrates. However, imbalances in epigenetic regulation contribute to the development of human diseases, such as fragile X syndrome, immunodeficiency, centromeric region instability, facial anomalies syndrome, Rett syndrome, α-thalassemia, and various cancers. Epigenetics encompasses several chromatin modulations at the molecular level that involve hierarchical covalent modifications of histone tails (including acetylation, phosphorylation, methylation, sumoylation, ribosylation, and ubiquitylation), DNA methylation (methylation of cytosine within CpG dinucleotide), and involvement of noncoding RNA functions. Hence, it may provide a major governing factor in tumor progression and cancer drug sensitivity. Since epigenetic marks, in contrast to genetic mutations, are reversible in nature, chemopreventive bioactive molecules are currently being evaluated as attractive drugs for cancer prevention. In view of this and to gain a better understanding and to design chemopreventive and therapeutic drugs from bioactive components (from food and spice products without any or less side effects), we have summarized the advances of applications of those drugs against cancer—notably, 5-aza-2′-deoxycytidine, epigallocatechin gallate, suberoylanilide, hydroxamic acid, trichostatin A, and others may be used as lead compounds for comparison of basic mechanisms of action while evaluating chemopreventive drugs without cytotoxicity from dietary sources.

Keywords

Cancer; Chemoprevention; Diet; Epigenetic; Herb products; Phytochemicals

INTRODUCTION

Cancer is a heterogeneous group of diseases evolved from a combinational crosstalk between genetic and epigenetic abnormalities, resulting in dysregulated gene expression and function (Cavalli, 2006; Dolinoy et al., 2007; Esteller, 2008; Hanahan and Weinberg, 2011). The initiation and progression of cancer is driven by

Epigenetics of Cancer Prevention
https://doi.org/10.1016/B978-0-12-812494-9.00002-0

changes in the series of clonally selected genetic processes that may lead to either oncogene activation or silencing of tumor-suppressor genes (Ellis et al., 2009). Genetic mutations play important roles in tumor progression, and abnormalities in epigenetic processes also play an equally vital role in the development of disease (Esteller, 2007). It is well documented that most cancers owe their causes to genetic and epigenetic changes acquired during the life span due to exposure to chemicals, radiation damage, and excess nutritional and dietary components that have the capability to modulate gene expression pattern and modify the susceptibility of individuals through changes in the epigenome (Anand et al., 2008; Herceg, 2007). Prospective studies of more than 500,000 participants from 10 European countries implicated that a heterogeneous diet and more consumption of red meat is associated with increased risk of cancer (Rohrmann et al., 2013). Cancer mortality is significantly higher for people associated with higher intake of processed meat compared with those with higher fruit and vegetable intake (Pan et al., 2012). Epidemiology and statistical surveys suggest that the migration of people from lower-risk areas of the world to higher-risk regions, resulting in a change from a normal diet to a Westernized one having lower fiber content, is a major contributor to cancer's causes. It has been estimated that in the United States more than 33% of deaths are caused by cancers that can be potentially avoided by modifying the composition of diet (Bingham and Riboli, 2004; Lin et al., 2009). The dynamic changes involved in epigenetics are influenced by a wide variety of social and environmental factors, including dietary intakes. Thus, observational studies have suggested that foods not only provide the energy and precursors for classic metabolism by undergoing chemical reaction processes, but also maintain a suitable environment that regulates the activity of the epigenome. Therefore, continuous exploration is suggested of nutritional factors and environmental toxicants in foods that may affect the functioning of genes via modulation of epigenetic marks. Unlike genetic changes (mutations), epigenetic modifications are reversible processes, which makes the epigenetic machinery an attractive therapeutic target for cancer prevention (Geutjes et al., 2012). In this review, we summarize the current advances on the interaction mechanisms between various bioactive components and specific epigenetic targets that can provide an optimum dietary solution against cancer.

EPIGENETIC REGULATION OF GENE ACTIVITY

DNA methylation at the cytosine 5-carbon position (hereafter, DNA methylation), and histone modifications by acetylation, methylation, and phosphorylation, play crucial roles in gene expression and chromatin dynamics. Changes in DNA methylation patterns may cause tumor development and cancer progression as well as affecting gene therapies, parthenogenesis, and cloning (Jones and Baylin, 2007). Histone modifications and DNA methylation control gene expression in association with various repressor proteins, such as methyl CpG domain-binding proteins, and through the remodeling of chromatin (Fig. 2.1).

Altered DNA Methylation and Abnormal Gene Expression

DNA methylation affects DNA packaging structurally. The process of the DNA methylation landscape differs between normal and tumor cells. Normally, cytosines in the context of CpG dinucleotides are methylated when scattered throughout the genome but are mostly unmethylated at the 5′ end of promoter regions of genes, where clustering of CpG dinucleotides occurs at the CpG islands (Kar et al., 2012). The

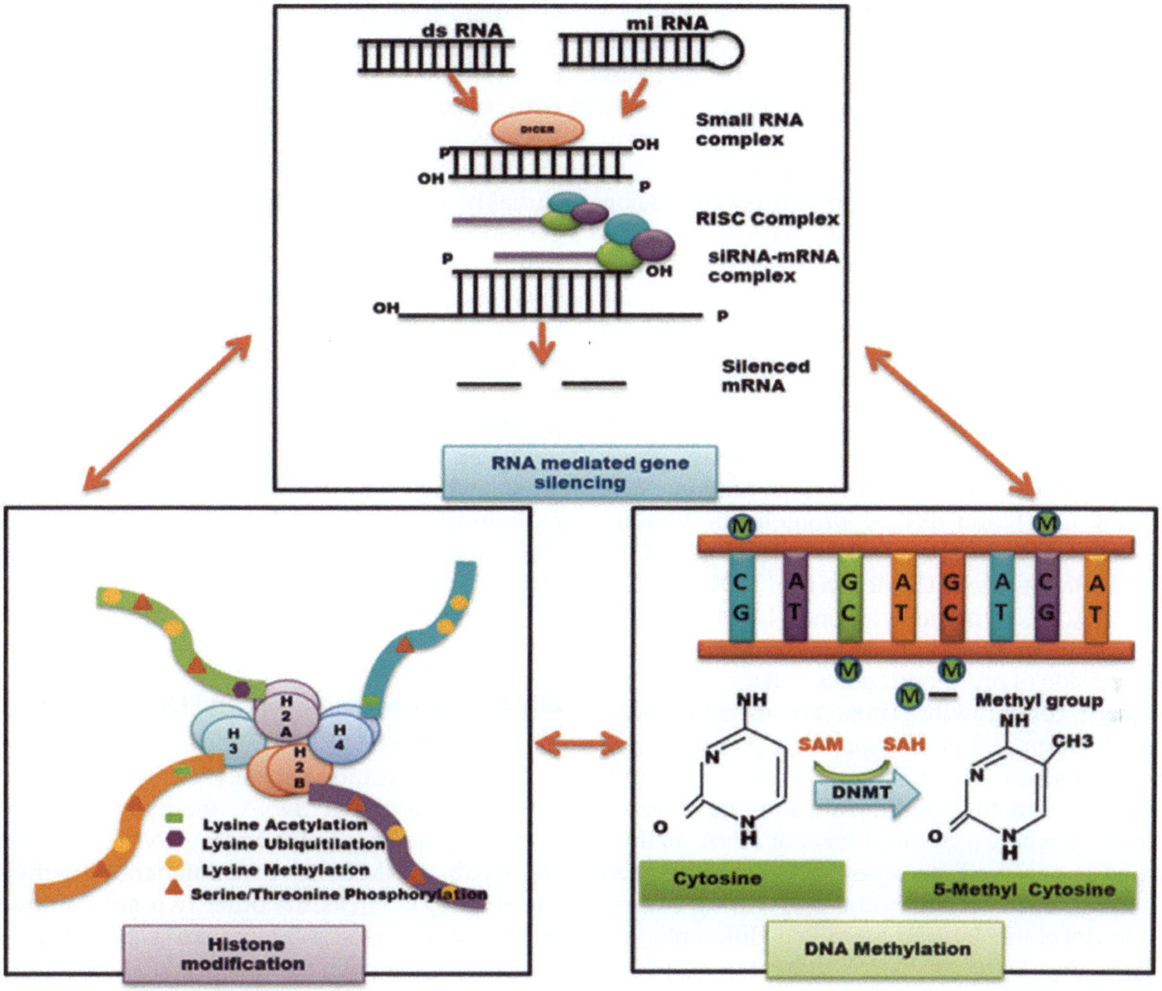

FIGURE 2.1 Different types of epigenetic modifications. DNA methylation, histone modifications, and RNA-mediated gene silencing are three different types of epigenetic regulatory mechanisms. DNA methylation involves the covalent modification of cytosine residue within the CpG dinucleotide in the presence of the DNMT enzyme. Histone modifications are the covalent posttranslational modifications in the N-terminal region of core histones, and the last mechanism of epigenetic process involves miRNA mediated gene regulation.

gene promoters, or any stretch of DNA having densely packed CpG-sequences, are considered to harbor CpG-islands, and expression of those genes might be regulated by DNA methylation. Genes that are associated with cell cycle regulation (p16INK4a, p15, and p14ARF), DNA repair (MLH1 and GST3), tumor progression (BRCA1 and VHL), cell–cell interaction and tissue integrity (TIMP3, CDH1, and CD44), and hormone reception (ESR1 and ESR2) harbor CpG rich promoters. Multiple evidence has suggested that promoter CpG islands remain hypermethylated in association with the inhibition of gene expression. In the oocyte, the CpG island having DNA methylation contains specific characteristic features of DNA sequences (Lienert et al., 2011; Quenneville et al., 2011). Methylation of DNA can inhibit transcriptional activation by directly preventing the binding of sequence-specific transcription factors, such as AP-2, E2F, and c-MyC (Patra et al., 2002, 2011). The indirect repression occurs by the binding of MBD proteins to the methylated DNA, along with the activity of histone deacetylase (HDAC) that deacetylates histone, to form repressive chromatin states (Patra et al., 2008; Kar et al., 2014). In the genome of cancer cells there is global hypomethylation and regional hypermethylation of CpG islands. As a result, DNA methylation status may either inhibit some transcription factors, such as E2F, CREB, CTCF and SF1, or promote the binding of transcription factor C/EBP2α (Burger et al., 2013; Marx et al., 2013; Mellen et al., 2012; Stadler et al., 2011). Hypomethylation of DNA is associated with an increased level of transcription and expression of proto-oncogenes that stimulate malignant cell growth. Thus, an aberrant DNA methylation and gene expression pattern might be causal in carcinogenesis when compared with DNA methylation during embryonic development, where a reduced level of DNA methylation is observed, because DNA demethylation is important in cellular processes during embryonic development and stem cell differentiation (Patra and Bettuzzi, 2007; Tahiliani et al., 2009). In the oocyte, the association of CGCGC motif in the CpG island in the absence of DNA methylation is independent of other factors; however, two transcription factors, such as E2f1 and E2f2, are present in the recognition site of CGCGC, which along with other cofactors of chromatin remodeling shows higher expression specifically in chromatin remodeling (Saadeh and Schulz, 2014). In mouse embryonic stem cells, the interaction between Cfp1 and CpG island is linked with hypomethylation, which occurs by the recruitment of a histone methyltransferase complex Set1 that deposits the H3K4me3 (Cedar and Bergman, 2009; Glastad et al., 2015). This trimethylation of H3K4 in turn inhibits the DNMT3a/1 complex that is directly associated with hypomethylation (Fillmore et al., 2015). Various mechanisms are suggested for removal of the −CH3 group from methylated DNA (Adjakly et al., 2011; Ruzov et al., 2011). In genomic DNA of various cells, it has been found that oxidation of 5hmC into 5-formylcytosine (5fC) and 5-carboxyl cytosine (5caC) is catalyzed by TET, and these 5fC and 5caC could be the additional intermediates of demethylation (Carroll et al., 2011; Zhenwei et al., 2015). Another possible mechanism of DNA demethylation mediated by TET is oxidation of derivatives of 5hmC, which is the main cause of passive demethylation because those derivatives are not recognized by methylation maintenance machinery. This is supported by the fact that in hemi-hydroxy-methylated DNA, DNMT1 is not in active form (Okashita et al., 2014).

DNA Methyltransferases and Demethylases During Cancer Progression

DNA methylation is catalyzed by three independently encoded DNA methyltransferases (DNMTs), such as DNMT1, DNMT3A, and DNMT3B. DNMT1 is a maintenance methyltransferase, whereas the other two are de novo methyltransferases. Mammalian DNMT1 is a multidomain enzyme (Zhang et al., 1999; Patra et al., 2003). Structural analyses has revealed how DNMT1 prefers hemimethylated DNA over unmethylated DNA (He et al., 2011; Song et al., 2012). During S-phase, normally DNA localizes at the site of DNA replication via interaction with proliferating cell nuclear antigen and ubiquitin-like PHD and ring-finger containing 1, which is a factor that binds the hemimethylated CGs and recruits DNMT1 at hemimethylation sites (Aasheim et al., 2015). Like DNMT1, DNMT3A, and DNMT3B, the protein DNMT3L is found in early embryos and germ cells and is related to DNMT3A and DNMT3B in sequence and in lacking a functional methyltransferase catalytic domain, resulting in the establishment of the DNA methylation pattern in both male and female germ line (Celik et al., 2015; Challen et al., 2012; Smallwood et al., 2011). Studies from biochemical and structural analysis revealed that DNMT can stimulate de novo methylation activity of DNMT3 by physically interacting with DNMT3 enzymes and thus promote their recruitment to chromatin.

Histone Modification

In addition to DNA methylation signatures, many histone modifications play important roles in chromatin dynamics, gene expression, and modulation. Acetylation and methylation are the two main histone signatures, and deregulation of their modifications is involved in tumor development and cancer progression (Chi et al., 2010; Choudhuri et al., 2010). Major posttranslational histone modifications include acetylation, phosphorylation, ubiquitylation, glycosylation, and sumoylation (Hebbes et al., 1988). Trimethylation of lysine 4 on histone H3 (H3K4me3) is enriched at transcriptionally active gene promoters, whereas trimethylation of lysine 9 on histone H3 (H3K9me3) and lysine 27 on histone H3 (H3K27me3) is present at gene promoters for transcriptional repression (Parbin et al., 2014). Thus, methylation of DNA along with H3K9me3 is associated with a repressive epigenetic signature, whereas histone acetylation and H3K4me3 are found to be associated with the active gene transcription process. It has also been reported that not all histone modifications are associated with methylation of DNA. For example, the polycomb target genes that play a crucial role in mammalian development and in the mechanism of silencing the epigenetic regulatory process are independent of DNMTs; rather, H3K27me3 is achieved by self-requirement of EZH2, which is a histone methyltransferase and a part of polycomb repressive complex 2 (Butler et al., 2012; Hansen and Helin, 2009). Various histone modifications are potentially reversible (Bird, 2002; Herman and Baylin, 2003; Ruter et al., 2004) and are catalyzed by many enzymes, such as histone acetyltransferases (HATs), HDACs, histone methyltransferases (HMTs), and histone demethylases (HDMs).

Histone Acetyltransferases and Deacetylases

The process of histone acetylation is catalyzed by HAT, which facilitates the transfer of acetyl group from acetyl-CoA to the NH_3^+ group of lysine residue, changing the overall charge of the histone tail from positive to neutral (Patra et al., 2001). Some of the major families of HATs include GNAT family, MYST family and P300/CBP family. The reverse process of histone acetylation is histone deacetylation catalyzed by the HDACs which involves the removal of acetyl group. The removal of acetyl group during deacetylation allows the DNA to become more tightly wrapped the histone cores hindering the binding of transcription factor to DNA and leads to decrease gene expression. There are four categories of HDAC, including class I, class II, class III, and class IV. Class I includes HDAC 1, 2, 3, and 8, which are the nuclear enzymes having a major focus as antitumor agents. The class II HDAC family is divided into two classes, class IIa (HDAC4, HDAC5, HDAC7, and HDAC9), which shuttle between the nuclear and cytoplasmic compartments, and class IIb, including HDAC6 and HDAC10, which are cytoplasmic and deacetylate nonhistone proteins. The class III HDAC family includes sirtuins, and HDAC 11 is a class IV HDAC.

Some cytosolic proteins are also modified along with nuclear histone acetylation–deacetylation modifications. The posttranslational modification of protein plays a crucial role in the stabilization of microtubule and intracellular transport. Recently, a cytosolic substrate of HDAC6, i.e., α-tubulin was identified, and HDAC6 inhibition is known to associate with the increased acetylation level of α-tubulin on Lys40, which was confirmed by intraperitoneal administration of a HDAC6 inhibitor—i.e., tubastatin (Selenica et al., 2014). Another recent study suggests that in human breast cancer cells (MCF-7), the inhibitory enzymatic activity of HDAC by tubastatin increases its direct binding to microtubule stability (Asthana et al., 2013).

Histone Methyltransferases and Demethylases

The covalent modification of histone involves two other enzymes, such as histone lysine methyltransferase and histone lysine demethylase, which add and remove the methyl group, respectively (Chaturvedi et al., 2011). The methylation of histone can occur either on lysine or on arginine residue, but lysine methylation is commonly observed on the histone tail of H3 and H4. The methylation sites associated with gene activation include H3K4, H3K9, H3K27, H3K48, and H3K79. Lysine may be mono-, di- or trimethylated with methyl groups by replacing the hydrogen of its NH_3 group, whereas arginine may be mono or dimethylated with a free NH_2 or NH_2^+ group. Generally, arginine methylation requires a complex known as protein arginine methyltransferase, and methylation of lysine requires a specific HMT containing a SET domain which harbors the enzymatic activity. The methylation of histone lysines depends on an abundance of S-adenosylmethionine (SAM), which acts as a methyl donor (Messina and Hilakivi-Clarke, 2009).

EPIGENETIC THERAPY OF CANCER

Epigenetic therapy of cancer is a novel therapy in the rapidly evolving field of cancer etiology (Boffetta et al., 2010). In response to integrated knowledge and views of epigenetic deregulations, and as one of the major causes of cancer, it is prime time for investigators to develop innovative new drugs and tools for treatment of cancers.

Epigenetic modulation is the imposition of a new mark or changes to existing marks and structural alternations associated with the genome function during tumor development and cancer progression. Studies on epigenetic modulations in a large cohort of samples considering epidemiology and clinical studies already recommended novel epigenetic therapy, including combinations of HDAC inhibitors and DNMT inhibitors.

Improving epigenetic therapy by discovering less cytotoxic drugs is currently considered as effective and valuable approach to correct epigenetic defects in the emerging field of cancer prevention and therapy. As epigenetic aberrations, in contrast to genetic defects, are potentially reversible, it offers exciting opportunities for discovering novel cancer drugs that can reactivate epigenetically silenced tumor-suppressor genes (Aasheim et al., 2015; Hathaway et al., 2012). As synthetic epigenetic drugs are associated with high levels of toxicity and lack specificity of action, the challenge is to determine the reversibility of adverse epigenetic marks in cancer inflammation that can be prevented by specific diets, natural photochemicals, or lifestyle changes (Heightman, 2011). Evidence from experimental data has suggested that diet and diet-derived polyphenols of plant origin have potent cancer chemopreventive effects in humans via modulation of functions of epigenetic machinery, such as DNMTs, HATs, HDACs, HMTs, and HDMs because of their anti-inflammatory, antioxidant, phytohormonal, and homeostatic effects within cancer cells (Fig. 2.2).

CANCER AND LIFESTYLE RISK FACTORS

Despite the enormous amount of research and rapid developments during the past few

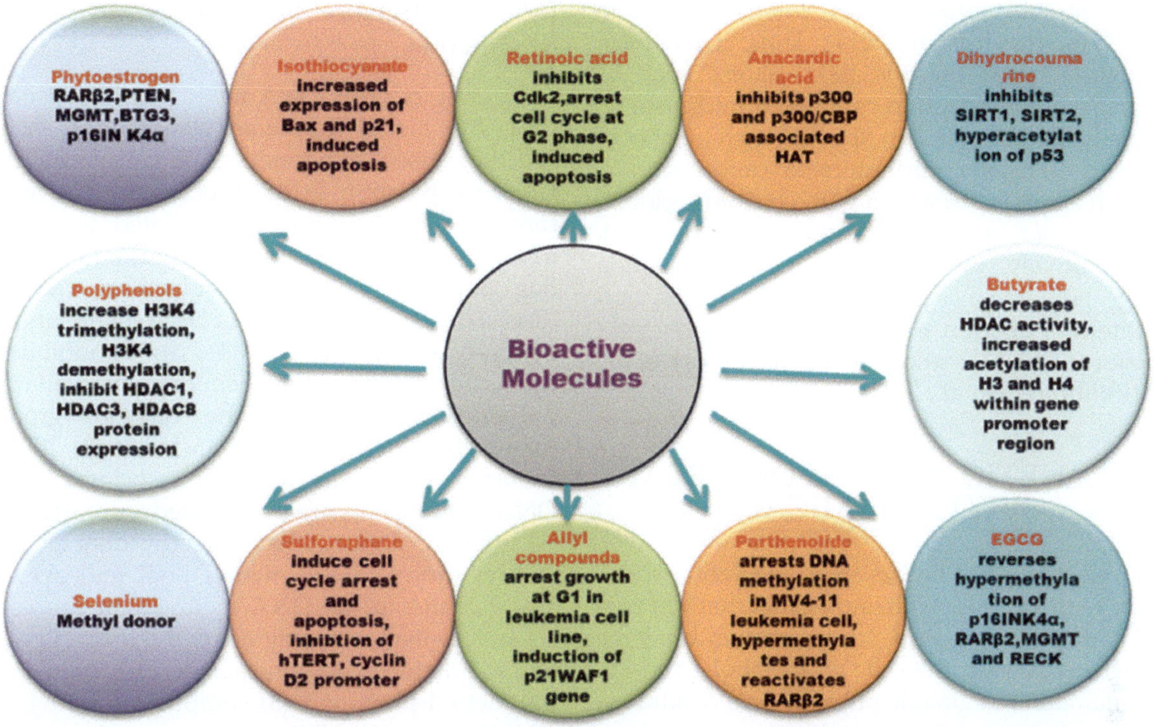

FIGURE 2.2 Role of different plant-derived bioactive molecules in cancer progression and epigenetics. Polyphenols derived from different plants play an important role in the inhibition of cancer progression by inhibiting different enzymes or by regulating the expression level of several genes or by inducing cell cycle arrest and apoptosis in several types of cancer.

decades, cancer is a worldwide killer and is believed to be a preventable disease (Baylin and Ohm, 2006). A recent survey of cancer shows that its incidence in American society is more than 1 million whereas worldwide it has been estimated that about 10 million people are diagnosed with cancer in the recent year. Epidemiological data has revealed that people of some cultures are more prone to certain cancers than others. For example, in Western countries the incidence of lung, colon, prostate, and breast cancer is more prevalent than they are in Eastern countries, such as India, where cancers of head, neck, and cervix are more common. Observational studies have suggested that worldwide, approximately 17% of new cancer is due to infectious disease—in low-to-middle resource regions, it is about 1.5 million (26%), while in high-resource countries it is about 7.2%. Of this, only 5%–10% of cancer cases are caused by genetic defects, whereas the remaining 90%–95% are linked to environmental and lifestyle factors. It is now expected that by the year 2020, approximately 55% of new cancer cases will be diagnosed and 12 million cancer patients will die (Scott and Weir, 1998).

In various countries there is large variation in rates of specific cancer, which may be linked to the incidence of cancer in migrating. This can be further emphasized by the migration of people from Asian countries, where the incidence of prostate and breast cancer is 25 times and 10 times lower respectively than it is for those in Western countries, and the rates for these cancers increase substantially when Asians migrate to the West (Rush et al., 2014). Thus, it has been suggested that lifestyle risk factors, such as increased consumption of tobacco and alcohol, obesity, and diets low in fiber but high in fat content, are major contributors to cancer causes, which again is associated with economic globalization and urbanization around the world. Alternatively, by modifying diet composition (e.g., polyphenols, phytochemicals, spices, and content of fiber), it will be possible to prevent a major fraction of cancer deaths (Haggarty et al., 2013; Ibrahim et al., 2011).

EPIGENETIC ACTIONS OF DIETARY PHYTOCHEMICALS

Cells of human beings are in constant association with various oxidizing agents that are present in air, food, and water, or they may be produced within the cells by metabolic activity to maintain sustainable physiological conditions with a balance between oxidants and antioxidants. The overproduction of oxidants results in oxidative stress, which may be associated with increased risk of cancer by enhancing oxidative damage to large biomolecules, such as membrane lipid, protein, and DNA (Bistulfi et al., 2010). Excess energy-rich diets and affluent lifestyles have major impacts on cancer progression and survival, with the inhibition of tumorigenesis. Thus, in order to reduce the risk of cancer, several dietary constituents, including fruits, vegetables, spices, and seeds that contain sufficient antioxidants—known as phytochemicals and retaining the pronounced chemopreventive effects, must be consumed. These agents include curcumin (turmeric), genistein (soybean), tea polyphenols (green tea), resveratrol (grapes), and sulforaphane (cruciferous vegetables), which can alter the methylation status of DNA and the histone modifications required for gene activation or silencing in cancer prevention and therapy (Fig. 2.3, Table 2.1).

DNA Methylation and Dietary Modulators

Overall in a cell, DNA methylation is visualized by global hypomethylation and site-specific regional hypermethylation., Particularly, DNA methylation in the promoters of several tumor-supressor genes is associated with the inactivation of various physiological and signalling pathways that are involved in various diseased state. This process of DNA methylation, catalyzed by specific DNMTs, can be inhibited by some dietary phytochemicals, particularly polyphenols containing the catechol group (Bhattacharya et al., 1999; Ghoshal et al., 2006). Dietary components that are associated with DNA methylation are categorized into 3 types: (1) methyl donors that directly donate the methyl group to the methyl pool and act as co-substrates in the DNA methylation process, (2) factors that indirectly affect the methyl pool by modulation of enzyme activity that regulates the methyl pool, and (3) compounds acting as DNMT enzyme inhibitors.

Role of Methyl Donors and Nutrients of One-Carbon Metabolism in DNA Methylation

Nutrients involved in one-carbon metabolism, such as folate, riboflavin, pyridoxine, and methionine, play an important role in maintaining the DNA methylation pattern. Folate, a water-soluble form of vitamin B9, acts as a methyl donor that transfers one carbon unit and is involved in multiple cellular processes, including specific amino acid synthesis, DNA synthesis, and DNA methylation. After dietary intake, folate is reduced to dihydrofolate and tetrahydrofolate (THF). Then through several metabolic pathways, THF results in the formation of 5-methyl-THF or 10-formyl THF, which act as cofactors for various enzymes. The cofactor 5-methyl-THF is involved in the remethylation of homocysteine by transferring the methyl group to homocysteine to produce

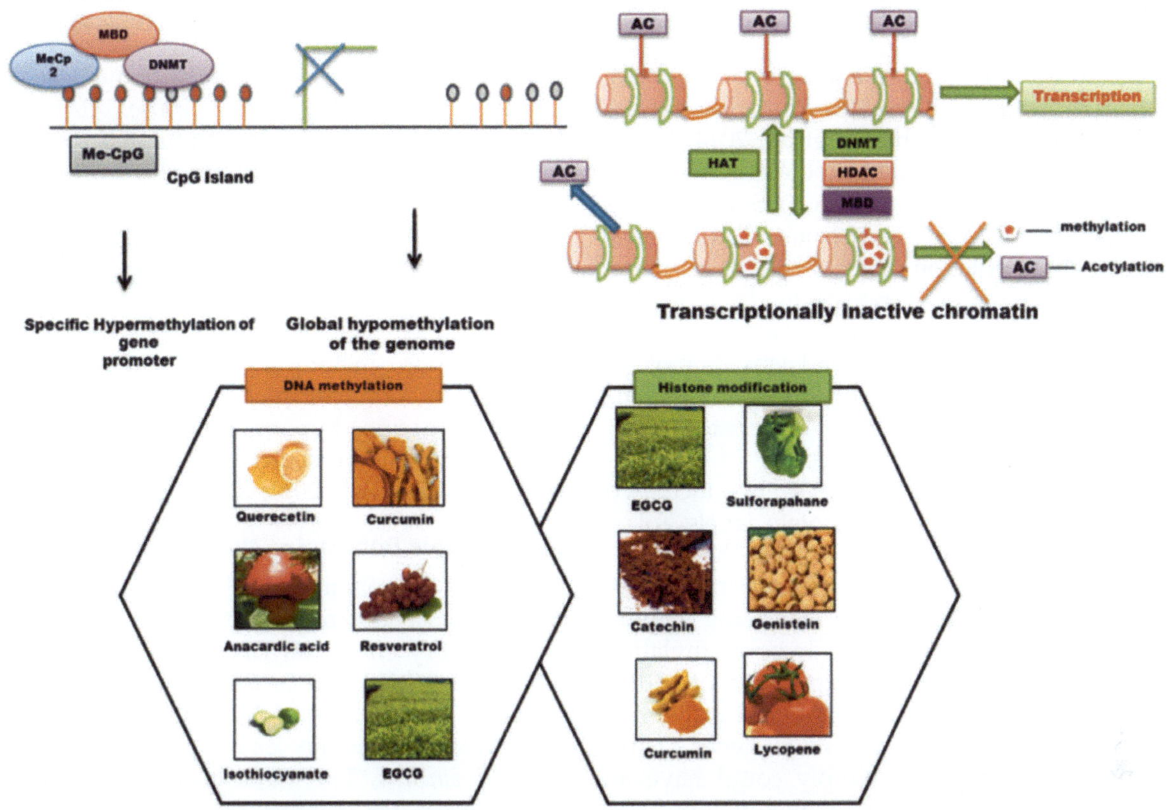

FIGURE 2.3 Effect of dietary polyphenols on DNA methylation and histone modifications. The diagram shows different types of epigenetic hypermethylation at the gene promoter that lead to the development of cancer with the silencing of the corresponding tumor-suppressor gene. Histone modifications are potentially reversible, and various histone modifying enzymes, such as histone acetyltransferases (HATs) and HDACs, play important roles in the acetylation and deacetylation of lysine residues present in the histone. Because of these modifications, conformational changes occur in the chromatin organization. Dietary phytochemicals having potent anticancer activity can have an impact on these enzymes and reverse epigenetic dysregulation by modulation of the epigenetic machinery.

TABLE 2.1 Dietary Ingredients With Various Epigenetic Mechanisms in Cancer

Nutrients	Source	Epigenetic Mechanisms	References
Soy isoflavones	Mulberries, peanuts	DNMT inhibitor	Pudenz et al. (2014)
Resveratrol	Grapes	DNMT inhibitor SIRT1 activator	Aldawsari et al. (2016)
Genistein	Soybeans	Both DNMT and HDAC inhibitor HAT activator	Adjakly et al. (2011) and Chen et al. (2011)
Curcumin	Turmeric	Both DNMT and HDAC inhibitor	Khor et al. (2011)
Sulforaphane	Cruciferous vegetables	DNMT and HDAC inhibitor	Fan et al. (2012)
Butyrate	Fruits and vegetables	HDAC inhibitor	Davie (2003)
Cambinol	Sweet clover	SIRT1 inhibitor	Lugrin et al. (2013)
Isothiocyanante	Broccoli	HDAC inhibitor	Nian et al. (2009)
Anacardic acid	Cashew nuts	HAT inhibitor	Yasutake et al. (2013)

methionine that acts as a precursor to SAM, which is the universal methylation donor (Kim et al., 2012; Nishihara et al., 2014; Ulrich et al., 2000). After transferring the methyl group, SAM gets converted to S-adenosyl-L-homocystein (SAH). Numerous studies have shown that depletion of SAM results in malignant transformation, gene-specific hypermethylation, global DNA hypomethylation, silencing of tumor-suppressor genes, and activation of oncogenes (Chandar et al., 1989; Pogribny et al., 2004). During the early stage of hepatocarcinogenesis, by increasing the activity of DNMT1, DNMT3B, MBD2, and MBD4 by increased expression of mRNA and protein levels, folate deficiency has been shown to alter the components of the DNA methylation machinery (Pogribny et al., 2006). The increasing activity of DNMT is associated with hypermethylation status, whereas increased expression of MBD2 and MBD4 is responsible for hypomethylation of oncogenes and prometastatic genes. Thus, factors that limit SAM supply could have an important impact on DNA methylation and histone marks. For this reason, folic acid supplementation is known to have a chemopreventive effect that has been demonstrated in some cancer modes, including hepatocarcinogenesis. Besides these, several other nutrients of one-carbon metabolism, such as riboflavin (B2), pyridoxine (B6), and cobalamin (B12), either by acting as direct methyl donors or cofactors in enzymes, also contribute to the methionine pool and DNA methylation in the cell. For example, riboflavin, a component of flavin adenine dinucleotide acts as a cofactor for methylene tetrahydrofolate (MTHF) reductase, which catalyzes the conversion of 5, 10-MTHF that is a substrate for homocysteine demethylation to methionine (Yang et al., 2008). The methionine level can also be modulated directly through the consumption of methionine in protein-rich diets. Vitamin B, also known as cobalamin, is involved in the regulation of DNA methylation by affecting one-carbon metabolism and is important for the methylation of homocysteine to methionine (Lee and Zhu, 2006). Also, the level of methionine can be modulated directly through consumption of methionine in methylation activity involving vitamin B-6, which is a cofactor for several enzymes involved in methionine synthesis, homocysteine regeneration, and metabolism of folate. It has been reported that a diet deficient in major dietary sources of the methyl group, including folate, vitamin B-12, and choline, induces global hypomethylation and site-specific hypermethylation and has been linked to increased risk of cancer development (Meeran et al., 2011; Zhang et al., 2010).

Polyphenols Possessing the Catechol Group

Certain polyphenols, such as bioflavonoids, tea catechins, and coffee polyphenols, exert major inhibitory effects on DNA methylation and DNMT activities. Tea is the the most popular beverage in the world, followed by water, and is produced from the leaves of the tea plant *Camellia sinensis*. The major chemical composition of green tea includes catechin, epicatechin (EC), (−) epicatechin-3-gallate (ECG), (−) epigallocatechin (EGC), and (−) epigallocatechin-3-gallate (EGCG). Among these phytochemicals, EGCG has been reported as a potential inhibitor of DNMTs, and hence causes passive demethylation. It has been observed that in animal models of lung, prostate, breast, and skin cancer, EGCG shows its anticancer activity by inhibiting cell proliferation, inducing apoptosis, and registering cell cycle arrest at the G0/G1 phase. It was also demonstrated that in a protein-binding assay, EGCG at a very low concentration and high specificity can bind to several proteins, including vimentin and insulin-like growth factor1 (IGF-1) receptor (Yang and Wang, 2011). EGCG can also induce apoptosis by inhibiting the PI3K/AKT/pBAD survival pathway, resulting in downregulation of Bcl-2 and upregulation of Bax (Gupta et al., 2012). It has also been observed that EGCG treatment can inhibit the process of angiogenesis, invasion, matrix metalloproteinases (MMPs) such as MMP2 and MMP9, and vascular endothelial growth factor. It can also inhibit the concentration of phosphorylated nuclear factor-κB (NF-κB)/p65 in HT-1080 cells in a dose-dependent manner (Lee et al., 2011). It has been found that EGCG has reversible activity on the hypermethylation status of p16INK4a, RARβ, MGMT, and hMLH1 genes by repressing the activity of DNMT1 in human esophageal cancer KYSE 510 cells. An in vitro study also suggested that EGCG is an effective inhibitor of DNMT. It has also been shown that with diverse potencies and in a concentration-dependent manner, tea catechins and bioflavonoids have inhibitory action on prokaryotic SssI methyltransferase-mediated DNA methylation and the human DNMT1-mediated DNA methylation reaction (Lee et al., 1994).

Because compounds in the catechol group serve as excellent substrates for methylation mediated by COMT, EGCG is also methylated, thus playing a crucial role in lowering the effectiveness of methyltransferase. This enzyme adds a methyl group to the catecholamine by using SAM as a methyl donor. This methylation reaction results in the demethylation of SAM and formation of SAH, which is a potent and selective inhibitor of DNA methylation. COMT and DMNT share a common core structure and belong to the same superfamily of SAM-dependent methyltransferases. EGCG can act as a mixed-type (competitive and noncompetitive) inhibitor for COMT (Zhu et al., 2010). Molecular docking studies indicate that EGCG can be well accommodated in a hydrophilic active pocket of DNMT1, and by forming four hydrogen bonds, it can be effectively tethered within the binding site of DNMT1, which is further stabilized by magnesium ions (Makela et al., 1998; Yoo et al., 2012). The methylation of EGCG occurs at both 4′ and 4″ positions to form 4′-*O*-MeEGCG and 4′, 4″-di-*O*-MeEGCG, respectively (Meng et al., 2002). It has been found that EGC and EC are the major catechins circulating within the blood; about 30% of EGC is in 4′-*O*-MeEGCG form (Renouf et al., 2011). From experimental study, it was observed that continuous 3-5 weeks of consumption of six cups of green tea daily, reduce the risk of prostate cancer. (Henning et al., 2013). Similarly, a different degree of EGCG methylation was also observed after consumption of green tea in kidney, lung, and xenografted prostate tissue (Henning et al., 2012). However, the methylation activity is organ-dependent, as the activity of COMT is higher in liver and kidney tissue as compared with lung and prostate cancer (Wang et al., 2012). In vitro study has suggested that EGCG can reverse the hypermethylation status of tumor-suppressor genes and reexpression of silenced genes in a concentration- and time-dependent manner. It has also been reported that EGCG acts as an efficient inhibitor of human dihydrofolate reductase. Through the interaction of EGCG with folic acid metabolism in cells, it can also inhibit the synthesis of DNA and RNA, altering the pattern of DNA methylation (Diaz-Chavez et al., 2013). It was also reported that EGCG can reactivate some methylation-silenced genes in human colon cancer and prostate cancer cells. Treatment of human oral cancer cells with EGCG partially reversed the hypermethylation status of the RECK gene and significantly enhanced the expression level of RECK mRNA, which is associated with the reduced expression of matrix metalloproteinases MMP-2 and MMP-9 and suppressed the invasive ability of cancer cells (Griffiths et al., 2014). Treatment of MCF-7 and MDA-MB-231 breast cancer cells with caffeic or chlorogenic acid decreased the promoter methylation of RARβ. Moreover, consumers of high doses of chlorogenic acid had an increased concentration of plasma to total homocysteine, which acts as an inhibitor of methylation reactions (Whitsett and Lamartiniere, 2006).

Soy Isoflavones and Phytoestrogens

Several soy isoflavones, commonly called phytoestrogens, act as both estrogen agonists and antagonists by differentially interacting with estrogen receptors α and β (Bowers et al., 2000) and regulate estrogen responsive genes, such as p21 (Wang et al., 2013). These observations suggest that soy or its derivatives exert anticancer effects.

Resveratrol (3,4′,5-trihydroxystilbene) is a type of natural phenol and a phytoalexin derived naturally from various plants, such as mulberries, peanuts, grapes, apricots, and pineapples (Casanova et al., 2012; Gupta et al., 2011). This polyphenol prevents against chemical-induced carcinogenesis by interfering with multiple signaling pathways that control cell division, cell growth, apoptosis, angiogenesis, and tumor metastasis in numerous types of human cancers (Vinod et al., 2013). Genistein, daidzein, and other isoflavones have attributed their anticarcinogenic effect by modulating DNMT activity and promoter methylation (Pudenz et al., 2014). Kinetic studies suggest the inhibitory activity of genistein on DNMT activity, which leads to silenced gene hypomethylation and reactivation, such as GSTp1, and tumor-suppressor gene BTG3 in prostate cancer. It has also been reported that dietary intake of genistein increases tumor latency in rats exposed to chemical carcinogens and decreases mammary tumor incidence (Majid et al., 2010). Upon treatment with these phytoestrogens, epigenomic changes have been reported in numerous studies. Exposure to resveratrol led to hypomethylation and reactivation of RARβ2 and phosphatase and tensin homolog (PTEN) tumor-suppressor genes with decreased expression of DNMT1 and p21 upregulation in noninvasive MCF-7. Treatment with resveratrol inhibits the expression of some enzymes such as cyclooxygenase, lipoxygense, NF-κB, Sirt1, and DNMTs (Aldawsari et al., 2016). The findings of the current study suggested that genistein plays an important role in the gene-transcription process by modifying epigenetic events. Genistein exposure resulted in the partial reversal of p16INK4a, RARβ, and MGMT promoter hypermethylation and reactivated methylation-silenced genes in human esophageal squamous KYSE 510 carcinoma cells associated with the inhibition of DNMT activity. Moreover, treatment with genistein resulted in a significant decrease in the methylation status of promoter region of *BTG3* tumor-suppressor gene in LNCaP and PC3 prostate carcinoma cell lines and in A498, ACHN, and HEK-293 renal cell carcinoma cell lines. Another epigenetic mechanism by which genistein can regulate the activity of androgen receptor (AR) is by inhibiting the activity of HDAC6, an Hsp90 deactylase that results in the inactivation of Hsp90 through increased acetylation and a consequent increased level of ubiquitylation and downregulation of AR. It has also been reported that genistein can effectively demethylate the CpG island of tumor-suppressor genes, such as GSTP1 and EPHB2, in the prostate cancer cell line (Adjakly et al., 2011). Recently, it has been proposed that genistein can suppress the growth of PC3 and DU-145 cell lines by inhibiting the oncogenic miRNA-151, resulting in upregulation of its target tumor-supressor genes (Rabiau et al., 2011). It has recently been observed that treatment of prostate cancer cells with isoflavones results in demethylation of the promoter sequences of miRNA-29a and miRNA-1256, leading to their upregulation and in turn leading to inhibition of their target genes and inhibition of prostate cancer cell growth and invasion (Li et al., 2012). Another report on the regulation of miRNA by phytoestrogens showed that in the PC-3 prostate cancer cell line, genistein and isoflavones can downregulate miRNA-221 and miRNA-222 and result in the induction of tumor-suppressor gene ARH1. In prostate cancer cell lines, increased expression of ARH1 is associated with inhibition of cellular proliferation, colony formation, and invasion (Chen et al., 2011; Chiyomaru et al., 2013). Multiple studies have suggested that daidzein, another soy isoflavone, may act as a growth modulator in various types of cancer, such as colon, prostate, breast, and pancreatic cancer (Bishop and Ferguson, 2015). Another bioflavonoid found in a wide variety of vegetables, leaves, fruits, and grains includes quercetin, which can demethylate and activate the tumor-suppressor gene p16 in RKO colon cancer cells (Jha et al., 2010).

Parthenolide and Curcumin

Parthenolide is a sesquiterpene lactone that occurs naturally in the plant feverfew and has been shown to have anticancer properties (Garcia-Pineres et al., 2004; Khor et al., 2011; Liu et al., 2011; Shu et al., 2011; Woynarowski and Konopa, 1981). Recently, it was reported that in MCF-7 breast cancer cells and MV4-11 leukemia cell lines, parthenolide shows inhibitory effect on DNMT1 activity linked to the alkylation of Cys1226 of the DNMT1 catalytic domain, and interferes with the binding of Sp1 to DNMT1 that results in the induction of sub-G1 cell cycle arrest (Guzman et al., 2005; Tahiliani et al., 2009). Another polyphenol curcumin, a diferuloylmethane that is present in the spice turmeric, was found to induce global methylation of DNA in MV4-11 leukemia cells (Costello and Franklin, 2007). It was also found to hypomethylate and subsequently reactivate the RARβ2 and WIF-1 tumor-suppressor genes in human cervical and non-small-cell lung cancer cells, respectively (Vermeulen et al., 2010). *In-silico* studies of interaction between curcumin and DNMT1 indicate that curcumin has an inhibitory effect on DNMT1 by covalently blocking the catalytic thiol group of the C1226 binding site. Furthermore, in human prostate cancer cells, curcumin exposure led to the demethylation of the promoter and re-expression of NEUROG1 and NRF1 (Wang et al., 2001). Treatment with curcumin inhbits HAT activity and is also a potential DNMT and HDAC inhibitor (Khor et al., 2011).

Selenium and Zinc

Several minerals, such as zinc and selenium, are involved in one-carbon metabolism and may act as methyl donors. It has been proposed that selenium reacts with homocysteine to form selenohomocysteine. In the methionine/transsulfuration pathway, zinc also acts as an important cofactor for various enzymes for the generation of methyl donor equivalents to SAM and betaine. In folate metabolism, the key enzyme serine hydroxymethyltransferase helps in the transfer of methyl groups into the methionine cycle, where serine acts as the methionine donor and is catalyzed by a zinc-dependent transcription factor. In rats, decreased turnover of SAM and depressed DNA and histone methylation in the liver are associated with zinc deficiency (Wallwork and Duerre, 1985). It has been reported that in malignant prostate tissues, the concentration of zinc is 10%–25% of that in a healthy prostate, suggesting that a high concentration of zinc is essential for the maintenance of prostate health (Wolk, 2007).

Sulforaphane

Sulforaphane, a biologically active phytonutrient within the isothiocyanate group of organosulfur compounds, is found in cruciferous vegetables, such as broccoli, broccoli sprouts, cabbage, and kale. It exhibits anticarcinogenic activity by induction of cell cycle arrest and apoptosis in various human cancer cells (Clarke et al., 2008). Recent studies demonstrate that sulforaphane may target the aberrant hypermethylation status of a gene by modulating DNMT expression. Treatment of breast cancer cells with sulforaphane results in decreased expression of DNMT1 and DNMT3a. Similarly, in LNCaP prostate cancer cells, sulforaphane downregulates DNMT1 and DNMT3b expression, but not that of DNMT3a (Arany et al., 2003). Treatment of

MCF-7 and MDA-MB-231 human breast cancer cell lines with sulforaphane resulted in the inhibition of hTERT, the catalytic regulatory subunit of telomerase. In human colon cancer CaCo-2 cells, sulforaphane downregulates DNMT1 activity. It prostate cancer cells, sulforaphane inhibited DNMT activity, caused demethylation of the cyclin D2 promoter, and enhanced transcription of cyclin D2 mRNA (Singh et al., 2004). In vitro study has suggested that in porcine satellite cells, sulforaphane acts as both an HDAC and a DNMT inhibitor (Fan et al., 2012).

Retinoic Acid

Contrary to other nutrients and vitamins, such as folate, zinc, and selenium that act directly on the synthesis of a methyl group donor, retinoic acid (RA) and vitamin D3 exert their epigenetic activity by interaction with their receptors. RA is produced by the oxidative conversion from β-carotene and acts by binding to the retinoic acid receptor (RAR) that is bound to DNA as a heterodimer with the retinoid X receptor (RXR). Moreover, in the promoters of target genes, RXR complex can further modulate the process of transcription by cognate response elements (Degos et al., 1995; Fujii et al., 2008). In cervical squamous carcinoma, RA activated p21 transcription, whereas it inhibits the AP-1 transcriptional activity in gastric cancer cells (Liu et al., 1996; Stefanska et al., 2010). This inhibitory activity is associated with decreased expression of DNA methylation with the downregulation of AP-1 responsive genes. The changes in the DNA methylation pattern affect the differentiating and proapoptotic actions of RA that have been observed in various types of cancers, such as leukemia as well as breast, ovarian, head, and neck cancers (Benkoussa et al., 2002; Daroqui and Augenlicht, 2010; Yu et al., 2008). It has also been reported that treatment of noninvasive MCF-7 cells with retinoic acid results in the reduction of methylation status of the promoter region and increases the expression level of RARβ2 and PTEN tumor-suppressor genes (Sowa et al., 1999). Additionally, treatment with RA results in the attenuation of DNMTs mediated by miRNA, and this constitutes another mechanism for the demethylating activity of this compound. RA can also upregulate glycine *N*-methyltransferase activity that catalyzes the conversion of SAM to SAH (Cheng et al., 1999). It has also been reported that in human head and neck squamous cell carcinoma cells, treatment of RA in combination with DNMT inhibitor, such as 5-aza-2′-deoxycytidine and valoproic acid, promoted growth arrest and cell differentiation (Gan et al., 2012). After treatment with all three drugs, the results was cell growth inhibition and increased granulocyte differentiation in promyelocytic leukemia cells (Gianni et al., 2012; Lengfelder et al., 2013; Lubbert et al., 2012; Podhorecka and Macheta, 2013).

Plant-Derived Natural Compound and Histone Modifications

Histone modification plays a vital role in epigenetic regulation, as it is a reversible process. It has been shown that several polyphenols possessing anticancer properties regulate the pattern of gene expression by altering the modifications of histone proteins (Myzak and Dashwood, 2006). Most plant-derived polyphenols possess HAT and HDAC activities and are being considered a group of agents for cancer treatment and prevention.

Dietary Fiber (Butyrate)

Butyrate is the smallest anticancer agent and is a known HDAC inhibitor with histone acetylation properties that contains a simple 3-carbon spacer attached to a carboxylic acid group. Most of its chemopreventive effect was observed in colorectal cancer. Butyrate decreases HDAC activity by entering the HDAC active pocket and forming a bidentate ligand of the carboxylate group with the buried zinc atom. The promoter region of butyrate-responsive genes consists of butyrate response elements where the action of butyrate is mediated through Sp1/Sp3 binding sites (Beishline and Azizkhan-Clifford, 2015; Nakano et al., 1997). Examples of such butyrate-responsive genes include p21WAF1 and Bax. For example, when Colo-320 human colon cancer cells are treated with 5 mM butyrate, the activation of the p21WAF1 gene occurs, which is associated with increased acetylation of H3 and H4 within its gene promoter region. It has been reported that inhibition of Sp1 and Sp3 binding sites with the activity of HDAC resulted in induction of histone acetylation and transcriptional activation of the p21WAF1/Cip1 gene, which inhibits cyclin-dependent kinase 2 activity and thus arrests cell cycling (Myzak et al., 2004). The potential therapeutic value of butyrate was also examined in the prostate cancer cell line by targeting the Sp1 gene (Sankpal et al., 2011).

In HCT116 human colon cancer cells, it has been observed that 1–10 mM of butyrate arrested cell growth at the G2 phase and induced apoptosis, which were associated with increased expression of p21WAF1 and Bax (Roovers and Assoian, 2000). More specifically, butyrate is known to associate with increased acetylation of Sp1, which results in decreased binding to these promoters, thus representing nonhistone proteins as relevant targets for the anticancer actions of this HDACi. Butyrate shows decreased HDAC activity and promotes cell cycle arrest (Davie, 2003). Because of HDAC-inhibitory effects of butyrate, it has also been considered a combinatorial anticancer intervention with several other agents, such as dietary and synthetic DNMT inhibitors and retinoid. Few studies have investigated the HDAC-inhibitory actions of butyrate in in vivo models. It has been demonstrated that a combination of sodium butyrate and folic acid has additive inhibitory actions in vivo on the development of colon cancer. Recently, it has been identified that butyrate not only acts as an HDAC inhibitor, but also can decrease the histone methylation process (Marinova et al., 2011). For example, butyrate treatment can significantly upregulated the expression of a specific histone demethylase—i.e., KDM5B. Another JSRID2 responsible for H3K27me3 is known to be upregulated by butyrate treatment (Pasini et al., 2010).

Cambinol and Dihydrocoumarin

Cambinol and dihydrocoumarin have been identified as SIRT inhibitors. SIRT1 has been shown to deacetylate the transcription factor p53. The inhibition of SIRT1 is associated with the inhibition of glycolysis and cell proliferation. In vitro and in vivo studies suggest that cambinol inhibits the deacetylase activity of SIRT1 and SIRT2 and also inhibits the inflammatory and innate immune response (Lugrin et al., 2013). Dihydrocoumarin is a dietary compound found in sweet clover that inhibits the deacetylase activities of yeast Sir2p and human SIRT1. A dose-dependent increased expression of ac-p53, cytotoxicity and apoptosis were also observed with the exposure of human lymphoblastoid cells to DHC 1 (Jiang et al., 2010). Another β-naphthol compound, cambinol, is known to inhibit SIRT1 and SIRT2. It has also been suggested that treatment of cancer cell lines with cambinol results in the hyperacetylation of p53 (Olaharski et al., 2005).

Isothiocyanates

In vitro and in vivo model studies suggest that sulforafane and related dietary isothiocynates, such as allyl isothiocyanate (allyl-ITC) (Wang et al., 2008), benzylisothiocyanate (BITC) (Batra et al., 2010), phenylhexyl isothiocyanate (PHITC) (Whitsett and Lamartiniere, 2006; Zeng et al., 2011), phenyl isothiocyanate (PEITC), and other longer-chain isothiocyanates (Do et al., 2010) inhibit the activity of HDAC with increased activity of histone acetylation. It has also been reported that in human leukemia cells, PHITC can inhibit the activity of HDAC by increasing the histone acetylation with an elevated level of H3K4 methylation and decreased H3K9 methylation marks. In acute leukemia cells, it can also reactivate the gene expression of the hypermethylated p16 gene via both DNA methylation and histone acetylation. In prostate cancer, PEITC depressed the promoter of P21WAF1 via HDAC inhibition and increased acetylation of histone with changes in the histone methylation level (Nian et al., 2009). It was also shown that in human leukemia cells, PEITC can overcome resistance to vorinostat, which can increase the inhibition of HDAC and restore gene re-expression (Rajendran et al., 2013).

In capan-2 pancreatic cells, it has been reported that BITC at a 10 μM concentration triggered apoptosis, resulting in decreased expression of HDAC1 and HDAC3. Sulforaphane treatment (15 μM) inhibits the activity of HDAC and increases expression of p21WAF1 and Bax, accompanied by H4 acetylation within p21WAF1 and Bax promoters in colon (HCT116) and prostate (LNCaP) cancer cells (Ruotolo et al., 2010). Recently, it has been identified that isothiocyanate plays an important role in the KEAP1/Nrf2/ARE pathway. When transgenic adenocarcinoma of mouse prostate cells (TRAMP C1) was treated with sulforaphane, it resulted in Nrf2 derepression by decreased DNMT1 and DNMT3a expression, specific CpG dinucleotide demethylation in the Nrf2 promoter, increased acetylation of histone H3, and elevated ac-H3 binding to the Nrf2 promoter, indicating active transcription (Chorley et al., 2012; Zhang et al., 2013). In hyperplastic (BPH1) and cancerous prostate cells (LNCaP and PC3), sulforaphane (at 15 μM) induced cell cycle arrest and apoptosis, decreased the activity of HDAC and class I and II HDAC proteins, and increased H3 acetylation within the p21WAF1 promoter region. It has been found that overexpression of HDAC3 and HDAC6 blocked sulforaphane-induced acetylation of respective substrates (Lea and Randolph, 2001; Zhu et al., 2010). Interestingly, it is noted that compared with normal cells, prostrate and colon cancer cells were more sensitive to sulforaphane (Balasubramanyam et al., 2003). In LNCaP prostate cancer cells, phenethyl isothiocyanate treatment (0.5–10 μM) results in cell growth arrest in association with inhibition of HDAC activity.

Anacardic Acid

Anacardic acids are the phenolic acids from the shell of the cashew nut and act as a potent inhibitor of HAT by inhibiting the activities of p300 and p300/CBP-associated HAT (Nian et al., 2008). In addition to its activity on p300, a structural analogue 6-pentadecyl salicylic acid was known to have inhibitory action on HAT in Tip60 and sensitize cancer cells (Balasubramanyam et al., 2004; Sun et al., 2006). It has also been observed that in pituitary adenoma cells, anacardic acid shows antiproliferative and cytotoxic effects that are associated with an increase expression of PARP, sub-G1-arrest, and apoptosis (Sukumari-Ramesh et al., 2011). In A549 cells, anacardic acid regulates LPS-stimulated IL-8 gene by modulation of histone acetylation (Yasutake et al., 2013).

Allyl Compounds

Allyl compounds, such as garlic, onions, and other compounds of the allium family, possess a complex range of water- and fat-soluble organosulfur compounds that have been known to possess anticancer properties (Iciek et al., 2009; Mai et al., 2006). An organosulfur compound known as diallyl disulfide, present in garlic and other allium vegetables, has shown HDAC-inhibitory actions in different cellular and in vivo systems (van Poppel et al., 1999). In human cancer cells, allyl mercaptan (AM), S-allylcysteine, S-allylmercaptocysteine, diallyldisulfide, and allicin increased histone acetylation (H3/H4), implicating HDACs as possible targets (Druesne-Pecollo and Latino-Martel, 2011). In human colon cancer cells, treatment with AM (0.5–2 mM) resulted in growth arrest at G1 phase and inhibition of HDAC activity with histone H3 hyperacetylation that facilitated

Sp3 and Sp3 binding on the promoter region of p21WAF1. (Pandey et al., 2010). In vitro and in vivo models have suggested that the growth inhibitory activity of diallyl disulfide on leukemia cell line involves the induction of p21WAF1 expression and acetylation of H3 and H4. It has been reported that in colonic epithelial cells and in prostate tissue, AM (20 μM) has been associated with HDAC inhibition.

Polyphenols

Natural polyphenols are present in many foods and beverages, and cancer prevention by these polyphenols has been investigated in light of modulation of histone modifications. The phytoestrogen genistein (at 50 μM) reactivated expression of silenced tumor-suppressor genes, including cylindromatosis (turban tumor syndrome, CYLD), PTEN, p53, and forkhead transcription factor 3a in LNCaP and PC3 prostate cancer cells (Nandakumar et al., 2011; Ye et al., 2001). In MCF-7 breast cancer cells, genistein (50–100 μM) inhibits the transcription of hTERT that involves increased H3K9 trimethylation and decreased H3K4 dimethylation (Li et al., 2009). In a recent report, it has been shown that in prostate cancer, genistein treatment results in an increased expression of HAT1 that may provide a mechanism for increased acetylation of H3K9 and induce the expression of SOX7, a Wnt inhibitory gene (Phillip et al., 2012). Curcuminoid, a polyphenol present in Indian spices, is known to inhibit the p300, HDAC1, HDAC3, and HDAC8 protein expression level (Kang et al., 2005). In multiple cancers, curcumin at a concentration of 20 μM or higher inhibits the activity of HAT by inducing proteasome-dependent degradation of p300 (Hatfield et al., 2009). Another study proposed that in medulloblastoma cells, curcumin induces apoptosis accompanied by reduced expression of HDAC4 with increased acetylation of tubulin (Carroll et al., 2011). EGCG, the major and most abundant polyphenolic catechin found in green tea, was also known to act as a histone modifier. In prostate, skin, and breast cancer cells, EGCG inhibited HDAC activity and increased histone acetylation (Batra et al., 2010; Khan and Mukhtar, 2008; Lee et al., 2012). When skin cancer cells are treated with EGCG, it reduces the expression level of BMI-1 and EZH2 proteins that lead to the decreased expression of histone H3K27me3 and reduced cell survival (Stefanska et al., 2012). It has also been identified that as an HAT inhibitor, EGCG suppressed the acetylation status of transcription factor p65 (ReIA) and thus inhibited NF-κB, interleukin-6, and downstream target genes.

Selenium

Selenium, a crucial microelement found in soil in inorganic forms, possesses anticarcinogenic properties that have been attributed to selenoproteins and organoselenium metabolites (Clark et al., 1996; de Miranda et al., 2014; Hatfield et al., 2014; Labbe et al., 2015). In different prostate cancer cells (LNCaP and PC3), metabolites of selenium are reported to induce H3 hyperacetylation (at a concentration of 10 and 50 μM) and inhibit activity of HDAC in nuclear fractions. Such amounts of selenium in the diet would be toxic and hence can not be recommended.

CONCLUSIONS

The incidence of colon, lung, breast, and prostate cancers is most prevalent in the Western world, such as Australia, New Zealand, USA, and Europe, whereas the lowest rates of mortality and morbidity are found in India, China, and Japan for these cancers. The cause of lower incidences of mortality and morbidity may be related to lifestyle and dietary factors that have shown preventive and curative potential in various types of cancers through different molecular mechanisms. Because of their low toxicity (both cytotoxicity and genotoxicity), dietary bioactive molecules are considered important chemopreventive agents. Some dietary phytochemicals and their epigenetic modes of action are described in this chapter; few phytochemicals are known to target two different epigenetic machineries (DNA methylation and histone modification), and many remain to be explored. Thus, upregulation of tumor-suppressor genes or inactivation of oncogenes by phytochemicals of diets may be effective for easier treatment of cancer. We also propose that molecular epigenetic events that are guided by reversible enzymatic actions, such as DNA methyltransferases/demethylases, HMTs/HDMs, and HATs/HDACs may be given better attention for dietary component-mediated therapeutic intervention of cancers.

Acknowledgments

SK, DS and MD are thankful to the National Institute of Technology, Rourkela for providing them fellowship. NP and SP are thankful to DST-INSPIRE scheme for granting fellowship to them. LD is a national postdoctoral fellow (DST-SERB, Grant No.: NPDF/2015/000757). We are extremely grateful to Professor Moshe Szyf, Department of Pharmacology and Therapeutics, McGill University, Montreal, Quebec, Canada, for reviewing this manuscript with corrections.

References

Aasheim, E.T., Sharp, S.J., Appleby, P.N., Shipley, M.J., Lentjes, M.A., Khaw, K.T., Brunner, E., Key, T.J., Wareham, N.J., 2015. Tinned fruit consumption and mortality in three prospective cohorts. PLoS One 10, e0117796.

Adjakly, M., Bosviel, R., Rabiau, N., Boiteux, J.P., Bignon, Y.J., Guy, L., Bernard-Gallon, D., 2011. DNA methylation and soy phytoestrogens: quantitative study in DU-145 and PC-3 human prostate cancer cell lines. Epigenomics 3, 795–803.

Aldawsari, F.S., Aguayo-Ortiz, R., Kapilashrami, K., Yoo, J., Luo, M., Medina-Franco, J.L., Velazquez-Martinez, C.A., 2016. Resveratrol-salicylate derivatives as selective DNMT3 inhibitors and anticancer agents. J. Enzym. Inhib. Med. Chem. 31, 695–703.

Anand, P., Kunnumakkara, A.B., Sundaram, C., Harikumar, K.B., Tharakan, S.T., Lai, O.S., Sung, B., Aggarwal, B.B., 2008. Cancer is a preventable disease that requires major lifestyle changes. Pharm. Res. 25, 2097–2116.

Arany, I., Whitehead, W.E., Ember, I.A., Tyring, S.K., 2003. Dose-dependent activation of p21WAF1 transcription by all-trans-acid in cervical squamous carcinoma cells. Anticancer Res. 23, 495–497.

Asthana, J., Kapoor, S., Mohan, R., Panda, D., 2013. Inhibition of HDAC6 deacetylase activity increases its binding with microtubules and suppresses microtubule dynamic instability in MCF-7 cells. J. Biol. Chem. 288, 22516–22526.

Balasubramanyam, K., Swaminathan, V., Ranganathan, A., Kundu, T.K., 2003. Small molecule modulators of histone acetyltransferase p300. J. Biol. Chem. 278, 19134–19140.

Balasubramanyam, K., Varier, R.A., Altaf, M., Swaminathan, V., Siddappa, N.B., Ranga, U., Kundu, T.K., 2004. Curcumin, a novel p300/CREB-binding protein-specific inhibitor of acetyltransferase, represses the acetylation of histone/nonhistone proteins and histone acetyltransferase-dependent chromatin transcription. J. Biol. Chem. 279, 51163–51171.

Batra, S., Sahu, R.P., Kandala, P.K., Srivastava, S.K., 2010. Benzyl isothiocyanate-mediated inhibition of histone deacetylase leads to NF-kappaB turnoff in human pancreatic carcinoma cells. Mol. Cancer Ther. 9, 1596–1608.

Baylin, S.B., Ohm, J.E., 2006. Epigenetic gene silencing in cancer - a mechanism for early oncogenic pathway addiction? Nat. Rev. Cancer 6, 107–116.

Beishline, K., Azizkhan-Clifford, J., 2015. Sp1 and the 'hallmarks of cancer'. FEBS J. 282, 224–258.

Benkoussa, M., Brand, C., Delmotte, M.H., Formstecher, P., Lefebvre, P., 2002. Retinoic acid receptors inhibit AP1 activation by regulating extracellular signal-regulated kinase and CBP recruitment to an AP1-responsive promoter. Mol. Cell. Biol. 22, 4522–4534.

Bhattacharya, S.K., Ramchandani, S., Cervoni, N., Szyf, M., 1999. A mammalian protein with specific demethylase activity for mCpG DNA. Nature 397, 579–583.

Bingham, S., Riboli, E., 2004. Diet and cancer—the European prospective investigation into cancer and nutrition. Nat. Rev. Cancer 4, 206–215.

Bird, A., 2002. DNA methylation patterns and epigenetic memory. Genes Dev. 16, 6–21.

Bishop, K.S., Ferguson, L.R., 2015. The interaction between epigenetics, nutrition and the development of cancer. Nutrients 7, 922–947.

Bistulfi, G., Vandette, E., Matsui, S., Smiraglia, D.J., 2010. Mild folate deficiency induces genetic and epigenetic instability and phenotype changes in prostate cancer cells. BMC Biol. 8, 6.

Boffetta, P., Couto, E., Wichmann, J., Ferrari, P., Trichopoulos, D., Bueno-de-Mesquita, H.B., van Duijnhoven, F.J., Buchner, F.L., Key, T., Boeing, H., Nothlings, U., Linseisen, J., Gonzalez, C.A., Overvad, K., Nielsen, M.R., Tjonneland, A., Olsen, A., Clavel-Chapelon, F., Boutron-Ruault, M.C., Morois, S.,

Lagiou, P., Naska, A., Benetou, V., Kaaks, R., Rohrmann, S., Panico, S., Sieri, S., Vineis, P., Palli, D., van Gils, C.H., Peeters, P.H., Lund, E., Brustad, M., Engeset, D., Huerta, J.M., Rodriguez, L., Sanchez, M.J., Dorronsoro, M., Barricarte, A., Hallmans, G., Johansson, I., Manjer, J., Sonestedt, E., Allen, N.E., Bingham, S., Khaw, K.T., Slimani, N., Jenab, M., Mouw, T., Norat, T., Riboli, E., Trichopoulou, A., 2010. Fruit and vegetable intake and overall cancer risk in the European prospective investigation into cancer and nutrition (EPIC). J. Natl. Cancer Inst. 102, 529–537.

Bowers, J.L., Tyulmenkov, V.V., Jernigan, S.C., Klinge, C.M., 2000. Resveratrol acts as a mixed agonist/antagonist for estrogen receptors alpha and beta. Endocrinology 141, 3657–3667.

Burger, L., Gaidatzis, D., Schubeler, D., Stadler, M.B., 2013. Identification of active regulatory regions from DNA methylation data. Nucleic Acids Res. 41, e155.

Butler, J.S., Koutelou, E., Schibler, A.C., Dent, S.Y., 2012. Histone-modifying enzymes: regulators of developmental decisions and drivers of human disease. Epigenomics 4, 163–177.

Carroll, R.E., Benya, R.V., Turgeon, D.K., Vareed, S., Neuman, M., Rodriguez, L., Kakarala, M., Carpenter, P.M., McLaren, C., Meyskens Jr., F.L., Brenner, D.E., 2011. Phase IIa clinical trial of curcumin for the prevention of colorectal neoplasia. Cancer Prev. Res. 4, 354–364.

Casanova, F., Quarti, J., da Costa, D.C., Ramos, C.A., da Silva, J.L., Fialho, E., 2012. Resveratrol chemosensitizes breast cancer cells to melphalan by cell cycle arrest. J. Cell. Biochem. 113, 2586–2596.

Cavalli, G., 2006. Chromatin and epigenetics in development: blending cellular memory with cell fate plasticity. Development (Camb. Engl.) 133, 2089–2094.

Cedar, H., Bergman, Y., 2009. Linking DNA methylation and histone modification: patterns and paradigms. Nat. Rev. Genet. 10, 295–304.

Celik, H., Mallaney, C., Kothari, A., Ostrander, E.L., Eultgen, E., Martens, A., Miller, C.A., Hundal, J., Klco, J.M., Challen, G.A., 2015. Enforced differentiation of Dnmt3a-null bone marrow leads to failure with c-Kit mutations driving leukemic transformation. Blood 125, 619–628.

Challen, G.A., Sun, D., Jeong, M., Luo, M., Jelinek, J., Berg, J.S., Bock, C., Vasanthakumar, A., Gu, H., Xi, Y., Liang, S., Lu, Y., Darlington, G.J., Meissner, A., Issa, J.P., Godley, L.A., Li, W., Goodell, M.A., 2012. Dnmt3a is essential for hematopoietic stem cell differentiation. Nat. Genet. 44, 23–31.

Chandar, N., Lombardi, B., Locker, J., 1989. c-myc gene amplification during hepatocarcinogenesis by a choline-devoid diet. Proc. Natl. Acad. Sci. U.S.A. 86, 2703–2707.

Chaturvedi, M.M., Sung, B., Yadav, V.R., Kannappan, R., Aggarwal, B.B., 2011. NF-kappaB addiction and its role in cancer: 'one size does not fit all'. Oncogene 30, 1615–1630.

Chen, Y., Zaman, M.S., Deng, G., Majid, S., Saini, S., Liu, J., Tanaka, Y., Dahiya, R., 2011. MicroRNAs 221/222 and genistein-mediated regulation of ARHI tumor suppressor gene in prostate cancer. Cancer Prev. Res. 4, 76–86.

Cheng, M., Olivier, P., Diehl, J.A., Fero, M., Roussel, M.F., Roberts, J.M., Sherr, C.J., 1999. The p21(Cip1) and p27(Kip1) CDK 'inhibitors' are essential activators of cyclin D-dependent kinases in murine fibroblasts. EMBO J. 18, 1571–1583.

Chi, P., Allis, C.D., et al., 2010. Covalent histone modifications—miswritten, misinterpreted and mis-erased in human cancers. Nature reviews. Cancer 10 (7), 457–469.

Chiyomaru, T., Yamamura, S., Fukuhara, S., Yoshino, H., Kinoshita, T., Majid, S., Saini, S., Chang, I., Tanaka, Y., Enokida, H., Seki, N., Nakagawa, M., Dahiya, R., 2013. Genistein inhibits prostate cancer cell growth by targeting miR-34a and oncogenic HOTAIR. PLoS One 8, e70372.

Chorley, B.N., Campbell, M.R., Wang, X., Karaca, M., Sambandan, D., Bangura, F., Xue, P., Pi, J., Kleeberger, S.R., Bell, D.A., 2012. Identification of novel NRF2-regulated genes by ChIP-Seq: influence on retinoid X receptor alpha. Nucleic Acids Res. 40, 7416–7429.

Choudhuri, S., Cui, Y., et al., 2010. Molecular targets of epigenetic regulation and effectors of environmental influences. Toxicology and applied pharmacology 245 (3), 378–393.

Clark, L.C., Combs Jr., G.F., Turnbull, B.W., Slate, E.H., Chalker, D.K., Chow, J., Davis, L.S., Glover, R.A., Graham, G.F., Gross, E.G., Krongrad, A., Lesher Jr., J.L., Park, H.K., Sanders Jr., B.B., Smith, C.L., Taylor, J.R., 1996. Effects of selenium supplementation for cancer prevention in patients with carcinoma of the skin. A randomized controlled trial. Nutritional Prevention of Cancer Study Group. JAMA 276, 1957–1963.

Clarke, J.D., Dashwood, R.H., Ho, E., 2008. Multi-targeted prevention of cancer by sulforaphane. Cancer Lett. 269, 291–304.

Costello, L.C., Franklin, R.B., 2007. Re: Silvano Gallus, Roberto Foschi, Eva Negri et al. Dietary zinc and prostate cancer risk: a case-control study from Italy, 2007. Eur. Urol.52, 1052–1057 Eur. Urol. 52, 1262–1263. author reply 1263-1264.

Daroqui, M.C., Augenlicht, L.H., 2010. Transcriptional attenuation in colon carcinoma cells in response to butyrate. Cancer Prev. Res. 3, 1292–1302.

Davie, J.R., 2003. Inhibition of histone deacetylase activity by butyrate. J. Nutr. 133, 2485S–2493S.

de Miranda, J.X., Andrade Fde, O., Conti, A., Dagli, M.L., Moreno, F.S., Ong, T.P., 2014. Effects of selenium compounds on proliferation and epigenetic marks of breast cancer cells. J. Trace Elem. Med. Biol. 28, 486–491.

Degos, L., Dombret, H., Chomienne, C., Daniel, M.T., Miclea, J.M., Chastang, C., Castaigne, S., Fenaux, P., 1995. All-trans-retinoic acid as a differentiating agent in the treatment of acute promyelocytic leukemia. Blood 85, 2643–2653.

Diaz-Chavez, J., Fonseca-Sanchez, M.A., Arechaga-Ocampo, E., Flores-Perez, A., Palacios-Rodriguez, Y., Dominguez-Gomez, G., Marchat, L.A., Fuentes-Mera, L., Mendoza-Hernandez, G., Gariglio, P., Lopez-Camarillo, C., 2013. Proteomic profiling reveals that resveratrol inhibits HSP27 expression and sensitizes breast cancer cells to doxorubicin therapy. PLoS One 8, e64378.

Do, D.P., Pai, S.B., Rizvi, S.A., D'Souza, M.J., 2010. Development of sulforaphane-encapsulated microspheres for cancer epigenetic therapy. Int. J. Pharm. 386, 114–121.

Dolinoy, D.C., Weidman, J.R., Jirtle, R.L., 2007. Epigenetic gene regulation: linking early developmental environment to adult disease. Reprod. Toxicol. 23, 297–307.

Druesne-Pecollo, N., Latino-Martel, P., 2011. Modulation of histone acetylation by garlic sulfur compounds. Anti Cancer Agents Med. Chem. 11, 254–259.

Ellis, L., Atadja, P.W., Johnstone, R.W., 2009. Epigenetics in cancer: targeting chromatin modifications. Mol. Cancer Ther. 8, 1409–1420.

Esteller, M., 2007. Cancer epigenomics: DNA methylomes and histone-modification maps. Nat. Rev. Genet. 8, 286–298.

Esteller, M., 2008. Epigenetics in cancer. N. Engl. J. Med. 358, 1148–1159.

Fan, H., Zhang, R., Tesfaye, D., Tholen, E., Looft, C., Holker, M., Schellander, K., Cinar, M.U., 2012. Sulforaphane causes a major epigenetic repression of myostatin in porcine satellite cells. Epigenetics 7, 1379–1390.

Fillmore, C.M., Xu, C., Desai, P.T., Berry, J.M., Rowbotham, S.P., Lin, Y.-J., Zhang, H., Marquez, V.E., Hammerman, P.S., Wong, K.-K., Kim, C.F., 2015. EZH2 inhibition sensitizes BRG1 and EGFR mutant lung tumours to TopoII inhibitors. Nature 520 (7546), 239–242.

Fujii, T., Yokoyama, G., Takahashi, H., Namoto, R., Nakagawa, S., Toh, U., Kage, M., Shirouzu, K., Kuwano, M., 2008. Preclinical studies of molecular-targeting diagnostic and therapeutic strategies against breast cancer. Breast Cancer (Tokyo Jpn.) 15, 73–78.

Gan, C.P., Hamid, S., Hor, S.Y., Zain, R.B., Ismail, S.M., Wan Mustafa, W.M., Teo, S.H., Saunders, N., Cheong, S.C., 2012. Valproic acid: growth inhibition of head and neck cancer by induction of terminal differentiation and senescence. Head Neck 34, 344–353.

Garcia-Pineres, A.J., Lindenmeyer, M.T., Merfort, I., 2004. Role of cysteine residues of p65/NF-kappaB on the inhibition by the sesquiterpene lactone parthenolide and N-ethyl maleimide, and on its transactivating potential. Life Sci. 75, 841–856.

Geutjes, E.J., Bajpe, P.K., Bernards, R., 2012. Targeting the epigenome for treatment of cancer. Oncogene 31, 3827–3844.

Ghoshal, K., Li, X., Datta, J., Bai, S., Pogribny, I., Pogribny, M., Huang, Y., Young, D., Jacob, S.T., 2006. A folate- and methyl-deficient diet alters the expression of DNA methyltransferases and methyl CpG binding proteins involved in epigenetic gene silencing in livers of F344 rats. J. Nutr. 136, 1522–1527.

Gianni, M., Peviani, M., Bruck, N., Rambaldi, A., Borleri, G., Terao, M., Kurosaki, M., Paroni, G., Rochette-Egly, C., Garattini, E., 2012. p38alphaMAPK interacts with and inhibits RARalpha: suppression of the kinase enhances the therapeutic activity of retinoids in acute myeloid leukemia cells. Leukemia 26, 1850–1861.

Glastad, K.M., Hunt, B.G., Goodisman, M.A., 2015. DNA methylation and chromatin organization in insects: insights from the ant Camponotus floridanus. Genome biology and evolution 27 (4), 931–942.

Griffiths, K., Wilson, D.W., Singh, R.B., De Meester, F., 2014. Effect of dietary phytoestrogens on human growth regulation: imprinting in health & disease. Indian J. Med. Res. 140 (Suppl), S82–S90.

Gupta, K., Thakur, V.S., Bhaskaran, N., Nawab, A., Babcook, M.A., Jackson, M.W., Gupta, S., 2012. Green tea polyphenols induce p53-dependent and p53-independent apoptosis in prostate cancer cells through two distinct mechanisms. PLoS One 7, e52572.

Gupta, S.C., Kannappan, R., Reuter, S., Kim, J.H., Aggarwal, B.B., 2011. Chemosensitization of tumors by resveratrol. Ann. N.Y. Acad. Sci. 1215, 150–160.

Guzman, M.L., Rossi, R.M., Karnischky, L., Li, X., Peterson, D.R., Howard, D.S., Jordan, C.T., 2005. The sesquiterpene lactone parthenolide induces apoptosis of human acute myelogenous leukemia stem and progenitor cells. Blood 105, 4163–4169.

Haggarty, P., Hoad, G., Campbell, D.M., Horgan, G.W., Piyathilake, C., McNeill, G., 2013. Folate in pregnancy and imprinted gene and repeat element methylation in the offspring. Am. J. Clin. Nutr. 97, 94–99.

Hanahan, D., Weinberg, R.A., 2011. Hallmarks of cancer: the next generation. Cell 144, 646–674.

Hansen, K.H., Helin, K., 2009. Epigenetic inheritance through self-recruitment of the polycomb repressive complex 2. Epigenetics 4, 133–138.

Hatfield, D.L., Tsuji, P.A., Carlson, B.A., Gladyshev, V.N., 2014. Selenium and selenocysteine: roles in cancer, health, and development. Trends Biochem. Sci. 39, 112–120.

Hatfield, D.L., Yoo, M.H., Carlson, B.A., Gladyshev, V.N., 2009. Selenoproteins that function in cancer prevention and promotion. Biochim. Biophys. Acta 1790, 1541–1545.

Hathaway, N.A., Bell, O., Hodges, C., Miller, E.L., Neel, D.S., Crabtree, G.R., 2012. Dynamics and memory of heterochromatin in living cells. Cell 149, 1447–1460.

He, Y.F., Li, B.Z., Li, Z., Liu, P., Wang, Y., Tang, Q., Ding, J., Jia, Y., Chen, Z., Li, L., Sun, Y., Li, X., Dai, Q., Song, C.X., Zhang, K., He, C., Xu, G.L., 2011. Tet-mediated formation of 5-carboxylcytosine and its excision by TDG in mammalian DNA. Science (New York N.Y.) 333, 1303–1307.

Hebbes, T.R., Thorne, A.W., Crane-Robinson, C., 1988. A direct link between core histone acetylation and transcriptionally active chromatin. EMBO J. 7, 1395–1402.

Heightman, T.D., 2011. Therapeutic prospects for epigenetic modulation. Expert Opin. Ther. Targets 15, 729–740.

Henning, S.M., Wang, P., Abgaryan, N., Vicinanza, R., de Oliveira, D.M., Zhang, Y., Lee, R.P., Carpenter, C.L., Aronson, W.J., Heber, D., 2013. Phenolic acid concentrations in plasma and urine from men consuming green or black tea and potential chemopreventive properties for colon cancer. Mol. Nutr. Food Res. 57, 483–493.

Henning, S.M., Wang, P., Said, J., Magyar, C., Castor, B., Doan, N., Tosity, C., Moro, A., Gao, K., Li, L., Heber, D., 2012. Polyphenols in brewed green tea inhibit prostate tumor xenograft growth by localizing to the tumor and decreasing oxidative stress and angiogenesis. J. Nutr. Biochem. 23, 1537–1542.

Herceg, Z., 2007. Epigenetics and cancer: towards an evaluation of the impact of environmental and dietary factors. Mutagenesis 22, 91–103.

Herman, J.G., Baylin, S.B., 2003. Gene silencing in cancer in association with promoter hypermethylation. N. Engl. J. Med. 349, 2042–2054.

Ibrahim, A.E., Arends, M.J., Silva, A.L., Wyllie, A.H., Greger, L., Ito, Y., Vowler, S.L., Huang, T.H., Tavare, S., Murrell, A., Brenton, J.D., 2011. Sequential DNA methylation changes are associated with DNMT3B overexpression in colorectal neoplastic progression. Gut 60, 499–508.

Iciek, M., Kwiecien, I., Wlodek, L., 2009. Biological properties of garlic and garlic-derived organosulfur compounds. Environ. Mol. Mutagen. 50, 247–265.

Jha, A.K., Nikbakht, M., Parashar, G., Shrivastava, A., Capalash, N., Kaur, J., 2010. Reversal of hypermethylation and reactivation of the RARbeta2 gene by natural compounds in cervical cancer cell lines. Folia Biol. 56, 195–200.

Jiang, S., Ma, X., Huang, Y., Xu, Y., Zheng, R., Chiao, J.W., 2010. Reactivating aberrantly hypermethylated p15 gene in leukemic T cells by a phenylhexyl isothiocyanate mediated inter-active mechanism on DNA and chromatin. J. Hematol. Oncol. 3, 48.

Jones, P.A., Baylin, S.B., 2007. The epigenomics of cancer. Cell 128, 683–692.

Kang, J., Chen, J., Shi, Y., Jia, J., Zhang, Y., 2005. Curcumin-induced histone hypoacetylation: the role of reactive oxygen species. Biochem. Pharmacol. 69, 1205–1213.

Kar, S., Deb, M., Sengupta, D., Shilpi, A., Parbin, S., Torrisani, J., Pradhan, S., Patra, S., 2012. An insight into the various regulatory mechanisms modulating human DNA methyltransferase 1 stability and function. Epigenetics 7, 994–1007.

Kar, S., Sengupta, D., Deb, M., Shilpi, A., Parbin, S., Rath, S.K., Pradhan, N., Rakshit, M., Patra, S.K., 2014. Expression profiling of DNA methylation-mediated epigenetic gene-silencing factors in breast cancer. Clin Epigenetics 6, 1186–1868.

Khan, N., Mukhtar, H., 2008. Multitargeted therapy of cancer by green tea polyphenols. Canc. Lett. 269, 269–280.

Khor, T.O., Huang, Y., Wu, T.Y., Shu, L., Lee, J., Kong, A.N., 2011. Pharmacodynamics of curcumin as DNA hypomethylation agent in restoring the expression of Nrf2 via promoter CpGs demethylation. Biochem. Pharmacol. 82, 1073–1078.

Kim, J., Cho, Y.A., Kim, D.H., Lee, B.H., Hwang, D.Y., Jeong, J., Lee, H.J., Matsuo, K., Tajima, K., Ahn, Y.O., 2012. Dietary intake of folate and alcohol, MTHFR C677T polymorphism, and colorectal cancer risk in Korea. Am. J. Clin. Nutr. 95, 405–412.

Labbe, D.P., Zadra, G., Ebot, E.M., Mucci, L.A., Kantoff, P.W., Loda, M., Brown, M., 2015. Role of diet in prostate cancer: the epigenetic link. Oncogene 34 (36), 4683–4691.

Lea, M.A., Randolph, V.M., 2001. Induction of histone acetylation in rat liver and hepatoma by organosulfur compounds including diallyl disulfide. Anticancer Res. 21, 2841–2845.

Lee, H., Wang, H.W., Su, H.Y., Hao, N.J., 1994. The structure-activity relationships of flavonoids as inhibitors of cytochrome P-450 enzymes in rat liver microsomes and the mutagenicity of 2-amino-3-methyl-imidazo[4,5-f] quinoline. Mutagenesis 9, 101–106.

Lee, M.H., Han, D.W., Hyon, S.H., Park, J.C., 2011. Apoptosis of human fibrosarcoma HT-1080 cells by epigallocatechin-3-O-gallate via induction of p53 and caspases as well as suppression of Bcl-2 and phosphorylated nuclear factor-kappaB. Apoptosis 16, 75–85.

Lee, W.J., Zhu, B.T., 2006. Inhibition of DNA methylation by caffeic acid and chlorogenic acid, two common catechol-containing coffee polyphenols. Carcinogenesis 27, 269–277.

Lee, Y.H., Kwak, J., Choi, H.K., Choi, K.C., Kim, S., Lee, J., Jun, W., Park, H.J., Yoon, H.G., 2012. EGCG suppresses prostate cancer cell growth modulating acetylation of androgen receptor by anti-histone acetyltransferase activity. Int. J. Mol. Med. 30, 69–74.

Lengfelder, E., Hofmann, W.K., Nowak, D., 2013. Treatment of acute promyelocytic leukemia with arsenic trioxide: clinical results and open questions. Expert Rev. Anticancer Ther. 13, 1035–1043.

Li, Y., Kong, D., Ahmad, A., Bao, B., Dyson, G., Sarkar, F.H., 2012. Epigenetic deregulation of miR-29a and miR-1256 by isoflavone contributes to the inhibition of prostate cancer cell growth and invasion. Epigenetics 7, 940–949.

Li, Y., Liu, L., Andrews, L.G., Tollefsbol, T.O., 2009. Genistein depletes telomerase activity through cross-talk between genetic and epigenetic mechanisms. Int. J. Cancer 125, 286–296. J. Int. du Cancer.

Lienert, F., Wirbelauer, C., Som, I., Dean, A., Mohn, F., Schubeler, D., 2011. Identification of genetic elements that autonomously determine DNA methylation states. Nat. Genet. 43, 1091–1097.

Lin, H.J., Zuo, T., Chao, J.R., Peng, Z., Asamoto, L.K., Yamashita, S.S., Huang, T.H., 2009. Seed in soil, with an epigenetic view. Biochim. Biophys. Acta 1790, 920–924.

Liu, M., Iavarone, A., Freedman, L.P., 1996. Transcriptional activation of the human p21(WAF1/CIP1) gene by retinoic acid receptor. Correlation with retinoid induction of U937 cell differentiation. J. Biol. Chem. 271, 31723–31728.

Liu, Y.L., Yang, H.P., Gong, L., Tang, C.L., Wang, H.J., 2011. Hypomethylation effects of curcumin, demethoxycurcumin and bisdemethoxycurcumin on WIF-1 promoter in non-small cell lung cancer cell lines. Mol. Med. Rep. 4, 675–679.

Lubbert, M., Ruter, B.H., Claus, R., Schmoor, C., Schmid, M., Germing, U., Kuendgen, A., Rethwisch, V., Ganser, A., Platzbecker, U., Galm, O., Brugger, W., Heil, G., Hackanson, B., Deschler, B., Dohner, K., Hagemeijer, A., Wijermans, P.W., Dohner, H., 2012. A multicenter phase II trial of decitabine as first-line treatment for older patients with acute myeloid leukemia judged unfit for induction chemotherapy. Haematologica 97, 393–401.

Lugrin, J., Ciarlo, E., Santos, A., Grandmaison, G., dos Santos, I., Le Roy, D., Roger, T., 2013. The sirtuin inhibitor cambinol impairs MAPK signaling, inhibits inflammatory and innate immune responses and protects from septic shock. Biochim. Biophys. Acta 1833, 1498–1510.

Mai, A., Rotili, D., Tarantino, D., Ornaghi, P., Tosi, F., Vicidomini, C., Sbardella, G., Nebbioso, A., Miceli, M., Altucci, L., Filetici, P., 2006. Small-molecule inhibitors of histone acetyltransferase activity: identification and biological properties. J. Med. Chem. 49, 6897–6907.

Majid, S., Dar, A.A., Shahryari, V., Hirata, H., Ahmad, A., Saini, S., Tanaka, Y., Dahiya, A.V., Dahiya, R., 2010. Genistein reverses hypermethylation and induces active histone modifications in tumor suppressor gene B-Cell translocation gene 3 in prostate cancer. Cancer 116, 66–76.

Makela, S., Poutanen, M., Kostian, M.L., Lehtimaki, N., Strauss, L., Santti, R., Vihko, R., 1998. Inhibition of 17beta-hydroxysteroid oxidoreductase by flavonoids in breast and prostate cancer cells. Proceedings of the Society for Experimental Biology and Medicine. Soc. Exp. Biol. Med. (New York N.Y.) 217, 310–316.

Marinova, Z., Leng, Y., Leeds, P., Chuang, D.M., 2011. Histone deacetylase inhibition alters histone methylation associated with heat shock protein 70 promoter modifications in astrocytes and neurons. Neuropharmacology 60, 1109–1115.

Marx, V.M., Herbert, M.B., Keitz, B.K., Grubbs, R.H., 2013. Stereoselective access to Z and E macrocycles by ruthenium-catalyzed Z-selective ring-closing metathesis and ethenolysis. J. Am. Chem. Soc. 135, 94–97.

Meeran, S.M., Patel, S.N., Chan, T.H., Tollefsbol, T.O., 2011. A novel prodrug of epigallocatechin-3-gallate: differential epigenetic hTERT repression in human breast cancer cells. Cancer Prev. Res. 4, 1243–1254.

Mellen, M., Ayata, P., Dewell, S., Kriaucionis, S., Heintz, N., 2012. MeCP2 binds to 5hmC enriched within active genes and accessible chromatin in the nervous system. Cell 151, 1417–1430.

Meng, X., Sang, S., Zhu, N., Lu, H., Sheng, S., Lee, M.J., Ho, C.T., Yang, C.S., 2002. Identification and characterization of methylated and ring-fission metabolites of tea catechins formed in humans, mice, and rats. Chem. Res. Toxicol. 15, 1042–1050.

Messina, M., Hilakivi-Clarke, L., 2009. Early intake appears to be the key to the proposed protective effects of soy intake against breast cancer. Nutr. Cancer 61, 792–798.

Myzak, M.C., Dashwood, R.H., 2006. Histone deacetylases as targets for dietary cancer preventive agents: lessons learned with butyrate, diallyl disulfide, and sulforaphane. Curr. Drug Targets 7, 443–452.

Myzak, M.C., Karplus, P.A., Chung, F.L., Dashwood, R.H., 2004. A novel mechanism of chemoprotection by sulforaphane: inhibition of histone deacetylase. Cancer Res. 64, 5767–5774.

Nakano, K., Mizuno, T., Sowa, Y., Orita, T., Yoshino, T., Okuyama, Y., Fujita, T., Ohtani-Fujita, N., Matsukawa, Y., Tokino, T., Yamagishi, H., Oka, T., Nomura, H., Sakai, T., 1997. Butyrate activates the WAF1/Cip1 gene promoter through Sp1 sites in a p53-negative human colon cancer cell line. J. Biol. Chem. 272, 22199–22206.

Nandakumar, V., Vaid, M., Katiyar, S.K., 2011. (-)-Epigallocatechin-3-gallate reactivates silenced tumor suppressor genes, Cip1/p21 and p16INK4a, by reducing DNA methylation and increasing histones acetylation in human skin cancer cells. Carcinogenesis 32, 537–544.

Nian, H., Delage, B., Ho, E., Dashwood, R.H., 2009. Modulation of histone deacetylase activity by dietary isothiocyanates and allyl sulfides: studies with sulforaphane and garlic organosulfur compounds. Environ. Mol. Mutagen. 50, 213–221.

Nian, H., Delage, B., Pinto, J.T., Dashwood, R.H., 2008. Allyl mercaptan, a garlic-derived organosulfur compound, inhibits histone deacetylase and enhances Sp3 binding on the P21WAF1 promoter. Carcinogenesis 29, 1816–1824.

Nishihara, R., Wang, M., Qian, Z.R., Baba, Y., Yamauchi, M., Mima, K., Sukawa, Y., Kim, S.A., Inamura, K., Zhang, X., Wu, K., Giovannucci, E.L., Chan, A.T., Fuchs, C.S., Ogino, S., Schernhammer, E.S., 2014. Alcohol, one-carbon nutrient intake, and risk of colorectal cancer according to tumor methylation level of IGF2 differentially methylated region. Am. J. Clin. Nutr. 100, 1479–1488.

Okashita, N., Kumaki, Y., Ebi, K., Nishi, M., Okamoto, Y., Nakayama, M., Hashimoto, S., Nakamura, T., Sugasawa, K., Kojima, N., Takada, T., Okano, M., Seki, Y., 2014. PRDM14 promotes active DNA demethylation through the ten-eleven translocation (TET)-mediated base excision repair pathway in embryonic stem cells. Development (Camb. Engl.) 141, 269–280.

Olaharski, A.J., Rine, J., Marshall, B.L., Babiarz, J., Zhang, L., Verdin, E., Smith, M.T., 2005. The flavoring agent dihydrocoumarin reverses epigenetic silencing and inhibits sirtuin deacetylases. PLoS Genet. 1, e77.

Pan, A., Sun, Q., Bernstein, A.M., Schulze, M.B., Manson, J.E., Stampfer, M.J., Willett, W.C., Hu, F.B., 2012. Red meat consumption and mortality: results from 2 prospective cohort studies. Arch. Intern. Med. 172, 555–563.

Pandey, M., Shukla, S., Gupta, S., 2010. Promoter demethylation and chromatin remodeling by green tea polyphenols leads to re-expression of GSTP1 in human prostate cancer cells. Int. J. Cancer 126, 2520–2533. J. Int. du Cancer.

Parbin, S., Kar, S., Shilpi, A., Sengupta, D., Deb, M., Rath, S.K., Patra, S.K., 2014. Histone deacetylases: a saga of perturbed acetylation homeostasis in cancer. J. Histochem. Cytochem. 62, 11–33.

Pasini, D., Cloos, P.A., Walfridsson, J., Olsson, L., Bukowski, J.P., Johansen, J.V., Bak, M., Tommerup, N., Rappsilber, J., Helin, K., 2010. JARID2 regulates binding of the Polycomb repressive complex 2 to target genes in ES cells. Nature 464, 306–310.

Patra, S.K., Bettuzzi, S., 2007. Epigenetic DNA-methylation regulation of genes coding for lipid raft-associated components: a role for raft proteins in cell transformation and cancer progression (review). Oncol. Rep. 17, 1279–1290.

Patra, S.K., Patra, A., Dahiya, R., 2001. Histone deacetylase and DNA methyltransferase in human prostate cancer. Biochem. Biophys. Res. Commun. 287, 705–713.

Patra, S.K., Patra, A., Zhao, H., Dahiya, R., 2002. DNA methyltransferase and demethylase in human prostate cancer. Mol. Carcinog. 33, 163–171.

Patra, S.K., Patra, A., Zhao, H., Carroll, P., Dahiya, R., 2003. Methyl-CpG-DNA binding proteins in human prostate cancer: expression of CXXC sequence containing MBD1 and repression of MBD2 and MeCP2. Biochem Biophys Res Commun 302, 759–766.

Patra, S.K., Patra, A., Rizzi, F., Ghosh, T.C., Bettuzzi, S., 2008. Demethylation of (Cytosine-5-C-methyl) DNA and regulation of transcription in the epigenetic pathways of cancer development. Cancer Metastasis Rev 27, 315–334.

Patra, S.K., Deb, M., Patra, A., 2011. Molecular marks for epigenetic identification of developmental and cancer stem cells. Clin. Epigenet. 2, 27–53.

Phillip, C.J., Giardina, C.K., Bilir, B., Cutler, D.J., Lai, Y.H., Kucuk, O., Moreno, C.S., 2012. Genistein cooperates with the histone deacetylase inhibitor vorinostat to induce cell death in prostate cancer cells. BMC Cancer 12, 145.

Podhorecka, M., Macheta, A., 2013. Acute promyelocytic leukemia—modern approach to disease pathogenesis and differentiation treatment. Postepy higieny i medycyny doswiadczalnej 67, 1083–1089 (Online).

Pogribny, I.P., James, S.J., Jernigan, S., Pogribna, M., 2004. Genomic hypomethylation is specific for preneoplastic liver in folate/methyl deficient rats and does not occur in non-target tissues. Mutat. Res. 548, 53–59.

Pogribny, I.P., Ross, S.A., Wise, C., Pogribna, M., Jones, E.A., Tryndyak, V.P., James, S.J., Dragan, Y.P., Poirier, L.A., 2006. Irreversible global DNA hypomethylation as a key step in hepatocarcinogenesis induced by dietary methyl deficiency. Mutat. Res. 593, 80–87.

Pudenz, M., Roth, K., Gerhauser, C., 2014. Impact of soy isoflavones on the epigenome in cancer prevention. Nutrients 6, 4218–4272.

Quenneville, S., Verde, G., Corsinotti, A., Kapopoulou, A., Jakobsson, J., Offner, S., Baglivo, I., Pedone, P.V., Grimaldi, G., Riccio, A., Trono, D., 2011. In embryonic stem cells, ZFP57/KAP1 recognize a methylated hexanucleotide to affect chromatin and DNA methylation of imprinting control regions. Mol. Cell 44, 361–372.

Rabiau, N., Trraf, H.K., Adjakly, M., Bosviel, R., Guy, L., Fontana, L., Bignon, Y.J., Bernard-Gallon, D.J., 2011. miRNAs differentially expressed in prostate cancer cell lines after soy treatment. In Vivo (Athens Greece) 25, 917–921.

Rajendran, P., Kidane, A.I., Yu, T.W., Dashwood, W.M., Bisson, W.H., Lohr, C.V., Ho, E., Williams, D.E., Dashwood, R.H., 2013. HDAC turnover, CtIP acetylation and dysregulated DNA damage signaling in colon cancer cells treated with sulforaphane and related dietary isothiocyanates. Epigenetics 8, 612–623.

Renouf, M., Redeuil, K., Longet, K., Marmet, C., Dionisi, F., Kussmann, M., Williamson, G., Nagy, K., 2011. Plasma pharmacokinetics of catechin metabolite 4′-O-Me-EGC in healthy humans. Eur. J. Nutr. 50, 575–580.

Rohrmann, S., Overvad, K., Bueno-de-Mesquita, H.B., Jakobsen, M.U., Egeberg, R., Tjonneland, A., Nailler, L., Boutron-Ruault, M.C., Clavel-Chapelon, F., Krogh, V., Palli, D., Panico, S., Tumino, R., Ricceri, F., Bergmann, M.M., Boeing, H., Li, K., Kaaks, R., Khaw, K.T., Wareham, N.J., Crowe, F.L., Key, T.J., Naska, A., Trichopoulou, A., Trichopoulos, D., Leenders, M., Peeters, P.H., Engeset, D., Parr, C.L., Skeie, G., Jakszyn, P., Sanchez, M.J., Huerta, J.M., Redondo, M.L., Barricarte, A., Amiano, P., Drake, I., Sonestedt, E., Hallmans, G., Johansson, I., Fedirko, V., Romieux, I., Ferrari, P., Norat, T., Vergnaud, A.C., Riboli, E., Linseisen, J., 2013. Meat consumption and mortality—results from the European prospective investigation into cancer and nutrition. BMC Med. 11, 63.

Roovers, K., Assoian, R.K., 2000. Integrating the MAP kinase signal into the G1 phase cell cycle machinery. BioEssays 22, 818–826.

Ruotolo, R., Tosi, F., Vernarecci, S., Ballario, P., Mai, A., Filetici, P., Ottonello, S., 2010. Chemogenomic profiling of the cellular effects associated with histone H3 acetylation impairment by a quinoline-derived compound. Genomics 96, 272–280.

Rush, E.C., Katre, P., Yajnik, C.S., 2014. Vitamin B12: one carbon metabolism, fetal growth and programming for chronic disease. Eur. J. Clin. Nutr. 68, 2–7.

Ruter, B., Wijermans, P.W., Lubbert, M., 2004. DNA methylation as a therapeutic target in hematologic disorders: recent results in older patients with myelodysplasia and acute myeloid leukemia. Int. J. Hematol. 80, 128–135.

Ruzov, A., Tsenkina, Y., Serio, A., Dudnakova, T., Fletcher, J., Bai, Y., Chebotareva, T., Pells, S., Hannoun, Z., Sullivan, G., Chandran, S., Hay, D.C., Bradley, M., Wilmut, I., De Sousa, P., 2011. Lineage-specific distribution of high levels of genomic 5-hydroxymethylcytosine in mammalian development. Cell Res. 21, 1332–1342.

Saadeh, H., Schulz, R., 2014. Protection of CpG islands against de novo DNA methylation during oogenesis is associated with the recognition site of E2f1 and E2f2. Epigenet. Chromatin 7, 26.

Sankpal, U.T., Goodison, S., Abdelrahim, M., Basha, R., 2011. Targeting Sp1 transcription factors in prostate cancer therapy. Med. Chem. (Shariqah (U.A.E.)) 7, 518–525.

Scott, J.M., Weir, D.G., 1998. Folic acid, homocysteine and one-carbon metabolism: a review of the essential biochemistry. J. Cardiovasc. Risk 5, 223–227.

Selenica, M.L., Benner, L., Housley, S.B., Manchec, B., Lee, D.C., Nash, K.R., Kalin, J., Bergman, J.A., Kozikowski, A., Gordon, M.N., Morgan, D., 2014. Histone deacetylase 6 inhibition improves memory and reduces total tau levels in a mouse model of tau deposition. Alzheimers Res. Ther. 6, 12.

Shu, L., Khor, T.O., Lee, J.H., Boyanapalli, S.S., Huang, Y., Wu, T.Y., Saw, C.L., Cheung, K.L., Kong, A.N., 2011. Epigenetic CpG demethylation of the promoter and reactivation of the expression of Neurog1 by curcumin in prostate LNCaP cells. AAPS J. 13, 606–614.

Singh, S.V., Herman-Antosiewicz, A., Singh, A.V., Lew, K.L., Srivastava, S.K., Kamath, R., Brown, K.D., Zhang, L., Baskaran, R., 2004. Sulforaphane-induced G2/M phase cell cycle arrest involves checkpoint kinase 2-mediated phosphorylation of cell division cycle 25C. J. Biol. Chem. 279, 25813–25822.

Smallwood, S.A., Tomizawa, S., Krueger, F., Ruf, N., Carli, N., Segonds-Pichon, A., Sato, S., Hata, K., Andrews, S.R., Kelsey, G., 2011. Dynamic CpG island methylation landscape in oocytes and preimplantation embryos. Nat. Genet. 43, 811–814.

Song, J., Teplova, M., Ishibe-Murakami, S., Patel, D.J., 2012. Structure-based mechanistic insights into DNMT1-mediated maintenance DNA methylation. Science (New York N.Y.) 335, 709–712.

Sowa, Y., Orita, T., Minamikawa-Hiranabe, S., Mizuno, T., Nomura, H., Sakai, T., 1999. Sp3, but not Sp1, mediates the transcriptional activation of the p21/WAF1/Cip1 gene promoter by histone deacetylase inhibitor. Cancer Res. 59, 4266–4270.

Stadler, M.B., Murr, R., Burger, L., Ivanek, R., Lienert, F., Scholer, A., van Nimwegen, E., Wirbelauer, C., Oakeley, E.J., Gaidatzis, D., Tiwari, V.K., Schubeler, D., 2011. DNA-binding factors shape the mouse methylome at distal regulatory regions. Nature 480, 490–495.

Stefanska, B., Karlic, H., Varga, F., Fabianowska-Majewska, K., Haslberger, A., 2012. Epigenetic mechanisms in anti-cancer actions of bioactive food components—the implications in cancer prevention. Br. J. Pharmacol. 167, 279–297.

Stefanska, B., Rudnicka, K., Bednarek, A., Fabianowska-Majewska, K., 2010. Hypomethylation and induction of retinoic acid receptor beta 2 by concurrent action of adenosine analogues and natural compounds in breast cancer cells. Eur. J. Pharmacol. 638, 47–53.

Sukumari-Ramesh, S., Singh, N., Jensen, M.A., Dhandapani, K.M., Vender, J.R., 2011. Anacardic acid induces caspase-independent apoptosis and radiosensitizes pituitary adenoma cells. J. Neurosurg. 114, 1681–1690.

Sun, Y., Jiang, X., Chen, S., Price, B.D., 2006. Inhibition of histone acetyltransferase activity by anacardic acid sensitizes tumor cells to ionizing radiation. FEBS Lett. 580, 4353–4356.

Tahiliani, M., Koh, K.P., Shen, Y., Pastor, W.A., Bandukwala, H., Brudno, Y., Agarwal, S., Iyer, L.M., Liu, D.R., Aravind, L., Rao, A., 2009. Conversion of

5-methylcytosine to 5-hydroxymethylcytosine in mammalian DNA by MLL partner TET1. Science (New York N.Y.) 324, 930–935.

Ulrich, C.M., Kampman, E., Bigler, J., Schwartz, S.M., Chen, C., Bostick, R., Fosdick, L., Beresford, S.A., Yasui, Y., Potter, J.D., 2000. Lack of association between the C677T MTHFR polymorphism and colorectal hyperplastic polyps. Cancer Epidemiol. Biomark. Prev. 9, 427–433 a publication of the American Association for Cancer Research, cosponsored by the American Society of Preventive Oncology.

van Poppel, G., Verhoeven, D.T., Verhagen, H., Goldbohm, R.A., 1999. Brassica vegetables and cancer prevention. Epidemiology and mechanisms. Adv. Exp. Med. Biol. 472, 159–168.

Vermeulen, L., De Sousa, E.M.F., van der Heijden, M., Cameron, K., de Jong, J.H., Borovski, T., Tuynman, J.B., Todaro, M., Merz, C., Rodermond, H., Sprick, M.R., Kemper, K., Richel, D.J., Stassi, G., Medema, J.P., 2010. Wnt activity defines colon cancer stem cells and is regulated by the microenvironment. Nat. Cell Biol. 12, 468–476.

Vinod, B.S., Maliekal, T.T., Anto, R.J., 2013. Phytochemicals as chemosensitizers: from molecular mechanism to clinical significance. Antioxidants Redox Signal. 18, 1307–1348.

Wallwork, J.C., Duerre, J.A., 1985. Effect of zinc deficiency on methionine metabolism, methylation reactions and protein synthesis in isolated perfused rat liver. J. Nutr. 115, 252–262.

Wang, L.G., Liu, X.M., Fang, Y., Dai, W., Chiao, F.B., Puccio, G.M., Feng, J., Liu, D., Chiao, J.W., 2008. De-repression of the p21 promoter in prostate cancer cells by an isothiocyanate via inhibition of HDACs and c-Myc. Int. J. Oncol. 33, 375–380.

Wang, P., Heber, D., Henning, S.M., 2012. Quercetin increased bioavailability and decreased methylation of green tea polyphenols in vitro and in vivo. Food Funct. 3, 635–642.

Wang, Q., Lee, D., Sysounthone, V., Chandraratna, R.A.S., Christakos, S., Korah, R., Wieder, R., 2001. 1,25-dihydroxyvitamin D3 and retonic acid analogues induce differentiation in breast cancer cells with function- and cell-specific additive effects. Breast Cancer Res. Treat. 67, 157–168.

Wang, Z., Nie, Z., Chen, W., Zhou, Z., Kong, Q., Seth, A.K., Liu, R., Chen, C., 2013. RNF115/BCA2 E3 ubiquitin ligase promotes breast cancer cell proliferation through targeting p21Waf1/Cip1 for ubiquitin-mediated degradation. Neoplasia (New York N.Y.) 15, 1028–1035.

Whitsett Jr., T.G., Lamartiniere, C.A., 2006. Genistein and resveratrol: mammary cancer chemoprevention and mechanisms of action in the rat. Expert Rev. Anticancer Ther. 6, 1699–1706.

Wolk, A., 2007. Editorial comment on: dietary zinc and prostate cancer risk: a case-control study from Italy. Eur. Urol. 52, 1056–1057.

Woynarowski, J.M., Konopa, J., 1981. Inhibition of DNA biosynthesis in HeLa cells by cytotoxic and antitumor sesquiterpene lactones. Mol. Pharmacol. 19, 97–102.

Yang, C.S., Fang, M., Lambert, J.D., Yan, P., Huang, T.H., 2008. Reversal of hypermethylation and reactivation of genes by dietary polyphenolic compounds. Nutr. Rev. 66 (Suppl. 1), S18–S20.

Yang, C.S., Wang, H., 2011. Mechanistic issues concerning cancer prevention by tea catechins. Mol. Nutr. Food Res. 55, 819–831.

Yasutake, T., Wada, H., Higaki, M., Nakamura, M., Honda, K., Watanabe, M., Ishii, H., Kamiya, S., Takizawa, H., Goto, H., 2013. Anacardic acid, a histone acetyltransferase inhibitor, modulates LPS-induced IL-8 expression in a human alveolar epithelial cell line A549. F1000Research 2, 78.

Ye, R., Bodero, A., Zhou, B.B., Khanna, K.K., Lavin, M.F., Lees-Miller, S.P., 2001. The plant isoflavenoid genistein activates p53 and Chk2 in an ATM-dependent manner. J. Biol. Chem. 276, 4828–4833.

Yoo, J., Kim, J.H., Robertson, K.D., Medina-Franco, J.L., 2012. Molecular modeling of inhibitors of human DNA methyltransferase with a crystal structure: discovery of a novel DNMT1 inhibitor. Adv. Protein Chem. Struct. Biol. 87, 219–247.

Yu, Z., Li, W., Lu, Q., Wang, L., Zhang, X., Han, P., Chen, P., Pei, Y., 2008. p21 is required for atRA-mediated growth inhibition of MEPM cells, which involves RAR. J. Cell. Biochem. 104, 2185–2192.

Zeng, H., Trujillo, O.N., Moyer, M.P., Botnen, J.H., 2011. Prolonged sulforaphane treatment activates survival signaling in nontumorigenic NCM460 colon cells but apoptotic signaling in tumorigenic HCT116 colon cells. Nutr. Cancer 63, 248–255.

Zhang, Y., Ng, H.H., Erdjument-Bromage, H., et al., 1999. Analysis of the NuRD subunits reveals a histone deacetylase core complex and a connection with DNA methylation. Genes & development 13 (15), 1924–1935.

Zhang, Y., Jurkowska, R., Soeroes, S., Rajavelu, A., Dhayalan, A., Bock, I., Rathert, P., Brandt, O., Reinhardt, R., Fischle, W., Jeltsch, A., 2010. Chromatin methylation activity of Dnmt3a and Dnmt3a/3L is guided by interaction of the ADD domain with the histone H3 tail. Nucleic Acids Res. 38, 4246–4253.

Zhang, C., Su, Z.Y., Khor, T.O., Shu, L., Kong, A.N., 2013. Sulforaphane enhances Nrf2 expression in prostate cancer TRAMP C1 cells through epigenetic regulation. Biochem. Pharmacol. 85, 1398–1404.

Zhenwei, J., Shuxin, G., Yongchun, Z., Xianhua, Z., 2015. Mechanisms of TET protein-mediated DNA demethylation and its role in the regulation of mouse development. Yi Chuan Hereditas 37, 34–40. Zhongguo yi chuan xue hui bian ji.

Zhu, B.T., Wu, K.Y., Wang, P., Cai, M.X., Conney, A.H., 2010. O-methylation of catechol estrogens by human placental catechol-o-methyltransferase: interindividual differences in sensitivity to heat inactivation and to inhibition by dietary polyphenols. Drug Metab. Dispos. 38, 1892–1899.

C H A P T E R

3

Natural Anticancer Agents: Modifying the Epigenome to Prevent and Treat Cancer

Kristina Andrijauskaite, Jay Morris, Michael J. Wargovich

Department of Molecular Medicine, UT Health San Antonio, San Antonio, TX, United States

Abstract

A growing body of research suggests that a diet rich in fruits and vegetables could not only reduce the risk of developing cancer, but also affect the treatment outcome. One common mechanism could involve epigenetic modulation. The precise mechanisms mediating epigenetics, however, are not well delineated. In this chapter, we provide a comprehensive overview of food-derived natural compounds with epigenetic activity such as alterations of DNA methylation, histone modifications, chromatin architecture, and small noncoding RNAs. Specifically, we discuss the chemopreventive mechanisms by which natural compounds alter the cancer epigenome and thereby reverse gene silencing. In addition, we present our perspective on natural epigenetic compounds as adjuvants in cancer prevention therapy. Finally, we conclude our chapter by proposing an epigenetic diet mainly designed for those subjects diagnosed with known cancers evolving from silenced tumor-suppressor genes.

Keywords

Cancer chemoprevention; Cancer treatment; Epigenetic diet; Epigenetic model systems; Epigenetic modifiers

INTRODUCTION: EPIGENETICS AND CANCER PREVENTION

Plants have long been used by humans as medicines. By some estimates, more than 60% of medicines in use have plant origins (Wachtel-Galor and Benzie, 2011). New to this inventory is a class of natural products that act to turn on or silence genes by epigenetic mechanisms. Research in recent years has identified epigenetic silencing of key regulatory genes as one hallmark of cancer initiation and progression (Corner and Cawley, 1976; Dong et al., 2016). Gene silencing can come about through a number of mechanisms, some well understood and others only now being queried as to mechanisms of action. At the most elementary level, it is in a tumor's best interest to disable genes whose products regulate drug metabolism, regulate proliferation, prime the immune system, induce programmed cell death, sustain growth signaling, and regulate tissue invasion and metastasis, among others, in order to survive

Epigenetics of Cancer Prevention
https://doi.org/10.1016/B978-0-12-812494-9.00003-2

(Bailón-Moscoso et al., 2014; Sonnenschein and Soto, 2013).

Cancers employ gene silencing by methylation of promoter sequences, histone modifications, micros and other noncoding RNAs, and remodeling of chromatin (Mei et al., 2016). On the surface this may seem to be another way that tumors subjugate normal cellular processes; unlike mutations, however, epigenetic silencing may be reversible, as it may well be possible to desilence regulatory genes to correct tumor behavior (Ahuja et al., 2016; Gołąbek et al., 2015).

Plant biodiversity has produced a myriad of natural products with the potential to prevent or treat cancer via epigenetic modulation (Gerhauser, 2013). The present review gives a comprehensive overview of the mechanisms by which tumor cell behavior could be realigned by reactivation of silenced regulatory genes, and it illustrates natural compounds that have been identified with a particular epigenetic event and in some cases illustrates their pluripotency in affecting multiple epigenetic pathways.

EPIGENETIC MECHANISMS FOR CANCER INTERVENTION

Epigenetic mechanisms play an instrumental role in the development and homeostasis of gene expression patterns. Furthermore, changes in the epigenetic landscape could significantly dysregulate the whole epigenome machinery. However, the reversible nature of epigenetic aberrations makes them attractive targets for identifying novel therapeutic interventions and investigating the utility of natural compounds to prevent and treat cancer.

Epigenetic changes, also called DNA modifications, refer to site-specific chemical alterations in DNA that regulate gene expression without altering the DNA sequence. In recent years, advances have been made in the understanding of epigenetic mechanisms, which include DNA methylation, histone modifications, chromatin architecture, and small noncoding RNAs.

Role of DNA Methylation

The most studied epigenetic change is DNA methylation. It occurs at the 5′ position of the cytosine ring within CpG dinucleotides via addition of the methyl group at the 5-carbon of the cytosine ring resulting in 5-methylcytosine. In normal cells, the CpG-rich regions remain usually unmethylated, allowing the normal transcription of genes (Rodríguez-Paredes and Esteller, 2011). By contrast, when the CpG islands in a promoter region are methylated, gene expression is suppressed. In cancer cells, this hypermethylation may lead to transcriptional silencing of tumor suppressor genes and therefore is considered to play a causal role in cancer development. In addition, hypomethylation is reportedly as prevalent as cancer-linked hypermethylation (Ehrlich, 2009), which results in chromosomal and genetic instability leading to further oncogenic events (Ellis et al., 2009). However, the mechanisms mediating hypomethylation are less understood. Both of these epigenetic changes are reported to be observed in neoplasias and early-stage tumors (Esteller, 2007). DNA methylation can suppress gene transcription through several mechanisms. One is by excluding binding proteins through the DNA-binding domains (Watt and Molloy, 1988). Another involves recruiting DNA-specialized proteins that bind to methylated CpG by forming repressor complexes with histone deacetylases (HDACs) and causing chromatin compaction (Feng and Zhang, 2001). DNA methylation is carried out by a family of enzymes called DNA methyltransferases (DNMTs). Three DNMTs (DNMT1, DNMT3a, and DNMT3b) are required for establishment and maintenance of DNA methylation patterns. Two additional enzymes (DNMT2 and DNMT3L) have more specialized but related functions. Specifically, DNMT1 is responsible for the maintenance of established patterns of DNA methylation, whereas DNMT3a and 3b affect de novo DNA methylation by preferably targeting unmethylated CpG sequences (Okano et al., 1999). DNA methylation plays a critical role in many cellular processes including chromosome inactivation, gene imprinting, genomic stability, and transcriptional regulation (Meier and Recillas-Targa, 2017). Abnormal patterns of DNA methylation are linked to poorer clinical outcomes in certain cancers (Bhalla, 2005; Herranz and Esteller, 2007).

Role of DNA Histone Modification

Another crucial epigenetic change involves histone modifications that modulate chromatin structure. Chromatin structure plays an important role in gene expression and is responsible for the storage of genetic information within the nucleus (Hauser and Jung, 2008). Chromatin can be either condensed ("heterochromatin"), leading to transcriptional repression, or open and accessible ("euchromatin"), resulting in active transcription (Hauser and Jung, 2008). Posttranslational modifications are carried out by the amino-terminal tails of histones and include acetylation, deacetylation, methylation, phosphorylation, ubiquitylation, and sumoylation. These modifications are catalyzed by histone-modifying enzymes such as histone methyltransferases (HMTs), histone demethylases (HDMs), histone acetyltransferases (HATs), and histone deacetylases (HDACs) (Yang and Seto, 2007). Histone acetylation is associated with an active state of the chromatin and is mediated by the opposing activities of HATs and HDACs (Ellis et al., 2009). Moreover, it has been reported that acetylation influences the translocation into nucleus of such transcription factors as NF-kb, STAT1-3 (Miceli et al., 2014). Hyperacetylation leads to the activation of the repressed genes, whereas hypoacetylation may result in silencing of gene transcription which in turn may promote carcinogenesis (Kondo, 2009). Histone deacetylation is the process where the role of HDACs is to oppose HAT activity by the removal of acetyl groups. Deregulation of HDAC activity has been associated with gene silencing and tumorigenesis (Ellis et al., 2009). Histone methylation involves transferring of the methyl groups on different lysine resides and is associated with either transcriptional activation or repression of gene expression (Jenuwein and Allis, 2001). This dual action depends on the residue (lysine or arginine), methylation status (mono, di or tri), and location (K4, K9, K27 in H3) (Kornberg and Lorch, 1999). Specifically, the methylation of H3K4, H3K4, H3K36 and H3K79 is known to activate gene transcription, whereas methylation of H3K9, H3K27, and H4K20 is associated with gene silencing or transcription repression (Mottet and Castronovo, 2008). Histone phosphorylation occurs on serine, threonine, and tyrosine resides and is mediated by protein kinases (PKs) and protein phosphatases (PPs). It plays a major role in DNA damage response and is associated with many other cellular responses such as transcriptional regulation, mitosis, cell cycle progression, and apoptosis (Cruickshank et al., 2010). Histone ubiquitination is a posttranslational modification in which the addition of ubiquitin from histones H2A and H2B is carried out by three enzymes (E1, E2, and E3). Ubiquitination of histones is associated with transcription initiation and elongation, silencing, and DNA repair (Weake and Workman, 2008). In addition, a small ubiquitin-like modifier, termed sumoylation, is essential for the maintenance of genomic integrity, the regulation of gene expression, and intracellular signaling (Seeler and Dejean, 2017). Taken together, aberrations in histone modifications broadly contribute to cancerous development (Leroy et al., 2013). In fact, it has been reported that genes encoding chromatin regulatory proteins are the most commonly mutated gene sets in cancer (Garraway and Lander, 2013). Therefore, understanding and targeting the cancer epigenome presents an opportunity for potential therapeutic intervention.

Role of Noncoding RNAs

Lastly, noncoding RNAs (ncRNAs) are considered important modulators of chromatin structure and gene expression. By definition, these RNA molecules do not encode for proteins but have important structural, catalytic, and regulatory functions (Ferdin et al., 2010). Based on transcript size, they can be divided into two major groups, small and long ncRNAs (Bartel and Chen, 2004; Harfe, 2005; Sana et al., 2012). Small noncoding RNAs are also called miRNAs and are involved in many biological processes such as development, differentiation, apoptosis, and proliferation (Calin and Croce, 2006). Extensive miRNA profiling reveals that their expression is completely different in tumors as compared with normal cells. Thus, miRNAs play an instrumental role in epigenetics because their phenotypic signatures greatly correlate with clinical and biological characteristics of tumors and more importantly, their response to therapy (Calin and Croce, 2006). In general, miRNAs are downregulated in cancer, but they can also be overexpressed depending on cell type and cellular differentiation status (Jansson and Lund, 2012).

EPIGENETICALLY SENSITIVE CANCERS AND TARGETS FOR INTERVENTIONS

The epigenetic landscape is altered in many cancer types as the transformation of cells from neoplastic to malignant development is accompanied by a misbalancing of the epigenetic orchestra, particularly increased DNA methylation in promoter regions and deacetylation of chromatin histones, resulting in epigenetic silencing of tumor-suppressor genes. Epigenetic changes affect the physiology of most cancers, but the most widely investigated cancers are prostate, breast, colon, ovarian, lung, and skin (Abbas and Gupta, 2008; Matkar et al., 2015). Because the silencing of tumor suppressor genes could be reactivated by DNMT and HDAC inhibitors, attempts have been made to use plant-derived compounds targeting epigenetic machinery (Table 3.1). Epigenetic alterations and their effect as biomarkers for cancer detection have been reported in a number of cancers. A well-studied example is hereditary nonpolyposis colon cancer (HNPCC), a form of colon cancer that shows early onset, predilection for right-sided etiology, and that includes extracolonic manifestations. HNPCC stems from epigenetic silencing of certain mismatch-repair genes, notably h*MLH*-1 and h*MSH*-2 among others. It has been reported that hypermethylation of these genes leads to their inactivation, resulting in a genome instability termed microsatellite instability. How this results in an early form of colon cancer is still under current investigation. Hypermethylation of the *GSTP1* gene is considered a way that the prostate becomes more vulnerable to the genomic damage that leads to cancer in that organ. CpG island methylation figures in the genesis of cervical cancer in as many as five genes. We will certainly discover that epigenetic alterations encompass a wider variety of human tumors (Fig. 3.1).

NATURAL PRODUCTS: SOURCE OF EPIGENETIC MODIFIERS

Natural products have been used as remedies for many diseases since ancient times. However, only during the past few decades have they taken on a new role as potential chemopreventive and chemotherapeutic agents to reverse cancer-related epigenetic aberrations. Cancer is a multistage process that may take years to develop before symptoms appear. Therefore, there is great interest in using natural compounds not only to treat cancer but also to

TABLE 3.1 Natural Compounds From Foods With Epigenetic Activity

Compound	Source	DNMT Inhibitor	HAT/HDAC Effects	Histone Marks	In Vitro	In Vivo	Structure	Authors
POLYPHENOLS								
Apigenin	Parsley	✓		↑ acetylation marks	✓		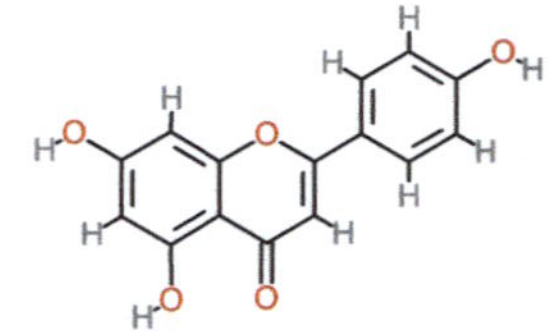	Kanwal et al. (2016) Tseng et al. (2017) Pandey et al. (2012) Fang et al. (2007) and Paredes-Gonzalez et al. (2014)
Curcumin	Turmeric	✓	↑ HAT ↓HDAC	↓ methylation marks	✓	✓	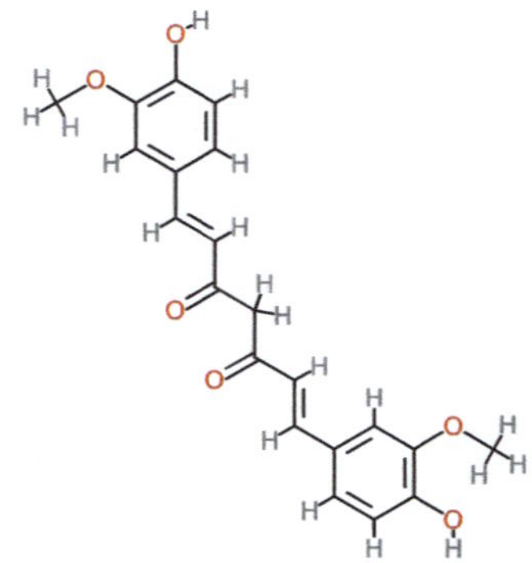	Jha et al. (2010) Lee et al. (2011)
Epigallocatechin-3-gallate (EGCG)	Tea	✓	↓ HDAC		✓	✓	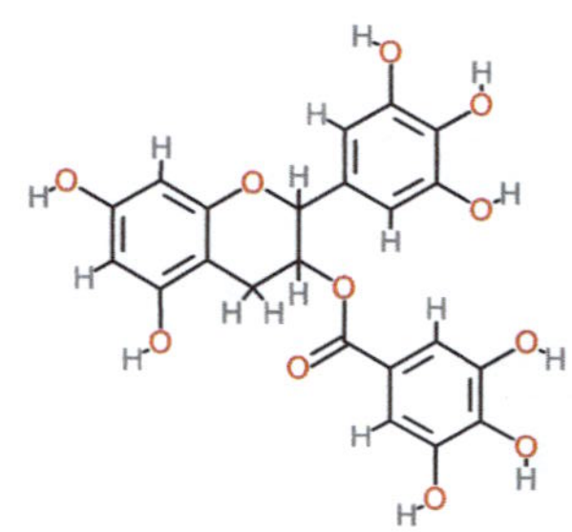	Khan et al. (2015) and Moseley et al. (2013) Nandakumar et al. (2011) Oya et al. (2017) Chang et al. (2015) and Pandey et al. (2010) Rajendran et al. (2011)

Continued

TABLE 3.1 Natural Compounds From Foods With Epigenetic Activity—cont'd

Compound	Source	DNMT Inhibitor	HAT/HDAC Effects	Histone Marks	In Vitro	In Vivo	Structure	Authors
Genistein	Soy	✓	↑ HAT ↓HDAC	↑ acetylation marks ↓ methylation marks	✓	✓	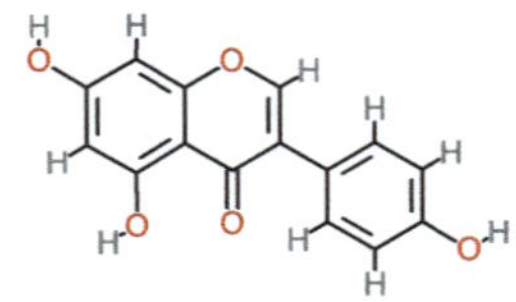	Vahid et al. (2015) Xie et al. (2014) Dagdemir et al. (2013)
Quercetin	Onions	✓	↑ HAT ↓HDAC	↑ acetylation marks	✓		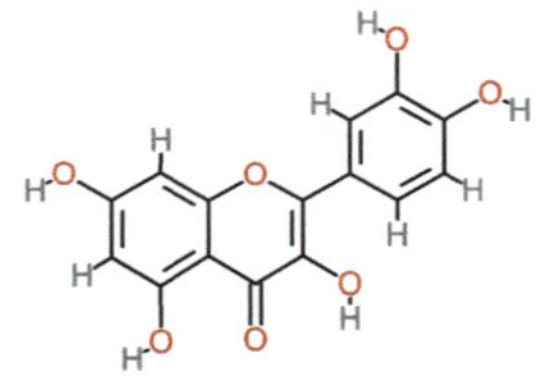	Sharma et al. (2016) Vargas et al. (2014)
Kaempferol	Kale, dill		↓HDAC		✓		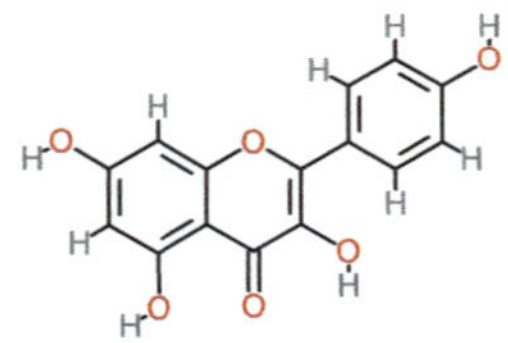	Berger et al. (2013)
Luteolin	Watercress	✓	↓HDAC		✓		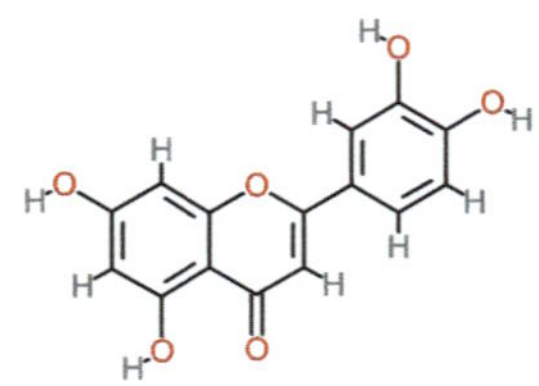	Attoub et al. (2011) and Klingstedt et al. (1989)

Resveratrol	Red fruits	✓	↓HDAC		✓	✓	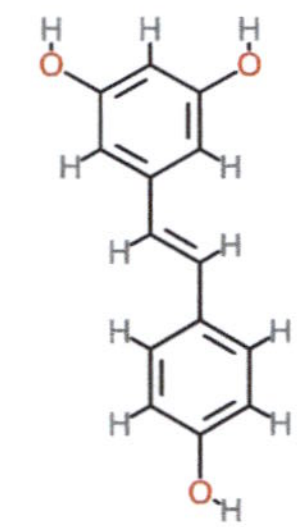	Qin et al. (2014) Kala and Tollefsbol (2016)
SULFUR-CONTAINING COMPOUNDS, ISOQUINOLINE ALKALOIDS, AND ISOTHIOCYANATES								
Diallyl disulfide (DADS)	Garlic		↓HDAC ↑ HAT	↑ acetylation marks	✓		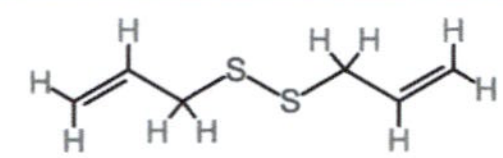	Druesne et al. (2004) and Myzak and Dashwood (2006) J. Huang et al. (2011)
3,3′-diindolylmethane (DIM)	Crucifers	✓	↓HDAC		✓		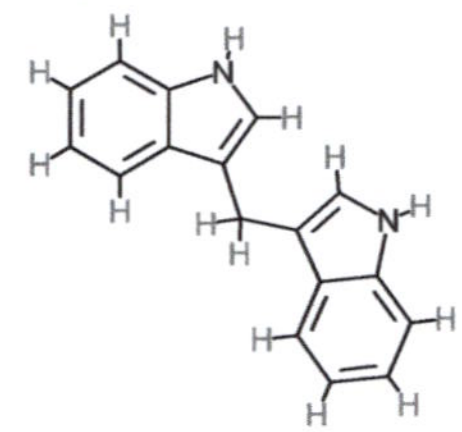	Fuentes et al. (2015) and Wu et al. (2013)
Berberine	Barberry	✓	↓HDAC		✓		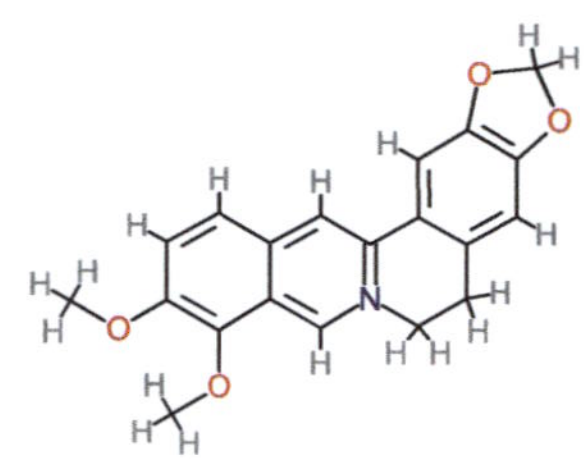	C. Huang et al. (2017) Kalaiarasi et al. (2016)
Sulphoraphane (SFN)	Crucifers	✓	↑ HAT ↓HDAC		✓		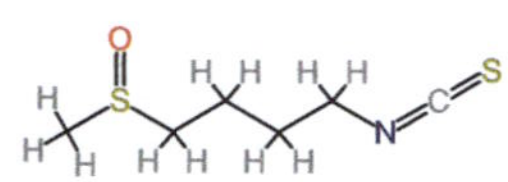	Fan et al. (2012) Jiang et al. (2016)

Continued

TABLE 3.1 Natural Compounds From Foods With Epigenetic Activity—cont'd

Compound	Source	DNMT Inhibitor	HAT/HDAC Effects	Histone Marks	In Vitro	In Vivo	Structure	Authors
OTHER COMPOUNDS OF NATURAL ORIGIN								
Psammplin A	Marine sponge	✓	↓HDAC		✓	✓	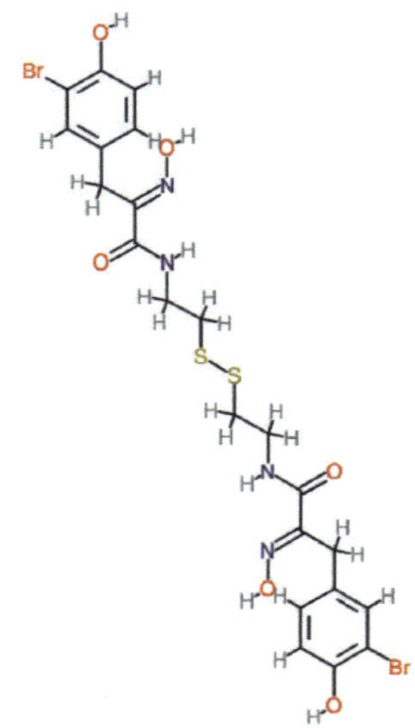	Pina et al. (2003) Kim et al. (2007)
Cyclostellamine	Marine sponge		↓HDAC		✓		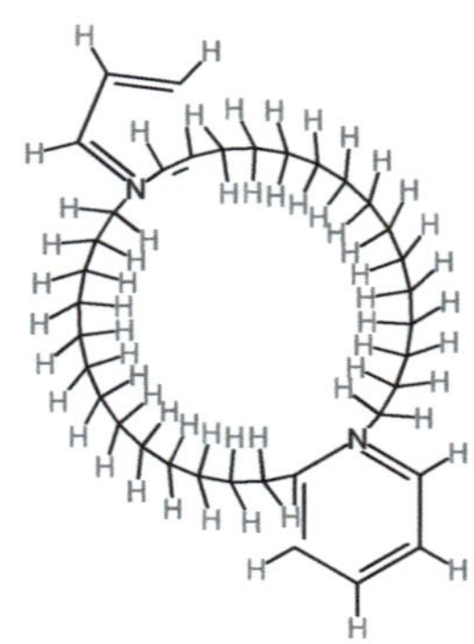	Oku et al. (2004)

Depudecin	Fungus	↓HDAC		✓	✓	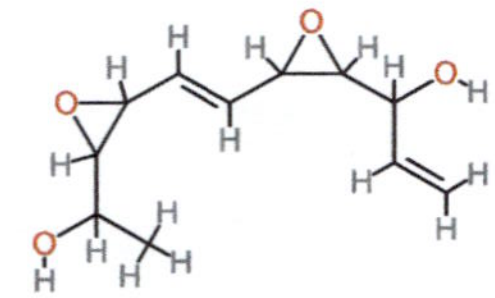	Kwon et al. (1998)
Dihydrocoumarin	Sweet clover	↓HDAC	↑ acetylation marks	✓	✓	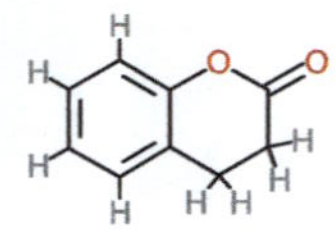	Olaharski et al. (2005)
Nicotinamide	Vitamin B3 metabolite	↓HDAC		✓		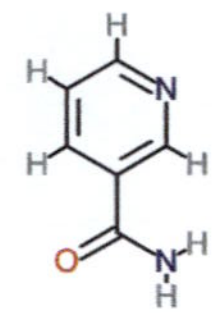	Zhang et al. (2011) Wang et al. (2013)
MCP30	Bitter melon seeds	↓HDAC	↑ acetylation marks	✓	✓	No structure available	Xiong et al. (2009)
Caffeine	Coffea arabica	↓HDAC	↑ acetylation marks	✓		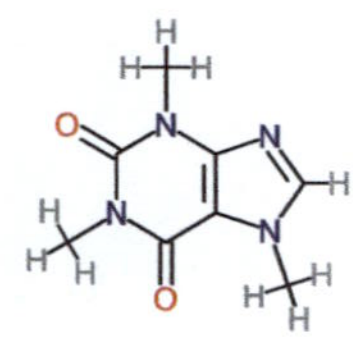	Mukwevho et al. (2008)
Withaferin A	Ashwagandha		↓methylation marks	✓	✓	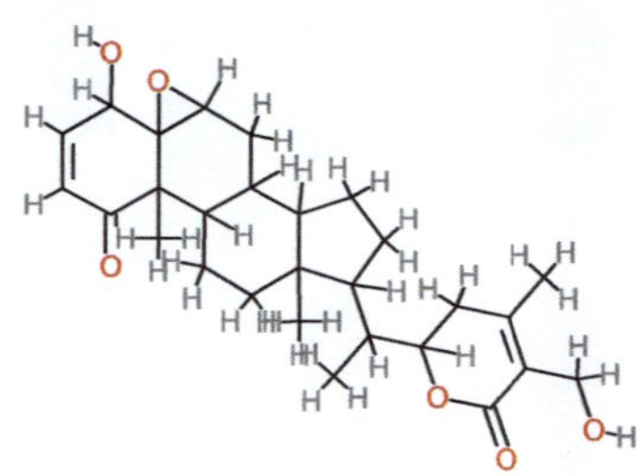	Szic et al. (2017)

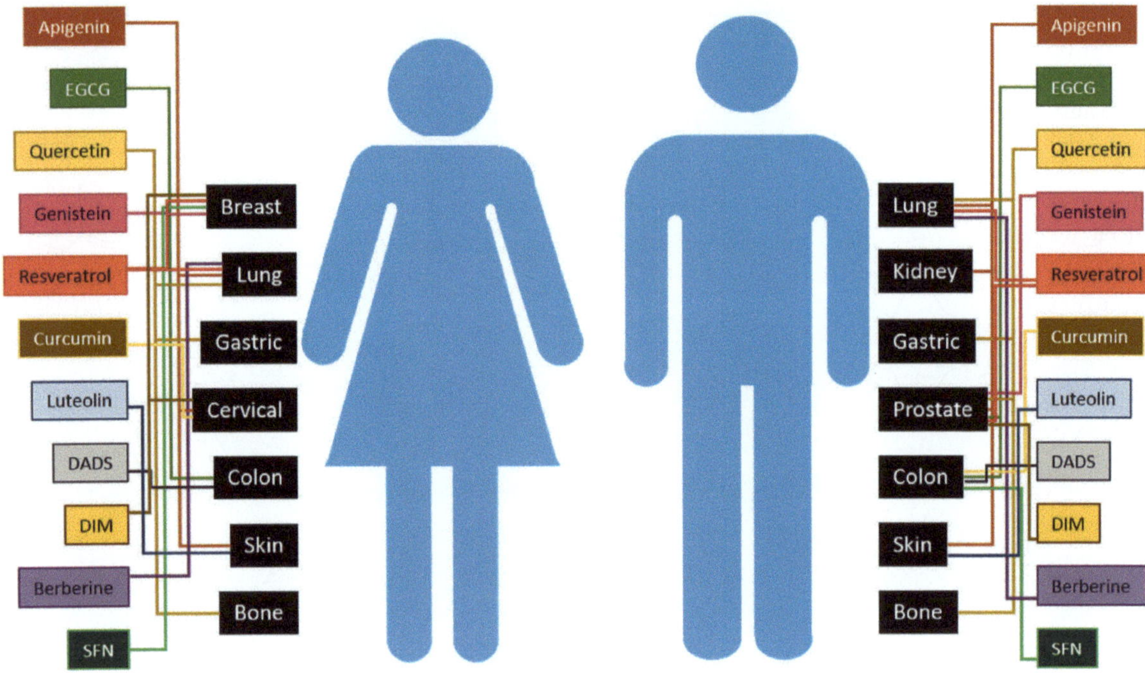

FIGURE 3.1 Natural anticancer agents targeting epigenetically sensitive cancers.

prevent it. Accumulated evidence from in vitro, in vivo, and clinical studies suggests that natural compounds derived from various vegetables and fruits can regulate epigenetic modifications via a number of mechanisms, such as apoptosis, silencing of cancer-related genes, reactivation of tumor-suppressor genes, and activation of cell-survival genes in different cancers. Several compounds belonging to different subclasses have been found to mediate these epigenetic alterations.

Polyphenolic Compounds

Polyphenols are plant secondary metabolites and abundant micronutrients, sometimes also called phytochemicals, that are mainly found in fruits, vegetables, cereals, and beverages (Pandey and Rizvi, 2009). Research indicates that regular consumption of a polyphenol-rich diet could offer some protection against developing cancer (Arts and Hollman, 2005; Graf et al., 2005). It is only conjecture at this point that polyphenols act epigenetically to prevent cancer, because many polyphenols have other modalities of action, such as antioxidant or anti-inflammatory activity. The polyphenol family comprises phenolic acids (hydroxybenzoic and hydroxycinamic), lignans, stilbenes, and flavonoids (Hardman, 2014; Manach et al., 2004). The latter is the largest group of plant phenolic compounds (Kumar and Pandey, 2013). There are six subclasses of flavonoids: flavan-3-ols (also known as flavanols or catechins), flavonols, flavones, flavanones, isoflavones, and anthocyanidins (Hardman, 2014). Flavonoids have been reported to exert a number of diverse biological activities including antibacterial, antiviral, analgesic, antiallergic, hepatoprotective, cytostatic, apoptotic, estrogenic, and antiestrogenic functions (Chung et al., 2010; Hodek et al., 2002; Malireddy et al., 2012). These effects are mediated by certain mechanisms including modulation of DNA methylation status and histone acetylation (Busch et al., 2015). Flavones are much less common than flavonols in fruits and vegetables (Manach et al., 2004).

Apigenin

One widely characterized flavone is **apigenin**, chemically known as 4′, 5, 7,-trihydroxyflavone, which is abundantly present in parsley, onions, oranges, tea, chamomile, wheat sprouts, and some seasonings (Patel et al., 2007). Apigenin has gained tremendous interest as a chemopreventive agent because of its low toxicity and reported anticancer effect in many cancers, such as breast, cervical, colon, lung, ovarian, prostate, skin, thyroid, gastric, and lung, to name a few (Shukla and Gupta, 2010). Specifically, apigenin mediated apoptosis due to HDAC1 and HDAC3 inhibition was observed in prostate cancer PC-3 and 22Rv1 cells (Pandey et al., 2012). (Paredes-Gonzales et al., 2014) demonstrated that apigenin can restore the silenced status of Nrf2 gene in skin epidermal JB6 P + cells by reducing the expression of the DNMT1, DNMT3a, and DNMT3b epigenetic proteins as well as the expression of some HDACs (Paredes-Gonzalez et al., 2014). In addition, treatment with apigenin led to cell cycle arrest with inhibited HDAC activity and H3 histone acetylation in MDA-MB-231 breast cancer cells and also delayed the tumor growth in a xenograft breast cancer model (Tseng et al., 2017). Finally, apigenin together with another flavone, luteolin, has been reported to exert inhibitory effects on 5-cytosine DNMT in KYSE 510 cells (Fang et al., 2007). Research suggests that a diet rich in flavones is related to a decreased risk of certain cancers, particularly cancers of the breast, digestive tract, skin, and prostate, as well as certain hematological malignancies (Shukla and Gupta, 2010).

Curcumin

Curcumin is a main component of turmeric (*Curcuma longa*). It possesses powerful anticancer activities (Reuter et al., 2011). Curcumin has been well investigated for its role in modulating the expression of DNMTs, HDACs, HATs, and miRNAs. Many studies have investigated the effect of curcumin on HDAC expression. Of these, Bora-Tatar et al. (2009) reported that among 33 carboxylic acid derivatives, curcumin was the most effective HDAC inhibitor. Other studies (Liu et al., 2005; Chen et al., 2007) revealed that treatment with curcumin led to significantly decreased levels of HDAC 1 and 3 in Raji cells. Curcumin has been shown to be an effective HDAC inhibitor in HeLa nuclear extracts (Bora-Tatar et al., 2009). It has also been reported to modulate miRNA expression in cancer cells (Sun et al., 2008; Ali et al., 2010).

Epigallocatechin Gallate

Epigallocatechin gallate (EGCG), also known as epigallocatechin-3-gallate, is the most abundant and powerful flavonoid in green tea (Gilbert and Liu, 2010). Other constituents of green tea include epicatechin-3-gallate, epigallocatechin, and epicatechin. Together they are called green tea polyphenols (GTP). It has been reported that EGCG inhibits DNMTs in a variety of cancer cells (Khan et al., 2015; Moseley et al., 2013) and in animal models (Chang et al., 2015; Yang et al., 2009). Thus, treatment with EGCG can lead to the reactivation of epigenetically silenced genes (Fang et al., 2003; Nandakumar et al., 2011). EGCG has also been defined as an effective histone-modifying agent (Balasubramanian et al., 2010; Deb et al., 2015; Nandakumar et al., 2011) and the modulator of miRNAs (Tsang and Kwok, 2010). Pandey et al. (2010) demonstrated that treatment with green tea polyphenols led to the reexpression of the epigenetically silenced glutathione-S transferase pi (GSTP1) gene in prostate cancer cells, which was correlated with DNMT inhibition (Pandey et al., 2010). In addition, EGCG modulates the activity of histone acetylation and in such way alters the chromatin structure (Rajendran et al., 2011). Thus, EGCG has been also shown to modulate polycomb proteins (Balasubramanian et al., 2010; Choudhury et al., 2011). Research indicates that EGCG alone or in combination with other compounds could be considered a potential agent for cancer prevention and treatment (Landis-Piwowar et al., 2007).

Genistein

Genistein belongs to the isoflavone polythenol group. It is found in beans, soy, and coffee (Miceli et al., 2014). Many studies indicate that a genistein-rich diet has been associated with decreased risk of hormone-dependent prostate and breast cancer (Banerjee et al., 2008; Jian, 2009). Genistein exerts epigenetic mechanisms through the modulation of DNMTs (Fang et al., 2005; Xie et al., 2014), HDACs (Majid et al., 2010; Vahid et al., 2015; Basak et al., 2008), and HATs (Vahid et al., 2015; Hong et al., 2004), and alterations of miRNAs (Parker et al., 2009). Although many in vitro studies show promising results, in-vivo studies (Qin et al., 2009; Zhang et al., 2016) are not conclusive and therefore there is an ongoing debate in the field regarding the effectiveness of genistein in certain cancers, especially breast and prostate cancer. Additionally, some reports indicate (Allred et al., 2001) that genistein may enhance the growth of breast cancer tumors in vivo. Flavonols are the most ubiquitous flavonoids (Manach et al., 2004). The main flavonols are **quercetin** and **kaempferol**. They can be found in onions, curly kale, leeks, broccoli, and blueberries (Manach et al., 2004). Other sources include tea, apples, berries, and wine (Busch et al., 2015).

Quercetin

Quercetin is the predominant flavonol (Aherne and O'Brien, 2002). It plays a significant role in cell-cycle regulation, survival/apoptotic signaling, and metastatic events (Aggarwal et al., 2015). It induces epigenetic changes by a number of mechanisms. Tan et al. (2009) reported that hypermethylation of the p16INK4a gene was successfully reversed after 120 h of treatment with quercetin. In addition, quercetin has been linked to inhibition of histone acetyl transferase activity (Ruiz et al., 2007), inhibition of HDAC and activation of HAT (Lee et al., 2011). Increased histone H3 acetylation has been reported after treatment with quercetin in leukemia HL60 cells (Rajendran et al., 2011). It has also been shown to be involved in DNMT inhibition (Gilbert and Liu, 2010; Lee et al., 2005; Priyadarsini et al., 2011). Together with other flavonols, it is also known to inhibit DNMT1-mediated DNA methylation in a concentration-dependent manner (Lee et al., 2005). Lastly, quercetin has been shown to modulate miRNAs in various human cancer cell lines including gastric cancer (Du et al., 2015), lung cancer (Sonoki et al., 2015), hepatocellular carcinoma (Lou et al., 2015), osteosarcoma (Zhang et al., 2015), and prostate cancer (Yang et al., 2015).

Kaempferol

Another important flavonol, **kaempferol,** is also known for its inhibitory effect for HDAC enzymes (Berger et al., 2013). Dietary plant sources include tomatoes, hop, red grapes, grapefruit, strawberries, and *Gingko biloba* (Busch et al., 2015).

Luteolin

Luteolin is a flavone widely distributed in sage, thyme, peppermint, carrot, broccoli, onion, and chili. It exerts similar biological activities to those of quercetin. Luetolin has been reported to have both chemopreventive and chemotherapeutic potential. Attoub (2011) demonstrated luetolin to be a potent HDAC inhibitor in lung cancer cells targeting the acetylation of histones H3 and H4.

Resveratrol

Resveratrol is an active ingredient in red grapes (wine), peanuts, and berries (Kumar et al., 2015). It has been linked to health and disease prevention because of its antiproliferative, antioxidative, anti-inflammatory, and anticancer properties. It belongs to the stilbenes family. It has been shown to exert an inhibitory effect on DNMTs (Kala and Tollefsbol, 2016) and also modulate the expression of miRNAs (Qin et al., 2014). Thus, it has also been shown to be a modulator of SIRT1 (Wang et al., 2008; Bishayee, 2009; Howitz et al., 2003).

Sulfur-Containing Compounds

Diallyl Disulfide and DIM

Diallyl Disulfide is an important garlic (*Allium sativum*) derivative. It has been reported to lead to HDAC inhibition and histone hyperacetylation in colon cancer cells (Druesne et al., 2004). It has also been described as HDACi (histone deacetylase inhibitor) (Huang et al., 2011). **DIM (3,3 − diindolylmethane),** another plant-derived compound known to have anticancer properties and is usually found in cruciferous vegetables, such as broccoli, cabbage, cauliflower, brussels sprouts, mustard, and radish (Aggarwal and Ichikawa, 2005). DIM has been reported to suppress DNMTs in prostate cancer in both in vitro and in vivo models (Wu et al., 2013) and to inhibit HDAC activity in prostate cancer cells (Beaver et al., 2012); it is also considered a chemopreventive agent in other cancers. Li et al. (2010) reported that DIM-altered miRNA expression led to inhibition of pancreatic cancer cell invasion (Y. Li et al., 2010). In addition, treatment with DIM resulted in the inhibition of breast cancer cell proliferation (Jin et al., 2010). In a mouse model designed to evaluate lung metastases, DIM consumption was found to cause a marked reduction in the number of lung metastasis nodules (Kim et al., 2009). Importantly, DIM has moved though preclinical studies into clinical trials targeting prostate, breast, and cervical cancers (Banerjee et al., 2011; Wu et al., 2013). Results from a Phase IIa clinical trial investigating the effectiveness of a new drug, Infemin, constituted of DIM solution excipient into hard gelatin capsules in patients diagnosed with a high-grade prostatic intraepithelial neoplasia (PIN), revealed that in the treatment group, the morphological index (MI) decreased from 0.50 to 0.08, while in the placebo group; it increased from 0.27 to 0.58. In addition, 45.5% of patients in the Infemin group had a complete regression of PIN (Paltsev et al., 2016). This study emphasizes the role of DIM, a natural epigenetic modifier, as a potential therapeutic agent.

Isoquinoline Alkaloids and Isothiocyanates

Berberine

Berberine is a bioactive isoquinoline alkaloid isolated from several herbal substances (Wang et al., 2017) that is mainly found in *Berberis aristata* (Kalaiarasi et al., 2016). Kalaiarasi and colleagues demonstrated that berberine caused epigenetic modifications mediated by HDAC inhibition, followed by histone hyperacetylation thereby, thereby inducing antineoplastic activity in the lung A549 cell line (Kalaiarasi et al., 2016). Berberine has also been reported to modulate DNMTs and miRNAs (Huang et al., 2017).

Sulforaphane

Sulforaphane is a natural compound derived from cruciferous vegetables. It has been associated with potential HDACi activity (Myzak et al., 2004) and downregulation of DNMTs in breast cancer cells (Meeran et al., 2010). It has also been shown to increase the level of histone acetylation (Meeran et al., 2010).

Other Compounds of Natural Origin With Epigenetic Properties

Other compounds isolated from foods and/or plants also demonstrate epigenetic properties. One of them, Psammaplin A, isolated from the marine sponge Aplysinella rhax, has been shown to inhibit HDAC activity in HeLa cells (Kim et al., 2007). Interestingly, Pina et al. (2003) reported that Psammaplin A, isolated from the sponge *Pseudoceratina purpurea*, inhibited both HDAC and DNMT and also suppressed tumor growth in the A549 lung xenograft mouse model while exhibiting low toxicity (Pina et al., 2003).

Other compounds of aqueous origin have also been investigated for their HDAC activities. Oku et al. (2004) reported three new cyclostellettamines (G,D,E), isolated from a marine sponge of the genus Xestospongia, to also have epigenetic properties. Together with cyclostellettamine A, they were found to inhibit HDAC activity in K562 human leukemia cells.

Depudecin, isolated from *Alternaria brassicicola*, due to its unique chemical structure, has been reported to inhibit HDAC activity both in vitro (HL60 cells) and in vivo (v-ras NIH 3T3 cells).

Dihydrocoumarin, a molecule isolated from *Melilotus officinalis* (sweet clover), has been shown to cause epigenetic silencing by inhibiting several human Sir2 family HDACs. In particular, SIRT1 and SIRT2 (also known as sirtuins) increased p53 acetylation leading to elevated levels of apoptosis (Olaharski et al., 2005).

Another known SIRT family inhibitor is nicotinamide. Avalos et al. (2005) demonstrated that nicotinamide inhibited deacetylation activity of sirtuins. In addition, nicotinamide has been shown to increase the formation of motoneurons from human embryonic stem cells (Zhang et al., 2011).

MCP30, isolated from the seeds of *Momordica charantia*, known as bitter melon, has been reported to induce apoptosis in PIN and PCa cell lines and suppress growth of PC-3 in vivo with no effect on normal prostate cells. This effect was shown to be mediated by the inhibition of HDAC-1 activity and by enhancing the acetylation of histones 3 and 4 (Xiong et al., 2009).

Caffeine has also been shown to play a role in epigenetics. Mukwevho et al. (2008) demonstrated that it reduced the activity of HDAC5 and increased the acetylation of H3, leading to enhanced expression of the GLUT4 glucose transporter involved in glucose disposal and protection against type II diabetes. Caffeic acid, a catechol-containing coffee polyphenol unrelated to caffeine, has been reported to inhibit DNMT activity (Lee and Zhu, 2006).

Lastly, there has been a huge interest in medicinal plant extracts as natural anticancer agents. They possess many as anti-inflammatory, antibacterial, and antioxidant biological properties, but little is known about their epigenetic properties. However, given their chemopreventive effects, it is very likely that future studies will focus on their epigenetic activities. Szic et al. (2017) demonstrated that withaferin A, the first described withanolide, derived from *Withania somnifera* (Ashwagandha in Ayurvedic medicine), induced DNA hypermethylation in breast cancer cells. Furthermore, it has been shown to silence receptor tyrosine-protein kinase erbB-2 (HER2)/progesterone receptor/estrogen receptor-—dependent gene expression in different clinical subtypes of breast cancer patients in the cancer genome atlas, and therefore may be considered a potential therapeutic agent to treat triple negative breast cancer.

In summary, epigenetic alterations are inevitably related to gene expression. DNMTs and HDACs are upregulated not only in advanced cancer, but also in the early phases of carcinogenesis. Therefore, finding compounds that could act as DNA demethylating agents and histone deacetylation inhibitors is of great importance. Some of these compounds have already been used in clinics, such as 5-Azacytidine (isolated from Streptoverticillium ladakanus), a DNMT inhibitor and Romidepsin (isolated from Chromobacterium violaceum), an HDAC inhibitor (Miceli et al., 2014). However, their efficacy is limited by toxicity and lack of effectiveness after the inhibitor is removed. Therefore, there is a great need to search for safer compounds, such as natural products, that are capable of modifying the epigenome. Given that plant-derived compounds target multiple epigenetic pathways, they have many advantages over the monotherapeutic agents that fail to be satisfying (Sarkar et al., 2009).

MODEL SYSTEMS: PRECLINICAL EVALUATION OF POTENTIAL EPIGENETIC MODIFIERS FROM NATURE

The search for naturally occurring epigenetic modifiers with potential utility in cancer chemoprevention has been aided by a number of in vitro and in vivo model systems that have enabled this research. Human cancer cell lines are often used as a prescreen for cancer-preventive activity and to distill the search down to epigenetically active candidates. Often these studies have employed the use of methylation-sensitive cell lines with those that are not amenable to epigenetic agents. For instance, in colon cancer the discovery of the CpG island methylator phenotype led to the identification of cancer cell lines with high levels of methylation in specific genes (Toyota et al., 1999). As well there are now identified several human breast cancer cell lines characterized by hypermethylation and gene silencing in a number of regulatory genes influencing the genesis of neoplasia (Roll et al., 2008). Other human cancer cell lines with methylation phenotypes exist for many common cancers. Screening for natural compounds with epigenetic-modifying activity in human cell lines can take place in several guises such as targeting the suppression of the epigenetic machinery and its components and reactivation of silenced genes (Perri et al., 2017). Reactivation of tumor-suppressor gene function is an important aspect of in vitro screening of natural products for epigenetic activity as restoring their functions often translates into easily measured effects of proliferation, induction of apoptosis, changes in the cell cycle, and reduction in motility along with measures of invasive potential (Boyanapalli et al., 2016; Datta et al., 2016; Garcia-Bloj et al., 2016).

While useful as initial screens, the potential for natural compounds to be epigenetic modifiers must pass through the gateway of animal model testing. In vivo assays tell a lot about the antitumor potential of a given agent, but also bring inherent pharmacological issues related to method of exposure, uptake, delivery, and metabolism of these compounds. Several approaches can be considered in the in vivo screen for epigenetic regulators for chemopreventive activity. These include mice in which certain genes have been knocked out for purely mechanistic studies, or wherein haploinsufficient mice are used in attempts to boost expression of a remaining allele (Bansal et al., 2016; Oka et al., 2005; Schemmer et al., 2013; Zagni et al., 2017). Some mouse models are useful in interrogating the effect of candidate natural product inhibitors on the expression and activity of epigenetic events. These could include gene and protein expression of DNMT, HDAC and sirtuin isoforms as well as effects on miRNA. In terms of functionality, the effect of selected natural products in the induction or suppression of histone marks has been researched. Other models for consideration include the use of xenografted human tumors in immunocompromised mice where the inoculated cells and resultant tumors are from well-characterized methylation-sensitive cell lines. The use of patient-derived xenografts is only now emerging in the study of mechanisms of action, where these xenografts stem from tissue samples for patients with epigenetically driven cancers (Maletzki et al., 2015a,b).

NATURAL EPIGENETIC COMPOUNDS AS ADJUVANTS FOR CANCER THERAPY

In principal, compounds that modify the epigenetic machinery could be employed as a new approach to improving the efficacy of cancer therapy. At least one major mechanism that has evolved from the research involves the release of silenced tumor-suppressor genes from epigenetic silencing by reducing or eliminating promoter hypermethylation. Several in vitro studies have examined the benefit of the DNMT inhibitor decitabine in combination with chemotherapy. For instance, use of L-asparaginase and decitabine in tandem evidenced synergistic cell killing effects on acute lymphoblastic leukemia cell lines (Serravalle et al., 2016). Decitabine was found to induce expression of the cancer immunotherapy target NY-ESO 1 in MCF-& breast cancer cells, leading to the potential for enhanced cytotoxicity. In hepatocellular cancer cell lines, decitabine in combination with hydroxymethyldibenzoyl methane induced p38 expression and was associated with enhanced cytotoxicity (Li et al., 2015). Initial enthusiasm for use of DNMT inhibitor (DNMTI) agents in hematological malignancies has been dampened due to unacceptable side effects in patients. Use of DNMT inhibitors in patients with and without standard chemotherapy has unveiled cotoxicity, unfortunately limiting the use of pharmaceutical DNMTIs in the clinic (Müller-Tidow et al., 2016; Prebet et al., 2016; Radsak et al., 2017). Natural sources of DNMTIs offer the tantalizing prospect of combinatorial efficacy with cancer therapeutics with far less toxicity due to the relatively nontoxic nature of natural products in general and the possible benefit of reducing the therapeutic doses of cancer drugs when combined with natural product DNMTIs.

Another approach has been to employ HDAC-inhibitor agents either alone or in combination with cancer therapy. HDAC inhibitors can augment the acetylation of cellular proteins. Pharmaceutical HDAC inhibitors such as hydroxamates (SAHAs), aliphatic acids (valproic acid), and benzamides either inhibit specific HDAC isoforms or have pan-HDAC inhibiting activity. HDAC inhibitors in their own right are anticancer agents but also have shown some promise in augmenting the action of standard therapeutics (Grabarska et al., 2017; Péneau et al., 2017). Cisplatin toxicity of rhabdomyosarcoma cells was shown to be enhanced in the presence of SAHAs or valproic acid (Jarząb et al., 2017). Panobinostat, an HDAC inhibitor, was shown to enhance the antitumor activity of trastuzumab in HER2+ xenografts (Medon et al., 2017). SAHAs and cisplatin worked better together in anticancer treatment of larynx cancer cells (Geng et al., 2017). In ovarian cancer cells, a combination of HDAC inhibition with 5-fluorouracil and paclitaxel caused G2 arrest associated with activation of the p38 signaling pathway (Akiyama et al., 2017). As with DNMT inhibitors, HDAC inhibition is not without discernible toxicity in patients; however, it would appear that refinement of dosages in Phase 1 clinical trials along with chemotherapeutics has led to lesser risk of thrombocytopenia and other off-target side effects (Iwahashi et al., 2014; Ngamphaiboon et al., 2015; Zibelman et al., 2015).

CONCLUSIONS AND FUTURE RESEARCH DIRECTIONS

In light of growing accumulated evidence suggesting that fruit-and-vegetable-rich diet is associated with reduced risk of cancer development, we aimed to provide a comprehensive overview of the natural anticancer agents. Specifically, we have chosen compounds from foods with reported epigenetic activity such as alteration of DNA methylation accompanied by reactivation of tumor-suppressor genes silenced by promoter hypermethylation, as well as histone modifications and miRNA alteration. The reversible nature of these epigenetic changes has led to increased demand for development of inhibitors targeting these process-mediating enzymes. A number of DNMT and HDAC pharmaceutical inhibitors are in clinical use or currently under investigation in clinical trials. The pyrimidine nucleoside analogues azacitidine (5-azacytidine,

Vidaza) and decitabine (5-aza-2′deoxycytidine, Dacogen) are approved by the US Food and Drug Administration for the treatment of myelodysplastic syndrome and acute myeloid leukemia. However, their effectiveness is limited by the associated toxicity, the possibility of activating oncogenes instead of the tumor-suppressor genes, and the lack of effect after treatment termination. Therefore, to overcome these hindrances, there is great demand for finding other sources of epigenetic modifiers that could alleviate the negative effects of currently available inhibitors. Plant-derived compounds are reported to be safe, could be taken continuously without severe side effects, and due to their bioactive components are capable of modulating the epigenome. Thus, because of chemical complexity and biodiversity, natural compounds are not limited to targeting only a single target. Although many

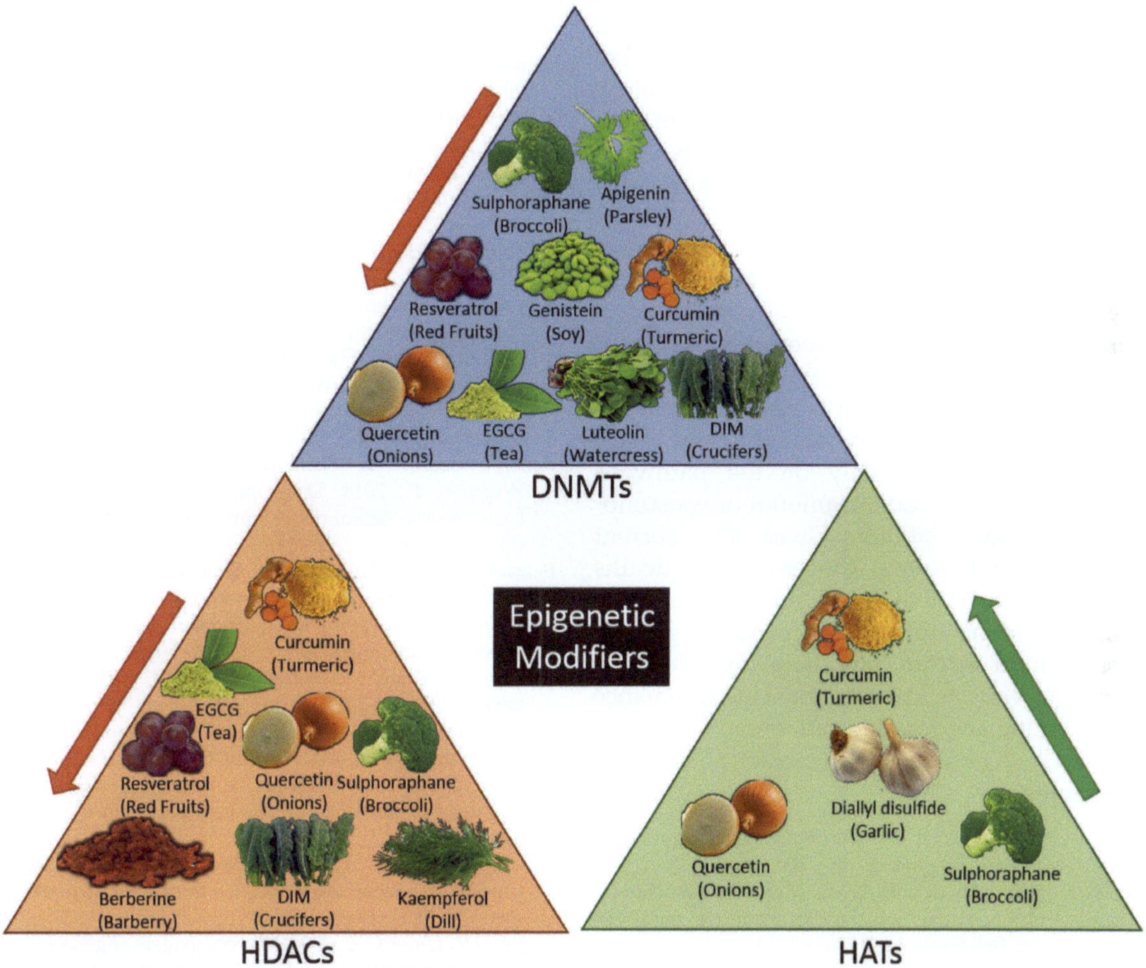

FIGURE 3.2 Foods as epigenetic modifiers. *Red arrows* represent decreased expression of DNMTs and HDACs, whereas green arrows represent increases in HATs. Food items shown in the figure may involve modulation of more than one epigenetic mechanism.

studies discussed in this chapter indicate the potential of these agents to reverse epigenetic aberrations, further investigations of the natural compounds are required in order to elucidate their true potential to prevent and treat cancer.

In addition, we can now consider that a broader approach to enhancing therapy might be envisioned. Aside from the use of specific natural compounds with specific epigenetic targets as their mechanism of action, it is reasonable to think of constructing an epigenetic diet enriched in foods with molecules that modify epigenetic events (Fig. 3.2). More needs to be discovered, but perhaps to sustain desilenced regulatory genes, an epigenetic diet may be of use in subjects diagnosed with known cancers evolving from silenced tumor-suppressor genes. As discussed previously, some natural compounds are reported to play a single epigenetic-modifier role, while others are known to be involved in multiple epigenetic events. In addition, some of them, such as quercetin, are found in nearly all plant products, whereas others are more specific to a particular type of food. Therefore, the efficacy of natural epigenetic compounds depends not only on the pathways they target, but also on the amount of consumption and bioavailability. Given the current perception that some cancer-related deaths could be prevented by a healthy lifestyle including proper nutrition, incorporating natural anticancer agents into daily diets may be the way forward to at least enhancing the chance of preventing and treating cancer.

References

Abbas, A., Gupta, S., 2008. The role of histone deacetylases in prostate cancer. Epigenetics 3 (6), 300–309.

Aggarwal, B.B., Ichikawa, H., 2005. Molecular targets and anticancer potential of indole-3-carbinol and its derivatives. Cell Cycle 4 (9), 1201–1215.

Aggarwal, R., Jha, M., Shrivastava, A., Jha, A.K., 2015. Natural compounds: role in reversal of epigenetic changes. Biochemistry (Mosc.) 80 (8), 972–989.

Aherne, S.A., O'Brien, N.M., 2002. Dietary flavonols: chemistry, food content, and metabolism. Nutrition 18 (1), 75–81.

Ahuja, N., Sharma, A.R., Baylin, S.B., 2016. Epigenetic therapeutics: a new weapon in the war against cancer. Annu. Rev. Med. 67, 73–89.

Akiyama, M., Sowa, Y., Taniguchi, T., Watanabe, M., Yogosawa, S., Kitawaki, J., Sakai, T., 2017. Three-combined treatment, a novel HDAC inhibitor OBP-801/YM753, 5-fluorouracil and paclitaxel, induces G2-phase arrest through the p38 pathway in human ovarian cancer cells. Oncol. Res. 1 (23).

Ali, S., Ahmad, A., Banerjee, S., Padhye, S., Dominiak, K., Schaffert, J.M., et al., 2010. Gemcitabine sensitivity can be induced in pancreatic cancer cells through modulation of miR-200 and miR-21 expression by curcumin or its analogue CDF. Cancer Res. 70 (9), 3606–3617.

Allred, C.D., Ju, Y.H., Allred, K.F., Chang, J., Helferich, W.G., 2001. Dietary genistin stimulates growth of estrogen-dependent breast cancer tumors similar to that observed with genistein. Carcinogenesis 22 (10), 1667–1673.

Arts, I.C., Hollman, P.C., 2005. Polyphenols and disease risk in epidemiologic studies. Am. J. Clin. Nutr. 81 (1 Suppl.), 317S–325S.

Attoub, S., Hassan, A.H., Vanhoecke, B., Iratni, R., Takahashi, T., Gaben, A.M., et al., 2011. Inhibition of cell survival, invasion, tumor growth and histone deacetylase activity by the dietary flavonoid luteolin in human epithelioid cancer cells. Eur. J. Pharmacol. 651 (1–3), 18–25.

Avalos, J.L., Bever, K.M., Wolberger, C., 2005. Mechanism of sirtuin inhibition by nicotinamide: altering the NAD(+) cosubstrate specificity of a Sir2 enzyme. Mol Cell 17 (6), 855–868.

Bailón-Moscoso, N., Romero-Benavides, J.C., Ostrosky-Wegman, P., 2014. Development of anticancer drugs based on the hallmarks of tumor cells. Tumour Biol. 35 (5), 3981–3995.

Balasubramanian, S., Adhikary, G., Eckert, R.L., 2010. The Bmi-1 polycomb protein antagonizes the (-)-epigallocatechin-3-gallate-dependent suppression of skin cancer cell survival. Carcinogenesis 31 (3), 496–503.

Banerjee, S., Kong, D., Wang, Z., Bao, B., Hillman, G.G., Sarkar, F.H., 2011. Attenuation of multi-targeted proliferation-linked signaling by 3,3'-diindolylmethane (DIM): from bench to clinic. Mutat. Res. 728 (1–2), 47–66.

Banerjee, S., Li, Y., Wang, Z., Sarkar, F.H., 2008. Multi-targeted therapy of cancer by genistein. Cancer Lett 269 (2), 226–242.

Bansal, N., David, G., Farias, E., Waxman, S., 2016. Emerging roles of epigenetic regulator Sin3 in cancer. Adv. Cancer Res. 130, 113–135.

Bartel, D.P., Chen, C.Z., 2004. Micromanagers of gene expression: the potentially widespread influence of metazoan microRNAs. Nat. Rev. Genet. 5 (5), 396–400. https://doi.org/10.1038/nrg1328.

Basak, S., Pookot, D., Noonan, E.J., Dahiya, R., 2008. Genistein down-regulates androgen receptor by modulating HDAC6-Hsp90 chaperone function. Mol. Cancer Ther. 7 (10), 3195–3202.

Beaver, L.M., Yu, T.W., Sokolowski, E.I., Williams, D.E., Dashwood, R.H., Ho, E., 2012. 3,3′-Diindolylmethane, but not indole-3-carbinol, inhibits histone deacetylase activity in prostate cancer cells. Toxicol. Appl. Pharmacol. 263 (3), 345–351.

Berger, A., Venturelli, S., Kallnischkies, M., Böcker, A., Busch, C., Weiland, T., et al., 2013. Kaempferol, a new nutrition-derived pan-inhibitor of human histone deacetylases. J. Nutr. Biochem. 24 (6), 977–985.

Bhalla, K.N., 2005. Epigenetic and chromatin modifiers as targeted therapy of hematologic malignancies. J. Clin. Oncol. 23 (17), 3971–3993.

Bishayee, A., 2009. Cancer prevention and treatment with resveratrol: from rodent studies to clinical trials. Cancer Prev. Res. (Phila.) 2 (5), 409–418.

Bora-Tatar, G., Dayangaç-Erden, D., Demir, A.S., Dalkara, S., Yelekçi, K., Erdem-Yurter, H., 2009. Molecular modifications on carboxylic acid derivatives as potent histone deacetylase inhibitors: activity and docking studies. Bioorg. Med. Chem. 17 (14), 5219–5228.

Boyanapalli, S.S., Li, W., Fuentes, F., Guo, Y., Ramirez, C.N., Gonzalez, X.P., et al., 2016. Epigenetic reactivation of RASSF1A by phenethyl isothiocyanate (PEITC) and promotion of apoptosis in LNCaP cells. Pharmacol. Res. 114, 175–184.

Busch, C., Burkard, M., Leischner, C., Lauer, U.M., Frank, J., Venturelli, S., 2015. Epigenetic activities of flavonoids in the prevention and treatment of cancer. Clin. Epigenet. 7, 64.

Calin, G.A., Croce, C.M., 2006. MicroRNA signatures in human cancers. Nat. Rev. Cancer 6 (11), 857–866.

Chang, X., Rong, C., Chen, Y., Yang, C., Hu, Q., Mo, Y., et al., 2015. (-)-Epigallocatechin-3-gallate attenuates cognitive deterioration in Alzheimer's disease model mice by upregulating neprilysin expression. Exp. Cell Res. 334 (1), 136–145.

Chen, Y., Shu, W., Chen, W., Wu, Q., Liu, H., Cui, G., 2007. Curcumin, both histone deacetylase and p300/CBP-specific inhibitor, represses the activity of nuclear factor kappa B and Notch 1 in Raji cells. Basic Clin. Pharmacol. Toxicol. 101 (6), 427–433.

Choudhury, S.R., Balasubramanian, S., Chew, Y.C., Han, B., Marquez, V.E., Eckert, R.L., 2011. (-)-Epigallocatechin-3-gallate and DZNep reduce polycomb protein level via a proteasome-dependent mechanism in skin cancer cells. Carcinogenesis 32 (10), 1525–1532.

Chung, S., Yao, H., Caito, S., Hwang, J.W., Arunachalam, G., Rahman, I., 2010. Regulation of SIRT1 in cellular functions: role of polyphenols. Arch. Biochem. Biophys. 501 (1), 79–90.

Corner, B.M., Cawley, P.F., 1976. Vibrio parahaemolyticus-food poisoning: case report. N.Z. Med. J. 83 (559), 155–156.

Cruickshank, M.N., Besant, P., Ulgiati, D., 2010. The impact of histone post-translational modifications on developmental gene regulation. Amino Acids 39 (5), 1087–1105.

Dagdemir, A., Durif, J., Ngollo, M., Bignon, Y.J., Bernard-Gallon, D., 2013. Histone lysine trimethylation or acetylation can be modulated by phytoestrogen, estrogen or anti-HDAC in breast cancer cell lines. Epigenomics 5 (1), 51–63.

Datta, J., Islam, M., Dutta, S., Roy, S., Pan, Q., Teknos, T.N., 2016. Suberoylanilide hydroxamic acid inhibits growth of head and neck cancer cell lines by reactivation of tumor suppressor microRNAs. Oral Oncol. 56, 32–39.

Deb, G., Thakur, V.S., Limaye, A.M., Gupta, S., 2015. Epigenetic induction of tissue inhibitor of matrix metalloproteinase-3 by green tea polyphenols in breast cancer cells. Mol. Carcinog. 54 (6), 485–499.

Dong, N., Shi, L., Wang, D.C., Chen, C., Wang, X., 2016. Role of epigenetics in lung cancer heterogeneity and clinical implication. Semin. Cell Dev. Biol. 64 (4), 18–25.

Druesne, N., Pagniez, A., Mayeur, C., Thomas, M., Cherbuy, C., Duée, P.H., et al., 2004. Diallyl disulfide (DADS) increases histone acetylation and p21(waf1/cip1) expression in human colon tumor cell lines. Carcinogenesis 25 (7), 1227–1236.

Du, F., Feng, Y., Fang, J., Yang, M., 2015. MicroRNA-143 enhances chemosensitivity of Quercetin through autophagy inhibition via target GABARAPL1 in gastric cancer cells. Biomed. Pharmacother. 74, 169–177. https://doi.org/10.1016/j.biopha.2015.08.005.

Ehrlich, M., 2009. DNA hypomethylation in cancer cells. Epigenomics 1 (2), 239–259.

Ellis, L., Atadja, P.W., Johnstone, R.W., 2009. Epigenetics in cancer: targeting chromatin modifications. Mol. Cancer Ther. 8 (6), 1409–1420.

Esteller, M., 2007. Cancer epigenomics: DNA methylomes and histone-modification maps. Nat. Rev. Genet. 8 (4), 286–298.

Fan, H., Zhang, R., Tesfaye, D., Tholen, E., Looft, C., Hölker, M., et al., 2012. Sulforaphane causes a major epigenetic repression of myostatin in porcine satellite cells. Epigenetics 7 (12), 1379–1390.

Fang, M., Chen, D., Yang, C.S., 2007. Dietary polyphenols may affect DNA methylation. J. Nutr. 137 (1 Suppl.), 223S–228S.

Fang, M.Z., Chen, D., Sun, Y., Jin, Z., Christman, J.K., Yang, C.S., 2005. Reversal of hypermethylation and reactivation of p16INK4a, RARbeta, and MGMT genes by genistein and other isoflavones from soy. Clin Cancer Res 11 (19 Pt 1), 7033–7041.

Fang, M.Z., Wang, Y., Ai, N., Hou, Z., Sun, Y., Lu, H., et al., 2003. Tea polyphenol (-)-epigallocatechin-3-gallate inhibits DNA methyltransferase and reactivates methylation-silenced genes in cancer cell lines. Cancer Res. 63 (22), 7563–7570.

Feng, Q., Zhang, Y., 2001. The MeCP1 complex represses transcription through preferential binding, remodeling, and deacetylating methylated nucleosomes. Genes Dev. 15 (7), 827–832.

Ferdin, J., Kunej, T., Calin, G.A., 2010. Non-coding RNAs: identification of cancer-associated microRNAs by gene profiling. Technol. Cancer Res. Treat. 9 (2), 123–138.

Fuentes, F., Paredes-Gonzalez, X., Kong, A.T., 2015. Dietary glucosinolates sulforaphane, phenethyl isothiocyanate, indole-3-carbinol/3,3′-diindolylmethane: anti-oxidative stress/inflammation, Nrf2, epigenetics/epigenomics and in vivo cancer chemopreventive efficacy. Curr. Pharmacol. Rep. 1 (3), 179–196.

Garcia-Bloj, B., Moses, C., Sgro, A., Plani-Lam, J., Arooj, M., Duffy, C., et al., 2016. Waking up dormant tumor suppressor genes with zinc fingers, TALEs and the CRISPR/dCas9 system. Oncotarget 7 (37), 60535–60554.

Garraway, L.A., Lander, E.S., 2013. Lessons from the cancer genome. Cell 153 (1), 17–37.

Geng, H., Zhao, L., Liang, Z., Zhang, Z., Xie, D., Bi, L., et al., 2017. Cigarette smoke extract-induced proliferation of normal human urothelial cells via the MAPK/AP-1 pathway. Oncol. Lett. 13 (1), 469–475.

Gerhauser, C., 2013. Cancer chemoprevention and nutriepigenetics: state of the art and future challenges. Top. Curr. Chem. 329, 73–132.

Gilbert, E.R., Liu, D., 2010. Flavonoids influence epigenetic-modifying enzyme activity: structure - function relationships and the therapeutic potential for cancer. Curr. Med. Chem. 17 (17), 1756–1768.

Gołąbek, K., Strzelczyk, J.K., Wiczkowski, A., Michalski, M., 2015. Potential use of histone deacetylase inhibitors in cancer therapy. Contemp. Oncol. (Pozn.) 19 (6), 436–440.

Grabarska, A., Łuszczki, J.J., Nowosadzka, E., Gumbarewicz, E., Jeleniewicz, W., Dmoszyńska-Graniczka, M., et al., 2017. Histone deacetylase inhibitor SAHA as potential targeted therapy agent for larynx cancer cells. J. Cancer 8 (1), 19–28.

Graf, B.A., Milbury, P.E., Blumberg, J.B., 2005. Flavonols, flavones, flavanones, and human health: epidemiological evidence. J. Med. Food 8 (3), 281–290.

Hardman, W.E., 2014. Diet components can suppress inflammation and reduce cancer risk. Nutr. Res. Pract. 8 (3), 233–240.

Harfe, B.D., 2005. MicroRNAs in vertebrate development. Curr. Opin. Genet. Dev. 15 (4), 410–415.

Hauser, A.T., Jung, M., 2008. Targeting epigenetic mechanisms: potential of natural products in cancer chemoprevention. Planta Med. 74 (13), 1593–1601.

Herranz, M., Esteller, M., 2007. DNA methylation and histone modifications in patients with cancer: potential prognostic and therapeutic targets. Methods Mol. Biol. 361, 25–62.

Hodek, P., Trefil, P., Stiborová, M., 2002. Flavonoids-potent and versatile biologically active compounds interacting with cytochromes P450. Chem. Biol. Interact. 139 (1), 1–21.

Hong, T., Nakagawa, T., Pan, W., Kim, M.Y., Kraus, W.L., Ikehara, T., et al., 2004. Isoflavones stimulate estrogen receptor-mediated core histone acetylation. Biochem. Biophys. Res. Commun. 317 (1), 259–264.

Howitz, K.T., Bitterman, K.J., Cohen, H.Y., Lamming, D.W., Lavu, S., Wood, J.G., et al., 2003. Small molecule activators of sirtuins extend Saccharomyces cerevisiae lifespan. Nature 425 (6954), 191–196.

Huang, C., Liu, H., Gong, X.L., Wu, L.Y., Wen, B., 2017. Effect of evodiamine and berberine on the interaction between DNMTs and target microRNAs during malignant transformation of the colon by TGF-β1. Oncol. Rep. 37 (3), 1637–1645.

Huang, J., Plass, C., Gerhauser, C., 2011. Cancer chemoprevention by targeting the epigenome. Curr. Drug Targets 12 (13), 1925–1956.

Iwahashi, S., Utsunomiya, T., Imura, S., Morine, Y., Ikemoto, T., Arakawa, Y., et al., 2014. Effects of valproic acid in combination with S-1 on advanced pancreatobiliary tract cancers: clinical study phases I/II. Anticancer Res. 34 (9), 5187–5191.

Jansson, M.D., Lund, A.H., 2012. MicroRNA and cancer. Mol. Oncol. 6 (6), 590–610.

Jarząb, A., Łuszczki, J.J., Guz, M., Gumbarewicz, E., Polberg, K., Stepulak, A., 2017. Additive interaction of cisplatin and histone deacetylase inhibitors combined treatment in rhabdomyosarcoma cells - an isobolographic analysis. Anticancer Res. 37 (3), 1067–1074.

Jenuwein, T., Allis, C.D., 2001. Translating the histone code. Science 293 (5532), 1074–1080.

Jha, A.K., Nikbakht, M., Parashar, G., Shrivastava, A., Capalash, N., Kaur, J., 2010. Reversal of hypermethylation and reactivation of the RARβ2 gene by natural compounds in cervical cancer cell lines. Folia Biol. (Praha) 56 (5), 195–200.

Jian, L., 2009. Soy, isoflavones, and prostate cancer. Mol. Nutr. Food Res. 53 (2), 217–226.

Jiang, L.L., Zhou, S.J., Zhang, X.M., Chen, H.Q., Liu, W., 2016. Sulforaphane suppresses in vitro and in vivo lung tumorigenesis through downregulation of HDAC activity. Biomed. Pharmacother. 78, 74–80.

Jin, Y., Zou, X., Feng, X., 2010. 3,3′-Diindolylmethane negatively regulates Cdc25A and induces a G2/M arrest by modulation of microRNA 21 in human breast cancer cells. Anticancer Drugs 21 (9), 814–822.

Kala, R., Tollefsbol, T.O., 2016. A novel combinatorial epigenetic therapy using resveratrol and pterostilbene for restoring estrogen receptor-α (ERα) expression in ERα--negative breast cancer cells. PLoS One 11 (5), e0155057.

Kalaiarasi, A., Anusha, C., Sankar, R., Rajasekaran, S., John Marshal, J., Muthusamy, K., Ravikumar, V., 2016. Plant isoquinoline alkaloid berberine exhibits chromatin remodeling by modulation of histone deacetylase to induce growth arrest and apoptosis in the A549 cell line. J. Agric. Food Chem. 64 (50), 9542–9550.

Kanwal, R., Datt, M., Liu, X., Gupta, S., 2016. Dietary Flavones as Dual Inhibitors of DNA Methyltransferases and Histone Methyltransferases. PLoS One 11 (9), e0162956.

Khan, M.A., Hussain, A., Sundaram, M.K., Alalami, U., Gunasekera, D., Ramesh, L., et al., 2015. (-)-Epigallocatechin-3-gallate reverses the expression of various tumor-suppressor genes by inhibiting DNA methyltransferases and histone deacetylases in human cervical cancer cells.

Kim, D.H., Shin, J., Kwon, H.J., 2007. Psammaplin A is a natural prodrug that inhibits class I histone deacetylase. Exp. Mol. Med. 39 (1), 47–55.

Kim, E.J., Shin, M., Park, H., Hong, J.E., Shin, H.K., Kim, J., et al., 2009. Oral administration of 3,3′-diindolylmethane inhibits lung metastasis of 4T1 murine mammary carcinoma cells in BALB/c mice. J. Nutr. 139 (12), 2373–2379.

Klingstedt, C., Ludwigs, U., Baehrendtz, S., Bokliden, A., Matell, G., 1989. A microcomputer system for on-line monitoring of pulmonary function during artificial ventilation. Int. J. Clin. Monit. Comput. 6 (2), 99–107.

Kondo, Y., 2009. Epigenetic cross-talk between DNA methylation and histone modifications in human cancers. Yonsei Med. J. 50 (4), 455–463.

Kornberg, R.D., Lorch, Y., 1999. Twenty-five years of the nucleosome, fundamental particle of the eukaryote chromosome. Cell 98 (3), 285–294.

Kumar, A., Dhar, S., Rimando, A.M., Lage, J.M., Lewin, J.R., Zhang, X., Levenson, A.S., 2015. Epigenetic potential of resveratrol and analogs in preclinical models of prostate cancer. Ann. N.Y. Acad. Sci. 1348 (1), 1–9.

Kumar, S., Pandey, A.K., 2013. Chemistry and biological activities of flavonoids: an overview. Sci. World J. 2013, 162750. https://doi.org/10.1155/2013/162750.

Kwon, H.J., Owa, T., Hassig, C.A., Shimada, J., Schreiber, S.L., 1998. Depudecin induces morphological reversion of transformed fibroblasts via the inhibition of histone deacetylase. Proc. Natl. Acad. Sci. U.S.A. 95 (7), 3356–3361.

Landis-Piwowar, K.R., Huo, C., Chen, D., Milacic, V., Shi, G., Chan, T.H., Dou, Q.P., 2007. A novel prodrug of the green tea polyphenol (-)-epigallocatechin-3-gallate as a potential anticancer agent. Cancer Res. 67 (9), 4303–4310.

Lee, W.J., Chen, Y.R., Tseng, T.H., 2011. Quercetin induces FasL-related apoptosis, in part, through promotion of histone H3 acetylation in human leukemia HL-60 cells. Oncol. Rep. 25 (2), 583–591.

Lee, S.J., Krauthauser, C., Maduskuie, V., Fawcett, P.T., Olson, J.M., Rajasekaran, S.A., 2011. Curcumin-induced HDAC inhibition and attenuation of medulloblastoma growth in vitro and in vivo. BMC Cancer 11, 144.

Lee, W.J., Shim, J.Y., Zhu, B.T., 2005. Mechanisms for the inhibition of DNA methyltransferases by tea catechins and bioflavonoids. Mol. Pharmacol. 68 (4), 1018–1030.

Lee, W.J., Zhu, B.T., 2006. Inhibition of DNA methylation by caffeic acid and chlorogenic acid, two common catechol-containing coffee polyphenols. Carcinogenesis 27 (2), 269–277.

Leroy, G., Dimaggio, P.A., Chan, E.Y., Zee, B.M., Blanco, M.A., Bryant, B., et al., 2013. A quantitative atlas of histone modification signatures from human cancer cells. Epigenet. Chromatin 6 (1), 20.

Li, C.F., Tsai, H.H., Ko, C.Y., Pan, Y.C., Yen, C.J., Lai, H.Y., et al., 2015. HMDB and 5-AzadC combination reverses tumor suppressor CCAAT/enhancer-binding protein delta to strengthen the death of liver cancer cells. Mol. Cancer Ther. 14 (11), 2623–2633.

Li, Y., Vandenboom, T.G., Wang, Z., Kong, D., Ali, S., Philip, P.A., Sarkar, F.H., 2010. miR-146a suppresses invasion of pancreatic cancer cells. Cancer Res. 70 (4), 1486–1495.

Liu, H.L., Chen, Y., Cui, G.H., Zhou, J.F., 2005. Curcumin, a potent anti-tumor reagent, is a novel histone deacetylase inhibitor regulating B-NHL cell line Raji proliferation. Acta Pharmacol. Sin. 26 (5), 603–609.

Lou, G., Liu, Y., Wu, S., Xue, J., Yang, F., Fu, H., Chen, Z., 2015. The p53/miR-34a/SIRT1 positive feedback loop in quercetin-induced apoptosis. Cell Physiol. Biochem. 35 (6), 2192–2202.

Majid, S., Dar, A.A., Shahryari, V., Hirata, H., Ahmad, A., Saini, S., Tanaka, Y., Dahiya, A.V., Dahiya, R., 2010. Genistein reverses hypermethylation and induces active histone modifications in tumor suppressor gene B-Cell translocation gene 3 in prostate cancer. Cancer 116 (1), 66–76.

Maletzki, C., Gock, M., Randow, M., Klar, E., Huehns, M., Prall, F., Linnebacher, M., 2015. Establishment and characterization of cell lines from chromosomal instable colorectal cancer. World J. Gastroenterol. 21 (1), 164–176.

Maletzki, C., Huehns, M., Knapp, P., Waukosin, N., Klar, E., Prall, F., Linnebacher, M., 2015. Functional characterization and drug response of freshly established patient-derived tumor models with CpG island methylator phenotype. PLoS One 10 (11), e0143194.

Malireddy, S., Kotha, S.R., Secor, J.D., Gurney, T.O., Abbott, J.L., Maulik, G., et al., 2012. Phytochemical antioxidants modulate mammalian cellular epigenome: implications in health and disease. Antioxid. Redox Signal. 17 (2), 327–339.

Manach, C., Scalbert, A., Morand, C., Rémésy, C., Jiménez, L., 2004. Polyphenols: food sources and bioavailability. Am. J. Clin. Nutr. 79 (5), 727–747.

Matkar, S., Sharma, P., Gao, S., Gurung, B., Katona, B.W., Liao, J., et al., 2015. An epigenetic pathway regulates sensitivity of breast cancer cells to HER2 inhibition via FOXO/c-Myc Axis. Cancer Cell 28 (4), 472–485.

Medon, M., Vidacs, E., Vervoort, S.J., Li, J., Jenkins, M.R., Ramsbottom, K.M., et al., 2017. HDAC inhibitor panobinostat engages host innate immune defenses to promote the tumoricidal effects of trastuzumab in HER2+ tumors. Cancer Res. 77 (10), 2594–2606. https://www.ncbi.nlm.nih.gov/pubmed/28249907.

Meeran, S.M., Ahmed, A., Tollefsbol, T.O., 2010. Epigenetic targets of bioactive dietary components for cancer prevention and therapy. Clin. Epigenet. 1 (3–4), 101–116.

Mei, Z., Zhang, X., Yi, J., Huang, J., He, J., Tao, Y., 2016. Sirtuins in metabolism, DNA repair and cancer. J. Exp. Clin. Cancer Res. 35 (1), 182.

Meier, K., Recillas-Targa, F., 2017. New insights on the role of DNA methylation from a global view. Front. Biosci. (Landmark Ed.) 22, 644–668.

Miceli, M., Bontempo, P., Nebbioso, A., Altucci, L., 2014. Natural compounds in epigenetics: a current view. Food Chem. Toxicol. 73, 71–83.

Moseley, V.R., Morris, J., Knackstedt, R.W., Wargovich, M.J., 2013. Green tea polyphenol epigallocatechin 3-gallate, contributes to the degradation of DNMT3A and HDAC3 in HCT 116 human colon cancer cells. Anticancer Res. 33 (12), 5325–5333.

Mottet, D., Castronovo, V., 2008. Histone deacetylases: target enzymes for cancer therapy. Clin. Exp. Metastasis 25 (2), 183–189.

Mukwevho, E., Kohn, T.A., Lang, D., Nyatia, E., Smith, J., Ojuka, E.O., 2008. Caffeine induces hyperacetylation of histones at the MEF2 site on the Glut4 promoter and increases MEF2A binding to the site via a CaMK-dependent mechanism. Am. J. Physiol. Endocrinol. Metab. 294 (3), E582–E588.

Müller-Tidow, C., Tschanter, P., Röllig, C., Thiede, C., Koschmieder, A., Stelljes, M., et al., 2016. Azacitidine in combination with intensive induction chemotherapy in older patients with acute myeloid leukemia: the AML-AZA trial of the Study Alliance Leukemia. Leukemia 30 (3), 555–561.

Myzak, M.C., Dashwood, R.H., 2006. Histone deacetylases as targets for dietary cancer preventive agents: lessons learned with butyrate, diallyl disulfide, and sulforaphane. Curr Drug Targets 7 (4), 443–452.

Myzak, M.C., Karplus, P.A., Chung, F.L., Dashwood, R.H., 2004. A novel mechanism of chemoprotection by sulforaphane: inhibition of histone deacetylase. Cancer Res. 64 (16), 5767–5774.

Nandakumar, V., Vaid, M., Katiyar, S.K., 2011. (-)-Epigallocatechin-3-gallate reactivates silenced tumor suppressor genes, Cip1/p21 and p16INK4a, by reducing DNA methylation and increasing histones acetylation in human skin cancer cells. Carcinogenesis 32 (4), 537–544.

Ngamphaiboon, N., Dy, G.K., Ma, W.W., Zhao, Y., Reungwetwattana, T., DePaolo, D., et al., 2015. A phase I study of the histone deacetylase (HDAC) inhibitor entinostat, in combination with sorafenib in patients with advanced solid tumors. Investig. New Drugs 33 (1), 225–232.

Oka, M., Meacham, A.M., Hamazaki, T., Rodić, N., Chang, L.J., Terada, N., 2005. De novo DNA methyltransferases Dnmt3a and Dnmt3b primarily mediate the cytotoxic effect of 5-aza-2′-deoxycytidine. Oncogene 24 (19), 3091–3099.

Okano, M., Bell, D.W., Haber, D.A., Li, E., 1999. DNA methyltransferases Dnmt3a and Dnmt3b are essential for de novo methylation and mammalian development. Cell 99 (3), 247–257.

Oku, N., Nagai, K., Shindoh, N., Terada, Y., van Soest, R.W., Matsunaga, S., Fusetani, N., 2004. Three new cyclostellettamines, which inhibit histone deacetylase, from a marine sponge of the genus Xestospongia. Bioorg. Med. Chem. Lett. 14 (10), 2617–2620.

Olaharski, A.J., Rine, J., Marshall, B.L., Babiarz, J., Zhang, L., Verdin, E., Smith, M.T., 2005. The flavoring agent dihydrocoumarin reverses epigenetic silencing and inhibits sirtuin deacetylases. PLoS Genet. 1 (6), e77.

Oya, Y., Mondal, A., Rawangkan, A., Umsumarng, S., Iida, K., Watanabe, T., et al., 2017. Down-regulation of histone deacetylase 4, -5 and -6 as a mechanism of synergistic enhancement of apoptosis in human lung cancer cells treated with the combination of a synthetic retinoid, Am80 and green tea catechin. J. Nutr. Biochem. 42, 7–16.

Paltsev, M., Kiselev, V., Drukh, V., Muyzhnek, E., Kuznetsov, I., Andrianova, E., Baranovskiy, P., 2016. First results of the double-blind randomized placebo-controlled multicenter clinical trial of DIM-based therapy designed as personalized approach to reverse prostatic intraepithelial neoplasia (PIN). EPMA J. 7, 5.

Pandey, M., Kaur, P., Shukla, S., Abbas, A., Fu, P., Gupta, S., 2012. Plant flavone apigenin inhibits HDAC and remodels chromatin to induce growth arrest and apoptosis in human prostate cancer cells: in vitro and in vivo study. Mol. Carcinog. 51 (12), 952–962.

Pandey, K.B., Rizvi, S.I., 2009. Plant polyphenols as dietary antioxidants in human health and disease. Oxid. Med. Cell. Longev. 2 (5), 270–278.

Pandey, M., Shukla, S., Gupta, S., 2010. Promoter demethylation and chromatin remodeling by green tea polyphenols leads to re-expression of GSTP1 in human prostate cancer cells. Int. J. Cancer 126 (11), 2520–2533.

Paredes-Gonzalez, X., Fuentes, F., Su, Z.Y., Kong, A.N., 2014. Apigenin reactivates Nrf2 anti-oxidative stress signaling in mouse skin epidermal JB6 P + cells through epigenetics modifications. AAPS J. 16 (4), 727–735.

Parker, L.P., Taylor, D.D., Kesterson, J., Metzinger, D.S., Gercel-Taylor, C., 2009. Modulation of microRNA associated with ovarian cancer cells by genistein. Eur. J. Gynaecol. Oncol. 30 (6), 616–621.

Patel, D., Shukla, S., Gupta, S., 2007. Apigenin and cancer chemoprevention: progress, potential and promise (review). Int. J. Oncol. 30 (1), 233–245.

Péneau, S., Giudici, K.V., Gusto, G., Goxe, D., Lantieri, O., Hercberg, S., Rolland-Cachera, M.F., 2017. Growth trajectories of body mass index during childhood: associated factors and health outcome at adulthood. J. Pediatr. 186, 64–71. https://www.ncbi.nlm.nih.gov/pubmed/28283258.

Perri, F., Longo, F., Giuliano, M., Sabbatino, F., Favia, G., Ionna, F., et al., 2017. Epigenetic control of gene expression: potential implications for cancer treatment. Crit. Rev. Oncol. Hematol. 111, 166–172.

Piña, I.C., Gautschi, J.T., Wang, G.Y., Sanders, M.L., Schmitz, F.J., France, D., et al., 2003. Psammaplins from the sponge Pseudoceratina purpurea: inhibition of both histone deacetylase and DNA methyltransferase. J. Org. Chem. 68 (10), 3866–3873.

Prebet, T., Sun, Z., Ketterling, R.P., Zeidan, A., Greenberg, P., Herman, J., et al., 2016. Azacitidine with or without Entinostat for the treatment of therapy-related myeloid neoplasm: further results of the E1905 North American Leukemia Intergroup study. Br. J. Haematol. 172 (3), 384–391.

Priyadarsini, R.V., Vinothini, G., Murugan, R.S., Manikandan, P., Nagini, S., 2011. The flavonoid quercetin modulates the hallmark capabilities of hamster buccal pouch tumors. Nutr. Cancer 63 (2), 218–226.

Qin, W., Zhang, K., Clarke, K., Weiland, T., Sauter, E.R., 2014. Methylation and miRNA effects of resveratrol on mammary tumors vs. normal tissue. Nutr. Cancer 66 (2), 270–277.

Qin, W., Zhu, W., Shi, H., Hewett, J.E., Ruhlen, R.L., MacDonald, R.S., et al., 2009. Soy isoflavones have an antiestrogenic effect and alter mammary promoter hypermethylation in healthy premenopausal women. Nutr. Cancer 61 (2), 238–244.

Radsak, M., Platzbecker, U., Schmidt, C.S., Hofmann, W.K., Nolte, F., 2017. Infectious complications in patients with myelodysplastic syndromes: a review of the literature with emphasis on patients treated with 5-azacitidine. Eur. J. Haematol. 99 (2), 112–118. https://www.ncbi.nlm.nih.gov/pubmed/28321924.

Rajendran, P., Ho, E., Williams, D.E., Dashwood, R.H., 2011. Dietary phytochemicals, HDAC inhibition, and DNA damage/repair defects in cancer cells. Clin. Epigenet. 3 (1), 4.

Reuter, S., Gupta, S.C., Park, B., Goel, A., Aggarwal, B.B., 2011. Epigenetic changes induced by curcumin and other natural compounds. Genes Nutr. 6 (2), 93–108.

Rodríguez-Paredes, M., Esteller, M., 2011. Cancer epigenetics reaches mainstream oncology. Nat. Med. 17 (3), 330–339.

Roll, J.D., Rivenbark, A.G., Jones, W.D., Coleman, W.B., 2008. DNMT3b overexpression contributes to a hypermethylator phenotype in human breast cancer cell lines. Mol. Cancer 7, 15.

Ruiz, P.A., Braune, A., Hölzlwimmer, G., Quintanilla-Fend, L., Haller, D., 2007. Quercetin inhibits TNF-induced NF-kappaB transcription factor recruitment to proinflammatory gene promoters in murine intestinal epithelial cells. J. Nutr. 137 (5), 1208–1215.

Sana, J., Faltejskova, P., Svoboda, M., Slaby, O., 2012. Novel classes of non-coding RNAs and cancer. J. Transl. Med. 10, 103.

Sarkar, F.H., Li, Y., Wang, Z., Kong, D., 2009. Cellular signaling perturbation by natural products. Cell Signal. 21 (11), 1541–1547.

Schemmer, J., Araúzo-Bravo, M.J., Haas, N., Schäfer, S., Weber, S.N., Becker, A., et al., 2013. Transcription factor TFAP2C regulates major programs required for murine fetal germ cell maintenance and haploinsufficiency predisposes to teratomas in male mice. PLoS One 8 (8), e71113.

Seeler, J.S., Dejean, A., 2017. SUMO and the robustness of cancer. Nat. Rev. Cancer 17 (3), 184–197.

Serravalle, S., Bertuccio, S.N., Astolfi, A., Melchionda, F., Pession, A., 2016. Synergistic cytotoxic effect of L-Asparaginase combined with decitabine as a demethylating agent in pediatric T-all, with specific epigenetic signature. Biomed. Res. Int. 2016, 1985750.

Sharma, V., Kumar, L., Mohanty, S.K., Maikhuri, J.P., Rajender, S., Gupta, G., 2016. Sensitization of androgen refractory prostate cancer cells to anti-androgens through re-expression of epigenetically repressed androgen receptor - synergistic action of quercetin and curcumin. Mol. Cell. Endocrinol. 431, 12–23.

Shukla, S., Gupta, S., 2010. Apigenin: a promising molecule for cancer prevention. Pharm. Res. 27 (6), 962–978.

Sonnenschein, C., Soto, A.M., 2013. The aging of the 2000 and 2011 Hallmarks of Cancer reviews: a critique. J. Biosci. 38 (3), 651–663.

Sonoki, H., Sato, T., Endo, S., Matsunaga, T., Yamaguchi, M., Yamazaki, Y., et al., 2015. Quercetin decreases Claudin-2 expression mediated by up-regulation of microRNA miR-16 in lung adenocarcinoma A549 cells. Nutrients 7 (6), 4578–4592.

Sun, M., Estrov, Z., Ji, Y., Coombes, K.R., Harris, D.H., Kurzrock, R., 2008. Curcumin (diferuloylmethane) alters the expression profiles of microRNAs in human pancreatic cancer cells. Mol. Cancer Ther. 7 (3), 464–473.

Szarc Vel Szic, K., Declerck, K., Crans, R.A.J., Diddens, J., Scherf, D.B., Gerhäuser, C., Vanden Berghe, W., 2017. Epigenetic silencing of triple negative breast cancer hallmarks by withaferin A. Oncotarget 8 (25), 40434–40453. https://www.ncbi.nlm.nih.gov/pubmed/28467815.

Tan, S., Wang, C., Lu, C., Zhao, B., Cui, Y., Shi, X., Ma, X., 2009. Quercetin is able to demethylate the p16INK4a gene promoter. Chemotherapy 55 (1), 6–10.

Toyota, M., Ahuja, N., Ohe-Toyota, M., Herman, J.G., Baylin, S.B., Issa, J.P., 1999. CpG island methylator phenotype in colorectal cancer. Proc. Natl. Acad. Sci. U.S.A. 96 (15), 8681–8686.

Tsang, W.P., Kwok, T.T., 2010. Epigallocatechin gallate up-regulation of miR-16 and induction of apoptosis in human cancer cells. J. Nutr. Biochem. 21 (2), 140–146.

Tseng, T.H., Chien, M.H., Lin, W.L., Wen, Y.C., Chow, J.M., Chen, C.K., et al., 2017. Inhibition of MDA-MB-231 breast cancer cell proliferation and tumor growth by apigenin through induction of G2/M arrest and histone H3 acetylation-mediated p21(WAF1/CIP1) expression. Environ. Toxicol. 32 (2), 434–444.

Vahid, F., Zand, H., Nosrat-Mirshekarlou, E., Najafi, R., Hekmatdoost, A., 2015. The role dietary of bioactive compounds on the regulation of histone acetylases and deacetylases: a review. Gene 562 (1), 8–15.

Vargas, J.E., Filippi-Chiela, E.C., Suhre, T., Kipper, F.C., Bonatto, D., Lenz, G., 2014. Inhibition of HDAC increases the senescence induced by natural polyphenols in glioma cells. Biochem. Cell. Biol. 92 (4), 297–304.

Wachtel-Galor, S., Benzie, I.F.F., 2011. Herbal medicine: an introduction to its history, usage, regulation, current trends, and research needs. In: Benzie, I.F.F., Wachtel-Galor, S. (Eds.), Biomolecular and Clinical Aspects. CRC Press/Taylor & Francis, Boca Raton (FL).

Wang, T., Cui, H., Ma, N., Jiang, Y., 2013. Nicotinamide-mediated inhibition of SIRT1 deacetylase is associated with the viability of cancer cells exposed to antitumor agents and apoptosis. Oncol. Lett. 6 (2), 600–604.

Wang, K., Feng, X., Chai, L., Cao, S., Qiu, F., 2017. The metabolism of berberine and its contribution to the pharmacological effects. Drug Metab. Rev. 1–54.

Wang, R.H., Sengupta, K., Li, C., Kim, H.S., Cao, L., Xiao, C., et al., 2008. Impaired DNA damage response, genome instability, and tumorigenesis in SIRT1 mutant mice. Cancer Cell 14 (4), 312–323.

Watt, F., Molloy, P.L., 1988. Cytosine methylation prevents binding to DNA of a HeLa cell transcription factor required for optimal expression of the adenovirus major late promoter. Genes Dev. 2 (9), 1136–1143.

Weake, V.M., Workman, J.L., 2008. Histone ubiquitination: triggering gene activity. Mol. Cell 29 (6), 653–663.

Wu, T.Y., Khor, T.O., Su, Z.Y., Saw, C.L., Shu, L., Cheung, K.L., et al., 2013. Epigenetic modifications of Nrf2 by 3,3′-diindolylmethane in vitro in TRAMP C1 cell line and in vivo TRAMP prostate tumors. AAPS J. 15 (3), 864–874.

Xie, Q., Bai, Q., Zou, L.Y., Zhang, Q.Y., Zhou, Y., Chang, H., Mi, M.T., 2014. Genistein inhibits DNA methylation and increases expression of tumor suppressor genes in human breast cancer cells. Genes Chromosomes Cancer 53 (5), 422–431.

Xiong, S.D., Yu, K., Liu, X.H., Yin, L.H., Kirschenbaum, A., Yao, S., Levine, A.C., 2009. Ribosome-inactivating proteins isolated from dietary bitter melon induce apoptosis and inhibit histone deacetylase-1 selectively in premalignant and malignant prostate cancer cells. Int. J. Cancer 125 (4), 774–782.

Yang, X.J., Seto, E., 2007. HATs and HDACs: from structure, function and regulation to novel strategies for therapy and prevention. Oncogene 26 (37), 5310–5318.

Yang, F., Song, L., Wang, H., Wang, J., Xu, Z., Xing, N., 2015. Combination of quercetin and 2-methoxyestradiol enhances inhibition of human prostate cancer LNCaP and PC-3 cells xenograft tumor growth. PLoS One 10 (5), e0128277.

Yang, C.S., Wang, X., Lu, G., Picinich, S.C., 2009. Cancer prevention by tea: animal studies, molecular mechanisms and human relevance. Nat. Rev. Cancer 9 (6), 429–439.

Zagni, C., Floresta, G., Monciino, G., Rescifina, A., 2017. The search for potent, small-molecule HDACIs in cancer treatment: a decade after vorinostat. Med. Res. Rev. 37 (6), 1373–1428. https://www.ncbi.nlm.nih.gov/pubmed/28181261.

Zhang, H.Y., Cui, J., Zhang, Y., Wang, Z.L., Chong, T., Wang, Z.M., 2016. Isoflavones and Prostate Cancer: A Review of Some Critical Issues. Chin Med J (Engl). 129 (3), 341–347.

Zhang, X., Guo, Q., Chen, J., Chen, Z., 2015. Quercetin enhances cisplatin sensitivity of human osteosarcoma cells by modulating microRNA-217-KRAS Axis. Mol. Cells 38 (7), 638–642.

Zhang, Y., Wang, J., Chen, G., Fan, D., Deng, M., 2011. Inhibition of Sirt1 promotes neural progenitors toward motoneuron differentiation from human embryonic stem cells. Biochem. Biophys. Res. Commun. 404 (2), 610–614.

Zibelman, M., Wong, Y.N., Devarajan, K., Malizzia, L., Corrigan, A., Olszanski, A.J., et al., 2015. Phase I study of the mTOR inhibitor ridaforolimus and the HDAC inhibitor vorinostat in advanced renal cell carcinoma and other solid tumors. Investig. New Drugs 33 (5), 1040–1047.

Further Reading

Chen, Q., Chen, Y., Bian, C., Fujiki, R., Yu, X., 2013. TET2 promotes histone O-GlcNAcylation during gene transcription. Nature 493 (7433), 561–564.

Kumar, S.V., Saravanan, D., Kumar, B., Jayakumar, A., 2014. An update on prodrugs from natural products. Asian Pac. J. Trop. Med. 7S1, S54–S59.

Ropero, S., Esteller, M., 2007. The role of histone deacetylases (HDACs) in human cancer. Mol. Oncol. 1 (1), 19–25.

Seidel, C., Florean, C., Schnekenburger, M., Dicato, M., Diederich, M., 2012a. Chromatin-modifying agents in anti-cancer therapy. Biochimie 94 (11), 2264–2279.

Seidel, C., Schnekenburger, M., Dicato, M., Diederich, M., 2012b. Histone deacetylase modulators provided by mother nature. Genes Nutr. 7 (3), 357–367.

CHAPTER

4

Histone Deacetylase Inhibitors in Cancer Prevention and Therapy

Shahrzad Movafagh, Amanda Munson

Department of Pharmacogenomics, Bernard J Dunn School of Pharmacy, Shenandoah University, Fairfax, VA, United States

Abstract

Histone deacetylases (HDACs) deacetylate histone and nonhistone proteins. In general, histone deacetylation leads to chromatin condensation and decreased gene expression, while the acetylation status of nonhistone proteins modulates a variety of cellular processes involved in survival and growth. In many cancers, alteration of HDAC function results in the altered expression and/or regulation of many proteins. Histone deacetylase inhibitors (HDACIs) allow for exposure of the chromatin structure, leading to reactivation of silenced genes as well as increased acetylation of nonhistone proteins, which together combat tumor cell survival and proliferation. HDACIs, as natural and synthetic products, have shown promise in cancer chemoprevention. In this chapter, we present an overview of HDAC function and association with cancers. We further discuss various HDACI classes and their preclinical and clinical efficacy as monotherapies or combination treatments for various cancers. We conclude with a discussion of novel HDACI development as a promising approach to chemoadjuvant therapy.

Keywords

Cancer treatment; Chemoprevention; Histone deacetylase; Histone deacetylase inhibitor; Histone modification

INTRODUCTION

Genome Organization

In total, the human genome is estimated to contain approximately three billion nucleotide pairs (Consortium, 2004). If all the DNA molecules in just one human cell were lined up end to end, their total length would be approximately 2 m (Chromosomes Fact Sheet, n.d.); however, this DNA is localized to the nucleus, an organelle with a 6 μm diameter (Alberts et al., 2014). Precisely condensing and compacting the DNA of each chromosome around special DNA-binding proteins addresses these intracellular space limitations (Andrews and Luger, 2011; Cutter and Hayes, 2015; Luger et al., 1997). There are two general classes of these proteins, histones and nonhistone chromosomal proteins (Alberts et al., 2014). Histones provide the first level of packaging and form the nucleosome. Nonhistone chromosomal proteins provide additional levels of DNA packaging. The combination of protein and DNA together is

Epigenetics of Cancer Prevention
https://doi.org/10.1016/B978-0-12-812494-9.00004-4

the chromatin of a cell (Alberts et al., 2014; Felsenfeld, 2014).

Several histone forms—i.e., H2A, H2B, H3, and H4—make up the histone core, while histone H1 links neighboring histones together (Cutter and Hayes, 2015). Each nucleosome core contains two each of the H2A, H2B, H3, and H4 histone monomers. These monomers have a conserved globular domain referred to as a histone fold, which interacts with the histone folds of the other monomers present in the nucleosome core (Felsenfeld, 2014). Each monomer has a divergent N-terminal tail that extends out from the histone core and serves as a binding site for the nonhistone chromosomal proteins and the location for many regulatory modifications (Cutter and Hayes, 2015; Tan et al., 2011).

Regulation of Gene Expression

All cells in a human body contain the same genetic information, but not all cells express the same genes, contributing to the wide variety of tissues and organs present in the human body. Similarly, different genes are expressed at different times in human development. These differences in tissue and developmental expression can in part be attributed to the expression and stability of transcription factors, DNA methylation, and/or chromatin modifications, including histone modifications (Felsenfeld, 2014). Transcription factors are proteins that bind to DNA in a sequence-specific manner and recruit other proteins, including histone-modifying enzymes. These histone-modifying enzymes can, among other things, reversibly modify histones via acetylation, methylation, and phosphorylation, influencing the ability of DNA to associate with histones and for the histones to associate with other proteins. These protein–protein interactions influence gene expression and ultimately the function of the cell.

Transcription of DNA into mRNA is influenced by the accessibility of the DNA to the transcriptional machinery (e.g., RNA polymerase and transcription factors); highly condensed DNA has relatively lower transcriptional activity than less tightly condensed DNA, providing one level of regulation (Li and Seto, 2016).

ACETYLATION OF HISTONE AND NONHISTONE PROTEINS REGULATES THEIR FUNCTIONS

Histones

Histone acetylation, performed by a collection of proteins referred to as histone acetyl transferases (HATs), occurs on specific lysines in the histone tail of each core histone. This often involves other histone modifications, such as trimethylation of specific other lysines in the vicinity, and a specific HAT or HAT complexes that recognize the histone tail and associated modifications of specific histones (Marmorstein and Zhou, 2014). The addition of an acetyl group to a lysine removes the positive charge from the R-group, thereby changing the structure and as a consequence the function of the histone. This results in loosening of the association between the negatively charged sugar–phosphate backbone of the DNA and the positively charged histone. This "looser" association makes the DNA more accessible to transcriptional machinery, resulting in increased gene expression. Furthermore, the acetylation of lysines on histones has an impact on the ability of other posttranslational modifications (e.g., phosphorylation, ubiquitination) (Caron et al., 2005; Tan et al., 2011) to occur on the lysines themselves as well as to other amino acids, further altering the DNA-spooling function of histones.

Conversely, the removal of these acetyl groups by proteins referred to as histone deacetylase complexes (HDACs) is thought to decrease gene expression by allowing DNA to associate more tightly with the histone, rendering the DNA less accessible to transcriptional machinery, and as a consequence, less transcriptionally active. However, this has been shown to be a rather oversimplified explanation, as acetylation of histones can also influence the binding of other transcription factors and regulatory proteins to chromatin, thus bringing these proteins into proximity of DNA (Seto and Yoshida, 2014).

Nonhistone Proteins

Acetylation of lysines can also occur on nonhistone transcriptional regulators (p53), cytoskeletal proteins (α-tubulin), and protein-folding chaperones (HSP90) (Caron et al., 2005; Kouzarides, 2000). Lysine acetylation can have wide-reaching effects. Acetylation of lysines can alter the three-dimensional structure of a protein, affecting its ability to bind other proteins or DNA. Furthermore, acetylation can have an impact on the subcellular localization, activity, and stability of a protein. Additionally, lysines are a common target for a variety of posttranslational modifications (Caron et al., 2005; Schwartz et al., 2011). Notably, lysine acetylation can influence lysine ubiquitination, thereby affecting ubiquitin-dependent protein degradation through the proteasome (Caron et al., 2005).

HISTONE DEACETYLASE COMPLEXES ARE RESPONSIBLE FOR THE DEACETYLATION OF HISTONES AND NONHISTONE PROTEINS

Two main families of proteins can deacetylate histones (and other substrates): "classical" HDACs, which utilize a Zn^{2+}-binding bromodomain in their catalytic site; and sirtuins (SIRTs), which utilize NAD^+ in their catalytic site (Imai et al., 2000). Table 4.1 lists various types of HDACs and their relation to certain cancers. The "classical" HDACs can be divided into three classes based on their structural similarities to yeast HDACs: class I (Rpd3p-like), class II (Hda1p-like), and class IV (Rpd3p-like and Hda1p-like). The SIRT family of proteins is the only member of class III (Sir2p-like) that is involved in protein deacetylation (Seto and Yoshida, 2014).

It is worth noting that class I, II, and IV HDACs are likely evolved from a prokaryotic acetylpolyamine amidohydrolase (Gregoretti et al., 2004; Leipe and Landsman, 1997). Since prokaryotes do not utilize histones to package DNA, these ancestors to modern "classical" HDACs likely targeted cellular proteins other than histones. In fact, acetylation has been demonstrated to be a regulatory mechanism used among both eukaryotes and prokaryotes (Kouzarides, 2000; Wang et al., 2010). Thus, while HATs and HDACs by name suggest that their primary mechanism of action is controlling the acetylation state of histones, acetylation/deacetylation is likely a far more ancient cellular regulatory mechanism targeting nonhistone proteins as well as histones (Kouzarides, 2000). Similarly, class III HDACs are also evolutionarily conserved (Grunstein and Gasser, 2013) in agreement with the observation that they have cellular targets in addition to histones. This is consistent with the fact that SIRTs have both an HDAC function and a mono−ADP−ribosyltransferase activity. Furthermore, some SIRTs have additional in vitro functionality (Du et al., 2011) as well as localization to different subcellular spaces (nucleus, cytoplasm, mitochondria, and nucleolus) (Seto and Yoshida, 2014).

It has been difficult to identify specific targets for each classical (classes I, II, and IV) HDAC, because when purified to homogeneity, these HDACs have low deacetylase function. Instead, many classical HDACs function as part of larger multiprotein complexes (e.g., CoREST, Sin3, NuRD, and NCoR) with greater activity than

TABLE 4.1 Histone Deacetylase Proteins Discussed in This Chapter[a]

Family	Class	Cellular Localization	Connection to Select Cancers
"Classical" Histone Deacetylase Family	**Class I**		
	HDAC1	nucleus	**Increased in:** various hematologic cancers; lung, gastric, liver, colorectal, and estrogen/progesterone receptor–positive breast cancers
	HDAC2	nucleus	**Decreased in:** some colorectal cancers **Increased in:** some colorectal cancers; various hematologic cancers; lung, gastric, liver, pancreatic, and estrogen/progesterone receptor–positive breast cancers
	HDAC3	nucleus	**Increased in:** various hematologic cancers; lung, gastric, liver, colorectal cancers
	HDAC8	nucleus	**Increased in:** ALL
"Classical" Histone Deacetylase Family	**Class IIa**		
	HDAC4	nucleus/cytoplasm	**Increased in:** T-ALL
	HDAC5	nucleus/cytoplasm	**Decreased in:** lung and liver cancers; AML **Increased in:** colorectal cancer
	HDAC7	nucleus/cytoplasm	**Increased in:** various hematologic cancers; pancreatic and colorectal cancers
	HDAC9	nucleus/cytoplasm	**Increased in:** various hematologic cancers
"Classical" Histone Deacetylase Family	**Class IIb**		
	HDAC6	cytoplasm (limited nucleus)	**Decreased in:** liver cancer; some hematologic cancers **Increased in:** pancreatic; estrogen-receptor-positive breast cancers; some hematologic cancers
	HDAC10	nucleus/cytoplasm	**Decreased in:** lung; gastric **Increased in:** CLL
Sirtuins	**Class III**		
	SIRT1	nucleus	**Decreased in:** breast and liver cancers **Increased in:** various hematologic cancers; colorectal cancer
	SIRT2	cytoplasm/nucleus	**Decreased in:** breast and liver cancer
	SIRT3	mitochondria/nucleus	**Decreased in:** breast, lung, liver, and pancreatic cancers; CLL
	SIRT4	mitochondria	**Decreased in:** gastric cancer, breast cancer, and various hematologic cancers
	SIRT5	mitochondria	**Increased in:** non-small-cell lung cancer
	SIRT6	nucleus	**Decreased in:** pancreatic, colorectal, and liver cancer Increased in: breast cancer; CLL
	SIRT7	nucleus/nucleolus	**Increased in:** colorectal cancer
"Classical" Histone Deacetylase Family	**Class IV**		
	HDAC11	nucleus/cytoplasm	**Increased in:** various hematologic cancers; colorectal and breast cancers

[a] *Expression of all HDACs has not been examined in all potential cancers, and thus the expression profiles of all HDACs in all cancers is not fully known.* *ALL*, acute lymphoblastic leukemia; *AML*, acute myelogenous leukemia; *CLL*, chronic lymphocytic leukemia; *T-ALL*, T-cell acute lymphoblastic leukemia.

do HDACs alone, and with different substrate specificity (Seto and Yoshida, 2014; West and Johnstone, 2014). Additionally, a high degree of sequence similarity exists within and between HDAC classes consistent with redundancy of function between classes.

In contrast, it has been possible to determine the histone substrate specificity of class III HDACs (SIRTs). SIRT1, the human SIRT with the strongest activity, has been demonstrated to deacetylate lysine 16 of histone 4 (H4K16) as well as H3K9 and H1K26, and to influence heterochromatin formation (Vaquero et al., 2004). SIRT2 normally localizes to the cytoplasm but can enter the nucleus and deacetylate H4K16 at the G_2/M transition (Bosch-Presegué and Vaquero, 2011; Michishita et al., 2005). SIRT3 deacetylates H4K16 but also functions in the mitochondria as a deacetylase during cellular stress (Finkel et al., 2009; Kim et al., 2010; Martínez-Redondo and Vaquero, 2013; Scher et al., 2007; Someya et al., 2010). SIRT4 and SIRT5 do not have in vivo HDAC function, as they have a mitochondrial localization. SIRT6 appears to have relatively week HDAC activity, but does deacetylate H3K9 and H3K56 (Michishita et al., 2008). SIRT7 deacetylates H3K18 (Barber et al., 2012), a key acetylated residue associated with promoters and transcriptional activation (Wang et al., 2008); depletion of H3K18 acetylation is involved in oncogenic transformation, aggressive cancers, and poor prognosis (Barber et al., 2012).

In addition to deacetylation of histones by HDACs and SIRTs, many other acetylated proteins, such as p53, can be deacetylated by HDAC1 (Luo et al., 2000) and SIRT1 (Vaziri et al., 2001). Large-scale screens examining protein acetylation have identified acetylated proteins in various subcellular localizations and involved in different cellular processes. Further, studies utilizing HDAC inhibitors (HDACIs) and SIRT1 knockouts have revealed increased (Choudhary et al., 2009; Peng et al., 2012) acetylation of nonhistone proteins with various cellular functions. Therefore, the regulatory functions of HDACs on both histones and nonhistone proteins are of great importance.

HISTONE DEACETYLASE COMPLEXES REGULATE MANY CELLULAR PROCESSES

Acetylation, and by extension deacetylation by HDACs, not only targets the lysines of histones, but also of other DNA-binding proteins, such as transcription factors, nuclear import factors, and cytoskeletal proteins. Much of what we know about HDAC function comes from studies performed with HDACIs and investigations utilizing HDAC deletions, mutations, and knockdown or overexpression studies.

HDAC inhibition results in acetylation of histones, which relaxes the chromatin structure and allows for transcriptional access to DNA strands and expression of various genes. HDACIs also enable acetylation of nonhistone proteins involved in transcriptional activity, intracellular signal transduction, and DNA repair (Lakshmaiah et al., 2014; Marks et al., 2004). Among the genes upregulated via HDACI activity are those that promote cell cycle arrest, apoptosis, DNA repair, immunogenicity, and reduced angiogenesis, all of which are antineoplastic in nature (Marks et al., 2004; Marks and Dokmanovic, 2005). Fig. 4.1 is an illustration of HDAC histone and nonhistone targets and subsequent cellular pathways that relate to antitumor activity of HDAC inhibition. In general, several specific mechanisms of action result from HDAC inhibition. Some of these known mechanisms are described below.

Cell Cycle

A specific result of HDAC inhibition is accumulation of acetylated histones intracellularly that are thought to affect centromere and kinetochore function. Increased acetylated histones during the S and G_2 phases of the cell cycle

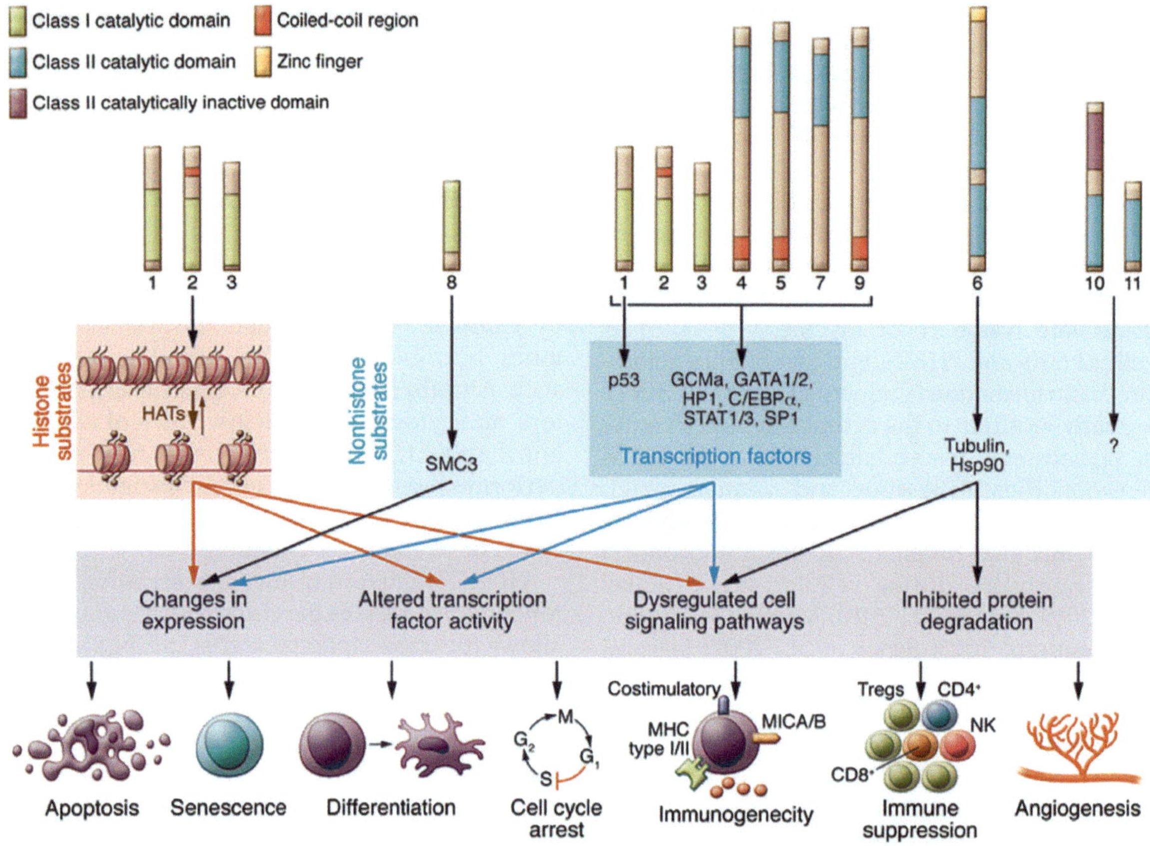

FIGURE 4.1 Illustration of HDAC histone and nonhistone targets and subsequent cellular pathways that relate to antitumor activity of HDAC inhibition. The various HDAC types and their classifications are indicated in the legend. The middle panel emphasizes the mechanisms of action, and the bottom tier outlines the biological outcomes of HDAC inhibition. *GCMa*, glial cells missing homolog 1; *HAT*, histone acetyltransferase; *HP1*, heterochromatin protein 1; *SMC3*, structural maintenance of chromosome 3. *Reproduced from West, A.C., Johnstone, R.W., 2014. New and emerging HDAC inhibitors for cancer treatment. J. Clin. Investig. 124, 30–39, with permission.*

trigger a G_2 checkpoint, which can lead to cell cycle arrest at the G_2 phase (Johnstone, 2002; Marks et al., 2004). Among the genes induced by histone deacetylation are cell cycle kinase inhibitor $p21^{WAF1}$, which leads to inhibition of cyclin E and cyclin A and eventual cell cycle arrest (Marks et al., 2004; Marks and Jiang, 2005; Sherr and Roberts, 1995). Additional genes involved in cell cycle progression are cyclin D and cyclin A, which are inhibited by use of HDACIs. Cyclins are subunits of cyclin-dependent kinases (CDKs), which serve as the regulatory controllers of cell cycle progression and facilitate promotion of the G_1 phase to the S phase (Sherr and Roberts, 1995). Repression of cyclin D and cyclin A leads to decreased activity of CDKs and results in cell cycle arrest.

HDACs regulate the G_1/S and the G_2/M checkpoints of the cell cycle both by regulation of transcription (repression of cell cycle genes such as cyclins and CDKs) and by stability of the cyclins, or the ability to complete mitosis.

HDAC1 and HDAC2 binding to the promoters of p21, p27, and p57 results in decreased expression of these CDK inhibitors (CDKIs), thereby allowing the cell cycle to proceed through the G_1/S checkpoint. Inhibition of HDAC1 was shown to increase the expression of p21 and p27 (Xie et al., 2012; Yamaguchi et al., 2010), which could stop progression through the cell cycle. Similarly, HDAC5 has also been demonstrated to play a role in the G_1/S transition, with knockdown of HDAC5 expression leading to upregulation of the CDKI p21 and a resultant decrease in the expression of cyclin D1 and CDKs 2, 4, and 6 (Fan et al., 2014), which could lead to cell cycle arrest.

Knockdown of HDAC1 has also been shown to inhibit the transition from G_2 to M in various tumor cell lines (Senese et al., 2007), again suggesting that HDAC1 is normally important for progression of the cell cycle. HDAC10 has also been demonstrated to play a role in the G_2/M transition by increasing the cyclin A2 transcription (Li et al., 2015).

The action of HDACs can also impact the progression through mitosis. Acetylated cyclin A is prone to degradation; however, during metaphase, HDAC3 deacetylates cyclin A, making it more stable and allowing mitosis to proceed. Secondly, HDAC3 in conjunction with the A-kinase-anchoring proteins AKAP95 and HA95 localizes to chromosomes in mitotic cells. These three proteins, acting in concert, remove acetyl groups from the H3 tail, providing a binding surface for the Aurora B kinase, which is responsible for chromosome condensation and segregation in mitosis (Li et al., 2006).

Apoptosis

Several mechanisms have been elucidated with regard to the proapoptotic effects of HDACIs. A key proapoptotic mechanism of HDACIs is upregulation of p53 activity via its hyperacetylation (Marks and Jiang, 2005). Additional mechanisms include activation of extrinsic apoptotic pathways via CD95, DR4, and DR5 death receptors and TRAIL as well as FasL coligands on tumor cells (Newbold et al., 2016; and West and Johnstone, 2014). Furthermore, activation of proapoptotic genes TNFSF10, BMF, Bax, and BaK and downregulation of antiapoptotic genes such as Mcl-1, BCl-2, and Bcl-X have been observed with HDACIs (Lakshmaiah et al., 2014; Lane and Chabner, 2009; West and Johnstone, 2014; Zhang et al., 2004). Increased expression of transcription factors such as E2F1 and FoxO1 and induction of proapoptotic BH3 family proteins such as Bim, Puma, Noxa, and SP1 have been seen with the use of HDACIs in various cancers (Inoue et al., 2007; Kang et al., 2014; Waby et al., 2010; Yang et al., 2009; Zhao et al., 2005). One proposed mechanism of increased expression of proapoptotic genes is through hyperacetylation of the promoter region of these various genes (Newbold et al., 2016).

Modified expression of certain microRNAs such as miR-15a, miR-16, and cluster miR-106b-93-25 also accounts for proapoptotic HDACI activity in different cancers (Sampath et al., 2012; Zhao et al., 2012). In chronic lymphocytic leukemia (CLL), miR-15a and miR-16 are silenced via chromosomal deletions or suppressed via histone deacetylation of chromatin on their promotor regions (Sampath et al., 2012). In endometrial cancer (EMC) cells, HDACI-induced miR-106b-93-25 suppression results from decreased Myc binding to the promotor region of miR-106b-93-25 host gene MCM7 (Zhao et al., 2012). Furthermore, in EMC cells, Myc protein levels are decreased under HDACI treatment, and low levels and Myc are found to be associated with increased proapoptotic activity (Zhao et al., 2012). On the contrary, in breast and lung cancer cells, Myc is expressed at low levels, and HDACIs seem to reactivate Myc. Myc in turn induces miR-15 and the Let-7 family, which inhibits Bcl-2 and Bcl-xL and leads to tumor cell apoptosis (Adams and Eischen, 2016).

Cytoplasmic FLICE-like inhibitory protein (c-FLIP), an inhibitor of the proapoptotic protein capsase-8, is regulated by HDAC1, HDAC2, and HDAC3; inhibition of these HDACs resulted in decreased c-FLIP expression and increased apoptosis in non-small-cell lung cancer (NSCLC) cells (Riley et al., 2013). Decreased c-FLIP expression in response to HDAC inhibition has been shown to be connected to increased acetylation of the Ku70 protein involved in nonhomologous end-joining (NHEJ) DNA repair, which results in an increase in c-FLIP polyubiquitination and resultant proteasomal degradation (Kerr et al., 2012). In another study, c-FLIP downregulation by HDACIs was inhibited by proteasome inhibitor MG132, further suggesting that ubiquitin-proteasome pathway is involved in c-FLIP downregulation (Riley et al., 2013).

HDAC2 is also generally antiapoptotic; knockdown of HDAC2 expression has been demonstrated to enhance TRAIL-induced apoptosis (Schüler et al., 2010), activate p53 and Bax expression and suppress Bcl2 expression leading to increased apoptosis (Jung et al., 2012), and induce proapoptotic factors (e.g., Bax, apoptosis-inducing factor, and Apaf-1) and repressed antiapoptotic factors (e.g., Bcl2) (Kim et al., 2013). HDAC8 is involved in regulating transcription of the proapoptotic protein BMF; inhibition of HDAC8 has been shown to activate transcription of BMF, leading to increased apoptosis (Kang et al., 2014).

Tumor Cell Immunogenicity

Another key function of HDACIs in cancer therapy is induced immunogenicity of tumor cells. Tumor growth in part involves downregulation of the immunogenic response (Setiadi et al., 2008). It has been shown that HDACIs increase expression of tumor antigens and antigen presentation by MHC I and II proteins (Chou et al., 2005; Manning et al., 2008; Setiadi et al., 2008; Tomasi et al., 2010). NKG2DL is a ligand that binds the activating NKG2D receptor expressed on natural killer (NK) cells and cytotoxic $CD8^{+}$ cells, leading to tumor cell lysis. Certain HDACIs such as sodium valproate have been shown to induce antigen presentation and tumor cytotoxicity via NKG2DL expression (Armeanu et al., 2005; Skov et al., 2005).

Differentiation

One aspect of tumorigenesis is inhibition of differentiation. Oncogenic fusion proteins such as PML−RARα and AML−1ETO formed by chromosomal translocation have been shown to partly induce tumor growth by inhibiting cellular differentiation (Ablain and de The, 2011). These fusion proteins recruit HDACs to bind to DNA and inhibit transcription of cellular differentiation genes. Thus, a key advantage of HDACIs in cancer therapy is reinduction of cellular differentiation (Leder and Leder, 1975). For example, treatment with the HDACI vorinostat in pediatric NUT midline carcinomas has led to restored differentiation and antitumor efficacy (Schwartz et al., 2011). Another effective approach to inducing differentiation in myeloid cells is via use of retinoic acid (RA). RA-induced differentiation is through binding to PML−RARα fusion protein. RA binding leads to dissociation of PML−RARα−associated HDAC from the chromatin structure, allowing for gene transcription (Minucci et al., 2001). Consequently, the combining of HDACIs with RA has been suggested as a promising approach to APL therapy.

DNA Damage and Repair

It has been shown that HDACIs inhibit DNA repair proteins by depressing the transcription of several repair genes such as Ku70, Ku80, RAD50, RAD51, MRE11, DNA-PK, BRCA1, EXO1, CHK1, and CHK2 (Adimoolam et al., 2007; Lee et al., 2010a,b; Robert and Rassool, 2012). Loss of DNA repair leads to tumor cell DNA damage and death. Furthermore, acetylation of histones and other proteins such as Ku70 leads to DNA exposure, which is another mechanism of DNA damage induced by HDACIs (Lee et al., 2010a,b). Other mechanisms include reactive oxygen species (ROS)–mediated DNA damage and activation of histone gammaH2AX, which results in replication-mediated DNA damage (Conti et al., 2010; Petruccelli et al., 2011).

HDAC1 and HDAC2 have been implicated in the regulation of NHEJ, an important aspect in the repair of double-strand breaks (Miller et al., 2010), whereas HDAC9 and HDAC10 have been implicated in DNA repair by homologous recombination (HR) (Kotian et al., 2011). Additionally, HDAC3 has been shown to be involved in both NHEJ and HR double-strand break-repair mechanisms (Bhaskara et al., 2010). HDAC4 and the p53 binding protein 53BP1 localize to sites of double-stranded DNA damage and play a role in progression through the G_2 checkpoint of the cell cycle (Kao et al., 2003). In addition to its function as a deacetylase, HDAC6 functions as a ubiquitin E3 ligase. In concert, these two activities regulate DNA mismatch repair (MMR) through deacetylation and ubiquitination of MSH2 (mutSα), a key protein in MMR (Zhang et al., 2014). SIRT1 and SIRT6 affect multiple aspects of DNA damage repair pathways by deacetylating proteins involved in both HR and NHEJ repair pathways (Li and Zhu, 2014), and SIRT3 has been demonstrated to contribute to the protection of cells from oxidative stress (Scher et al., 2007; Someya et al., 2010).

Generation of Reactive Oxygen Species

Another mechanism by which HDACIs induce tumor cell death is via generation of ROS (Ruefli et al., 2001). The mechanism of generation of ROS by HDACIs is thought to involve activation of mitochondrial proapoptotic Bcl-2 family member Bid (Ruefli et al., 2001). ROS in turn increases the release of the mitochondrial cytochrome *c*, resulting in tumor cell death. Along these lines, the use of antioxidants and free radical scavengers have been shown to reverse HDACI-induced cell death (Ruefli et al., 2001; Newbold et al., 2016). Another suggested mechanism of ROS generation in HDACI therapy is the increased expression of thioredoxin binding protein-2, which undermines the antioxidant effect of thioredoxin, leading to ROS accumulation and subsequent cell death in transformed cells (Marks and Jiang, 2005). ROS can further induce cell death via DNA damage and activation of histone gammaH2AX, which results in replication-mediated DNA damage (Conti et al., 2010; Petruccelli et al., 2011).

Autophagy

A controversial aspect of HDACI therapy is its effect on autophagy-related cell death. On the one hand, it has been shown that HDACIs can induce autophagy in tumor cells via key autophagy genes such as Atg5 and Atg7 (Robert et al., 2011; Newbold et al., 2016). More specifically, in Atg5 and Atg7 deleted cells, HDACIs cannot induce autophagy (Newbold et al., 2016). On the other hand, the induction of autophagy does not always induce cell death and may be used as a defense mechanism by tumor cells to prevent HDACI-induced apoptosis (Dupéré-Richer et al., 2013; Newbold et al., 2016). This has been shown in HDACI-resistant cells that have high basal levels of autophagy to protect the cells from apoptosis and preserve HDACI-resistant cell lines. Along these lines, inhibitors of autophagy have been shown to restore sensitivity to HDAC inhibition (Dupéré-Richer et al., 2013).

Autophagy has multiple roles within a cell. It regulates the cellular stress/starvation response, organelle recycling, degradation of protein aggregates, cell differentiation, and cell death/survival, processes that can be both pro- and anti-oncogenic (Ouyang et al., 2012). One role of HDAC1 and HDAC2 in autophagy is to increase the transcription of genes involved in the autophagic pathway. Deletion of these two HDAC genes decreases the expression of proteins involved in autophagy and thus autophagic flux (Moresi et al., 2012). Further, HDAC1 inhibits autophagy; reduction of HDAC1 activity, either by HDAC1 siRNA/shRNA knockdown or HDAC1 inhibition, leads to increased formation of autophagic vacuoles/autophagosomes and autophagy (Jeong et al., 2009; Oh et al., 2008).

Another major player in the regulation of autophagy is class IIb HDAC6, which connects the autophagic pathway with proteasomal degradation. Unlike HDAC1 and HDAC2, which play a role in the formation of autophagosomes, HDAC6 is involved in the fusion of autophagosomes to lysosomes via the actin cytoskeleton (Lee et al., 2010a,b). Additionally, HDAC6 deacetylates proteins such as salt-inducible kinase 2, thereby increasing its activity and autophagosome processing (Yang et al., 2013). HDAC10, another class IIb HDAC, also functions in the fusion of autophagosomes and lysosomes, notably via deacetylation of heat shock protein Hsp70 (Oehme et al., 2013).

SIRT1 plays a role in autophagy via regulation of the acetylation state of proteins involved in the autophagy pathway and has also been shown to have a role in oxidative-stress-induced autophagy (Huang et al., 2015; Lee et al., 2008; Ou et al., 2014). SIRT3, which has a mitochondrial localization (Lombard et al., 2007; Michishita et al., 2005), regulates cellular responses to oxidative stress by inducing mitophagy by deacetylation of FOXO3 (Tseng et al., 2013). SIRT5, which also localizes to the mitochondria, inhibits autophagy and mitophagy in a glutamine-dependent manner (Li and Seto, 2016; Polletta et al., 2015), and SIRT6, which localizes to the nucleus, has been demonstrated to promote autophagy in primary neonatal rat cardiomyocytes (Lu et al., 2016).

Gene Transcription

HDACIs modulate transcriptional activity of a number of genes highly involved in the activity and proliferation of tumor cells. This modulation occurs through acetylation of nonhistone proteins. HDACIs increase transcriptional activity of p53, E2F, and nuclear factor-κB, prevent the expression and activity of c-Myc, and induce the degradation of hypoxia-inducible factor 1 (HIF-1) (Lakshmaiah et al., 2014.; Lane and Chabner, 2009; Newbold et al., 2016). HDACIs can also indirectly affect the oncogenic process by increasing the activity of proteins that affect oncoprotein stability or function. One suggested mechanism for this process is hyperacetylation of heat shock protein 90 (Hsp90). Hsp90 serves as a key chaperone to a number of oncoproteins such as AKT, c-KIT, Her-2, Bcr-Abl, c-RAF, and BRAF (Newbold et al., 2016). Decreased association of Hsp90 with target proteins leads to their proteasomal degradation and subsequent antitumor activity. An example of this type of inhibition is decreased association of Hsp90 with Bcr-Abl, leading to Bcr-Abl degradation and subsequent tumor cell apoptosis in chronic myeloid leukemia (Bali et al., 2005; Newbold et al., 2016).

Angiogenesis

A number of HDACIs are known to prevent angiogenesis by repressing expression and activity of vascular endothelial growth factor (VEGF) (Deroanne et al., 2002; Liang et al., 2006). It has been shown that VEGF-induced angiogenesis is reversed by HDACIs through induction of VEGF competitor semaphorin III (Deroanne et al., 2002).

Another key antiangiogenic mechanism of HDACIs is suppression of HIF-1. HIF-1 is a master regulator of hypoxic signaling and induces adaptive responses to hypoxemia by inducing transcription of many genes, including proangiogenic and prometabolic genes. HIF-1 activity and levels are increased in many cancers. An effective therapeutic aspect of HDACIs in cancer is thought to be suppression of HIF-1. Repression of HIF-1 by HDACIs has been suggested to involve ubiquitination of the HIF-1α subunit through increased expression of Von Hippel–Lindau (VHL) and direct HIF-1α acetylation leading to its ubiquitination (Liang et al., 2006). Later studies have found VHL-independent mechanisms of HIF-1α degradation, including hyperacetylation of chaperone protein Hsp90. Decreased activity of Hsp90 is thought to cause misfolding of the HIF-1α subunit, leading to its proteasomal degradation (Liang et al., 2006).

Under normoxic conditions, HIF-1α is acetylated, which allows for binding of the E3 ubiquitin ligase VHL to HIF-1α. This leads to HIF-1α ubiquitination and proteasomal degradation (Jeong et al., 2002; Lee et al., 2004). HDAC1 deacetylates HIF-1α, resulting in the persistence of HIF-1α (Yoo et al., 2006).

Like HDAC1, HDAC4 also deacetylates HIF-1α, resulting in persistence of HIF-1α (Geng et al., 2011). HDAC5 deacetylates Hsp70, resulting in decreased association with HIF-1α and increased association with Hsp90 (Chen et al., 2015). HDAC6 deacetylates Hsp90, increasing the interaction between this chaperone protein and HIF-1α, resulting in increased HIF-1α stability (Kong et al., 2006).

The effect of HDACs on angiogenesis is variable and HDAC specific. HDAC5 represses the expression of genes involved in angiogenesis by binding to their promoters (Urbich et al., 2009), whereas HDAC6 regulates angiogenesis in endothelial cells by deacetylating cortactin, permitting cell migration and vascular growth (Kaluza et al., 2011). Interestingly, in hepatocellular carcinoma (HCC), HDAC6 has antiangiogenic effects, in part through HIF-1α and VEGFA (Lv et al., 2016). Additionally, HDAC2 and HDAC3, in conjunction with KLF-4, repress VEGF transcription by binding to the VEGF promoter (Ray et al., 2013).

HDAC4 and HDAC5 enhance recruitment of the transcriptional coactivator p300 to HIF-1α, resulting in increased HIF-1 activity (Seo et al., 2009). Similarly, HDAC7 increases the transcriptional activity of HIF-1α via p300 recruitment (Kato et al., 2004) but also promotes angiogenesis and vascular maintenance by regulating histone acetylation and by extension gene expression of the promoter of the antiangiogenic gene AKAP12 (Turtoi et al., 2012).

HIF-1α can also be deacetylated by SIRT1, and under hypoxic conditions SIRT1 is downregulated, resulting in sustained HIF-1α acetylation and by extension activation (Lim et al., 2010). SIRT3, a mitochondrially localized deacetylase, regulates the cellular response to ROS by reducing ROS levels and destabilizing HIF-1α (Chalkiadaki and Guarente, 2015). SIRT6 can function as a regulator of HIF-1α transcriptional activity by altering the acetylation of H3 near HIF-1's target gene promoters, particularly those involved in glycolytic pathways (Zhong et al., 2010). SIRT7 has been shown to be a negative regulator of HIF function independent of its deacetylase activity (Hubbi et al., 2013).

Natural and Synthetic Histone Deacetylase Complex Inhibitors in Cancer Therapy

HDACIs are a variety of natural and synthetic compounds that target different HDACs and have a wide range of activities both mechanistically and in treating various types of cancers. The key activity of HDACIs is via interactions with the catalytic zinc ions of HDAC enzymes. The classification of current HDACIs is listed in Table 4.2. In general, these drugs are classified into groups based on their chemical structure as well as their ability to block one or more HDACs. This classification further determines the mechanism of action of different HDACIs.

Short-Chain Fatty Acids

Short-chain fatty acids inhibit HDACs by chelating zinc ions via their carboxylate groups (Folmer et al., 2010). Sodium butyrate is a prototype HDACI that primarily targets HDAC types I, IIa, and IV (Seidel et al., 2012). Sodium butyrate is formed in the gastrointestinal tract by

TABLE 4.2 HDAC Inhibitors in Cancer Therapy

Class	HDAC Class Affected	Source
SHORT-CHAIN FATTY ACIDS		
Butyrates	I, II, IIa, IV	Natural
Sodium propionate	I, II, IIa, IV	Natural
Valproic acid	I, II	Synthetic
AR-42	I, IIb	Synthetic
HYDROXAMIC ACIDS		
Trichostatin A	I, II, IV	Natural
SAHA	I, II, IV	Synthetic
Abexinostat	I, II	Synthetic
Belinostat	I, II, IV	Synthetic
CHR-3996	I, II, IV	Synthetic
CUDC-101	I, II, EGFR, HER2	Synthetic
Givinostat	PI3K[a]	Synthetic
Panobinostat	I, II, IV	Synthetic
Pracinostat	I, II, IV	Synthetic
Quisinostat	I, II, IV	Synthetic
Resminostat	I, II	Synthetic
	I, II	
CYCLIC PEPTIDES		
Apicidin D	I	Natural
FK235222	I	Natural
DEPSIPEPTIDES		
Largazole	I	Synthetic
Romidepsin	I, II	Synthetic
Spiruchostatin A	I, HDAC10	Synthetic
BENZAMIDES		
Entinostat	HDAC1, HDAC3	Synthetic
Mocetinostat	HDAC1, HDAC2	Synthetic
Ricolinostat	Class IIb (HDAC6)	Synthetic
OTHER		
BRD8430	HDAC1, HDAC2	Synthetic
Compound 60	HDAC1, HDAC2	Synthetic
MRL-223	HDAC1, HDAC2	Synthetic
RGFP699	HDAC 3	Synthetic
BG45	HDAC 3	Synthetic
T247 and T436	HDAC 3	Synthetic
TM269	Class IIa	Synthetic
MC1575 and MC1568	Class IIa	Synthetic
YK-4-272	Class IIa	Synthetic
ACY-1215 and ACY-241	Class IIb	Synthetic
Compound 12	Class IIb	Synthetic
Tubacin, tubastatin A	Class IIb	Synthetic
CIA	Class IIb	Synthetic

[a] *Hybrid molecule mode of action.*

microbial fermentation of dietary fiber. The key microbial producers of butyrate are *Faecalibacterium prausnitzii* and *Eubacterium rectale/Roseburia* spp. (Berni Canani et al., 2012).

In terms of anticancer therapy, butyrate is thought to activate the $P21^{WAF1}$ cell cycle arrest inhibitor and the proapoptotic BAK protein of the Bcl-2 family (Berni Canani et al., 2012). Butyrate is also thought to induce autophagic cell death in B-cell lymphoma cell lines (Shao et al., 2004) and prevent angiogenesis (Deroanne et al., 2002).

The beneficial effects of dietary fiber against colon cancer are thought to be partly due to production of HDACIs in the colon. Additional studies have shown a high-fiber diet to be protective against lymphomas and NSCLC (Berni Canani et al., 2012; Wei et al., 2016). The chemopreventive effects of butyrate include induction of detoxifying enzymes such as glutathione S-transferase that serve as ROS scavengers (Berni Canani et al., 2012).

Two other short-chain fatty acid HDACIs are sodium propionate and valproic acid (VPA), each of which has a similar anticancer profile to that of butyrate. Sodium propionate has limited clinical efficacy due to low stability. VPA is an antiepileptic agent that seems to inhibit HDAC classes I and II (West and Johnstone, 2014). There have been reports that VPA can induce autophagy and apoptosis, reduce expression of c-Myc and Mcl-1, increase levels of ^{WAF1}, Puma, and Bim, and increase ROS production in pancreatic tumor cells (Gilardini Montani et al., 2017). VPA is also shown to decrease β-catenin and VEGF levels in colon cancer cell lines (Akbarzadeh et al., 2016). Additionally, both sodium butyrate and VPA has been shown to depress the growth of breast cancer cell lines (Damaskos et al., 2017).

Hydroxamic Acids

HDACIs in the hydroxamic acid family exert their effects by chelating zinc ions via their hydroxamate moieties (Folmer et al., 2010). A key member of the hydroxamic acid family is trichostatin A (TSA), which is produced by actinomycete *Streptomyces hygroscopicus*. TSA is an antifungal agent with antitumor activity. It is known to be a potent pan-HDACI that induces cell cycle arrest and differentiation (Vigushin et al., 2001). Furthermore, TSA has been shown to induce apoptosis in tumor cells via activation of Bax and caspase 3 (Seidel et al., 2012). The other members of the hydroxamic family have similar actions on tumor cell death and differentiation. Interestingly, this class of HDACIs seems to induce protective autophagy by tumor cells and therefore leads to HDACI resistance (Dupéré-Richer et al., 2013). A synthetic analog of TSA, suberoylanilide hydroxamic acid (SAHA) or vorinostat, was approved by the United States Food and Drug Administration (FDA) for cutaneous T-cell lymphoma (CTCL). Aside from vorinostat, two other hydroxamic acids; belinostat and panobinostat, have been approved for peripheral T-cell lymphoma (PTCL) and multiple myeloma, respectively. Similar to TSA, the three listed agents block HDAC types I, II, and IV. Additionally, givinostat and pracinostat of this family are being tested in leukemias (Finazzi et al., 2013; Garcia-Manero et al., 2017; Li et al., 2016). Other agents in this category are quisinostat and abexinostat, which are in clinical trials for various types of lymphomas (Clinicaltrials.gov #NCT01486277, #NCT00724984) (Child et al., 2016). Abexinostat has also been in trials for treatment of sarcomas and other solid tumors (West and Johnstone, 2014). Some other hydroxamic acid−type molecules currently under investigation for various tumor types have narrower ranges of HDAC inhibition while having additional effects, such as HER2 and EGFR blockade (Li and Seto, 2016).

Cyclic Peptides

Another class of HDACIs studied is cyclic peptides. Cyclic peptides structurally contain epoxide, acetamide, and ketone groups, which allow for chelation of zinc ions (Folmer et al., 2010). Several natural products exist in this category. FK235222 is a product isolated from fermentation of *Acremonium* sp. It is a cyclic peptide with HDACI activity shown to induce apoptosis and cell cycle arrest in both leukemia and prostate cancers (D'Acunto et al., 2010; Petrella et al., 2008). Apicidin D is another member of the cyclic peptide family, originating from *Fusarium* spp., and has antiparasitic activity mediated by HDAC inhibition (Han et al., 2000). Apicidin's HDAC-inhibitory activity has been tested in a number of cancer cell lines, including HeLa cervical cancer cell lines, mouse colon carcinoma, human osteosarcoma, and MCF-7 breast cancer cell lines. Apicidin has been shown to be antiproliferative in a number of listed cell lines. It is further shown to induce G_0/G_1 cell cycle arrest by activating $P21^{WAF1/Cip1}$, Cyclin D, and Cdk2, and to change cellular morphology in HeLa cells (Han et al., 2000).

Depsipeptides

Depsipeptides, another class of HDACIs, exert their activity by chelating Zn^{2+} ions via their exposed thiol groups. A depsipeptide isolated from *Chromobacterium violaceum* is a class I and II HDACI and has been FDA-approved as romidepsin for the treatment of CTCL and PTCL. Romidepsin has proapoptotic and autophagic effects in leukemia cell lines (Seidel et al., 2012). Other depsipeptides are largazole, isolated from marine cyanobacterium *Symploca* sp., and spiruchostatin A, isolated from *Pseudomonas*, both of which have class I HDAC-inhibitory activity. Largazole has shown antiproliferative activity in transformed colorectal, breast, neuroblastoma, and osteosarcoma cell lines in nanomolar concentrations (Hong and Luesch, 2012). A more specific X-ray crystal structure analysis reveals that largazole has selectivity for class I HDACs as well as for HDAC10 of class II (Hong and Luesch, 2012). Spiruchostatin A, with structural similarity to romidepsin, has been shown to have antiproliferative activity against several cell lines, including breast, colon, and ovarian cancers. In breast cancer cell lines, spiruchostatin A has been shown to cause G_1/G_2 cell cycle arrest, and in comparison with Trichostatin A of the hydroxamic acid family, spiruchostatin A seems to inhibit HDAC1 with greater potency (Crabb et al., 2008).

Benzamides

Another class of HDACIs is benzamides, which have slightly different pharmacological properties than those of other classes. Benzamides have been shown to be more selective and more potent for specific HDACs rather than for entire classes of HDACs. Specifically, benzamides seem to inhibit HDAC1, HDAC2, and HDAC3 of class I with high potency (Beckers et al., 2007; Marson et al., 2013). Benzamides have been shown to induce expression of $P21^{WAF1}$, cause cell cycle arrest, activate several proapoptotic genes, and be cytotoxic in various tumor cell types at micromolar concentrations. Additionally, specific benzamide analogs have been shown to induce breast cancer cell differentiation that is comparable to that of SAHA (Beckers et al., 2007). This family of HDACIs has no natural products; however, specific synthetic benzamides have been promising in clinical trials of breast cancer and lymphomas. Among these, entinostat and mocetinostat have received a great deal of attention.

Dietary Polyphenols

Flavonoids

Flavonoids and their precursor chalcones have been identified as potent HDAC and SIRT inhibitors (Folmer et al., 2010.; Orlikova et al., 2012; Seidel et al., 2012). Flavone isolated from the southern African fruiting pineapple guava tree (*Feijoa sellowiana*) has been shown to have HDACI activity. Additionally, prenylated isoflavone pomiferin isolated from horse-apple fruit (*Maclura pomifera*) is another class I HDAC-inhibitory flavonoid (Folmer et al., 2010; Seidel et al., 2012). Chalcones are precursors of flavonoids found in many natural sources, such as plants and vegetables, that also induce HDAC inhibition. A specific molecule in this category is butenin, derived from *Rhus verniciflua*, which has both HDAC and SIRT-inhibitory activity (Orlikova et al., 2012; Seidel et al., 2012). Genistein derived from soybeans is another isoflavone and has been shown to decrease HDAC activity in squamous esophageal and renal cancer cells. Other members of this group with reported HDACI or SIRTI activity are luteolin, quercetin, and silibinin (Link et al., 2010).

Ally Derivatives

A dietary source of allyl derivatives or organosulfurs is garlic. Allyl derivatives such as allyl mercaptan (AM), diallyl disulfide,

S-allylcysteine, and S-allylmecaptocysteine induce increased histone acetylation in cancer cells and in rat cancer colonocyte models (Link et al., 2010). AM isolated from garlic has been shown to be a potent inhibitor of HDAC activity in human colon adenocarcinoma cell lines. The specific HDAC8-inhibitory activity of AM was shown via docking studies and suggests that zinc–sulfur interactions may mediate organosulfur inhibition of the HDAC8 catalytic site (Link et al., 2010). Furthermore, increased acetylation of the CKDKN1A promoter, increased $P21^{WAF1}$ expression, cell cycle arrest, and apoptosis secondary to organosulfur exposure have been reported (Druesne et al., 2004; Nian et al., 2008).

Isothiocyanates

Another group of natural compounds, derived from cruciferous vegetables such as broccoli, is isothiocyanates. Sulforaphane derivatives of isothiocyanates have shown HDACI activity in colon, prostate, and breast cancer cell lines (Link et al., 2010; Meeran et al., 2010; Myzak et al., 2006, 2004). The effects of isothiocyanates seem to be through increased acetylation of bax and $P21^{WAF1}$ promoters (Link et al., 2010). Similar HDACI activities have been reported in humans peripheral blood mononuclear cells following intake of broccoli (Myzak et al., 2007).

Other Polyphenols

Another polyphenol is 3,3′-diinolylmethane (DIM), identified in vegetables such as broccoli, which has class I HDAC-inhibitory effects and specifically blocks HDAC1, HDAC2, and HDAC3. The efficacy of DIM has been tested in colon and breast cancers (Link et al., 2010). Green tea polyphenol epigallocatechin-3-gallate (EGCG) has also shown some activity against HDAC1, HDAC2, HDAC3, and SIRT1 (Choi et al., 2009). Resveratrol, a polyphenol found in red grapes, has been shown to be a pan-HDACI with HDAC class I and II binding affinity comparable to those of SAHA and TSA (Venturelli et al., 2013). Resveratrol was found to be antiproliferative in hepatic, colon, prostate, pancreatic, breast, and lung cancer cell lines (Venturelli et al., 2013). Other polyphenols include quercetin and dihydrocoumarin, which seem to have activity against SIRTs and class III HDACs (Link et al., 2010; Seidel et al., 2012).

HDACs in Cancer and Current and Emerging HDAC Inhibitors in Cancer Therapy

To date, most of the approved HDACI therapy has been in the area of hematological malignancies. The four FDA-approved HDACIs are used in lymphomas and multiple myeloma. Table 4.4 describes the current FDA-approved HDACIs and their clinical applications. However, a number of HDACIs are being studied in various solid tumors and are currently in clinical trials for these additional indications. Table 4.3 provides a list of HDACIs currently under study for various cancers. A discussion of preclinical and clinical findings with various investigational HDACIs for breast, colorectal, pancreatic, lung, gastric, HCC, and hematological cancers is included in this section along with a brief summary of some HDACs commonly associated with each of these cancers (Table 4.1). It is important to note that HDACs may have different activities in different tissues and different cancers, with different prognoses.

Breast Cancer

Several HDACs and SIRTs have altered expression in breast cancer. Among class I HDACs, HDAC1 is highly expressed in ER/progesterone receptor (PgR)-positive breast cancers, and elevated expression of HDAC1 has been associated with better outcomes. HDAC2 and HDAC3 have elevated expression levels in ER/PgR-negative breast cancers. Elevated expression of HDAC6 has been shown to be associated

TABLE 4.3 HDAC Inhibitors Currently in Clinical Trials

HDACI	Clinical Trial Phase	Cancer	Clinicaltrials.gov#
Abexinostat (PCI-24781)	I/II	sarcoma, lymphoma, advanced cancer	NCT01027910 NCT00724984 NCT00562224
AR42	I	multiple myeloma, lymphoma	NCT01129193 NCT00562224
CHR-3996	I	advanced solid tumors	NCT00697879
CUDC-101	I	advanced solid tumors, multiple cancers	NCT00728793 NCT01171924
Entinostat (MS-2750)	I/II	multiple solid tumors, breast, lung, melanoma	Multiple[a]
Givinostat (ITF-2357)	II	lymphoma, multiple myeloma	NCT00702506 NCT00792467
Mocetinostat (SGN-35)	II	lymphoma, non-small-cell lung cancer, MDS, B-cell lymphoma, leukemia, pancreatic cancer	Multiple[a]
Pracinostat (SB-939)	I/II	acute myelogenous leukemia, MDS, prostate cancer, solid tumors	NCT01912274 NCT01873703 NCT01075308 NCT01184274
Quisinostat (JNJ-26481585)	I/II	cutaneous T-cell lymphoma, multiple myeloma, solid tumors, lymphoma	NCT01486277 NCT01464112 NCT00677105
Resminostat (4SC-201, RAS2410)	I/II	Non-Hodgkin lymphoma, advanced colorectal cancer, hepatocellular carcinoma	NCT01037474 NCT00943449 NCT01277406
Ricolinostat (ACY-1215)	I/II	multiple myeloma, lymphoid malignancies, chronic lymphocytic leukemia, metastatic breast cancer	NCT01997840 NCT02091063 NCT01583283 NCT02632071

[a] *See www.clinicaltrials.gov.*
MDS, myelodysplastic disorders.

with better outcomes in ER-positive breast cancers (Li and Seto, 2016). HDAC11 has been shown to be overexpressed in breast cancer cell lines (Deubzer et al., 2013). Among the SIRTs, SIRT1, SIRT2, SIRT3, and SIRT4 have decreased expression in breast cancer, while SIRT6 has increased expression (Chalkiadaki and Guarente, 2015).

Great emphasis has been placed on combination HDACI therapy for breast cancer (Damaskos et al., 2017). Current studies show a promising place in therapy for several HDACIs in the treatment of various types of breast cancer. Among the approved HDACIs, vorinostat has been extensively studied and although monotherapy with vorinostat has not been conclusive,

TABLE 4.4 Current FDA-Approved HDAC Inhibitors

Drug	HDAC Class	Type of Cancer
Istodax (romidepsin)	I, II	cutaneous T-cell lymphoma peripheral T-cell lymphoma
Farydak (panobinostat)	I, II	multiple myeloma
Beleodaq (belinostat)	I, II, IV	refractory peripheral T-cell lymphoma
Zolinza (vorinostat)	I, II, IV	cutaneous T-cell lymphoma

in ER-positive breast cancer, a combination of vorinostat with chemotherapy or hormone therapy has been promising (Damaskos et al., 2017). Panobinostat has also been investigated in several studies, and a combination therapy of panobinostat with aromatase inhibitors in aromatase inhibitor-resistant cells has shown to be effective in suppressing the growth of these cells (Damaskos et al., 2017). A natural HDACI, santacruzamate A, which is produced by the cyanobacterium *Symploca* sp. and has similar structural motifs to those of SAHA, has shown promise in growth suppression of MCF-7 breast cancer cell lines (Damaskos et al., 2017). Other HDACIs tested with efficacy in breast cancer are entinostat, TSA, sodium butyrate, VPA, and SAHA. Treatment with entinostat; a benzamide HDACI, has been promising in combination with aromatase-inhibitor exemestane in metastatic cancer. A phase III clinical trial investigating this combination therapy in advanced breast cancer is currently underway and nearing its primary completion date (www.clinicaltrials.gov #NCT02115282). Ricolinostat is another HDACI that is currently in a phase I trial for metastatic breast cancer (www.clinicaltrials.gov #NCT02632071). Overall, treatment of breast cancer with HDACIs in combination with other chemotherapy agents has received a great deal of attention.

Colorectal Cancer

Among HDAC class I family members, HDAC2 plays a major role in cancer of the colon. A lack of expression of HDAC2 has been found in unstable sporadic colorectal cancer. HDAC2 also exhibits elevated expression levels in other forms of colorectal cancer and can serve as both a biomarker and a prognostic factor. HDAC1, HDAC3, HDAC5, and HDAC7 also show increased expression in colorectal cancers (Li and Seto, 2016). HDAC11 is highly expressed in colon cancer cell lines (Deubzer et al., 2013). Among the SIRTs, increased expression of SIRT1 and SIRT7 and decreased expression of SIRT6 are found in colon cancer (Chalkiadaki and Guarente, 2015).

Microarray studies indicate that the HDACIs vorinostat and panobinostat lead to growth inhibition and cell cycle arrest in several colon cancer cells (LaBonte et al., 2009). VPA has also been tested in colon cancer cells with proapoptotic activity (Akbarzadeh et al., 2016). A number of natural products including allyl derivatives, isothiocyanates, DIM, spiruchostatin A, and apicidin have also shown efficacy against colon cancer cell lines in preclinical studies (Crabb et al., 2008; Han et al., 2000; Link et al., 2010). In terms of clinical trials, only combination therapy of HDACIs with other chemotherapy agents has been studied. Phase II trials of entinostat and 5-azacitidine have shown drug tolerability with minimal clinical efficacy in metastatic colorectal cancer (Azad et al., 2017). Vorinostat with fluorouracil, leucovorin, and oxaliplatin has been tested in metastatic colorectal cancer with no significant clinical efficacy (Wilson et al., 2010). Vorinostat together with 5-fluorouracil has been found effective in 21 of 38 patients with refractory colorectal cancer (Fakih et al., 2010). A new trial is currently underway that compares the efficacy of vorinostat plus hydroxychloroquine versus regorafenib in colorectal cancer (www.clinicaltrials.gov #NCT02316340). Overall, despite evidence of the involvement of epigenetic variation in colon cancer development, clinical efficacy still needs to be established for HDACIs in combination with other therapies.

Pancreatic Cancer

In pancreatic cancer, HDAC2, HDAC3, and HDAC7 have increased expression, with elevated HDAC7 shown to be associated with a poor prognosis (Li and Seto, 2016). Among the SIRTs, both SIRT3 and SIRT6 have decreased expression in pancreatic cancers (Chalkiadaki and Guarente, 2015).

Among HDACIs, TSA has shown efficacy against various pancreatic cell lines individually and in combination with other chemotherapy agents (Koutsounas et al., 2013). Other members of the hydroxamic acid family, including SAHA (vorinostat), have shown antiproliferative and proapoptotic activity in various pancreatic cell lines (Koutsounas et al., 2013). Cyclic peptides such as romidepsin have been shown to be antiproliferative and to cause cell cycle arrest in several pancreatic cell lines. Additionally, romidepsin has been shown to induce tumor immunogenicity to NK cells. Short-chain fatty acids and benzamides show antiproliferative activity and induce cell cycle arrest in pancreatic cells (Koutsounas et al., 2013). Several clinical trials have investigated HDACIs in pancreatic ductal adenocarcinomas (PDACs). Combination therapies of vorinostat and gemcitabine or proteasome inhibitors have shown little promise in halting the progression of PDAC (Arlt and Schäfer, 2016). Panobinostat combination therapy with bortezomib was shown to be effective in stabilizing one patient, whereas romidepsin in combination with gemcitabine was shown to induce remission in five patients with PDAC, and belinostat combined with carboplatin retained partial remission in one patient (Arlt and Schäfer, 2016). Although there seems to be some degree of efficacy with HDACs in pancreatic cell lines, the clinical studies to date fail to show robust efficacy.

Lung Cancer

Among the classical HDACs, HDAC1, HDAC2, and HDAC3 have elevated levels in lung cancers, with HDAC1 and HDAC3 overexpression associated with poor outcomes in lung adenocarcinoma patients. Conversely, HDAC5 and HDAC10 have decreased expression levels, which is associated with poor outcomes in lung cancer patients (Li and Seto, 2016). Among the SIRTs, reduced expression of SIRT3 and SIRT4 is seen in lung cancer, and increased expression of SIRT5 is associated with NSCLC (Chalkiadaki and Guarente, 2015).

HDACI therapy in lung cancer treatment has been explored mostly in terms of combination therapy. A specific area of interest is combination therapy with RAs. As mentioned previously, in some cancer cells, RAR complexes with coreceptors to form fusion proteins that bind DNA and suppress genes such as those involved in cellular proliferation and apoptosis. HDACs are components of RAR coreceptor complexes. HDACIs inhibit RAR complex formation and therefore prevent gene suppression. It has been shown that combination therapy with RAs and HDACIs improves antitumor efficacy. Two HDACIs, SL142 and SL325, have been shown to increase apoptosis in lung cancer cells (Han et al., 2010). Additionally, combination therapy with RA increases the activity of the RA receptor element in lung cancer cells (Han et al., 2010). Other agents that have shown efficacy in lung cancer cells are TSA and vorinostat (Ansari et al., 2016). TSA seems to induce apoptosis and cell cycle arrest in H157 lung cancer cells, whereas vorinostat has been shown to increase expression of $p21^{WAF1}$ and cell cycle arrest (Ansari et al., 2016). The combination of HDACIs with traditional chemotherapy agents such as cisplatin has shown antiproliferative and proapoptotic effects in adenocarcinoma and squamous cell carcinomas of the lung (Gumbarewicz et al., 2016). A phase II clinical study investigating the better efficacy of an erlotinib-with-entinostat combination compared with erlotinib alone in non-small-cell lung carcinoma in underway (www.clinicaltrials.gov #NCT00750698).

Gastric Cancer

Microarray studies have shown high expression of HDACs in gastric tumor tissue, which can serve as an independent prognostic factor in gastric cancer (Weichert et al., 2008). Specifically, elevation of HDAC1, HDAC2, and HDAC3 levels are related to tumor spread to the lymph nodes and decreased survival. Additionally, HDAC4 levels are elevated in gastric cancer tumors. Conversely, decreased HDAC10 levels are associated with a poor prognosis in gastric cancer (Li and Seto, 2016). Among the SIRTs, low levels of SIRT4 are associated with gastric cancer (Chalkiadaki and Guarente, 2015).

In term of therapeutics, microarray studies have identified several genes downregulated by vorinostat treatment (Claerhout et al., 2011), and a combination therapy of vorinostat with taxanes has shown cytotoxic effects in gastric cancer cells (Abdelfatah et al., 2016; Chang et al., 2010). However, clinical trials with combination HDACI therapy have failed to show robust response thus far. A Phase I/II study of vorinostat plus capecitabine, and cisplatin for metastatic or recurrent gastric cancer, was limited by adverse events and did not reach clinical significance (Yoo et al., 2016). Other clinical trials have also been completed with variable tolerability and efficacy results (www.clinicaltrials.gov #NCT01249443, NCT00537121).

Liver Cancer/Hepatocellular Carcinoma

In HCC, HDAC1, HDAC2, and HDAC3 levels have been shown to be elevated and are associated with a poor prognosis, especially in early-stage liver cancers. By contrast, decreased HDAC6 expression is associated with poor outcomes in liver cancer. HDAC5 also has increased expression in HCC (Li and Seto, 2016). Among the SIRTs, decreased expression of SIRT2, SIRT3, and SIRT6 is observed in liver cancer (Chalkiadaki and Guarente, 2015).

Several studies have shown both in vitro and in vivo responses of HCCs to HDACIs (Ma et al., 2014; Wahid et al., 2017; Zhang et al., 2012). A natural product, resveratrol derived from the polyphenols found in red grapes, has shown promise in inhibiting the proliferation of HCC (Venturelli et al., 2013). Pracinostat in combination with sorafenib has been shown to reduce tumor volume and prevent vessel density in HCC xenografts (Lachenmayer et al., 2012). In terms of clinical trials, belinostat in a phase II trial was found to induce stable disease in 45.2% patients with HCC (Yeo et al., 2012). Furthermore, in the SHELTER study, resminostat and sorafenib showed safety and potential for greater progression-free survival in patients receiving resminostat plus sorafenib compared with resminostat monotherapy (Bitzer et al., 2016). Overall, HDACIs in HCC seem to be relatively effective, especially in combination with sorafenib, which is promising given the shortage of treatments available for this cancer.

Hematological Cancers

The role of HDACs has most extensively been studied in hematologic cancers, with the resulting altered expression of HDACs in classes I, II, and III being associated with various types of cancers. Among the classical HDACs, high expression of HDACs 1–9 is associated with acute lymphoblastic leukemia (ALL), and high expression of HDACs 1, 3, 7, 9, and 10 is associated with advanced CLL (Wang et al., 2011). In CLL, the prognostic power of HDAC6 is unclear, with one set of studies showing an association between elevated HDAC6 levels and poor prognosis (Wang et al., 2011) and another showing a correlation between low HDAC6 levels and poor prognosis (Van Damme et al., 2012). Interestingly, in acute myelogenous leukemia (AML), HDAC5 levels are decreased and HDAC6 levels are increased (Li and Seto, 2016).

Increased HDAC1 and HDAC2 levels are found in diffuse large B-cell lymphoma (DLBCL) and are associated with decreased survival, and HDAC3 is also found to be elevated in DLBCL. Elevated HDAC6 is associated with positive outcomes in DLBCL but poor outcomes in PTCL (Li and Seto, 2016). In CTCL, high expression of HDAC2 is found in aggressive cancers and HDAC6 has decreased expression; however, HDAC6 expression in CTCL was predictive of better survival (Li and Seto, 2016; Marquard et al., 2008).

Hodgkin lymphoma (HL) is associated with elevated levels of HDAC1, HDAC2, and HDAC3, and these high HDAC levels are associated with poorer outcomes (Li and Seto, 2016). HDAC11 is overexpressed in HL and in Philadelphia-negative chronic myeloproliferative neoplasms (Barneda-Zahonero and Parra, 2012). HDAC1 is overexpressed in myeloma and is associated with a poor outcome (Li and Seto, 2016).

Among the SIRTs, overexpression of SIRT1 is associated with AML, chronic myelogenous leukemia, and CLL (Chalkiadaki and Guarente, 2015; Li and Seto, 2016). Reduced expression of SIRT4 is associated with adult T-cell leukemia and AML. Further, decreased expression of SIRT3 is associated with CLL and a poor prognosis (Chalkiadaki and Guarente, 2015; Li and Seto, 2016).

The use of HDACIs in hematological malignancies has been the most promising among cancers. Among the actions of HDACIs in hematologic cancer cells are induction of $p21^{WAF1}$ and cell cycle arrest, accumulation of aggresomes and misfolded proteins, and inhibition of AML1−ETO/HDAC1 fusion protein complexes in AML (Imai et al., 2016). All FDA-approved HDACIs have hematological cancer indications. Additionally, a number of newer HDACIs currently in trials have indications for lymphomas, multiple myeloma, and MDS. Among the newer agents, resminostat is currently in a phase II trial for non-Hodgkin lymphoma (www.clinicaltrials.gov #NCT01037474). Resminostat has also been shown to be antiproliferative in multiple myeloma cell lines and induces proapoptotic pathways and cell cycle arrest in these cells (Mandl-Weber et al., 2010). The abexinostat phase I trial for AML, ALL, and MDS has not shown particular clinical effectiveness (Vey et al., 2017). However, a phase II trial of abexinostat has shown significant tumor size reduction in follicular lymphomas (Evens et al., 2016). Mocetinostat has been tested with and has demonstrated efficacy and tolerability in CLL (Mottamal et al., 2015). In vivo studies of givinostat have shown an antileukemic response associated with induction of apoptosis and differentiation in ALL (Pinazza et al., 2016). A phase II study of givinostat in polycythemia vera in combination with hydroxycarbamide led to complete or partial response in up to 50% of patients (Finazzi et al., 2013). In addition to the above, entinostat, mocetinostat, ricolinostat, and AR-42 are in various trials for lymphomas, multiple myeloma, and MDS. Overall, HDACIs seem to be a promising approach for the treatment of hematological malignancies both as monotherapy and in combination with traditional chemotherapy or newer agents.

Histone Deacetylase Complex Inhibitors in Development

The new direction of HDACI therapy is toward the development of selective HDACIs and hybrid molecules targeting several proteins involved in tumor growth or development. Selective HDACIs may have a promising profile in select cancers and present fewer limiting side effects in comparison with pan-HDACIs. Hybrid molecules, on the other hand, target several molecules, which leads to better efficacy and avoids the use of combination therapy.

Currently in development are selective HDAC1/HDAC2, HDAC3, HDAC4/5, HDAC6, and HDAC8 inhibitors as well as several hybrid molecules.

Histone Deacetylase Complex Selective Inhibitors

HDAC1/HDAC2

Several agents developed with HDAC1 and HDAC2 specificity have shown antitumor activity. BRD8430 and compound 60 are among these agents, with antitumor activity in neuroblastoma cell lines. Another agent; MRLB-223 has been shown to induce apoptosis in the brachial lymph nodes of C57BL/6 mice (Li and Seto, 2016; Newbold et al., 2013).

HDAC3

HDAC3-specific inhibitors have shown antitumor efficacy in various cancers. This category has several agents: RGFP699, studied in CTCL cells, BG45, studied in MM; and T247 and T436, studied in prostate and colon cancer cells (Li and Seto, 2016). Among the mechanisms of HDAC3 inhibitors are apoptosis and growth retardation, which seem to be independent of HDAC1 and HDAC6 inhibition. HDAC3 is also recruited by fusion protein complexes in AML, and the expression of LPS-induced proinflammatory genes seems to be HDAC3 dependent (West and Johnstone, 2014). Therefore, specific inhibition of HDAC3 may have multifaceted activity in some cancers.

HDAC 4,5,7,9 (Class IIa HDAC)

This class of drugs has been harder to develop due to a current lack of knowledge about the exact mechanism of action of class IIa HDACs. Several compounds have been identified via high-throughput screening as having selectivity for class IIa HDACs. These include MC1568 and MC1575, with antiproliferative effects in melanoma and breast cancer cells, and TMP269, with modest growth-inhibitory effects in multiple myeloma. MC1568, in conjunction with other agents, has shown efficacy in pancreatic, colorectal, and gastric cancers (Li and Seto, 2016). A selective inhibitor of HDAC4/5, LMK235, has shown promise against some solid tumors (Li and Seto, 2016). Two other agents, YK-4-272 and tasquinimod, seem to have antitumor efficacy in specific prostate cancer cell lines. The latter agent seems to inhibit HDAC4 activity by disrupting its nucleus–cytoplasmic shuttling (Kong et al., 2012; Li and Seto, 2016).

HDAC6 (Class IIb HDAC)

HDAC6 is a special class of HDACs that catalyzes primarily cytoplasmic nonhistone proteins. HDAC6 inhibition leads to the inhibition of proteasomes, aggresomes, and chaperone proteins such as HSP90. As a result, proper degradation of proteins is affected by HDAC6 inhibitors, which leads to accumulation of polyubiquitinated proteins and activation of apoptotic and autophagy pathways (Li and Seto, 2016; Newbold et al., 2013). A number of HDAC6 inhibitors have been identified, among which ricolinostat (ACY-1215) is currently in clinical trial for multiple myeloma, lymphomas, and breast cancer (Yee et al., 2016). Another promising anticancer application of ricolinostat is in glioblastoma multiforme (Li and Seto, 2016). A similar molecule to ricolinostat, ACY-241, with potential better tolerability is also being tested in combination with immunomodulatory agents in multiple myeloma in clinical trials (www.clinicaltrials.gov #NCT02400242) (Niesvizky et al., 2015). Other agents reported in this group as tubastatin A, tubacin, CIA, HPOB, nexturastat A, and compound 12 have shown preclinical efficacy in various cancer cells (Li and Seto, 2016).

HDAC8

Although the role of HDAC8 in cancer has been harder to elucidate, developing HDAC8-

specific inhibitors has been promising due to its molecular structure. Two agents with HDAC8 selectivity are PCI-34,051 and PCI-48,012, which have been shown to induce cell death in T-cell leukemia and lymphoma cell lines (Li and Seto, 2016). In combination with RA, both PCI-34,051 and PCI-48,012 and another agent, Cpd2, have shown to induce differentiation and cell cycle arrest in neuroblastomas in both in vitro and in vivo mouse models (Rettig et al., 2015).

Hybrid Molecules

Another direction of emerging HDACI therapy is through the development of hybrid molecules that target HDACs and other related proteins leading to enhanced antitumor activity. An example of these molecules is CUDC-101, which has a multimodal mechanism of action inhibiting HDACs, EGFR, HER-2, and phosphoinositide 3-kinase. CUDC-101 is currently in clinical trial for treatment of various advanced solid tumors (www.clinicaltrials.gov NCT01171924). Several other agents with various dual activities are in the pipeline as well and may show promise in targeting specific pathways within tumors (Li and Seto, 2016).

CONCLUSION

HDACI therapy is a unique approach to cancer treatment with potential benefits as adjuvant therapy or monotherapy. Use of HDACIs has been tested in many cancers with variable results. The most promising evidence of HDACIs' anticancer activity has been noted in vitro, with partial efficacy in clinical trials. In general, HDACI therapy is most effective in combination with traditional chemotherapeutic agents. The advantages of using HDACIs as adjunct agents are decreased exposure to toxicity and increased efficacy. HDACIs can augment the effects of existing therapies by enhancing the action of the acetylated proteins involved in the pharmacology of chemotherapy agents. Furthermore, HDACIs can enhance tumor cell death by increasing apoptosis, autophagy, and other cell damage processes. As there is not a global "HDAC expression profile" for cancer, specific HDAC expression profiles for each cancer are critical for selecting the optimal HDACI monotherapy or combination therapy and considering dose-dependent toxicity and tolerability. Optimizing HDACI therapy requires a better understanding of class-specific HDAC mechanisms in various cancers and development of more-potent HDACIs with less toxicity. Along these lines, class-specific and hybrid molecules targeting multiple pathways seem to be promising directions in HDACI therapy.

References

Abdelfatah, E., Kerner, Z., Nanda, N., Ahuja, N., 2016. Epigenetic therapy in gastrointestinal cancer: the right combination. Ther. Adv. Gastroenterol. 9, 560–579.

Ablain, J., de The, H., 2011. Revisiting the differentiation paradigm in acute promyelocytic leukemia. Blood 117, 5795–5802.

Adams, C.M., Eischen, C.M., 2016. Molecular underpinnings of HDAC inhibition revealed. Cell Cycle 15, 1943–1944.

Adimoolam, S., Sirisawad, M., Chen, J., Thiemann, P., Ford, J.M., Buggy, J.J., 2007. HDAC inhibitor PCI-24781 decreases RAD51 expression and inhibits homologous recombination. Proc. Natl. Acad. Sci. U.S.A. 104, 19482–19487.

Akbarzadeh, L., Moini Zanjani, T., Sabetkasaei, M., 2016. Comparison of anticancer effects of carbamazepine and valproic acid. Iran. Red Crescent Med. J. 18, e37230.

Alberts, B., Johnson, A., Lewis, J., Morgan, D., Raff, M., Roberts, K., Walter, P., 2014. Molecular Biology of the Cell, sixth ed. Garland Science.

Andrews, A.J., Luger, K., 2011. Nucleosome structure(s) and stability: variations on a theme. Annu. Rev. Biophys. 40, 99–117.

Ansari, J., Shackelford, R.E., El-Osta, H., 2016. Epigenetics in non-small cell lung cancer: from basics to therapeutics. Transl. Lung Cancer Res. 5, 155–171.

Arlt, A., Schäfer, H., 2016. Investigational histone deacetylase inhibitors for treating pancreatic adenocarcinoma. Expert Opin. Investig. Drugs 25, 1251–1254.

Armeanu, S., Bitzer, M., Lauer, U.M., Venturelli, S., Pathil, A., Krusch, M., Kaiser, S., Jobst, J., Smirnow, I., Wagner, A., Steinle, A., Salih, H.R., 2005. Natural killer cell–mediated lysis of hepatoma cells via specific induction of NKG2D ligands by the histone deacetylase inhibitor sodium valproate. Cancer Res. 65, 6321–6329.

Azad, N.S., El-Khoueiry, A., Yin, J., Oberg, A.L., Flynn, P., Adkins, D., Sharma, A., Weisenberger, D.J., Brown, T., Medvari, P., Jones, P.A., Easwaran, H., Kamel, I., Bahary, N., Kim, G., Picus, J., Pitot, H.C., Erlichman, C., Donehower, R., Shen, H., Laird, P.W., Piekarz, R., Baylin, S., Ahuja, N., May 23 2017. Combination epigenetic therapy in metastatic colorectal cancer (mCRC) with subcutaneous 5-azacitidine and entinostat; a phase 2 consortium/stand up 2 cancer study. Oncotarget 8 (21), 35326–35338.

Bali, P., Pranpat, M., Bradner, J., Balasis, M., Fiskus, W., Guo, F., Rocha, K., Kumaraswamy, S., Boyapalle, S., Atadja, P., Seto, E., Bhalla, K., 2005. Inhibition of histone deacetylase 6 acetylates and disrupts the chaperone function of heat shock protein 90 a novel BASIS for Antileukemia activity of histone deacetylase inhibitors. J. Biol. Chem. 280, 26729–26734.

Barber, M.F., Michishita-Kioi, E., Xi, Y., Tasselli, L., Kioi, M., Moqtaderi, Z., Tennen, R.I., Paredes, S., Young, N.L., Chen, K., Struhl, K., Garcia, B.A., Gozani, O., Li, W., Chua, K.F., 2012. SIRT7 links H3K18 deacetylation to maintenance of oncogenic transformation. Nature 487, 114–118.

Barneda-Zahonero, B., Parra, M., 2012. Histone deacetylases and cancer. Mol. Oncol. 6, 579–589.

Beckers, T., Burkhardt, C., Wieland, H., Gimmnich, P., Ciossek, T., Maier, T., Sanders, K., 2007. Distinct pharmacological properties of second generation HDAC inhibitors with the benzamide or hydroxamate head group. Int. J. Cancer 121, 1138–1148.

Berni Canani, R., Di Costanzo, M., Leone, L., 2012. The epigenetic effects of butyrate: potential therapeutic implications for clinical practice. Clin. Epigenet. 4, 4.

Bhaskara, S., Knutson, S.K., Jiang, G., Chandrasekharan, M.B., Wilson, A.J., Zheng, S., Yenamandra, A., Locke, K., Yuan, J., Bonine-Summers, A.R., Wells, C.E., Kaiser, J.F., Washington, M.K., Zhao, Z., Wagner, F.F., Sun, Z.-W., Xia, F., Holson, E.B., Khabele, D., Hiebert, S.W., 2010. Hdac3 is essential for the maintenance of chromatin structure and genome stability. Cancer Cell 18, 436–447.

Bitzer, M., Horger, M., Giannini, E.G., Ganten, T.M., Wörns, M.A., Siveke, J.T., Dollinger, M.M., Gerken, G., Scheulen, M.E., Wege, H., Zagonel, V., Cillo, U., Trevisani, F., Santoro, A., Montesarchio, V., Malek, N.P., Holzapfel, J., Herz, T., Ammendola, A.S., Pegoraro, S., Hauns, B., Mais, A., Lauer, U.M., Henning, S.W., Hentsch, B., 2016. Resminostat plus sorafenib as second-line therapy of advanced hepatocellular carcinoma - the SHELTER study. J. Hepatol. 65, 280–288.

Bosch-Presegué, L., Vaquero, A., 2011. The dual role of sirtuins in cancer. Genes Cancer 2, 648–662.

Caron, C., Boyault, C., Khochbin, S., 2005. Regulatory cross-talk between lysine acetylation and ubiquitination: role in the control of protein stability. BioEssays 27, 408–415.

Chalkiadaki, A., Guarente, L., 2015. The multifaceted functions of sirtuins in cancer. Nat. Rev. Cancer 15, 608–624.

Chang, H., Rha, S.Y., Jeung, H.-C., Jung, J.-J., Kim, T.S., Kwon, H.J., Kim, B.S., Chung, H.C., 2010. Identification of genes related to a synergistic effect of taxane and suberoylanilide hydroxamic acid combination treatment in gastric cancer cells. J. Cancer Res. Clin. Oncol. 136, 1901–1913.

Chen, S., Yin, C., Lao, T., Liang, D., He, D., Wang, C., Sang, N., 2015. AMPK-HDAC5 pathway facilitates nuclear accumulation of HIF-1α and functional activation of HIF-1 by deacetylating Hsp70 in the cytosol. Cell Cycle 14, 2520–2536.

Child, F., Ortiz-Romero, P. l., Alvarez, R., Bagot, M., Stadler, R., Weichenthal, M., Alves, R., Quaglino, P., Beylot-Barry, M., Cowan, R., Geskin, L. j., Pérez-Ferriols, A., Hellemans, P., Elsayed, Y., Phelps, C., Forslund, A., Kamida, M., Zinzani, P. l, 2016. Phase II multicentre trial of oral quisinostat, a histone deacetylase inhibitor, in patients with previously treated stage IB–IVA mycosis fungoides/Sézary syndrome. Br. J. Dermatol. 175, 80–88.

Choi, K.-C., Jung, M.G., Lee, Y.-H., Yoon, J.C., Kwon, S.H., Kang, H.-B., Kim, M.-J., Cha, J.-H., Kim, Y.J., Jun, W.J., Lee, J.M., Yoon, H.-G., 2009. Epigallocatechin-3-gallate, a histone acetyltransferase inhibitor, inhibits EBV-induced B lymphocyte transformation via suppression of RelA acetylation. Cancer Res. 69, 583–592.

Chou, S.-D., Khan, A.N.H., Magner, W.J., Tomasi, T.B., 2005. Histone acetylation regulates the cell type specific CIITA promoters, MHC class II expression and antigen presentation in tumor cells. Int. Immunol. 17, 1483–1494.

Choudhary, C., Kumar, C., Gnad, F., Nielsen, M.L., Rehman, M., Walther, T.C., Olsen, J.V., Mann, M., 2009. Lysine acetylation targets protein complexes and Co-Regulates major cellular functions. Science 325, 834–840.

Chromosomes Fact Sheet [WWW Document], n.d. Natl. Hum. Genome Res. Inst. NHGRI. https://www.genome.gov/26524120/Chromosomes-Fact-Sheet.

Claerhout, S., Lim, J.Y., Choi, W., Park, Y.-Y., Kim, K., Kim, S.-B., Lee, J.-S., Mills, G.B., Cho, J.Y., 2011. Gene expression signature analysis identifies vorinostat as a candidate therapy for gastric cancer. PLoS One 6.

International Human Genome Sequencing Consortium, 2004. Finishing the euchromatic sequence of the human genome. Nature 431, 931–945.

Conti, C., Leo, E., Eichler, G.S., Sordet, O., Martin, M.M., Fan, A., Aladjem, M.I., Pommier, Y., 2010. Inhibition of histone deacetylase in cancer cells slows down replication forks, activates dormant origins, and induces DNA damage. Cancer Res. 70, 4470–4480.

Crabb, S.J., Howell, M., Rogers, H., Ishfaq, M., Yurek-George, A., Carey, K., Pickering, B.M., East, P., Mitter, R., Maeda, S., Johnson, P.W.M., Townsend, P., Shin-ya, K., Yoshida, M., Ganesan, A., Packham, G., 2008. Characterisation of the in vitro activity of the depsipeptide histone deacetylase inhibitor spiruchostatin A. Biochem. Pharmacol. 76, 463–475.

Cutter, A.R., Hayes, J.J., 2015. A brief review of nucleosome structure. FEBS Lett. 589, 2914–2922.

D'Acunto, C.W., Fontanella, B., Rodriquez, M., Taddei, M., Parente, L., Petrella, A., 2010. Histone deacetylase inhibitor FR235222 sensitizes human prostate adenocarcinoma cells to apoptosis through up-regulation of Annexin A1. Cancer Lett. 295, 85–91.

Damaskos, C., Valsami, S., Kontos, M., Spartalis, E., Kalampokas, T., Kalampokas, E., Athanasiou, A., Moris, D., Daskalopoulou, A., Davakis, S., Tsourouflis, G., Kontzoglou, K., Perrea, D., Nikiteas, N., Dimitroulis, D., 2017. Histone deacetylase inhibitors: an attractive therapeutic strategy against breast cancer. Anticancer Res. 37, 35–46.

Deroanne, C.F., Bonjean, K., Servotte, S., Devy, L., Colige, A., Clausse, N., Blacher, S., Verdin, E., Foidart, J.-M., Nusgens, B.V., Castronovo, V., 2002. Histone deacetylases inhibitors as anti-angiogenic agents altering vascular endothelial growth factor signaling. Oncogene 21, 427–436.

Deubzer, H.E., Schier, M.C., Oehme, I., Lodrini, M., Haendler, B., Sommer, A., Witt, O., 2013. HDAC11 is a novel drug target in carcinomas. Int. J. Cancer 132, 2200–2208.

Druesne, N., Pagniez, A., Mayeur, C., Thomas, M., Cherbuy, C., Duée, P.-H., Martel, P., Chaumontet, C., 2004. Diallyl disulfide (DADS) increases histone acetylation and p21 waf1/cip1 expression in human colon tumor cell lines. Carcinogenesis 25, 1227–1236.

Du, J., Zhou, Y., Su, X., Yu, J.J., Khan, S., Jiang, H., Kim, J., Woo, J., Kim, J.H., Choi, B.H., He, B., Chen, W., Zhang, S., Cerione, R.A., Auwerx, J., Hao, Q., Lin, H., 2011. Sirt5 is a NAD-dependent protein lysine demalonylase and desuccinylase. Science 334, 806–809.

Dupéré-Richer, D., Kinal, M., Ménasché, V., Nielsen, T.H., del Rincon, S., Pettersson, F., Miller, W.H., 2013. Vorinostat-induced autophagy switches from a death-promoting to a cytoprotective signal to drive acquired resistance. Cell Death Dis. 4, e486.

Evens, A.M., Balasubramanian, S., Vose, J.M., Harb, W., Gordon, L.I., Langdon, R., Sprague, J., Sirisawad, M., Mani, C., Yue, J., Luan, Y., Horton, S., Graef, T., Bartlett, N.L., 2016. A phase I/II multicenter, open-label study of the oral histone deacetylase inhibitor abexinostat in relapsed/refractory lymphoma. Clin. Cancer Res. 22, 1059–1066.

Fakih, M.G., Fetterly, G., Egorin, M.J., Muindi, J.R., Espinoza-Delgado, I., Zwiebel, J.A., Litwin, A., Holleran, J.L., Wang, K., Diasio, R.B., 2010. A phase I, pharmacokinetic, and pharmacodynamic study of two schedules of vorinostat in combination with 5-fluorouracil and leucovorin in patients with refractory solid tumors. Clin. Cancer Res. 16, 3786–3794.

Fan, J., Lou, B., Chen, W., Zhang, J., Lin, S., Lv, F., Chen, Y., 2014. Down-regulation of HDAC5 inhibits growth of human hepatocellular carcinoma by induction of apoptosis and cell cycle arrest. Tumor Biol. 35, 11523–11532.

Felsenfeld, G., 2014. A brief history of epigenetics. Cold Spring Harb. Perspect. Biol. 6.

Finazzi, G., Vannucchi, A.M., Martinelli, V., Ruggeri, M., Nobile, F., Specchia, G., Pogliani, E.M., Olimpieri, O.M., Fioritoni, G., Musolino, C., Cilloni, D., Sivera, P., Barosi, G., Finazzi, M.C., Di Tollo, S., Demuth, T., Barbui, T., Rambaldi, A., 2013. A phase II study of Givinostat in combination with hydroxycarbamide in patients with polycythaemia vera unresponsive to hydroxycarbamide monotherapy. Br. J. Haematol. 161, 688–694.

Finkel, T., Deng, C.-X., Mostoslavsky, R., 2009. Recent progress in the biology and physiology of sirtuins. Nature 460, 587–591.

Folmer, F., Orlikova, B., Schnekenburger, M., Dicato, M., 2010. Naturally occurring regulators of histone acetylation/deacetylation. Curr. Nutr. Food Sci. 6, 78–99.

Garcia-Manero, G., Montalban-Bravo, G., Berdeja, J.G., Abaza, Y., Jabbour, E., Essell, J., Lyons, R.M., Ravandi, F., Maris, M., Heller, B., DeZern, A.E., Babu, S., Wright, D., Anz, B., Boccia, R., Komrokji, R.S., Kuriakose, P., Reeves, J., Sekeres, M.A., Kantarjian, H.M., Ghalie, R., Roboz, G.J., 2017. Phase 2, randomized, double-blind study of pracinostat in combination with azacitidine in patients with untreated, higher-risk myelodysplastic syndromes. Cancer 123, 994–1002.

Geng, H., Harvey, C.T., Pittsenbarger, J., Liu, Q., Beer, T.M., Xue, C., Qian, D.Z., 2011. HDAC4 protein regulates HIF1α protein lysine acetylation and cancer cell response to hypoxia. J. Biol. Chem. 286, 38095–38102.

Gilardini Montani, M.S., Granato, M., Santoni, C., Del Porto, P., Merendino, N., D'Orazi, G., Faggioni, A., Cirone, M., 2017. Histone deacetylase inhibitors VPA and TSA induce apoptosis and autophagy in pancreatic cancer cells. Cell. Oncol. Dordr.

Gregoretti, I., Lee, Y.-M., Goodson, H.V., 2004. Molecular evolution of the histone deacetylase family: functional implications of phylogenetic analysis. J. Mol. Biol. 338, 17–31.

Grunstein, M., Gasser, S.M., 2013. Epigenetics in *Saccharomyces cerevisiae*. Cold Spring Harb. Perspect. Biol. 5, a017491.

Gumbarewicz, E., Luszczki, J.J., Wawruszak, A., Dmoszynska-Graniczka, M., Grabarska, A.J., Jarząb, A.M., Polberg, K., Stepulak, A., 2016. Isobolographic analysis demonstrates additive effect of cisplatin and HDIs combined treatment augmenting their anticancer activity in lung cancer cell lines. Am. J. Cancer Res. 6, 2831–2845.

Han, J.-W., Ahn, S.H., Park, S.H., Wang, S.Y., Bae, G.-U., Seo, D.-W., Kwon, H.-K., Hong, S., Lee, H.Y., Lee, Y.-W., Lee, H.-W., 2000. Apicidin, a histone deacetylase inhibitor, inhibits proliferation of tumor cells via induction of p21WAF1/cip1 and gelsolin. Cancer Res. 60, 6068–6074.

Han, S., Fukazawa, T., Yamatsuji, T., Matsuoka, J., Miyachi, H., Maeda, Y., Durbin, M., Naomoto, Y., 2010. Anti-tumor effect in human lung cancer by a combination treatment of novel histone deacetylase inhibitors: SL142 or SL325 and retinoic acids. PLoS One 5.

Hong, J., Luesch, H., 2012. Largazole: from discovery to broad-spectrum therapy. Nat. Prod. Rep. 29, 449–456.

Huang, R., Xu, Y., Wan, W., Shou, X., Qian, J., You, Z., Liu, B., Chang, C., Zhou, T., Lippincott-Schwartz, J., Liu, W., 2015. Deacetylation of nuclear LC3 drives autophagy initiation under starvation. Mol. Cell 57, 456–466.

Hubbi, M.E., Hu, H., Kshitiz, Gilkes, D.M., Semenza, G.L., 2013. Sirtuin-7 inhibits the activity of hypoxia-inducible factors. J. Biol. Chem. 288, 20768–20775.

Imai, S., Armstrong, C.M., Kaeberlein, M., Guarente, L., 2000. Transcriptional silencing and longevity protein Sir2 is an NAD-dependent histone deacetylase. Nature 403, 795–800.

Imai, Y., Maru, Y., Tanaka, J., 2016. Action mechanisms of histone deacetylase inhibitors in the treatment of hematological malignancies. Cancer Sci. 107, 1543–1549.

Inoue, S., Riley, J., Gant, T.W., Dyer, M.J.S., Cohen, G.M., 2007. Apoptosis induced by histone deacetylase inhibitors in leukemic cells is mediated by Bim and Noxa. Leukemia 21, 1773–1782.

Jeong, J.-W., Bae, M.-K., Ahn, M.-Y., Kim, S.-H., Sohn, T.-K., Bae, M.-H., Yoo, M.-A., Song, E.J., Lee, K.-J., Kim, K.-W., 2002. Regulation and destabilization of HIF-1α by ARD1-mediated acetylation. Cell 111, 709–720.

Jeong, H., Then, F., Melia Jr., T.J., Mazzulli, J.R., Cui, L., Savas, J.N., Voisine, C., Paganetti, P., Tanese, N., Hart, A.C., Yamamoto, A., Krainc, D., 2009. Acetylation targets mutant huntingtin to autophagosomes for degradation. Cell 137, 60–72.

Johnstone, R.W., 2002. Histone-deacetylase inhibitors: novel drugs for the treatment of cancer. Nat. Rev. Drug Discov. 1, 287–299.

Jung, K.H., Noh, J.H., Kim, J.K., Eun, J.W., Bae, H.J., Xie, H.J., Chang, Y.G., Kim, M.G., Park, H., Lee, J.Y., Nam, S.W., 2012. HDAC2 overexpression confers oncogenic potential to human lung cancer cells by deregulating expression of apoptosis and cell cycle proteins. J. Cell. Biochem. 113, 2167–2177.

Kaluza, D., Kroll, J., Gesierich, S., Yao, T.-P., Boon, R.A., Hergenreider, E., Tjwa, M., Rössig, L., Seto, E., Augustin, H.G., Zeiher, A.M., Dimmeler, S., Urbich, C., 2011. Class IIb HDAC6 regulates endothelial cell migration and angiogenesis by deacetylation of cortactin. EMBO J. 30, 4142–4156.

Kang, Y., Nian, H., Rajendran, P., Kim, E., Dashwood, W.M., Pinto, J.T., Boardman, L.A., Thibodeau, S.N., Limburg, P.J., Löhr, C.V., Bisson, W.H., Williams, D.E., Ho, E., Dashwood, R.H., 2014. HDAC8 and STAT3 repress BMF gene activity in colon cancer cells. Cell Death Dis. 5, e1476.

Kao, G.D., McKenna, W.G., Guenther, M.G., Muschel, R.J., Lazar, M.A., Yen, T.J., 2003. Histone deacetylase 4 interacts with 53BP1 to mediate the DNA damage response. J. Cell Biol. 160, 1017–1027.

Kato, H., Tamamizu-Kato, S., Shibasaki, F., 2004. Histone deacetylase 7 associates with hypoxia-inducible factor 1α and increases transcriptional activity. J. Biol. Chem. 279, 41966–41974.

Kerr, E., Holohan, C., McLaughlin, K.M., Majkut, J., Dolan, S., Redmond, K., Riley, J., McLaughlin, K., Stasik, I., Crudden, M., Van Schaeybroeck, S., Fenning, C., O'Connor, R., Kiely, P., Sgobba, M., Haigh, D., Johnston, P.G., Longley, D.B., 2012. Identification of an acetylation-dependant Ku70/FLIP complex that regulates FLIP expression and HDAC inhibitor-induced apoptosis. Cell Death Differ. 19, 1317–1327.

Kim, H.-S., Patel, K., Muldoon-Jacobs, K., Bisht, K.S., Aykin-Burns, N., Pennington, J.D., van der Meer, R., Nguyen, P., Savage, J., Owens, K.M., Vassilopoulos, A., Ozden, O., Park, S.-H., Singh, K.K., Abdulkadir, S.A., Spitz, D.R., Deng, C.-X., Gius, D., 2010. SIRT3 is a mitochondria-localized tumor suppressor required for maintenance of mitochondrial integrity and metabolism during stress. Cancer Cell 17, 41–52.

Kim, J.K., Noh, J.H., Eun, J.W., Jung, K.H., Bae, H.J., Shen, Q., Kim, M.G., Chang, Y.G., Kim, S.-J., Park, W.S., Lee, J.Y., Borlak, J., Nam, S.W., 2013. Targeted inactivation of HDAC2 restores p16INK4a activity and exerts antitumor effects on human gastric cancer. Mol. Cancer Res. 11, 62–73.

Kong, X., Lin, Z., Liang, D., Fath, D., Sang, N., Caro, J., 2006. Histone deacetylase inhibitors induce VHL and ubiquitin-independent proteasomal degradation of hypoxia-inducible factor 1α. Mol. Cell Biol. 26, 2019–2028.

Kong, H.-S., Tian, S., Kong, Y., Du, G., Zhang, L., Jung, M., Dritschilo, A., Brown, M.L., 2012. Preclinical studies of YK-4-272, an inhibitor of class II histone deacetylases by disruption of nucleocytoplasmic shuttling. Pharm. Res. 29, 3373–3383.

Kotian, S., Liyanarachchi, S., Zelent, A., Parvin, J.D., 2011. Histone deacetylases 9 and 10 are required for homologous recombination. J. Biol. Chem. 286, 7722–7726.

Koutsounas, I., Giaginis, C., Patsouris, E., Theocharis, S., 2013. Current evidence for histone deacetylase inhibitors in pancreatic cancer. World J. Gastroenterol. 19, 813–828.

Kouzarides, T., 2000. Acetylation: a regulatory modification to rival phosphorylation? EMBO J. 19, 1176–1179.

LaBonte, M.J., Wilson, P.M., Fazzone, W., Groshen, S., Lenz, H.-J., Ladner, R.D., 2009. DNA microarray profiling of genes differentially regulated by the histone deacetylase inhibitors vorinostat and LBH589 in colon cancer cell lines. BMC Med. Genomics 2, 67.

Lachenmayer, A., Toffanin, S., Cabellos, L., Alsinet, C., Hoshida, Y., Villanueva, A., Minguez, B., Tsai, H.-W., Ward, S.C., Thung, S., Friedman, S.L., Llovet, J.M., 2012. Combination therapy for hepatocellular carcinoma: additive preclinical efficacy of the HDAC inhibitor panobinostat with sorafenib. J. Hepatol. 56, 1343–1350.

Lakshmaiah, K.C., Jacob, L.A., Aparna, S., Lakanatha, D., Saldanha, S.C., 2014. Epigenetic therapy of cancer with histone deacetylase inhibitors. J. Cancer Res. Ther. 10, 469–478.

Lane, A.A., Chabner, B.A., 2009. Histone deacetylase inhibitors in cancer therapy. J. Clin. Oncol. 27, 5459–5468.

Leder, A., Leder, P., 1975. Butyric acid, a potent inducer of erythroid differentiation in cultured erythroleukemic cells. Cell 5, 319–322.

Lee, J., Bae, S., Jeong, J., Kim, S., Kim, K., 2004. Hypoxia-inducible factor (HIF-1)[alpha]: its protein stability and biological functions. Exp. Mol. Med. Seoul 36, 1–12.

Lee, I.H., Cao, L., Mostoslavsky, R., Lombard, D.B., Liu, J., Bruns, N.E., Tsokos, M., Alt, F.W., Finkel, T., 2008. A role for the NAD-dependent deacetylase Sirt1 in the regulation of autophagy. Proc. Natl. Acad. Sci. U.S.A. 105, 3374–3379.

Lee, J.-H., Choy, M.L., Ngo, L., Foster, S.S., Marks, P.A., 2010a. Histone deacetylase inhibitor induces DNA damage, which normal but not transformed cells can repair. Proc. Natl. Acad. Sci. U.S.A. 107, 14639–14644.

Lee, J.-Y., Koga, H., Kawaguchi, Y., Tang, W., Wong, E., Gao, Y.-S., Pandey, U.B., Kaushik, S., Tresse, E., Lu, J., Taylor, J.P., Cuervo, A.M., Yao, T.-P., 2010b. HDAC6 controls autophagosome maturation essential for ubiquitin-selective quality-control autophagy. EMBO J. 29, 969–980.

Leipe, D.D., Landsman, D., 1997. Histone deacetylases, acetoin utilization proteins and acetylpolyamine amidohydrolases are members of an ancient protein superfamily. Nucleic Acids Res. 25, 3693–3697.

Li, Y., Seto, E., 2016. HDACs and HDAC inhibitors in cancer development and therapy. Cold Spring Harb. Perspect. Med. 6, a026831.

Li, Z., Zhu, W.-G., 2014. Targeting histone deacetylases for cancer therapy: from molecular mechanisms to clinical implications. Int. J. Biol. Sci. 10, 757–770.

Li, Y., Kao, G.D., Garcia, B.A., Shabanowitz, J., Hunt, D.F., Qin, J., Phelan, C., Lazar, M.A., 2006. A novel histone deacetylase pathway regulates mitosis by modulating Aurora B kinase activity. Genes Dev. 20, 2566–2579.

Li, Y., Peng, L., Seto, E., 2015. Histone deacetylase 10 regulates the cell cycle G2/M phase transition via a novel Let-7–HMGA2–cyclin A2 pathway. Mol. Cell Biol. 35, 3547–3565.

Li, Y., Zhao, K., Yao, C., Kahwash, S., Tang, Y., Zhang, G., Patterson, K., Wang, Q.-E., Zhao, W., 2016. Givinostat, a type II histone deacetylase inhibitor, induces potent caspase-dependent apoptosis in human lymphoblastic leukemia. Genes Cancer 7, 292–300.

Liang, D., Kong, X., Sang, N., 2006. Effects of histone deacetylase inhibitors on HIF-1. Cell Cycle Georget. Tex. 5, 2430–2435.

Lim, J.-H., Lee, Y.-M., Chun, Y.-S., Chen, J., Kim, J.-E., Park, J.-W., 2010. Sirtuin 1 modulates cellular responses to hypoxia by deacetylating hypoxia-inducible factor 1α. Mol. Cell 38, 864–878.

Link, A., Balaguer, F., Goel, A., 2010. Cancer chemoprevention by dietary polyphenols: promising role for epigenetics. Biochem. Pharmacol. 80, 1771–1792.

Lombard, D.B., Alt, F.W., Cheng, H.-L., Bunkenborg, J., Streeper, R.S., Mostoslavsky, R., Kim, J., Yancopoulos, G., Valenzuela, D., Murphy, A., Yang, Y., Chen, Y., Hirschey, M.D., Bronson, R.T., Haigis, M., Guarente, L.P., Farese, R.V., Weissman, S., Verdin, E., Schwer, B., 2007. Mammalian Sir2 homolog SIRT3 regulates global mitochondrial lysine acetylation. Mol. Cell Biol. 27, 8807–8814.

Lu, J., Sun, D., Liu, Z., Li, M., Hong, H., Liu, C., Gao, S., Li, H., Cai, Y., Chen, S., Li, Z., Ye, J., Liu, P., 2016. SIRT6 suppresses isoproterenol-induced cardiac hypertrophy through activation of autophagy. Transl. Res. 172, 96–112 e6.

Luger, K., Mäder, A.W., Richmond, R.K., Sargent, D.F., Richmond, T.J., 1997. Crystal structure of the nucleosome core particle at 2.8 Å resolution. Nature 389, 251–260.

Luo, J., Su, F., Chen, D., Shiloh, A., Gu, W., 2000. Deacetylation of p53 modulates its effect on cell growth and apoptosis. Nature 408, 377–381.

Lv, Z., Weng, X., Du, C., Zhang, C., Xiao, H., Cai, X., Ye, S., Cheng, J., Ding, C., Xie, H., Zhou, L., Wu, J., Zheng, S., 2016. Downregulation of HDAC6 promotes angiogenesis in hepatocellular carcinoma cells and predicts poor prognosis in liver transplantation patients. Mol. Carcinog. 55, 1024–1033.

Ma, L., Chua, M.-S., Andrisani, O., So, S., 2014. Epigenetics in hepatocellular carcinoma: an update and future therapy perspectives. World J. Gastroenterol. 20, 333–345.

Mandl-Weber, S., Meinel, F.G., Jankowsky, R., Oduncu, F., Schmidmaier, R., Baumann, P., 2010. The novel inhibitor of histone deacetylase resminostat (RAS2410) inhibits proliferation and induces apoptosis in multiple myeloma (MM) cells. Br. J. Haematol. 149, 518–528.

Manning, J., Indrova, M., Lubyova, B., Pribylova, H., Bieblova, J., Hejnar, J., Simova, J., Jandlova, T., Bubenik, J., Reinis, M., 2008. Induction of MHC class I molecule cell surface expression and epigenetic activation of antigen-processing machinery components in a murine model for human papilloma virus 16-associated tumours. Immunology 123, 218–227.

Marks, P.A., Dokmanovic, M., 2005. Histone deacetylase inhibitors: discovery and development as anticancer agents. Expert Opin. Investig. Drugs 14, 1497–1511.

Marks, P.A., Jiang, X., 2005. Histone deacetylase inhibitors in programmed cell death and cancer therapy. Cell Cycle 4, 549–551.

Marks, P.A., Richon, V.M., Miller, T., Kelly, W.K., 2004. Histone deacetylase inhibitors. B.-A. in C. In: Research. Academic Press, pp. 137–168.

Marmorstein, R., Zhou, M.-M., 2014. Writers and readers of histone acetylation: structure, mechanism, and inhibition. Cold Spring Harb. Perspect. Biol. 6.

Marquard, L., Gjerdrum, L.M., Christensen, I.J., Jensen, P.B., Sehested, M., Ralfkiaer, E., 2008. Prognostic significance of the therapeutic targets histone deacetylase 1, 2, 6 and acetylated histone H4 in cutaneous T-cell lymphoma. Histopathology 53, 267–277.

Marson, C.M., Matthews, C.J., Yiannaki, E., Atkinson, S.J., Soden, P.E., Shukla, L., Lamadema, N., Thomas, N.S.B., 2013. Discovery of potent, isoform-selective inhibitors of histone deacetylase containing chiral heterocyclic capping groups and a N-(2-aminophenyl)benzamide binding unit. J. Med. Chem. 56, 6156–6174.

Martínez-Redondo, P., Vaquero, A., 2013. The diversity of histone versus nonhistone sirtuin substrates. Genes Cancer 4, 148–163.

Meeran, S.M., Patel, S.N., Tollefsbol, T.O., 2010. Sulforaphane causes epigenetic repression of hTERT expression in human breast cancer cell lines. PLoS One 5, e11457.

Michishita, E., Park, J.Y., Burneskis, J.M., Barrett, J.C., Horikawa, I., 2005. Evolutionarily conserved and nonconserved cellular localizations and functions of human SIRT proteins. Mol. Biol. Cell 16, 4623–4635.

Michishita, E., McCord, R.A., Berber, E., Kioi, M., Padilla-Nash, H., Damian, M., Cheung, P., Kusumoto, R., Kawahara, T.L.A., Barrett, J.C., Chang, H.Y., Bohr, V.A., Ried, T., Gozani, O., Chua, K.F., 2008. SIRT6 is a histone H3 lysine 9 deacetylase that modulates telomeric chromatin. Nature 452, 492–496.

Miller, K.M., Tjeertes, J.V., Coates, J., Legube, G., Polo, S.E., Britton, S., Jackson, S.P., 2010. Human HDAC1 and HDAC2 function in the DNA-damage response to promote DNA nonhomologous end-joining. Nat. Struct. Mol. Biol. 17, 1144–1151.

Minucci, S., Nervi, C., Lo Coco, F., Pelicci, P.G., 2001. Histone deacetylases: a common molecular target for differentiation treatment of acute myeloid leukemias? Oncogene 20, 3110–3115.

Moresi, V., Carrer, M., Grueter, C.E., Rifki, O.F., Shelton, J.M., Richardson, J.A., Bassel-Duby, R., Olson, E.N., 2012. Histone deacetylases 1 and 2 regulate autophagy flux and skeletal muscle homeostasis in mice. Proc. Natl. Acad. Sci. U.S.A. 109, 1649–1654.

Mottamal, M., Zheng, S., Huang, T.L., Wang, G., 2015. Histone deacetylase inhibitors in clinical studies as templates for new anticancer agents. Mol. Basel Switz. 20, 3898–3941.

Myzak, M.C., Karplus, P.A., Chung, F.-L., Dashwood, R.H., 2004. A novel mechanism of chemoprotection by sulforaphane: inhibition of histone deacetylase. Cancer Res. 64, 5767–5774.

Myzak, M.C., Dashwood, W.M., Orner, G.A., Ho, E., Dashwood, R.H., 2006. Sulforaphane inhibits histone deacetylase in vivo and suppresses tumorigenesis in Apc-minus mice. FASEB J. 20, 506–508.

Myzak, M.C., Tong, P., Dashwood, W.-M., Dashwood, R.H., Ho, E., 2007. Sulforaphane retards the growth of human PC-3 xenografts and inhibits HDAC activity in human subjects. Exp. Biol. Med. Maywood NJ 232, 227–234.

Newbold, A., Matthews, G.M., Bots, M., Cluse, L.A., Clarke, C.J.P., Banks, K.-M., Cullinane, C., Bolden, J.E., Christiansen, A.J., Dickins, R.A., Miccolo, C., Chiocca, S., Kral, A.M., Ozerova, N.D., Miller, T.A., Methot, J.L., Richon, V.M., Secrist, J.P., Minucci, S., Johnstone, R.W., 2013. Molecular and biologic analysis of histone deacetylase inhibitors with diverse specificities. Mol. Cancer Ther. 12, 2709–2721.

Newbold, A., Falkenberg, K.J., Prince, H.M., Johnstone, R.W., 2016. How do tumor cells respond to HDAC inhibition? FEBS J. 283, 4032–4046.

Nian, H., Delage, B., Pinto, J.T., Dashwood, R.H., 2008. Allyl mercaptan, a garlic-derived organosulfur compound, inhibits histone deacetylase and enhances Sp3 binding on the P21WAF1 promoter. Carcinogenesis 29, 1816–1824.

Niesvizky, R., Richardson, P.G., Gabrail, N.Y., Madan, S., Yee, A.J., Quayle, S.N., Almeciga-Pinto, I., Jones, S.S., Houston, L., Hayes, D., Duzer, J.V., Wheeler, C., Trede, N.S., Raje, N.S., 2015. ACY-241, a novel, HDAC6 selective inhibitor: synergy with immunomodulatory (IMiD®) drugs in multiple myeloma (MM) cells and early clinical results (ACE-MM-200 study). Blood 126, 3040.

Oehme, I., Linke, J.-P., Böck, B.C., Milde, T., Lodrini, M., Hartenstein, B., Wiegand, I., Eckert, C., Roth, W., Kool, M., Kaden, S., Gröne, H.-J., Schulte, J.H., Lindner, S., Hamacher-Brady, A., Brady, N.R., Deubzer, H.E., Witt, O., 2013. Histone deacetylase 10 promotes autophagy-mediated cell survival. Proc. Natl. Acad. Sci. U.S.A. 110, E2592–E2601.

Oh, M., Choi, I.-K., Kwon, H.J., 2008. Inhibition of histone deacetylase1 induces autophagy. Biochem. Biophys. Res. Commun. 369, 1179–1183.

Orlikova, B., Orlikova, B., Schnekenburger, M., Schnekenburger, M., Zloh, M., Zloh, M., Golais, F., Golais, F., Diederich, M., Diederich, M., Tasdemir, D., Tasdemir, D., 2012. Natural chalcones as dual inhibitors of HDACs and NF-κB. Oncol. Rep. 28, 797–805.

Ou, X., Lee, M.R., Huang, X., Messina-Graham, S., Broxmeyer, H.E., 2014. SIRT1 positively regulates autophagy and mitochondria function in embryonic stem cells under oxidative stress. Stem Cells 32, 1183–1194.

Ouyang, L., Shi, Z., Zhao, S., Wang, F.-T., Zhou, T.-T., Liu, B., Bao, J.-K., 2012. Programmed cell death pathways in cancer: a review of apoptosis, autophagy and programmed necrosis. Cell Prolif. 45, 487–498.

Peng, L., Ling, H., Yuan, Z., Fang, B., Bloom, G., Fukasawa, K., Koomen, J., Chen, J., Lane, W.S., Seto, E., 2012. SIRT1 negatively regulates the activities, functions, and protein levels of hMOF and TIP60. Mol. Cell. Biol. 32, 2823–2836.

Petrella, A., D'Acunto, C.W., Rodriquez, M., Festa, M., Tosco, A., Bruno, I., Terracciano, S., Taddei, M., Paloma, L.G., Parente, L., 2008. Effects of FR235222, a novel HDAC inhibitor, in proliferation and apoptosis of human leukaemia cell lines: role of Annexin A1. Eur. J. Cancer 44, 740–749.

Petruccelli, L.A., Dupéré-Richer, D., Pettersson, F., Retrouvey, H., Skoulikas, S., Miller, W.H., 2011. Vorinostat induces reactive oxygen species and DNA damage in acute myeloid leukemia cells. PLoS One 6, e20987.

Pinazza, M., Borga, C., Agnusdei, V., Minuzzo, S., Fossati, G., Paganin, M., Michielotto, B., De Paoli, A., Basso, G., Amadori, A., te Kronnie, G., Indraccolo, S., 2016. An immediate transcriptional signature associated with response to the histone deacetylase inhibitor Givinostat in T acute lymphoblastic leukemia xenografts. Cell Death Dis. 6, e2047.

Polletta, L., Vernucci, E., Carnevale, I., Arcangeli, T., Rotili, D., Palmerio, S., Steegborn, C., Nowak, T., Schutkowski, M., Pellegrini, L., Sansone, L., Villanova, L., Runci, A., Pucci, B., Morgante, E., Fini, M., Mai, A., Russo, M.A., Tafani, M., 2015. SIRT5 regulation of ammonia-induced autophagy and mitophagy. Autophagy 11, 253–270.

Ray, A., Alalem, M., Ray, B.K., 2013. Loss of epigenetic Kruppel-like factor 4 histone deacetylase (KLF-4-HDAC)-mediated transcriptional suppression is crucial in increasing vascular endothelial growth factor (VEGF) expression in breast cancer. J. Biol. Chem. 288, 27232–27242.

Rettig, I., Koeneke, E., Trippel, F., Mueller, W.C., Burhenne, J., Kopp-Schneider, A., Fabian, J., Schober, A., Fernekorn, U., von Deimling, A., Deubzer, H.E., Milde, T., Witt, O., Oehme, I., 2015. Selective inhibition of HDAC8 decreases neuroblastoma growth in vitro and in vivo and enhances retinoic acid-mediated differentiation. Cell Death Dis. 6, e1657.

Riley, J.S., Hutchinson, R., McArt, D.G., Crawford, N., Holohan, C., Paul, I., Van Schaeybroeck, S., Salto-Tellez, M., Johnston, P.G., Fennell, D.A., Gately, K., O'Byrne, K., Cummins, R., Kay, E., Hamilton, P., Stasik, I., Longley, D.B., 2013. Prognostic and therapeutic relevance of FLIP and procaspase-8 overexpression in non-small cell lung cancer. Cell Death Dis. 4, e951.

Robert, C., Rassool, F.V., 2012. HDAC inhibitors: roles of DNA damage and repair. Adv. Cancer Res. 116, 87–129.

Robert, T., Vanoli, F., Chiolo, I., Shubassi, G., Bernstein, K.A., Rothstein, R., Botrugno, O.A., Parazzoli, D., Oldani, A., Minucci, S., Foiani, M., 2011. HDACs link the DNA damage response, processing of double-strand breaks and autophagy. Nature 471, 74–79.

Ruefli, A.A., Ausserlechner, M.J., Bernhard, D., Sutton, V.R., Tainton, K.M., Kofler, R., Smyth, M.J., Johnstone, R.W., 2001. The histone deacetylase inhibitor and chemotherapeutic agent suberoylanilide hydroxamic acid (SAHA) induces a cell-death pathway characterized by cleavage of Bid and production of reactive oxygen species. Proc. Natl. Acad. Sci. U.S.A. 98, 10833–10838.

Sampath, D., Liu, C., Vasan, K., Sulda, M., Puduvalli, V.K., Wierda, W.G., Keating, M.J., 2012. Histone deacetylases mediate the silencing of miR-15a, miR-16, and miR-29b in chronic lymphocytic leukemia. Blood 119, 1162–1172.

Scher, M.B., Vaquero, A., Reinberg, D., 2007. SirT3 is a nuclear NAD+-dependent histone deacetylase that translocates to the mitochondria upon cellular stress. Genes Dev. 21, 920–928.

Schüler, S., Fritsche, P., Diersch, S., Arlt, A., Schmid, R.M., Saur, D., Schneider, G., 2010. HDAC2 attenuates TRAIL-induced apoptosis of pancreatic cancer cells. Mol. Cancer 9, 80.

Schwartz, B.E., Hofer, M.D., Lemieux, M.E., Bauer, D.E., Cameron, M.J., West, N.H., Agoston, E.S., Reynoird, N., Khochbin, S., Ince, T.A., Christie, A., Janeway, K.A., Vargas, S.O., Perez-Atayde, A.R., Aster, J.C., Sallan, S.E., Kung, A.L., Bradner, J.E., French, C.A., 2011. Differentiation of NUT midline carcinoma by epigenomic reprogramming. Cancer Res. 71, 2686–2696.

Seidel, C., Schnekenburger, M., Dicato, M., Diederich, M., 2012. Histone deacetylase modulators provided by mother nature. Genes Nutr. 7, 357–367.

Senese, S., Zaragoza, K., Minardi, S., Muradore, I., Ronzoni, S., Passafaro, A., Bernard, L., Draetta, G.F., Alcalay, M., Seiser, C., Chiocca, S., 2007. Role for histone deacetylase 1 in human tumor cell proliferation. Mol. Cell Biol. 27, 4784–4795.

Seo, H.-W., Kim, E.-J., Na, H., Lee, M.-O., 2009. Transcriptional activation of hypoxia-inducible factor-1α by HDAC4 and HDAC5 involves differential recruitment of p300 and FIH-1. FEBS Lett. 583, 55–60.

Setiadi, A.F., Omilusik, K., David, M.D., Seipp, R.P., Hartikainen, J., Gopaul, R., Choi, K.B., Jefferies, W.A., 2008. Epigenetic enhancement of antigen processing and presentation promotes immune recognition of tumors. Cancer Res. 68, 9601–9607.

Seto, E., Yoshida, M., 2014. Erasers of histone acetylation: the histone deacetylase enzymes. Cold Spring Harb. Perspect. Biol. 6, a018713.

Shao, Y., Gao, Z., Marks, P.A., Jiang, X., 2004. Apoptotic and autophagic cell death induced by histone deacetylase inhibitors. Proc. Natl. Acad. Sci. U.S.A. 101, 18030–18035.

Sherr, C.J., Roberts, J.M., 1995. Inhibitors of mammalian G1 cyclin-dependent kinases. Genes Dev. 9, 1149–1163.

Skov, S., Pedersen, M.T., Andresen, L., Straten, P.T., Woetmann, A., Ødum, N., 2005. Cancer cells become susceptible to natural killer cell killing after exposure to histone deacetylase inhibitors due to glycogen synthase kinase-3–dependent expression of MHC class I–Related chain a and B. Cancer Res. 65, 11136–11145.

Someya, S., Yu, W., Hallows, W.C., Xu, J., Vann, J.M., Leeuwenburgh, C., Tanokura, M., Denu, J.M., Prolla, T.A., 2010. Sirt3 mediates reduction of oxidative damage and prevention of age-related hearing loss under caloric restriction. Cell 143, 802–812.

Tan, M., Luo, H., Lee, S., Jin, F., Yang, J.S., Montellier, E., Buchou, T., Cheng, Z., Rousseaux, S., Rajagopal, N., Lu, Z., Ye, Z., Zhu, Q., Wysocka, J., Ye, Y., Khochbin, S., Ren, B., Zhao, Y., 2011. Identification of 67 histone marks and histone lysine crotonylation as a new type of histone modification. Cell 146, 1016–1028.

Tomasi, T.B., Magner, W.J., Wiesen, J.L., Oshlag, J.Z., Cao, F., Pontikos, A.N., Gregorie, C.J., 2010. MHC class II regulation by epigenetic agents and microRNAs. Immunol. Res. 46, 45–58.

Tseng, A.H.H., Shieh, S.-S., Wang, D.L., 2013. SIRT3 deacetylates FOXO3 to protect mitochondria against oxidative damage. Free Radic. Biol. Med. 63, 222–234.

Turtoi, A., Mottet, D., Matheus, N., Dumont, B., Peixoto, P., Hennequière, V., Deroanne, C., Colige, A., Pauw, E.D., Bellahcène, A., Castronovo, V., 2012. The angiogenesis suppressor gene AKAP12 is under the epigenetic control of HDAC7 in endothelial cells. Angiogenesis 15, 543–554.

Urbich, C., Rössig, L., Kaluza, D., Potente, M., Boeckel, J.-N., Knau, A., Diehl, F., Geng, J.-G., Hofmann, W.-K., Zeiher, A.M., Dimmeler, S., 2009. HDAC5 is a repressor of angiogenesis and determines the angiogenic gene expression pattern of endothelial cells. Blood 113, 5669–5679.

Van Damme, M., Crompot, E., Meuleman, N., Mineur, P., Bron, D., Lagneaux, L., Stamatopoulos, B., 2012. HDAC isoenzyme expression is deregulated in chronic lymphocytic leukemia B-cells and has a complex prognostic significance. Epigenetics 7 (12), 1403–1412.

Vaquero, A., Scher, M., Lee, D., Erdjument-Bromage, H., Tempst, P., Reinberg, D., 2004. Human SirT1 interacts with histone H1 and promotes formation of facultative heterochromatin. Mol. Cell 16, 93–105.

Vaziri, H., Dessain, S.K., Eaton, E.N., Imai, S.-I., Frye, R.A., Pandita, T.K., Guarente, L., Weinberg, R.A., 2001. hSIR2-SIRT1 Functions as an NAD-Dependent p53 Deacetylase. Cell 107, 149–159.

Venturelli, S., Berger, A., Böcker, A., Busch, C., Weiland, T., Noor, S., Leischner, C., Schleicher, S., Mayer, M., Weiss, T.S., Bischoff, S.C., Lauer, U.M., Bitzer, M., 2013. Resveratrol as a Pan-HDAC inhibitor alters the acetylation status of jistone proteins in human-derived hepatoblastoma cells. PLoS One 8.

Vey, N., Prebet, T., Thalamas, C., Charbonnier, A., Rey, J., Kloos, I., Liu, E., Luan, Y., Vezan, R., Graef, T., Recher, C., 2017. Phase 1 dose-escalation study of oral abexinostat for the treatment of patients with relapsed/refractory higher-risk myelodysplastic syndromes, acute myeloid leukemia, or acute lymphoblastic leukemia. Leuk. Lymphoma 58, 1880–1886.

Vigushin, D.M., Ali, S., Pace, P.E., Mirsaidi, N., Ito, K., Adcock, I., Coombes, R.C., 2001. Trichostatin a is a histone deacetylase inhibitor with potent antitumor activity against breast cancer in vivo. Clin. Cancer Res. 7, 971–976.

Waby, J.S., Chirakkal, H., Yu, C., Griffiths, G.J., Benson, R.S., Bingle, C.D., Corfe, B.M., 2010. Sp1 acetylation is associated with loss of DNA binding at promoters associated with cell cycle arrest and cell death in a colon cell line. Mol. Cancer 9, 275.

Wahid, B., Ali, A., Rafique, S., Idrees, M., 2017. New Insights into the epigenetics of hepatocellular carcinoma. BioMed Res. Int. 2017.

Wang, Z., Zang, C., Rosenfeld, J.A., Schones, D.E., Barski, A., Cuddapah, S., Cui, K., Roh, T.-Y., Peng, W., Zhang, M.Q., Zhao, K., 2008. Combinatorial patterns of histone acetylations and methylations in the human genome. Nat. Genet. 40, 897–903.

Wang, Q., Zhang, Y., Yang, C., Xiong, H., Lin, Y., Yao, J., Li, H., Xie, L., Zhao, W., Yao, Y., Ning, Z.-B., Zeng, R., Xiong, Y., Guan, K.-L., Zhao, S., Zhao, G.-P., 2010. Acetylation of metabolic enzymes coordinates carbon source utilization and metabolic flux. Science 327, 1004–1007.

Wang, J.C., Kafeel, M.I., Avezbakiyev, B., Chen, C., Sun, Y., Rathnasabapathy, C., Kalavar, M., He, Z., Burton, J., Lichter, S., 2011. Histone deacetylase in chronic lymphocytic leukemia. Oncology 81, 325–329.

Wei, W., Sun, W., Yu, S., Yang, Y., Ai, L., 2016. Butyrate production from high-fiber diet protects against lymphoma tumor. Leuk. Lymphoma 57, 2401–2408.

Weichert, W., Röske, A., Gekeler, V., Beckers, T., Ebert, M.P.A., Pross, M., Dietel, M., Denkert, C., Röcken, C., 2008. Association of patterns of class I histone deacetylase expression with patient prognosis in gastric cancer: a retrospective analysis. Lancet Oncol. 9, 139–148.

West, A.C., Johnstone, R.W., 2014. New and emerging HDAC inhibitors for cancer treatment. J. Clin. Investig. 124, 30–39.

Wilson, P.M., El-Khoueiry, A., Iqbal, S., Fazzone, W., LaBonte, M.J., Groshen, S., Yang, D., Danenberg, K.D., Cole, S., Kornacki, M., Ladner, R.D., Lenz, H.-J., 2010. A phase I/II trial of vorinostat in combination with 5-fluorouracil in patients with metastatic colorectal cancer who previously failed 5-FU-based chemotherapy. Cancer Chemother. Pharmacol. 65, 979–988.

Xie, H.J., Noh, J.H., Kim, J.K., Jung, K.H., Eun, J.W., Bae, H.J., Kim, M.G., Chang, Y.G., Lee, J.Y., Park, H., Nam, S.W., 2012. HDAC1 inactivation induces mitotic defect and caspase-independent autophagic cell death in liver cancer. PLoS One 7, e34265.

Yamaguchi, T., Cubizolles, F., Zhang, Y., Reichert, N., Kohler, H., Seiser, C., Matthias, P., 2010. Histone deacetylases 1 and 2 act in concert to promote the G1-to-S progression. Genes Dev. 24, 455–469.

Yang, Y., Zhao, Y., Liao, W., Yang, J., Wu, L., Zheng, Z., Yu, Y., Zhou, W., Li, L., Feng, J., Wang, H., Zhu, W.-G., 2009. Acetylation of FoxO1 activates Bim expression to induce apoptosis in response to histone deacetylase inhibitor depsipeptide treatment. Neoplasia 11, 313–324.

Yang, F.-C., Tan, B.C.-M., Chen, W.-H., Lin, Y.-H., Huang, J.-Y., Chang, H.-Y., Sun, H.-Y., Hsu, P.-H., Liou, G.-G., Shen, J., Chang, C.-J., Han, C.-C., Tsai, M.-D., Lee, S.-C., 2013. Reversible acetylation regulates salt-inducible kinase (SIK2) and its function in autophagy. J. Biol. Chem. 288, 6227–6237.

Yee, A.J., Bensinger, W.I., Supko, J.G., Voorhees, P.M., Berdeja, J.G., Richardson, P.G., Libby, E.N., Wallace, E.E., Birrer, N.E., Burke, J.N., Tamang, D.L., Yang, M., Jones, S.S., Wheeler, C.A., Markelewicz, R.J., Raje, N.S., 2016. Ricolinostat plus lenalidomide, and dexamethasone in relapsed or refractory multiple myeloma: a multicentre phase 1b trial. Lancet Oncol. 17, 1569–1578.

Yeo, W., Chung, H.C., Chan, S.L., Wang, L.Z., Lim, R., Picus, J., Boyer, M., Mo, F.K.F., Koh, J., Rha, S.Y., Hui, E.P., Jeung, H.C., Roh, J.K., Yu, S.C.H., To, K.F., Tao, Q., Ma, B.B., Chan, A.W.H., Tong, J.H.M., Erlichman, C., Chan, A.T.C., Goh, B.C., 2012. Epigenetic therapy using belinostat for patients with unresectable hepatocellular carcinoma: a multicenter phase I/II study with biomarker and pharmacokinetic analysis of tumors from patients in the Mayo phase II consortium and the cancer therapeutics research group. J. Clin. Oncol. 30, 3361–3367.

Yoo, Y.-G., Kong, G., Lee, M.-O., 2006. Metastasis-associated protein 1 enhances stability of hypoxia-inducible factor-1α protein by recruiting histone deacetylase 1. EMBO J. 25, 1231–1241.

Yoo, C., Ryu, M.-H., Na, Y.-S., Ryoo, B.-Y., Lee, C.-W., Kang, Y.-K., 2016. Vorinostat in combination with capecitabine plus cisplatin as a first-line chemotherapy for patients with metastatic or unresectable gastric cancer: phase II study and biomarker analysis. Br. J. Cancer 114, 1185–1190.

Zhang, X.D., Gillespie, S.K., Borrow, J.M., Hersey, P., 2004. The histone deacetylase inhibitor suberic bishydroxamate regulates the expression of multiple apoptotic mediators and induces mitochondria-dependent apoptosis of melanoma cells. Mol. Cancer Ther. 3, 425–435.

Zhang, C.Z., Pan, Y., Cao, Y., Lai, P.B.S., Liu, L., Chen, G.G., Yun, J., 2012. Histone deacetylase inhibitors facilitate dihydroartemisinin-induced apoptosis in liver cancer in vitro and in vivo. PLoS One 7.

Zhang, M., Xiang, S., Joo, H.-Y., Wang, L., Williams, K.A., Liu, W., Hu, C., Tong, D., Haakenson, J., Wang, C., Zhang, S., Pavlovicz, R.E., Jones, A., Schmidt, K.H., Tang, J., Dong, H., Shan, B., Fang, B., Radhakrishnan, R., Glazer, P.M., Matthias, P., Koomen, J., Seto, E., Bepler, G., Nicosia, S.V., Chen, J., Li, C., Gu, L., Li, G.-M., Bai, W., Wang, H., Zhang, X., 2014. HDAC6 deacetylates and ubiquitinates MSH2 to maintain proper levels of MutSα. Mol. Cell 55, 31–46.

Zhao, Y., Tan, J., Zhuang, L., Jiang, X., Liu, E.T., Yu, Q., 2005. Inhibitors of histone deacetylases target the Rb-E2F1 pathway for apoptosis induction through activation of proapoptotic protein Bim. Proc. Natl. Acad. Sci. U.S.A 102, 16090–16095.

Zhao, Z.-N., Bai, J.-X., Zhou, Q., Yan, B., Qin, W.-W., Jia, L.-T., Meng, Y.-L., Jin, B.-Q., Yao, L.-B., Wang, T., Yang, A.-G., 2012. TSA suppresses miR-106b-93-25 cluster expression through downregulation of MYC and inhibits proliferation and induces apoptosis in human EMC. PLoS One 7.

Zhong, L., D'Urso, A., Toiber, D., Sebastian, C., Henry, R.E., Vadysirisack, D.D., Guimaraes, A., Marinelli, B., Wikstrom, J.D., Nir, T., Clish, C.B., Vaitheesvaran, B., Iliopoulos, O., Kurland, I., Dor, Y., Weissleder, R., Shirihai, O.S., Ellisen, L.W., Espinosa, J.M., Mostoslavsky, R., 2010. The histone deacetylase Sirt6 regulates glucose homeostasis via Hif1α. Cell 140, 280–293.

CHAPTER

5

Epigenetic Effects of Curcumin in Cancer Prevention

Muthu K. Shanmugam[1], Frank Arfuso[2], Judy C. Sng[1], Anupam Bishayee[3], Alan Prem Kumar[1,4,5,6,7], Gautam Sethi[1,8]

[1]Department of Pharmacology, Yong Loo Lin School of Medicine, National University of Singapore, Singapore, Singapore; [2]Stem Cell and Cancer Biology Laboratory, School of Biomedical Sciences, Curtin Health Innovation Research Institute, Curtin University, Perth, WA, Australia; [3]Lake Erie College of Osteopathic Medicine, Bradenton, FL, United States; [4]Cancer Science Institute of Singapore, National University of Singapore, Singapore, Singapore; [5]Curtin Medical School, Faculty of Health Sciences, Curtin University, Perth, WA, Australia; [6]National University Cancer Institute, National University Health System, Singapore, Singapore; [7]Department of Biological Sciences, University of North Texas, Denton, TX, United States; [8]School of Pharmacy and Biomedical Sciences, Curtin Health Innovation Research Institute, Curtin University, Perth, WA, Australia

Abstract

Curcumin, a polyphenolic compound derived from turmeric (*Curcuma longa* Linn), possesses remarkable medicinal properties and has been extensively studied over four decades for its potential anti-inflammatory and/or anticancer effects. The epigenetic impact of curcumin in cancer prevention is yet largely an unexplored area of research, with sporadic reports of its epigenetic modulatory properties. Epigenetics represents an evolving strategy for cancer chemoprevention and has been established as a promising approach in human clinical trials. Epigenetic studies include DNA methylation, histone lysine acetylation/deacetylation, histone methylation/demethylation, and arginine methylation. These epigenetic changes are mediated by specific enzymes, such as lysine acetyltransferases/deacetylases and lysine methyltransferases/demethylases, which regulate this reversible posttranslational modification as well as nuclear transcriptional processes. Curcumin has been reported to modulate epigenetic changes in cancer cells; however, this preventive aspect has not been much reported. Curcumin has been shown to be a DNA hypomethylation agent in colon, prostate, and breast cancer, thus serving as a chemopreventive agent. Among these curcumin-induced epigenetic alterations is modulation of the expression of several oncogenic and tumor-suppressor microRNAs (miRs). Suppression of oncomiRs, such as miR-21, miR-17-5p, miR-20a, and miR-27a, and overexpression of miR-34 a/c and epithelial–mesenchymal transition-suppressor miRs, are among the most important effects of curcumin on miRs homeostasis. The present chapter is designed to collate the existing literature

Epigenetics of Cancer Prevention
https://doi.org/10.1016/B978-0-12-812494-9.00005-6

and comprehensively discuss the current understanding of potential epigenetic targets of curcumin and its effects on cancer chemoprevention.

Keywords
Curcumin; Deacetylases; DNA methyltransferases; Epigenetics; Histone acetyltransferases; microRNA.

INTRODUCTION

The word epigenetics was first introduced to the scientific community by Conrad H. Waddington. It was known as the Waddington's equation, the combination of two words, epigenesis + genetics = epigenetics (Van Speybroeck, 2002; Waddington 1942, 1952; Waddington and Carter, 1952). Epigenetics is defined as heritable changes in gene expression patterns that occur without any changes to DNA but involve changes that are adequate to regulate activation or repression of gene expression. In the human body, cells contain identical genetic materials; nevertheless, all cells have the ability to maintain their specific phenotype and biological functions. It has been found that these specific capabilities are guaranteed by the chromatin-associated histones on the DNA sequence that make transient and heritable chemical changes that occur during gene expression (Berger, 2007; Berger et al., 2009; Hirst and Marra, 2009; Shanmugam et al., 2016). Decades of research have identified the pivotal role of epigenetics in all physiological processes, such as during development, in the differentiation of tissues and organs, for maintaining cell identity, in cell cycle regulation and proliferation, and in genome integrity (Feinberg, 2007; Kouzarides, 2007). Several factors have been identified as regulating epigenetic gene processes through changes in the chromatin structure; however, the covalent modifications on histones and the DNA itself are the most decisive events that regulate these processes (Segal and Widom, 2009; Simo-Riudalbas and Esteller, 2014, 2015).

Chromatin modification is the pivotal event that creates an interaction between the DNA sequence and histones and provides accessibility to specific loci on the DNA sequence, such as creation of docking sites for transcription regulators. These covalent modifications are now known as the histone code, which is defined by four different DNA modifications (Baylin and Jones, 2011; Pfaffeneder et al., 2014; Wu and Zhang, 2011). These events often dictate key cellular events such as mitosis, meiosis, DNA damage response, gene expression, cell cycle, cell signaling pathways, energy, and metabolic pathways. Several posttranslational modifications have been observed on histone and non-histone proteins, such as phosphorylation, N-terminal acetylation, methylation, sumoylation, ubiquitination, propionylation, butyrylation, carbonylation, neddylation, proline isomerization, citrullination, and ADP ribosylation, all of which regulate diverse protein functions (Kouzarides, 2007; Lee et al., 2010; Shanmugam et al., 2016; Shanmugam and Sethi, 2012).

In the eukaryotic cell, a 147-base pair of DNA is wrapped around the core octamural globular histone proteins, the histone octamer, containing pairs of H2A, H2B, H3, and H4; thus they form the fundamental unit, the nucleosome, and then chromatin and chromosomes (Luger et al., 1997; Tessarz and Kouzarides, 2014). The structure and function of the epigenome is regulated by the covalent marks, which are catalyzed by specific enzymes, such as histone lysine methyltransferases, histone acetyltransferases (HATs), and DNA methyltransferases (DNMTs) that transfer methyl or acetyl groups to the specific amino acids on the histones or DNA and are known as the "writers." The proteins that recognize and interpret these covalent marks on the DNA or histones are known as "readers," including the methyl-CpG binding domain protein on the DNA, and histone mark readers such as chromo, bromo, tudor, pro-trp-trp-pro, malignant brain tumor, and plant homeodomain (PHD) finger domain proteins. These posttranslational modifications are found to be reversible because these marks can be removed, and the enzymes that are involved in this process are called "erasers," such as histone demethylases, histone deacetylases (HDACs), and the ten-eleven translocation family of 5-methylcytosine oxidases. Therefore, this tightly regulated process enables accessibility to the chromatin and thereby enhances the ability of transcription factors to bind to DNA and initiate transcription (Baylin and Jones, 2016; Berger, 2007; Shanmugam et al., 2016).

Several regions in the chromatin may assume two opposing conformations based on the degree of DNA coiling. First is the euchromatin, which is characterized by a relaxed DNA that enables transcription factors to bind to specific regions of the DNA and initiates transcription, and the second is the heterochromatin, which is characterized by super-coiled compact DNA that does not allow gene transcription. Thus, the state of the chromatin determines gene expression patterns, resulting in significant changes in their phenotypes and biological functions (Baylin and Jones, 2011, 2016; Jones et al., 2016; Venkatesh and Workman, 2015; Yan et al., 2016).

In addition to their pivotal role in the development of an organism, deregulated epigenetic mechanisms are often encountered in the development and severity of disease. It is now increasingly apparent that chronic inflammation-driven diseases, including cancer, occur due to both genetic and epigenetic alterations (Baylin and Jones, 2011; Egger et al., 2004; Esteller, 2008; Shanmugam et al., 2016; Shanmugam and Sethi, 2012). Cancer primarily occurs due to heritable gene mutations in somatic and reproductive cells. These mutations are often found in the genes regulating cell cycle, oncogenes, and tumor-suppressor genes, as well as genes regulating apoptosis (Mummaneni and Shord, 2014). These mutations occur in such a fashion that they enhance cell division and prevent cells from undergoing apoptosis, and inactivate/silence tumor-suppressor genes (Hanahan and Weinberg, 2011). A gene can also be silenced or switched off completely by heritable epigenetic changes rather than by mutation on the DNA sequence (Baylin and Jones, 2016). Epigenetic silencing can occur by deregulated methylation of the cytosine (C) residue in the CpG sequence motifs that reside within the regions that govern gene transcription, repetitive sequence, and oncogene promoters (Esteller, 2008). In addition, genes encoding epigenetic regulators often have several point mutations, translocations, amplifications, and deletions (Simo-Riudalbas and Esteller, 2014).

The most important aspect of epigenetic alterations is that posttranslational modifications on histones and DNA are reversible, in contrast with the irreversible genomic mutations that completely inactivate tumor-suppressor genes or activate oncogenes. Consequently, epigenetic modification provides an opportunity to correct the aberrant methylation or acetylation of histones and DNA by therapeutic manipulation (Yan et al., 2016). The United States Food and Drug Administration has already approved several epigenetic regulators that reverse epigenetic changes and restore gene activity in cancer cells (Perri et al., 2017). Vorinostat and romidepsin are HDACs that have been approved for refractory cutaneous T-cell lymphoma (Dhanak and Jackson, 2014; Guo et al., 2015b). Irreversible DNA methyltransferase DNMT1 and DNMT3B inhibitor 5-azacytidine and 5-aza-2′-deoxycytidine nucleoside analogues are currently used for the treatment of patients with myelodysplastic syndrome (Garcia-Manero and Fenaux, 2011; Kantarjian et al., 2010; Wells et al., 2014). It has been shown that epigenetic silencing of genes has implications in cancer detection, prevention, and therapy (Baylin and Jones, 2016).

Changes in the DNA methylation pattern have served to detect cancer early, and it is also of note, that DNA methylation can be detected with a great degree of accuracy (Baylin and Jones, 2016).

Histone lysine acetyltransferases activate inflammatory genes, whereas histone lysine deacetylases (KDACs) repress inflammatory gene expression. Various cytokines, such as interleukin 1β (IL-1β), tumor necrosis factor α, lipopolysaccharides, and other stimulants can promote histone acetylation. Promoters of pro-inflammatory cytokines (such as IL-1, IL-2, IL-8, and IL-12) are rapidly acetylated and become transcriptionally active. KDACs regulate transcription of both pro-inflammatory and anti-inflammatory cytokines via their co-repressor complexes and transcription factors such as forkhead box P3, signal transducer and activator of transcription, GATA transcription factor, zinc finger E-Box binding homeobox 1, and nuclear factor-κB (NF-κB) (Cantley and Haynes, 2013; Shanmugam et al., 2016; Shanmugam and Sethi, 2012). The levels of circulating histones and nucleosomes are increased in cancer-related inflammation and infection, which suggests that histones are therapeutic targets for infectious and inflammatory disorders and cancer (Allam et al., 2014; Chen et al., 2014). In addition to novel synthetic compounds, natural product compounds have shown tremendous potential as modulators of deregulated epigenetic mechanism(s). Thus, changes to the epigenome mediated by various enzymes are now being recognized as important mediators in the activation or silencing of genes, the impact of which are as significant as heritable permanent genetic mutations in DNA (Portela and Esteller, 2010). The impact of epigenetic alterations in disease progression has led pharmaceutical companies to target these histone- and nonhistone-modifying enzymes for anticancer drug development (Eliseeva et al., 2007; Selvi et al., 2010; Shanmugam et al. 2011, 2016).

CURCUMIN

Indian traditional medicine, known as Ayurvedic medicine, clearly describes turmeric, the rhizome from *Curcuma longa Linn* (family Zingiberaceae), as an effective medicine for various ailments, such as asthma, bronchial hyperactivity, allergy, anorexia, coryza, cough, sinusitis, and hepatic disease (Ammon and Wahl, 1991; Goel et al., 2008; Gupta et al. 2012, 2013; Hatcher et al., 2008; Reuter et al., 2011; Shanmugam et al., 2015) (Fig. 5.1). Curcumin and its analogues have been reported from other Curcuma species such as *Curcuma mangga, Curcuma zedoaria, Costus speciosus, Curcuma xanthorrhiza, Curcuma aromatic, Curcuma phaeocaulis, Etlingera elatior*, and *Zingiber cassumunar* (Aggarwal et al., 2007; Shanmugam et al., 2015). Approximately 77% of turmeric contains the bioactive compound curcumin (Shanmugam et al., 2015). Several beneficial effects of curcumin have been reported over the years, highlighting it as a protective and medicinal agent. Curcumin has been reported to have anti-infective (Chan et al., 2005), antioxidant (Sreejayan and Rao, 1997), anti-inflammatory (Brouet and Ohshima, 1995; Dikshit et al., 1995), hepatoprotective (Kiso et al., 1983), cardioprotective (Venkatesan, 1998), thrombosuppressive (Srivastava et al., 1985), antiarthritic (Deodhar et al., 1980), chemopreventive, and anticarcinogenic activity (Aggarwal et al., 2013; Chen et al., 2006; Chen et al., 2006; Darvesh et al., 2012a; Divya and Pillai, 2006; Kasi et al., 2016; Shanmugam et al., 2015; Shishodia et al., 2007) as well as helping in the treatment of neurodegenerative disorders (Darvesh et al., 2012b; Frautschy et al.,

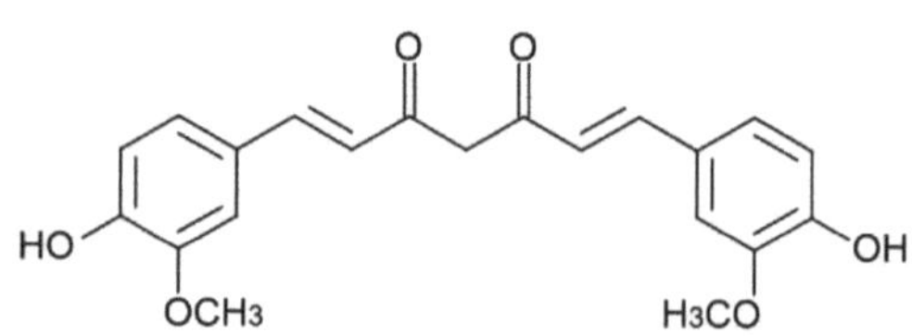

FIGURE 5.1 The chemical structure of curcumin.

2001; Lim et al., 2001), metabolic disorders (Arun and Nalini, 2002; Babu and Srinivasan, 1995), stomachache, postpartum abdominal and menstrual pain, and wound healing (Phan et al., 2001). Curcumin has also been shown to modulate multiple intracellular molecular targets in several preclinical disease models, including cancer (Kunnumakkara et al., 2008; Mukherjee et al., 2015; Shanmugam et al. 2011, 2015; Shishodia et al., 2007).

MODULATION OF EPIGENETIC MECHANISMS BY DIETARY CURCUMIN

Environmental and several dietary factors have been shown to influence the pathological progression of inflammation-driven diseases, including cancer (Remely et al., 2015). Several naturally occurring dietary compounds found in fruits and vegetables have been demonstrated to alleviate oxidative stress, inhibit nitrosamine formation, bind and dilute carcinogens in the digestive track, and prevent malignant transformation of cells in the body (Guo et al., 2015b; Remely et al., 2015). Recent studies have indicated that curcumin is also an epigenetic modulator (Balasubramanyam et al., 2004; Boyanapalli and Tony Kong, 2015; Fu and Kurzrock, 2010; Reuter et al., 2011; Selvi et al., 2010; Shanmugam et al., 2015; Teiten et al., 2013) (Fig. 5.2).

DNA METHYLATION

DNA methylation is a process whereby a methyl group is transferred by DNMTs to the cytosine bases located 5′ to a guanosine in the CpG dinucleotide (Bird, 2002). CpG dinucleotides are found to be clustered in short regions known as CpG islands that are 0.5–4 kb in length (Bird, 2002). In approximately 60% of genes, the CpG islands are found near the proximal promoter end and are generally unmethylated in normal cells (Suzuki and Bird, 2008; Wang and Leung, 2004). Unmethylated DNA has a euchromatin formation that allows transcription factor binding to initiate active transcription of genes (Baylin and Jones, 2016). In cancer cells, the CpG islands are hypermethylated and subsequently there is silencing of the transcription of several tumor-suppressor genes (Baylin and Herman, 2000; Jones and Baylin, 2002). Curcumin has been shown to be a DNMT inhibitor in cancer cells (Guo et al., 2015b). In a human androgen-dependent prostate cancer cell line, curcumin suppressed DNA methylation in the promoter region and reactivated the expression of neurogenin1 (Shu et al., 2011). Similarly, curcumin was shown to hypomethylate transgenic adenocarcinoma of mice prostate (TRAMP) C1 cell line, and restored the activity of nuclear factor erythroid 2-related factor 2 (Nrf2) (Khor et al., 2011). In a follow-up study, a synthetic curcumin analogue FN1 showed stronger anticancer activity than that of curcumin. In the TRAMP-C1 cell line, FN1 hypomethylated promoter regions of the Nrf2 gene, inactivated DNMTs, and activated an antioxidant defense pathway (Li et al., 2016).

The hypomethylation activity of curcumin was also observed in novel synthetic curcumin analogues, such as EF31 and UBS109, where they activated silenced genes, including epithelial cadherin, p16, and secreted protein acidic and rich in cysteine in MIA PaCa-2 and PANC-1 pancreatic cancer cell lines (Nagaraju et al., 2013). In MCF-7 breast cancer cells, downregulation of DNMT1 levels by curcumin was associated with the reactivation of ras-association domain family protein 1A (Du et al., 2012). In a systematic genome-wide approach, the effect of curcumin was evaluated in three colorectal cancer cell lines, HCT116, HT29, and TKO, and it was found that curcumin treatment resulted in methylation changes at selected partially methylated loci instead of fully methylated CpG sites (Link et al., 2013). Deleted in lung and

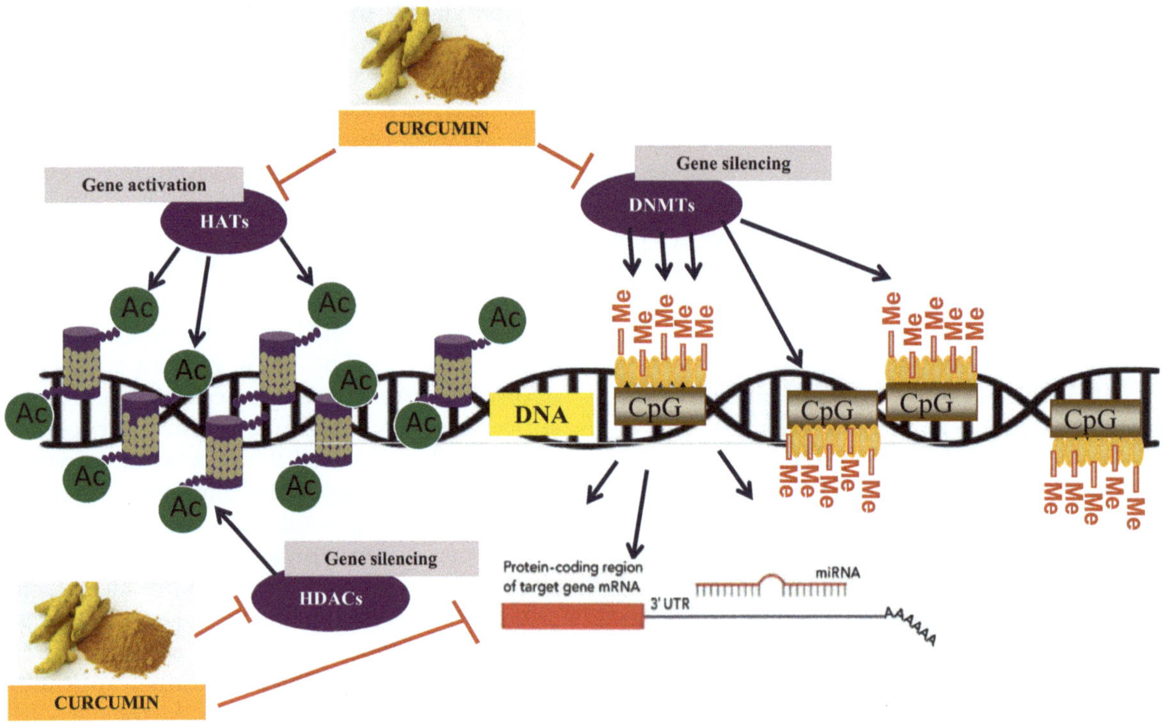

FIGURE 5.2 Effect of curcumin on DNA and histone epigenetic modifications. Curcumin regulates epigenetic mechanisms such as DNA methylation, histone acetylation modifying enzymes, and miRs that induce and/or suppress transcription and gene expression, leading to cancer progression and/or apoptosis.

esophageal cancer 1 (DLEC1) is a tumor-suppressor gene that is hypermethylated in cancer cells, thus rendering it inactive. When HT29 colorectal cancer cells were treated with curcumin for 5 days, it inhibited colony formation of HT29 cells, and a decrease in CpG methylation and DNMTs was observed with a concomitant increase in the mRNA expression of DLEC1, suggesting reversal of DLEC1 silencing (Guo et al., 2015a). In a recent study, it has been shown that curcumin inhibited the growth of breast cancer cells in vitro and in vivo by demethylating tumor suppressor deleted in liver cancer (DLC1), downregulating DNMT1 through the downregulation of sp1 transcription factor, and inducing DLC1 expression (Liu et al., 2017). In SiHa cervical cancer cell lines, hypermethylation of retinoic acid receptor β2 gene was reversed by curcumin when treated with a dose of 20 μM curcumin for 6 days (Jha et al., 2010). In another study, it was shown that curcumin treatment hypomethylated the Fanconi anemia (FANCF) promoter region and increased the expression of FANCF genes in SiHa cervical cancer cells (Parashar et al., 2012). Molecular docking studies have shown that curcumin covalently blocks the catalytic thiolate of DNMT1 in MV4-11 leukemic cell lines and serves as a global hypomethylating agent (Liu et al., 2009). Ubiquitin-like PHD and RING finger domain-containing 1 (UHRF1) mitotic chromatin association is required for DNA methylation (Rothbart et al., 2012) and plays an important role in tumor cell proliferation (Jenkins et al., 2005). In melanoma cells, the expression of phosphodiesterase 1A, cyclin A, the epigenetic integrator UHRF1, and DNMT1 were decreased by curcumin (Abusnina et al., 2011). Other curcumin derivatives, such as dimethoxycurcumin and bisdemethoxycurcumin, were also reported to exhibit demethylation activity as well as restoring tumor-suppressor gene Wnt inhibitor factor-1 expression in non-small-cell lung cancer cells (Liu et al., 2011). In another study, dimethoxycurcumin was reported to induce the expression of promoter methylated genes without reversing DNA methylation, and upregulated H3K36me3 marks in leukemia cells (Hassan et al., 2015). Table 5.1 summarizes the effects of curcumin in various cancer models.

HISTONE ACETYLTRANSFERASES AND HISTONE DEACETYLASES

Acetylation of histones and nonhistone proteins has been identified as a key posttranslational modification and is actively involved in the regulation of gene transcription in eukaryotes as well as in viral DNA that integrates to the human genome, such as human immunodeficiency virus. Dysregulated histone acetylation is often observed in inflammation-driven diseases, including cancers (Choudhary et al., 2009; Kaypee et al., 2016). Alterations in the normal acetylation pattern often leads to tumorigenesis by both inducing gene activation through HATs and by inducing gene silencing by HDACs (Hornbeck et al., 2012; Kaypee et al., 2016). Thus, a perfect balance between acetylation and deacetylation is essential in maintaining regulated cell functions. Several reports have highlighted that the activity of HATs and HDACs can be modulated by several natural and synthetic compounds (Delage and Dashwood, 2008; Kaypee et al., 2016; Seidel et al. 2012a,b; Selvi et al., 2010). Clinical trials with curcumin have either been completed or are in various phases of the clinical trial process, and several novel formulations are also being tested, such as nanoparticles, capsules, and solutions (Di Costanzo et al., 2014; Gupta et al., 2013; Shanmugam et al., 2015). Dietary curcumin has also been characterized as a natural HAT and HDAC inhibitor. In a study by Balasubramanyam et al. (2004), the investigators reported that curcumin is the first cell-permeable inhibitor of p300/CREB-binding protein (CBP) HAT activity but not of p300/CBP-associated factor in vitro and in vivo. Marcu et al. (2006) also reported that curcumin is a p300/CBP inhibitor (Marcu et al., 2006). In their study, they show that curcumin inhibited the acetyltransferase activity of purified p300 using either histone H3 or p53 as substrate. In addition, radiolabeled curcumin formed a covalent association with p300 compared with tetrahydrocurcumin, where there was no inhibitory activity, following a Michael reaction–dependent mechanism. This effect was further confirmed in PC3-M prostate cancer cells where curcumin inhibited MS-275 (an HDAC inhibitor)-induced hyperacetylation (Marcu et al., 2006). In human acute promyelocytic leukemia NB4 cells, curcumin induced inhibition of proliferation, and HDAC activity was associated with a concomitant increase in H3 acetylation and the expression of tumor-suppressor protein p53 (Li et al., 2005). Similarly, curcumin was shown to acetylate histone H3 at the site of p21(WAF1/CIP1), and induced cell cycle arrest in the human lymphoma cell line Raji (Li et al., 2005). Another study showed that curcumin inhibited HepG2 cell growth by inhibiting HDAC1 and upregulating p21(WAF1/CIP1). In Raji cells, curcumin inhibited HDAC1, HDAC3, and HDAC8, increased histone H4 acetylation, inhibited cell proliferation, and induced apoptosis (Liu et al., 2005; Wu et al., 2006). Similarly, curcumin inhibition of Raji cell proliferation was associated with a reduction in p300, HDAC1, and HDAC3 expression. In addition, curcumin suppressed the phosphorylation of IκB-α and Notch-1 expression, thereby inhibiting nuclear translocation of NF-κB (Chen et al., 2007). Curcumin was shown to restore the expression of

TABLE 5.1 Curcumin-Modulated Epigenetic Events in Various Cancer Models

Cancer Type(s)	Cell Lines	Curcumin and Its Derivatives	Epigenetic Events	Effect(s)	References
Brain cancer	Glioma cells	Curcumin	Histone modification	Inhibition of HAT	Kang et al. (2006)
	Medullo-blastoma DAOY, D283 Med, and D341 cells	Curcumin	Histone modification	Inhibition of HDAC 4 activity, tubulin acetylation	Lee et al. (2011)
Breast cancer	MCF-7 cells	Curcumin	DNA methylation	DNA hypomethylation	Du et al. (2012)
Burkitt lymphoma	Raji cells	Curcumin	Histone modification	Inhibition of p300-mediated acetylation of RelA isoform Suppression in l evels of HDAC1, HDAC3, and HDAC8	Chen et al. (2007), Liu et al. (2005) and Wu et al. (2006)
Cervical cancer	SiHa cells	Curcumin	DNA methylation	DNA hypomethylation Hypomethylation of the Fanconi anemia (FANCF) promoter Hypomethylation of RAR2 gene	Jha et al. (2010) and Parashar et al. (2012)
	HeLa cells	Curcumin	Histone modification	Inhibition of p300/CREB-binding protein (CBP) HAT activity Molecular docking inhibition of curcumin with HDAC8	Balasubramanyam et al. (2004) and Bora-Tatar et al. (2009)
Leukemia	MV4–11 cells	Curcumin	DNA methylation	DNA hypomethylation	Liu et al. (2009)
	THP-1 cells	Curcumin	Histone modification	Suppression of HAT activity, hypoacetylation of p65 isoform of NF-κB	Yun et al. (2011)
Lung cancer	A549 cells, H460 cells, SPC-A-1 cells	Demethoxy curcumin/bisdem ethoxycurcumin	DNA methylation	Hypomethylation of WIF-1 promoter	Liu et al. (2011)
Melanoma	B16F10 murine cells	Curcumin	DNA methylation	DNA hypomethylation	Abusnina et al. (2011)

TABLE 5.1 Curcumin-Modulated Epigenetic Events in Various Cancer Models—cont'd

Cancer Type(s)	Cell Lines	Curcumin and Its Derivatives	Epigenetic Events	Effect(s)	References
Prostate cancer	LNCaP cells	Curcumin	DNA methylation; histone modification	Reduction in promoter CpG methylation of Neurog1 Induction of HDAC 1, 4, 5, and 8 expression, inhibition of HDAC3 Decrease in total HDAC activity	Shu et al. (2011)
	LNCaP cells	Curcumin	Histone modification	Stimulation of histone H3 and H4 and p53 acetylation	Shankar and Srivastava (2007)
	PC3 cells	Curcumin	Histone modification	Inhibition of histone hyperacetylation and proteasome-dependent degradation of p300	Marcu et al. (2006)
	TRAMP C1 cells	Curcumin	DNA methylation	DNA Hypomethylation of Nrf2 gene	Khor et al. (2011)

suppressor of cytokine signaling 1 (SOCS1) and SOCS3 by inhibiting HDAC8 activity in myeloproliferative tumor cell lines K562 and HEL (Chen et al., 2013). In this study, inhibition of HDAC8 increased histone acetylation in the promoter regions of SOCS1 and SOCS3 (Chen et al., 2013). One of the best characterized natural HAT inhibitors is curcumin, and it exhibits high efficacy in the prevention and treatment of several tumor types, such as those of head, neck, and lung cancers (Kumar et al., 2014; Malhotra et al., 2014). In contrast, however, curcumin treatment resulted in increased global levels of acetylated H3K18 and H4K16 and was less effective in inducing DNA damage markers (Collins et al., 2013). Curcumin, in combination with trichostatin (TSA), was found to have a synergistic antiproliferative effect and induced apoptosis in SKBR3 and 435eB breast cancer cells. Combination treatment decreased phospho-ERK and phospho-AKT, and induced caspase 3 activation (Yan et al., 2013). *In silico* studies with curcumin and human HDAC8 demonstrated that curcumin has higher free-binding energy and had an inhibition constant comparable to those of TSA and vorinostat, suggesting that curcumin is a more potent inhibitor than sodium butyrate and valproic acid (Reuter et al., 2011). In another study, curcumin was found to be a potent HDAC inhibitor in HeLa cell nuclear extracts when compared with 33 carboxylic acid derivatives (Bora-Tatar et al., 2009). In LNCaP prostate cancer cells, curcumin acetylated histone H3 and H4 in a dose-dependent manner, upregulated p53, and induced apoptosis through the global regulation of acetylation on genes related to survival or apoptosis (Shankar and Srivastava, 2007). In brain glioma cells, curcumin induced histone hypoacetylation, activated caspase 3 and PARP, and induced apoptosis-mediated cell death (Kang et al., 2006). Triggering receptor expressed on myeloid cells 1 (TREM-1) is a potent inducer of the toll-like receptor initiated inflammatory response and is overexpressed in tumor-associated macrophages. TREM-1 is associated with cancer progression and recurrence, and is a poor prognostic indicator in patients with non-small-cell lung cancer (Nguyen et al., 2015; Yuan et al. 2012, 2014). In a recent study by Yuan Z et al. (2012), it has been reported that curcumin modulated TREM-1 expression by inhibiting p300 acetylation of the TREM-1 promoter region, especially on the H3 and H4 lysine residues (Yuan et al., 2012). Curcumin was reported to inhibit HAT activity in the THP-1 human monocytic cell line (derived from an acute monocytic leukemia patient) and prevented acetylation of NF-κB at Lys310 and subsequent nuclear transcription of pro-inflammatory cytokines (Yun et al., 2011). Medulloblastoma is the most common brain tumor in children. Curcumin dose-dependently inhibited cell cycle progression and induced apoptosis in medulloblastoma cells. These effects were associated with a reduction in HDAC4 expression and an increase in tubulin acetylation, thus disrupting mitotic regulation (Lee et al., 2011). Royt et al. (2011) demonstrated that curcumin in combination with either cyclophosphamide or paclitaxel downregulated HDACs in MCF-7 (ER-positive), MDA-MB-231 (ER-negative), MCF- 12F (control) breast cell lines, and also in mice mammary tumors (Royt et al., 2011). In another study, when HDAC inhibitors were administered alone, they resulted in an increase in tumor-associated metastasis proteins, and promoted tumor growth in a variety of cancers. However, when HDAC inhibitors were combined with protein kinase C inhibitors curcumin or tamoxifen, they suppressed expression of tumor metastasis proteins and tumor growth in vivo (Kao et al., 2011; Lin, 2004; Lin et al., 2012). It is of note, that combination therapy of curcumin with other HDAC inhibitors provides an opportunity to reduce the dosage of HDAC inhibitors and thus enables the development of low-cost treatment options for cancer patients.

CURCUMIN AS AN EPIGENETIC MODULATOR IN MICRORNAS

MicroRNAs (miRs) are endogenous RNAs that belong to the family of small noncoding RNAs, about 22 nucleotides in length (Bartel, 2009; Vidigal and Ventura, 2015). They are single-stranded regulatory RNAs. More than 2000 miRNAs have been discovered in humans and are involved in several developmental and regulatory processes both physiologically and pathologically (Ventura and Jacks, 2009). It has been shown that miRs regulate gene expression through multiple mechanisms (Fendler et al., 2016; Hammond, 2015; Masliah-Planchon et al., 2016; Wagner et al., 2014). The genome has been reported to express more noncoding RNAs than the ones that code for proteins (Smith et al., 2017). The conserved noncoding miR let-7 sequence, from *Caenorhabditis elegans* to humans, is downregulated in multiple cancers and highlights the importance of the 3′UTR region during gene regulation (Pasquinelli et al., 2000; Vella et al., 2004). Overall, about 30%–50% of protein-coding genes are probably regulated by miRs, which play a key role in disease progression (Lagos-Quintana et al., 2001). In a study by Calin et al. (2002), it was reported that downregulation of specific miRs was associated with increased incidence of chronic B-cell lymphocytic leukemia (Calin and Croce, 2006; Calin et al., 2002).

Several recent studies have clearly indicated that miRs serve as robust biomarkers of cancer progression and therapy response. Briefly, miRs are transcribed as pri-miRNAs that are cleaved by the Drosha/DGCR8 complex in the nucleus. The resulting hairpin-shaped pre-miRNA is further processed by dicer in the cytoplasm. The mature miRs employ the RNA-induced silencing complex in the two mechanisms of gene regulation through mRNA degradation and translation inhibition. The degree of pairing between the miR and target mRNA seems to be a determinant of the mechanism of regulation (Brighenti, 2015; Fendler et al., 2016; Masliah-Planchon et al., 2016; Smith et al., 2017; Wagner et al., 2014). In addition, several let-7 family miRs, including miR-48, miR-84, and miR-241 were shown to have tumor-suppressor activity. They have been found to bind to human RAS 3′UTR. In addition, miR-34 has been found to have tumor-suppressor properties (Chang et al., 2007). Interestingly, in colorectal and breast cancers, upregulation of miR-21 downregulates PTEN activity and is associated with tumor initiation, progression, and metastasis (Asangani et al., 2008). Thus, regulation of miR expression is now considered a novel strategy for the detection, prevention, and treatment of aggressive cancers (Li et al., 2010; Negrini et al., 2007). Moreover, an miR's signature in cancer chemoprevention is now widely identified and could be exploited for therapy (Karius et al., 2012; Li et al., 2010; Neelakandan et al., 2012). Recent evidence has suggested that epigenetic alterations are also involved in the antitumor properties of curcumin. Curcumin abrogates cell proliferation and the expression miR-21, and upregulates PTEN tumor-suppressor gene expression (Asangani et al., 2008). Curcumin was also shown to induce the expression of miR-200, thereby suppressing the expression of polycomb repressive complexes such as EZH2, SUZ12, and epithelial-to-mesenchymal transition (EMT). Curcumin also inhibits Ras activity through upregulation of tumor-suppressive miRs such as let-7a/b/c/d that target cancer stem cell marker genes such as Nanog, Sox2, and Oct4 (Momtazi et al., 2016).

COLON CANCER

In colon cancer, several miRs have been shown to be regulated by curcumin, such as miR-17-5p, miR-20a, miR-21, miR-27a, miR-34a/c, miR-101, miR-141, miR-200b/c, and miR-429. A semisynthetic derivative of curcumin, di-fluorinated curcumin (CDF), was shown to prevent the growth of chemoresistant colon cancer stemlike cells (Kanwar et al., 2011). In a similar study with chemoresistant colon cancer cells HCT116, HT29 cells, and metastatic SW620 colon cells, CDF at 100 nM inhibited miR-21 and concomitantly increased PTEN expression in all cell lines tested (Roy et al., 2013). In an earlier study, CDF was very effective in the re-expression of miR-34a and miR-34c, which was consistent with inhibition of cell growth of both chemosensitive and chemoresistant colon cancer cells (Roy et al., 2012). Curcumin was also shown to dose-dependently inhibit miR-21 expression in RKo and HCT116 colon cancer cells in vitro and in vivo using a chick chorioallantoic membrane assay, and it stabilized the expression of tumor-suppressor PDCD4 in colorectal cancer (Mudduluru et al., 2011). Curcuminoids' dose-dependent inhibition of miR-27a in SW480 colon cancer cells was associated with disruption of miR-27a-ZBTB10-Sp axis (Noratto et al., 2013). EMT has been implicated in the progression and metastasis of cancer and the development of chemoresistance.

Toden et al. (2015) evaluated the effect of curcumin on the expression pattern of EMT-suppressive miRs in HCT116 and SW480 5-fluorouracil (5-FU)-labeled cells and in an in vivo xenograft mouse model. Curcumin treatment was found to upregulate the expression of miR-200b, miR-200c, miR-141, miR-429, and miR-101 in 5-FU resistant cell lines. In contrast, 5-FU treatment did not affect expression of these

EMT-suppressive miRNAs in 5-FU resistant cells. Only miR-34a expression was upregulated in HCT116-5FU cells but not in the SW480−5FU cell line. In the murine xenograft tumor, curcumin treatment (50 mg/kg day) for 40 days overexpressed miR-200c, a key suppressive miR, and decreased tumor growth (Toden et al., 2015). In another study, curcumin (at a dose of 10 μM) decreased the expression of miR-27a, miR-20a, and miR-17-5p, and inhibited the expression of ZBTB10 and ZBTB4 in RKO colon cancer cells (Gandhy et al., 2012).

PANCREATIC CANCER

miR-22 regulates estrogen receptor 1 (ESR1) and SP1 transcription factor expression. In human BxPC-3 pancreatic cancer cells, curcumin upregulated miR-22 and miR-199a* and suppressed its target genes ESR1 and SP1, while inhibiting miR-22 expression by antisense enhanced ESR1 and SP1 expression (Sun et al., 2008). In the same study, liposomal-encapsulated curcumin enhanced bioavailability, and at a dose of 10 μmol/L, upregulated miR-22 by 68.6%, whereas miR-199a* was downregulated by 36.4% (Sun et al., 2008). Oncogenic miR-21 has been shown to have antiapoptotic activity in various cancer cell lines and targets the PTEN tumor-suppressor gene. In contrast, miR-200b/c reverses the oncogenic EMT phenotype in cancer cells. Ali et al. (2010) showed that curcumin, CDF, or gemcitabine treatment can modulate MIA PaCa-E (relatively resistant to gemcitabine), MIA PaCa-M (highly resistant to gemcitabine), and BxPC-3 (gemcitabine-sensitive) cells. A significant decrease in miR-21 was observed in both MIA PaCa-E and MIA PaCa-M cells compared with BxPC-3 cells after treatment with CDF alone or in combination with gemcitabine (Ali et al., 2010). miR-200b and miR-200c expression was also increased, but the effect was greater in CDF-treated cells than it was with curcumin or gemcitabine alone (Ali et al., 2010). In another study, Bao et al. (2011) reported the in vivo effect of curcumin and CDF in a pancreatic cancer xenograft tumor model. Curcumin (5 mg/kg day) or CDF (5 mg/kg day) administered intragastrically once a day for 12 days increased the expression of both miR-200b and miR-200c and suppressed tumor growth compared with curcumin alone, where the expression of miRs was minimal and tumor growth was not inhibited (Bao et al., 2011). Several miRs are found to be tumor-suppressive miRs, such as let-7 family, miR-26a, miR-101, miR-146a, and miR-200b that target the cancer stem cell marker genes EZH2, EpCAM, Nanog, Sox2, and Oct4.

AsPC-1 and MIA PaCa-2 pancreatic cancer cells are characterized by the absence of tumor-suppressive miRs. CDF treatment induced the expression of tumor-suppressive miRs in these two cell lines. In contrast, expression of the miR-21 oncomir, which is overexpressed in AsPC-1 and MIA PaCa-2 cells, was also inhibited by CDF (Bao et al., 2012). In an orthotopically implanted pancreatic tumor model, CDF (2.5 mg/kg day) administered intragastrically for 3 weeks reduced the expression of EZH2, Notch-1, CD44, EpCAM, and Nanog, increased the expression of let-7, miR-26a, and miR-101, and inhibited tumor growth and aggressiveness by targeting an EZH2-miRNA regulatory circuit responsible for epigenetically controlled gene expression (Bao et al., 2012). Curcumin treatment (6 μM for 72 h) of AsPC-1 and BxPC-3 pancreatic cancer cells resulted in suppressed cell proliferation, induced cell apoptosis, and inhibited cell migration in pancreatic cancer cells partly through upregulation of miR-7 and subsequent downregulation of SET8 and its downstream effectors, including p53 (Ma et al., 2014). High miR-221 expression is a poor prognostic indicator associated with a shorter life span in pancreatic cancer patients. A study by Sarkar et al. (2013) investigated the effect of CDF on the levels of miR-221 in MIA PaCa-2 and Panc-1 cells. CDF was found to downregulate the expression of miR-221 and upregulate the expression of PTEN, p27(kip1), p57(kip2), p53, and modulator of apoptosis, leading to the inhibition of cell proliferation and migration of pancreatic cancer cells (Sarkar et al., 2013).

LUNG CANCER

In A549 lung cancer cells, curcumin at doses of 20–40 μM suppressed miR-21 expression and induced apoptosis of A549 cells (Zhang et al., 2014). In non-small-cell lung cancer cells, miR-192-5p and miR-215 were investigated in the p53 wild type H460, A427, and A549 cell lines as well as the p53-null H1299 cell line (Ye et al., 2015). Ye et al. (2015) showed that miR-192-5p/215 expression was upregulated in curcumin (15 μM) treated A549 cells (p53 wild type), but not in H1299 cells (p53-null). The curcumin-induced apoptosis in these cells was associated with overexpression of miR-192-5p/215. Interestingly, it was observed that X-linked inhibitor of apoptosis (XIAP) is a target of miR-192-5p/215, and upon curcumin treatment, induced apoptosis by activating the p53-miR-192-5p/215-XIAP pathway in non-small-cell lung cancer cells (Ye et al., 2015). However, curcumin was reported to modulate the PI3K/AKT signaling pathway by upregulating miR-192-5p and induced apoptosis in non-small-cell lung cancer cells (Jin et al., 2015). The oncogenic microRNA miR-186* suppresses proapoptotic genes in lung cancer cells. Curcumin downregulated the expression of miR-186* in A549/DDP lung cancer cells, which was associated with inhibition of cell proliferation and induction of apoptosis (Tang et al., 2010; Zhang et al. 2010a,b). miR-874 is another tumor-suppressive miR and has been shown to target matrix metalloproteases (MMP). Ahamed et al. (2015) demonstrated that CDF upregulated miR-874 in A549 and H1299 non-small-cell lung cancer cells and inhibited MMP-2 expression (Ahmad et al., 2015).

BREAST CANCER

In MCF-7 breast cancer cells, curcumin-mediated upregulation of miR-15a and miR-16 was associated with inhibition of antiapoptotic protein Bcl2 in MCF-7 breast cancer cells and induced apoptosis (Yang et al., 2010). Overexpression of the miR-17-92 cluster is a key oncogenic event in a variety of cancers. The miR-17-92 cluster functions by targeting and inactivating the tumor-suppressor gene PTEN. Bisphenol A (BPA) a known endocrine disrupter, is associated with the development of breast cancer. In another study using BPA induced-MCF-7 breast cancer cells, curcumin decreased oncogenic miR-19a/b expression. Curcumin was able to increase the expression of miR-19-related target proteins, including PTEN, p-AKT, p-MDM2, p53, and proliferating cell nuclear antigen (PCNA). By modulating the miR-19/PTEN/AKT/p53 axis, curcumin inhibited the proliferative effects of BPA on MCF-7 cells (Li et al., 2014). Using an miR microarray, Kronski et al. (2014) showed that curcumin induces miR-181b expression in metastatic triple-negative MDA-MB-231 breast cancer cells and downregulates the prometastatic chemokine ligands CXCL1 and 2 by binding to their 3′UTR (Kronski et al., 2014). In another study using a triple-negative breast cancer cell line, curcumin induced the expression of miR-34a, a tumor-suppressor miR, with concomitant downregulation of Bcl-2 and Bmi-1 (Guo et al., 2013).

PROSTATE CANCER

miR-205 is a tumor suppressor. In prostate cancer cells C4-2, DU145, and PC-3, poly(lactic-coglycolic acid)-curcumin nanoparticles induced the expression of miR-205 and inhibited expression of the oncogenic miR-21 in a dose- and time-dependent manner and suppressed tumor growth in a xenograft mouse model

(Yallapu et al., 2014). EF24 (diphenyl difluoroketone) is a novel curcumin analogue that was tested for its anti-miR and antitumor properties (Adams et al., 2004; Reid et al., 2014). In androgen independent DU145 prostate cancer cells, EF24 was found to downregulate miR-21 expression both in vitro and in vivo. EF24 increased the expression of the miR-21 target genes PTEN and PDCD4 in DU145 tumor tissue, and suppressed the expression of cyclin D1 and Ki67 (Yang et al., 2013). Bioinformatics analysis revealed that 3 miRs, namely miR-141, miR-183 and miR-152, may be regulated by curcumin in prostate cancer cells. miR-141 is an androgen regulated circulating miR and is antiapoptotic. miR-183 was reported to be upregulated in prostate cancer, and Dkk-3/SMAD4 was identified as potential target genes of miR-183 (Ueno et al., 2013). miR-152 was also found to be a tumor suppressor. Recently, Guo et al. (2015) reported that a dose-dependent inhibition of prostate cell proliferation was associated with miR-208 mediated CDKN1A activation (Guo et al., 2015c).

MELANOMA

The novel curcumin analogue EF24 targets miR-21 in B16 melanoma cells, induces apoptosis, and also results in the upregulation of PTEN (Yang et al., 2013). The miR-33b target is the high-mobility group AT-hook 2 (HMGA2), a key mediator of tumor cell EMT. Zhang et al. (2015) reported the overexpression of miR-33b in Lu1205 and A375 melanoma cells. The miR-33a target, HMGA2, was predicted by the use software programs such as TargetScan, PicTar, and miRandam (Zhang et al., 2015). In melanoma cells, mmu-miR-205-5p targets are Bcl-2 and PCNA proteins. The most upregulated miR was mmu-mir-205-5p, which was expressed 135-fold greater in dietary curcumin-treated melanoma murine B78H1, human SK-MEL-28, and MeWo cells (Dahmke et al., 2013). The miR targets Bcl-2 and PCNA were predicted by the online analysis-tool GeneTrail (Dahmke et al., 2013).

OTHER CANCERS

Oncogenic miR-31 was found to be upregulated in oral squamous-cell carcinoma (OSCC). Curcumin at doses of 6, 12, and 24 μM attenuated phosphorylation of AKT and downregulated C/EBPβ in response to miR-31 inhibition in SAS, OECM-1, and HSC-3 oral cancer cells (Lu et al., 2014). In another study, curcumin-mediated inhibition of OSCC cells was associated with elevated miR-9 expression and suppression of Wnt/β-catenin signaling in a dose-dependent manner (Xiao et al., 2014). In bladder cancer cell lines T24, J82, and TCCSUP, curcumin upregulated the tumor suppressor miR-203 nearly twofold. miRANDA and TargetScan software predicted the miR-203 targets AKT2 and Src (Saini et al., 2011). Curcumin directly induces hypomethylation of the miR-203 promoter, and subsequently suppresses cellular proliferation, migration, and invasion of bladder cancer (Saini et al., 2011). An miR microarray analysis of curcumin-treated Y79 retinoblastoma cells showed significant upregulation of four tumor-suppressor miRs (let-7g*, miR-22, 503, and 200c) as well as significant downregulation of eight oncomiRs (miR-135b, 210, 25*, 95, 514, 106b*, 34c-3p, and 92a-1*) (Sreenivasan et al., 2012). In SKOV3 ovarian cancer cells, curcumin (60 μM) upregulated miR-9 and induced apoptosis (Zhao et al., 2014).

Dendrosomal curcumin (DNC) was prepared as a 142-nm spherical structure. DNC-treated U87MG glioblastoma cells induced the expression of miR-145 and decreased the expression of OCT4 (octamer binding protein 4), SOX-2 (SRY [sex determining region Y]-box 2), and Nanog in a miR-145-dependent manner (Tahmasebi Mirgani et al., 2014). Curcumin caused inhibitory effects on HONE1 nasopharyngeal carcinoma cells by inhibiting the expression of hsa-miR-125a-5p and subsequently enhancing expression of TP53 (Gao et al., 2014). In another study, Gao et al. (2012) found that miR-15a/16-1 mediates the downregulation of Wilms tumor 1 gene, which plays an important role in the antiproliferative effect of curcumin in leukemic cells (Gao et al., 2012). Recently, it has been found that addition of curcumin to chronic myelogenous leukemia cells caused a dose-dependent increase of PTEN, a target of miR-21, and also downregulated Bcr-Abl expression through the cellular increase of miR-196b in vitro and in vivo (Taverna et al., 2015). Subramaniam et al. (2012) showed that curcumin treatment downregulated the expression of Notch-1 specific miRs, miR-21 and miR-34a, and upregulated tumor suppressor let-7a miR in esophageal cancer cells TE-7, TE-10, and Eso-1 (Subramaniam et al., 2012). DNC increased the expression of the tumor-suppressor long-noncoding RNA gene MEG3 via up-DNC-dependent overexpression of miR-29a and miR-185 and could downregulate the expression of DNMT1 and subsequently overexpress MEG3 (Zamani et al., 2015). Additionally, the expression levels of miR-200a/b might determine the therapeutic efficacy of curcumin in hepatocellular carcinoma cells (Liang et al., 2013).

CONCLUSIONS

In recent years, epigenetic alterations, such as aberrant activation or inactivation, have been demonstrated to lead to the upregulation or silencing of genes. Epigenetic regulation has gained much attention, especially in the area of DNA methylation, where histone modifications, such as acetylation or deacetylation occurring without any changes in the actual DNA sequence, establish an important mechanism by which dietary natural compounds, such as curcumin, can selectively hypomethylate or inhibit deacetylases to activate gene expression. The role of curcumin in negating various inflammation-associated diseases, including cancer, has long been reported. Several lines of evidence suggest that curcumin can epigenetically regulate the expression of key transcription factors by reversing DNA methylation and modifying histone acetylation patterns that play key roles in the development of diseases. We highlighted various miRs that are modulated by curcumin in various types of cancer. Activation or suppression of miRs offers a novel strategy to inhibit tumor growth. Suppression of miR-21, miR-17-5p, miR-20a, and miR-27a, and the overexpression of miR-34a/c and EMT-suppressive miRs are some of the notable strategies available. The translational relevance of these findings on curcumin is primarily based on its nontoxicity (even up to doses of 8000 mg/day), tolerability, and efficacy properties in healthy volunteers as well as some clinical benefits observed in several clinical trials. However, curcumin activity is limited by its poor bioavailability and some possible adverse effects. These studies provide a solid foundation for more well-controlled studies in larger cohorts as well as opening avenues for future drug development. Several new formulations of curcumin, such as liposomes, micelles, nanoparticles, and phospholipid complexes that could enhance its bioavailability and improve its efficacy, are still in the early stages of development. Curcumin has also been shown to be chemosensitizer in several models of cancer. In this chapter, we have discussed several potential roles of curcumin in regulating various epigenetic markers in the prevention and therapy of cancer. Therefore, the anticancer potential of curcumin could be linked to its epigenetic alterations and the subsequent regulation of gene expression. However, further in vivo studies and clinical trials are required to further validate the epigenome potential of curcumin as a novel dietary agent and also as an adjuvant when administered with conventional chemotherapeutics.

ABBREVIATIONS

DNMT DNA methyltransferases
FANCF Fanconi anemia
FOXP3 Forkhead box P3
HAT Histone acetyltransferases
HDAC Histone deacetylases
IL Interleukin
KAT Histone lysine acetyltransferases
KDAC Histone lysine deacetylases
KDM Istone lysine demethylases
KMT Histone lysine methyltransferases
MBD Methyl-CpG binding domain
MBT Malignant brain tumor
NF-κB Nuclear factor-κB
PHD Plant homeodomain
PWWP Pro-trp-trp-pro
STAT Signal transducer and activator of transcription
TET Ten-eleven translocation
TNF Tumor necrosis factor
ZEB1 Zinc finger E-Box binding homeobox 1

Acknowledgments

This work was supported by NUHS Basic seed grant [T1-BSRG 2015-02] and Ministry of Education Tier 1 grant to G. Sethi. APK was supported by grants from National Medical Research Council of Singapore, NCIS Yong Siew Yoon Research Grant through donations from the Yong Loo Lin Trust and by the National Research Foundation Singapore and the Singapore Ministry of Education under its Research Centers of Excellence initiative to Cancer Science Institute of Singapore, National University of Singapore.

References

Abusnina, A., Keravis, T., Yougbare, I., Bronner, C., Lugnier, C., 2011. Anti-proliferative effect of curcumin on melanoma cells is mediated by PDE1A inhibition that regulates the epigenetic integrator UHRF1. Mol. Nutr. Food Res. 55 (11), 1677–1689.

Adams, B.K., et al., 2004. Synthesis and biological evaluation of novel curcumin analogs as anti-cancer and anti-angiogenesis agents. Bioorg. Med. Chem. 12 (14), 3871–3883.

Aggarwal, B., Prasad, S., Sung, B., Krishnan, S., Guha, S., 2013. Prevention and treatment of colorectal cancer by natural agents from mother nature. Curr. Colorectal Cancer Rep. 9 (1), 37–56.

Aggarwal, B.B., Sundaram, C., Malani, N., Ichikawa, H., 2007. Curcumin: the Indian solid gold. Adv. Exp. Med. Biol. 595, 1–75.

Ahmad, A., et al., 2015. Molecular docking and inhibition of matrix metalloproteinase-2 by novel difluorinated-benzylidene curcumin analog. Am. J. Transl. Res. 7 (2), 298–308.

Ali, S., et al., 2010. Gemcitabine sensitivity can be induced in pancreatic cancer cells through modulation of miR-200 and miR-21 expression by curcumin or its analogue CDF. Cancer Res. 70 (9), 3606–3617.

Allam, R., Kumar, S.V., Darisipudi, M.N., Anders, H.J., 2014. Extracellular histones in tissue injury and inflammation. J. Mol. Med. (Berl.) 92 (5), 465–472.

Ammon, H.P., Wahl, M.A., 1991. Pharmacology of Curcuma longa. Planta Med. 57 (1), 1–7.

Arun, N., Nalini, N., 2002. Efficacy of turmeric on blood sugar and polyol pathway in diabetic albino rats. Plant Foods Hum. Nutr. 57 (1), 41–52.

Asangani, I.A., et al., 2008. MicroRNA-21 (miR-21) post-transcriptionally downregulates tumor suppressor Pdcd4 and stimulates invasion, intravasation and metastasis in colorectal cancer. Oncogene 27 (15), 2128–2136.

Babu, P.S., Srinivasan, K., 1995. Influence of dietary curcumin and cholesterol on the progression of experimentally induced diabetes in albino rat. Mol. Cell. Biochem. 152 (1), 13–21.

Balasubramanyam, K., et al., 2004. Curcumin, a novel p300/CREB-binding protein-specific inhibitor of acetyltransferase, represses the acetylation of histone/nonhistone proteins and histone acetyltransferase-dependent chromatin transcription. J. Biol. Chem. 279 (49), 51163–51171.

Bao, B., et al., 2012. Curcumin analogue CDF inhibits pancreatic tumor growth by switching on suppressor microRNAs and attenuating EZH2 expression. Cancer Res. 72 (1), 335–345.

Bao, B., et al., 2011. Anti-tumor activity of a novel compound-CDF is mediated by regulating miR-21, miR-200, and PTEN in pancreatic cancer. PLoS One 6 (3), e17850.

Bartel, D.P., 2009. MicroRNAs: target recognition and regulatory functions. Cell 136 (2), 215–233.

Baylin, S.B., Herman, J.G., 2000. DNA hypermethylation in tumorigenesis: epigenetics joins genetics. Trends Genet. 16 (4), 168–174.

Baylin, S.B., Jones, P.A., 2011. A decade of exploring the cancer epigenome - biological and translational implications. Nat. Rev. Cancer 11 (10), 726–734.

Baylin, S.B., Jones, P.A., 2016. Epigenetic determinants of cancer. Cold Spring Harb. Perspect. Biol. 8 (9), a019505.

Berger, S.L., 2007. The complex language of chromatin regulation during transcription. Nature 447 (7143), 407–412.

Berger, S.L., Kouzarides, T., Shiekhattar, R., Shilatifard, A., 2009. An operational definition of epigenetics. Genes Dev. 23 (7), 781–783.

Bird, A., 2002. DNA methylation patterns and epigenetic memory. Genes Dev. 16 (1), 6–21.

Bora-Tatar, G., et al., 2009. Molecular modifications on carboxylic acid derivatives as potent histone deacetylase inhibitors: activity and docking studies. Bioorg. Med. Chem. 17 (14), 5219–5228.

Boyanapalli, S.S., Tony Kong, A.N., 2015. "Curcumin, the king of spices": epigenetic regulatory mechanisms in the prevention of cancer, neurological, and inflammatory diseases. Curr. Pharmacol. Rep. 1 (2), 129–139.

Brighenti, M., 2015. MicroRNA and MET in lung cancer. Ann. Transl. Med. 3 (5), 68.

Brouet, I., Ohshima, H., 1995. Curcumin, an anti-tumour promoter and anti-inflammatory agent, inhibits induction of nitric oxide synthase in activated macrophages. Biochem. Biophys. Res. Commun. 206 (2), 533–540.

Calin, G.A., Croce, C.M., 2006. MicroRNA signatures in human cancers. Nat. Rev. Cancer 6 (11), 857–866.

Calin, G.A., et al., 2002. Frequent deletions and down-regulation of micro- RNA genes miR15 and miR16 at 13q14 in chronic lymphocytic leukemia. Proc. Natl. Acad. Sci. U.S.A. 99 (24), 15524–15529.

Cantley, M.D., Haynes, D.R., 2013. Epigenetic regulation of inflammation: progressing from broad acting histone deacetylase (HDAC) inhibitors to targeting specific HDACs. Inflammopharmacology 21 (4), 301–307.

Chan, M.M., Adapala, N.S., Fong, D., 2005. Curcumin overcomes the inhibitory effect of nitric oxide on Leishmania. Parasitol. Res. 96 (1), 49–56.

Chang, T.C., et al., 2007. Transactivation of miR-34a by p53 broadly influences gene expression and promotes apoptosis. Mol. Cell. 26 (5), 745–752.

Chen, A., Xu, J., Johnson, A.C., 2006. Curcumin inhibits human colon cancer cell growth by suppressing gene expression of epidermal growth factor receptor through reducing the activity of the transcription factor Egr-1. Oncogene 25 (2), 278–287.

Chen, C.Q., et al., 2013. Pure curcumin increases the expression of SOCS1 and SOCS3 in myeloproliferative neoplasms through suppressing class I histone deacetylases. Carcinogenesis 34 (7), 1442–1449.

Chen, J., et al., 2006. Curcumin protects PC12 cells against 1-methyl-4-phenylpyridinium ion-induced apoptosis by bcl-2-mitochondria-ROS-iNOS pathway. Apoptosis 11 (6), 943–953.

Chen, R., Kang, R., Fan, X.G., Tang, D., 2014. Release and activity of histone in diseases. Cell Death Dis. 5, e1370.

Chen, Y., et al., 2007. Curcumin, both histone deacetylase and p300/CBP-specific inhibitor, represses the activity of nuclear factor kappa B and Notch 1 in Raji cells. Basic Clin. Pharmacol. Toxicol. 101 (6), 427–433.

Choudhary, C., et al., 2009. Lysine acetylation targets protein complexes and co-regulates major cellular functions. Science 325 (5942), 834–840.

Collins, H.M., et al., 2013. Differential effects of garcinol and curcumin on histone and p53 modifications in tumour cells. BMC Cancer 13, 37.

Dahmke, I.N., et al., 2013. Curcumin intake affects miRNA signature in murine melanoma with mmu-miR-205-5p most significantly altered. PLoS One 8 (12), e81122.

Darvesh, A.S., Aggarwal, B.B., Bishayee, A., 2012a. Curcumin and liver cancer: a review. Curr. Pharm. Biotechnol. 13 (1), 218–228.

Darvesh, A.S., et al., 2012b. Curcumin and neurodegenerative diseases: a perspective. Expert Opin. Investig. Drugs. 21 (8), 1123–1140.

Delage, B., Dashwood, R.H., 2008. Dietary manipulation of histone structure and function. Annu. Rev. Nutr. 28, 347–366.

Deodhar, S.D., Sethi, R., Srimal, R.C., 1980. Preliminary study on antirheumatic activity of curcumin (diferuloyl methane). Indian J. Med. Res. 71, 632–634.

Dhanak, D., Jackson, P., 2014. Development and classes of epigenetic drugs for cancer. Biochem. Biophys. Res. Commun. 455 (1–2), 58–69.

Di Costanzo, A., Del Gaudio, N., Migliaccio, A., Altucci, L., 2014. Epigenetic drugs against cancer: an evolving landscape. Arch. Toxicol. 88 (9), 1651–1668.

Dikshit, M., Rastogi, L., Shukla, R., Srimal, R.C., 1995. Prevention of ischaemia-induced biochemical changes by curcumin & quinidine in the cat heart. Indian J. Med. Res. 101, 31–35.

Divya, C.S., Pillai, M.R., 2006. Antitumor action of curcumin in human papillomavirus associated cells involves down-regulation of viral oncogenes, prevention of NFkB and AP-1 translocation, and modulation of apoptosis. Mol. Carcinog. 45 (5), 320–332.

Du, L., et al., 2012. Reactivation of RASSF1A in breast cancer cells by curcumin. Nutr. Cancer 64 (8), 1228–1235.

Egger, G., Liang, G., Aparicio, A., Jones, P.A., 2004. Epigenetics in human disease and prospects for epigenetic therapy. Nature 429 (6990), 457–463.

Eliseeva, E.D., Valkov, V., Jung, M., Jung, M.O., 2007. Characterization of novel inhibitors of histone acetyltransferases. Mol. Cancer Ther. 6 (9), 2391–2398.

Esteller, M., 2008. Epigenetics in cancer. N. Engl. J. Med. 358 (11), 1148–1159.

Feinberg, A.P., 2007. Phenotypic plasticity and the epigenetics of human disease. Nature 447 (7143), 433–440.

Fendler, A., Stephan, C., Yousef, G.M., Kristiansen, G., Jung, K., 2016. The translational potential of microRNAs as biofluid markers of urological tumours. Nat. Rev. Urol. 13 (12), 734–752.

Frautschy, S.A., et al., 2001. Phenolic anti-inflammatory antioxidant reversal of Abeta-induced cognitive deficits and neuropathology. Neurobiol. Aging 22 (6), 993–1005.

Fu, S., Kurzrock, R., 2010. Development of curcumin as an epigenetic agent. Cancer 116 (20), 4670–4676.

Gandhy, S.U., Kim, K., Larsen, L., Rosengren, R.J., Safe, S., 2012. Curcumin and synthetic analogs induce reactive oxygen species and decreases specificity protein (Sp) transcription factors by targeting microRNAs. BMC Cancer 12, 564.

Gao, S.M., et al., 2012. Pure curcumin decreases the expression of WT1 by upregulation of miR-15a and miR-16-1 in leukemic cells. J. Exp. Clin. Cancer Res. 31 (1), 27.

Gao, W., Chan, J.Y., Wong, T.S., 2014. Curcumin exerts inhibitory effects on undifferentiated nasopharyngeal carcinoma by inhibiting the expression of miR-125a-5p. Clin. Sci. (Lond.) 127 (9), 571–579.

Garcia-Manero, G., Fenaux, P., 2011. Hypomethylating agents and other novel strategies in myelodysplastic syndromes. J. Clin. Oncol. 29 (5), 516–523.

Goel, A., Kunnumakkara, A.B., Aggarwal, B.B., 2008. Curcumin as "Curecumin": from kitchen to clinic. Biochem. Pharmacol. 75 (4), 787–809.

Guo, Shu, L., Zhang, C., Su, Z.Y., Kong, A.N., 2015a. Curcumin inhibits anchorage-independent growth of HT29 human colon cancer cells by targeting epigenetic restoration of the tumor suppressor gene DLEC1. Biochem. Pharmacol. 94 (2), 69–78.

Guo, Su, Z.Y., Kong, A.N., 2015b. Current perspectives on epigenetic modifications by dietary chemopreventive and herbal phytochemicals. Curr. Pharmacol. Rep. 1 (4), 245–257.

Guo, Xu, Y., Fu, Q., 2015c. Curcumin inhibits growth of prostate carcinoma via miR-208-mediated CDKN1A activation. Tumour Biol. 36 (11), 8511–8517.

Guo, J., et al., 2013. Synergistic effects of curcumin with emodin against the proliferation and invasion of breast cancer cells through upregulation of miR-34a. Mol. Cell. Biochem. 382 (1–2), 103–111.

Gupta, S.C., Patchva, S., Aggarwal, B.B., 2013. Therapeutic roles of curcumin: lessons learned from clinical trials. AAPS J. 15 (1), 195–218.

Gupta, S.C., Patchva, S., Koh, W., Aggarwal, B.B., 2012. Discovery of curcumin, a component of golden spice, and its miraculous biological activities. Clin. Exp. Pharmacol. Physiol. 39 (3), 283–299.

Hammond, S.M., 2015. An overview of microRNAs. Adv. Drug Deliv. Rev. 87, 3–14.

Hanahan, D., Weinberg, R.A., 2011. Hallmarks of cancer: the next generation. Cell 144 (5), 646–674.

Hassan, H.E., et al., 2015. Curcumin and dimethoxycurcumin induced epigenetic changes in leukemia cells. Pharm. Res. 32 (3), 863–875.

Hatcher, H., Planalp, R., Cho, J., Torti, F.M., Torti, S.V., 2008. Curcumin: from ancient medicine to current clinical trials. Cell Mol. Life Sci. 65 (11), 1631–1652.

Hirst, M., Marra, M.A., 2009. Epigenetics and human disease. Int. J. Biochem. Cell Biol. 41 (1), 136–146.

Hornbeck, P.V., et al., 2012. PhosphoSitePlus: a comprehensive resource for investigating the structure and function of experimentally determined post-translational modifications in man and mouse. Nucleic Acids Res. 40 (Database Issue), D261–D270.

Jenkins, Y., et al., 2005. Critical role of the ubiquitin ligase activity of UHRF1, a nuclear RING finger protein, in tumor cell growth. Mol. Biol. Cell 16 (12), 5621–5629.

Jha, A.K., et al., 2010. Reversal of hypermethylation and reactivation of the RARbeta2 gene by natural compounds in cervical cancer cell lines. Folia Biol. (Praha) 56 (5), 195–200.

Jin, H., Qiao, F., Wang, Y., Xu, Y., Shang, Y., 2015. Curcumin inhibits cell proliferation and induces apoptosis of human non-small cell lung cancer cells through the upregulation of miR-192-5p and suppression of PI3K/Akt signaling pathway. Oncol. Rep. 34 (5), 2782–2789.

Jones, P.A., Baylin, S.B., 2002. The fundamental role of epigenetic events in cancer. Nat. Rev. Genet. 3 (6), 415–428.

Jones, P.A., Issa, J.P., Baylin, S., 2016. Targeting the cancer epigenome for therapy. Nat. Rev. Genet. 17 (10), 630–641.

Kang, S.K., Cha, S.H., Jeon, H.G., 2006. Curcumin-induced histone hypoacetylation enhances caspase-3-dependent glioma cell death and neurogenesis of neural progenitor cells. Stem Cells Dev. 15 (2), 165–174.

Kantarjian, H.M., et al., 2010. Phase 2 study of romiplostim in patients with low- or intermediate-risk myelodysplastic syndrome receiving azacitidine therapy. Blood 116 (17), 3163–3170.

Kanwar, S.S., et al., 2011. Difluorinated-curcumin (CDF): a novel curcumin analog is a potent inhibitor of colon cancer stem-like cells. Pharm. Res. 28 (4), 827–838.

Kao, H.H., et al., 2011. Kinase gene expression and subcellular protein expression pattern of protein kinase C isoforms in curcumin-treated human hepatocellular carcinoma Hep 3B cells. Plant Foods Hum. Nutr. 66 (2), 136–142.

Karius, T., Schnekenburger, M., Dicato, M., Diederich, M., 2012. MicroRNAs in cancer management and their modulation by dietary agents. Biochem. Pharmacol. 83 (12), 1591–1601.

Kasi, P.D., et al., 2016. Molecular targets of curcumin for cancer therapy: an updated review. Tumour Biol. 37 (10), 13017–13028.

Kaypee, S., et al., 2016. Aberrant lysine acetylation in tumorigenesis: implications in the development of therapeutics. Pharmacol. Ther. 162, 98–119.

Khor, T.O., et al., 2011. Pharmacodynamics of curcumin as DNA hypomethylation agent in restoring the expression of Nrf2 via promoter CpGs demethylation. Biochem. Pharmacol. 82 (9), 1073–1078.

Kiso, Y., Suzuki, Y., Watanabe, N., Oshima, Y., Hikino, H., 1983. Antihepatotoxic principles of Curcuma longa rhizomes. Planta Med. 49 (3), 185–187.

Kouzarides, T., 2007. Chromatin modifications and their function. Cell 128 (4), 693–705.

Kronski, E., et al., 2014. miR181b is induced by the chemopreventive polyphenol curcumin and inhibits breast cancer metastasis via down-regulation of the inflammatory cytokines CXCL1 and -2. Mol. Oncol. 8 (3), 581–595.

Kumar, B., et al., 2014. A novel curcumin analog (H-4073) enhances the therapeutic efficacy of cisplatin treatment in head and neck cancer. PLoS One 9 (3), e93208.

Kunnumakkara, A.B., Anand, P., Aggarwal, B.B., 2008. Curcumin inhibits proliferation, invasion, angiogenesis and metastasis of different cancers through interaction with multiple cell signaling proteins. Cancer Lett. 269 (2), 199–225.

Lagos-Quintana, M., Rauhut, R., Lendeckel, W., Tuschl, T., 2001. Identification of novel genes coding for small expressed RNAs. Science 294 (5543), 853–858.

Lee, J.S., Smith, E., Shilatifard, A., 2010. The language of histone crosstalk. Cell 142 (5), 682–685.

Lee, S.J., et al., 2011. Curcumin-induced HDAC inhibition and attenuation of medulloblastoma growth in vitro and in vivo. BMC Cancer 11, 144.

Li, W., et al., 2016. Epigenetics reactivation of Nrf2 in prostate TRAMP C1 cells by curcumin analogue FN1. Chem. Res. Toxicol. 29 (4), 694–703.

Li, X., et al., 2014. Curcumin modulates miR-19/PTEN/AKT/p53 axis to suppress bisphenol A-induced MCF-7 breast cancer cell proliferation. Phytother. Res. 28 (10), 1553–1560.

Li, X.G., Chen, Y., Wu, Q., Liu, H.L., 2005. [Effects of curcumin on the acetylation of histone H3, P53 and the proliferation of NB4 cells]. Zhonghua Xue Ye Xue Za Zhi 26 (9), 551–553.

Li, Y., Kong, D., Wang, Z., Sarkar, F.H., 2010. Regulation of microRNAs by natural agents: an emerging field in chemoprevention and chemotherapy research. Pharm. Res. 27 (6), 1027–1041.

Liang, H.H., et al., 2013. MicroRNA-200a/b influenced the therapeutic effects of curcumin in hepatocellular carcinoma (HCC) cells. Tumour Biol. 34 (5), 3209–3218.

Lim, G.P., et al., 2001. The curry spice curcumin reduces oxidative damage and amyloid pathology in an Alzheimer transgenic mouse. J. Neurosci. 21 (21), 8370–8377.

Lin, J.K., 2004. Suppression of protein kinase C and nuclear oncogene expression as possible action mechanisms of cancer chemoprevention by Curcumin. Arch. Pharm. Res. 27 (7), 683–692.

Lin, K.T., et al., 2012. HDAC inhibitors augmented cell migration and metastasis through induction of PKCs leading to identification of low toxicity modalities for combination cancer therapy. Clin. Cancer Res. 18 (17), 4691–4701.

Link, A., et al., 2013. Curcumin modulates DNA methylation in colorectal cancer cells. PLoS One 8 (2), e57709.

Liu, H.L., Chen, Y., Cui, G.H., Zhou, J.F., 2005. Curcumin, a potent anti-tumor reagent, is a novel histone deacetylase inhibitor regulating B-NHL cell line Raji proliferation. Acta Pharmacol. Sin. 26 (5), 603–609.

Liu, Y., Zhou, J., Hu, Y., Wang, J., Yuan, C., 2017. Curcumin inhibits growth of human breast cancer cells through demethylation of DLC1 promoter. Mol. Cell Biochem. 425 (1–2), 47–58.

Liu, Y.L., Yang, H.P., Gong, L., Tang, C.L., Wang, H.J., 2011. Hypomethylation effects of curcumin, demethoxycurcumin and bisdemethoxycurcumin on WIF-1 promoter in non-small cell lung cancer cell lines. Mol. Med. Rep. 4 (4), 675–679.

Liu, Z., et al., 2009. Curcumin is a potent DNA hypomethylation agent. Bioorg. Med. Chem. Lett. 19 (3), 706–709.

Lu, W.C., et al., 2014. EGF up-regulates miR-31 through the C/EBPbeta signal cascade in oral carcinoma. PLoS One 9 (9), e108049.

Luger, K., Mader, A.W., Richmond, R.K., Sargent, D.F., Richmond, T.J., 1997. Crystal structure of the nucleosome core particle at 2.8 A resolution. Nature 389 (6648), 251–260.

Ma, J., et al., 2014. Curcumin inhibits cell growth and invasion through up-regulation of miR-7 in pancreatic cancer cells. Toxicol. Lett. 231 (1), 82–91.

Malhotra, A., Nair, P., Dhawan, D.K., 2014. Study to evaluate molecular mechanics behind synergistic chemopreventive effects of curcumin and resveratrol during lung carcinogenesis. PLoS One 9 (4), e93820.

Marcu, M.G., et al., 2006. Curcumin is an inhibitor of p300 histone acetylatransferase. Med. Chem. 2 (2), 169–174.

Masliah-Planchon, J., Garinet, S., Pasmant, E., 2016. RAS-MAPK pathway epigenetic activation in cancer: miRNAs in action. Oncotarget 7 (25), 38892–38907.

Momtazi, A.A., et al., 2016. Curcumin as a MicroRNA regulator in cancer: a review. Rev. Physiol. Biochem. Pharmacol. 171, 1–8.

Mudduluru, G., et al., 2011. Curcumin regulates miR-21 expression and inhibits invasion and metastasis in colorectal cancer. Biosci. Rep. 31 (3), 185–197.

Mukherjee, N., Kumar, A.P., Ghosh, R., 2015. DNA methylation and flavonoids in genitourinary cancers. Curr. Pharmacol. Rep. 1 (2), 112–120.

Mummaneni, P., Shord, S.S., 2014. Epigenetics and oncology. Pharmacotherapy 34 (5), 495–505.

Nagaraju, G.P., et al., 2013. Novel synthetic curcumin analogues EF31 and UBS109 are potent DNA hypomethylating agents in pancreatic cancer. Cancer Lett. 341 (2), 195–203.

Neelakandan, K., Babu, P., Nair, S., 2012. Emerging roles for modulation of microRNA signatures in cancer chemoprevention. Curr. Cancer Drug Targets 12 (6), 716–740.

Negrini, M., Ferracin, M., Sabbioni, S., Croce, C.M., 2007. MicroRNAs in human cancer: from research to therapy. J. Cell Sci. 120 (Pt 11), 1833–1840.

Nguyen, A.H., Berim, I.G., Agrawal, D.K., 2015. Chronic inflammation and cancer: emerging roles of triggering receptors expressed on myeloid cells. Expert Rev. Clin. Immunol. 11 (7), 849–857.

Noratto, G.D., Jutooru, I., Safe, S., Angel-Morales, G., Mertens-Talcott, S.U., 2013. The drug resistance suppression induced by curcuminoids in colon cancer SW-480 cells is mediated by reactive oxygen species-induced disruption of the microRNA-27a-ZBTB10-Sp axis. Mol. Nutr. Food Res. 57 (9), 1638–1648.

Parashar, G., Parashar, N.C., Capalash, N., 2012. Curcumin causes promoter hypomethylation and increased expression of FANCF gene in SiHa cell line. Mol. Cell. Biochem. 365 (1–2), 29–35.

Pasquinelli, A.E., et al., 2000. Conservation of the sequence and temporal expression of let-7 heterochronic regulatory RNA. Nature 408 (6808), 86–89.

Perri, F., et al., 2017. Epigenetic control of gene expression: potential implications for cancer treatment. Crit. Rev. Oncol. Hematol. 111, 166–172.

Pfaffeneder, T., et al., 2014. Tet oxidizes thymine to 5-hydroxymethyluracil in mouse embryonic stem cell DNA. Nat. Chem. Biol. 10 (7), 574–581.

Phan, T.T., See, P., Lee, S.T., Chan, S.Y., 2001. Protective effects of curcumin against oxidative damage on skin cells in vitro: its implication for wound healing. J. Trauma. 51 (5), 927–931.

Portela, A., Esteller, M., 2010. Epigenetic modifications and human disease. Nat. Biotechnol. 28 (10), 1057–1068.

Reid, J.M., et al., 2014. Mouse pharmacokinetics and metabolism of the curcumin analog, 4-piperidinone,3,5-bis[(2-fluorophenyl)methylene]-acetate(3E,5E) (EF-24; NSC 716993). Cancer Chemother. Pharmacol. 73 (6), 1137–1146.

Remely, M., et al., 2015. Therapeutic perspectives of epigenetically active nutrients. Br. J. Pharmacol. 172 (11), 2756–2768.

Reuter, S., Gupta, S.C., Park, B., Goel, A., Aggarwal, B.B., 2011. Epigenetic changes induced by curcumin and other natural compounds. Genes Nutr. 6 (2), 93–108.

Rothbart, S.B., et al., 2012. Association of UHRF1 with methylated H3K9 directs the maintenance of DNA methylation. Nat. Struct. Mol. Biol. 19 (11), 1155–1160.

Roy, S., Levi, E., Majumdar, A.P., Sarkar, F.H., 2012. Expression of miR-34 is lost in colon cancer which can be re-expressed by a novel agent CDF. J. Hematol. Oncol. 5, 58.

Roy, S., Yu, Y., Padhye, S.B., Sarkar, F.H., Majumdar, A.P., 2013. Difluorinated-curcumin (CDF) restores PTEN expression in colon cancer cells by down-regulating miR-21. PLoS One 8 (7), e68543.

Royt, M., Mukherjee, S., Sarkar, R., Biswas, J., 2011. Curcumin sensitizes chemotherapeutic drugs via modulation of PKC, telomerase, NF-kappaB and HDAC in breast cancer. Ther. Deliv. 2 (10), 1275–1293.

Saini, S., et al., 2011. Curcumin modulates microRNA-203-mediated regulation of the Src-Akt axis in bladder cancer. Cancer Prev. Res. (Phila) 4 (10), 1698–1709.

Sarkar, S., et al., 2013. Down-regulation of miR-221 inhibits proliferation of pancreatic cancer cells through up-regulation of PTEN, p27(kip1), p57(kip2), and PUMA. Am. J. Cancer Res. 3 (5), 465–477.

Segal, E., Widom, J., 2009. What controls nucleosome positions? Trends Genet. 25 (8), 335–343.

Seidel, C., Florean, C., Schnekenburger, M., Dicato, M., Diederich, M., 2012a. Chromatin-modifying agents in anti-cancer therapy. Biochimie 94 (11), 2264–2279.

Seidel, C., Schnekenburger, M., Dicato, M., Diederich, M., 2012b. Histone deacetylase modulators provided by mother nature. Genes Nutr. 7 (3), 357–367.

Selvi, B.R., Mohankrishna, D.V., Ostwal, Y.B., Kundu, T.K., 2010. Small molecule modulators of histone acetylation and methylation: a disease perspective. Biochim. Biophys. Acta 1799 (10–12), 810–828.

Shankar, S., Srivastava, R.K., 2007. Involvement of Bcl-2 family members, phosphatidylinositol 3′-kinase/AKT and mitochondrial p53 in curcumin (diferulolylmethane)-induced apoptosis in prostate cancer. Int. J. Oncol. 30 (4), 905–918.

Shanmugam, M.K., Arfuso, F., Sethi, G., 2016. Epigenetic modifications and their potential role in tumorigenesis. In: Dincer, Y. (Ed.), Epigenetics: Mechanisms and Clinical Perspectives. Nova Science Publishers, Inc., New York, pp. 31–82.

Shanmugam, M.K., Kannaiyan, R., Sethi, G., 2011. Targeting cell signaling and apoptotic pathways by dietary agents: role in the prevention and treatment of cancer. Nutr. Cancer 63 (2), 161–173.

Shanmugam, M.K., et al., 2015. The multifaceted role of curcumin in cancer prevention and treatment. Molecules 20 (2), 2728–2769.

Shanmugam, M.K., Sethi, G., 2012. Role of epigenetics in inflammation-associated diseases. In: Kundu, T.K. (Ed.), Epigenetics: Development and Disease. Springer Dordrecht, Heidelberg New York London, pp. 627–658.

Shishodia, S., Chaturvedi, M.M., Aggarwal, B.B., 2007. Role of curcumin in cancer therapy. Curr .Probl. Cancer. 31 (4), 243–305.

Shu, L., et al., 2011. Epigenetic CpG demethylation of the promoter and reactivation of the expression of Neurog1 by curcumin in prostate LNCaP cells. AAPS J. 13 (4), 606–614.

Simo-Riudalbas, L., Esteller, M., 2014. Cancer genomics identifies disrupted epigenetic genes. Hum. Genet. 133 (6), 713–725.

Simo-Riudalbas, L., Esteller, M., 2015. Targeting the histone orthography of cancer: drugs for writers, erasers and readers. Br. J. Pharmacol. 172 (11), 2716–2732.

Smith, B., Agarwal, P., Bhowmick, N.A., 2017. MicroRNA applications for prostate, ovarian and breast cancer in the era of precision medicine. Endocr. Relat. Cancer 24 (5), R157–R172.

Sreejayan, Rao, M.N., 1997. Nitric oxide scavenging by curcuminoids. J. Pharm. Pharmacol. 49 (1), 105–107.

Sreenivasan, S., Thirumalai, K., Danda, R., Krishnakumar, S., 2012. Effect of curcumin on miRNA expression in human Y79 retinoblastoma cells. Curr. Eye Res. 37 (5), 421–428.

Srivastava, R., Dikshit, M., Srimal, R.C., Dhawan, B.N., 1985. Anti-thrombotic effect of curcumin. Thromb. Res. 40 (3), 413–417.

Subramaniam, D., et al., 2012. Curcumin induces cell death in esophageal cancer cells through modulating Notch signaling. PLoS One 7 (2), e30590.

Sun, M., et al., 2008. Curcumin (diferuloylmethane) alters the expression profiles of microRNAs in human pancreatic cancer cells. Mol. Cancer Ther. 7 (3), 464–473.

Suzuki, M.M., Bird, A., 2008. DNA methylation landscapes: provocative insights from epigenomics. Nat. Rev. Genet. 9 (6), 465–476.

Tahmasebi Mirgani, M., et al., 2014. Dendrosomal curcumin nanoformulation downregulates pluripotency genes via miR-145 activation in U87MG glioblastoma cells. Int. J. Nanomed. 9, 403–417.

Tang, N., Zhang, J., Du, Y., 2010. Curcumin promoted the apoptosis of cisplain-resistant human lung carcinoma cells A549/DDP through down-regulating miR-186*. Zhongguo Fei Ai Za Zhi 13 (4), 301–306.

Taverna, S., et al., 2015. Curcumin inhibits in vitro and in vivo chronic myelogenous leukemia cells growth: a possible role for exosomal disposal of miR-21. Oncotarget 6 (26), 21918–21933.

Teiten, M.H., Dicato, M., Diederich, M., 2013. Curcumin as a regulator of epigenetic events. Mol. Nutr. Food Res. 57 (9), 1619–1629.

Tessarz, P., Kouzarides, T., 2014. Histone core modifications regulating nucleosome structure and dynamics. Nat. Rev. Mol. Cell Biol. 15 (11), 703–708.

Toden, S., et al., 2015. Curcumin mediates chemosensitization to 5-fluorouracil through miRNA-induced suppression of epithelial-to-mesenchymal transition in chemoresistant colorectal cancer. Carcinogenesis 36 (3), 355–367.

Ueno, K., et al., 2013. microRNA-183 is an oncogene targeting Dkk-3 and SMAD4 in prostate cancer. Br. J. Cancer 108 (8), 1659–1667.

Van Speybroeck, L., 2002. From epigenesis to epigenetics: the case of C. H. Waddington. Ann. N.Y. Acad. Sci. 981, 61–81.

Vella, M.C., Reinert, K., Slack, F.J., 2004. Architecture of a validated microRNA::target interaction. Chem. Biol. 11 (12), 1619–1623.

Venkatesan, N., 1998. Curcumin attenuation of acute adriamycin myocardial toxicity in rats. Br. J. Pharmacol. 124 (3), 425–427.

Venkatesh, S., Workman, J.L., 2015. Histone exchange, chromatin structure and the regulation of transcription. Nat. Rev. Mol. Cell Biol. 16 (3), 178–189.

Ventura, A., Jacks, T., 2009. MicroRNAs and cancer: short RNAs go a long way. Cell 136 (4), 586–591.

Vidigal, J.A., Ventura, A., 2015. The biological functions of miRNAs: lessons from in vivo studies. Trends Cell Biol. 25 (3), 137–147.

Waddington, C.H., 1942. Canalization of development and the inheritance of acquired characters. Nature 150 (3811), 563–565.

Waddington, C.H., 1952. Selection of the genetic basis for an acquired character. Nature 169 (4294), 278.

Waddington, C.H., Carter, T.C., 1952. Malformations in mouse embryos induced by trypan blue. Nature 169 (4288), 27–28.

Wagner, S., Ngezahayo, A., Murua Escobar, H., Nolte, I., 2014. Role of miRNA let-7 and its major targets in prostate cancer. Biomed. Res. Int. 2014, 376326.

Wang, Y., Leung, F.C., 2004. An evaluation of new criteria for CpG islands in the human genome as gene markers. Bioinformatics 20 (7), 1170–1177.

Wells, R.A., Leber, B., Zhu, N.Y., Storring, J.M., 2014. Optimizing outcomes with azacitidine: recommendations from Canadian centres of excellence. Curr. Oncol. 21 (1), 44–50.

Wu, H., Zhang, Y., 2011. Mechanisms and functions of Tet protein-mediated 5-methylcytosine oxidation. Genes Dev. 25 (23), 2436–2452.

Wu, Q., Chen, Y., Li, X., 2006. HDAC1 expression and effect of curcumin on proliferation of Raji cells. J. Huazhong Univ. Sci. Technol. Med. Sci. 26 (2), 199–201, 210.

Xiao, C., Wang, L., Zhu, L., Zhang, C., Zhou, J., 2014. Curcumin inhibits oral squamous cell carcinoma SCC-9 cells proliferation by regulating miR-9 expression. Biochem. Biophys. Res. Commun. 454 (4), 576–580.

Yallapu, M.M., et al., 2014. Anti-cancer activity of curcumin loaded nanoparticles in prostate cancer. Biomaterials 35 (30), 8635–8648.

Yan, G., Graham, K., Lanza-Jacoby, S., 2013. Curcumin enhances the anticancer effects of trichostatin a in breast cancer cells. Mol. Carcinog. 52 (5), 404–411.

Yan, W., Herman, J.G., Guo, M., 2016. Epigenome-based personalized medicine in human cancer. Epigenomics 8 (1), 119–133.

Yang, C.H., Yue, J., Sims, M., Pfeffer, L.M., 2013. The curcumin analog EF24 targets NF-kappaB and miRNA-21, and has potent anticancer activity in vitro and in vivo. PLoS One 8 (8), e71130.

Yang, J., Cao, Y., Sun, J., Zhang, Y., 2010. Curcumin reduces the expression of Bcl-2 by upregulating miR-15a and miR-16 in MCF-7 cells. Med. Oncol. 27 (4), 1114–1118.

Ye, M., et al., 2015. Curcumin promotes apoptosis by activating the p53-miR-192-5p/215-XIAP pathway in non-small cell lung cancer. Cancer Lett. 357 (1), 196–205.

Yuan, Z., et al., 2014. TREM-1 is induced in tumor associated macrophages by cyclo-oxygenase pathway in human non-small cell lung cancer. PLoS One 9 (5), e94241.

Yuan, Z., et al., 2012. Curcumin mediated epigenetic modulation inhibits TREM-1 expression in response to lipopolysaccharide. Int. J. Biochem. Cell. Biol. 44 (11), 2032–2043.

Yun, J.M., Jialal, I., Devaraj, S., 2011. Epigenetic regulation of high glucose-induced proinflammatory cytokine production in monocytes by curcumin. J. Nutr. Biochem. 22 (5), 450–458.

Zamani, M., Sadeghizadeh, M., Behmanesh, M., Najafi, F., 2015. Dendrosomal curcumin increases expression of the long non-coding RNA gene MEG3 via up-regulation of epi-miRs in hepatocellular cancer. Phytomedicine 22 (10), 961–967.

Zhang, J., et al., 2010a. Curcumin promotes apoptosis in human lung adenocarcinoma cells through miR-186* signaling pathway. Oncol. Rep. 24 (5), 1217–1223.

Zhang, J., et al., 2010b. Curcumin promotes apoptosis in A549/DDP multidrug-resistant human lung adenocarcinoma cells through an miRNA signaling pathway. Biochem. Biophys. Res. Commun. 399 (1), 1–6.

Zhang, P., et al., 2015. MicroRNA-33b, upregulated by EF24, a curcumin analog, suppresses the epithelial-to-mesenchymal transition (EMT) and migratory potential of melanoma cells by targeting HMGA2. Toxicol. Lett. 234 (3), 151–161.

Zhang, W., Bai, W., Zhang, W., 2014. MiR-21 suppresses the anticancer activities of curcumin by targeting PTEN gene in human non-small cell lung cancer A549 cells. Clin. Transl. Oncol. 16 (8), 708–713.

Zhao, S.F., et al., 2014. Induction of microRNA-9 mediates cytotoxicity of curcumin against SKOV3 ovarian cancer cells. Asian Pac. J. Cancer Prev. 15 (8), 3363–3368.

CHAPTER

6

The Epigenetic Targets of Berry Anthocyanins in Cancer Prevention

Dina Moustafa Abo El-Ella[1], *Anupam Bishayee*[2]

[1]Department of Pharmacology and Toxicology, Faculty of Pharmacy, October 6 University, 6th October City, Egypt; [2]Lake Erie College of Osteopathic Medicine, Bradenton, FL, United States

Abstract

The most common forms of epigenetic regulation of gene expression are DNA methylation, histone modifications, and miRNA-related mechanisms. Irregularities of these epigenetic modifications are observed in an increasing number of pathological disorders, including cancer. The reversibility of these epigenetic changes makes them an attractive and encouraging target in the treatment of cancer. Thus, a growing number of epigenetically active compounds, such as berry anthocyanins, are suggested for the alteration of these epigenetic cellular mechanisms. It has been documented that berry anthocyanins are potent anticarcinogenic agents and are protective against genomic instability in the oncogenic pathway. It remains to be established whether berry anthocyanins can influence these epigenetic alteration mechanisms in cancer. Accordingly, the aim of this review is to shed light on the molecular mechanisms of berry anthocyanins in cancer prevention through the targeting of epigenetic modifications.

Keywords

Anthocyanins; Berry; Cancer; DNA methylation; Epigenetic targets; Histone modification

INTRODUCTION

In the last few years, increased interest has emerged for the use of fruits, especially berries, as chemopreventive and protective agents against a variety of chronic diseases (Kristo et al., 2016). Berries are popularly consumed in North America both in fresh and frozen forms and in a variety of processed foods, such as yogurts, beverages, jam, canned fruits, and jellies (Seeram, 2008). Some important berry genera contain many pharmacologically active phytochemicals. These berry genera are the *Vaccinium* genus (blueberry, lingonberry, cranberry, and bilberry), *Rubus* genus (blackberry, red raspberry, black raspberry, and cloudberry), *Fragaria* genus (strawberry), and *Sambucus* genus (elderberry and red elderberry) (Stoner et al., 2010). Berries also contain some of the best dietary sources of bioactive compounds (De Souza et al., 2014; Slatnar et al., 2012; Namiesnik et al., 2014). These bioactive compounds include

Epigenetics of Cancer Prevention
https://doi.org/10.1016/B978-0-12-812494-9.00006-8

phenolic acids, benzoic acid as hydroxybenzoic, derivatives of cinnamic acid, such as hydroxycinnamic acids (Mattila et al., 2006), and stilbenes, such as resveratrol (D'Archivio et al., 2007). In addition, berries contain lignans, such as secoisolariciresinol (Mazur et al., 2000), flavonoids including anthocyanins, such as cyanidin, pelargonidin, delphinidin, peonidin, and malvidin (Clifford, 2000), flavonols, such as quercetin, myricetin, kaempferol and flavanols, such as catechins (Crozier et al., 2009). Berries additionally contain condensed tannins, such as proanthocyanidins (Santos-Buelga and Scalbert, 2000), and hydrolyzable tannins, such as ellagitannins and gallotannins (Clifford and Scalbert, 2000), vitamins A, C, and E (Pantelidis et al., 2007; Szajdek and Borowska, 2008), folate (Subar et al., 1989), and alkaloids, such as berberine, berbamine and palmatine (Ivanovska and Philipov, 1996). Moreover, berries contain carotenoids (Ovaskainen et al., 2008), xanthones, such as α-mangostin, β-mangostin, γ-mangostin, and methoxy-β-mangostin (Akao et al., 2008) as well as polysaccharides (Hilz et al., 2005).

BIOLOGICAL, PHARMACOLOGICAL, AND ANTICANCER ACTIVITIES OF BERRIES

Berries are a rich source of phenolic compounds, particularly phenolic acids, anthocyanins, proanthocyanidins, and other flavonoids. These compounds provide the pigmentation of fruits and have been found to be beneficial to human health (Bagchi et al., 2004). Recently, lingonberries induced hypermethylation globally and also specifically to nuclear receptor corepressor 2 (NCoR), which encodes the NCoR/silencing mediator for retinoid (SMRT) that regulates the metabolic homeostasis and inflammatory pathways. NCoR (known as SMRT) is implicated in the regulation of lipid, inflammation, and oxidative stress pathways (Heyman-Lindén et al., 2016). Moreover, berry bioactive components induce anticancer activity through various pharmacological activities, including the induction of xenobiotic-metabolizing enzymes, modulation of gene expression and subcellular signaling pathways, reduction of cell proliferation, and stimulation of apoptotic cell death and anti-inflammatory properties (Bagchi et al., 2004; Seeram, 2008). In the following sections, we present several laboratory and animal studies showing the anticancer activity of berries.

A group of researchers found that black raspberry extract inhibited several signal transduction pathways, such as mitogen-activated protein kinases (MAPKs), phosphatidylinositol-3-kinase (PI3K)/protein kinase B(Akt)/ c-Jun N-terminal kinase (JNK), and p38/extracellular signal-regulated kinase 1/2 that control several transcription factors, such as activator protein 1 (AP-1) and nuclear factor-κB (NF-κB), as well as the expression of their target genes, such as cyclooxygenase-2 (COX-2) and vascular endothelial growth factor (VEGF), in benzo[a]pyrene-7,8-diol-9,10-epoxide—exposed JB-6 Cl 41 cells (Lu et al., 2006). Moreover, black raspberry reversed oxidative stress and suppressed NF-κB and MAPK pathways in N-nitrosomethylbenzylamine (NMBA)—induced esophageal squamous cell carcinogenesis in rats (Shi et al., 2017). Additionally, berry phytochemicals can interfere with both estrogen receptor (ESR) and tyrosine kinase receptor (TKR) signaling (Aiyer et al., 2012). They can also induce apoptotic as well as autophagic cell death by modulating the kinase signaling involved in ESR and TKR pathways (Aiyer et al., 2012). Furthermore, black raspberry topical gel suppressed genes associated with RNA processing, growth factor recycling, and the inhibition of apoptosis (Mallery et al., 2008). Moreover, Wang et al. (2014) showed that treating a familial adenomatous polyposis patient with black raspberries resulted in the reduction of cell-proliferation rates as measured by Ki-67 expression, decreased DNA methyltransferase 1, and increased cyclin-dependent kinase inhibitor 2A (P16) levels, while the expression of c-Myc and the extent of apoptosis were unchanged. Additionally, black raspberries induced anti-inflammatory activity by several mechanisms, such as suppressing the messenger RNA (mRNA) expression of pro-inflammatory cytokines, tumor necrosis factor-α, and interleukin-1β, and the protein expression of p-IκBα, COX-2, and plasma prostaglandin E2 (Montrose et al., 2011). Chen et al. (2006) showed that a diet supplemented with five percent black raspberries and given to rats subjected to NMBA-induced esophageal tumorigenesis resulted in reduced mRNA and protein expression levels of COX-2, iNOS, c-Jun, and prostaglandin E2 in preneoplastic lesions of the esophagus. Also, pretreatment of an A/J mouse with 10% lyophilized strawberries one - week before carcinogen treatment failed to inhibit methylnitrosamino-1-3-pyridyl-1-butanone and B[a]PD-induced mouse lung cancer (Carlton et al., 2000).

Knobloch et al. (2016) showed that treating oral squamous cell carcinoma patients with black raspberries enhanced the expression of pro4-survival genes—such as aurora kinase A, baculoviral IAP repeat containing 5, and epidermal growth factor (EGF)—and reduces other pro-inflammatory genes, such as NF-κB1 and prostaglandin-endoperoxide synthase 2. Also, cyanidin-3-glucoside (C3G) extracted from mulberry showed anticancer activity against MDA-MB-453 breast cancer cells and xenografted MDA-MB-453 cells in nude mice (Cho et al., 2017). Treatment with C3G (a member of the anthocyanin family) showed active apoptosis by caspase-3 cleavage and DNA fragmentation through the alteration of B-cell lymphoma 2 (Bcl-2) and the Bcl2-associated X protein (Bax) pathway. Indeed, C3G inhibited tumor growth in xenograft mice (Cho et al., 2017). Studies included in this review refer mostly to freezing fruit as berries (Table 6.1).

ANTHOCYANIN

Berries are rich in anthocyanins, which account for approximately 5%—10% of their dry weight (Stoner et al., 2008). Anthocyanins are responsible for much of the red, blue, and purple colors of fruits, vegetables, grains, flowers, and herbs. The Greek word "anthocyanins" is derived from "*anthos*," meaning flower, and "*kyanos*," meaning blue. Anthocyanins are the polyphenolic compounds present in most colorful berries (Delgado-Vargas et al., 2000; Fossen et al., 2002). Anthocyanins are secondary metabolites of plants and are the most important subclass of flavonoids (He and Giusti, 2010). Moreover, these phytochemicals are of great interest to nutritionists and food processors because of their health benefits and their colorful characteristics, for which they are considered as a natural alternative to synthetic dyes (Jing and Giusti, 2011).

CHEMICAL STRUCTURE OF ANTHOCYANINS

Anthocyanins are water-soluble pigments that belong to the flavonoid group. Their basic structure is three molecules, such as a flavylium ring or aglycone base (as anthocyanidins), one or more sugar molecules attached at C3, C5, or C7, and sometimes acylating groups (Kong et al., 2003). The most common sugar molecule found attached to the aglycone is glucose, followed by galactose, rhamnose, xylose, and arabinose. The sugar moieties bound to anthocyanidins (anthocyanin aglycones) are mono-, di-, and trisaccharide (Fang, 2014). Some anthocyanins may be rarely acylated by acids attached to the sugar moiety as p-coumaric, caffeic, ferulic, sinapic, p-hydroxybenzoic, malonic, oxalic, malic, succinic, or acetic acid (De Pascual-Teresa et al., 2013). Anthocyanins are predominantly found in nature as water-soluble glycosides of polyhydroxy and polymethoxy derivatives of

TABLE 6.1 Laboratory and Animal Studies Showing Anticancer Activity of Different Berries

Berries	Cell Culture, In Vivo or Clinical Study	Biological Activity	References
Black raspberry	Mouse epidermal JB-6 Cl 41 cells	↓ MAPKs, PI-3K/Akt/JNKs, and p38/ERK1/2 ↓ AP-1 and NF-κB ↓ COX-2 and VEGF	Lu et al. (2006)
	Oral squamous cell carcinoma patients	↓ NF-κB1and PTGS2 ↑ AURKA, BIRC5, and EGF	Knobloch et al. (2016)
	Esophageal squamous cancer cells in rats	↓ MAPKs ↓ NF-κB	Shi et al. (2016)
	Human premalignant oral lesions	↓ Genes for RNA processing, signal transduction, and inflammatory pathways ↓ COX-2 and NF-κB	Mallery et al. (2008)
Mulberry	MDA-MB-453 breast cancer cells, Xenograft nude mice	↑ Caspase-3 cleavage ↓ Tumor growth in mice	Cho et al. (2017)
Lingonberry Blackcurrant Bilberry	C57BL/6JBomTac mice receiving high fat diets (45 kcal% fat)	↓ STAT3, mTOR and NF-κB signaling; Hypermethylation of NCoR-2 ↓NF-κB translocation to nucleus ↓ Lipid synthesis by ↓ SREBP1c transcriptional activity ↓NF-κB translocation to nucleus ↓ Intestinal lipid absorption	Heyman-Lindén et al. (2016)
Strawberry	NNK- and B[a]PD- induced lung cancer in A/J mice	No protection against lung tumor	Carlton et al. (2000)
Black raspberry	Familial adenomatous polyposis patient	↓ Ki-67 ↓ DNMT1 ↑ P16	Wang et al. (2014)
	Ulcerative colitis in C57BL/6Jmice	↓RNA expression of the pro-inflammatory cytokines TNF-α and IL-1β ↓p-IκBα, COX-2 and prostaglandin E2	Montrose et al. (2011)
	Esophageal cancer in rats	↓mRNA and protein expression of COX-2, iNOS, c-Jun, and prostaglandin E2	Chen et al. (2006)

2-phenyl-benzopyrylium or flavylium salts (Kong et al., 2003).

Various combinations of types and numbers of sugars and acids attached to different types of aglycons lead to the wide range of anthocyanins that are found in berries. However, the predominant berry anthocyanin is C3G (Castaneda-Ovando et al., 2009). The major anthocyanin compounds found in different berries are listed in Table 6.2.

More than 635 anthocyanins have been identified according to the number and position of hydroxyl and methoxyl moieties at the B ring and by the nature and number of aliphatic and aromatic acids attached to the sugar residues. The six most mentioned anthocyanidins found in

TABLE 6.2 Anthocyanin Composition of Various Berries

Berries	Anthocyanins	References
Raspberry	Cyanidin glycosides, cyanidin-3-arabinose, cyanidin-3-sophoroside, cyanidin-3-rutinoside, and pelargonidin-3- glycosides	Probst (2015) and Skrovankova et al. (2015)
Blueberry	Malvidin glycosides, cyanidin glycosides, delphinidin glycosides, and petunidin glycosides	Giovanelli and Buratti (2009) and Nile and Park (2014)
Strawberry	Cyanidin-3-glucoside, pelargonidin, and pelargonidin-3-rutinoside	Giampieri et al. (2012) and Wang et al. (2007)
Cranberry	Cyanidin glycosides, peonidin glycosides, pelargonidin glycosides, malvidin glycosides, and delphinidin glycosides	Neto. (2007) and Skrovankova et al. (2015)
Bilberry	Cyanidin-3-galactoside, cyanidin-3-glucoside, and cyanidin-3-arabinoside	Cocetta et al. (2012) and Madhavi et al. (1998)
Blackberry	Cyanidin glycosides, pelargonidin glycosides, and peonidin glycosides	Probst (2015)
Blackcurrant	Delphinidin-3-O-rutinoside and cyanidin-3-O-rutinoside	Gopalan et al. (2012)
Chokeberry	Cyanidin-3-galactoside and cyanidin-3-arabinoside	Bernabé et al. (2015), Denev et al. (2012) and Kulling and Rawel (2008)
Acai berry	Cyanidin, delphinidin, malvidin, pelargonidin, and peonidin	Poulose et al. (2012)
Lingonberry	Cyanidin-3-O-sambubioside, delphinidin-3-O-galactoside, and peonidin-3-O-galactoside	Brown et al. (2014), Lehtonen et al. (2013) and Ogawa et al. (2014)
Arctic bramble	Cyanidin-3-glucoside and cyanidin-3-rutinoside	Määttä-Riihinen et al. (2004)
Jamun berry	3,5-diglucosides of delphinidin, petunidin, and malvidin.	Faria et al. (2011)
Elderberry	Cyanidin-3,5-diglucoside and cyanidin-3-glucoside	González-Molina et al. (2012)

berries are pelargonidin, cyanidin, peonidin, delphinidin, petunidin, and malvidin (Fig. 6.1) (de Pascual-Teresa and Sanchez-Ballesta, 2008; Jing and Giusti, 2007).

Anthocyanins are poorly absorbed in the gastrointestinal tract (He et al., 2005) and largely metabolized by gut microflora to several metabolites, including megal, gallic acid, syringic acid, vanillic acid, protocatechuic acid, and 2,4,6-trihydroxybenzaldehyde (THBA) (Aura et al., 2005; Forester and Waterhouse, 2008; Keppler and Humpf, 2005).

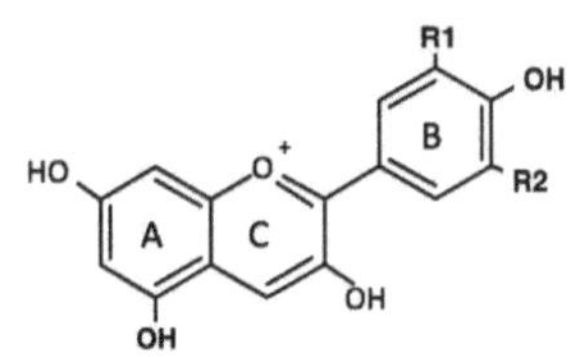

FIGURE 6.1 Chemical structures of anthocyanidins and their related compounds: cyanidin (R1 = OH, R2 = H), delphinidin (R1 = R2 = OH), peonidin (R1 = OCH_3, R2 = H), petunidin (R1 = OCH_3, R2 = OH), pelargonidin (R1 = R2 = H), and malvidin (R1 = R2 = OCH_3).

ANTICANCER EFFECTS OF BERRY ANTHOCYANINS

Numerous studies have reported the antiproliferative activity of anthocyanin-rich extract from blackberries on colon cancer HT-29, breast cancer MCF7, lung cancer A549, and leukemia HL-60 cells (Aqil et al., 2012; Dai et al., 2009). Also, blackcurrant anthocyanins reduced oxidative DNA damage in mononuclear blood cells of humans as determined by strand breaks as well as endonuclease III and formamidopyrimidine DNA glycosylase–sensitive sites by the comet assay (Møller et al., 2004).

An anthocyanin extract from black raspberries suppressed NMBA-induced esophageal carcinogenesis and downregulated Ki-67, COX-2, and CD45 in esophageal cancer (Wang et al., 2009). Concurrently, the anthocyanin-rich extract decreased cell proliferation, inflammation, and angiogenesis together with increased apoptosis (Wang et al., 2009). In addition, anthocyanins derived from black raspberries and protocatechuic acid (anthocyanin metabolites) are documented to inhibit inflammation and tumorigenesis in N- NMBA-induced esophageal cancer in rats (Peiffer et al., 2014). Moreover, treatment of esophageal squamous cancer cells by cyanidin-3-rutinoside, an anthocyanin friction of black raspberries, resulted in downregulation of COX-2 and iNOS protein and suppressed the activation of extracellular receptor kinase (ERK) and Akt (Shi et al., 2016).

Other studies showed that blueberry anthocyanins exhibited strong antioxidant properties (Schantz et al., 2010), inhibited tumor cell growth (Liu et al., 2010; Seeram et al., 2006), and induced apoptosis (Srivastava et al., 2007). A bilberry anthocyanin-rich extract decreased Caco-2 and Hep-G2 cancer cell proliferation and increased the accumulation of sub-G1 cells and lactate dehydrogenase activity in the Caco-2 and Hep-G2 cell lines (Liu et al., 2010). It was shown that the anthocyanin content of black raspberry, strawberry, and blueberry extracts stimulated the incidence of apoptosis in an HT-29 colon cancer cell line (Seeram et al., 2006). Also, chokeberry anthocyanin extracts containing high levels of monoglycosylated cyanidin derivatives showed stronger cytotoxic activity against colon cancer HT-29 cells than that of bilberry anthocyanin extracts (Zhao et al., 2004).

Another study found that berry anthocyanins induced autophagy as characterized by upregulation of eIF2a and downregulation of mammalian target of rapamycin and Bcl-2 (Longo et al., 2008). Inhibition of autophagy by either 3-methyladenine or Atg5 small interfering RNA–enhanced anthocyanin-triggered apoptosis in liver cancer cells (Longo et al., 2008). In prostate cancer, anthocyanin-enriched fractions from blueberries downregulate matrix metalloproteinases (MMPs) MMP-2/MMP-9 and upregulate the tissue inhibitor of metalloproteinases 1/2 in DU145 cells by modulating protein kinase C and MAPK pathways (Matchett et al., 2005, 2006).

It was found that chokeberry anthocyanin-rich extract upregulated the CDK p21WAF1 and cyclin-dependent kinase inhibitor 1B ($p27^{Kip1}$), both binding to cyclin-CDK complexes and blocking the G1/G0 and G2/M phases of cell cycle in colon cancer HT-29 cells (Malik et al., 2003). Additionally, the extract downregulated cyclin A and cyclin B1 to inhibit colon carcinoma cell growth without any effect on the growth of normal colon cells (Malik et al., 2003). Furthermore, Ding et al. (2006) demonstrated that C3G derived from blackberries exhibits chemotherapeutic activities against different cancer models—in vitro using the JB6 and A546 cell line, and in vivo using mice bearing lung or skin tumors. Moreover, C3G inhibited ERK, JNK, and MAPK p38 signaling cascades in addition to AP-1, MAPK, NF-κB, COX-2, and TNF-α expression in different cancer models (Ding et al., 2006).

Anthocyanin-rich mixtures from bilberries and grapes inhibited the receptor protein-tyrosine kinases associated with EGF and VEGF receptor-2 more than they did those linked to epidermal growth factor 2 (ErbB2), VEGF receptor-w3, and insulin-like growth factor receptor 1 using in vitro kinase assays in A431 human vulva carcinoma cells. Also, both mixtures inhibited phosphorylation of the ErbB3-receptor slightly more potently than their inhibition of the EGF and ErbB2-receptor in A431 human vulva carcinoma cells (Teller et al., 2009). These TKRs are important in controlling the oncogenic signaling cascade that engages with MAPK and PI3K (Gerber et al., 1998; Marko et al., 2004). Furthermore, a bilberry extract containing 15 different anthocyanins showed significant reduction in the intestinal adenoma burden in an Apc^{Min} mouse model, which is a model of human familial adenomatous polyposis (Cooke et al., 2006). Moreover, anthocyanin-rich blackberry extracts potently suppressed the DNA-damaging properties of the topoisomerase poisons camptothecin and doxorubicin in HT29 colon cancer cells (Esselen et al., 2011).

Interestingly, another study showed that anthocyanins extracted from raspberries downregulated COX-2 and iNOS gene expression in the rat esophageal epithelial cell line RE-149 DHD (Zikri et al., 2009). Magnuson et al. (2003) found that anthocyanin-rich extracts from chokeberries and bilberries reduced the number of aberrant crypt foci in rats by inhibition of COX-2 gene expression. More recently, dried raspberries reduced the activity of MMP-2 and urokinase, upregulated epithelial markers, such as E-cadherin and α-catenin, and inhibited mesenchymal markers, such as N-cadherin, fibronectin, snail-1, and vimentin in the A549 cancer cell line (Hsieh et al., 2013). Also, dried raspberry inhibited p-FAK, p-paxillin, and AP-1, indicating its antiepithelial mesenchymal transition in lung cancer cells (Hsieh et al., 2013). A BALB/c nude mice xenograft model showed a reduction in lung tumor volume and weight upon treatment with dried raspberry (Hsieh et al., 2013). Xu et al. (2010) investigated the effect of C3G (a member of the anthocyanin family) purified from blackberries on ethanol-induced breast cancer cell line ($MCF7^{ErbB2}$ and MDA-MB231) migration and invasion. The investigators found that C3G inhibited ethanol-stimulated phosphorylation of ErbB2, cSrc, focal adhesion kinase, and p130 (Cas) as well as interactions among these proteins and the p130(Cas)/JNK interaction (Xu et al., 2010). Moreover, Misikangas et al. (2007) showed that the anthocyanins, proanthocyanidins, and ellagic acid content in lingonberries and cloudberries resulted in a significant reduction in the number of intestinal tumors in mice. Also, they showed that lingonberries and cloudberries markedly inhibited the growth of adenomas and the accumulation of nuclear β-catenin and cyclin D1 in the mucosa. Table 6.3 summarizes the effects of various berry anthocyanins on various cancer types.

C3G (an anthocyanin member) showed significant inhibition of UVB-induced epidermal thickening and decreased mRNA and protein expression of proliferating cell nuclear antigen and cyclin D1 (Pratheeshkumar et al., 2014). On the other hand, the metabolites of the berry anthocyanins megal, gallic acid, and THBA are also effective in reducing colon cancer cell (Caco-2) viability (Forester and Waterhouse, 2010). In addition, a recent study by Forester et al. (2014) has documented that these anthocyanin metabolites activated the apoptotic pathway through activation of caspase-3. Furthermore, megal and gal metabolites caused DNA fragmentation and nuclear condensation. Also, megal, gal, and THBA inhibited the transcription factors NF-κB, AP-1, signal transducer and activator of transcription 1, and octamer transcription factor activated in colorectal cancer. Accordingly, these anthocyanin metabolites could exert anticarcinogenic actions through the modulation of multiple signaling pathways, especially NF-κB (Forester et al., 2014).

TABLE 6.3 Effects of Berry Anthocyanins on Different Cancer Types

Berry Anthocyanins	Cancer Type	Effects	References
Blackberry	Colon cancer cell HT-29 Lung cancer cell A549 Leukemia cancer cell HL-60 Breast cancer cell MCF-7	↓ Cell viability ↑ H_2O_2 ↑ ROS ↑ Cytotoxicity ↓ DPPH ↑ ORAC ↓ DNA damage and mutation	Dai et al. (2009) Aqil et al. (2012)
Blackcurrant	MNBC human cells	↓ Oxidative DNA damage (Fdp) site	Møller et al. (2004)
Black raspberry	Esophageal cancer in Fischer 344 rats	↓ Esophageal tumorigenesis ↓ Ki-67, COX-2, and CD45 ↓ VEGF and HIF-1α ↓ ERK1/2 ↓ Bcl2 ↑ BAX	Wang et al. (2009)
		↓ Esophageal tumorigenesis ↓ Pentraxin-3 ↓ COX-2, iNOS, NF-κB, and sEH	Peiffer et al. (2014)
Blackberry	Colon cancer cell HT-29	↓ DNA-damaging effects	Esselen et al. (2011)
Bilberry	Colorectal cancer cell Caco-2 Liver cancer cell line Hep-G2	↑ Accumulation of sub-G1 cells ↑ LDH activity	Liu et al. (2010)
Black raspberry, strawberry, and blueberry	Colon cancer cell HT-29 Oral cancer cells (KB, CAL-27) Breast (MCF-7) Prostate (LNCaP)	↓ Cancer cell growth and proliferation ↑ Apoptosis of COX-2 expressing in HT-29	Seeram et al. (2006)
Chokeberry and bilberry	Colon cancer cell HT-29	↓ Cancer cell growth	Zhao et al. (2004)
Wild berries	Human HCC cell lines PLC/PRF/5	↑ Bax, cytochrome *c* and caspase 3 ↑ Autophagy ↓ Phosphorylation of mTOR ↓ Bcl2 ↑ Phosphorylation of eIF2α ↓ Cancer cell growth ↑ Nuclear fragmentation	Longo et al. (2008)
Blueberry	Prostate cancer cell line DU145	↓ MMP-2/MMP-9 ↑ TIMP-1/TIMP-2	Matchett et al. (2005, 2006)
Chokeberry	Colon cancer cell HT-29	↑ CDK $p21^{WAF1}$ and $p27^{Kip1}$ Blockade of G1/G0 and G2/M ↓ Cyclin A and cyclin B1	Malik et al. (2003)
Bilberry	Intestinal adenoma in Apc^{Min} mouse	↓ Number of intestinal adenomas	Cooke et al. (2006)

TABLE 6.3 Effects of Berry Anthocyanins on Different Cancer Types—cont'd

Berry Anthocyanins	Cancer Type	Effects	References
Blackberry	Lung cancer cell A549 Mouse epidermal JB6 in vivo model of lung and skin tumor in mice	↓ERK, JNK, and p38 ↓AP-1, MAPK, NF-κB, COX-2, and TNFα	Ding et al. (2006)
Bilberries and Grapes	Human vulva carcinoma cells A431	↓ EGF and VEGFR-2 receptor ↓Phosphorylation of ErbB3-receptor	Teller et al. (2009)
Chokeberry and Bilberry	Colon carcinogen azoxymethane in rats	↓Number of aberrant crypt foci ↓Gene expression of COX-2	Magnuson et al. (2003)
	Intestinal C57BL/6J Min/1 mice	↓ Adenoma growth ↓ Nuclear β-catenin accumulation ↓Cyclin D1 ↓Gene expression of adenosine deaminase, 5′ectonucleoditase, and prostaglandin E2	Misikangas et al. (2007)
Strawberry	Rat esophageal epithelial cell lines RE-149 DHD	↓ Gene expression of COX-2 and iNOS	Zikri et al. (2009)
Raspberry	Lung cancer cell A549 BALB/c nude mice xenograft model	↓MMP-2 and u-PA ↑E-cadherin and α-catenin ↓N-cadherin, fibronectin, snail-1, and vimentin ↓Lung tumor volume and weight	Hsieh et al. (2013)
Blackberry	Breast cancer cells ($MCF7^{ErbB2}$ and MDA-MB231)	↓Phosphorylation of ErbB2, cSrc, FAK, and p130 (Cas) ↓ p130(Cas)/JNK interaction	Xu et al. (2010)
Black raspberry	Esophageal squamous cancer cell in rats	↓COX-2 and iNOS ↓Activation of ERK ↓AKT expression	Shi et al. (2016)

EPIGENETIC REGULATION IN CANCER

Epigenetics is defined as hereditary changes in gene expression without changes in the DNA sequence (Wolffe and Matzke, 1999). DNA methylation, covalent modifications of histone tails, and alteration of miRNA expression are all components of epigenetic regulations and are involved in ruling gene expression (Winter et al., 2009; Yoo and Jones, 2006). Recently, it was shown that epigenetic alterations could largely contribute to the development of cancer. Epigenetic disorders of gene expression can induce the initiation and progression of cancer (Esteller, 2008). In addition, several factors—either endogenous, such as nutrients, infections, and physical activity, or exogenous, such as the environment—can influence these epigenetic mechanisms (Dolinoy et al., 2007). In the present review, we focus mainly on the three epigenetic mechanisms (DNA methylation, histone modification, and miRNA expression) that can potentially yield insights into cancer chemoprevention using berry anthocyanins.

DNA Methylation

DNA methylation is a transmissible alteration of the DNA structure that inhibits gene expression directly. This methylation occurs without any sequence alteration (Jones and Baylin, 2002). In normal cells, DNA is methylated by DNA methyltransferases (DNMTs) including DNMT1, DNMT3a, and DNMT3b via the addition of a methyl group that forms 5-methylcytosine (Bird, 2002; Esteller, 2002). In cancer cells, an imbalance in DNA methylation has been reported, and DNA undergoing either hypomethylation or hypermethylation is a key characteristic of human neoplasia (Baylin et al., 1991). DNA hypomethylation can facilitate the expression of quiescent proto-oncogenes and prometastatic genes as well as promote tumor progression. Localized DNA hypermethylation can increase methylation at CpG islands within the gene-promoter regions of specific genes (such as tumor-suppressor genes), and this can result in transcriptional silencing and an inability to control tumorigenesis (Ehrlich, 2009).

DNA methylation in cancer cells is now believed to be a major cause of transcriptional gene silencing in carcinogenesis. Many cancer-related genes, such as human MutL homolog 1, cyclin-dependent kinase inhibitor 2A ($p16^{INK4a}$), cyclin-dependent kinase 4 inhibitor B ($p15^{INK4b}$), ARF tumor suppressor ($p14^{ARF}$), death-associated protein kinase, cadherin-1, cadherin-13, and glutathione S-transferase P1 are inactivated by this epigenetic alteration (Esteller, 2002). Other studies have shown that a chronic methyl-deficient diet in rodents caused an increase in oxidative stress and DNA hypomethylation that resulted in liver tumorigenesis (Pogribny et al., 2004, 2012). On the other hand, hypermethylation is known to cause gene silencing by preventing the recruitment of transcription factors to the promoters of the gene (Bird, 2002; Esteller, 2008).

Histone Modifications

Histones are basic proteins that bind to DNA in the nucleus, forming nucleosome, the basic subunit of chromatin, which further condenses chromatin to form the chromosome. Histone is an octamer consisting of an H3/H4 tetramer and two H2A/H2B dimers, and wrapped with 146 base pairs of DNA, forms the nucleosome. Moreover, histones' function is to preserve the stability of the folded chromatin structure and to undergo posttranslational modifications that are linked directly to gene expression (Berlowitz and Pallotta, 1972; Luger et al., 1997; Tremethick, 2007).

The majority of these modifications happen at the lysine, arginine, and serine residues of the N-terminal tails extending from the histones via acetylation, methylation, phosphorylation, ubiquitination, and sumoylation (Berger, 2007; Bhaumik et al., 2007; Ellis et al., 2009; Kouzarides, 2007; Luger et al., 1997). Concurrently, this modification can either activate or dominate the gene transcription depending on the location and type of histone modifications involved (Kouzarides, 2007). For example, euchromatin modifications such as acetylation of histone 3 and histone 4 (H3 and H4), and di- and trimethylation of H3K4, are accompanied by active gene transcription. Also, di- and trimethylation of H3K9, and trimethylation of H3K27, are accompanied by gene suppression (Barski et al., 2007; Li et al., 2007). On the other hand, various histone-modifying enzymes such as HATs and HMTs are involved in these modifications, adding acetyl groups and methyl groups respectively to the histone tails. Furthermore, histone deacetylases (HDACs) and histone demethylases remove acetyl and methyl groups respectively from histones (Haberland et al., 2009; Shi, 2007).

Fraga et al. (2005) documented that disturbances in histone modification, such as trimethylation of H4K20 and monoacetylation of H4K16, are found in cancer cells. Similarly, Bolden et al. (2006) acknowledged that overexpression of HDACs, such as HDAC1, HDAC2, and HDAC6, has been observed in cancer cells. Hence, global disruption by these histone modifications and their modifying enzymes is a cause for cancer incidence, and this point is also of interest for cancer prevention using berry anthocyanins.

Noncoding RNA (Microribonucleic Acid miRNAs)

In addition to DNA methylation and histone modification, miRNAs are considered another factor in the epigenetic regulation of gene expression . They are short RNA molecules that exert their regulatory effects either through mRNA destabilization or translation inhibition (Lagos-Quintana et al., 2001; Lujambio and Lowe, 2012; Jing et al., 2005). Moreover, they are important factors in the regulation of gene expression by either inhibiting translation or triggering degradation of target mRNAs (Croce, 2009). miRNAs are convoluted in variant biological incidents, such as cell proliferation, apoptosis, tumor growth, cell cycle, and immunity (Baltimore et al., 2008; Carleton et al., 2007).

In cancer, it was found that miRNAs have a strong effect on the epigenetic mechanisms and epigenetic regulations of miRNA expression (Esteller, 2007; Jones and Baylin, 2002; Kai and Pasquinelli, 2010). Accordingly, miRNAs can be tumor suppressors or oncogenes depending on their target transcripts and cellular contexts. Oncogenic miRNAs target tumor-suppressor mRNAs, whereas tumor-suppressor miRNAs target oncogenic mRNAs (Jing et al., 2005). On the other hand, miRNAs affect the epigenome by targeting enzymes of the epigenetic machinery (Iorio et al., 2010; Zhou et al., 2010). Subsequently, DNMT1 and DNMT3B are targets of miR-148a in cholangiocarcinoma and cervical cancer, respectively. Also, HDAC1 is targeted by miR-449a and miR-449b in prostate cancer, and HDAC4 by miR-1 in hepatocellular carcinoma (Stefanska et al., 2012). Therefore, targeting specific miRNAs with berry anthocyanins can be a novel therapeutic approach for cancer prevention.

EPIGENETIC TARGETS OF BERRY ANTHOCYANINS IN CANCER PREVENTION

Documented data from various cell culture and animal studies reveal that berry anthocyanins are potent anticarcinogenic agents through various mechanisms of action (Duthie, 2007). Moreover, the alterations observed in all aspects of epigenetic regulation, starting from DNA methylation to histone modifications and anomalous miRNA expression, provides an interesting theme to study the effects of berry anthocyanins on these targets. Therefore, we will discuss what is known about the effects of berry anthocyanins on various epigenetic targets in cancer, because several studies have investigated how berry anthocyanins can modulate epigenetic alterations in cancer.

Wang et al. (2013a) showed that black raspberries administered to knockout mice resulted in an inhibitory effect on ulcerative colitis. The knockout mouse model recapitulates what occurs in human ulcerative colitis. This inhibitory effect is due to the black raspberry–mediated maintenance of DNA methylation homeostasis of several genes associated with regulation of the Wnt pathway in knockout mice. Also, black raspberries prevented epigenetic dysregulation of the Wnt signaling pathway, in part by deterring DNA hypermethylation, inappropriate histone modifications, and condensation of chromatin (Wang et al., 2013a).

Moreover, a dietary supplement of freeze-dried black raspberries given to C57BL/6J mice resulted in reduction of the colonic ulceration induced by dextran sodium sulfate administration. Black raspberry administration suppresses colonic ulceration by decreasing NF-κB p65 protein expression, leading to decreased DNMT3B and thus decreased promoter methylation of tumor-suppressor genes in the Wnt pathway and decreased translocation of β-catenin to the nucleus (Wang et al., 2013b). In another study, black raspberry powder protectively modulated both genetic and epigenetic biomarkers in tissues from colorectal cancer patients (Wang et al., 2011). Black raspberries suppressed DNMT1 and reactivated tumor-suppressor genes by demethylating their promoters (Wang et al., 2011). Clinical studies using black raspberries found an alteration in the methylation of DNMT1, SFRP2, and PAX6a in normal adjacent tissues and DNMT1, SFRP2, PAX6a, and WIF1 in colorectal tumor tissues (Wang et al., 2011). These studies explain well the epigenetic targets of berries, but not whether inhibitory effects are due to black raspberry anthocyanins or other phytochemicals. Subsequently, treating colon cancer HCT-116, Caco-2, and SW480 cells with black raspberry–derived anthocyanins resulted in the suppression of cancer cell proliferation and increased the incidence of apoptosis due to the decreased mRNA expression of β-catenin and c-Myc (Wang et al., 2013c). Anthocyanins also suppressed DNMT1 and DNMT3B expression in colon cancer cells, in addition to demethylating the promoters of tumor-suppressor genes (CDKN2A, and SFRP2, SFRP5, and WIF1) in colon cancer cells (Wang et al., 2013c).

Delphinidin is an anthocyanidin with potent antioxidant, anti-inflammatory, and anticancer properties (Afaq et al., 2007). Another study showed that the treatment of LNCaP human prostate cancer cells with delphinidin resulted in activation of initiator and effector caspases. Also, this treatment resulted in cleavage of PARP-1 and HDAC3 as well as increased expression of the proapoptotic proteins Bax, PUMA, and p21 (Jeong et al., 2016). HDACs are transcriptional corepressors that regulate cell cycle arrest, cell differentiation, and apoptosis in cancer cell lines (Wilson et al., 2006). Moreover, HDAC3 knockdown LNCaP cells treated with delphinidin resulted in apoptosis, which in turn led to p53-dependent activation of the expression of p21 and Bax (Jeong et al., 2016). Indeed, this study observed that delphinidin treatment led to apoptosis in caspase-mediated HDAC3 cleavage that resulted in the hyperacetylation of p53 (Jeong et al., 2016). Subsequently, it was shown that protocatechuic acid (anthocyanin metabolite) and cyanidin inhibited the DNMT activity of an enzyme involved in the methylation of DNA and hence caused epigenetic modifications in MCF7 cells (Paluszczak et al., 2010).

Ko et al. (2015) documented that treating prostate cancer cells with a combination of delphinidin and TNF-related apoptosis-inducing ligand resulted in apoptosis. This apoptotic effect results from the activation of effector caspases via induction of the death receptor 5 pathway, leading to HDAC3 cleavage-dependent mitochondrial apoptosis. However, molecular studies that deal with berry anthocyanins as epigenetic effectors are still limited. The documented studies are presented in Table 6.4.

CONCLUSION AND FUTURE PERSPECTIVE

Although DNA methylation, histone modifications, and miRNA-related mechanisms are epigenetic regulations involved in ruling gene expression, modifications of these epigenetic targets contribute to the development of cancer. Particular interest has been placed on edible berries, because they are rich in a variety of compounds shown to exert favorable effects on several types of cancers. Data from numerous cell culture and animal models suggest that berry extracts and berry-derived phytochemicals, such as the anthocyanins, are potential cancer chemotherapeutic agents. Anthocyanins and

TABLE 6.4 The Epigenetic Effects of Berry Anthocyanins and Their Related Bioactive Compounds

Cell Culture, In Vivo Model or Clinical Study	Berry Anthocyanins	Epigenetic Effect	References
Colorectal cancer patient	Black raspberry powder	↓DNMT1, β-catenin, Ki-67, and CD105 ↑E-cadherin, TUNEL, and p16 ↓DNA methylation of tumor-suppressor genes (SFRP2, p16, SFRP5, and WIF1, PAX6a)	Wang et al. (2011)
Ulcerative colitis in interleukin-10 knockout mice		↓ β-catenin nuclear translocation ↓Promoter methylation of wnt3a ↑mRNA expression of Wif1, sox17, and qki ↑mRNA expression of dkk2 and dkk3 ↓DNA hypermethylation of wif1, sox17, qki, dkk2, and wnt3a ↓Epigenetic dysregulation of the Wnt signaling pathway ↓ DNMT3B, MBD2, HDAC1, and HDAC2 ↓DNA hypermethylation of dkk2 ↑ dkk2 ↓Condensation of chromatin	Wang et al. (2013a)
DSS-induced ulcerative colitis	Black raspberry powder	↓ Ulceration in mucosa, submucosa and muscularis ↓Nuclear localization of NF-κB p65 ↓DNMT3B, HDAC1, HDAC2, and MBD2 ↓DNA methylation of dkk3, dkk2, apc, sfrp1, and sox17 ↓β-Catenin nuclear translocation ↓c-Myc ↑mRNA expression of apc gene	Wang et al. (2013b)
Colon cancer HCT-116, Caco-2, and SW480 cells	Black raspberry–derived anthocyanin	↓DNMT1 and DNMT3B ↓mRNA expression of c-Myc and β-catenin ↑Promoter demethylation of CDKN2A, SFRP5, SFRP2, and WIF1	Wang et al. (2013c)
Human prostate cancer cell LNCaP	Delphinidin	↑Caspases-8, caspase-3, caspase-7, and PARP-1↑Bax, PUMA, and p21 ↓HDAC3 ↑PARP-1 cleavage ↑Hyperacetylation of p53	Jeong et al. (2016)
Breast cancer cell MCF7	Protocatechuic acid (anthocyanin metabolite)	No change in methylation of lysines 9 and 27 in histone H3	Paluszczak et al. (2010)
Human prostate cancer cells LNCaP, Du145	Delphinidin + TRAIL	↓ mRNA expression of XIAP, cIAP-2, Bcl-2, survivin, and MCL-1 ↑Cleavage of caspase-3 and caspase-7 ↑ p53 acetylation ↑ DR5, BAX, and p21 ↓HDAC3	Ko et al. (2015)

their metabolites, as well as anthocyanidins, possess antioxidant, anti-inflammatory, and antiproliferative effects that seem to offer a notable arsenal for reducing the risk of cancer incidence. Despite a lack of evidence for a clear protective effect of berry anthocyanins on epigenetic modulations, several studies have investigated the efficacy of berries, especially black raspberries and their derived anthocyanin extracts, in inhibiting the epigenetic modulations involved in cancer. These modifications involved reactivating tumor-suppressor genes by demethylating their promoters and the suppression of DNA methyl transferase 1. We observed that the anthocyanin metabolites megal, gal, and THBA exert anticarcinogenic effects through modulation of multiple signaling pathways, but it remains to be established whether these metabolites can alter the epigenetic modulations in cancer. A greater understanding of the exact mechanisms by which different types of berry anthocyanins and their metabolites influence epigenetic modulations in cancer will facilitate the design of the preliminary protocol for using berry anthocyanins as chemopreventive agents.

ABBREVIATIONS

A431 Human vulva carcinoma
A549 Adenocarcinoma human alveolar basal epithelial cells
Akt Protein kinase B
AP-1 Activator protein 1
Apc Adenomatous polyposis coli
Atg5 Autophagy-related gene 5
AURKA Aurora kinase A
B(a)PDE Benzo[a]pyrene-7,8-diol-9,10-epoxide
BAX BCL2-associated X protein
Bcl-2 B-cell lymphoma 2
BIRC5 Baculoviral IAP repeat containing 5
CD105 Endoglin
CDH1 Cadherin −1
CDH13 Cadherin 13
CDK Cyclin-dependent kinases
CDKN2A Cyclin-dependent kinase inhibitor 2A
COX-2 Cyclooxygenase-2
DAPK Death-associated protein kinase
dkk2 Dickkopf-related protein 2
dkk3 Dickkopf-related protein 3
DNMT DNA methyltransferases
DPPH 1,1-diphenyl-2-picrylhydrazyl
EGF Epidermal growth factor
ErbB Epidermal growth factor 2
eIF2a Eukaryotic initiation factor 2asubunit
ESR Estrogen receptor
ERK1/2 Extracellular signal-regulated kinase 1/2
Gal Gallic acid
GSTP1 Glutathione S-transferase P1
HATs Histone acetyltransferases
HDACs Histone deacetylases
HDMs Histone demethylases
hMLH1 human MutL homolog 1
HMTs Histone methyltransferases
IGFR1 Insulin like growth factor receptor 1
iNOS Inducible nitric oxide synthase
JNK c-Jun N-terminal kinase
LDH Lactate dehydrogenase
MAPK Mitogen-activated protein kinase
MCF7ErbB2 MCF7 cells overexpressing ErbB2
Megal 3-O-methylgallic acid
miRNA Microribonucleic acids
MMP-2 Matrix metalloproteinase-2
MMP-9 Matrix metallopeptidase 9
MRC-5 Normal human fetal lung fibroblast
NCoR Nuclear receptor corepressor 2
NF-κB Nuclear factor kappa-light-chain-enhancer of activated B cells
NMBA N-nitrosomethylbenzylamine
NNK Methylnitrosamino-1-3-pyridyl-1-butanone
OCT-1 Octamer transcription factor
ORAC Oxygen radical-absorbing capacity
P1−3K Phosphatidylinositol-3-kinase
p14ARF ARF tumor suppressor
p15INK4b Cyclin-dependent kinase 4 inhibitor B
p16 Cyclin-dependent kinase inhibitor 2A
p16INK4a Cyclin-dependent kinase inhibitor 2A
p21Cip1 CDK-interacting protein 1
p27Kip1 Cyclin-dependent kinase inhibitor 1B
p38 Mitogen-activated protein kinases
p53 Tumor protein
PARP-1 Poly (ADP-ribose) polymerase 1
pax6a Paired box 6a
PKC Protein kinase C
PTGS2 Prostaglandin-endoperoxide synthase 2
PUMA p53 upregulated modulator of apoptosis
Qki RNA-binding protein gene quaking
She Soluble epoxide hydrolase
SFRP2 Secreted Frizzled Related Protein 2
SFRP5 Secreted frizzled-related protein 5

SMRT Silencing mediator for retinoid
STAT-1 Signal transducer and activator of transcription 1
THBA 2,4,6-trihydroxybenzaldehyde
TIMP-1 Tissue inhibitors of metalloproteinases-1
TIMP-2 Tissue inhibitors of metalloproteinases-2
TKR Tyrosine kinase receptor
TRAIL TNF-related apoptosis-inducing ligand
u-PA Urokinase-type plasminogen activator
VEGF Vascular endothelial growth factor
VEGFR-2 Vascular endothelial growth factor receptor
WIF1 Wnt inhibitory factor 1
Wnt Wingless-related integration site
XIAP X-linked inhibitor of apoptosis protein

References

Afaq, F., Syed, D.N., Malik, A., Hadi, N., Sarfaraz, S., Kweon, M.H., Khan, N., Abu Zaid, M., Mukhtar, H., 2007. Delphinidin, an anthocyanidin in pigmented fruits and vegetables, protects human HaCaT keratinocytes and mouse skin against UVB-mediated oxidative stress and apoptosis. J. Investig. Dermatol. 127, 222–232.

Aiyer, H.S., Warri, A.M., Woode, D.R., Hilakivi-Clarke, L., Clarke, R., 2012. Influence of berry polyphenols on receptor signaling and cell-death pathways: implications for breast cancer prevention. J. Agric. Food Chem. 60, 5693–5708.

Akao, Y., Nakagawa, Y., Nozawa, Y., 2008. Anti-cancer effects of xanthones from pericarps of mangosteen. Int. J. Mol. Sci. 9, 355–370.

Aqil, F., Gupta, A., Munagala, R., Jeyabalan, J., Kausar, H., Sharma, R.J., Singh, I.P., Gupta, R.C., 2012. Antioxidant and antiproliferative activities of anthocyanin/ellagitannin-enriched extracts from Syzygium cumini L. (Jamun, the Indian Blackberry). Nutr. Cancer 64, 428–438.

Aura, A.M., Martin-Lopez, P., O'Leary, K.A., Williamson, G., Oksman-Caldentey, K.M., Poutanen, K., Santos-Buelga, C., 2005. In vitro metabolism of anthocyanins by human gut microflora. Eur. J. Nutr. 44, 133–142.

Bagchi, D., Sen, C.K., Bagchi, M., Atalay, M., 2004. Anti-angiogenic, antioxidant, and anti-carcinogenic properties of a novel anthocyanin-rich berry extract formula. Biochemistry (Mosc.) 69, 75–80.

Baltimore, D., Boldin, M.P., O'Connell, R.M., Rao, D.S., Taganov, K.D., 2008. MicroRNAs: new regulators of immune cell development and function. Nat. Immunol. 9, 839–845.

Barski, A., Cuddapah, S., Cui, K., Roh, T.Y., Schones, D.E., Wang, Z., Wei, G., Chepelev, I., Zhao, K., 2007. High-resolution profiling of histone methylations in the human genome. Cell 129, 823–837.

Baylin, S.B., Makos, M., Wu, J.J., Yen, R.W., de Bustros, A., Vertino, P., Nelkin, B.D., 1991. Abnormal patterns of DNA methylation in human neoplasia: potential consequences for tumor progression. Cancer Cells 3, 383–390.

Berger, S.L., 2007. The complex language of chromatin regulation during transcription. Nature 447, 407–412.

Berlowitz, L., Pallotta, D., 1972. Acetylation of nuclear protein in the heterochromatin and euchromatin of mealy bugs. Exp. Cell Res. 71, 45–48.

Bernabé, J., Mulero, J., Marhuenda, J., Cerdá, B., Avilés, F., 2015. Changes in antioxidant enzymes in metabolic syndrome patients after consumption a citrus-based juice enriched with Aronia melanocarpa. J. Nutr. Disord. Ther. 5, 4.

Bhaumik, S.R., Smith, E., Shilatifard, A., 2007. Covalent modifications of histones during development and disease pathogenesis. Nat. Struct. Mol. Biol. 14, 1008–1016.

Bird, A., 2002. DNA methylation patterns and epigenetic memory. Genes Dev. 16, 6–21.

Bolden, J.E., Peart, M.J., Johnstone, R.W., 2006. Anticancer activities of histone deacetylase inhibitors. Nat. Rev. Drug Discov. 5, 769–784.

Brown, E.M., Nitecki, S., Pereira-Caro, G., McDougall, G.J., Stewart, D., Rowland, I., Crozier, A., Gill, C.I.R., 2014. Comparison of in vivo and in vitro digestion on polyphenol composition in lingonberries: potential impact on colonic health. BioFactors 40, 611–623.

Carleton, M., Cleary, M.A., Linsley, P.S., 2007. MicroRNAs and cell cycle regulation. Cell Cycle 6, 2127–2132.

Carlton, P.S., Kresty, L.A., Stoner, G.D., 2000. Failure of dietary lyophilized strawberries to inhibit 4-(methylnitrosamino)-1-(3-pyridyl)-1-butanone-and benzo[a]pyrene-induced lung tumorigenesis in strain A/J mice. Cancer Lett. 159, 113–117.

Castaneda-Ovando, A., de Lourdes Pacheco-Hernández, M., Páez-Hernández, M.E., Rodríguez, J.A., Galán-Vidal, C.A., 2009. Chemical studies of anthocyanins: a review. Food Chem. 113, 859–871.

Chen, T., Hwang, H., Rose, M.E., Nines, R.G., Stoner, G.D., 2006. Chemopreventive properties of black raspberries in N-nitrosomethylbenzylamine-induced rat esophageal tumorigenesis: down-regulation of cyclooxygenase-2, inducible nitric oxide synthase, and c-Jun. Cancer Res. 66, 2853–2859.

Cho, E., Chung, E.Y., Jang, H.Y., Hong, O.Y., Chae, H.S., Jeong, Y.J., Kim, S.Y., Kim, B.S., Kim, J.S., Park, K.H., March 27, 2017. Anti-cancer effect of cyanidin-3-glucoside from mulberry via caspase-3 cleavage and DNA fragmentation in vitro and in vivo. Anticancer Agents Med. Chem. 17, 1519–1525.

Clifford, M.N., 2000. Anthocyanins–nature, occurrence and dietary burden. J. Sci. Food Agric. 80, 1063–1072.

Clifford, M.N., Scalbert, A., 2000. Ellagitannins, occurrence in food, bioavailability and cancer prevention. J. Food Sci. Agric. 80, 1118–1125.

Cocetta, G., Karppinen, K., Suokas, M., Hohtola, A., Häggman, H., Spinardi, A., Mignani, I., Jaakola, L., 2012. Ascorbic acid metabolism during bilberry (Vaccinium myrtillus L.) fruit development. Plant Physiol. 169 (11), 1059–1065.

Cooke, D., Schwarz, M., Boocock, D., Winterhalter, P., Steward, W.P., Gescher, A.J., Marczylo, T.H., 2006. Effect of cyanidin-3-glucoside and an anthocyanin mixture from bilberry on adenoma development in the ApcMin mouse model of intestinal carcinogenesis-relationship with tissue anthocyanin levels. Int. J. Cancer 119, 2213–2220.

Croce, C.M., 2009. Causes and consequences of microRNA dysregulation in cancer. Nat. Rev. Genet. 10, 704–714.

Crozier, A., Jaganath, I.B., Clifford, M.N., 2009. Dietary phenolics: chemistry, bioavailability and effects on health. Nat. Prod. Rep. 26, 1001–1043.

D'Archivio, M., Filesi, C., di Benedetto, R., Gargiulo, R., Giovannini, C., Masella, R., 2007. Polyphenols, dietary sources and bioavailability. Ann. Ist. Super. Sanit. 43, 348–361.

Dai, J., Gupte, A., Gates, L., Mumper, R.J., 2009. A comprehensive study of anthocyanin containing extracts from selected blackberry cultivars: extraction methods, stability, anticancer properties and mechanisms. Food Chem. Toxicol. 47, 837–847.

De Pascual-Teresa, S., Sanchez-Ballesta, M.T., 2008. Anthocyanins: from plant to health. Phytochem. Rev. 7, 281–299.

De Pascual-Teresa, S., Sanchez-Ballesta, M.T., García-Viguera, C., 2013. Anthocyanins. In: Ramawat, K.G., Merillon, J.M. (Eds.), Natural Products: Phytochemistry, Botany and Metabolism of Alkaloids, Phenolics and Terpenes. Springer, Berlin-Heidelberg, Germany, pp. 1803–1819.

De Souza, V.R., Pereira, P.A., da Silva, T.L., de Oliveira Lima, L.C., Pio, R., Queiroz, F., 2014. Determination of the bioactive compounds, antioxidant activity and chemical composition of Brazilian blackberry, red raspberry, strawberry, blueberry and sweet cherry fruits. Food Chem. 156, 362–368.

Delgado-Vargas, F., Jimenez, A.R., Paredes-Lopez, O., 2000. Natural pigments: carotenoids, anthocyanins, and betalains—characteristics, biosynthesis, processing, and stability. Crit. Rev. Food Sci. Nutr. 40, 173.

Denev, P.N., Kratchanov, C.G., Ciz, M., Lojek, A., Kratchanova, M.G., 2012. Bioavailability and antioxidant activity of black chokeberry (Aronia melanocarpa) polyphenols: in vitro and in vivo evidences and possible mechanisms of action: a review. Compr. Rev. Food Sci. Food Saf. 11, 471–489.

Ding, M., Feng, R., Wang, S.Y., Bowman, L., Lu, Y., Qian, Y., Castranova, V., Jiang, B.H., Shi, X., 2006. Cyanidin-3-glucoside, a natural product derived from blackberry, exhibits chemopreventive and chemotherapeutic activity. J. Biol. Chem. 281, 17359–17368.

Dolinoy, D.C., Weidman, J.R., Jirtle, R.L., 2007. Epigenetic gene regulation: linking early developmental environment to adult disease. Reprod. Toxicol. 23, 297–307.

Duthie, S.J., 2007. Berry phytochemicals, genomic stability and cancer: evidence for chemoprotection at several stages in the carcinogenic process. Mol. Nutr. Food Res. 665, 665–674.

Ehrlich, M., 2009. DNA hypomethylation in cancer cells. Epigenomics 1, 239–259.

Ellis, L., Atadja, P.W., Johnstone, R.W., 2009. Epigenetics in cancer: targeting chromatin modifications. Mol. Cancer Ther. 8, 1409–1420.

Esselen, M., Fritz, J., Hutter, M., Teller, N., Baechler, S., Boettler, U., Marczylo, T.H., Gescher, A.J., Marko, D., 2011. Anthocyanin-rich extracts suppress the DNA-damaging effects of topoisomerase poisons in human colon cancer cells. Mol. Nutr. Food Res. 55, S143–S153.

Esteller, M., 2002. CpG island hypermethylation and tumor suppressor genes: a booming present, a brighter future. Oncogene 21, 5427–5440.

Esteller, M., 2007. Epigenetic gene silencing in cancer: the DNA hypermethylome. Hum. Mol. Genet. 16, R50–R59.

Esteller, M., 2008. Epigenetics in cancer. N. Engl. J. Med. 358, 1148–1159.

Fang, J., 2014. Bioavailability of anthocyanins. Drug Metab. Rev. 46, 508–520.

Faria, A.F., Marques, M.C., Mercadante, A.Z., 2011. Identification of bioactive compounds from jambolão (Syzygium cumini) and antioxidant capacity evaluation in different pH conditions. Food Chem. 126, 1571–1578.

Forester, S.C., Choy, Y.Y., Waterhouse, A.L., Oteiza, P.I., 2014. The anthocyanin metabolites gallic acid, 3-O-methylgallic acid, and 2,4,6-trihydroxybenzaldehyde decrease human colon cancer cell viability by regulating pro-oncogenic signals. Mol. Carcinog. 53, 432–439.

Forester, S.C., Waterhouse, A.L., 2008. Identification of cabernet sauvignon anthocyanin gut microflora metabolites. J. Agric. Food Chem. 56, 9299–9304.

Forester, S.C., Waterhouse, A.L., 2010. Gut metabolites of anthocyanins,gallic acid, 3-O-methylgallic acid, and 2,4,6-trihydroxybenzaldehyde, inhibit cell proliferation of Caco-2 cells. J. Agric. Food Chem. 58, 5320–5327.

Fossen, T., Slimestad, R., Ovstedal, D.O., Andersen, O.M., 2002. Anthocyanin of grasses. Biochem. Syst. Ecol. 30, 855.

Fraga, M.F., Ballestar, E., Villar-Garea, A., Boix-Chornet, M., Espada, J., Schotta, G., Haydon, C., Ropero, S., Petrie, K., Iyer, N.G., Pérez-Rosado, A., Calvo, E., Lopez, J.A., Cano, A., Calasanz, M.J., Colomer, D., Piris, M.A.,

Ahn, N., Imhof, A., Caldas, C., Jenuwein, T., Esteller, M., 2005. Loss of acetylation at Lys16 and trimethylation at Lys20 of histone H4 is a common hallmark of human cancer. Nat. Genet. 37, 391–400.

Gerber, H.P., McMurtrey, A., Kowalski, J., Yan, M., Keyt, B.A., Dixit, V., Ferrara, N., 1998. Vascular endothelial growth factor regulates endothelial cell survival through the phosphatidylinositol 3-kinase/Akt signal transduction pathway. Requirement for Flk-1/KDR activation. J. Biol. Chem. 273, 30336–30343.

Giampieri, F., Tulipani, S., Alvarez-Suarez, J.M.L., Quiles, J., Mezzetti, B., Battino, M., 2012. The strawberry: composition, nutritional quality, and impact on human health. Nutrition 28, 9–19.

Giovanelli, G., Buratti, S., 2009. Comparison of polyphenolic composition and antioxidant activity of wild Italian blueberries and some cultivated varieties. Food Chem. 112, 903–908.

González-Molina, E., Gironés-Vilaplana, A., Mena, P., Moreno, D.A., García-Viguera, C., 2012. New beverages of lemon juice with elderberry and grape concentrates as a source of bioactive compounds. J. Food Sci. 77, C727–C733.

Gopalan, A., Reuben, S.C., Ahmed, S., Darvesh, A.S., Hohmann, J., Bishayee, A., 2012. The health benefits of blackcurrants. Food Funct. 3, 795–809.

Haberland, M., Montgomery, R.L., Olson, E.N., 2009. The many roles of histone deacetylases in development and physiology: implications for disease and therapy. Nat. Rev. Genet. 10, 32–42.

He, J., Giusti, M.M., 2010. Anthocyanins: natural colorants with health-promoting properties. Annu. Rev. Food Sci. Technol. 1, 163–187.

He, J., Magnuson, B.A., Giusti, M.M., 2005. Analysis of anthocyanins in rat intestinal contents—impact of anthocyanin chemical structure on fecal excretion. J. Agric. Food Chem. 53, 2859–2866.

Heyman-Lindén, L., Seki, Y., Storm, P., Jones, H.A., Charron, M.J., Berger, K., Holm, C., 2016. Berry intake changes hepatic gene expression and DNA methylation patterns associated with high-fat diet. J. Nutr. Biochem. 27, 79–95.

Hilz, H., Bakx, E.J., Schols, H.A., Voragen, A.G.J., 2005. Cell wall polysaccharides in black currants and bilberries-characterisation in berries, juice, and press cake. Carbohydr. Polym. 59, 477–488.

Hsieh, Y.S., Chu, S.C., Hsu, L.S., Chen, K.S., Lai, M.T., Yeh, C.H., Chen, P.N., Rubus idaeus, L., 2013. Reverses epithelial-to-mesenchymal transition and suppresses cell invasion and protease activities by targeting ERK1/2 and FAK pathways in human lung cancer cells. Food Chem. Toxicol. 62, 908–918.

Iorio, M.V., Piovan, C., Croce, C.M., 2010. Interplay between microRNAs and the epigenetic machinery: an intricate network. Biochim. Biophys. Acta 1799, 694–701.

Ivanovska, N., Philipov, S., 1996. Study on the anti-inflammatory action of berberis vulgaris root extract, alkaloid fractions and pure alkaloids. Int. J. Immunopharmacol. 18, 553–561.

Jeong, M.H., Ko, H., Jeon, H., Sung, G.J., Park, S.Y., Jun, W.J., Lee, Y.H., Lee, J., Lee, S.W., Yoon, H.G., Choi, K.C., 2016. Delphinidin induces apoptosis via cleaved HDAC3-mediated p53 acetylation and oligomerization in prostate cancer cells. Oncotarget 7, 56767–56780.

Jing, P., Giusti, M.M., 2007. Effects of extraction conditions on improving the yield and quality of an anthocyanin-rich purple corn (Zea mays L.) color extract. J. Food Sci. 72, C363–C638.

Jing, P., Giusti, M.M., 2011. In: Seeram, N.P., Stoner, G.D. (Eds.), Contribution of Berry Anthocyanins to Their Chemopreventive Properties, in Berries and Cancer Prevention. Springer, New York, pp. 3–40.

Jing, Q., Huang, S., Guth, S., Zarubin, T., Motoyama, A., Chen, J., Di Padova, F., Lin, S.C., Gram, H., Han, J., 2005. Involvement of microRNA in AU-rich element-mediated mRNA instability. Cell 120, 623–634.

Jones, P.A., Baylin, S.B., 2002. The fundamental role of epigenetic events in cancer. Nat. Rev. Genet. 3, 415–428.

Kai, Z.S., Pasquinelli, A.E., 2010. MicroRNA assassins: factors that regulate the disappearance of miRNAs. Nat. Struct. Mol. Biol. 17, 5–10.

Keppler, K., Humpf, H.U., 2005. Metabolism of anthocyanins and their phenolic degradation products by the intestinal micro- flora. Bioorg. Med. Chem. 13, 5195–5205.

Knobloch, T.J., Uhrig, L.K., Pearl, D.K., Casto, B.C., Warner, B.M., Clinton, S.K., Sardo-Molmenti, C.L., Ferguson, J.M., Daly, B.T., Riedl, K., Schwartz, S.J., Vodovotz, Y., Buchta, A.J., Schuller, D.E., Ozer, E., Agrawal, A., Weghorst, C.M., 2016. Suppression of pro-inflammatory and pro-survival biomarkers in oral cancer patients consuming a black raspberry phytochemical-rich troche. Cancer Prev. Res. 9, 159–171.

Ko, H., Jeong, M.H., Jeon, H., Sung, G.J., So, Y., Kim, I., Son, J., Lee, S.W., Yoon, H.G., Choi, K.C., 2015. Delphinidin sensitizes prostate cancer cells to TRAIL-induced apoptosis, by inducing DR5 and causing caspase-mediated HDAC3 cleavage. Oncotarget 6, 9970–9984.

Kong, J.M., Chia, L.S., Goh, N.K., Chia, T.F., Brouillard, R., 2003. Analysis and biological activities of anthocyanins. Phytochemistry 64, 923–933.

Kouzarides, T., 2007. Chromatin modifications and their function. Cell 128, 693–705.

Kristo, A.S., Klimis-Zacas, D., Sikalidis, A.K., 2016. Protective role of dietary berries in cancer. Antioxidants (Basel) 5, E37.

Kulling, S.E., Rawel, H.M., 2008. Chokeberry (Aronia melanocarpa)-A review on the characteristic components and potential health effects. Planta Med. 74, 1625–1634.

Lagos-Quintana, M., Rauhut, R., Lendeckel, W., Tuschl, T., 2001. Identification of novel genes coding for small expressed RNAs. Science 294, 853–858.

Lehtonen, H.M., Lindstedt, A., Järvinen, R., Sinkkonen, J., Graça, G., Viitanen, M., Kallio, H., Gil, A.M., 2013. 1H-NMR-based metabolic fingerprinting of urine metabolites after consumption of lingonberries (Vaccinium vitis-idaea) with a high-fat meal. Food Chem. 138, 982–990.

Li, B., Carey, M., Workman, J.L., 2007. The role of chromatin during transcription. Cell 128, 707–719.

Liu, J., Zhang, W., Jing, H., Popovich, D.G., 2010. Bog bilberry (*Vaccinium uliginosum* L.) extract reduces cultured Hep-G2, Caco-2, and 3T3-L1 cell viability, affects cell cycle progression, and has variable effects on membrane permeability. J. Food Sci. 75, H103–H107.

Longo, L., Platini, F., Scardino, A., Alabiso, O., Vasapollo, G., Tessitore, L., 2008. Autophagy inhibition enhances anthocyanin-induced apoptosis in hepatocellular carcinoma. Mol. Cancer Ther. 7, 2476–2485.

Lu, H., Li, H., Zhang, D., Stoner, G.D., Chuanshu, H., 2006. Molecular mechanisms involved in chemoprevention of black raspberry extracts: from transcription factors to their target genes. Nutr. Cancer 54, 69–78.

Luger, K., Mader, A.W., Richmond, R.K., Sargent, D.F., Richmond, T.J., 1997. Crystal structure of the nucleosome core particle at 2.8 A resolution. Nature 389, 251–260.

Lujambio, A., Lowe, S.W., 2012. The microcosmos of cancer. Nature 482, 347–355.

Määttä-Riihinen, K.R., Kamal-Eldin, A., Törrönen, A.R., 2004. Identification and quantification of phenoliccompounds in berries of Fragaria and Rubus species (family Rosaceae). J. Agric. Food Chem. 52, 6178–6187.

Madhavi, D.L., Bomser, J., Smith, M.A.L., Singletary, K., 1998. Isolation of bioactive constituents from Vaccinium myrtillus (bilberry) fruits and cell cultures. Plant Sci. 131, 95–103.

Magnuson, B.A., Lala, G., Kwon, Y.J., 2003. Anthocyanin-rich extracts inhibit growth of human colon cancer cells and azoxymethane-induced colonaberrant crypts in rats: implications for colon cancer chemoprevention. Cancer Epidemiol. Biomark. Prev. 12, 1323s–1324s.

Malik, M., Zhao, C., Schoene, N., Guisti, M.M., Moyer, M.P., Magnuson, B.A., 2003. Anthocyanin-rich extract from Aronia meloncarpa E induces a cell cycle block in colon cancer but not normal colonic cells. Nutr. Cancer 46, 186–196.

Mallery, S.R., Zwick, J.C., Pei, P., Tong, M., Larsen, P.E., Shumway, B.S., Lu, B., Fields, H.W., Mumper, R.J., Stoner, G.D., 2008. Topical application of a bioadhesive black raspberry gel modulates gene expression and reduces cyclooxygenase 2 protein in human premalignant oral lesions. Cancer Res. 68, 4945–4957.

Marko, D., Puppel, N., Tjaden, Z., Jakobs, S., Pahlke, G., 2004. The substitution pattern of anthocyanidins affects different cellular signalling cascades regulating cell proliferation. Mol. Nutr. Food Res. 48, 318–325.

Matchett, M.D., MacKinnon, S.L., Sweeney, M.I., Gottschall-Pass, K.T., Hurta, R.A., 2005. Blueberry flavonoids inhibit matrix metalloproteinase activity in DU145 human prostate cancer cells. Biochem. Cell. Biol. 83, 637–643.

Matchett, M.D., MacKinnon, S.L., Sweeney, M.I., Gottschall-Pass, K.T., Hurta, R.A., 2006. Inhibition of matrix metalloproteinase activity in DU145 human prostate cancer cells by flavonoids from lowbush blueberry (Vaccinium angustifolium): possible roles for protein kinase C and mitogen-activated protein-kinase-mediated events. J. Nutr. Biochem. 17, 117–125.

Mattila, P., Hellström, J., Törrönen, R., 2006. Phenolic acids in berries, fruits, and beverages. J. Agric. Food Chem. 54, 7193–7199.

Mazur, W.M., Uehara, M., Wähälä, K., Adlercreutz, H., 2000. Phyto-oestrogen content of berries, and plasma concentrations and urinary excretion of enterolactone after a single strawberry-meal in human subjects. Br. J. Nutr. 83, 381–387.

Misikangas, M., Pajari, A.M., Paivarinta, E., Oikarinen, S.I., Rajakangas, J., Marttinen, M., Tanayama, H., Torronen, R., Mutanen, M., 2007. Three Nordic berries inhibit intestinal tumorigenesis in multiple intestinal neoplasia/+ mice by modulating beta-catenin signaling in the tumor and transcription in the mucosa. J. Nutr. 137, 2285–2290.

Møller, P., Loft, S., Alfthan, G., Freese, R., 2004. Oxidative DNA damage in circulating mononuclear blood cells after ingestion of blackcurrant juice or anthocyanin-rich drink. Mutat. Res. 551, 119–126.

Montrose, D.C., Horelik, N.A., Madigan, J.P., Stoner, G.D., Wang, L.S., Bruno, R.S., Park, H.J., Giardina, C., Rosenberg, D.W., 2011. Anti-inflammatory effects of freeze-dried black raspberry powder in ulcerative colitis. Carcinogenesis 32, 343–350.

Namiesnik, J., Vearasilp, K., Nemirovski, A., Leontowicz, H., Leontowicz, M., Pasko, P., Martinez-Ayala, A.L., González-Aguilar, G.A., Suhaj, M., Gorinstein, S., 2014. In vitro studies on the relationship between the antioxidant activities of some berry extracts and their binding properties to serum albumin. Appl. Biochem. Biotechnol. 172, 2849–2865.

Neto, C.C., 2007. Cranberry and its phytochemicals: a review of in vitro anticancer studies. J. Nutr. 137, 186S–193S.

Nile, S.H., Park, S.W., 2014. Edible berries: bioactive components and their effect on human health. Nutrition 30, 134–144.

Ogawa, K., Kuse, Y., Tsuruma, K., Kobayashi, S., Shimazawa, M., Hara, H., 2014. Protective effects of bilberry and lingonberry extracts against blue light-emitting diode light-induced retinal photoreceptor cell damage in vitro. BMC Complement. Altern. Med. 14, 120.

Ovaskainen, M.L., Törrönen, R., Koponen, J.M., Sinkko, H., Hellström, J., Reinivuo, H., Mattila, P., 2008. Dietary intake and major food sources of polyphenols in Finnish adults. J. Nutr. 138, 562–566.

Paluszczak, J., Krajka-Kuzniak, V., Baer-Dubowska, W., 2010. The effect of dietary polyphenols on the epigenetic regulation of gene expression in MCF7 breast cancer cells. Toxicol. Lett. 192, 119–125.

Pantelidis, G.E., Vasilakakis, M., Manganaris, G.A., Diamantidis, G., 2007. Antioxidant capacity, phenol, anthocyanin and ascorbic acid contents in raspberries, blackberries, red currants, gooseberries and cornelian cherries. Food Chem. 102, 777–783.

Peiffer, D.S., Zimmerman, N.P., Wang, L.S., Ransom, B.W., Carmella, S.G., Kuo, C.T., Siddiqui, J., Chen, J.H., Oshima, K., Huang, Y.W., Hecht, S.S., Stoner, G.D., 2014. Chemoprevention of esophageal cancer with black raspberries, their component anthocyanins, and a major anthocyanin metabolite, protocatechuic acid. Cancer Prev. Res. (Phila) 7, 574–584.

Pogribny, I.P., James, S.J., Beland, F.A., 2012. Molecular alterations in hepatocarcinogenesis induced by dietary methyl deficiency. Mol. Nutr. Food 56, 116–125.

Pogribny, I.P., James, S.J., Jernigan, S., Pogribna, M., 2004. Genomic hypomethylation is specific for preneoplastic liver in folate/methyl deficient rats and does not occur in non-target tissues. Mutat. Res. Fundam. Mol. Mech. Mutagen. 548, 53–59.

Poulose, S.M., Fisher, D.R., Larson, J., Bielinski, D.F., Rimando, A.M., Carey, A.N., Schauss, A.G., Shukitt-Hale, B., 2012. Anthocyanin-rich açai (Euterpe oleracea Mart.) fruit pulp fractions attenuate inflammatory stress signaling in mouse brain BV-2 microglial cells. J. Agric. Food Chem. 60, 1084–1093.

Pratheeshkumar, P., Son, Y.O., Wang, X., Divya, S.P., Joseph, B., Hitron, J.A., Wang, L., Kim, D., Yin, Y., Roy, R.V., Lu, J., Zhang, Z., Wang, Y., Shi, X., 2014. Cyanidin-3-glucoside inhibits UVB-induced oxidative damage and inflammation by regulating MAP kinase and NF-κB signaling pathways in SKH-1 hairless mice skin. Toxicol. Appl. Pharmacol. 280, 127–137.

Probst, Y., 2015. A review of the nutrient composition of selected Rubus berries. Nutr. Food Sci. 45, 242–254.

Santos-Buelga, C., Scalbert, A., 2000. Proanthocyanidins and tannin-like compounds—nature, occurrence, dietary intake and effects on nutrition and health. J. Sci. Food Agric. 80, 1094–1117.

Schantz, M., Mohn, C., Baum, M., Richling, E., 2010. Antioxidative efficiency of an anthocyanin rich bilberry extract in the human colon tumor cell lines Caco-2 and HT-29. J. Berry Res. 1, 25–33.

Seeram, N.P., 2008. Berry fruits for cancer prevention: current status and future prospects. J. Agric. Food Chem. 56, 630–635.

Seeram, N.P., Adams, L.S., Zhang, Y., Lee, R., Sand, D., Scheuller, H.S., Heber, D., 2006. Blackberry, black raspberry, blueberry, cranberry, red raspberry, and strawberry extracts inhibit growth and stimulate apoptosis of human cancer cells in vitro. J. Agric. Food Chem. 54, 9329–9339.

Shi, N., Chen, F., Zhang, X., Clinton, S.K., Tang, X., Sun, Z., Chen, T., 2017. Suppression of oxidative stress and NFκB/MAPK signaling by lyophilized black raspberries for esophageal cancer prevention in rats. Nutrients 9, E413.

Shi, N., Riedl, K.M., Schwartz, S.K., Zhang, X., Clinton, S.K., Chen, T., 2016. Efficacy comparison of lyophilised black raspberries and combination of celecoxib and PBIT in prevention of carcinogen-induced oesophageal cancer in rats. J. Funct. Food 27, 11.

Shi, Y., 2007. Histone lysine demethylases: emerging roles in development, physiology and disease. Nat. Rev. Genet. 8, 829–833.

Skrovankova, S., Sumczynski, D., Mlcek, J., Jurikova, T., Sochor, J., 2015. Bioactive compounds and antioxidant activity in different types of berries. Int. J. Mol. Sci. 16, 24673–24706.

Slatnar, A., Jakopic, J., Stampar, F., Veberic, R., Jamnik, P., 2012. The effect of bioactive compounds on in vitro and in vivo antioxidant activity of different berry juices. PLoS One 7, e47880.

Srivastava, A., Akoh, C.C., Fischer, J., Krewer, G., 2007. Effect of anthocyanin fractions from selected cultivars of Georgia-grown blueberries on apoptosis and phase II enzymes. J. Agric. Food Chem. 55, 3180–3185.

Stefanska, B., Karlic, H., Varga, F., Fabianowska-Majewska, K., Haslberger, A., 2012. Epigenetic mechanisms in anti-cancer actions of bioactive food components—the implications in cancer prevention. Br. J. Pharmacol. 167, 279–297.

Stoner, G.D., Wang, L., Sardo, C., Zikri, N., Hecht, S.S., Mallery, S.R., 2010. Cancer prevention with berries: role of anthocyanins. Nutr. Health 703–723. Part of the series Nutrition and Health.

Stoner, G.D., Wang, L.S., Casto, B.C., 2008. Laboratory and clinical studies of cancer chemoprevention by antioxidants in berries. Carcinogenesis 29, 1665–1674.

Subar, A.F., Block, G., James, L.D., 1989. Folate intake and food sources in the US population. Am. J. Clin. Nutr. 50, 508–516.

Szajdek, A., Borowska, E.J., 2008. Bioactive compounds and health-promoting properties of berry fruits: a review. Plant Foods Hum. Nutr. 63, 147–156.

Teller, N., Thiele, W., Marczylo, T.H., Gescher, A.J., Boettler, U., Sleeman, J., Marko, D., 2009. Suppression of the kinase activity of receptor tyrosine kinases by anthocyanin-rich mixtures extracted from bilberries and grapes. J. Agric. Food Chem. 57, 3094–3101.

Tremethick, D.J., 2007. Higher-order structures of chromatin: the elusive 30 nm fiber. Cell 128, 651–654.

Wang, L.S., Arnold, M., Huang, Y.W., Sardo, C., Seguin, C., Martin, E., Huang, T.H., Riedl, K., Schwartz, S., Frankel, W., Pearl, D., Xu, Y., Winston, J., Yang, G.Y., Stoner, G., 2011. Modulation of genetic and epigenetic biomarkers of colorectal cancer in humans by black raspberries: a phase I pilot study. Clin. Cancer Res. 17, 598–610.

Wang, L.S., Burke, C.A., Hasson, H., Kuo, C.T., Molmenti, C.L., Seguin, C., Liu, P., Huang, T.H., Frankel, W.L., Stoner, G.D., 2014. A phase Ib study of the effects of black raspberries on rectal polyps in patients with familial adenomatous polyposis. Cancer Prev. Res. (Phila) 7, 666–674.

Wang, L.S., Hecht, S.S., Carmella, S.G., Yu, N., Larue, B., Henry, C., McIntyre, C., Rocha, C., Lechner, J.F., Stoner, G.D., 2009. Anthocyanins in black raspberries prevent esophageal tumors in rats. Cancer Prev. Res. (Phila) 2, 84–93.

Wang, L.S., Kuo, C.T., Cho, S.J., Seguin, C., Siddiqui, J., Stoner, K., Weng, Y.I., Huang, T.H.M., Tichelaar, J., Yearsley, M., 2013c. Black raspberry-derived anthocyanins demethylate tumor suppressor genes through the inhibition of DNMT1 and DNMT3B in colon cancer cells. Nutr. Cancer 65, 118–125.

Wang, L.S., Kuo, C.T., Huang, T.H.M., Yearsley, M., Oshima, K., Stoner, G.D., Yu, J., Lechner, J.F., Huang, Y.W., 2013a. Black raspberries protectively regulate methylation of wnt pathway genes in precancerous colon tissue. Cancer Prev. Res. 6, 1317–1327.

Wang, L.S., Kuo, C.T., Stoner, K., Yearsley, M., Oshima, K., Yu, J., Huang, T.H.M., Rosenberg, D., Peiffer, D., Stoner, G., 2013b. Dietary black raspberries modulate DNA methylation in dextran sodium sulfate (DSS)-induced ulcerative colitis. Carcinogenesis 34, 2842–2850.

Wang, S.Y., Chen, C.T., Wang, C.Y., Chen, P., 2007. Resveratrol content in strawberry fruit is affected by preharvest conditions. J. Agric. Food Chem. 55, 8269–8274.

Wilson, A.J., Byun, D.S., Popova, N., Murray, L.B., L'Italien, K., Sowa, Y., Arango, D., Velcich, A., Augenlicht, L.H., Mariadason, J.M., 2006. Histone deacetylase 3 (HDAC3) and other class I HDACs regulate colon cell maturation and p21 expression and are deregulated in human colon cancer. J. Biol. Chem. 281, 13548–13558.

Winter, J., Jung, S., Keller, S., Gregory, R.I., Diederichs, S., 2009. Many roads to maturity: microRNA biogenesis pathways and their regulation. Nat. Cell Biol. 11, 228–234.

Wolffe, A.P., Matzke, M.A., 1999. Epigenetics: regulation through repression. Science 286, 481–486.

Xu, M., Bower, K.A., Wang, S., Frank, J.A., Chen, G., Ding, M., Wang, S., Shi, X., Ke, Z., Luo, J., 2010. Cyanidin-3-glucoside inhibits ethanol-induced invasion of breast cancer cells overexpressing ErbB2. Mol. Cancer 29, 285.

Yoo, C.B., Jones, P.A., 2006. Epigenetic therapy of cancer: past, present and future. Nat. Rev. Drug Discov. 5, 37–50.

Zhao, C., Giusti, M.M., Malik, M., Moyer, M.P., Magnuson, B.A., 2004. Effects of commercial anthocyanin-rich extracts on colonic cancer and nontumorigenic colonic cell growth. J. Agric. Food Chem. 52, 6122–6128.

Zhou, H., Hu, H., Lai, M., 2010. Non-coding RNAs and their epigenetic regulatory mechanisms. Biol. Cell 102, 645–655.

Zikri, N.N., Riedl, K.M., Wang, L.S., Lechner, J., Schwartz, S.J., Stoner, G.D., 2009. Black raspberry components inhibit proliferation, induce apoptosis, and modulate gene expression in rat esophageal epithelial cells. Nutr. Cancer 61, 816–826.

CHAPTER

7

Isothiocyanates, Epigenetics, and Cancer Prevention

Silvia Novío, María Jesús Núñez-Iglesias, Manuel Freire-Garabal

SNL Laboratory, School of Medicine and Dentistry, University of Santiago de Compostela, Santiago de Compostela, Spain

Abstract

Isothiocyanates (ITCs) are the most studied glucosinolate-derived bioactive food components in terms of bioactivity. Various research has emphasized the chemopreventive properties of ITCs on breast, lung, colorectal, and prostate cancer, the four most common cancers worldwide. Although multiple mechanisms are involved in the anticancer effects of ITCs, until quite recently little attention has focused on their ability to influence the epigenetic processes that lead to correction of aberrant epigenetic changes in cancer cells. This chapter is intended to review evidence of the impact of ITCs on epigenetic regulatory mechanisms (DNA methylation, histone posttranslational modifications, and microRNA expression) and how their effects regulate the expression of mediators involved in various signaling pathways, including those involved in apoptosis and cell cycle regulation of breast, lung, colorectal, and prostate cancer cells.

Keywords

DNA methylation; Epigenetic; Histone modifications; Isothiocyanate; microRNA

INTRODUCTION

Cancer remains a public health concern worldwide. It is the second-leading cause of death globally. Breast, colorectal, and lung cancer are the most frequent types of cancer in women, while lung, prostate, and colorectal cancer are the most common among men. The incidence of cancer is expected to rise by about 70% over the next two decades. This global increase in the cancer burden is mainly attributed to, among other factors, aging and exposure to modifiable risk factors, such as diet (Thun et al., 2010). In fact, one-third of cancers are attributable to nutrition, hence rendering the disease highly preventable (30%–50%) by avoiding or modifying risk factors and implementing prevention strategies.

Carcinogenesis develops in three stages: initiation, promotion, and progression (Anisimov, 2009) (Fig. 7.1). The first stage results from

Epigenetics of Cancer Prevention
https://doi.org/10.1016/B978-0-12-812494-9.00007-X

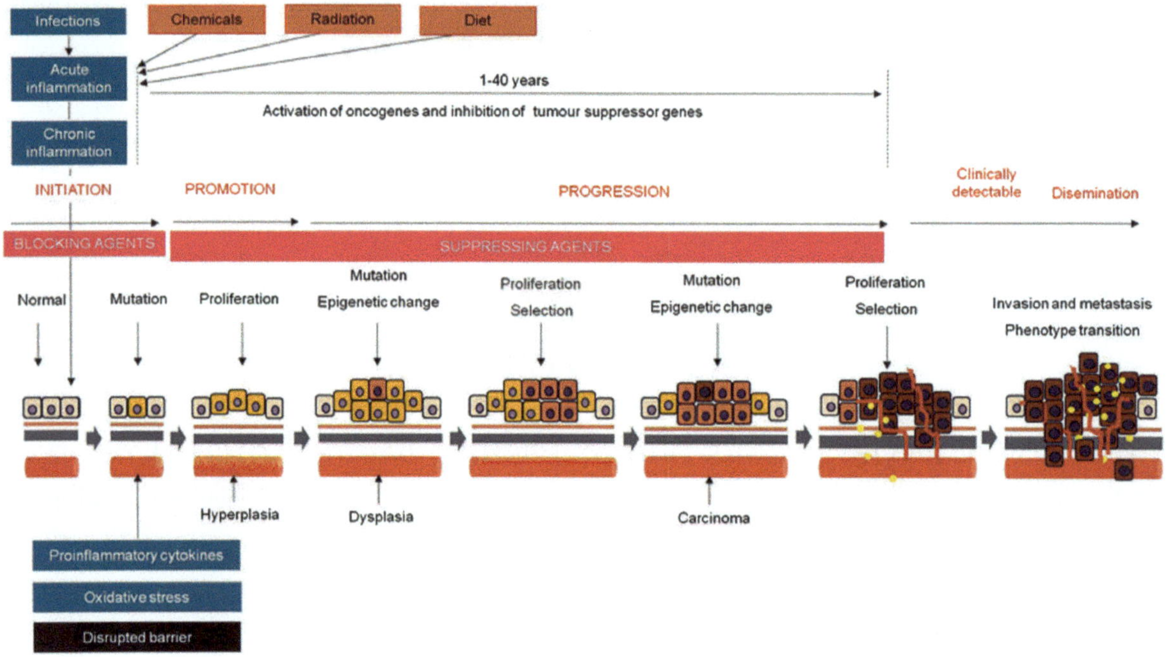

FIGURE 7.1 Various stages of cancer. Carcinogenesis is a multistep process that includes initiation, promotion, and progression. Genetic alterations related to activation of oncogenes and inhibition of tumor-suppressor genes, together with inflammation and oxidative stress, lead to genomic instability that results in the formation of cancer. Bioactive food components, such as isothiocyanates, can prevent or reverse the changes in each step of this process, acting as "blocking agents" during initiation and "suppressing agents" during promotion and progression.

irreversible genetic damage induced by chemical, physical, and biological carcinogens that leads to the activation of oncogenes and/or inactivation of tumor-suppressor genes. The second stage, promotion, is a reversible process that involves changes in the expression of the genome mediated through sustained interactions between carcinogens and receptors. Finally, progression is an irreversible stage characterized by malignant growth, with invasion and/or metastasis. The reversibility of promotion supports the implication of epigenetic mechanisms in the etiology of cancer. Thus, nowadays it is known that genetic alterations (e.g., deletions, aneuploidy, and chromosome rearrangements) cooperate with epigenetics throughout the carcinogenesis process (Fuentes et al., 2015; Kaufman-Szymczyk et al., 2015). This finding has given rise to the development of therapeutic and preventive strategies targeting the epigenome (Basse and Arock, 2015).

Many attempts have been made to discover epigenetic drugs (epi-drugs). Synthetic epi-drugs, such as 5-aza-2′-deoxycytidine (5′-Aza, decitabine), are used in cancer patients to reverse abnormal epigenetic changes (Halpern et al., 2017). However, its toxicity, lack of specificity, and delivery-related problems (Sinha et al., 2015), together with the fact that certain foods seem to have a protective role against cancer (Gao and Tollefsbol, 2015; Hardy and Tollefsbol, 2011), have aroused great interest in natural compounds. In this sense, bioactive food components (BFCs) have been shown to modulate molecular targets involved in the initiation, promotion, and progression of cancer by epigenetic mechanisms. Accordingly, BFCs with epigenetic properties are known as "epigenetic diets" (Li et al., 2016) or "nutri-epigenetics" (Gianfredi et al., 2017). Among BFCs, cruciferous vegetables, which are rich sources of isothiocyanates (ITCs), seem promising as dietary phytochemicals for improving current cancer treatment (Fuentes et al., 2015; Gao and Tollefsbol, 2015; Hardy and Tollefsbol, 2011).

In this chapter, we focus on the chemopreventive activity of ITCs as epigenome modifiers in breast, prostate, lung, and colorectal cancers.

CANCER EPIGENETICS

The field of epigenetics has rapidly expanded since the 1940s. It is defined as heritable but potentially reversible changes in gene expression and phenotype that are not accompanied by changes in underlying nucleotide sequence (Momparler, 2003). Since epigenetic modifications are vital in regulating translation and degradation of messenger RNA (mRNA), transcription of genes, DNA repair, and replication (Clarke et al., 2011; Ho et al., 2009), abnormal epigenetic changes may lead to the induction and maintenance of cancer (Dawson and Kouzarides, 2012), even with phenotypes very similar to those of hereditary cancers (Lips et al., 2013).

The major epigenetic processes are aberrations in DNA methylation (hypermethylation and hypomethylation), histone posttranslational modifications (HPMs)—e.g., histone acetylation/deacetylation and methylation—and microRNA (miRNA) expression (Esteller, 2008).

Aberrations in DNA Methylation

Aberrations in DNA methylation were the first epigenetic modification to be observed in cancer cells, and nowadays they are the best characterized epigenetic hallmark evidenced in cancer (Robertson, 2005). In general, methylation changes in DNA include hypermethylation and loss of methylation, which cause silencing of tumor-suppressor genes and the activation of oncogenes, respectively (Esteller, 2008). Hypermethylation has been far more deeply studied in cancer chemoprevention than other DNA aberrations.

DNA methylation refers to the addition of a methyl group (CH_3) at the 5′ cytosine residues within CpG dinucleotides. This reaction is catalyzed by DNA methyltransferases (DNMTs), which include DNMT1 (also referred as the "maintenance" DNMT), DNMT3a, and DNMT3b (Li and Tollefsbol, 2010; Meeran et al., 2010a). Whereas DNMT1 maintains methylation through cell division, DNMT3a and DNMT3b are responsible for de novo methylation of DNA during development (Law and Jacobsen, 2010). Overexpression of DNMTs has been observed in many cancers (Lin and Wang, 2014).

Short regions of 0.5–3 kb in length, known as CpG islands (Antequera and Bird, 1993), are rich in CpG content. In normal cells, the majority of CpG dinucleotides in the genome are methylated, whereas CpG islands often remain unmethylated. On the contrary, in cancer cells, hypermethylation of CpG dinucleotides and islands occurs and usually results in transcriptional gene silencing and gene inactivation, including tumor-suppressor genes, genes encoding hormone receptors, and DNA repair enzymes (Esteller, 2008; Hun Lee et al., 2013).

DNA methylation changes have already been thoroughly described for breast (Tang et al., 2016; Widschwendter et al., 2017), lung (Ansari et al., 2016; Dong et al., 2017; Duruisseaux and Esteller, 2017), colorectal (Savio and Bapat, 2015; Tse et al., 2017; Weisenberger et al., 2018), and prostate (Chiam et al., 2014; Savio and Bapat, 2015) cancers. This deep knowledge has led to the identification of cancer type-specific hypermethylation profiles that can be used like biomarkers for the diagnosis, prognosis, and prediction of response to therapy (Basse and Arock, 2015; Chiam et al., 2014; Duruisseaux and Esteller, 2017; Pan et al., 2013; Puccini et al., 2017).

Histone Posttranslational Modifications

Histone proteins (core histones, such as H2A, H2B, H3, and H4, and linker histones, such as H1) pack DNA into nucleosomes, fundamental units of the chromatin. H2A, H2B, H3, and H4 have N-terminal tails that are the sites of posttranslational modifications, such as acetylation, methylation, phosphorylation, biotinylation, and ubiquitination (Kouzarides, 2007; Strahl and Allis, 2000). Although most of these modifications are reversible (Davis and Ross, 2007), all of them alter the accessibility of DNA to transcriptional activity (see next paragraph). The most important HPMs that affect gene expression are those located on the lysine and arginine residues of H3 and H4 (Bronner et al., 2010).

Acetylation and methylation are by far the most thoroughly studied modifications. Acetylation/deacetylation and methylation/demethylation are phenomena regulated by the enzymes histone acetyltransferase (HAT)/histone deacetylase (HDAC) and histone methyltransferase (HMT)/histone demethylase (HDM) (Kouzarides, 2007), respectively. Whereas acetylation of lysine residues of the N-terminal tail leads to a loosely packed chromatin that is accessible to transcription factors and is transcriptionally active, deacetylation leads to a tightly packed chromatin that is inaccessible to transcription factors and transcriptionally inactive (Li et al., 2017b; Meeran et al., 2012). Similarly, histone methylation has been shown to be another mechanism of activation (e.g., methylation of H3 at lysine 4 (K4) or methylation of histones at arginine residues) and repression (e.g., methylation of H3 at lysine 9 (K9)) of gene expression (Esteller, 2008).

A disrupted balance of histone acetylation and/or methylation regulated by HATs/HDACs and HMTs/HDMs, respectively, can lead to increased risk of cancer. Thus, HPMs linked to cancer are also known as "histone onco-modifications" (Füllgrabe et al., 2011). These modifications have been shown to dysregulate pathways of cell differentiation, cell cycle, and apoptosis in different types of cancer, such as those of the breast (Cang et al., 2014; Lewinska et al., 2017; Li et al., 2016, 2017b; Liu et al., 2013b; Meeran et al., 2010b, 2012; Royston et al., 2017; Sinha et al., 2015), colorectal (Liu et al., 2013c; Myzak et al., 2004; Park et al., 2017; Rajendran et al., 2013), lung (Berry and Janknecht, 2013; El-Khoury et al., 2004), and prostate (Beklemisheva et al., 2006; Boyanapalli et al., 2016; Clarke et al., 2011; Myzak et al., 2006a,b; Wang et al., 2007; Zhang et al., 2013).

microRNAs

miRNAs are small noncoding RNAs (18–25 nucleotides in length) that regulate the expression of at least one-third of all protein-coding genes in humans (Rajewsky, 2006) by modulating protein translation or enhancing degradation of mRNAs (Maia et al., 2014). Their biological actions are cell stage-dependent. Although in general, miRNAs are translational repressors in dividing cells (Bueno and Malumbres, 2011), under specific conditions they can promote the translation of their target (Vasudevan et al., 2007), especially in quiescent cells (Zhao and Liu, 2009).

The regulatory effects of miRNAs on human gene expression have aroused interest in studying their application in the treatment of human diseases, including cancer. Nowadays it is known that miRNAs play important roles in the development and progression of cancers (Tomasetti et al., 2014) by modulating the activity of cancer stem cells (CSCs) (Bao et al., 2012) and differentiated cells (Zhang et al., 2016). Dysregulated miRNA levels have been linked to nearly every type of human cancer (Calin and Croce, 2006). Thus, high-quality articles have been published on the functions of miRNAs in colorectal (Dacosta and Bao, 2017), breast (Lewinska et al., 2017; Li et al., 2014b), lung (Li et al., 2015b, 2017a; Zhu et al., 2017), and prostate (Yu et al., 2013; Zhang et al., 2016) cancers.

Considering the foregoing, compounds with the capability to demethylate CpG (e.g., inhibiting DNMT function), target HAT/HDAC and HMT/HDM activities, or modulate miRNA expression might be useful in cancer treatment.

ISOTHIOCYANATES AS EPIGENETIC MODIFIERS IN CANCER

Glucosinolates are secondary metabolites found in Brassicaceae and related families. Brassicaceae vegetables are widely cultivated (Ishida et al., 2014). The content of glucosinolate in these vegetables depends on many factors, such as species and cultivar, growing conditions, climate, physiological age, and tissue type (Cartea et al., 2008; Fenwick et al., 1983; Mithen et al., 2000).

Depending on the amino acid precursor (methionine, tryptophan, tyrosine, or phenylalanine), glucosinolates can be grouped into three chemical classes: aliphatic, indole, and aromatic. Glucosinolates themselves are biologically inert compounds; however, in the presence of myrosinase (b-thioglucosidase, and enzyme present in plants and intestinal microbes), they undergo hydrolysis and chemical rearrangement to generate bioactive compounds, such as ITCs, the most studied of known glucosinolate-derived BFCs in terms of bioactivity (Fahey et al., 2001). In recent decades, ITCs have received growing attention as therapeutic compounds to be used in medicine because in contrast to synthetic drugs, they have low or no toxicity and side effects, are readily available, and can be taken continuously (Saha et al., 2013).

ITCs have been shown effective in treating a wide range of conditions, including cancer. Concretely, ITCs have been studied extensively in experimental in vitro and in vivo carcinogenesis models for their chemopreventive properties on breast, lung, colorectal, and prostate cancers, with the conclusion that the ITC sulforaphane (SFN) is among the most important BFCs that have anticancer properties (Hardy and Tollefsbol, 2011). In addition, several authors emphasize the role of ITCs at the initiation stage (Li et al., 2016; Wang et al., 2007). Different in vitro studies have shown that benzyl ITC (BITC) induces apoptosis in human breast (Kim et al., 2011), prostate (Lin et al., 2017), and lung (Liu et al., 2013a) cancer cells and inhibits migration and invasion in colon cancer cells (Lai et al., 2010). Likewise, in vivo studies have provided evidence that ITC treatment prolongs tumor-free survival times (Aras et al., 2013), decreases tumor incidence (Aras et al., 2013; Myzak et al., 2006a), and inhibits the growth of tumors (Park et al., 2017).

Multiple interactive mechanisms are involved in the anticancer effects of ITCs, with the majority focusing on ITCs as regulators of antioxidant (Atwell et al., 2015) and anti-inflammatory (Atwell et al., 2015) pathways. Until quite recently, however, little attention has been focused on ITCs' ability to influence epigenetic processes that lead to the correction of aberrant epigenetic changes in cancer cells. The ability of ITCs to reverse abnormal epigenetic alterations that are fundamental to the initiation, promotion, and progression of carcinogenesis suggests that consumption of a diet rich in cruciferous vegetables could become a pertinent, nontoxic, and economical option for preventing the occurrence of cancer.

Available evidence indicates that the chemopreventive activities of ITCs in cancer involve the epigenetic regulation (through the three main mechanisms highlighted in the previous section) of several processes, including cell cycle control (Boyanapalli et al., 2016; Liu et al., 2013b), proliferation (Royston et al., 2017), apoptosis (Boyanapalli et al., 2016), autophagy (Lewinska et al., 2017), cell–cell interaction (Wong et al., 2014), migration (Wong et al., 2014), cell adhesion (Wong et al., 2014), DNA repair (Barrera et al., 2013), redox (Wang et al.,

2007), and inflammation (Fuentes et al., 2015; Liu and Dey, 2017) under both in vitro and in vivo conditions (Table 7.1).

DNA Methylation

Silencing of tumor-suppressor genes mediated by DNA methylation is a phenomenon commonly observed in different cancers (Hsu et al., 2011; Kaufman-Szymczyk et al., 2015; Li and Tollefsbol, 2010), and thus their reactivation has gained much attention in cancer research. Boyanapalli et al. (2016) observed that when LNCaP cells are treated with phenethyl ITC (PEITC, 2.5 or 5 μM), ras association domain family 1 isoform A (RASSF1A) expression, which is frequently inactivated in advanced stages of prostate cancer (Kuzmin et al., 2002), is reactivated at least partly by inducing demethylation. The restored expression of RASSF1A causes G2/M cell cycle arrest and apoptosis (Boyanapalli et al., 2016) and could even improve the response to standard treatments (Gupta and Srivastava, 2012), because it is known that human epidermal growth factor receptor 2 (i.e., HER2) overexpression is linked to hypermethylation of RASSF1A (Cho et al., 2012).

One mechanism involved in tumor-suppressor gene silencing is attributed to overexpression of DNMTs. Different approaches to modulate DNMT expression have been described (Li et al., 2013); however, the most successful DNMT inhibitors to date have important limitations: they are not selective toward different DNMTs, are unstable, have limited activity, and have been associated with serious side effects (Schiffmann et al., 2016; Wijermans et al., 2005). Therefore, it is necessary to identify more specific and selective inhibitors.

ITCs inhibit DNMT enzyme activity as shown in different models of breast (Meeran et al., 2010b; Shankar et al., 2013), colorectal (Liu et al., 2013c; Traka et al., 2005), and prostate (Chen et al., 2016; Hsu et al., 2011; Wong et al., 2014) cancer (no evidence for lung cancer). This inhibition seems to be responsible for at least part of cell cycle arrest (Myzak et al., 2004), apoptosis induction (Royston et al., 2017; Boyanapalli et al., 2016), and stimulation of antioxidant (Zhang et al., 2013) and anti-inflammatory (Liu et al., 2013c) defense mechanisms. Unlike those of other dietary phytochemicals, these effects are caused by low micromolar concentrations of ITCs that do not affect normal cells (Lewinska et al., 2017), and thus their effects are specific to cancer cells.

Among the anticancer effects of ITCs mediated by modulation of DNA methylation, antioxidant effects have probably received the most concentrated attention. These effects have been observed with all ITCs tested [SFN, PEITC, BITC, phenyl hexyl ITC, iberin (IB), and allyl ITC]. For example, SFN (1–2.5 μM) has been shown to restore the expression of nuclear factor erythroid-2 related factor-2 (Nrf2), a key regulator in the cellular defense system against oxidative stress (Zhang et al., 2013); as well, it has restored the expression of its downstream gene NAD(P)H quinone dehydrogenase 1 in TRAMP C1 cells through the suppression of DNMT1 and DNMT3a. This SFN-mediated detoxifying activity has also been observed with PETIC, which reactivates antioxidant enzymes that are frequently underexpressed in cancer. Wang et al. (2007) observed that after treating both androgen-dependent and androgen-independent human prostate cancer cells for 5 days with PETIC, the methylation of the CpG island of GSTP1 reached the level found in normal prostatic cells, the effect being similar to or higher than that caused by 5′-Aza.

Sometimes the ability of ITCs to reverse cancer-associated DNA methylation alterations is limited. Aberrantly methylated genes involved in cell adhesion and chemotaxis, as well as in immune functions and inflammatory pathways (Wong et al., 2014), have been associated with cancer progression and recurrence (D'Antonio et al., 2010) and are a finding of bad prognosis (Terada et al., 2012). When cancer cells with this methylation profile are treated

TABLE 7.1 In Vitro and In Vivo Effects of Isothiocyanates on DNA Methylation, Histone Modifications, and miRNA Mechanisms in Breast, Colorectal, Lung, and Prostate Cancer

Epigenetic Mechanism	Epigenetic Modification (Observed or Analyzed)	Isothiocyanates	Target	Experimental Model	References
Inhibition of DNMT1 and DNMT3a Inhibition of HDAC Increase of RBP2	Hyperacetylation and demethylation in hTERT promoter region	SFN	hTERT, MAD1, c-MYC, CTCF	In vitro: MCF-7 and MDA-MB-231 cells	Meeran et al. (2010b)
Inhibition in global HDAC activity	No significant changes in the acetylation of H3, H4 or α-tubulin	SFN	ERα, EGFR, EGFR-2 (HER-2), cyclin B1, cyclin D1, caspase-3 caspase-8, caspase-9, PARP, Bcl-2, cytochrome *c*	In vitro: MDA-MB-231, MDA-MB-468, MCF-7 and T47D cells	Pledgie-Tracy et al. (2007)
Inhibition of DNMT1 and HDAC1	Increase in acetylation of H3 Hypermethylation of promoter regions of MAML2, PON1, PBX1, MAGED1, SLCO3A1, CCKBR, SEPT9, GRB10, IPO7, CCNY, EPHB1, RDH10 Hypomethylation of promoter regions of PSMA7, PRKCA, MIR196B, MEM198, ALCAM, CD166, DCBLD2, MSI2, RPS19, TRMT12	SFN (alone or in combination with EGCG)	MAML2, PON1, PBX1, MAGED1, SLCO3A1, CCKBR, SEPT9, GRB10, IPO7, CCNY, EPHB1, RDH10, PSMA7, PRKCA, MIR196B, MEM198, ALCAM, CD166, DCBLD2, MSI2, RPS19, TRMT12	In vitro: precancerous breast cells and breast cancer cells transformed by transfection In vivo: subcutaneous xenograft (breast cancer cells transformed by transfection)	Li et al. (2016)
Inhibition of DNMT1 Inhibition of HDAC1	Hyperacetylation and hypermethylation of promoter regions of ERα Alteration of the binding of transcriptional corepressors/coactivators to the ERα promoter	SFN (alone or in combination with EGCG and/or tamoxifen)	ERα	In vitro: MCF-7, MDA-MB-231 and MDA-MB-157 cells In vivo: Orthotopic assay (MDA-MB-231 cells)	Li et al. (2017b)
Inhibition of DNMT1	Decrease of methylation in promoter regions of PTEN and RARβ2	SFN (alone or in combination with clofarabine)	PTEN, RARβ2, p53, p21	In vitro: MDA-MB-231 and MCF-7 cells	Lubecka-Pietruszewska et al. (2015)

Continued

TABLE 7.1 In Vitro and In Vivo Effects of Isothiocyanates on DNA Methylation, Histone Modifications, and miRNA Mechanisms in Breast, Colorectal, Lung, and Prostate Cancer—cont'd

Epigenetic Mechanism	Epigenetic Modification (Observed or Analyzed)	Isothiocyanates	Target	Experimental Model	References
Upregulation of miR-140	Demethylation in WWP2 promoter regions	SFN	SOX9, ALDH1, WWP2	In vitro: MCF10DCIS, SUM225CWN, SUM102PT, MCF-10A and breast tumor samples from patients with ductal carcinoma in situ In vivo: Nude mice (MCF10DCIS stemlike cells)	Li et al. (2014b)
Inhibition of HAT (only green tea polyphenols) Inhibition of HDAC1, HDAC4, HDAC6 Inhibition of SUV39H1 Inhibition of DNMTs	Increased acetylation and decreased methylation of ERα promoter Global DNA hypomethylation	SFN (alone or in combination with green tea polyphenols and/or tamoxifen)	ERα	In vitro: MDA-MB-231, MDA-MB-453 and MCF-7 cells	Meeran et al. (2012)
	Demethylation on p21 and KLOTHO promoters Acetylation at promoter regions of KLOTHO and p21	SFN (alone or in combination with green tea polyphenols)	Cyclin B1, CDK1, CDC25C, ER, p21, KLOTHO, IGF-1 receptor, ER1/2, Akt	In vitro: MDA-MB-231 and MDA-MB-453 cells	Sinha et al. (2015)
Inhibition of HDAC1 Inhibition of DNMT1, DNMT3a, DNMT3b		SFN (alone or in combination with withaferin A)	Bax, Bcl-2	In vitro: MDA-MB-231, MCF-7 and T47D cells	Royston et al. (2017)
Probably inhibition of HDACs		PEITC (alone or in combination with paclitaxel)	N.S.	In vitro: MDA-MB-231 and MCF-7 cells	Liu et al. (2013b)
Probably inhibition of HDAC6	Increased acetylation of α-tubulin	PEITC (alone or in combination with paclitaxel)	Cyclin B1, CDK1, Bax, Bcl-2, PARP	In vitro: MDA-MB-231 and MCF-7 cells	Cang et al. (2014)
Possibly inhibition of HDACs	Demethylation of CAV1 promoter region	SFN	CAV1	In vitro: MCF-7 and MDA-MB-231 cells	Deb et al. (2014)
Increased exosomal miR-140 Decreased exosomal miR-21 and miR-29		SFN	ALDH1	In vitro: MCF10DCIS, MCF-7 and MDA-MB-231 cells	Li et al. (2014a)

Downregulation of miR-23b, miR-92b, miR-381, miR-382 Inhibition of DNMT1, DNMT3a and DNMT3b Upregulation of HDAC5 and downregulation of HDAC2, HDAC3, HDAC4, HDAC6, HDAC7, HDAC8, HDAC9, HDAC10, SIN3A, SAP18, SAP30, RBBP4, RBBP7, MECP2, MBD2, MBD3, MBD4, MBD7	Global DNA hypomethylation	SFN	Cyclins (A2, B1, B2, D1, D2, D3, E1, H), TGFβ2, TGFβ3, p53, p21, p27, ERK1/2	In vitro: SK-BR-3, MCF-7 and MDA-MB-231 cells	Lewinska et al. (2017)
	Hypermethylation of promoter regions of IL8, CCL2 and MMP7 Hypoacetylation of promoter regions of IL8 and MMP7	PEITC	MMP7, MMP9, CCL2, CD40, CSF2, CXCL10, IL8, REL, STAT1, NFkB, TNFαip3	In vitro: SW480 and HT-29 cells	Liu et al. (2013c)
Inhibition of HDAC	Global histone (H3 and H4) acetylation	SFN	p21, Bax	In vivo: Apc^{min} mice colon cancer cells	Myzak et al. (2006a)
Inhibition of DNMT1		SFN	N.S.	In vitro: Caco-2 cells	Traka et al. (2005)
Downregulation of DNMT3a and DNMT3b	Unaffected methylation of CpG islands of p16 and ERα	SFN (alone or in combination with selenium)	p16, ERα	In vitro: Caco-2 and HCT116 cells	Barrera et al. (2013)
Transient upregulation of DNMT1 and downregulation of DNMT3a	Unaffected methylation of CpG islands of p16, ERα, APC and MGMT	IB (alone or in combination with selenium)	p16, ERα	In vitro: Caco-2 and HCT116 cells	Barrera et al. (2013)
Inhibition of HDAC1	Enhanced acetylation of H3 and H4 Increased acetylation of p21 promoter regions	SFN (alone or in combination with trichostatin A)	p21	In vitro: HCT116 cells	Myzak et al. (2004)
Inhibition of HDAC3 and HDAC6	Increased acetylation of CtIP	SFN	N.S.	In vitro: HCT116, HT-29, SW48 and SW480 cells	Rajendran et al. (2013)
No effect on HDAC3, HDAC6, HDAC1, HDAC2	No increased acetylation of CtIP	AITC	N.S.	In vitro: HCT116, HT-29, SW48 and SW480 cells	Rajendran et al. (2013)

Continued

TABLE 7.1 In Vitro and In Vivo Effects of Isothiocyanates on DNA Methylation, Histone Modifications, and miRNA Mechanisms in Breast, Colorectal, Lung, and Prostate Cancer—cont'd

Epigenetic Mechanism	Epigenetic Modification (Observed or Analyzed)	Isothiocyanates	Target	Experimental Model	References
Modulation of PRC, HMTs, HATs, HDACs and LSDs	Hypomethylation of PRCS target genes (SPG20, PAK7, PCDH10, HNF4A, VWC2, CDH6, RASSF5, SOX3)	PEITC	Bad, Bim, Blk, Bmf, Bak, Ap-1 Cre/Creb, NfκB, BMI-1	In vitro: SW620, SW480 and HCT116 cells In vivo: subcutaneous xenograft (SW620 cells)	Park et al. (2017)
Upregulation of miR-214		PEITC (alone or in combination with cisplatin or doxorubicin)	c-MYC, Cyclin E1, β-catenin, survivin, Ezh2	In vitro: H460, H1299 and A549 cells In vivo: subcutaneous xenograft (H460 cells)	Li et al. (2017a)
Upregulation of let-7a, let-7b, let-7c, let-7f, miR-10a, miR-26a, miR-30a, miR-30c, miR-34b, miR-34c, miR-99b, miR-122a, miR-123-prec, miR-124a-prec, miR-125a-prec, miR-125b, miR-140s, miR-145-prec, miR-146-prec, miR-191-prec, miR192, miR-219-prec, miR-222-prec, miR-223-prec		PEITC (alone or in combination with oltipraz and I3C)	N.S.	In vivo: cigarette smoke-induced lung cancer in rats	Izzotti et al. (2010)
Downregulation of miR-19		SFN	β-catenin, Wnt/β-catenin, c-Myc, cyclin D1, PCNA,, bcl-2, caspase 8, caspase 9, caspase 3, Bax	In vitro: A549 and H1299 cells	Zhu et al. (2017)
Upregulation of miR-125a-3p		SFN	p53 and MDM-2	In vitro: A549, NCI–H460 and SPCA-1 cells	Li et al. (2015b)
Inhibition of DNMT1 and DNMT3a Inhibition of HDAC1, HDAC4, HDAC5 and HDAC7	Decreased methylation in Nrf2 promoter regions Global DNA acetylation Increased acetylation of Nrf2 promoter	SFN	Nrf2, NQO1	In vitro: TRAMP0 C1 cells	Zhang et al. (2013)

Inhibition of HDAC1	Demethylation in GSTP1 promoter Increased acetylation of H3	PEITC	GSTP1	In vitro: PC3 and LNCaP cells	Wang et al. (2007)
Inhibition of DNMT1 and DNMT3b	Decreased methylation in cyclin D2 promoter regions containing c-Myc and multiple Sp1 binding sites	SFN	Cyclin D2	In vitro: BHP-1, PC3 and LNCaP cells	Hsu et al. (2011)
Inhibition of DNMT1 and DNMT3b (LNCaP cells) Inhibition of DNMT1, DNMT3a, and DNMT3b (PC3 cells)	Increased methylation in TGFBR1, CCR4, CXCR4 and CYR61 promoter regions	SFN (or 3,3′-diindolylmethane)	CCR4, CYR61, TGFBR1, CXCR4	In vitro: LNCaP and PC3 cells	Wong et al. (2014)
Inhibition of DNMT1	Acetylation of H3K4 increased Demethylation in GSTP1 promoter	PEITC	GSTP1	In vitro: LNCaP cells	Chen et al. (2016)
Inhibition of HDAC	Enhanced acetylation of H3 and H4 Increase in acetylation of p21 promoter regions	SFN	Bax, p21, p53 (not in PC3 cells), Bcl2 (not in LNCaP cells)	In vitro: BPH-1, PC3 and LNCaP cells	Myzak et al. (2006b)
Inhibition of HDAC	Increased acetylation of histones H3 and H4 Probably increased histone acetylation at the Bax promoter	SFN	Bax	In vivo: subcutaneous xenograft (PC3 cells)	Myzak et al. (2007)
	Enhancement of acetylation of histone H3 Demethylation of promoter regions of p21	PEITC	p21, p27, p53, c-Myc	In vitro: LNCaP cells	Wang et al. (2008)
Inhibition of HDAC2, HDAC3, HDAC4, and HDAC6	Increase in acetylated histone H3 at the p21 promoter Increase in acetylated α-tubulin	SFN	p21, GSTP1, HO1, NQO1	In vitro: BPH-1, PC3 and LNCaP cells	Clarke et al. (2011)
Inhibition of HDAC1 and HDAC2	Enhanced acetylation of H3 and H4	PHI	P21	In vitro: LNCaP cells	Beklemisheva et al. (2006)

Continued

TABLE 7.1 In Vitro and In Vivo Effects of Isothiocyanates on DNA Methylation, Histone Modifications, and miRNA Mechanisms in Breast, Colorectal, Lung, and Prostate Cancer—cont'd

Epigenetic Mechanism	Epigenetic Modification (Observed or Analyzed)	Isothiocyanates	Target	Experimental Model	References
Inhibition of DNMT1, DNMT3a, and DNMT3b Inhibition of HDAC1, HDAC2, HDAC4, HDAC6	RASSF1A promoter demethylation	PEITC	RASSF1A, p21, caspase 3, caspase 7, Bax, cyclin B1	In vitro: LNCaP cells	Boyanapalli et al. (2016)
Induction of miR-17		PEITC	PCAF	In vitro: LNCaP, PC3, DU145, C4−2B and ALVA-31 cells	Yu et al. (2013)
Upregulation of miR-194 Downregulation of miR-106a/b and miR-695		PEITC	BMP1, MMP2, MMP9	In vitro: LNCaP and PC3 cells	Zhang et al. (2016)

AITC, allyl isothiocyanate; *ALDH1*, aldehyde dehydrogenase 1; *APC*, adenomatous polyposis coli; *BHP-1*, benign prostate hyperplasia; *CAV1*, caveolin-1; *CCR4*, C−C chemokine receptor type 4; *CDH6*, cadherin 6; *CtIP*, c-terminal binding protein (CtBP) interacting protein; *CXCR4*, C-X-C chemokine receptor type 4; *CYR61*, cysteine-rich angiogenic inducer 61; *DNMT*, DNA methyltransferases; *EGCG*, epigallocatechin-3-gallate; *EGFR*, epidermal growth factor receptor; *ERα*, estrogen receptor-α; *Ezh2*, enhancer of zeste homologue 2 gene; *GSTP1*, glutathione-S-transferase-π1; *H3*, core histone H3; *H4*, core histone H4; *HAT*, histone acetyltransferase; *HDAC*, histone deacetylase; *HTLM*, histone lysine methyltransferase; *HNF4A*, hepatocyte nuclear factor 4; *HO1*, heme oxygenase; *hTERT*, human telomerase reverse transcriptase; *I3C*, indole-3-carbinol; *IB*, iberin; *IL*, interleukin; *LSD*, lysine demethylases; *MBD2*, methyl-CpG binding domain protein 2; *MBD3*, methyl-CpG binding domain protein 3; *MBD4*, methyl-CpG binding domain protein 4; *MBD7*, methyl-CpG binding domain protein 7; *MECP2*, methyl-CpG binding protein 2; *MGMT*, O6-methylguanine-DNA methyltransferase; *MMP*, matrix metalloproteinase; *NQO1*, NAD(P)H: quinone reductase; *Nrf2*, nuclear factor erythroid-2 related factor-2; *N.S.*, not studied; *PAK7*, p21 protein activated kinase 7; *PCDH10*, protocadherin 10; *PEITC*, phenethyl isothiocyanate; *PHI*, phenyl hexyl isothiocyanate; *PRC*, polycomb repressive complexes; *PTEN*, phosphatase and tensin homologue; *RARβ2*, retinoic acid receptor-β 2; *RASSF5*, Ras association domain-containing protein 5; *RBBP4*, RB binding protein 4; *RBBP7*, RB binding protein 7; *RBP2*, histone demethylase RBP2; *REL*, reticuloendotheliosis oncogene; *SAP18*, histone deacetylase complex subunit SAP18; *SAP30*, histone deacetylase complex subunit SAP30; *SEPT9*, septin 9; *SFN*, sulforaphane; *SIN3A*, histone deacetylase complex subunit Sin3A; *SOX3*, SRY-related HMG-box3; *SPG20*, spastic paraplegia 20; *STAT1*, signal transducer and activator of transcription 1; *SUV39H1*, histone H3K9 methyltransferase; *TGFBR1*, transforming growth factor receptor β 1; *TGFβ2*, transforming growth factor β 2; *TGFβ3*, transforming growth factor β 3; *TNFαip3*, tumor necrosis factor α-induced protein 3; *VWC2*, Von Willebrand factor C domain containing 2; *WWP2*, E3 ubiquitin ligase with SFN, it can reverse the methylated profile of only 15% of aberrantly methylated genes, increasing the methylation in genes that had decreased methylation and decreasing the methylation in genes that had increased methylation (Wong et al., 2014). This emphasizes the necessity of administering combinatorial treatments in order to increase the efficacy of therapeutic interventions. In this way, Royston et al. (2017) observed greater inhibition of cellular viability and induction of apoptosis of cancer cells with combinatorial approaches (SFN and withaferin A) than with either compound administered alone, which could be explained by their effects on multiple epigenetic processes.

Hormone-dependent and hormone-independent cancer cells can have different DNA methylation patterns that may determine their degree of response to ITC treatment. According to this, it has been demonstrated that the number of genes associated with inflammation and defense functions having altered promoter methylation profiles that are affected by treatment with SFN is higher in LNCaP cells than it is in PC3 cells (Wong et al., 2014). This effect has not been verified in breast cancer cells, however, where three different cell lines responded to SFN treatment in a similar manner (Lewinska et al., 2017).

The anticancer action of a particular ITC on a specific cell line cannot be extrapolated to other cell lines because a priori, some cells seem more resistant to treatment. For example, for Caco-2 cells treated with IB or SFN, no permanent effect was observed with regard to the methylation of genes involved in cell growth control, differentiation, migration, apoptosis, and DNA damage repair (Barrera et al., 2013). Although causes for the lack of effect are not known, it has been demonstrated that SFN can downregulate DNMT1 in this cell line (Traka et al., 2005), which could be attributed to the culture conditions used because it is known that methylation patterns can be modified by serial passages (Antequera et al., 1990; Shmookler Reis and Goldstein, 1982).

Histone Posttranslational Modifications

HPM and DNA CpG island methylation are the two major epigenetic changes involved in gene silencing (Jones and Baylin, 2002). The most frequent HPM is that mediated by the regulation of HDAC activity (Meeran et al., 2010b). Unlike other natural compounds, such as polyphenol epigallocatechin-3-gallate (Choi et al., 2009), ITCs inhibit HDAC without causing any significant alteration in HAT activity (Meeran et al., 2010b). This effect might be mediated by a direct interaction between ITC and the HDAC active site (Myzak et al., 2004).

The inhibition of HDAC leads to apoptosis induction (Liu et al., 2013b; Meeran et al., 2012; Myzak et al., 2006b; Royston et al., 2017), cellular proliferation inhibition (Liu et al., 2013b; Meeran et al., 2010b, 2012; Royston et al., 2017; Sinha et al., 2015), and cell cycle arrest (Clarke et al., 2011; Liu et al., 2013b; Myzak et al., 2006b; Sinha et al., 2015). When these effects are evaluated, experimental conditions must be carefully optimized. While prolonged exposure of SW620 cells to 2–5 μM of PEITC results in decreased viability (by 40% vs. control), the effect is not seen in cells treated for only 3 days because this ITC is not an epigenetic modifier in short-term cultures (Park et al., 2017).

The epigenetic activity of ITCs can be attributed to dual mechanisms (Meeran et al., 2010b; Wang et al., 2007). For example, the reactivation of GSTP1, frequently inactivated in prostate tumors, is mediated by at least two epigenetic mechanisms (DNA demethylation and HPM) (Wang et al., 2007). Likewise, Meeran et al. (2010b) have observed that changes in the HPMs of the hTERT promoter, and DNA demethylation of hTERT exon 1 in response to treatment with SFN, lead to inhibition of cellular growth and induction of apoptosis in human breast cancer cells. This evidence suggests a close interdependence between DNA methylation and histone deacetylation, as pointed out by Kaufman-Szymczyk et al. (2015).

As noted previously, combinatorial approaches with epigenetic compounds (Li et al., 2016; Royston et al., 2017; Sinha et al., 2015) or epigenetic and chemotherapeutic agents (Cang et al., 2014; Li et al., 2017a; Liu et al., 2013b; Meeran et al., 2012) can be necessary for enhancing therapeutic efficacy. Combined treatments can lead to synergistic (Cang et al., 2014; Li et al., 2017b; Liu et al., 2013b; Meeran et al., 2012; Royston et al., 2017; Sinha et al., 2015) or additive (Royston et al., 2017) effects. These strategies are useful because (1) achieving effective concentrations for two different compounds is more feasible than the uptake of higher concentrations of a single agent (Sinha et al., 2015); (2) using low doses of two or more drugs, instead of high doses of one drug, reduces the probability of side effects (Li et al., 2017b); (3) the cells of certain types of cancer can have higher susceptibility to one epigenetic mechanism than to another, as shown in studies with breast cancer cells (MDA-MB-231 and MDA-MB-157) (Li et al., 2017b); and (4) they can resensitize refractory cancer cells to conventional therapies (Li et al., 2017a,b; Sinha et al., 2015). In this way, it has been observed that the combination of green tea polyphenols with SFN causes reactivation of tumor-suppressor genes ($p21^{CIP1/WAF1}$ and KLOTHO) (Sinha et al., 2015) in estrogen receptor-α (ERα)-negative MDA-MB-231 cells through epigenetic mechanisms and sensitizes them to tamoxifen therapy via epigenetic reactivation of ERα at a minimum (Meeran et al., 2012). The therapeutic efficacy of this combinatorial treatment is similar to that observed with inhibitors of DNMTs and HDACs (5-azacytidine and trichostatin A, respectively) but without the toxicity and side effects (Li et al., 2017b) associated with azacytidine and trichostatin A (Sinha et al., 2015; Suzuki and Miyata, 2005). Thus, these findings are a new window of hope for patients with breast cancer that is refractory to hormonal therapy, and it demonstrates the importance of diet during chemotherapy.

In general, effective doses of BFCs for inducing apoptosis, cell cycle arrest, and inhibition of cellular growth against hormone-independent cancer cells (e.g., MDA-MB-231 and PC3) are often higher than those needed against hormone-dependent cancer cells (MCF-7 and LNCaP) and benign hyperplastic cells (Deb et al., 2014; Liu et al., 2013b). Accounting for the fact that PC3 prostate cancer cells have higher HDAC activity, which premaligns cells (Patra et al., 2001), HDAC inhibitors might be a good therapeutic option in hormone-refractory prostate cancer. In fact, it has been shown that SFN inhibits HDAC activity more efficiently in PC-3 cells than it does in LNCaP cells (Myzak et al., 2006b).

microRNAs

The evidence that ITCs can suppress proliferation (Li et al., 2015b; Yu et al., 2013; Zhu et al., 2017), induce apoptosis (Zhu et al., 2017), arrest cell cycle (Lewinska et al., 2017), and inhibit invasion (Zhang et al., 2016) of breast, lung, and prostate cancer cells (no evidence for colorectal cancer cells) through the regulation of miRNAs is quite recent. Their effects on lung cancer cells are especially interesting.

ITC-mediated changes in the profile of miRNAs in breast cancer have been thoroughly studied. The miRNA profile of ER-negative breast cancer cells is more affected by SFN treatment than is the miRNA profile of those that are ER-positive (Lewinska et al., 2017). The expression of 352 miRNAs was evaluated in MDA-MB-231, MCF-7, and SK-BR-3 cells, and 92 miRNAs in the three cell lines showed changes when treated with SFN (32 were downregulated and 60 were upregulated). Intriguing interactions between miRNAs and cell signaling pathways are responsible for ITC-mediated effects. The downregulation of miR-23b, miR-92b, and miR-381 in SK-BR-3 cells by SFN was associated with elevated levels of transforming growth factor-β, which in turn increases the levels of p21 (Lewinska et al., 2017). This finding would explain, at least in part, how SFN can induce cell cycle arrest by modulation of epigenetic mechanisms.

Regulation of the expression of oncogenic and tumor-suppressor miRNAs (e.g., miR-19 and miR-140, respectively) and their target genes is an emerging field of interest in cancer treatment. miR-19 (Zhu et al., 2017) and miR-140 (Li et al., 2014b) are up- and downregulated respectively in CSCs, which are responsible for the progress, heterogeneity, metastasis, and relapse of tumors (Lundin and Driscoll, 2013; Li et al., 2014a) because of their ability to self-renew, resistance to treatment, regulation of neovascularization, and evasion of growth-control mechanisms (Chen et al., 2016; Farnie and Clarke, 2007; Takebe and Ivy, 2010). Recently, it has been shown that under in vitro conditions, SFN (1−15 μM) inhibits cell proliferation and induces apoptosis in lung CSCs, its suppressive effects being mediated by downregulation of miR-19 (Zhu et al., 2017), which in turn causes the inhibition of Wnt/β-catenin pathway activation (Xu et al., 2016; Zhu et al., 2017). SFN's modulation of the expression of miRNAs in CSCs has also been demonstrated in in vivo studies. SFN treatment can inhibit the growth of breast CSCs when they are injected into the mammary glands of nude mice through upregulation of miR-140, which in turn downregulates SOX9 and ALDH1 (Li et al., 2014b). Thus, the miR-19/Wnt/β-catenin and miR-140/ALDH1/SOX9 axes are potential targets for the design of new anticancer drugs.

miRNA genes play an important role in the formation of metastasis in lung (Li et al., 2015a) and prostate (Zhang et al., 2016) cancers because they can trigger the epithelial-to-mesenchymal transition (Li et al., 2015a), regulate coactivators of the androgen receptor (Yu et al., 2013), and regulate the expression of matrix metalloproteinases (MMPs) (Zhang et al., 2016). Considering that metastasis is a main cause for the high mortality of lung cancer patients, intervention strategies that target miRNAs might be promising approaches for treatment. In this way, Li et al. (2015b) have shown that SFN treatment inhibits the invasion and migration abilities of lung cancer cells by inducing overexpression of miRNA-125a-3p. Similar effects have been observed in advanced stages of prostate cancer. PEITC attenuates the invasive capacity of prostate cancer cells (demonstrated by transwell migration assay) because it decreases the levels of oncogenic MMPs (MMP2 and MMP9) by suppressing BMP1 expression (Zhang et al., 2016). In this case, however, the effect is mediated through upregulation of the tumor suppressor miR-194, which is known to have many targets besides BMP1 (Das et al., 2017; Kong et al., 2017). Therefore, modulation of miRNAs with therapeutic aims must be done with caution because it could have a broader functional impact on cells than expected.

Another Achilles' heel in the treatment of lung cancer is long-term loss of response. Although chemotherapy in non-small-cell lung cancer (NSCLC) is generally successful in the beginning, over time patients usually display resistance so that relapses are frequent. Since c-MYC is involved in cancer initiation and progression and its upregulation is commonly found in patients with aggressive and poorly differentiated NSCLC (Vita and Henriksson, 2006), inhibiting c-MYC could be a promising therapeutic approach for NSCLC. Li et al. (2017a) observed that SFN suppresses c-MYC expression through upregulating miR-214, which in turn increases the therapeutic efficacy of cisplatin, a drug frequently used in the treatment of NSCLC. These findings emphasize the necessity of clinical studies to evaluate the application of ITCs in the treatment of patients with recurrent NSCLC.

CONCLUSIONS

We can conclude, in light of the evidence available, that ITCs may (1) act as epigenetic modulators of transcriptional activity in cancer cells, especially those of the breast and prostate (evidence is limited for lung and colorectal cancers) and (2) reverse the abnormal epigenetic alterations that are fundamental to the initiation, promotion, and progression of these cancers. The above-mentioned results about the protective effects of ITCs are encouraging, but they could be only the tip of the iceberg. More studies are needed in order to evaluate the chemopreventive properties of ITCs besides those that have been most studied, SFN and PEITC.

References

Anisimov, V.N., 2009. Carcinogenesis and aging 20 years after: escaping horizon. Mech. Ageing Dev. 130, 105–121.

Ansari, J., Shackelford, R.E., El-Osta, H., 2016. Epigenetics in non-small cell lung cancer: from basics to therapeutics. Transl. Lung Cancer Res. 5, 155–171.

Antequera, F., Bird, A., 1993. Number of CpG islands and genes in human and mouse. Proc. Natl. Acad. Sci. U.S.A 90, 11995–11999.

Antequera, F., Boyes, J., Bird, A., 1990. High levels of de novo methylation and altered chromatin structure at CpG islands in cell lines. Cell 62, 503–514.

Aras, U., Gandhi, Y.A., Masso-Welch, P.A., Morris, M.E., 2013. Chemopreventive and anti-angiogenic effects of dietary phenethyl isothiocyanate in an N-methyl nitrosourea-induced breast cancer animal model. Biopharm. Drug Dispos. 34, 98–106.

Atwell, L.L., Beaver, L.M., Shannon, J., Williams, D.E., Dashwood, R.H., Ho, E., 2015. Epigenetic regulation by sulforaphane: opportunities for breast and prostate cancer chemoprevention. Curr. Pharmacol. Rep. 1, 102–111.

Bao, B., Li, Y., Ahmad, A., Azmi, A.S., Bao, G., Ali, S., Banerjee, S., Kong, D., Sarkar, F.H., 2012. Targeting CSC-related miRNAs for cancer therapy by natural agents. Curr. Drug Targets 13, 1858–1868.

Barrera, L.N., Johnson, I.T., Bao, Y., Cassidy, A., Belshaw, N.J., 2013. Colorectal cancer cells Caco-2 and HCT116 resist epigenetic effects of isothiocyanates and selenium in vitro. Eur. J. Nutr. 52, 1327–1341.

Basse, C., Arock, M., 2015. The increasing roles of epigenetics in breast cancer: implications for pathogenicity, biomarkers, prevention and treatment. Int. J. Cancer 137, 2785–2794.

Beklemisheva, A.A., Fang, Y., Feng, J., Ma, X., Dai, W., Chiao, J.W., 2006. Epigenetic mechanism of growth inhibition induced by phenylhexyl isothiocyanate in prostate cancer cells. Anticancer Res. 26, 1225–1230.

Berry, W.L., Janknecht, R., 2013. KDM4/JMJD2 histone demethylases: epigenetic regulators in cancer cells. Cancer Res. 73, 2936–2942.

Boyanapalli, S.S., Li, W., Fuentes, F., Guo, Y., Ramirez, C.N., Gonzalez, X.P., Pung, D., Kong, A.N., 2016. Epigenetic reactivation of RASSF1A by phenethyl isothiocyanate (PEITC) and promotion of apoptosis in LNCaP cells. Pharmacol. Res. 114, 175–184.

Bronner, C., Fuhrmann, G., Chédin, F.L., Macaluso, M., Dhe-Paganon, S., 2010. UHRF1 links the histone code and DNA methylation to ensure faithful epigenetic memory inheritance. Genet. Epigenet. 2009, 29–36.

Bueno, M.J., Malumbres, M., 2011. MicroRNAs and the cell cycle. Biochim. Biophys. Acta 1812, 592–601.

Calin, G.A., Croce, C.M., 2006. MicroRNA signatures in human cancers. Nat. Rev. Cancer 6, 857–866.

Cang, S., Ma, Y., Chiao, J.W., Liu, D., 2014. Phenethyl isothiocyanate and paclitaxel synergistically enhanced apoptosis and alpha-tubulin hyperacetylation in breast cancer cells. Exp. Hematol. Oncol. 3, 5.

Cartea, M.E., Rodríguez, V.M., de Haro, A., Velasco, P., Ordás, A., 2008. Variation of glucosinolates and nutritional value in nabicol (Brassica napus pabularia group). Euphytica 159, 111–122.

Chen, Y., Cang, S., Han, L., Liu, C., Yang, P., Solangi, Z., Lu, Q., Liu, D., Chiao, J.W., 2016. Establishment of prostate cancer spheres from a prostate cancer cell line after phenethyl isothiocyanate treatment and discovery of androgen-dependent reversible differentiation between sphere and neuroendocrine cells. Oncotarget 7, 26567–26579.

Chiam, K., Ricciardelli, C., Bianco-Miotto, T., 2014. Epigenetic biomarkers in prostate cancer: current and future uses. Cancer Lett. 342, 248–256.

Cho, Y.H., Shen, J., Gammon, M.D., Zhang, Y.J., Wang, Q., Gonzalez, K., Xu, X., Bradshaw, P.T., Teitelbaum, S.L., Garbowski, G., Hibshoosh, H., Neugut, A.I., Chen, J., Santella, R.M., 2012. Prognostic significance of gene-specific promoter hypermethylation in breast cancer patients. Breast Cancer Res. Treat. 131, 197–205.

Choi, K.C., Jung, M.G., Lee, Y.H., Yoon, J.C., Kwon, S.H., Kang, H.B., Kim, M.J., Cha, J.H., Kim, Y.J., Jun, W.J., Lee, J.M., Yoon, H.G., 2009. Epigallocatechin-3-gallate, a histone acetyltransferase inhibitor, inhibits EBV-induced B lymphocyte transformation via suppression of RelA acetylation. Cancer Res. 69, 583–592.

Clarke, J.D., Hsu, A., Yu, Z., Dashwood, R.H., Ho, E., 2011. Differential effects of sulforaphane on histone deacetylases, cell cycle arrest and apoptosis in normal prostate cells versus hyperplastic and cancerous prostate cells. Mol. Nutr. Food Res. 55, 999–1009.

Dacosta, C., Bao, Y., 2017. The role of microRNAs in the chemopreventive activity of sulforaphane from cruciferous vegetables. Nutrients 9, E902.

D'Antonio, K.B., Schultz, L., Albadine, R., Mondul, A.M., Platz, E.A., Netto, G.J., Getzenberg, R.H., 2010. Decreased expression of Cyr61 is associated with prostate cancer recurrence after surgical treatment. Clin. Cancer Res. 16, 5908–5913.

Das, R., Gregory, P.A., Fernandes, R.C., Denis, I., Wang, Q., Townley, S.L., Zhao, S.G., Hanson, A.R., Pickering, M.A., Armstrong, H.K., Lokman, N.A., Ebrahimie, E., Davicioni, E., Jenkins, R.B., Karnes, R.J., Ross, A.E., Den, R.B., Klein, E.A., Chi, K.N., Ramshaw, H.S., Williams, E.D., Zoubeidi, A., Goodall, G.J., Feng, F.Y., Butler, L.M., Tilley, W.D., Selth, L.A., 2017. MicroRNA-194 promotes prostate cancer metastasis by inhibiting SOCS2. Cancer Res. 77, 1021–1034.

Davis, C.D., Ross, S.A., 2007. Dietary components impact histone modifications and cancer risk. Nutr. Rev. 65, 88–94.

Dawson, M.A., Kouzarides, T., 2012. Cancer epigenetics: from mechanism to therapy. Cell 150, 12–27.

Deb, M., Sengupta, D., Kar, S., Rath, S.K., Parbin, S., Shilpi, A., Roy, S., Das, G., Patra, S.K., 2014. Elucidation of caveolin 1 both as a tumor suppressor and metastasis promoter in light of epigenetic modulators. Tumour Biol. 35, 12031–12047.

Dong, N., Shi, L., Wang, D.C., Chen, C., Wang, X., 2017. Role of epigenetics in lung cancer heterogeneity and clinical implication. Semin. Cell Dev. Biol. 64, 18–25.

Duruisseaux, M., Esteller, M., 2017. Lung cancer epigenetics: from knowledge to applications. Semin. Cancer Biol. S1044–S1579X, 30166-30169.

El-Khoury, V., Gomez, D., Liautaud-Roger, F., Trussardi-Régnier, A., Dufer, J., 2004. Effects of the histone deacetylase inhibitor trichostatin A on nuclear texture and c-jun gene expression in drug-sensitive and drug-resistant human H69 lung carcinoma cells. Cytometry A 62, 109–117.

Esteller, M., 2008. Epigenetics in cancer. N. Engl. J. Med. 358, 1148–1159.

Fahey, J.W., Zalcmann, A.T., Talalay, P., 2001. The chemical diversity and distribution of glucosinolates and isothiocyanates among plants. Phytochemistry 56, 5–51.

Farnie, G., Clarke, R.B., 2007. Mammary stem cells and breast cancer–role of Notch signalling. Stem Cell Rev. 3, 169–175.

Fenwick, G.R., Heaney, R.K., Mullin, W.J., 1983. Glucosinolates and their breakdown products in food and food plants. Crit. Rev. Food Sci. Nutr. 18, 123–201.

Fuentes, F., Paredes-Gonzalez, X., Kong, A.N., 2015. Dietary glucosinolates sulforaphane, phenethyl isothiocyanate, indole-3-carbinol/3,3'-diindolylmethane: anti-oxidative stress/inflammation, Nrf2, epigenetics/epigenomics and in vivo cancer chemopreventive efficacy. Curr. Pharmacol. Rep. 1, 179–196.

Füllgrabe, J., Kavanagh, E., Joseph, B., 2011. Histone onco-modifications. Oncogene 30, 3391–3403.

Gao, Y., Tollefsbol, T.O., 2015. Impact of epigenetic dietary components on cancer through histone modifications. Curr. Med. Chem. 22, 2051–2064.

Gianfredi, V., Vannini, S., Moretti, M., Villarini, M., Bragazzi, N.L., Izzotti, A., Nucci, D., 2017. Sulforaphane and epigallocatechin gallate restore estrogen receptor expression by modulating epigenetic events in the breast cancer cell line MDA-MB-231: a systematic review and meta-analysis. J. Nutrigenet. Nutrigenom. 10, 126–135.

Gupta, P., Srivastava, S.K., 2012. Antitumor activity of phenethyl isothiocyanate in HER2-positive breast cancer models. BMC Med. 10, 80.

Halpern, A.B., Othus, M., Huebner, E.M., Buckley, S.A., Pogosova-Agadjanyan, E.L., Orlowski, K.F., Scott, B.L., Becker, P.S., Hendrie, P.C., Chen, T.L., Percival, M.M., Estey, E.H., Stirewalt, D.L., Walter, R.B., 2017. Mitoxantrone, etoposide and cytarabine following epigenetic priming with decitabine in adults with relapsed/refractory acute myeloid leukemia or other high-grade myeloid neoplasms: a phase 1/2 study. Leukemia 31, 2560–2567.

Hardy, T.M., Tollefsbol, T.O., 2011. Epigenetic diet: impact on the epigenome and cancer. Epigenomics 3, 503–518.

Ho, E., Clarke, J.D., Dashwood, R.H., 2009. Dietary sulforaphane, a histone deacetylase inhibitor for cancer prevention. J. Nutr. 139, 2393–2396.

Hsu, A., Wong, C.P., Yu, Z., Williams, D.E., Dashwood, R.H., Ho, E., 2011. Promoter de-methylation of cyclin D2 by sulforaphane in prostate cancer cells. Clin. Epigenet. 3, 3.

Hun Lee, J., Shu, L., Fuentes, F., Su, Z.Y., Tony Kong, A.N., 2013. Cancer chemoprevention by traditional Chinese herbal medicine and dietary phytochemicals: targeting nrf2-mediated oxidative stress/anti-inflammatory responses, epigenetics, and cancer stem cells. J. Tradit. Complement. Med. 3, 69–79.

Ishida, M., Hara, M., Fukino, N., Kakizaki, T., Morimitsu, Y., 2014. Glucosinolate metabolism, functionality and breeding for the improvement of Brassicaceae vegetables. Breed Sci. 64, 48–59.

Izzotti, A., Calin, G.A., Steele, V.E., Cartiglia, C., Longobardi, M., Croce, C.M., De Flora, S., 2010. Chemoprevention of cigarette smoke-induced alterations of microRNA expression in rat lungs. Cancer Prev. Res. (Phila) 3, 62–72.

Jones, P.A., Baylin, S.B., 2002. The fundamental role of epigenetic events in cancer. Nat. Rev. Genet. 3, 415–428.

Kaufman-Szymczyk, A., Majewski, G., Lubecka-Pietruszewska, K., Fabianowska-Majewska, K., 2015. The role of sulforaphane in epigenetic mechanisms, including interdependence between histone modification and DNA methylation. Int. J. Mol. Sci. 16, 29732–29743.

Kim, S.H., Nagalingam, A., Saxena, N.K., Singh, S.V., Sharma, D., 2011. Benzyl isothiocyanate inhibits oncogenic actions of leptin in human breast cancer cells by suppressing activation of signal transducer and activator of transcription 3. Carcinogenesis 32, 359–367.

Kong, Q., Chen, X.S., Tian, T., Xia, X.Y., Xu, P., 2017. MicroRNA-194 suppresses prostate cancer migration and invasion by downregulating human nuclear distribution protein. Oncol. Rep. 37, 803–812.

Kouzarides, T., 2007. Chromatin modifications and their function. Cell 128, 693–705.

Kuzmin, I., Gillespie, J.W., Protopopov, A., Geil, L., Dreijerink, K., Yang, Y., Vocke, C.D., Duh, F.M., Zabarovsky, E., Minna, J.D., Rhim, J.S., Emmert-Buck, M.R., Linehan, W.M., Lerman, M.I., 2002. The RASSF1A tumor suppressor gene is inactivated in prostate tumors and suppresses growth of prostate carcinoma cells. Cancer Res. 62, 3498–3502.

Lai, K.C., Huang, A.C., Hsu, S.C., Kuo, C.L., Yang, J.S., Wu, S.H., Chung, J.G., 2010. Benzyl isothiocyanate (BITC) inhibits migration and invasion of human colon cancer HT29 cells by inhibiting matrix metalloproteinase-2/-9 and urokinase plasminogen (uPA) through PKC and MAPK signaling pathway. J. Agric. Food Chem. 58, 2935–2942.

Law, J.A., Jacobsen, S.E., 2010. Establishing, maintaining and modifying DNA methylation patterns in plants and animals. Nat. Rev. Genet. 11, 204–220.

Lewinska, A., Adamczyk-Grochala, J., Deregowska, A., Wnuk, M., 2017. Sulforaphane-induced cell cycle arrest and senescence are accompanied by DNA hypomethylation and changes in microRNA profile in breast cancer cells. Theranostics 7, 3461–3477.

Li, J., Yang, S., Yan, W., Yang, J., Qin, Y.J., Lin, X.L., Xie, R.Y., Wang, S.C., Jin, W., Gao, F., Shi, J.W., Zhao, W.T., Jia, J.S., Shen, H.F., Ke, J.R., Liu, B., Zhao, Y.Q., Huang, W.H., Yao, K.T., Li, D.J., Xiao, D., 2015a. MicroRNA-19 triggers epithelial-mesenchymal transition of lung cancer cells accompanied by growth inhibition. Lab. Investig. 95, 1056–1070.

Li, K.K., Li, F., Li, Q.S., Yang, K., Jin, B., 2013. DNA methylation as a target of epigenetic therapeutics in cancer. Anticancer Agents Med. Chem. 13, 242–247.

Li, Q., Eades, G., Yao, Y., Zhang, Y., Zhou, Q., 2014a. Characterization of a stem-like subpopulation in basal-like ductal carcinoma in situ (DCIS) lesions. J. Biol. Chem. 289, 1303–1312.

Li, Q., Yao, Y., Eades, G., Liu, Z., Zhang, Y., Zhou, Q., 2014b. Downregulation of miR-140 promotes cancer stem cell formation in basal-like early stage breast cancer. Oncogene 33, 2589–2600.

Li, Q.Q., Xie, Y.K., Wu, Y., Li, L.L., Liu, Y., Miao, X.B., Liu, Q.Z., Yao, K.T., Xiao, G.H., 2017a. Sulforaphane inhibits cancer stem-like cell properties and cisplatin resistance through miR-214-mediated downregulation of c-MYC in non-small cell lung cancer. Oncotarget 8, 12067–12080.

Li, S., Li, X., Zhao, H., Gao, M., Wang, F., Li, W., 2015b. Overexpression of microRNA-125a-3p effectively inhibits the cell growth and invasion of lung cancer cells by regulating the mouse double minute 2 homolog/p53 signaling pathway. Mol. Med. Rep. 12, 5482–5486.

Li, Y., Buckhaults, P., Cui, X., Tollefsbol, T.O., 2016. Combinatorial epigenetic mechanisms and efficacy of early breast cancer inhibition by nutritive botanicals. Epigenomics 8, 1019–1037.

Li, Y., Meeran, S.M., Tollefsbol, T.O., 2017b. Combinatorial bioactive botanicals re-sensitize tamoxifen treatment in ER-negative breast cancer via epigenetic reactivation of ERα expression. Sci. Rep. 7, 9345.

Li, Y., Tollefsbol, T.O., 2010. Impact on DNA methylation in cancer prevention and therapy by bioactive dietary components. Curr. Med. Chem. 17, 2141–2151.

Lin, J.F., Tsai, T.F., Yang, S.C., Lin, Y.C., Chen, H.E., Chou, K.Y., Hwang, T.I., 2017. Benzyl isothiocyanate induces reactive oxygen species-initiated autophagy and apoptosis in human prostate cancer cells. Oncotarget 8, 20220–20234.

Lin, R.K., Wang, Y.C., 2014. Dysregulated transcriptional and post-translational control of DNA methyltransferases in cancer. Cell Biosci. 4, 46.

Lips, E.H., Mulder, L., Oonk, A., van der Kolk, L.E., Hogervorst, F.B., Imholz, A.L., Wesseling, J., Rodenhuis, S., Nederlof, P.M., 2013. Triple-negative breast cancer: BRCAness and concordance of clinical features with BRCA1-mutation carriers. Br. J. Cancer 108, 2172–2177.

Liu, B.N., Yan, H.Q., Wu, X., Pan, Z.H., Zhu, Y., Meng, Z.W., Zhou, Q.H., Xu, K., 2013a. Apoptosis induced by benzyl isothiocyanate in gefitinib-resistant lung cancer cells is associated with Akt/MAPK pathways and generation of reactive oxygen species. Cell Biochem. Biophys. 66, 81–92.

Liu, K., Cang, S., Ma, Y., Chiao, J.W., 2013b. Synergistic effect of paclitaxel and epigenetic agent phenethyl isothiocyanate on growth inhibition, cell cycle arrest and apoptosis in breast cancer cells. Cancer Cell Int. 13, 10.

Liu, Y., Chakravarty, S., Dey, M., 2013c. Phenethylisothiocyanate alters site- and promoter-specific histone tail modifications in cancer cells. PLoS One 8, e64535.

Liu, Y., Dey, M., 2017. Dietary phenethyl isothiocyanate protects mice from colitis associated colon cancer. Int. J. Mol. Sci. 18, E1908.

Lubecka-Pietruszewska, K., Kaufman-Szymczyk, A., Stefanska, B., Cebula-Obrzut, B., Smolewski, P., Fabianowska-Majewska, K., 2015. Sulforaphane alone and in combination with clofarabine epigenetically regulates the expression of DNA methylation-silenced tumour suppressor genes in human breast cancer cells. J. Nutrigenet. Nutrigenom. 8, 91–101.

Lundin, A., Driscoll, B., 2013. Lung cancer stem cells: progress and prospects. Cancer Lett. 338, 89–93.

Maia, B.M., Rocha, R.M., Calin, G.A., 2014. Clinical significance of the interaction between non-coding RNAs and the epigenetics machinery: challenges and opportunities in oncology. Epigenetics 9, 75–80.

Meeran, S.M., Ahmed, A., Tollefsbol, T.O., 2010a. Epigenetic targets of bioactive dietary components for cancer prevention and therapy. Clin. Epigenet. 1, 101–116.

Meeran, S.M., Patel, S.N., Li, Y., Shukla, S., Tollefsbol, T.O., 2012. Bioactive dietary supplements reactivate ER expression in ER-negative breast cancer cells by active chromatin modifications. PLoS One 7, e37748.

Meeran, S.M., Patel, S.N., Tollefsbol, T.O., 2010b. Sulforaphane causes epigenetic repression of hTERT expression in human breast cancer cell lines. PLoS One 5, e11457.

Mithen, R.F., Dekker, M., Verkerk, R., Rabot, S., Johnson, I.T., 2000. The nutritional significance, biosynthesis and bioavailability of glucosinolates in human foods. J. Sci. Food Agric. 80, 967–984.

Momparler, R.L., 2003. Cancer epigenetics. Oncogene 22, 6479–6483.

Myzak, M.C., Dashwood, W.M., Orner, G.A., Ho, E., Dashwood, R.H., 2006a. Sulforaphane inhibits histone deacetylase in vivo and suppresses tumorigenesis in Apc-minus mice. FASEB J. 20, 506–508.

Myzak, M.C., Hardin, K., Wang, R., Dashwood, R.H., Ho, E., 2006b. Sulforaphane inhibits histone deacetylase activity in BPH-1, LnCaP and PC-3 prostate epithelial cells. Carcinogenesis 27, 811–819.

Myzak, M.C., Karplus, P.A., Chung, F.L., Dashwood, R.H., 2004. A novel mechanism of chemoprotection by sulforaphane: inhibition of histone deacetylase. Cancer Res. 64, 5767–5774.

Myzak, M.C., Tong, P., Dashwood, W.M., Dashwood, R.H., Ho, E., 2007. Sulforaphane retards the growth of human PC-3 xenografts and inhibits HDAC activity in human subjects. Exp. Biol. Med. (Maywood) 232, 227–234.

Pan, J., Chen, J., Zhang, B., Chen, X., Huang, B., Zhuang, J., Mo, C., Qiu, S., 2013. Association between RASSF1A promoter methylation and prostate cancer: a systematic review and meta-analysis. PLoS One 8, e75283.

Park, J.E., Sun, Y., Lim, S.K., Tam, J.P., Dekker, M., Chen, H., Sze, S.K., 2017. Dietary phytochemical PEITC restricts tumor development via modulation of epigenetic writers and erasers. Sci. Rep. 7, 40569.

Patra, S.K., Patra, A., Dahiya, R., 2001. Histone deacetylase and DNA methyltransferase in human prostate cancer. Biochem. Biophys. Res. Commun. 287, 705–713.

Pledgie-Tracy, A., Sobolewski, M.D., Davidson, N.E., 2007. Sulforaphane induces cell type-specific apoptosis in human breast cancer cell lines. Mol. Cancer Ther. 6, 1013–1021.

Puccini, A., Berger, M.D., Naseem, M., Tokunaga, R., Battaglin, F., Cao, S., Hanna, D.L., McSkane, M., Soni, S., Zhang, W., Lenz, H.J., 2017. Colorectal cancer: epigenetic alterations and their clinical implications. Biochim. Biophys. Acta 1868, 439–448.

Rajendran, P., Kidane, A.I., Yu, T.W., Dashwood, W.M., Bisson, W.H., Löhr, C.V., Ho, E., Williams, D.E., Dashwood, R.H., 2013. HDAC turnover, CtIP acetylation and dysregulated DNA damage signaling in colon cancer cells treated with sulforaphane and related dietary isothiocyanates. Epigenetics 8, 612–623.

Rajewsky, N., 2006. microRNA target predictions in animals. Nat. Genet. 38, S8–S13.

Robertson, K.D., 2005. DNA methylation and human disease. Nat. Rev. Genet. 6, 597–610.

Royston, K.J., Udayakumar, N., Lewis, K., Tollefsbol, T.O., 2017. A novel combination of withaferin A and sulforaphane inhibits epigenetic machinery, cellular viability and induces apoptosis of breast cancer cells. Int. J. Mol. Sci. 18, E1092.

Saha, K., Hornyak, T.J., Eckert, R.L., 2013. Epigenetic cancer prevention mechanisms in skin cancer. AAPS J. 15, 1064–1071.

Savio, A.J., Bapat, B., 2015. Beyond the island: epigenetic biomarkers of colorectal and prostate cancer. Methods Mol. Biol. 1238, 103–124.

Schiffmann, I., Greve, G., Jung, M., Lübbert, M., 2016. Epigenetic therapy approaches in non-small cell lung cancer: update and perspectives. Epigenetics 11, 858–870.

Shankar, S., Kumar, D., Srivastava, R.K., 2013. Epigenetic modifications by dietary phytochemicals: implications for personalized nutrition. Pharmacol. Ther. 138, 1–17.

Shmookler Reis, R.J., Goldstein, S., 1982. Variability of DNA methylation patterns during serial passage of human diploid fibroblasts. Proc. Natl. Acad. Sci. U.S.A. 79, 3949–3953.

Sinha, S., Shukla, S., Khan, S., Tollefsbol, T.O., Meeran, S.M., 2015. Epigenetic reactivation of p21CIP1/WAF1 and KLOTHO by a combination of bioactive dietary supplements is partially ERα-dependent in ERα-negative human breast cancer cells. Mol. Cell. Endocrinol. 406, 102–114.

Strahl, B.D., Allis, C.D., 2000. The language of covalent histone modifications. Nature 403, 41–45.

Suzuki, T., Miyata, N., 2005. Non-hydroxamate histone deacetylase inhibitors. Curr. Med. Chem. 12, 2867–2880.

Takebe, N., Ivy, S.P., 2010. Controversies in cancer stem cells: targeting embryonic signaling pathways. Clin. Cancer Res. 16, 3106–3112.

Tang, Q., Cheng, J., Cao, X., Surowy, H., Burwinkel, B., 2016. Blood-based DNA methylation as biomarker for breast cancer: a systematic review. Clin. Epigenet. 8, 115.

Terada, N., Kulkarni, P., Getzenberg, R.H., 2012. Cyr61 is a potential prognostic marker for prostate cancer. Asian J. Androl. 14, 405–408.

Thun, M.J., DeLancey, J.O., Center, M.M., Jemal, A., Ward, E.M., 2010. The global burden of cancer: priorities for prevention. Carcinogenesis 31, 100–110.

Tomasetti, M., Neuzil, J., Dong, L., 2014. MicroRNAs as regulators of mitochondrial function: role in cancer suppression. Biochim. Biophys. Acta 1840, 1441–1453.

Traka, M., Gasper, A.V., Smith, J.A., Hawkey, C.J., Bao, Y., Mithen, R.F., 2005. Transcriptome analysis of human colon Caco-2 cells exposed to sulforaphane. J. Nutr. 135, 1865–1872.

Tse, J.W.T., Jenkins, L.J., Chionh, F., Mariadason, J.M., 2017. Aberrant DNA methylation in colorectal cancer: what should we target? Trends Cancer 3, 698–712.

Vasudevan, S., Tong, Y., Steitz, J.A., 2007. Switching from repression to activation: microRNAs can up-regulate translation. Science 318, 1931–1934.

Vita, M., Henriksson, M., 2006. The Myc oncoprotein as a therapeutic target for human cancer. Semin. Cancer Biol. 16, 318–330.

Wang, L.G., Beklemisheva, A., Liu, X.M., Ferrari, A.C., Feng, J., Chiao, J.W., 2007. Dual action on promoter demethylation and chromatin by an isothiocyanate restored GSTP1 silenced in prostate cancer. Mol. Carcinog. 46, 24–31.

Wang, L.G., Liu, X.M., Fang, Y., Dai, W., Chiao, F.B., Puccio, G.M., Feng, J., Liu, D., Chiao, J.W., 2008. Derepression of the p21 promoter in prostate cancer cells by an isothiocyanate via inhibition of HDACs and c-Myc. Int. J. Oncol. 33, 375–380.

Weisenberger, D.J., Liang, G., Lenz, H.J., 2018. DNA methylation aberrancies delineate clinically distinct subsets of colorectal cancer and provide novel targets for epigenetic therapies. Oncogene 37, 566–577.

Widschwendter, M., Evans, I., Jones, A., Ghazali, S., Reisel, D., Ryan, A., Gentry-Maharaj, A., Zikan, M., Cibula, D., Eichner, J., Alunni-Fabbroni, M., Koch, J., Janni, W.J., Paprotka, T., Wittenberger, T., Menon, U., Wahl, B., Rack, B., Lempiäinen, H., 2017. Methylation patterns in serum DNA for early identification of disseminated breast cancer. Genome Med. 9, 115.

Wijermans, P.W., Lübbert, M., Verhoef, G., Klimek, V., Bosly, A., 2005. An epigenetic approach to the treatment of advanced MDS; the experience with the DNA demethylating agent 5-aza-2′-deoxycytidine (decitabine) in 177 patients. Ann. Hematol. 84 (Suppl. 1), 9–17.

Wong, C.P., Hsu, A., Buchanan, A., Palomera-Sanchez, Z., Beaver, L.M., Houseman, E.A., Williams, D.E., Dashwood, R.H., Ho, E., 2014. Effects of sulforaphane and 3,3′-diindolylmethane on genome-wide promoter methylation in normal prostate epithelial cells and prostate cancer cells. PLoS One 9, e86787.

Xu, K., Zhao, Y.C., 2016. MEF2D/Wnt/β-catenin pathway regulates the proliferation of gastric cancer cells and is regulated by microRNA-19. Tumour Biol. 37, 9059–9069.

Yu, C., Gong, A.Y., Chen, D., Solelo Leon, D., Young, C.Y., Chen, X.M., 2013. Phenethyl isothiocyanate inhibits androgen receptor-regulated transcriptional activity in prostate cancer cells through suppressing PCAF. Mol. Nutr. Food Res. 57, 1825–1833.

Zhang, C., Shu, L., Kim, H., Khor, T.O., Wu, R., Li, W., Kong, A.N., 2016. Phenethyl isothiocyanate (PEITC) suppresses prostate cancer cell invasion epigenetically through regulating microRNA-194. Mol. Nutr. Food Res. 60, 1427–1436.

Zhang, C., Su, Z.Y., Khor, T.O., Shu, L., Kong, A.N., 2013. Sulforaphane enhances Nrf2 expression in prostate cancer TRAMP C1 cells through epigenetic regulation. Biochem. Pharmacol. 85, 1398–1404.

Zhao, S., Liu, M.F., 2009. Mechanisms of microRNA-mediated gene regulation. Sci. China C Life Sci. 52, 1111–1116.

Zhu, J., Wang, S., Chen, Y., Li, X., Jiang, Y., Yang, X., Li, Y., Wang, X., Meng, Y., Zhu, M., Ma, X., Huang, C., Wu, R., Xie, C., Geng, S., Wu, J., Zhong, C., Han, H., 2017. miR-19 targeting of GSK3β mediates sulforaphane suppression of lung cancer stem cells. J. Nutr. Biochem. 44, 80–91.

CHAPTER

8

Epigenetic Mechanisms of Resveratrol and Its Analogs in Cancer Prevention and Treatment

Avinash Kumar[1], *Anait S. Levenson*[1,2]

[1]Arnold and Marie Schwartz College of Pharmacy and Health Sciences, Long Island University, Brooklyn, NY, United States; [2]College of Veterinary Medicine, Long Island University, Brookville, NY, United States

Abstract

Resveratrol is a dietary polyphenol that has attracted immense attention from biomedical researchers due to its numerous biological properties implicated for human health and disease. The molecular mechanisms that contribute to the beneficiary nature of resveratrol are multilayered and include epigenetic regulation. This chapter focuses on the effects of resveratrol, and its two natural analogs, on anticancer activity mediated by DNA methylation, histone modification, and microRNAs. Studies on resveratrol's epigenetic action, specifically histone deacetylase activity, add a new dimension to the expanding roles of resveratrol and its analogs as potential epichemopreventive and chemosensitizing agents. Clinical trials examining epigenetic changes facilitated by resveratrol and its analogs for cancer chemoprevention and therapy are urgently needed.

Keywords

Cancer; Chemoprevention; DNA methylation; Epigenetics; Histone modifications; microRNAs; Piceatannol; Pterostilbene; Resveratrol.

INTRODUCTION

A growing body of molecular and clinical evidence indicates that epigenetic machinery plays a critical role in the development and progression of cancer. While genetic information is constant, epigenetic modifications are extraordinarily complex and dynamic in their responses to environmental signals. The most well-studied epigenetic mechanisms are DNA methylation, histone modifications, and noncoding RNAs (ncRNAs). The interrelated key epigenetic processes that affect chromatin structure and conformation are regulated by the interplay of several enzymes such as DNA methyltransferases (DNMTs), histone acetyltransferases (HATs), and histone deacetylases (HDACs) as well as histone methyltransferases (HMTs) and histone demethylases (HDMs). In addition to these chromatin "writers" and "erasers,"

Epigenetics of Cancer Prevention
https://doi.org/10.1016/B978-0-12-812494-9.00008-1

other chromatin-modifying complexes called "readers" (i.e., methyl-CpG-binding domain and bromodomain proteins) play a major role in coordinating the cross talk between DNA methylation, histone modifications, and chromatin organization (Arrowsmith et al., 2012; Du et al., 2015). Finally, the most widely studied class of ncRNAs, microRNAs (miRNAs), can directly interact with genes to abrogate gene transcription and prevent translation as well as modify chromatin structure by mediating the recruitment of DNMTs and HMTs (Holoch and Moazed, 2015). Epigenetic modifications lead to chromatin remodeling that results in changes to cellular processes, such as cell cycle progression, DNA repair, cellular differentiation and growth, apoptosis, and angiogenesis. Therefore, abundant evidence highlights the contribution of epigenetics to the initiation and progression of cancer, generation of metastasis, and prediction of drug response (Giorgio et al., 2016; Kristiansen and Soletormos, 2016; Wang et al., 2016; Xiao and Li, 2016).

EPIGENETICS AND DIET

Epidemiological studies have linked environmental factors and diet to changes in an individual's epigenome and consequent genomic instability that trigger cancer. Proper nutrition, including diets rich in fresh fruits and vegetables, have been found to lower cancer rates. Dietary phytochemicals represented by different classes of polyphenols such as stilbenes, flavonoids, lignans, and others have been widely studied for their beneficiary cardioprotective, anti-inflammatory, antioxidative, and anticancer properties. The molecular mechanisms that contribute to the beneficiary nature of dietary polyphenols are multilayered and include regulation of epigenetic mechanisms that affect the DNA methylation, histone modifications, and ncRNAs that eventually regulate gene expression (Link et al., 2010; Vanden Berghe, 2012; Abbas et al., 2013; Kumar et al., 2016). The significant contribution of diet to epigenetic mechanisms is being fast realized and extensively studied in the emerging field of nutriepigenomics. Moreover, a great potential is being realized for nutritional intervention through natural epigenetic agents for cancer chemoprevention and possibly cancer therapeutics through their immuno-, radio-, and chemosensitization properties.

RESVERATROL AND ITS ANALOGS

Resveratrol (*trans*-3,4′,5-trihydroxystilbene) Fig. 8.1) is a natural phytoalexin synthesized in plants as a defensive response against various environmental stressors (Hasan and Bae, 2017). Resveratrol is present in a wide variety of plant

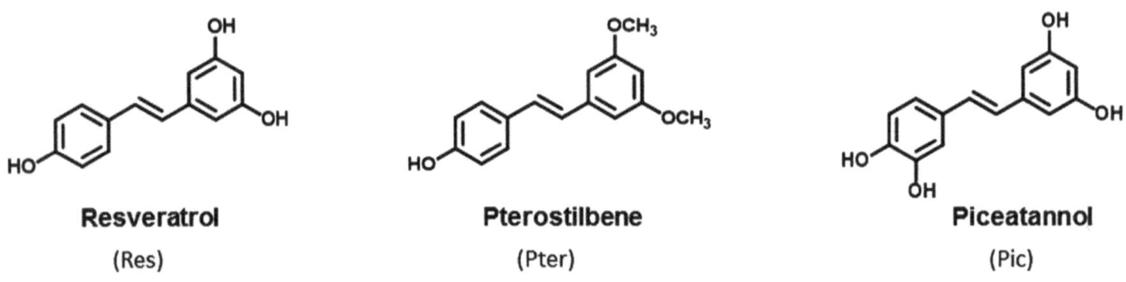

FIGURE 8.1 Chemical structures of resveratrol (*trans*-3,4′,5-trihydroxystilbene), pterostilbene (*trans*-3,5-dimethoxystilbene), and piceatannol (*trans*-3,3′,4′,5-tetrahydroxystilbene) known to exert anticancer activity through epigenetic mechanisms.

species including grapes, peanuts, and mulberries, and thus is a constituent of the human diet. The epidemiological finding of an inverse relationship between consumption of red wine and a low incidence of cardiovascular disease led to the phenomenon known as the "French paradox" when the cardioprotective properties of red wine were attributed to resveratrol (Constant, 1997; Sato et al., 2000). Since publication of the pioneer work on resveratrol's potential for cancer chemoprevention by Pezzuto and colleagues (Jang et al., 1997), numerous studies have shown the anticancer effects of resveratrol and its analogs in vitro and in vivo in various cancer types (Gusman et al., 2001; Bishayee, 2009). Resveratrol has been shown to inhibit growth, induce cell cycle arrest and apoptosis, inhibit angiogenesis, and modulate various signaling pathways in cancer cells. As a phytoestrogen, resveratrol binds and activates the estrogen receptor (ERs) ERα and ERβ, thus triggering biological responses (Gehm et al., 1997; Bowers et al., 2000; Gehm and Levenson, 2006). The role of ER isoforms that mediate resveratrol's effects is contradictory, however, and the nature of resveratrol as a selective ER modulator (SERM) remains complex because it depends on tissue microenvironment factors including the levels of ER isoforms, availability of certain cellular proteins, and balance of coactivators and corepressors (Chakraborty et al., 2013). The mixed agonist/antagonist nature of resveratrol is of great interest considering its potential for affecting human health. As a natural SERM, the partial estrogenic activity of resveratrol could provide health benefits in bone tissues, the cardiovascular system, Alzheimer disease, clinical depression, and menopausal symptoms, to mention a few. However, concerns remain regarding its adverse effects if it is used as a preventive or therapeutic agent for breast and endometrial cancers (Le et al., 1984; Levenson and Jordan, 1999; Jain et al., 2000; Sengupta and Jordan, 2008). Understanding resveratrol-mediated epigenetic protective signaling during estrogen (E2)-induced carcinogenesis and progression may have a tremendous impact on the lives of thousands of women.

Substantial efforts have been directed toward studies of natural and synthetic analogs of resveratrol with the goal of enhancing their bioavailability and efficacy in vivo. Among them, pterostilbene (Pter) (*trans*-3, 5-dimethoxystilbene) and piceatannol (Pic) (*trans*-3,3′,4′,5-tetrahydroxystilbene) (Fig. 8.1) are the most studied, particularly in terms of their epigenetic mechanisms of action. Pter, found in grapes and blueberries, is a natural analog of resveratrol with decreased metabolism, higher bioavailability, and proven anti-inflammatory and anticancer activity (Pan et al., 2008, 2009; Remsberg et al., 2008; Chakraborty et al., 2010; Chiou et al., 2011; Kapetanovic et al., 2011). Pter, along with resveratrol, was suggested as a natural epigenetic chemopreventive and therapeutic agent in prostate cancer on the basis of epigenetic reader metastasis-associated 1 (MTA1)-mediated and miRNA-mediated tumor reduction in vitro and in various animal models (Kai et al., 2010; Dhar et al., 2011, 2015a,b, 2016; Li et al., 2013). Pic is found in grapes and red wine and has been reported as a cancer-preventive agent (Thakkar et al., 1993; Fleming et al., 1995; Su and David, 2000; Vo et al., 2010; Kang et al., 2011). As suggested by many, the potential of resveratrol and its analogs for chemoprevention and anticancer therapy may largely reside in their ability to cause or sensitize tumor cells for cell death (Fulda and Debatin, 2004; Opipari et al., 2004; Pervaiz and Holme, 2006). However, the exact mechanisms by which resveratrol and its analogs/derivatives exert these anticarcinogenic effects are not completely elucidated. During development of the concept of epigenetic therapy over the last two decades (Ahuja et al., 2014; Jones, 2014), research efforts have been driven toward an understanding of the epigenetic mechanisms of anticancer activity by dietary polyphenols (Link et al., 2010). In the current chapter, we will deliberate the regulation of epigenetic mechanisms by resveratrol and two of its natural analogs, Pter and Pic, in cancer.

RESVERATROL AND ITS ANALOGS AS REGULATORS OF DNA METHYLATION

DNA methylation is the most studied of the major epigenetic mechanisms. It is well known that even in normal tissues, a highly dynamic turnover in DNA methylation patterns dictates large epigenetic heterogeneity. Still, in contrast to methylation patterns in normal tissues, cancer cells typically display global DNA hypomethylation combined with hypermethylation of promoter-associated cytosine-phosphate-guanine (CpG) dinucleotide islands, which is linked to gene silencing. Unique methylation signatures in tumor cells can be used as diagnostic and predictive biomarkers for chemotherapeutic drugs (Shen et al., 2007; Heyn and Esteller, 2012). Altered DNA methylation events contribute to tumorigenesis and cancer progression in at least two ways: (1) deregulated expression of genes that encode members of DNA methylation machinery such as DNMTs, and (2) inactivation of tumor-suppressor genes by hypermethylation of CpG islands. The potential reversibility of DNA methylation suggests that these are viable targets for cancer prevention and treatment. Clearly, if resveratrol can affect the mechanisms of DNA methylation and restore normal DNA methylation patterns, it may have health benefits. Surprisingly, studies on resveratrol's effects on DNA methylation are limited when compared with other dietary bioactive molecules such as curcumin, genistein, and epigallocatechin-3-gallate (EGCG) (Kumar et al., 2016). Nevertheless, here we will summarize available data on the role of resveratrol and its analogs in DNA methylation (Table 8.1).

A number of studies have reported the effects of resveratrol and its analogs on the expression and activity of DNMTs as well as global and gene-specific methylation/demethylation. Resveratrol has been shown to decrease the expression of various DNMTs in breast cancer (Papoutsis et al., 2012; Stefanska et al., 2012; Mirza et al., 2013; Qin et al., 2014). Interestingly, resveratrol displayed differential effects in tumorous versus normal tissue by decreasing DNMT3b expression in breast tumors and increasing its expression in normal breast tissue (Qin et al., 2014). Combinatorial resveratrol and Pter treatment of breast cancer cells resulted not only in significant downregulation of DNMT1, DNMT3a, and DNMT3b, but also in the inhibition of their activity (Kala et al., 2015). On the other hand, resveratrol and Pter increased DNMT3b-mediated DNA methylation at mastermind like transcriptional coactivator 2 enhancer, a coactivator of Notch targets, resulting in transcriptional silencing of oncogenic signaling in breast cancer cells (Lubecka et al., 2016). Several studies have reported on resveratrol's ability to play a role in decreasing gene-promoter hypermethylation and modulating DNA hypomethylation (Berner et al., 2010; Lee et al., 2012; Zhu et al., 2012; Singh et al., 2014; Medina-Aguilar et al., 2016). The genome-wide DNA methylation signatures were identified in triple-negative breast cancer cells treated with resveratrol, and decreased gene-promoter hypermethylation and increased DNA methylation were found (Medina-Aguilar et al., 2016). Importantly, resveratrol's effects were very specific—it restored the hypomethylated status of key tumor suppressors and the hypermethylated status of oncogenes. The authors concluded that resveratrol as an epi-drug may affect the methylome of cancer cells, revealing novel potential targets for epigenetic therapy. In addition, a couple studies reported on "resveratrol as a phytoestrogen" in epigenetic mechanisms of action. While the role of resveratrol in breast cancer remains controversial (Gehm and Levenson, 2006), it has been recently reported that resveratrol mediated epigenetic protective signaling during E2-induced breast carcinogenesis in

TABLE 8.1 Regulation of DNA Methylation by Resveratrol and Its Analogs

Resveratrol and Analogs	Dose	Epigenetic Mechanism[a]	Genes/Histones	Cells/Organ	References
Resveratrol (Res)	100 μM	DNA methylation and demethylation	*AURKA, CCNB1, DDIT4, DLGAP5, EYS, FAM83D, HIST1H2BM, IL24, LPXN, NFIL3, PFKFB3, SLC14A1, STC1, AMY2A, IL18 and SLIT3*	MDA-MB-231 breast cancer cells	Medina-Aguilar et al. (2016)
	50 μM	Reversal of estrogen-induced promoter methylation	*NRF2*	MCF-10A breast epithelial cells	Singh et al. (2014)
	5, 25 mg/kg	↓ DNMT3b	Differential effect in normal tissue	Rat breast tumor tissue	Qin et al. (2014)
	5–50 μM	↓ DNMT1, 3a, 3b, MeCP2	–	Patient breast tumors	Mirza et al. (2013)
	0.5–20 μM	↓ DNMT1	*BRCA1*	MCF-7 breast cancer cells	Papoutsis et al. (2012)
	50 μM	DNA demethylation, promoter demethylation	ERα	A2058 melanoma cells; MCF-7, MDA-MB-231, MDA-MB-468 breast cancer cells	Lee at al. (2012)
	5, 50 mg/kg	Promoter demethylation	*RASSF1A*	High-risk breast cancer patients	Zhu et al. (2012)
	10, 14, 30 μM	↓ DNMTs	*PTEN*, p21	MCF-7, MDA-MB-231 breast cancer cells	Stefanska et al. (2012)
	10, 50 μM	Promoter methylation	ERα, p16	Caco-2 colon cancer cells	Berner et al. (2010)
	10, 50 μM	Promoter demethylation	p15	Caco-2 colon cancer cells	Berner et al. (2010)
Res Pterostilbene (Pter)	5–20 μM 5–15 μM	DNA methylation, enhancer methylation	*MAML2*, NOTCH target genes	Breast cancer cells	Lubecka et al. (2016)
Res + Pter	5–15 μM	↓ DNMT1, 3a, 3b DNMT activity inhibition	γH2AX, *hTERT*	HCC1806, MDA-MB-157, breast cancer cells; MCF-10A breast epithelial cells	Kala et al. (2015)

[a] ↓, *downregulation.*

female August Copenhagen Irish rats (Singh et al., 2014). The authors found that resveratrol treatment inhibited E2-mediated alteration in nuclear factor erythroid 2-related factor 2 (Nrf2)-promoter methylation, which resulted in an increase in Nrf2-regulated antioxidant genes that are involved in protection against oxidative DNA damage. Interestingly, resveratrol treatment led to increased methylation of the ERα promoter in colon cancer cells (Berner et al., 2010) but demethylated and activated ERα in breast and skin cancer cells, sensitizing tumor cells to antiestrogens (Lee et al., 2012). In MCF-7 breast cancer cells, resveratrol prevented epigenetic silencing of the breast cancer 1 gene (Papoutsis et al., 2012) and restored phosphatase and tensin homolog (PTEN) tumor-suppressor expression by hypomethylating its promoter (Stefanska et al., 2012). Finally, resveratrol demonstrated a dose-related effect on gene-specific DNA methylation of Ras-association domain family 1 isoform A (RASSF-1α) in women at high risk for breast cancer after 12 weeks of consumption at 5 and 50 mg twice daily (Zhu et al., 2012).

RESVERATROL AND ITS ANALOGS AS REGULATORS OF HISTONE MODIFICATIONS

Chromatin remodeling is the dynamic modification of the chromatin architecture partially carried out by histone posttranslational modifications, which include acetylation, methylation, phosphorylation, ubiquitylation, glycosylation, and sumoylation (Esteller, 2008; Ellis et al., 2009). The final composition and conformation of chromatin as either relaxed transcriptionally active euchromatin or condensed inactive heterochromatin, associated with gene silencing, is defined by a variety of possible combinations of histone modifications together with tightly bound DNA, transcriptional machinery, and chromatin remodeling complexes. The most well-studied histone modifications are acetylation/deacetylation and methylation/demethylation, which are catalyzed by HATs/HDACs and HMTs/HDMs, respectively. Generally, acetylation has been positively associated with gene expression, and inhibitors of HDACs have been found to induce growth arrest and apoptosis in cancer cells. The molecular mechanisms that contribute to the chemopreventive and anticancer nature of resveratrol and its analogs include modulation of processes affecting histone modifications (Table 8.2).

A large number of structurally diverse HDAC inhibitors have been identified from natural sources and developed synthetically. Five HDAC inhibitors have been approved by the United States Food and Drug Administration (FDA), and many others are undergoing clinical trials either as monotherapies or in combinations with other drugs. The underlying mechanisms of the anticancer actions of HDAC inhibitors include the regulation of gene expression, induction of apoptosis and cell cycle arrest, inhibition of angiogenesis, and regulation of DNA damage and repair (Chun, 2016). Based on the available literature, resveratrol's actions as an HDAC inhibitor seem very promising. Most studies have reported downregulation of class 1 and 2 HDACs by resveratrol, implying that it promotes hyperacetylation rather than hypoacetylation. Resveratrol inhibited 11 human HDACs of class I, II, and IV in a dose-dependent manner in hepatocellular carcinoma cell lines with detectable histone hyperacetylation in HepG2 cells (Venturelli et al., 2013). Moreover, resveratrol demonstrated pan-HDAC inhibitory activity in human blood samples (Venturelli et al., 2013). Along the same lines, resveratrol has been shown to inhibit the expression levels of the MTA1/HDAC1,2 units of the nucleosome remodeling and deacetylase corepressor complex associated with deacetylation of histones and nonhistone proteins. It has been demonstrated that resveratrol, through downregulation

TABLE 8.2 Regulation of Histone Modifications by Resveratrol and Its Analogs

Resveratrol and Analogs	Dose	Epigenetic Mechanism[a]	Genes/Histones	Cells/Organ	References
Resveratrol (Res)	1–10 μM	↓ SIRT1	*NF-κB*, *MMP9*, *CXCR4*	HCT116, SW480 colorectal cancer cells	Buhrmann et al. (2016)
	5–100 μM 50 mg/kg	↓ MTA1, 2, 3, HDAC1, 2	*PTEN*	DU145, 22Rv1, PC3M, PTEN knockout prostate cancer cells and xenografts	Dhar et al. (2015)
	5–100 μM 0.1%–0.2% diet	↑ SIRT1	–	RWPE1 prostate epithelial cells, LNCaP, C42B, DU145, PC3 prostate cancer cells and prostate-specific PTEN knockout mice	Li et al. (2013)
SRT501 (proprietary formulation of Res)	100 μM	↑ SIRT1	*GADD45G*	ALL Reh and NALM-6 pre-B cells, MOLT-4 T-cells, Daudi, Raji and 2F7 Burkitt lymphoma cells, U266 multiple myeloma cells, REC-1 mantle cell lymphoma cells, DLBCL DHL-8, DLBCL Ly3 and DLBCL DHL-6 B-cell lymphoma cells	Scuto et al. (2013)
	5–100 μM	Class I, II and IV HDACs inhibition	–	HepG2, Hep3B, HuH7 hepatoma cells	Venturelli et al. (2013)
	20, 50, 100 μM	LSD-1 histone methyltransferase inhibition	–	HepG2, FAO hepatoma cells, HEK293 embryonic kidney cells and C2C12 myoblasts	Abdulla et al. (2013)
	40 μM	Histone phosphorylation	γH2AX, *ATM*, *BCL-2*, *BAX*, cleaved caspase-3	Peripheral blood and bone marrow mononuclear cells from CLL patients	Podhorecka et al. (2011)
	1–100 μM	↓ MTA1, HDAC1	p53	LNCaP, DU145 prostate cancer cells	Kai et al. (2010)

(Continued)

TABLE 8.2 Regulation of Histone Modifications by Resveratrol and Its Analogs—cont'd

Resveratrol and Analogs	Dose	Epigenetic Mechanism[a]	Genes/Histones	Cells/Organ	References
	5–50 μM	Histone phosphorylation	–	PANC-1 pancreatic cancer cells and MDA-MB-231 breast cancer cells	Hong et al. (2009)
	20 μM	↑ SIRT1	*MDR1, BCL-xL*	MCF-7 breast cancer cells	Bourguignon et al. (2009)
	7.5 μg/mL	SIRT1 activation	–	SIRT1 mutant mice	Wang et al. (2008)
Pter	50 μM 10 mg/kg	↓ MTA1	*PTEN, AKT, AR, MYC, CCND1, TGFB1, NOTCH2, ETS2, HSP90, NF-κB, IL1B, CDH1, VIM, VEGFC*, p53, p21, p27, CD31, Ki-67	LNCaP, DU145 prostate cancer cells and prostate-specific PTEN knockout mice	Dhar et al. (2016)
Res + Pter	5–15 μM	↓ SIRT1	γH2AX, *hTERT*	MCF10A breast epithelial cells and HCC1806, MDA-MB-157 breast cancer cells	Kala et al. (2015)
Res, Pter, Piceatannol (Pic), Trimethoxy resveratrol	5–100 μM 50 mg/kg	↓ MTA1	p53	RWPE1 prostate epithelial cells, LNCaP, DU145, PC3M prostate cancer cells	Li et al. (2013)
LYR71 (Trimeric resveratrol derivative)	5–80 μM 20, 40 mg/kg	Histone deacetylation	H3/H4 hyperacetylation, *MMP9*	MDA-MB-231 breast cancer cells and xenografts	Kim et al. (2008)
Pter + SAHA	50 μM Pter + 10 μM SAHA, 10 mg/kg Pter + 50 mg/kg SAHA	↓ MTA1	*HIF1A, VEGF, IL1B*	LNCaP and PC3M prostate cancer cells and prostate-specific PTEN knockout mice	Butt et al. (2017)

[a] ↑, *upregulation;* ↓, *downregulation.*

of MTA1, disrupted the MTA1/HDAC1 complex, resulting in increased p53 acetylation and apoptosis in prostate cancer cells (Kai et al., 2010). In addition, resveratrol reversed acetylation and activation of PTEN by inhibition of MTA1/HDAC1/2 in prostate cancer (Dhar et al., 2015a). Inhibition of MTA1 was further demonstrated by natural and synthetic analogs of resveratrol, particularly by Pter, which showed potent MTA1 inhibition and increased MTA1-dependent p53 acetylation in prostate cancer xenografts (Li et al., 2013).

Moreover, preclinical proof for Pter as a promising MTA1-targeted chemopreventive and therapeutic strategy has been recently demonstrated in prostate-specific PTEN-heterozygous and PTEN-null mouse models (Dhar et al., 2016). Furthermore, a recent report provided preclinical evidence that a combination treatment with Pter and an approved HDAC inhibitor, suberoylanilide hydroxamic acid (SAHA), may result in potent inhibition of prostate tumor growth and progression partially due to the ability of Pter to sensitize tumor cells to SAHA treatment. The effects were associated with a reduction in MTA1-regulated proangiogenic factors, hypoxia-induced factor 1 alpha, vascular endothelial growth factor, and interleukin 1β leading to decreased angiogenesis (Butt et al., 2017). Although resveratrol has been called a "natural HDAC inhibitor" by some (Kai et al., 2010; Dhar et al., 2015a; Kumar et al., 2015), others have reported histone H3 and H4 deacetylation leading to matrix metalloproteinase 9 (MMP9) silencing in breast cancer cells treated with LYR71, a trimeric resveratrol derivative (Kim et al., 2008). In fact, the activity of resveratrol as a histone acetylation/deacetylation modulator remains controversial, mainly because of its known effect on sirtuins (SIRTs), particularly SIRT1 expression and activity. SIRT1 is a nicotinamide adenine dinucleotide–dependent class III HDAC that possesses both oncogenic and tumor-suppressor potential, and it remains controversial whether inhibition or activation of SIRT1 is more efficacious in anticancer therapy (Bosch-Presegue and Vaquero, 2011). It has been shown that many human cancers exhibit reduced levels of SIRT1 compared with those of normal tissues, suggesting the tumor-suppressor nature of SIRT1 (Wang et al., 2008). When SIRT1 mutant mice that developed tumors in multiple tissues were treated with resveratrol, activation of SIRT1 resulted in reduced tumorigenesis (Wang et al., 2008). Other studies also have found increased expression and activation of SIRT1 in cancer cells and mouse models by resveratrol (Wang et al., 2008; Bourguignon et al., 2009; Li et al., 2013; Scuto et al., 2013) that are associated with beneficial physiological effects, such as significant reduction in prostate weight and the incidence of high-grade prostatic intraepithelial neoplasia lesions in prostate cancer (Li et al., 2013). SRT501, a SIRT1 activator that is a proprietary formulation of resveratrol with enhanced pharmacokinetic properties and improved bioavailability, has been tested in combination with the HDAC inhibitor panobinostat in human malignant lymphoid cell lines and shown to have SIRT1-promoted nuclear factor kappa B (NF-κB)/signal transducer-mediated and activator of transcription 3 (STAT3)/growth arrest DNA damage gene 45 gamma–mediated beneficial effects (Scuto et al., 2013). The antileukemia, antilymphoma, and antimyeloma effects associated with the inhibition of STAT3 and NF-κB signaling by resveratrol have been previously shown in several studies (Bhardwaj et al., 2007; Li et al., 2010; Singh et al., 2011). The reduction of chemoresistance and induction of apoptosis by resveratrol has been associated with resveratrol-induced SIRT1-p300 association and acetyltransferase inactivation leading to the deacetylation of hyaluronic acid/cluster of differentiation 44–induced β-catenin and NF-κB p65, inhibition of β-catenin-T-cell factor/lymphocyte enhancer factor and NF-κB-specific transcriptional activation, and the impairment of multidrug resistance 1 and B-cell lymphoma x gene expression in MCF-7 breast cancer cells (Bourguignon et al., 2009). Alternatively, a recent study has reported the anticancer effects of resveratrol in colorectal cancer cells through the downregulation of SIRT1 leading to NF-κB activation-dependent suppression of C-X-C chemokine receptor type 4 and MMP9 (Buhrmann et al., 2016). Moreover, recent combination treatments with resveratrol and Pter resulted in the downregulation of SIRT1 expression in three breast cancer cell lines, accompanied by decreases in DNA damage-associated gamma histone 2A variant X and telomerase expression. The authors concluded that combinatorial resveratrol and Pter epigenetic regulation might contribute to triple-negative breast cancer prevention and/or therapy (Kala et al., 2015). Noticeably, only a few studies have reported the effects of resveratrol on histone methylation (Abdulla et al., 2013) and histone phosphorylation (Hong et al., 2009; Podhorecka et al., 2011). Of note, while resveratrol predominantly regulates HDACs, other natural bioactive molecules such as curcumin, genistein, and EGCG very often demonstrate inhibitory effects on HATs in addition to HDACs (Kumar et al., 2016).

RESVERATROL AND ITS ANALOGS AS REGULATORS OF MICRORNAS

A subset of small ncRNAs, miRNAs are involved in posttranscriptional gene silencing. They inhibit the translation of about 60% of protein-coding genes by recognizing complementary sequences in the 3′UTR of the targeted messenger RNA (mRNA). Dietary polyphenols can regulate miRNA biosynthesis and expression, which in turn can affect miRNA interaction and the targeting of specific mRNAs (Ross and Davis, 2011). In cancer, miRNAs that target oncogenic mRNAs are called tumor-suppressor miRNAs, whereas miRNAs that target the mRNAs of tumor-suppressor genes are called oncomiRs. Numerous studies have investigated changes in miRNA expression by resveratrol in cancer, though only a few have directly linked miRNA modulation to the outcome. Yet accumulated data indicate that resveratrol and its analogs are attractive miRNA-mediated chemopreventive and therapeutic strategy options together with responsive circulating miRNAs, which are considered potential chemopreventive and predictive biomarkers for clinical development in cancer (Brase et al., 2011; Tili and Michaille, 2011; Dhar et al., 2015b; Su et al., 2015). An overview of the miRNA-mediated effects of resveratrol and its analogs in cancer is summarized in Table 8.3.

The miRNA-mediated role of resveratrol has been studied in various cancer types. Resveratrol treatment of T-cell and B-cell acute lymphoblastic leukemia cells induced levels of miR-196b and miR-1290 leading to insulin growth factor binding protein 3–mediated promotion of apoptosis and inhibition of cell proliferation and migration (Zhou et al., 2017). Another study has reported on the upregulation of miR-15a and miR-16-1 upon resveratrol treatment of acute lymphoblastic leukemia cells resulting in apoptosis (Azimi et al., 2015). Pic inhibited the tumor necrosis factor alpha (TNFα)/NF-κB pathway in human leukemia cells by downregulating v-akt murine thymoma viral oncogene/forkhead box 3–mediated miR-183 expression (Liu and Chang, 2012). In various cancer cells, resveratrol increased the levels of miR-326 leading to pyruvate kinase muscle 2–mediated promotion of apoptosis (Wu et al., 2016). Treatment of MCF-7 and MDA-MB-231 breast cancer cells with resveratrol resulted in the downregulation of miR-125b-5p, miR-200c-3p, and miR-542-3p and the upregulation of miR-409-3p and miR-122-5p leading to B-cell lymphoma 2 (Bcl-2), X-linked inhibitor of apoptosis, and cyclin-dependent kinase–mediated apoptosis and cell cycle arrest (Venkatadri et al., 2016). In the E2-dependent mammary carcinoma rodent model, resveratrol demonstrated differential regulation of miR-21, miR-129, miR-204, and miR-489 in tumorous versus normal tissues (Qin et al., 2014). A recent study demonstrated that Pter increased the levels of miR-205, resulting in proto-oncogene tyrosine-protein kinase/focal adhesion kinase–mediated inhibition of epithelial-to-mesenchymal transition in triple-negative breast cancer cells and xenografts (Su et al., 2015). Moreover, Pter treatment of endometrial cancer cells caused downregulation of miR-663b, which resulted in B-cell lymphoma-like 14–mediated decreased cell viability and increased apoptosis (Wang et al., 2017).

TABLE 8.3 Regulation of MicroRNAs by Resveratrol and Its Analogs

Resveratrol and Analogs	Dose	microRNAs[a]	Target MicroRNA/Function	Cells/Organs	References
Resveratrol (Res)	25–100 μM	↑ miR-196b, miR-1290	IGFBP3/apoptosis, migration, proliferation, cell cycle	TALL-104 T-cell acute lymphoblastic leukemia cells and SUP-B15 B-cell acute lymphoblastic leukemia cells	Zhou et al. (2017)
	25–200 μM	↑ miR-326	PKM2/cell viability, apoptosis, endoplasmic reticulum stress	DLD-1 colon cancer cells, HeLa cervical cancer cells, MCF-7 breast cancer cells and HepG2 liver cancer cells	Wu et al. (2016)
	50–300 μM	↓ miR-125b-5p, miR-200c-3p ↑ miR-409-3p, miR-122-5p and ↓ miR-542-3p	Bcl-2, XIAP, CDKs/ apoptosis, cell cycle	MCF-7 and MDA-MB-231 breast cancer cells	Venkatadri et al. (2016)
	12.5–400 μM	↑ miR-34c	KITLG/cell viability, proliferation, migration, invasion and apoptosis	HT-29 and HCT-116 colorectal cancer cells and xenografts	Yang et al. (2015a)
	15–100 μM	↑ miR-15a, miR-16-1	apoptosis	CCRF-CEM acute lymphoblastic leukemia cells	Azimi et al. (2015)
	25–100 μM 40, 100 mg/kg	↑ miR-328	MMP2/cell migration	HOS, MG-63, U2OS, Saos-2, and 143B osteosarcoma cells and xenografts	Yang et al. (2015b)
	50–150 μM 8 mg/kg	↓ miR-21, miR-30a-5p ↓ miR-19	P53, PTEN, EGFR, STAT3, COX-2, NF-κB and PI3K/ Akt/mTOR pathway; cell cycle, apoptosis	Rat C6 glioma cells, rats bearing C6 glioma cells	Wang et al. (2015)
	5, 25 mg/kg	↑ miR-21, miR-129 ↑ miR-204, miR-489	Differential effect in normal tissue	Rat mammary carcinoma	Qin et al. (2014)
	10 μM	↑ miR-34a	E2F3/Sirt1, growth inhibition, apoptosis, MAPK/Erk1/2, chemosensitizer to 5-fluorouracil	DLD-1 colon cancer cells	Kumazaki et al. (2013)
	10–100 μM	↓ miR-21	pIκB, nuclear NF-κB decrease/apoptosis	U251 glioma cells	Li et al. (2013)

Continued

TABLE 8.3 Regulation of MicroRNAs by Resveratrol and Its Analogs—cont'd

Resveratrol and Analogs	Dose	microRNAs[a]	Target MicroRNA/Function	Cells/Organs	References
	10–100 μM	↓ miR-21	BCL-2/apoptosis	PANC-1, CFPAC-1, MIA Paca-2 pancreatic cancer cells	Liu et al. (2013)
	12.5–50 μM	↑ miR-622	K-Ras/cell growth, cell cycle	16HBE-T bronchial epithelial cells, H460 l ung cancer cells	Han et al. (2012)
	50 μM	↑ miR-663 ↓ miR-21, miR-17, miR-25 ↓ miR-26a, miR-92a	TGFβ1/TGFβR1 and TGFβR2, PTEN, PDCD4, E-cad, SMAD7 increase	SW480 colon cancer cells	Tili et al. (2010b)
	50 μM	↑ miR-663	miR-155, JunB, JunD	THP-1 monocytic cells and human blood monocytes	Tili et al. (2010a)
	50 μM	↓ 23 miRs, ↑ 28 miRs	miRNA microarray study	LNCaP prostate cancer cells	Dhar et al. (2011)
Combination with quercetin (1:1)	0–60 μg/mL	↓ miR-27a	ZBTB10 –mediated inhibition of Sp1, Sp3, Sp4 mRNA and protein, surviving	HT-29 colon cancer cells	Del Follo-Martinez et al. (2013)
Pterostilbene (Pter)	25–100 μM	↓ miR-663b	BCL2L14/inhibition of cell viability and induction of apoptosis	HTB-111 and Ishikawa endometrial cancer cells	Wang et al. (2017)
	2.5–10 μM 10 mg/kg	↑ miR-205	Src/epithelial-to-mesenchymal transition	MCF-7, MDA-MB-231, Hs578 t breast cancer cells and xenografts	Su et al. (2015)
	0.1–5 μM 2 mg/kg	↑ miR-205	GRP78-mediated inhibition of stem cell renewal and maintenance	CD133 + stem cells from GBM8401 and U87MG glioblastoma cells and xenografts	Huynh et al. (2015)
Res, Pter	25–100 μM 50 mg/kg	↓ miR-17-5p, miR-20a ↓ miR-106a, miR-106b	PTEN/apoptosis, growth inhibition, treatment-response miRs	DU145, LNCaP prostate cancer cells and xenografts, serum	Dhar et al. (2015b)
Piceatannol (Pic)	30 μM	↑ miR-129	BCL-2/growth, apoptosis	HCT116 and HT-29 colorectal cancer	Zhang et al. (2014)
	1–10 μM	↓ miR-183	B-TrCP/TNFα/NF-κB; SP1, ADAM17	U937 leukemia cells	Liu et al. (2012)

[a] ↑, *upregulation* ↓, *downregulation.*

In HT-29 and HCT-116 colorectal cancer cells, treatment with resveratrol induced miR-34c/KIT proto-oncogene receptor tyrosine kinase ligand—mediated apoptosis and inhibited cell proliferation, invasion, and migration (Yang et al., 2015a), whereas Pic induced apoptosis by increasing miR-129-mediated downregulation of Bcl-2 (Zhang et al., 2014). Resveratrol treatment of colon cancer cells resulted in decreased levels of several oncomiRs and increased levels of the putative tumor suppressor miR-663 that targets transforming growth factor β (TGFβ) and downstream small worm phenotype mothers against decapentaplegic homolog effectors (Tili et al., 2010b). Interestingly, upregulation of miR-663 was also linked to resveratrol-mediated anti-inflammatory effects through the activator protein 1 signaling pathway (Tili et al., 2010a). The combination of resveratrol with 5-fluorouracil induced a synergistic increase in growth inhibition and apoptosis in colon cancer cells through the suppression of the miR-34a-mediated mitogen-activated kinase-like protein/extracellular signal-regulated kinase 1 and 2 signaling pathway, specifying resveratrol as a chemosensitizer (Kumazaki et al., 2013). Resveratrol in combination with quercetin also demonstrated miR-27a/zinc finger and BTB domain-containing 10—mediated inhibition of colon cancer cell survival (Del Follo-Martinez et al., 2013). Resveratrol upregulated tumor suppressor miR-622, affecting the expression of the Kirsten rat sarcoma viral oncogene homolog in human bronchial epithelial cells and lung cancer cells (Han et al., 2012). Resveratrol's effects on miR-21-mediated apoptosis were also seen in pancreatic cancer cells (Liu et al., 2013). The miRNA-mediated antiglioma effect of resveratrol and Pter has been reported, suggesting miR-21 and miR-205 involvement in apoptosis (Li et al., 2013) and stem cell renewal and maintenance (Huynh et al., 2015), respectively. In the study with allografts, mediation has been demonstrated by miR-21, miR-30a-5p, and miR-19 in the activation of tumor suppressors and the inhibition of oncogenic signaling upon treatment with resveratrol in vitro and in vivo (Wang et al., 2015). Resveratrol inhibited the migration of osteosarcoma cells through miR-328-mediated repression of MMP2 (Yang et al., 2015b). Lastly, it was observed that resveratrol modulated 51 miRNAs in LNCaP prostate cancer cells (Dhar et al., 2011), and subsequently it was demonstrated that resveratrol and Pter directly regulate miR-17—92 cluster-mediated rescue of the PTEN tumor suppressor in DU145 and LNCaP prostate cancer cells (Dhar et al., 2015b). The investigators also demonstrated that Pter treatment resulted in significantly diminished tumor growth in miRNA-overexpressing prostate cancer xenografts by downregulating miR-17 and miR-106a while simultaneously increasing PTEN levels. Moreover, Pter treatment reversed the increase in cell proliferation and the decrease in apoptosis caused by miR-17/106a overexpression in tumors (Dhar et al., 2015b).

CONCLUSIONS

Here, we aimed to give an overview of the role of resveratrol and its natural analogs as novel potential epigenetic agents for cancer chemoprevention and treatment. Although many epi-drugs are in clinical development, to date only limited epigenetic therapies have been approved by the FDA for use in cancer. Growing evidence demonstrates the epigenetic mechanisms of action by dietary polyphenols in the alteration of gene expression, apoptosis, the cell cycle, and angiogenesis. However, it has not yet been determined which natural bioactive molecules will be therapeutically effective against cancer. Particularly, although resveratrol and its analogs have been shown to regulate interrelated key epigenetic mechanisms that involve DNA methylation, histone and nonhistone protein modifications, and miRNAs, further exploration of their efficacy in preclinical studies is necessary. At the same time, there is obvious potential for greater effects from resveratrol and its analogs as HDAC inhibitors, especially when combined with other chemotherapeutic agents. Chemosensitization of tumor cells by resveratrol and analogs in combinatorial settings may become the most promising strategy for better outcomes in cancer. In addition, available experimental data suggest the miRNA-mediated potential for resveratrol and analogs as chemopreventive and therapeutic agents. Utilization of resveratrol-regulated circulating miRNAs as noninvasive prognostic and predictive biomarkers promises to be a breakthrough for evaluating and monitoring clinical trials. In summary, "drugging" the epigenome with natural dietary compounds, such as resveratrol and its analogs, promises new chemopreventive options for a wide range of diseases including cancer. We believe that natural compounds with epigenetic mechanisms of action will likely play a fundamental role in forthcoming "diet-precision" medicine. Finally, the characteristic structural simplicity of the resveratrol molecule allows for the rational design of a number of its synthetic analogs, derivatives, adducts, and conjugates as new possibilities for epi-drug development.

Acknowledgments

The writing of this chapter was supported by the Department of Defense Prostate Cancer Research Program under Award Number W81XWH-13-1-0370 and partially by the National Cancer Institute of the National Institutes of Health under Award Number R15CA216070 to Dr. Anait S. Levenson. Views and opinions of, and endorsements by the authors do not reflect those of the US Army of the Department of Defense or the National Institutes of Health. We are grateful to Ms. Elena V. Levenson for critically editing the manuscript.

References

Abbas, A., Patterson 3rd, W., Georgel, P.T., 2013. The epigenetic potentials of dietary polyphenols in prostate cancer management. Biochem. Cell. Biol. 91 (6), 361–368.

Abdulla, A., Zhao, X., Yang, F., 2013. Natural polyphenols inhibit lysine-specific Demethylase-1. J. Biochem. Pharmacol. Res. 1 (1), 56–63.

Ahuja, N., Easwaran, H., Baylin, S.B., 2014. Harnessing the potential of epigenetic therapy to target solid tumors. J. Clin. Investig. 124 (1), 56–63.

Arrowsmith, C.H., Bountra, C., Fish, P.V., Lee, K., Schapira, M., 2012. Epigenetic protein families: a new frontier for drug discovery. Nat. Rev. Drug Discov. 11 (5), 384–400.

Azimi, A., Hagh, M.F., Talebi, M., Yousefi, B., Hossein pour feizi, A.A., Baradaran, B., et al., 2015. Time–and concentration–dependent effects of resveratrol on miR15a and miR16-1 expression and apoptosis in the CCRF-CEM acute lymphoblastic leukemia cell line. Asian Pac. J. Cancer Prev. 16 (15), 6463–6468.

Berner, C., Aumuller, E., Gnauck, A., Nestelberger, M., Just, A., Haslberger, A.G., 2010. Epigenetic control of estrogen receptor expression and tumor suppressor genes is modulated by bioactive food compounds. Ann. Nutr. Metab. 57 (3–4), 183–189.

Bhardwaj, A., Sethi, G., Vadhan-Raj, S., Bueso-Ramos, C., Takada, Y., Gaur, U., et al., 2007. Resveratrol inhibits proliferation, induces apoptosis, and overcomes chemoresistance through down-regulation of STAT3 and nuclear factor-kappaB-regulated antiapoptotic and cell survival gene products in human multiple myeloma cells. Blood 109 (6), 2293–2302.

Bishayee, A., 2009. Cancer prevention and treatment with resveratrol: from rodent studies to clinical trials. Cancer Prev. Res. (Phila.) 2 (5), 409–418.

Bosch-Presegue, L., Vaquero, A., 2011. The dual role of sirtuins in cancer. Genes Cancer 2 (6), 648–662.

Bourguignon, L.Y., Xia, W., Wong, G., 2009. Hyaluronan-mediated CD44 interaction with p300 and SIRT1 regulates beta-catenin signaling and NFkappaB-specific transcription activity leading to MDR1 and Bcl-xL gene expression and chemoresistance in breast tumor cells. J. Biol. Chem. 284 (5), 2657–2671.

Bowers, J.L., Tyulmenkov, V.V., Jernigan, S.C., Klinge, C.M., 2000. Resveratrol acts as a mixed agonist/antagonist for estrogen receptors alpha and beta. Endocrinology 141 (10), 3657–3667.

Brase, J.C., Johannes, M., Schlomm, T., Falth, M., Haese, A., Steuber, T., et al., 2011. Circulating miRNAs are correlated with tumor progression in prostate cancer. Int. J. Cancer 128 (3), 608–616.

Buhrmann, C., Shayan, P., Popper, B., Goel, A., Shakibaei, M., 2016. Sirt1 is required for resveratrol-mediated chemopreventive effects in colorectal cancer cells. Nutrients 8 (3), 145.

Butt, N.A., Kumar, A., Dhar, S., Rimando, A.M., Akhtar, I., Hancock, J.C., et al., 2017. Targeting MTA1/HIF-1alpha signaling by pterostilbene in combination with histone deacetylase inhibitor attenuates prostate cancer progression. Cancer Med. 6 (11), 2673–2685.

Chakraborty, A., Gupta, N., Ghosh, K., Roy, P., 2010. In vitro evaluation of the cytotoxic, anti-proliferative and anti-oxidant properties of pterostilbene isolated from Pterocarpus marsupium. Toxicol. In Vitro 24 (4), 1215–1228.

Chakraborty, S., Levenson, A.S., Biswas, P.K., 2013. Structural insights into Resveratrol's antagonist and partial agonist actions on estrogen receptor alpha. BMC Struct. Biol. 13, 27.

Chiou, Y.S., Tsai, M.L., Nagabhushanam, K., Wang, Y.J., Wu, C.H., Ho, C.T., et al., 2011. Pterostilbene is more potent than resveratrol in preventing azoxymethane (AOM)-induced colon tumorigenesis via activation of the NF-E2-related factor 2 (Nrf2)-mediated antioxidant signaling pathway. J. Agric. Food Chem. 59 (6), 2725–2733.

Chun, P., 2016. Histone deacetylase inhibitors in medical therapeutics. In: Tollefsbol, T.O. (Ed.), Medical Epigenetics. Elsevier Inc, United Kingdom.

Constant, J., 1997. Alcohol, ischemic heart disease, and the French paradox. Coron. Artery Dis. 8 (10), 645–649.

Del Follo-Martinez, A., Banerjee, N., Li, X., Safe, S., Mertens-Talcott, S., 2013. Resveratrol and quercetin in combination have anticancer activity in colon cancer cells and repress oncogenic microRNA-27a. Nutr. Cancer 65 (3), 494–504.

Dhar, S., Hicks, C., Levenson, A.S., 2011. Resveratrol and prostate cancer: promising role for microRNAs. Mol. Nutr. Food Res. 55 (8), 1219–1229.

Dhar, S., Kumar, A., Li, K., Tzivion, G., Levenson, A.S., 2015a. Resveratrol regulates PTEN/Akt pathway through inhibition of MTA1/HDAC unit of the NuRD complex in prostate cancer. Biochim. Biophys. Acta 1853 (2), 265–275.

Dhar, S., Kumar, A., Rimando, A.M., Zhang, X., Levenson, A.S., 2015b. Resveratrol and pterostilbene epigenetically restore PTEN expression by targeting oncomiRs of the miR-17 family in prostate cancer. Oncotarget 6 (29), 27214–27226.

Dhar, S., Kumar, A., Zhang, L., Rimando, A.M., Lage, J.M., Lewin, J.R., et al., 2016. Dietary pterostilbene is a novel MTA1-targeted chemopreventive and therapeutic agent in prostate cancer. Oncotarget 7 (14), 18469–18484.

Du, Q., Luu, P.L., Stirzaker, C., Clark, S.J., 2015. Methyl-CpG-binding domain proteins: readers of the epigenome. Epigenomics 7 (6), 1051–1073.

Ellis, L., Atadja, P.W., Johnstone, R.W., 2009. Epigenetics in cancer: targeting chromatin modifications. Mol. Cancer Ther. 8 (6), 1409–1420.

Esteller, M., 2008. Epigenetics in cancer. N. Engl. J. Med. 358 (11), 1148–1159.

Fleming, I., Fisslthaler, B., Busse, R., 1995. Calcium signaling in endothelial cells involves activation of tyrosine kinases and leads to activation of mitogen-activated protein kinases. Circ. Res. 76 (4), 522–529.

Fulda, S., Debatin, K.M., 2004. Sensitization for tumor necrosis factor-related apoptosis-inducing ligand-induced apoptosis by the chemopreventive agent resveratrol. Cancer Res. 64 (1), 337–346.

Gehm, B.D., McAndrews, J.M., Chien, P.Y., Jameson, J.L., 1997. Resveratrol, a polyphenolic compound found in grapes and wine, is an agonist for the estrogen receptor. Proc. Natl. Acad. Sci. U.S.A. 94 (25), 14138–14143.

Gehm, B.D., Levenson, A.S., 2006. Resveratrol as a phytoestrogen. In: Aggarwal, B.B., Shishodia, S. (Eds.), Resveratrol in Health and Disease. Taylor & Francis Group, LLC, United States of America.

Giorgio, C., Dell'Aversana, C., Altucci, L., 2016. Epigenetics in acute myeloid leukemia. In: Tollefsbol, T.O. (Ed.), Medical Epigenetics. Elsevier Inc., United Kingdom.

Gusman, J., Malonne, H., Atassi, G., 2001. A reappraisal of the potential chemopreventive and chemotherapeutic properties of resveratrol. Carcinogenesis 22 (8), 1111–1117.

Han, Z., Yang, Q., Liu, B., Wu, J., Li, Y., Yang, C., et al., 2012. MicroRNA-622 functions as a tumor suppressor by targeting K-Ras and enhancing the anticarcinogenic effect of resveratrol. Carcinogenesis 33 (1), 131–139.

Hasan, M., Bae, H., 2017. An overview of stress-induced resveratrol synthesis in grapes: perspectives for resveratrol-enriched grape products. Molecules 22 (2), E294.

Heyn, H., Esteller, M., 2012. DNA methylation profiling in the clinic: applications and challenges. Nat. Rev. Genet. 13 (10), 679–692.

Holoch, D., Moazed, D., 2015. RNA-mediated epigenetic regulation of gene expression. Nat. Rev. Genet. 16 (2), 71–84.

Hong, Y.B., Kang, H.J., Kim, H.J., Rosen, E.M., Dakshanamurthy, S., Rondanin, R., et al., 2009. Inhibition of cell proliferation by a resveratrol analog in human pancreatic and breast cancer cells. Exp. Mol. Med. 41 (3), 151–160.

Huynh, T.T., Lin, C.M., Lee, W.H., Wu, A.T., Lin, Y.K., Lin, Y.F., et al., 2015. Pterostilbene suppressed irradiation-resistant glioma stem cells by modulating GRP78/miR-205 axis. J. Nutr. Biochem. 26 (5), 466–475.

Jain, M.G., Ferrenc, R.G., Rehm, J.T., Bondy, S.J., Rohan, T.E., Ashley, M.J., et al., 2000. Alcohol and breast cancer mortality in a cohort study. Breast Cancer Res. Treat. 64 (2), 201–209.

Jang, M., Cai, L., Udeani, G.O., Slowing, K.V., Thomas, C.F., Beecher, C.W., et al., 1997. Cancer chemopreventive activity of resveratrol, a natural product derived from grapes. Science 275 (5297), 218–220.

Jones, P.A., 2014. At the tipping point for epigenetic therapies in cancer. J. Clin. Investig. 124 (1), 14–16.

Kai, L., Samuel, S.K., Levenson, A.S., 2010. Resveratrol enhances p53 acetylation and apoptosis in prostate cancer by inhibiting MTA1/NuRD complex. Int. J. Cancer 126 (7), 1538–1548.

Kala, R., Shah, H.N., Martin, S.L., Tollefsbol, T.O., 2015. Epigenetic-based combinatorial resveratrol and pterostilbene alters DNA damage response by affecting SIRT1 and DNMT enzyme expression, including SIRT1-dependent gamma-H2AX and telomerase regulation in triple-negative breast cancer. BMC Canc. 15, 672.

Kang, C.H., Moon, D.O., Choi, Y.H., Choi, I.W., Moon, S.K., Kim, W.J., et al., 2011. Piceatannol enhances TRAIL-induced apoptosis in human leukemia THP-1 cells through Sp1- and ERK-dependent DR5 up-regulation. Toxicol. In Vitro 25 (3), 605–612.

Kapetanovic, I.M., Muzzio, M., Huang, Z., Thompson, T.N., McCormick, D.L., 2011. Pharmacokinetics, oral bioavailability, and metabolic profile of resveratrol and its dimethylether analog, pterostilbene, in rats. Cancer Chemother. Pharmacol. 68 (3), 593–601.

Kim, J.E., Kim, H.S., Shin, Y.J., Lee, C.S., Won, C., Lee, S.A., et al., 2008. LYR71, a derivative of trimeric resveratrol, inhibits tumorigenesis by blocking STAT3-mediated matrix metalloproteinase 9 expression. Exp. Mol. Med. 40 (5), 514–522.

Kristiansen, S., Soletormos, G., 2016. Clinical utility of solid tumor epigenetics. In: Tollefsbol, T.O. (Ed.), Medical Epigenetics. Elsevier Inc, United Kingdom.

Kumar, A., Dhar, S., Rimando, A.M., Lage, J.M., Lewin, J.R., Zhang, X., et al., 2015. Epigenetic potential of resveratrol and analogs in preclinical models of prostate cancer. Ann. N.Y. Acad. Sci 1348 (1), 1–9.

Kumar, A., Butt, N.A., Levenson, A.S., 2016. Natural epigenetic-modifying molecules in medical therapy. In: Tollefsbol, T.O. (Ed.), Medical Epigenetics. Elsevier Inc, United Kingdom.

Kumazaki, M., Noguchi, S., Yasui, Y., Iwasaki, J., Shinohara, H., Yamada, N., et al., 2013. Anti-cancer effects of naturally occurring compounds through modulation of signal transduction and miRNA expression in human colon cancer cells. J. Nutr. Biochem. 24 (11), 1849–1858.

Le, M.G., Hill, C., Kramar, A., Flamanti, R., 1984. Alcoholic beverage consumption and breast cancer in a French case-control study. Am. J. Epidemiol. 120 (3), 350–357.

Lee, H., Zhang, P., Herrmann, A., Yang, C., Xin, H., Wang, Z., et al., 2012. Acetylated STAT3 is crucial for methylation of tumor-suppressor gene promoters and inhibition by resveratrol results in demethylation. Proc. Natl. Acad. Sci. U.S.A. 109 (20), 7765–7769.

Levenson, A.S., Jordan, V.C., 1999. Selective oestrogen receptor modulation: molecular pharmacology for the millennium. Eur. J. Cancer 35 (14), 1974–1985.

Li, G., Rivas, P., Bedolla, R., Thapa, D., Reddick, R.L., Ghosh, R., et al., 2013. Dietary resveratrol prevents development of high-grade prostatic intraepithelial neoplastic lesions: involvement of SIRT1/S6K axis. Cancer Prev. Res. (Phila.) 6 (1), 27–39.

Li, H., Jia, Z., Li, A., Jenkins, G., Yang, X., Hu, J., et al., 2013. Resveratrol repressed viability of U251 cells by miR-21 inhibiting of NF-kappaB pathway. Mol. Cell. Biochem. 382 (1–2), 137–143.

Li, K., Dias, S.J., Rimando, A.M., Dhar, S., Mizuno, C.S., Penman, A.D., et al., 2013. Pterostilbene acts through metastasis-associated protein 1 to inhibit tumor growth, progression and metastasis in prostate cancer. PLoS One 8 (3), e57542.

Li, T., Wang, W., Chen, H., Li, T., Ye, L., 2010. Evaluation of anti-leukemia effect of resveratrol by modulating STAT3 signaling. Int. Immunopharmacol. 10 (1), 18–25.

Link, A., Balaguer, F., Goel, A., 2010. Cancer chemoprevention by dietary polyphenols: promising role for epigenetics. Biochem. Pharmacol. 80 (12), 1771–1792.

Liu, P., Liang, H., Xia, Q., Li, P., Kong, H., Lei, P., et al., 2013. Resveratrol induces apoptosis of pancreatic cancers cells by inhibiting miR-21 regulation of BCL-2 expression. Clin. Transl. Oncol. 15 (9), 741–746.

Liu, W.H., Chang, L.S., 2012. Suppression of Akt/Foxp3-mediated miR-183 expression blocks Sp1-mediated ADAM17 expression and TNFalpha-mediated NFkappaB activation in piceatannol-treated human leukemia U937 cells. Biochem. Pharmacol. 84 (5), 670–680.

Lubecka, K., Kurzava, L., Flower, K., Buvala, H., Zhang, H., Teegarden, D., et al., 2016. Stilbenoids remodel the DNA methylation patterns in breast cancer cells and inhibit oncogenic NOTCH signaling through epigenetic regulation of MAML2 transcriptional activity. Carcinogenesis 37 (7), 656–668.

Medina-Aguilar, R., Perez-Plasencia, C., Marchat, L.A., Gariglio, P., Garcia Mena, J., Rodriguez Cuevas, S., et al., 2016. Methylation landscape of human breast cancer cells in response to dietary compound resveratrol. PLoS One 11 (6), e0157866.

Mirza, S., Sharma, G., Parshad, R., Gupta, S.D., Pandya, P., Ralhan, R., 2013. Expression of DNA methyltransferases in breast cancer patients and to analyze the effect of natural compounds on DNA methyltransferases and associated proteins. J. Breast Cancer 16 (1), 23–31.

Opipari Jr., A.W., Tan, L., Boitano, A.E., Sorenson, D.R., Aurora, A., Liu, J.R., 2004. Resveratrol-induced autophagocytosis in ovarian cancer cells. Cancer Res. 64 (2), 696–703.

Pan, M.H., Chang, Y.H., Tsai, M.L., Lai, C.S., Ho, S.Y., Badmaev, V., et al., 2008. Pterostilbene suppressed lipopolysaccharide-induced up-expression of iNOS and COX-2 in murine macrophages. J. Agric. Food Chem. 56 (16), 7502–7509.

Pan, M.H., Chiou, Y.S., Chen, W.J., Wang, J.M., Badmaev, V., Ho, C.T., 2009. Pterostilbene inhibited tumor invasion via suppressing multiple signal transduction pathways in human hepatocellular carcinoma cells. Carcinogenesis 30 (7), 1234–1242.

Papoutsis, A.J., Borg, J.L., Selmin, O.I., Romagnolo, D.F., 2012. BRCA-1 promoter hypermethylation and silencing induced by the aromatic hydrocarbon receptor-ligand TCDD are prevented by resveratrol in MCF-7 cells. J. Nutr. Biochem. 23 (10), 1324–1332.

Pervaiz, S., Holme, A.L., 2006. Mechanism of apoptosis by resveratrol. In: Aggarwal, B.B., Shishodia, S. (Eds.), Resveratrol in Health and Disease. Taylor & Francis Group, LLC, United States of America.

Podhorecka, M., Halicka, D., Klimek, P., Kowal, M., Chocholska, S., Dmoszynska, A., 2011. Resveratrol increases rate of apoptosis caused by purine analogues in malignant lymphocytes of chronic lymphocytic leukemia. Ann. Hematol. 90 (2), 173–183.

Qin, W., Zhang, K., Clarke, K., Weiland, T., Sauter, E.R., 2014. Methylation and miRNA effects of resveratrol on mammary tumors vs. normal tissue. Nutr. Cancer 66 (2), 270–277.

Remsberg, C.M., Yanez, J.A., Ohgami, Y., Vega-Villa, K.R., Rimando, A.M., Davies, N.M., 2008. Pharmacometrics of pterostilbene: preclinical pharmacokinetics and metabolism, anticancer, antiinflammatory, antioxidant and analgesic activity. Phytother. Res. 22 (2), 169–179.

Ross, S.A., Davis, C.D., 2011. MicroRNA, nutrition, and cancer prevention. Adv. Nutr. 2 (6), 472–485.

Sato, M., Maulik, G., Bagchi, D., Das, D.K., 2000. Myocardial protection by protykin, a novel extract of trans-resveratrol and emodin. Free Radic. Res. 32 (2), 135–144.

Scuto, A., Kirschbaum, M., Buettner, R., Kujawski, M., Cermak, J.M., Atadja, P., et al., 2013. SIRT1 activation enhances HDAC inhibition-mediated upregulation of GADD45G by repressing the binding of NF-kappaB/STAT3 complex to its promoter in malignant lymphoid cells. Cell Death Dis. 4, e635.

Sengupta, S., Jordan, V.C., 2008. Selective estrogen modulators as an anticancer tool: mechanisms of efficiency and resistance. Adv. Exp. Med. Biol. 630, 206–219.

Shen, L., Kondo, Y., Ahmed, S., Boumber, Y., Konishi, K., Guo, Y., et al., 2007. Drug sensitivity prediction by CpG island methylation profile in the NCI-60 cancer cell line panel. Cancer Res. 67 (23), 11335–11343.

Singh, B., Shoulson, R., Chatterjee, A., Ronghe, A., Bhat, N.K., Dim, D.C., et al., 2014. Resveratrol inhibits estrogen-induced breast carcinogenesis through induction of NRF2-mediated protective pathways. Carcinogenesis 35 (8), 1872–1880.

Singh, N.P., Singh, U.P., Hegde, V.L., Guan, H., Hofseth, L., Nagarkatti, M., et al., 2011. Resveratrol (trans-3,5,4′-trihydroxystilbene) suppresses EL4 tumor growth by induction of apoptosis involving reciprocal regulation of SIRT1 and NF-kappaB. Mol. Nutr. Food Res. 55 (8), 1207–1218.

Stefanska, B., Salame, P., Bednarek, A., Fabianowska-Majewska, K., 2012. Comparative effects of retinoic acid, vitamin D and resveratrol alone and in combination with adenosine analogues on methylation and expression of phosphatase and tensin homologue tumour suppressor gene in breast cancer cells. Br. J. Nutr. 107 (6), 781–790.

Su, C.M., Lee, W.H., Wu, A.T., Lin, Y.K., Wang, L.S., Wu, C.H., et al., 2015. Pterostilbene inhibits triple-negative breast cancer metastasis via inducing microRNA-205 expression and negatively modulates epithelial-to-mesenchymal transition. J. Nutr. Biochem. 26 (6), 675–685.

Su, L., David, M., 2000. Distinct mechanisms of STAT phosphorylation via the interferon-alpha/beta receptor. Selective inhibition of STAT3 and STAT5 by piceatannol. J. Biol. Chem. 275 (17), 12661–12666.

Thakkar, K., Geahlen, R.L., Cushman, M., 1993. Synthesis and protein-tyrosine kinase inhibitory activity of polyhydroxylated stilbene analogues of piceatannol. J. Med. Chem. 36 (20), 2950–2955.

Tili, E., Michaille, J.J., Adair, B., Alder, H., Limagne, E., Taccioli, C., et al., 2010a. Resveratrol decreases the levels of miR-155 by upregulating miR-663, a microRNA targeting JunB and JunD. Carcinogenesis 31 (9), 1561–1566.

Tili, E., Michaille, J.J., Alder, H., Volinia, S., Delmas, D., Latruffe, N., et al., 2010b. Resveratrol modulates the levels of microRNAs targeting genes encoding tumor-suppressors and effectors of TGFbeta signaling pathway in SW480 cells. Biochem. Pharmacol. 80 (12), 2057–2065.

Tili, E., Michaille, J.J., 2011. Resveratrol, MicroRNAs, inflammation, and cancer. J. Nucleic Acids 2011, 102431.

Vanden Berghe, W., 2012. Epigenetic impact of dietary polyphenols in cancer chemoprevention: lifelong remodeling of our epigenomes. Pharmacol. Res. 65 (6), 565–576.

Venkatadri, R., Muni, T., Iyer, A.K., Yakisich, J.S., Azad, N., 2016. Role of apoptosis-related miRNAs in resveratrol-induced breast cancer cell death. Cell Death Dis. 7, e2104.

Venturelli, S., Berger, A., Bocker, A., Busch, C., Weiland, T., Noor, S., et al., 2013. Resveratrol as a pan-HDAC inhibitor alters the acetylation status of histone [corrected] proteins in human-derived hepatoblastoma cells. PLoS One 8 (8), e73097.

Vo, N.T., Madlener, S., Bago-Horvath, Z., Herbacek, I., Stark, N., Gridling, M., et al., 2010. Pro- and anticarcinogenic mechanisms of piceatannol are activated dose dependently in MCF-7 breast cancer cells. Carcinogenesis 31 (12), 2074–2081.

Wang, G., Dai, F., Yu, K., Jia, Z., Zhang, A., Huang, Q., et al., 2015. Resveratrol inhibits glioma cell growth via targeting oncogenic microRNAs and multiple signaling pathways. Int. J. Oncol. 46 (4), 1739–1747.

Wang, H., Zhang, Y., Kriska, A., Chen, H., 2016. Epigenetic regulation in cancer metastasis. In: Tollefsbol, T.O. (Ed.), Medical Epigenetics. Elsevier Inc, United Kingdom.

Wang, R.H., Sengupta, K., Li, C., Kim, H.S., Cao, L., Xiao, C., et al., 2008. Impaired DNA damage response, genome instability, and tumorigenesis in SIRT1 mutant mice. Cancer Cell 14 (4), 312–323.

Wang, Y.L., Shen, Y., Xu, J.P., Han, K., Zhou, Y., Yang, S., et al., 2017. Pterostilbene suppresses human endometrial cancer cells in vitro by down-regulating miR-663b. Acta Pharmacol. Sin. 38 (10), 1394–1400.

Wu, H., Wang, Y., Wu, C., Yang, P., Li, H., Li, Z., 2016. Resveratrol induces cancer cell apoptosis through MiR-326/PKM2-mediated ER stress and mitochondrial fission. J. Agric. Food Chem. 64 (49), 9356–9367.

Xiao, J., Li, G., 2016. Prospective advances in medical epigenetics. In: Tollefsbol, T.O. (Ed.), Medical Epigenetics. Elsevier Inc, United Kingdom.

Yang, S., Li, W., Sun, H., Wu, B., Ji, F., Sun, T., et al., 2015a. Resveratrol elicits anti-colorectal cancer effect by activating miR-34c-KITLG in vitro and in vivo. BMC Cancer 15, 969.

Yang, S.F., Lee, W.J., Tan, P., Tang, C.H., Hsiao, M., Hsieh, F.K., et al., 2015b. Upregulation of miR-328 and inhibition of CREB-DNA-binding activity are critical for resveratrol-mediated suppression of matrix metalloproteinase-2 and subsequent metastatic ability in human osteosarcomas. Oncotarget 6 (5), 2736–2753.

Zhang, H., Jia, R., Wang, C., Hu, T., Wang, F., 2014. Piceatannol promotes apoptosis via up-regulation of microRNA-129 expression in colorectal cancer cell lines. Biochem. Biophys. Res. Commun. 452 (3), 775–781.

Zhou, W., Wang, S., Ying, Y., Zhou, R., Mao, P., 2017. miR-196b/miR-1290 participate in the antitumor effect of resveratrol via regulation of IGFBP3 expression in acute lymphoblastic leukemia. Oncol. Rep. 37 (2), 1075–1083.

Zhu, W., Qin, W., Zhang, K., Rottinghaus, G.E., Chen, Y.C., Kliethermes, B., et al., 2012. Trans-resveratrol alters mammary promoter hypermethylation in women at increased risk for breast cancer. Nutr. Cancer 64 (3), 393–400.

CHAPTER

9

Epigenetic Mechanisms of Quercetin and Other Flavonoids in Cancer Therapy and Prevention

Gian Luigi Russo[1], *Paola Ungaro*[2]

[1]National Research Council, Institute of Food Sciences, Avellino, Italy; [2]National Research Council, Institute of Experimental Endocrinology and Oncology 'G. Salvatore', Naples, Italy

Abstract

Flavonoids are the most abundant polyphenols in fruits, vegetables, and plant-derived beverages, and their presence in food has been associated with many positive outcomes in terms of protection against overall mortality and degenerative diseases. Recent findings suggest that quercetin and other flavonoids may possess the capacity to counteract the adverse epigenetic regulation involved in various forms of cancer. Epigenetics is the study of heritable changes in gene expression independent from those occurring in the genome and includes histone modifications, DNA methylation, and posttranscriptional gene regulation by noncoding RNAs.

In general, hypomethylation leads to activation of genes that cause cancer, whereas global DNA hypomethylation is common in the metastatic phase. In several cases, it has been shown that hypermethylation of tumor-suppressive genes determines the silencing of these genes that is commonly associated with cancer development. In recent years, several studies have demonstrated that dietary factors may influence the structure of chromatin and contribute to regulating both gene transcription and translation. Here, we review the role of quercetin and other flavonoids in regulating epigenetic processes in cancer cells. Being that a large part of the studies analyzed are based on preclinical models, we conclude that although promising, the impact of flavonoids in the regulation of key epigenetic mechanisms controlling cancer requires validation from clinical studies.

Keywords

DNA methylation; Flavonoids; Histone modifications; microRNA; Quercetin.

INTRODUCTION

Polyphenols are part of the large family of phytochemicals, a growing group (>10,000) of nonnutritional compounds often associated with the prevention of degenerative pathologies (Bilotto et al., 2013; Kim and Kim, 2013; Teiten et al., 2013). The presence of polyphenols in fruits and vegetables has been associated with recent positive outcomes of "protective" diets against overall mortality in observational

Epigenetics of Cancer Prevention
https://doi.org/10.1016/B978-0-12-812494-9.00009-3

studies. In fact, participants in the PREDIMED study (Prevention with Mediterranean Diet) showed that those at high risk who reported high polyphenol intake (mainly stilbenes and lignans) presented a reduced risk of overall mortality compared with subjects who had lower intakes (Tresserra-Rimbau et al., 2014). In a more recent systematic review and dose-response meta-analysis of prospective studies, fruit and vegetable intakes were associated with reduced risk of all-cause mortality, including cancer and cardiovascular diseases (Aune et al., 2017). According to Phenol-Explorer, one of the most cited and accredited polyphenol databases (Neveu et al., 2010; Rothwell et al., 2012, 2013), polyphenols include the following classes: flavonoids, lignans, phenolic acids, stilbenes, nonphenolic metabolites, other polyphenols. Flavonoids, the topic of the present chapter, are divided into the following subclasses: anthocyanins, chalcones, dihydrochalcones, dihydroflavonols, flavanols, flavanones, flavones, flavonols, and isoflavonoids. Table 9.1 reports the chemical structure of the most representative compounds in each flavonoid subclass, the epigenetic activity of which will be analyzed in the present chapter.

During the last decade, a flourish of studies have described how polyphenols can play a significant role in the prevention of degenerative pathologies, including cancer as well as cardiovascular, neurodegenerative, and metabolic diseases acting on multiple mechanisms controlling cell proliferation, gene expression, and redox homeostasis (Watson et al., 2014). Starting from evidence indicating that epigenetic mechanisms, throughout the diet and other environmental conditions, can influence later health outcomes on the exposed individuals (Russo et al., 2017b; Zeisel, 2009), experimental data have been accumulated on the capacity of polyphenols to interfere with the regulation of the epigenome in such a way as to modulate the progression of common complex diseases. A significant number of interesting reviews have been published on these topics in the last 5 years (Busch et al., 2015; Huang et al., 2016; Petronis, 2010; Remely et al., 2015; Russo et al., 2017b; Vahid et al., 2015). Considering that epigenetic modifications that include DNA methylation, histone modification, and other patterns are associated with carcinogenesis and cancer progression, consistent efforts have been devoted to exploring the possibility that dietary polyphenols could contribute to cancer prevention and therapy by modulating the activity of enzymes, such as histone methyltransferases (HMTs), DNA methyltransferases (DNMTs), histone acetyltransferases (HATs), histone demethylases (HDMs), histone deacetylases (HDACs), and other related proteins—e.g., methyl-CpG binding proteins (MBPs) and DNMT1-associated protein (Gao and Tollefsbol, 2015; George et al., 2017; Hardy and Tollefsbol, 2011; Mukherjee et al., 2015; Vanden Berghe, 2012; Verma, 2013).

In this chapter, after reviewing the basic concepts of epigenetics, we select a specific class of polyphenols, the flavonoids, and critically analyze the most significant experimental observations, showing if and how they can participate in the control of cancer development and its progression via modulation of epigenetic pathways, such as histone modifications, DNA methylation, and microRNA (miRNA) expression. Considering the abundant and complex literature on this topic, we apologize in advance to those authors whose works are not cited owing to space limitations, and we hope that our critical analysis may contribute to inspiring future investigations in this area.

EPIGENETIC MECHANISMS: A GENERAL DESCRIPTION

The term "epigenetics" indicates mechanisms that regulate reversible heritable changes in gene expression not caused by changes in the DNA sequence and that are transmitted through meiosis and mitotic cell division. This means

TABLE 9.1 Chemical Structures of Selected Flavonoids Affecting Epigenetic Mechanisms in Cancer

Flavonoid Subclass	Name	Chemical Structure[a]	Epigenetic Mechanisms
Flavonols	Quercetin	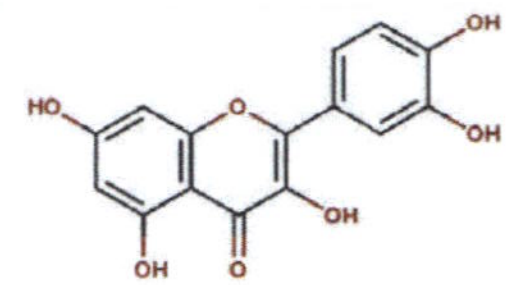	Inhibition of DNA methylation by DNMTs Reduced expression of DNMT1 Demethylation of gene promoter Suppression of HAT activity Activation of HAT Inhibition of HDAC
Flavanols	(−)-Epigallocatechin 3-O-gallate (EGCG)	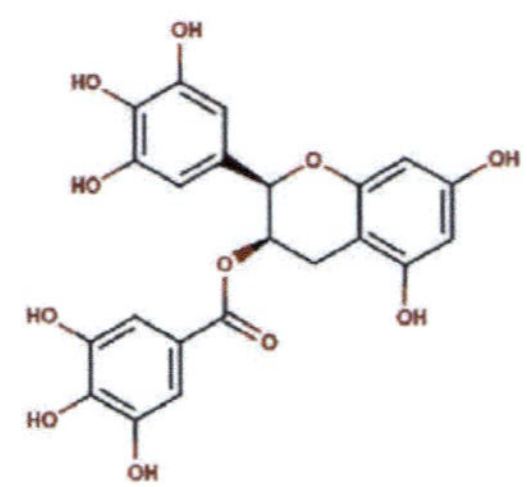	Reduction of DNMT and HDAC activities
	3, 3′-di-O-gallate (procyanidin B2)	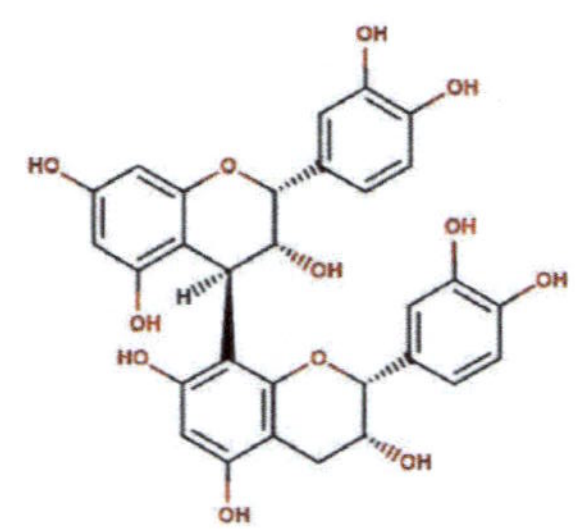	Reduction of DNMT activity

Continued

TABLE 9.1 Chemical Structures of Selected Flavonoids Affecting Epigenetic Mechanisms in Cancer—cont'd

Flavonoid Subclass	Name	Chemical Structure[a]	Epigenetic Mechanisms
Isoflavonoids	Genistein	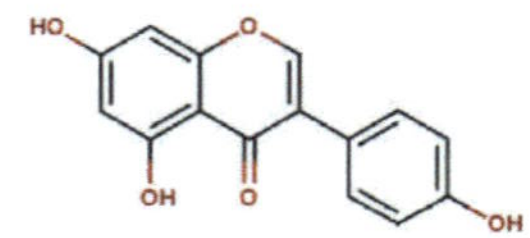	Amplification of 180, 170, and 150 miRNA Reverse DNA hypermethylation of BRCA1 and BRCA2 Demethylation promoter sequence of miR-29a and miR-1256
	Daidzein	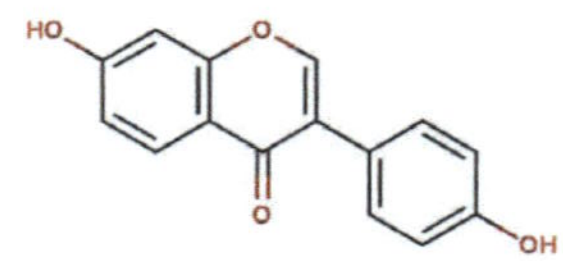	
Dihydroflavonols	3,6-Dihydroxyflavone	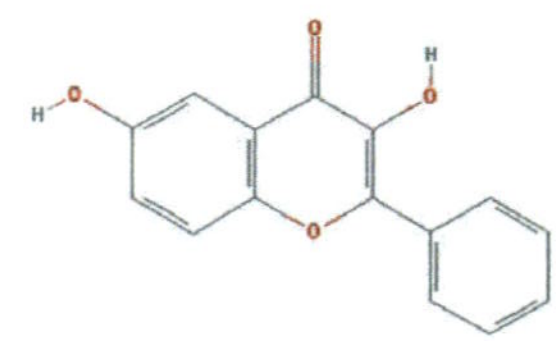	Downregulation of miR34a and upregulation of miR21 Inhibition of DNMT
Flavones	Apigenin	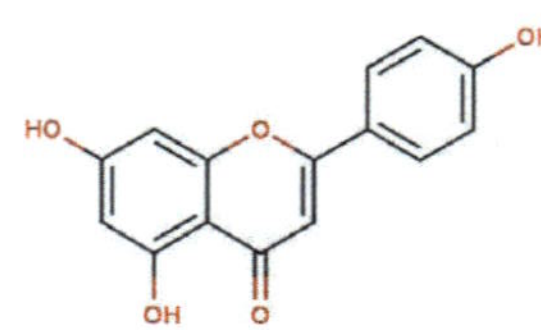	Demethylation of Nrf2 promoter Downregulation of expression of DNMTs

[a] *Structures were retrieved from the Phenol-Explorer database (Neveu et al., 2010; Rothwell et al., 2012, 2013).*

that the phenotype is determined by both the genotype and the epigenotype, the latter changing during development due to environmental influences. Epigenetic mechanisms are capable of mediating the effects of environmental conditions on exposed individuals, and for this reason they represent a link between the environment and gene expression and might act as determinants of later health outcomes (Zeisel, 2009). Nowadays, growing evidence suggests that epigenetic marks are involved not only in cancer that is being extensively studied (Vanden Berghe, 2012), but also in the development of chronic noncommunicable metabolic diseases such as obesity, diabetes, neurodegeneration, and cardiovascular disease (Petronis, 2010). Because epigenetic phenomena can be modulated and reverted by external or internal environmental factors, epigenetics is now considered a starting point for clinical intervention, and especially, nutrients and bioactive food components can modify the expression of the genes involved in physiologic and pathologic processes by modulating their epigenetic patterns.

Different epigenetic modifications have been identified and described so far: DNA methylation of cytosine within cytosine-phosphate-guanine (CpG) dinucleotides, posttranslational modifications of histones that include methylation, acetylation, ubiquitination, and sumoylation of lysine, phosphorylation of serine and threonine, methylation of arginine, and a variety of noncoding RNAs (ncRNAs), including miRNAs, that regulate gene expression at multiple levels (Kim et al., 2009) (Fig. 9.1).

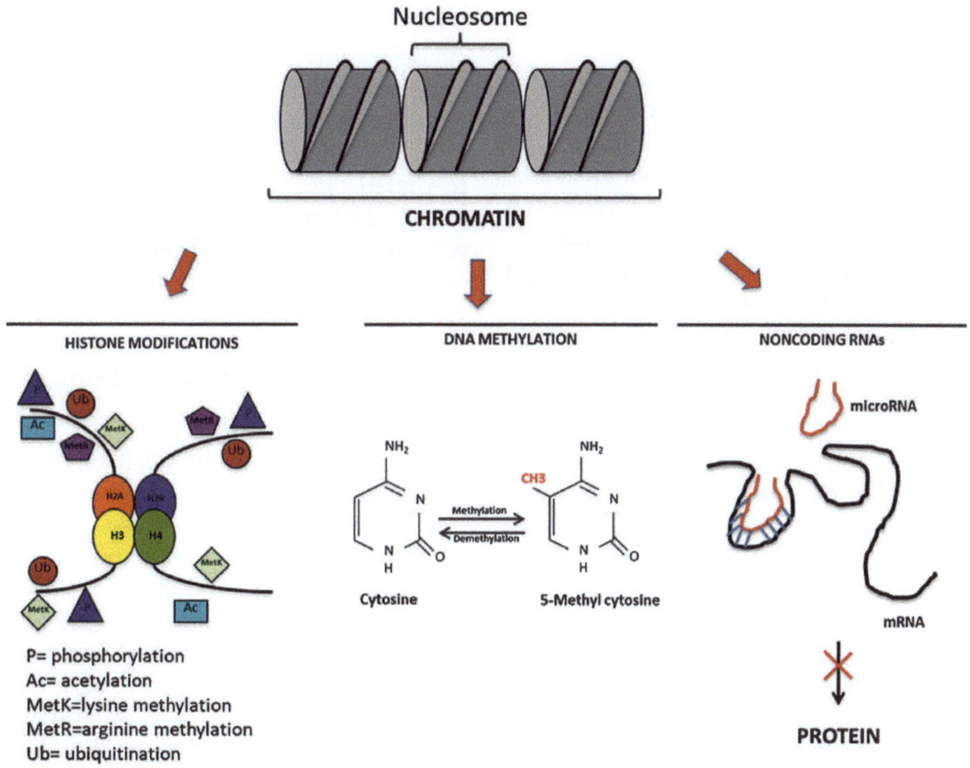

FIGURE 9.1 A scheme of the classical epigenetic mechanisms involved in gene expression regulation (see text for details).

DNA Methylation

DNA methylation, the best-known epigenetic mark (Chanet et al., 2012), consists of the addition of a methyl group to the 5′ position of the cytosine ring within CpG dinucleotides to form 5-methylcytosine. CpG dinucleotides are especially enriched in the promoter regions of genes and in regions of large repetitive sequences (e.g., LINE and ALU retrotransposon elements, and centromeric repeats) (Bird, 2002). Short CpG-rich regions are clustered at 5′ ends of many genes and constitute CpG islands that are present in more than 50% of human gene promoters (Wang and Leung, 2004). Normally, CpG islands remain unmethylated during development and in undifferentiated normal cells (Suzuki and Bird, 2008). Hypermethylation and hypomethylation of CpG islands within gene promoters can result in the silencing and activation of gene expression, respectively (Suzuki and Bird, 2008).

DNA methylation influences several key physiological processes, such as X chromosome inactivation, imprinting, and the silencing of germline-specific genes and repetitive elements. The modification at 5-methylcytosine is catalyzed by two types of DNMTs. DNMT1 is a maintenance methyltransferase that adds methyl groups to hemimethylated CpG sites following DNA replication (Plass, 2002; Plass and Soloway, 2002); DNMT3A and DNMT3B are de novo methyltransferases acting on both unmethylated and hemimethylated DNA (Jonesand and Liang, 2009; Lawand and Jacobsen, 2010). S-adenosylmethionine (SAM) is the universal methyl donor for methyltransferases, including DNMTs and protein methyltransferases (Loenen, 2006). SAM is synthesized in the methionine pathway from several precursors present in the diet, such as methionine, folate, choline, betaine, and vitamins B2, B6, and B12 (Feil and Fraga, 2012; McKay and Mathers, 2011). In rodents, it has been shown that diets deficient in methyl donors (i.e., low levels of folate, choline, and methionine) produce global DNA hypomethylation (Craciunescu et al., 2010; Mehedint et al., 2010; Niculescu et al., 2006; Pogribny et al., 2008, 2009). Conversely, an increased availability of methyl donors increases DNA methylation in specific loci (Ba et al., 2011; Waterland et al., 2006, 2008; Waterland and Jirtle, 2003).

Although in most cases DNA methylation is a stable epigenetic mark, DNA methylation recently has been shown to undergo active demethylation during development. In mammals, a loss of methylation yields multiple intermediate forms of CpG modification, including hydroxymethylation, formylation, and carboxylation (Ito et al., 2010). Indeed, a role for 5-hydroxymethylcytosine modification in mammalian DNA demethylation has been observed, in terms of acting as an intermediate in the active DNA demethylation pathway involving DNA repair and 5-hydroxymethylcytosine-specific DNA glycosylase activity (Lawand and Jacobsen, 2010). Of particular interest, reactive oxygen species and oxidative stress may affect DNA demethylation by DNA oxidation or ten-eleven translocation–mediated hydroxymethylation (Chia et al., 2011). Alternatively, demethylation can also occur passively by replication in the absence of functional maintenance of methylation pathways (Vanden Berghe, 2012).

Posttranslational Modifications of Histones

Within the eukaryotic cell nucleus, DNA is wound around a core of histone proteins with the resulting structure referred to as chromatin, a highly conserved structural polymer that plays a central role in regulated gene expression. The functional unit of chromatin is the nucleosome, which consists of 147 bp DNA wrapped around an octamer of histones comprising two H3/H4 dimers and two H2A/H2B dimers (Zhang et al., 1999). Short stretches of linker DNA enable the primary conformation to coil into 30 nm fibers or a higher-order chromatin structure as seen in the nucleus and have an appearance of beads on a string. Chromatin has regions of transcriptionally active euchromatin and inactive heterochromatin (Lund and van Lohuizen, 2004). The interconversion of these two states of the chromatin for DNA accessibility to transcription factors are dynamically regulated by groups of enzymes that remodel nucleosomes along the DNA or reversibly modify histones on their amino-terminal histone "tails," which extend from the core octamer. These modifications, including methylation, acetylation, phosphorylation, ribosylation, ubiquitination, sumoylation, and biotinylation (Kouzarides, 2007), take place at lysine, arginine, and serine residues and constitute the so-called "histone code" that is read and translated into transcriptional changes (Kouzarides, 2007). Histone modifications can be associated with either transcriptional activation or transcriptional repression and can regulate one another, providing regulatory cross talk. Methylation and acetylation of lysine residues are the best-characterized histone modifications. Lysine acetylation associates with transcriptional activation, since its major action is to neutralize the basic charge of the lysine residue and unfolded chromatin (Jenuwein and Allis, 2001). This function is not easy to observe in vivo, but biophysical analysis indicates that internucleosomal contacts play an important role in determining the higher-order chromatin structure. Conversely, histone methylation does not influence nucleosomal contacts and can be associated with both gene activation and gene repression depending on which specific lysine is modified. There is good evidence that trimethylation of lysine 4 on histone H3 is enriched at transcriptionally active gene promoters (Liang et al., 2004), whereas trimethylation of lysines 9 and 27 on histone H3 is present at transcriptionally repressed promoters. Similar to DNA methylation changes, various histone modifications are potentially reversible and are determined by chromatin writer, reader, and eraser enzyme complexes that add or remove covalent modifications to and from histone proteins (Ellis et al., 2009; Iacobuzio-Donahue, 2009). A number of histone-modifying enzymes, including various HATs and HDACs that add and remove acetyl groups from histone proteins, and HMTs and HDMs responsible for adding and removing methyl groups, respectively, have been identified in recent years (Haberland et al., 2009; Shi, 2007; Wang et al., 2008). Evidence now shows that histone- and DNA-modifying enzymes utilize intermediary metabolic products as substrates and cofactors for their activities. These metabolites include acetyl-CoA, uridine diphosphate-glucose, α-ketoglutarate, nicotinamide adenine dinucleotide (NAD+), flavin adenine dinucleotide, ATP, and SAM, whose concentrations are tightly linked to global cellular metabolism and energy levels (Luoand and Kuo, 2009; Teperino et al., 2010). These coregulators could act as metabolic sensors and can translate change in metabolism into change in gene expression by affecting the activity of transcription factors as well as changing the structure of the epigenome by modulating the activity of chromatin modifiers. In particular, nutritional status may influence the epigenome by directly affecting the activities of chromatin-modifying enzymes. Observations in mammalian cell culture experiments suggest that histone acetylation also depends on intracellular acetyl-CoA pools (Wellen et al., 2009). HMTs and HDMs also require metabolites for their enzymatic activity. For example, HMTs use methyl group donors present in the diet, including methionine, folate choline, betaine, and vitamins B2, B6, and B12. The idea that gene transcription is influenced by intermediary metabolic products through epigenetic mechanisms was suggested several years ago (Ernstand and Kells, 2010). The ability of specific foods and dietary regimens to prevent diseases and maintain health by epigenetic modulation has been taken into consideration and investigated in more detail. A current hypothesis implies that the response to either food intake and individual compounds present in or deriving from foods may affect the chromatin structure.

Noncoding RNA

In addition to DNA methylation and histone modifications, ncRNAs regulate chromatin architecture (Rouhi et al., 2008). The term ncRNA is commonly employed for RNA that does not encode a protein. Initially, ncRNAs were noted to perform catalytic functions in facilitating RNA splicing (Link et al., 2010); more recently, it has been recognized that they contribute to the posttranscriptional regulation of genes (Rouhi et al., 2008). These ncRNAs include long ncRNAs, miRNAs, and tiny RNAs that control various levels of gene expression in physiology and development, including chromatin architecture/epigenetic memory, transcription (enhancer function), and RNA splicing, editing, translation, and turnover (De Santa et al., 2010; Guiland and Esteller, 2009; Lujambio et al., 2008). The best-characterized family of small ncRNAs, miRNAs, are small single-stranded RNAs about 19–24 nucleotides in length that regulate gene expression through sequence-specific base pairing with 3′ untranslated regions of the target messenger RNA (mRNA), resulting in the degradation or inhibition of translation (He and Hannon, 2004). In metazoans, one miRNA can regulate several genes because miRNAs do not need to form a perfect base-pair match to their target site. They function by recruiting the RNA-induced silencing complex, a multiple-protein nuclease complex constituting at least TRBP, DICER1, and Argonaute 2 to target gene transcripts that are silenced via degradation of the mRNA or by locking the translation through complementary binding to target mRNA (Macfarlane and Murphy, 2010). The miRNAs control various levels of gene expression and are involved in cell growth and proliferation, differentiation, development, organogenesis, metabolism, and immunity. With such a large spectrum of biological processes, miRNA can influence multiple diseases including obesity, cancer, cardiovascular disease, and diabetes (Iorio and Croce, 2012). For example, miRNA profiles are now being used to classify human cancers (Nicoloso et al., 2009; Peter, 2009), and the development of miRNA-based therapies is currently under investigation.

EPIGENETIC MECHANISMS OF FLAVONOIDS

Quercetin

The flavonol quercetin (3,3′,4′,5,7-pentahydroxyflavone) (Table 9.1) is the most abundant flavonoid in fruits and vegetables, with a daily intake in the diet ranging between 5 and 40 mg/day, although these values can scale up to greater than 500 mg/day in subjects who normally consume high quantities of foods rich in flavonols, such as apples, onions, red grapes, and green tea. We mostly review the capacity of quercetin to interfere with different well-established "hallmarks of cancer," although these studies are largely based on cell lines (Russo et al., 2012, 2014a,b).

Circumstantial evidence exists about the capacity of quercetin to exert part of its anticancer effects by acting on epigenetic-controlling mechanisms. The molecule can inhibit DNA methylation catalyzed by prokaryotic Sssl DNMT and human DNMT1 in a concentration-dependent manner with an IC_{50} of 1.6 μM (Lee et al., 2005). The capacity of quercetin to inhibit DNMT was also confirmed in PC3 and DU145 prostate cancer (CaP) cell lines. Resistance in CaP to androgen-deprivation therapy with anti-androgens has often been associated with hypermethylation of the androgen receptor (AR) gene promoter. Both quercetin and curcumin synergistically restored AR mRNA and protein levels in AR-negative cell lines by global hypomethylation resulting in apoptotic induction via mitochondrial depolarization. The IC_{50} for quercetin in PC3 and DU145 was approximately 20–23 μM, and the combined treatment was more effective than a single monotreatment (Sharma et al., 2016). Quercetin can also inhibit about 40% the growth in RKO human colon cancer cells by inducing demethylation of the *CDKN2A* gene promoter and enhancing gene expression at a concentration of 80 μM (Tan et al., 2009). Similarly, in three bladder cancer cell lines (EJ, J82, and T24), quercetin (100 μM) strongly decreased (by 35%–70%) the DNA methylation of estrogen receptor-β -(ER-β), *CDKN2A,* and *RASSF1A* genes (Ma et al., 2006). In human breast cancer MDA-MB-231 cells, quercetin (100–200 μM) suppresses p300 HAT activity, resulting in the attenuation of p300-mediated acetylation of nuclear factor-κB (NF-κB) p50 (Xiao et al., 2011). In human esophageal 9706 cancer cells (Eca9706), a nanoliposomal formulation of quercetin (20–60 μM) administered alone or in combination with butyrate (1–4 mM) suppressed Eca9706 cell proliferation. This effect was associated with reduced global expression of DNMT1, NF-κB p65, HDAC1, and cyclin D1 as well as increased expression of caspase-3 and $p16^{INK4\alpha}$. The efficacy of quercetin and butyrate to reduce the level of DNA methylation on the $p16^{INK4\alpha}$ gene promoter was demonstrated by the order of methylation-specific PCR of $p16^{INK4\alpha}$, which resulted in the following for specific treatments: control > butyrate or quercetin > quercetin + butyrate. Overall, this result suggests that a combination of two epigenetic interfering agents—e.g., quercetin and butyrate—can target altered DNA methylation and histone acetylation via NF-κB cascade signaling (Zheng et al., 2014). In human leukemia HL60 cells, quercetin (25–100 μM administered for 12 h) increased histone H3 acetylation, which resulted in enhanced expression of the Fas ligand. In this cellular model, quercetin activated HAT and inhibited HADC, contributing to the modulation of histone acetylation (Lee et al., 2011). Unfortunately, the relationship between the concentrations of quercetin nanoparticles applied and the corresponding values of active quercetin is not clear.

An interesting approach derives from the work of Bishayee et al. (2015), who prepared synthetic poly(lactic-co-glycolic acid)-encapsulated gold-quercetin nanoparticles (NQs). The biological activity of this formulation (24 μg/mL) was tested on HepG2 hepatocarcinoma cells to measure HDAC activity and the expression levels of HDAC1 and HDAC2. NQ treatment for 24 h suppressed the HDAC enzymatic activity associated with decreased HDAC1 and HDAC2 expression levels, indicating NQ's capacity to act as a repressor of transcription and proliferation (Bishayee et al., 2015).

In a hamster model of 7,12-dimethylbenz[a]anthracene–induced buccal pouch carcinomas, quercetin, after intragastric administration of 25 mg/kg body weight for 1–8 weeks, ameliorated several markers associated with cell proliferation and survival, and this effect was positively correlated to the inhibition of HDAC1 and DNMT1 (Priyadarsini et al., 2011). An animal model was also employed to demonstrate that a prenatal diet enriched with quercetin can induce persistent alterations in phase I and II enzymes that may result in protection against cancer risk in offspring. Pregnant mice supplemented with 1 mmol/kg (338 mg/kg) quercetin until delivery generated adult females with altered capacity to metabolize the environmental pollutant benzo[a]pyrene (B[a]P). In fact, although the quercetin supplementation was stopped in the newborn mice, CYP1A1/CYP1B1 (phase I enzymes) and GSTP1/NQO1/UGT1A6 (phase II enzymes) were upregulated in the liver and lung tissues of females. Following an ex vivo analysis, it was shown that the amount of unmetabolized B[a]P was lower in the lung microsomes of offspring that had received quercetin during gestation. Overall, this study demonstrates that exposure to quercetin triggers epigenetic mechanisms in a tissue- and gender-dependent manner that reduces the formation of B[a]P-induced DNA adducts in offspring and consequently reduces cancer risk (Vanhees et al., 2012). It is worthwhile to note that the same authors demonstrated in a previous work that adult mice prenatally exposed to quercetin (302 mg/kg starting from 3 days before conception until the end of gestation) showed increased iron storage in the liver by upregulation of iron-associated cytokine expression of interleukin-6 (I6), IL10, hepcidin, and IL1β, but despite this increase, measurement of 8oxodG levels indicated that liver oxidative stress was significantly lowered in these animals. The authors suggest that prenatal quercetin exposure resulted in hypomethylation of repetitive elements (short interspersed nuclear element B1, or SINEB1), an epigenetic modification that persisted in offspring (Vanhees et al., 2011).

Searching PubMed for studies where the epigenetic effects of quercetin were investigated in clinical trials, no results were retrieved. The same search was performed on https://www.ClinicalTrials.gov, a service of the United States National Institutes of Health, filtering the search for all type of trials (e.g., recruiting, not yet recruiting, active, not recruiting, completed, etc.), with the same result.

Other Flavonoids

In an in silico analysis, the capacity of different flavonoids to bind and inhibit DNMTs in both humans and mice was compared. Although docking and simulation analyses indicated that the most energetically favorable complex was ((−)-Epigallocatechin 3-O-gallate (EGCG))−DNMT, in vitro assays on triple-negative breast cancer cells (MDA-MB 231) showed that a different flavanol, 3,3′-di-O-gallate (procyanidin B2), reduced DNMT activity with an IC_{50} of 6.88 ± 0.647 μM, compared with a value of 9.36 ± 1.0 μM for EGCG, and this inhibitory effect resulted in enhanced expression of DNMT target genes, such as E-cadherin, maspin, and BRCA1 (Shilpi et al., 2015). Considering the broad spectrum of action of EGCG in different diseases, these authors proposed procyanidin B2 as a more promising anticancer agent for medicinal applications.

In a different study involving the flavanol EGCG (1−100 μM) and butyrate (0.1−6 mM), the authors described that the combined treatment triggered apoptosis and cell cycle arrest in HT29, RKO, and HCT116 colorectal cancer cells. The epigenetic capacity of EGCG and butyrate individually and in combination was demonstrated by measuring the levels of DNMT1 and HDAC1, which were significantly reduced in the three cell lines investigated, with the combination being more effective than single treatments and thus indicating a combinatorial epigenetic effect (Saldanha et al., 2014). Using the same compound, a different group suggested that EGCG (50−150 μM) maintained its capacity to reduce HDAC and DNMT protein expression, but only in methylation-sensitive colon cell lines, such as HCT 116, whereas in others (e.g., HT29) HDAC and DNMT remained relatively stable (Moseley et al., 2013). To confirm the epigenetic potential of green tea components, it is worthwhile to cite a review published a few years ago in which the authors reported that in seven out of nine types of cell lines and tissues derived from different forms of cancer, green tea or EGCG decreased DNMT1 and reversed hypermethylation, whereas in three out of four mouse models of cancer development, the treatment confirmed the inhibitory effect of EGCG on DNA methylation (Henning et al., 2013).

In a subclass of isoflavonoids, genistein (40 μM) and daidzein (110 μM) showed demethylation/methylation effects on the CaP cell lines DU145 and LNCaP. In fact, using a genome-wide DNA-methylation analysis, a significant change in methylation patterns was observed in both CaP cell lines following genistein and daidzein treatments, and this effect was even more pronounced when compared with that of 5-azacytidine (a demethylating agent) (Karsli-Ceppioglu et al., 2015). In fact, in three different CaP cell lines (PC-3, DU145 and LNCaP) treated for 48 h with genistein and daidzein at the same concentrations, it was reported that these compounds induced amplification (2% variation) of 180, 170, and 150 miRNA in PC-3, DU145, and LNCaP cells, respectively. These changes in miRNA expression were comparable to those induced by 2 μM 5-azacytidine. The results suggest that both soy phytoestrogens interfere with miRNA methylation status and may possess potential protective effects against CaP (Adjakly et al., 2015; Rabiau et al., 2011). In a different study, treatment of the breast cancer cell lines MCF7, MDAMB231, and MCF10a with 8.5 μM genistein and 78.5 μM daidzein reversed DNA hypermethylation and rescued the expression of BRCA1 and BRCA2 genes. Considering that hypermethylation of tumor suppressors BRCA1 and BRCA2 has been measured in sporadic breast cancer cases, the epigenetic effects of genistein and daidzein seem to restore the functionality of these oncosuppressor genes (Bosviel et al., 2012). These results were only partially confirmed in the CaP cell lines DU145 and PC3, where 40 μM genistein and 110 μM daidzein decreased the methylation of BRCA1, but not BRCA2 that was unmethylated in control cells (Adjakly et al., 2011). A mixture of isoflavones (70.5% genistein, 26.3% daidzein, and 0.31% glycitein) applied to PCa cells at a concentration equivalent to 20 μM genistein resulted in demethylation of the methylation sites in the promoter sequences of miR-29a and miR-1256, increasing their expression. This upregulation was linked to the capacities of miR-29a and miR-1256 to inhibit the expression of TRIM68 and phosphoglycerate kinase 1 (PGK1). TRIM68 is significantly upregulated in CaP tissues and contributes to the activation of the AR. High expression of PGK1 is correlated with tumor progression in multiple cancer types. Acting on miR-29a and miR-1256, this mixture of isoflavone may represent a potential nontoxic demethylating agent for the prevention of CaP (Li et al., 2012).

Dihydroflavonols are a subclass of flavonoids including dihydroquercetin 3-*O*-rhamnoside, dihydroquercetin, and dihydromyricetin 3-*O*-rhamnoside. It has been reported that 3,6-dihydroxyflavone (3,6-DHF; 5–20 μM) effectively suppressed carcinogens NNK and B[a]P-induced carcinogenic transformation in MCF10A cells, and this effect was associated with the downregulation of miR34a and upregulation of miR21. Further, 3,6-DHF inhibited hypermethylation of the miR34a promoter, and additional docking studies suggest that the molecule acts as an effective DNMT1 inhibitor (Peng et al., 2015).

Apigenin belongs to the flavone subclass and has been proposed as a demethylation compound acting on the nuclear factor erythroid 2–related factor 2 (Nrf2) promoter at 15 CpG sites in mouse skin epidermal JB6 P+ cells in a concentration range of 1.56–6.25 μM. The molecule dose-dependently reversed the hypermethylated status of the 15 CpG sites in the Nrf2 promoter, resulting in increased translocation of Nrf2 and enhanced expression of NAD(P)H: quinone oxidoreductase 1 at the protein and mRNA levels. Apigenin was also effective in downregulating the expressions of DNMT1, DNMT3a, and DNMT3b as well as the activity of class I HDAC members (Paredes-Gonzalez et al., 2014).

FUTURE PERSPECTIVES

The studies reviewed in this chapter convincingly show that specific dietary flavonoids may directly or indirectly modulate specific epigenetic processes and lead to changes in chromatin modification patterns, and in some cases may correct aberrant expression of miRNA. Table 9.1 summarizes the key epigenetic mechanisms triggered by quercetin and other bioactive flavonoids. However, important weaknesses emerge that may limit the potential applicative outcomes of these compounds. Firstly, in the previous paragraphs, we paid great attention to indicate the working concentrations of flavonoids applied in different experimental models. It is easy to verify that these values are always in the micromolar range, sometimes above 100 μM. As we have reviewed elsewhere (Russo et al., 2014a,b, 2017b), these pharmacological concentrations are far from those acceptable for chemopreventive interventions. Secondly, the flavonoids used in the cited studies are always in the aglycone form. We know that after oral administration, circulating aglycones are almost absent in the blood; therefore, it is difficult to associate the encouraging anticancer effects of these classes of compounds observed in vitro with the potential outcomes in vivo, due to low bioavailability and the high metabolic transformation that they undergo before reaching target tissues (Russo et al., 2014a,b). Thirdly, the large majority of studies discussed above have been performed on cell lines, with only a few in animal models and none in humans. Even among ongoing clinical trials (https://www.clinicaltrials.gov database), we did not retrieve any clinical study designed to explore the epigenetic potential of flavonoids in cancer. We recently discussed the controversial and contradictory issues related to the efficacy of polyphenol administration in cancer treatment and prevention based on preclinical models (Russo et al., 2017a). We believe that adequate and specifically selected cellular models are essential for identifying the epigenetic molecular mechanisms triggered by quercetin and other flavonoids in cancer; however, we disagree with the easy extrapolation that because these compounds are present in a diet rich in fruits and vegetables, they can therefore prevent those forms of cancer that depend on uncontrolled epigenetic mechanisms. At the opposite end, the use of cellular models is essential to helping scientists understand how flavonoids can eventually be employed in the treatment of specific types of cancer, possibly in association with canonical chemotherapy/radiotherapy and other epigenetic interfering agents, such as butyrate, mentioned above.

CONCLUSIONS

The data summarized in Table 9.1 suggest the existence of features common to several flavonoids reviewed in this chapter. As an example, many show the ability to inhibit DNMT activity. This goal may be achieved directly by binding and inhibiting the enzyme, or indirectly by reducing its expression, for example. In silico analysis suggests that the former mechanism is most suitable, at least for flavanols. The existence of structural similarities among different flavonoids probably explains their common functional ability to interact with DNMTs. Future studies with ad hoc chemically modified analogs designed on the basis of in silico outcomes will confirm this possibility.

In conclusion, we believe that the study of the epigenetic potential of quercetin and other flavonoids is still in its infancy. Until now, specific mechanisms of actions have only been hypothesized based on data obtained from cell lines. The field remains strongly promising and of great interest for future research, but requires two key contributions: (1) a robust demonstration of a cause–effect relationship between the single agent applied and its direct interaction with one or more specific epigenetic markers, and (2) validation in clinical studies of the capability of specific flavonoids to influence the epigenetic processes involved in cancer development.

References

Adjakly, M., Bosviel, R., Rabiau, N., Boiteux, J.P., Bignon, Y.J., Guy, L., Bernard-Gallon, D., 2011. DNA methylation and soy phytoestrogens: quantitative study in DU-145 and PC-3 human prostate cancer cell lines. Epigenomics 3 (6), 795–803.

Adjakly, M., Ngollo, M., Dagdemir, A., Judes, G., Pajon, A., Karsli-Ceppioglu, S., Penault-Llorca, F., Boiteux, J.P., Bignon, Y.J., Guy, L., Bernard-Gallon, D., 2015. Prostate cancer: the main risk and protective factors-Epigenetic modifications. Ann. Endocrinol. 76 (1), 25–41.

Aune, D., Giovannucci, E., Boffetta, P., Fadnes, L.T., Keum, N., Norat, T., Greenwood, D.C., Riboli, E., Vatten, L.J., Tonstad, S., 2017. Fruit and vegetable intake and the risk of cardiovascular disease, total cancer and all-cause mortality-a systematic review and dose-response meta-analysis of prospective studies. Int. J. Epidemiol. 46 (3), 1029–1056.

Ba, Y., Yu, H., Liu, F., Geng, X., Zhu, C., Zhu, Q., Zheng, T., Ma, S., Wang, G., Li, Z., Zhang, Y., 2011. Relationship of folate, vitamin B12 and methylation of insulin-like growth factor-II in maternal and cord blood. Eur. J. Clin. Nutr. 65, 480–485.

Bilotto, S., Spagnuolo, C., Russo, M., Tedesco, I., Laratta, B., Russo, G.L., 2013. Dietary phytochemicals in chemoprevention of cancer: an update. Immunol. Endocr. Metab. Agents Med. Chem. 13, 2–24.

Bird, A., 2002. DNA methylation patterns and epigenetic memory. Genes Dev. 16, 6–21.

Bishayee, K., Khuda-Bukhsh, A.R., Huh, S.O., 2015. PLGA-loaded gold-nanoparticles precipitated with quercetin downregulate HDAC-Akt activities controlling proliferation and activate p53-ROS crosstalk to induce apoptosis in hepatocarcinoma cells. Mol. Cells 38 (6), 518–527.

Bosviel, R., Dumollard, E., Dechelotte, P., Bignon, Y.J., Bernard-Gallon, D., 2012. Can soy phytoestrogens decrease DNA methylation in BRCA1 and BRCA2 oncosuppressor genes in breast cancer? OMICS 16 (5), 235–244.

Busch, C., Burkard, M., Leischner, C., Lauer, U.M., Frank, J., Venturelli, S., 2015. Epigenetic activities of flavonoids in the prevention and treatment of cancer. Clin. Epigenet. 7, 64.

Chanet, A., Milenkovic, D., Manach, C., Mazur, A., Morand, C., 2012. Citrus flavanones: what is their role in cardiovascular protection? J. Agric. Food Chem. 60, 8809–8822.

Chia, N., Wang, L., Lu, X., Senut, M.C., Brenner, C., Ruden, D.M., 2011. Hypothesis: environmental regulation of 5-hydroxymethylcytosine by oxidative stress. Epigenetics 6, 853–856.

Craciunescu, C.N., Johnson, A.R., Zeisel, S.H., 2010. Dietary choline reverses some, but not all, effects of folate deficiency on neurogenesis and apoptosis in fetal mouse brain. J. Nutr. 140, 1162–1166.

De Santa, F., Barozzi, I., Mietton, F., Ghisletti, S., Polletti, S., Tusi, B.K., Muller, H., Ragoussis, J., Wei, C.L., Natoli, G., 2010. A large fraction of extragenic RNA pol II transcription sites overlap enhancers. PLoS Biol. 8, e1000384.

Ellis, L., Atadja, P.W., Johnstone, R.W., 2009. Epigenetics in cancer: targeting chromatin modifications. Mol. Cancer Ther. 8, 1409–1420.

Ernstand, J., Kells, M., 2010. Discovery and characterization of chromatin states for systematic annotation of the human genome. Nat. Biotechnol. 28, 817–825.

Feil, R., Fraga, M.F., 2012. Epigenetics and the environment: emerging patterns and implications. Nat. Rev. Genet. 13, 97–109.

Gao, Y., Tollefsbol, T.O., 2015. Impact of epigenetic dietary components on cancer through histone modifications. Curr. Med. Chem. 22 (17), 2051–2064.

George, V.C., Dellaire, G., Rupasinghe, H.P.V., 2017. Plant flavonoids in cancer chemoprevention: role in genome stability. J. Nutr. Biochem. 45, 1–14.

Guiland, S., Esteller, M., 2009. DNA methylomes, histone codesand miRNAs: tying it all together. Int. J. Biochem. Cell Biol. 41, 87–95.

Haberland, M., Montgomery, R.L., Olson, E.N., 2009. The many roles of histone deacetylases in development and physiology: implications for disease and therapy. Nat. Rev. Genet. 10 (1), 32–42.

Hardy, T.M., Tollefsbol, T.O., 2011. Epigenetic diet: impact on the epigenome and cancer. Epigenomics 3 (4), 503–518.

He, L., Hannon, G.J., 2004. MicroRNAs: small RNAs with a big role in gene regulation. Nat. Rev. Genet. 5, 522–531.

Henning, S.M., Wang, P., Carpenter, C.L., Heber, D., 2013. Epigenetic effects of green tea polyphenols in cancer. Epigenomics 5 (6), 729–741.

Huang, Z., Huang, Q., Ji, L., Wang, Y., Qi, X., Liu, L., Liu, Z., Lu, L., 2016. Epigenetic regulation of active Chinese herbal components for cancer prevention and treatment: a follow-up review. Pharmacol. Res. 114, 1–12.

Iacobuzio-Donahue, C.A., 2009. Epigenetic changes in cancer. Annu. Rev. Pathol. 4, 229–249.

Iorio, M.V., Croce, C.M., 2012. Causes and consequences of microRNA dysregulation. Cancer J. 18 (3), 215–222.

Ito, S., D'Alessio, A.C., Taranova, O.V., Hong, K., Sowers, L.C., Zhang, Y., 2010. Role of Tet proteins in 5mC to 5hmC conversion, ES-cell self-renewal and inner cell mass specification. Nature 466 (7310), 1129–1133.

Jenuwein, T., Allis, C.D., 2001. Translating the histone code. Science 293, 1074–1080.

Jonesand, P., Liang, G., 2009. Rethinking how DNA methylation patterns are maintained. Nat. Rev. Genet. 10, 805–811.

Karsli-Ceppioglu, S., Ngollo, M., Adjakly, M., Dagdemir, A., Judes, G., Lebert, A., Boiteux, J.P., Penault, L.F., Bignon, Y.J., Guy, L., Bernard-Gallon, D., 2015. Genome-wide DNA methylation modified by soy phytoestrogens: role for epigenetic therapeutics in prostate cancer? OMICS 19 (4), 209–219.

Kim, J.K., Samaranayake, M., Pradhan, S., 2009. Epigenetic mechanisms in mammals. Cell. Mol. Life Sci. 66, 596–612.

Kim, S.O., Kim, M.R., 2013. (-)-Epigallocatechin 3-gallate inhibits invasion by inducing the expression of Raf kinase inhibitor protein in AsPC1 human pancreatic adenocarcinoma cells through the modulation of histone deacetylase activity. Int. J. Oncol. 42 (1), 349–358.

Kouzarides, T., 2007. Chromatin modifications and their function. Cell 128 (4), 693–705.

Lawand, J., Jacobsen, S., 2010. Establishing, maintaining and modifying DNA methylation patterns in plants and animals. Nat. Rev. Genet. 11, 204–220.

Lee, W.J., Chen, Y.R., Tseng, T.H., 2011. Quercetin induces FasL-related apoptosis, in part, through promotion of histone H3 acetylation in human leukemia HL-60 cells. Oncol. Rep. 25 (2), 583–591.

Lee, W.J., Shim, J.Y., Zhu, B.T., 2005. Mechanisms for the inhibition of DNA methyltransferases by tea catechins and bioflavonoids. Mol. Pharmacol. 68 (4), 1018–1030.

Li, Y., Kong, D., Ahmad, A., Bao, B., Dyson, G., Sarkar, F.H., 2012. Epigenetic deregulation of miR-29a and miR-1256 by isoflavone contributes to the inhibition of prostate cancer cell growth and invasion. Epigenetics 7 (8), 940–949.

Liang, G., Lin, J.C., Wei, V., Yoo, C., Cheng, J.C., Nguyen, C.T., Weisenberger, D.J., Egger, G., Takai, D., Gonzales, F.A., Jones, P.A., 2004. Distinct localization of histone H3 acetylation and H3-K4 methylation to the transcription start sites in the human genome. Proc. Natl. Acad. Sci. U.S.A. 101, 7357–7362.

Link, A., Balaguer, F., Goel, A., 2010. Cancer chemoprevention by dietary polyphenols: promising role for epigenetics. Biochem. Pharmacol. 80 (12), 1771–1792.

Loenen, W., 2006. S-Adenosylmethionine: jack of all trades and master of everything? Biochem. Soc. Trans. 34, 330–333.

Lujambio, A., Calin, G.A., Villanueva, A., Ropero, S., Sánchez-Céspedes, M., Blanco, D., Montuenga, L.M., Rossi, S., Nicoloso, M.S., Faller, W.J., Gallagher, W.M., Eccles, S.A., Croce, C.M., Esteller, M., 2008. A microRNA DNA methylation signature for human cancer metastasis. Proc. Natl. Acad. Sci. U.S.A. 105, 13556–13561.

Lund, A.H., van Lohuizen, M., 2004. Epigenetics and cancer. Genes Dev. 18 (19), 2315–2335.

Luoand, J., Kuo, M.H., 2009. Linking nutrient metabolism to epigenetics. Cell Sci. Rev. 6, 49–54.

Ma, L., Feugang, J.M., Konarski, P., Wang, J., Lu, J., Fu, S., Ma, B., Tian, B., Zou, C., Wang, Z., 2006. Growth inhibitory effects of quercetin on bladder cancer cell. Front. Biosci. 11, 2275–2285.

Macfarlane, L.A., Murphy, P.R., 2010. MicroRNA: biogenesis, function and role in cancer. Curr.Genomics 11, 537–561.

McKay, J.A., Mathers, J.C., 2011. Diet induced epigenetic changes and their implications for health. Acta Physiol (Oxf.) 202 (2), 103–118.

Mehedint, M.G., Craciunescu, C.N., Zeisel, S.H., 2010. Maternal dietary choline deficiency alters angiogenesis in fetal mouse hippocampus. Proc. Natl. Acad. Sci. U.S.A. 107 (29), 12834–12839.

Moseley, V.R., Morris, J., Knackstedt, R.W., Wargovich, M.J., 2013. Green tea polyphenol epigallocatechin 3-gallate, contributes to the degradation of DNMT3A and HDAC3 in HCT 116 human colon cancer cells. Anticancer Res. 33 (12), 5325–5333.

Mukherjee, N., Kumar, A.P., Ghosh, R., 2015. DNA methylation and flavonoids in genitourinary cancers. Current Pharmacol. Rep. 1 (2), 112–120.

Neveu, V., Perez-Jimenez, J., Vos, F., Crespy, V., du Chaffaut, L., Mennen, L., Knox, C., Eisner, R., Cruz, J., Wishart, D., Scalbert, A., 2010. Phenol-Explorer: an online comprehensive database on polyphenol contents in foods. Database (Oxf.) 2010, bap024.

Nicoloso, M.S., Spizzo, R., Shimizu, M., Rossi, S., Calin, G.A., 2009. MicroRNAs-the micro steering wheel of tumour metastases. Nat. Rev. Cancer 9, 293–302.

Niculescu, M.D., Craciunescu, C.N., Zeisel, S.H., 2006. Dietary choline deficiency alters global and gene-specific DNA methylation in the developing hippocampus of mouse fetal brains. FASEB J. 20 (1), 43–49.

Paredes-Gonzalez, X., Fuentes, F., Su, Z.Y., Kong, A.N., 2014. Apigenin reactivates Nrf2 anti-oxidative stress signaling in mouse skin epidermal JB6 P + cells through epigenetics modifications. AAPS J. 16 (4), 727–735.

Peng, X., Chang, H., Gu, Y., Chen, J., Yi, L., Xie, Q., Zhu, J., Zhang, Q., Mi, M., 2015. 3,6-Dihydroxyflavone suppresses breast carcinogenesis by epigenetically regulating miR-34a and miR-21. Cancer Prev. Res. (Phila.) 8 (6), 509–517.

Peter, M.E., 2009. Let-7 nd miR-200 microRNAs: guardians against pluripotency and cancer progression. Cell Cycle 8, 843–852.

Petronis, A., 2010. Epigenetics as a unifying principle in the aetiology of complex traits and diseases. Nature 465 (7299), 721–727.

Plass, C., 2002. Cancer epigenomics. Hum. Mol. Genet. 11, 2479–2488.

Plass, C., Soloway, P., 2002. DNA methylation, imprinting and cancer. Eur. J. Hum. Genet. 10, 6–16.

Pogribny, I.P., Karpf, A.R., James, S.R., Melnyk, S., Han, T., Tryndyak, V.P., 2008. Epigenetic alterations in the brains of Fisher 344 rats induced by long-term administration of folate/methyl-deficient diet. Brain Res. 1237, 25–34.

Pogribny, I.P., Tryndyak, V.P., Bagnyukova, T.V., Melnyk, S., Montgomery, B., Ross, S.A., Latendresse, J.R., Rusyn, I., Beland, F.A., 2009. Hepatic epigenetic phenotype predetermines individual susceptibility to hepatic steatosis in mice fed a lipogenic methyl-deficient diet. J. Hepatol. 51, 176–186.

Priyadarsini, R.V., Vinothini, G., Murugan, R.S., Manikandan, P., Nagini, S., 2011. The flavonoid quercetin modulates the hallmark capabilities of hamster buccal pouch tumors. Nutr. Cancer 63 (2), 218–226.

Rabiau, N., Trraf, H.K., Adjakly, M., Bosviel, R., Guy, L., Fontana, L., Bignon, Y.J., Bernard-Gallon, D.J., 2011. miRNAs differentially expressed in prostate cancer cell lines after soy treatment. In Vivo 25 (6), 917–921.

Remely, M., Lovrecic, L., de la Garza, A.L., Migliore, L., Peterlin, B., Milagro, F.I., Martinez, A.J., Haslberger, A.G., 2015. Therapeutic perspectives of epigenetically active nutrients. Br. J. Pharmacol. 172 (11), 2756–2768.

Rothwell, J.A., Perez-Jimenez, J., Neveu, V., Medina-Remon, A., M'Hiri, N., Garcia-Lobato, P., Manach, C., Knox, C., Eisner, R., Wishart, D.S., Scalbert, A., 2013. Phenol-Explorer 3.0: a major update of the Phenol-Explorer database to incorporate data on the effects of food processing on polyphenol content. Database (Oxf.) 2013, bat070.

Rothwell, J.A., Urpi-Sarda, M., Boto-Ordonez, M., Knox, C., Llorach, R., Eisner, R., Cruz, J., Neveu, V., Wishart, D., Manach, C., Andres-Lacueva, C., Scalbert, A., 2012. Phenol-Explorer 2.0: a major update of the Phenol-Explorer database integrating data on polyphenol metabolism and pharmacokinetics in humans and experimental animals. Database (Oxf.) 2012, bas031.

Rouhi, A., Mager, D.L., Humphries, R.K., Kuchenbauer, F., 2008. MiRNAs, epigenetics, and cancer. Mamm. Genome 19, 517–525.

Russo, G.L., Russo, M., Spagnuolo, C., 2014a. The pleiotropic flavonoid quercetin: from its metabolism to the inhibition of protein kinases in chronic lymphocytic leukemia. Food Funct. 5 (10), 2393–2401.

Russo, G.L., Russo, M., Spagnuolo, C., Tedesco, I., Bilotto, S., Iannitti, R., Palumbo, R., 2014b. Quercetin: a pleiotropic kinase inhibitor against cancer. Cancer Treat Res. 159, 185–205.

Russo, G.L., Tedesco, I., Spagnuolo, C., Russo, M., 2017a. Antioxidant polyphenols in cancer treatment: friend, foe or foil? Semin. Cancer Biol. (Epub ahead of print.).

Russo, G.L., Vastolo, V., Ciccarelli, M., Albano, L., Macchia, P.E., Ungaro, P., 2017b. Dietary polyphenols and chromatin remodeling. Crit. Rev. Food Sci. Nutr. 57 (12), 2589–2599.

Russo, M., Spagnuolo, C., Tedesco, I., Bilotto, S., Russo, G.L., 2012. The flavonoid quercetin in disease prevention and therapy: facts and fancies. Biochem. Pharmacol. 83 (1), 6–15.

Saldanha, S.N., Kala, R., Tollefsbol, T.O., 2014. Molecular mechanisms for inhibition of colon cancer cells by combined epigenetic-modulating epigallocatechin gallate and sodium butyrate. Exp. Cell Res. 324 (1), 40–53.

Sharma, V., Kumar, L., Mohanty, S.K., Maikhuri, J.P., Rajender, S., Gupta, G., 2016. Sensitization of androgen refractory prostate cancer cells to anti-androgens through re-expression of epigenetically repressed androgen receptor - synergistic action of quercetin and curcumin. Mol. Cell. Endocrinol. 431, 12–23.

Shi, Y., 2007. Histone lysine demethylases: emerging roles in development, physiology and disease. Nat. Rev. Genet. 8 (11), 829–833.

Shilpi, A., Parbin, S., Sengupta, D., Kar, S., Deb, M., Rath, S.K., Pradhan, N., Rakshit, M., Patra, S.K., 2015. Mechanisms of DNA methyltransferase-inhibitor interactions: procyanidin B2 shows new promise for therapeutic intervention of cancer. Chem. Biol. Interact. 233, 122–138.

Suzuki, M.M., Bird, A., 2008. DNA methylation landscapes: provocative insights from epigenomics. Nat. Rev. Genet. 465–476.

Tan, S., Wang, C., Lu, C., Zhao, B., Cui, Y., Shi, X., Ma, X., 2009. Quercetin is able to demethylate the p16INK4a gene promoter. Chemotherapy 55 (1), 6–10.

Teiten, M.H., Dicato, M., Diederich, M., 2013. Curcumin as a regulator of epigenetic events. Mol. Nutr. Food Res. 57, 1619–1629.

Teperino, R., Schoonjans, K., Auwerx, J., 2010. Histone methyl transferases and demethylases: can they link metabolism and transcription? Cell Metab. 12, 321–327.

Tresserra-Rimbau, A., Rimm, E.B., Medina-Remon, A., Martinez-Gonzalez, M.A., Lopez-Sabater, M.C., Covas, M.I., Corella, D., Salas-Salvado, J., Gomez-Gracia, E., Lapetra, J., Aros, F., Fiol, M., Ros, E., Serra-Majem, L., Pinto, X., Munoz, M.A., Gea, A., Ruiz-Gutierrez, V., Estruch, R., Lamuela-Raventos, R.M., Investigators, P.S., 2014. Polyphenol intake and mortality risk: a re-analysis of the PREDIMED trial. BMC Med. 12, 77.

Vahid, F., Zand, H., Nosrat-Mirshekarlou, E., Najafi, R., Hekmatdoost, A., 2015. The role dietary of bioactive compounds on the regulation of histone acetylases and deacetylases: a review. Gene 562 (1), 8–15.

Vanden Berghe, W., 2012. Epigenetic impact of dietary polyphenols in cancer chemoprevention: lifelong remodeling of our epigenomes. Pharmacol. Res. 65 (6), 565–576.

Vanhees, K., Godschalk, R.W., Sanders, A., van Waalwijk van Doorn-Khosrovani, S.B., van Schooten, F.J., 2011. Maternal quercetin intake during pregnancy results in an adapted iron homeostasis at adulthood. Toxicology 290 (2–3), 350–358.

Vanhees, K., van Schooten, F.J., Moonen, E.J., Maas, L.M., van Waalwijk van Doorn-Khosrovani, S.B., Godschalk, R.W., 2012. Maternal intake of quercetin during gestation alters ex vivo benzo[a]pyrene metabolism and DNA adduct formation in adult offspring. Mutagenesis 27 (4), 445–451.

Verma, M., 2013. Cancer control and prevention: nutrition and epigenetics. Curr. Opin. Clin. Nutr. Metab. Care 16 (4), 376–384.

Wang, Y., Leung, F., 2004. An evaluation of new criteria for CpG islands in the human genome as gene markers. Bioinformatics 20, 1170–1177.

Wang, Z., Zang, C., Rosenfeld, j.A., Schones, D.E., Barski, A., Cuddapah, S., Cui, K., Roh, T.Y., Peng, W., Zhang, M.Q., Zhao, K., 2008. Combinatorial patterns of histone acetylations and methylations in the human genome. Nat. Genet. 40 (7), 897–903.

Waterland, R.A., Dolinoy, D.C., Lin, J.R., Smith, C.A., Shi, X., Tahiliani, K.G., 2006. Maternal methyl supplements increase offspring DNA methylation at Axin Fused. Genesis 44, 401–406.

Waterland, R.A., Jirtle, R.L., 2003. Transposable elements: targets for early nutritional effects on epigenetic gene regulation. Mol. Cell Biol. 23, 5293–5300.

Waterland, R.A., Travisano, M., Tahiliani, K.G., Rached, M.T., Mirza, S., 2008. Methyl donor supplementation prevents transgenerational amplification of obesity. Int. J. Obes. (Lond.) 32, 1373–1379.

Watson, R.R., Preedy, V.R., Zibadi, S., 2014. Polyphenols in Human Health and Disease, 2 vols. Academic Press, Tokio.

Wellen, K., Hatzivassiliou, G., Sachdeva, U., Bui, T., Cross, J., Thompson, C., 2009. ATP-citrate lyase links cellular metabolism to histone acetylation. Science 324 (5930), 1076–1080.

Xiao, X., Shi, D., Liu, L., Wang, J., Xie, X., Kang, T., Deng, W., 2011. Quercetin suppresses cyclooxygenase-2 expression and angiogenesis through inactivation of P300 signaling. PLoS One 6 (8), e22934.

Zeisel, S.H., 2009. Epigenetic mechanisms for nutrition determinants of later health outcomes. Am. J. Clin. Nutr. 89 (5), 1488S–1493S.

Zhang, Y., Ng, H.H., Erdjument-Bromage, H., Tempst, P., Bird, A., Reinberg, D., 1999. Analysis of the NuRD subunits reveals a histone deacetylase core complex and a connection with DNA methylation. Genes Dev. 13 (15), 1924–1935.

Zheng, N.G., Wang, J.L., Yang, S.L., Wu, J.L., 2014. Aberrant epigenetic alteration in Eca9706 cells modulated by nanoliposomal quercetin combined with butyrate mediated via epigenetic-NF-kappaB signaling. Asian Pac. J. Cancer Prev. 15 (11), 4539–4543.

PART II

ORGAN SPECIFIC EPIGENETIC TARGETS OF NATURAL PRODUCTS

CHAPTER

10

Dietary Phytochemicals, Epigenetics, and Colon Cancer Chemoprevention

Ritikraj Arya[1], *Sabita N. Saldanha*[2]

[1]Loveless Academic Magnet Program High School, Montgomery, AL, United States; [2]Department of Biological Sciences, Alabama State University, Montgomery, AL, United States

Abstract

Ranking as the third most commonly diagnosed cancer, colorectal cancer (CRC) is a high-mortality-rate disease with a distinguished prevalence in worldwide populations. Socioeconomic status often determines an individual's access to health care and preventative screening; moreover, such factors also correlate with an increased exposure to poor diet and inadequate food habits. CRCs are typically more predominant in western culture where consumption of meats and fats is widespread. In contrast, diets around the world, most notably the Asian diet, have been found to correlate with lower cancer risk due to increased consumption of phytochemical-rich fruits and vegetables. Interestingly, consumption of these phytocompound-rich foods often favors genetic and epigenetic homeostasis in cellular environments. Many studies have demonstrated the efficacy of using these isolated compounds as therapeutic regimens in cancers, including those of the colorectal phenotype, and many further studies have confirmed the epigenetic roles of these compounds. This chapter will address the chemopreventive and therapeutic efficacy of these molecules via modulation of epigenetic signatures commonly known to influence CRC development and progression.

Keywords

Chemoprevention; Colon cancer; DNA methylation; Epigenetics; Histone modification; microRNA; Phytochemicals

INTRODUCTION

Colorectal cancer (CRC) is a term that collectively describes cancers of the colon and rectal regions. Although incidence rates of CRCs have been on the decline in some areas, most notably Asian populations, the disease is still highly prevalent in Western populations. Lifestyle and dietary habits are considered major contributing factors in the development of CRCs (Hughes et al., 2017). The typical Western diet is characterized by a high intake of red meat (pork, beef, veal, lamb, and mutton), processed meats, unhealthy fats (fried foods, cakes, and pastries), and alcohol (Table 10.1) (Slattery

Epigenetics of Cancer Prevention
https://doi.org/10.1016/B978-0-12-812494-9.00010-X

TABLE 10.1 Carcinogenic Compounds Found in Dietary Sources That May Pose a Risk for Colorectal Cancer

Diet	Cancer-Promoting Factor	Cancer-Causing Factors	Compound Type/Effect	References
Red meat	Rich in heme iron	Formation of N-nitroso compound	Carcinogen	Ananthakrishnan et al. (2015)
		Aldehydes	Cytotoxic and genotoxic	
Processed meats	Rich in nitrites and sodium	Cooking produces HCAs and PAHs	Carcinogens	Tuan and Chen (2016)
Fats	Fats as energy source	Intestinal stem cells	Increased proliferation of intestinal stem cells	Motawi et al. (2017) and Park and Kwak (2012)
			Nutrient sensing pathway hyperactivated (PPAR)-delta Allows to burn fat instead of carbohydrates	
Alcohol	Ethanol	Acetaldehyde	Carcinogenic	Tuan and Chen (2016)
		ROS	Damages DNA	

HCA, heterocyclic amines; *PAH*, polycyclic aromatic hydrocarbons; *PPAR*, peroxisome proliferator-activated receptor; *ROS*, reactive oxygen species.

et al., 2000). Meta-analyses have shown a strong association between the consumption of red meat and CRC onset; a correlation has also been drawn between red meat consumption and gene polymorphisms of cytochrome P450 that are associated with sporadic CRCs (Alexander et al., 2011; Kury et al., 2007).

Diets rich in vegetables and fruits are considered healthy, primarily as a consequence of the many resulting health benefits following digestion. Included among these health benefits are antioxidant, anti-obesity, and anti-inflammatory effects. Collectively, biologically active compounds with the attributes of these health-promoting effects are termed phytochemicals. Based on their chemical composition, phytochemicals are broadly classified under five categories: terpenoids, polyphenols, organosulfur compounds, phytosterols, and alkaloids (Upadhyay and Dixit, 2015; Saldanha and Tollefsbol, 2012) Often, the term bioactive molecules is also used to describe them. Bioactive molecules belonging to the phytochemical groups of terpenoids, polyphenols, and organosulfur compounds are well studied for their health-promoting benefits and cancer-preventive roles.

Based on the molecular structure and functional groups attached, these molecules exhibit the cellular effects that promote apoptosis, inhibit cell growth and proliferation, and promote pro-anti-inflammatory responses that collectively help alleviate tumor development. The many beneficial biological effects are based on in vitro studies and are cell-type specific. Some groups of phytochemicals are more potent in inhibiting the initiation of tumor formation, as shown in Fig. 10.1, with others having benefits over the spectrum of CRC disease pathogenesis with many roles. The type of CRC, familial versus sporadic, is also affected by a multitude of genetic and epigenetic factors that interplay to intensify the mutagenic effect. For the many cell-beneficial functions shown in the figure, these groups of phytochemicals can be manipulated for use in chemoprevention as well as treatment strategies where antimetastatic, anti-invasive, and anti-inflammatory responses are

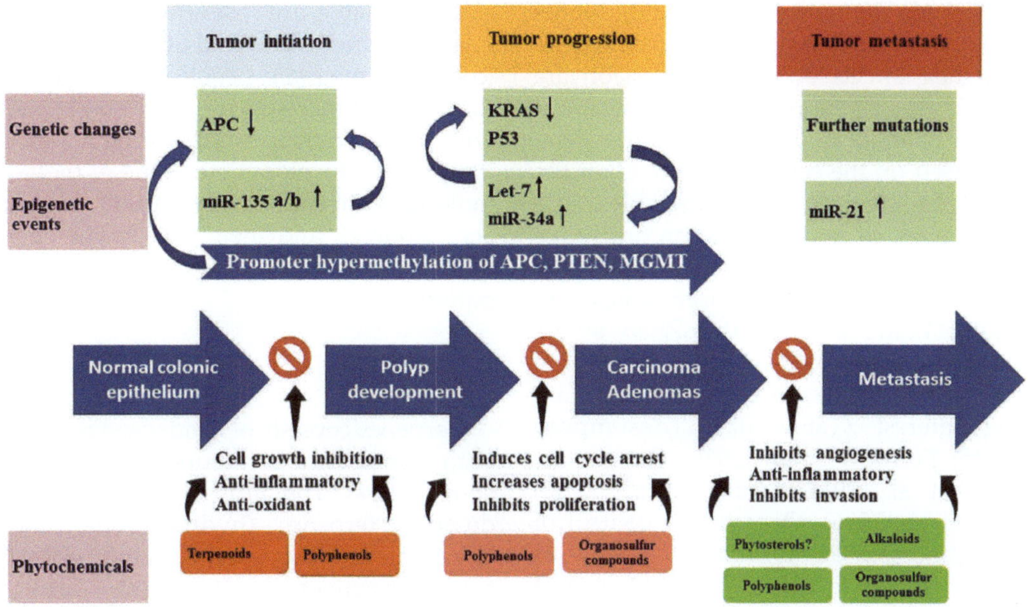

FIGURE 10.1 Chemopreventive and therapeutic effects of phytochemicals. Phytochemicals are classified into major groups of terpenoids, polyphenols, organosulfur compounds, alkaloids, and phytosterols. Genetic factors in colorectal cancer cells (CRCs) include mutations of key genes such as adenomatous polyposis coli (APC), KRAS, and BRAF, which propels a mutagenic and further metastatic cellular phenotype. In many cases epigenetic factors such as CPG island methylator phenotype, a common epigenetic trigger, are observed in the pathogenesis of CRCs. Included among them is the hypermethylation of promoters of tumor suppressors such as APC, phosphatase and tensin homolog (PTEN), and mismatch repair gene human mutL homolog 1. Additionally, miRNAs to specific genes and pathways can compound the development and progression of the disease. In some cases, APC may be mutated, but if normal, the overexpression of miR-135, which inhibits APC, can affect the phenotypic outcome of cellular normalcy. Hypermethylation of APC could also lead to the silencing of its expression without mutations. Similarly, miR-34a, a downstream target of p53 that mediates many p53 biological functions, can alter cell population balance. Let-7 inhibits KRAS expression, and mutations in KRAS contribute to the progression of CRCs.

needed to subdue the rapid growth of the tumor mass and cause its regression in CRCs (Fig. 10.1) (Saldanha and Tollefsbol, 2012; Wang et al., 2012).

Besides genetic regulation and control, phytochemicals also affect epigenetic processes, which are reversible and contribute to changes in gene function or expression without alterations to gene sequences. The primary epigenetic mechanisms known to regulate gene expression are DNA methylation, histone modifications, and microRNAs (miRNAs) (Lee et al., 2013; Li et al., 2016b). Of the three overarching epigenetic processes, DNA methylation is well studied and involves regulation of gene expression by the addition of methyl groups to cytosine residues in the CpG dinucleotide configuration (Sahnane et al., 2015; Sakai et al., 2014; Tse et al., 2017). These processes are defined by DNA methyltransferases (DNMTs) and involve DNA methylation patterns that are either de novo, or their maintenance. Hypermethylation of genes results in gene silencing in most cases, with some exceptions, whereas hypomethylation primarily contributes to increases in gene expression (Chandler and Jones, 1988; Guilleret et al., 2002; Kato et al., 2009; Takasawa et al., 2018). Histone modifications are rather dynamic, and of the various histone modifiers, the roles of acetylation and deacetylation on gene expression are well established (Dong and Weng, 2013). Nonetheless, the epigenetic profiles of histone residues and their effects on gene expression are difficult to assess, as the resulting modifications are quite dynamic and therefore transient, changing based on the cellular milieu. Over the years, miRNAs have gained considerable attention as modifiers of gene regulation and have been shown to influence various cellular processes in both normal and diseased states (Liu et al., 2011; Wang et al., 2016; Zhao et al., 2012). Further, long noncoding RNAs (lncRNAs) and picoRNAs (piRNAs) in gene regulation are garnering interest (Table 10.2) (Assumpcao et al., 2015; Ferracin et al., 2016; Gao et al., 2017; Han et al., 2014; Muller et al., 2015; Thorenoor et al., 2016). Overall, the control of epigenetic processes stems from the ability of bioactive molecules to affect the function and enzymatic activity of enzymes that critically affect DNA methylation and histone modifications. In addition, some of these molecules can directly inhibit or overexpress miRNAs, lncRNAs, and piRNAs in diseased states. In this chapter, we aim to disseminate current information on the chemopreventive and epigenetic effects of phytochemicals on genes and the key metabolic pathways they affect within the colonic microenvironment. To date, several bioactive molecules with potent chemotherapeutic effects are known, and their biological relevance has been determined. However, to fit within the scope of the chapter, the focus will be primarily on the most well-studied molecules directly relevant to CRC.

PLANT PHYTOCHEMICALS AND MEDICINAL VALUE

Alternate medicine that involves the use of plant-based derivative compounds is tending toward becoming the preferred route for the prevention and treatment of many diseases, including cancer, because such compounds are easily available, are cost-effective, and pose minimal side effects. The goal of chemoprevention is to alter or maintain beneficial changes in genes that affect either specific or multiple pathways with related functionalities in the hope of inhibiting tumor development. Some of these natural molecules are abundantly present in various dietary sources, and their roles in the chemoprevention of CRCs are discussed in subsequent paragraphs.

Terpenoids

Terpenes consists of condensed isoprene units with a molecular structure of $(C5H8)n$ (Yadav et al., 2014). The term terpenoids is used to describe terpenes (hydrocarbons) and their oxygenated, hydrogenated, and dehydrogenated derivatives (Yadav et al., 2014). These organic compounds are found in plants and account for the flavor and scents of plant structures (flowers and fruits) (Yadav et al., 2014). Terpenoids are classified as mono- ($C = 10$), di- ($C = 20$), tri- ($C = 30$), and tetraterpenoids ($C = 40$) based on the total number of carbon atoms and are further subclassified according to the number of rings in the structure (Yadav et al., 2014). These molecules have a diverse array of functions, among them anti-inflammatory, antifungal, and antiviral effects (Huang et al., 2012; Sharma et al., 2017). Thus, the use of terpenoids in medicine has invoked tremendous interest, and studies have consistently shown that this group of phytochemicals is highly useful in the prevention and therapy of several diseases including cancers (Huang et al., 2012; Sharma et al., 2017).

Monoterpenes

Monoterpenes and triterpenoids have been shown to have preventive and therapeutic effects in colon carcinomas (Zhang et al., 2005). In their therapeutic roles against cancers, these molecules have been shown to inhibit the

TABLE 10.2 Effect of Phytochemicals on Long Noncoding RNA Expression in Various Cancers

Dietary Compound	Long Noncoding RNA	Cell Line/Tumor Type	Effect	References
Resveratrol	LncRNA MALAT1	Colon tumor samples from patients	Downregulation of LncRNA MALAT1 Lowered Wnt/β-catenin signaling, causing the inhibition of invasion and metastasis	Ji et al. (2013)
Curcumin	Silencing of lncRNA PANDAR	Colon tumor samples from patients at Second affiliated Hospital, Nanjing Medical University	Switch from senescence to apoptosis due to lncRNA PANDAR silencing	Chen et al. (2017)
Sulforaphane	LINC01116	PC-3 prostate cancer cells	Decrease in expression of LINC01116 and cellular proliferation Upregulation of several genes, including GAPDH, MAP1LC3B2, and H2AFY LINC01116 is typically overexpressed in colon cancers	Beaver et al. (2017)
Genistein	LncRNA HOTAIR	786-O and ACHN renal cancer cell lines	Genistein treatment in renal cancer significantly reduced the expression of HOTAIR, which is typically overexpressed in renal cancers. HOTAIR is also overexpressed in colon cancers	Chiyomaru et al. (2014)
EGCG	NEAT1	A459, H460, and H1299 NSCLC cell lines	Upregulation of NEAT1, which decreased drug resistance responses and increased sensitivity to cisplatin treatment, a platinum based drug commonly used to treat NSCLC Cisplatin treatment is plagued with drug resistance in NSCLC Cisplatin can also be used to treat colon cancer, but drug resistance diminishes the drug's efficacy; hence, treatment with cisplatin in combination with EGCG in colon cancer seems promising	Jiang et al. (2016)

GAPDH, glyceraldehyde3-phosphate dehydrogenase; *H2AFY*, H2A histone family member Y; *HOTAIR*, HOX transcript antisense RNA; *lncRNA*, long noncoding RNA; *MALAT1*, metastasis-associated lung adenocarcinoma transcript 1; *MAP1LC3B2*, microtubule-associated protein 1 light-chain 3 beta 2; *NEAT1*, nuclear paraspeckle assembly transcript 1, *NSCLC*, non-small-cell lung cancer; *PANDAR*, Promoter of CDKN1A antisense DNA damage-activated RNA.

initiation and progression of tumorigenesis, encourage the differentiation and apoptosis of tumor cells, inhibit the process of angiogenesis, inhibit tumor invasion, and inhibit metastasis by regulating transcription factors and signaling mechanisms (Bardon et al., 2002; Sobral et al., 2014). Perillyl alcohol (POH) and its metabolite, a monoterpene, have been shown effective against colon cancer cells by upregulating p21 and downregulating cyclin D1 and CDK4 (Bardon et al., 2002). However, a phase II clinical study of POH at doses of 1200 mg/m^2 and advanced acceptable doses at 1600 mg/m^2 failed to have any appreciable antitumor effects in advanced metastatic CRCs (Meadows et al., 2002). POH appears to be a suitable chemopreventive agent as a result of its cell cycle regulatory role, and it may be particularly useful in mitigating early CRC tumor development through the prevention of polyps. However, its role in polyp development and regulation has not been examined or analyzed to date. Although its clinical significance is still questionable, POH may be useful in combination therapies as shown with other cancers (Yeruva et al., 2010). D-limonene, a monoterpene found in the essential oils of citrus fruits, and its metabolite, POH, have been shown to induce the activation of detoxifying phase II enzymes and to activate genes involved in apoptosis (Belanger, 1998). D-limonene induces apoptosis in different cancers by targeting specific pathways that mediate apoptosis. In colon cancer cells, D-limonene inactivates the serine threonine kinase (Akt) pathway (Jia et al., 2013); in leukemic cells, it activates the caspase-dependent mitochondrial death pathway (Ji et al., 2006); and in prostate cancer cells, it induces the generation of reactive oxygen species (Rabi and Bishayee, 2009). The metabolites of D-limonene—POH, perillic acid, and limonene 1,2-diol—are effective inhibitors of colon cancer cell growth and have also been shown to inhibit angiogenesis and induce apoptosis (Sun, 2007; Chidambara Murthy et al., 2012a). In CRCs, fatty-acid synthase (FASN), an enzyme that regulates de novo lipogenesis, is highly expressed and has been investigated for its potential role as a marker for the disease (Lee et al., 2017). The increased expression of the enzyme is linked to colon cell proliferation. Because this enzyme is differentially regulated in normal cells compared with colon cancer cells, it serves as a therapeutic target against the disease (Lee et al., 2017). In addition to monotremes, including D-limonene and POH, polyphenols such as epigallocatechin-3-gallate (EGCG) are also known to be effective in inhibiting FASN in vitro and in vivo. Nonetheless, their clinical significance remains to be determined.

Diterpenes

Carnosol, a polyphenolic diterpene, has a myriad of chemopreventive effects. Carnosol prevents the activation of carcinogens, enhances the activities of detoxifying enzymes, inhibits inflammatory processes that promote tumor formation, selectively induces apoptosis, inhibits cell proliferation, and hinders tumor angiogenesis and invasion (Al Dhaheri et al., 2014; Chun et al., 2014; Johnson et al., 2008; Johnson, 2011; Park et al., 2014; Sanchez-Camargo et al., 2016; Vergara et al., 2014). Epidemiological studies have supported the hypothesis that diterpenes from coffee, cafestol, and kahweol lower the risk of CRC (Cavin et al., 2002; Lee et al., 2012b). They have been shown to regulate specificity protein 1 expression and induce apoptosis in human malignant pleural mesothelioma (Lee et al., 2012b). These natural coffee-based diterpenes affect xenobiotic metabolism and reduce the activation of carcinogens by inhibiting cytochrome P450 enzymes (Huber et al., 2003). They also exhibit detoxifying activities through the induction of detoxifying enzyme systems, such as glutathione S-transferase and UDP-glucuronosyltransferase (Huber et al., 2003). Additionally, these molecules suppress the expression of cyclooxygenase-2 (COX-2), a molecule that strongly regulates the inflammatory process that promotes carcinogenesis (Kim et al., 2006).

Triterpenes

Fungal foods, including several types of mushrooms, are commonly used in traditional Chinese medicine. Ganodermanontriol (GDNT), a triterpene from the extract of the Chinese medicinal mushroom *Ganoderma lucidum,* has been found to suppress the proliferation of colon cancer cells (Jedinak et al., 2011). These changes were mediated by the downregulation of β-catenin, CDK4, and proliferating cell nuclear antigen. Additionally, GDNT decreased tumor growth in a xenograft model in which cyclin D1 inhibition was also observed (Jedinak et al., 2011). Lanostane (2, 3, 4, 6, and 8) and ergostane (1, 5, and **7**) triterpenes extracted from the medicinal fungus *Antrodia camphorata* exhibit similar antiproliferative and cytotoxic efficacies, albeit through different molecular mechanisms (Yeh et al., 2009). Lanostane 3 was effective against colon cells (HT-29, HCT-116, and SW480 cell lines). However, lanostanes 3 and 8 were specific to triple-negative MDA-MB 231 breast cancer cells (Yeh et al., 2009). Additionally, these compounds are nontoxic to normal cells, which is a pertinent characteristic in treatment regimens of the disease. Ergostanes 1, 5, and 7 exhibit synergistic cytotoxicity against HT29 cells and are proapoptotic molecules. Apoptotic induction is mediated by the decrease in antiapoptotic proteins Bcl2 and procaspase-3 and the expression of apoptosis-associated (ADP-ribose) polymerase cleavage (Yeh et al., 2009). Triterpene glycosides also have antitumor effects. Mogroside IVe, a triterpenoid glycoside obtained from the monk fruit, has been found to be effective in inhibiting CRC cell proliferation through the induction of p53, the downregulation of matrix metalloproteinase (MMP) 9, and the phosphorylation of extracellular signaling protein kinases (ERKs) 1 and 2 (Liu et al., 2016). Similar results have been observed in throat cancer cells. The therapeutic effects of mogroside IVe have also been observed in vivo in xenograft models (Liu et al., 2016).

Oleanolic acid (OA) and maslinic acid (MA) are triterpenes found in the skins of olives that are made available in the oil through the extraction process (Pozo et al., 2017). These triterpenes have been found to have health benefits that include antitumor and anti-inflammatory activities (Juan et al., 2008). However, the quality of the oil consumed and the concentration of these molecules dictate the resulting health benefits. A high concentration of these molecules can result in biphasic effects as well. MA has been shown to induce apoptosis in colon cancer cells via both the intrinsic and extrinsic mitochondrial pathways based on cell type (Reyes-Zurita et al., 2009). In Caco-2 (p53-deficient) cells, MA's apoptotic effects include increased caspase-9 activation with no change in Bax expression via c-Jun N-terminal protein kinase (JNK) activation in a p53-independent manner (Reyes-Zurita et al., 2013). Since MA can induce both p53-dependent and p53-independent apoptosis, it can serve as a candidate compound in treating CRCs regardless of the p53 genotype and phenotype.

The survival and invasive capacity of tumors stem from their ability to sustain growth and proliferation through the generation of blood vessels via a process known as angiogenesis. Targeting this process has been a long-standing chemotherapeutic goal. Intracellular signaling pathways, including signal transducer and activator of transcription 3 (STAT3) and sonic hedgehog (SHH), are implicated in these processes. In a CRC mouse xenograft model, the effect of OA on angiogenesis was evaluated, and the regulation of STAT3 and SHH was assessed (Li et al., 2016a). Following experimentation, it was found that the STAT3 and SHH signaling pathways were inhibited, resulting in altered expression of their downstream targets, vascular endothelial growth factor (VEGF) A and fibroblast growth factor (FGF); both VEGF and FGF are considered proangiogenic molecules. This study also demonstrated a steep decrease in microvessel density, which greatly affected tumor growth. In another study, both in vitro (HT-29 cells) and in vivo (mouse xenograft model) assessments of OA's effect on tumor growth was investigated (Li et al., 2015). The study found that OA significantly modulated multiple targets to inhibit tumor growth by both volume and weight in the mouse model. In vitro studies showed that the molecule effectively induced apoptosis by Bax and p21 upregulation along with the suppression of Akt, protein S6 kinase, and mitogen-activated protein kinase signaling activation. Additionally, Bcl-2, cyclin D1, and CDK4 were significantly reduced and p53 pathway activation was promoted (Li et al., 2015). Ursolic acid, another type of triterpene compound found in olive oil, has been shown to inhibit the growth and metastasis of human colon cancer both in vitro and in an orthotopic mouse model of CRCs (Kim et al., 2014; Shan et al., 2009; Wang et al., 2013a,b).

Tetraterpenes

Lycopene, β-carotene, and lutein are natural carotenoids classified as tetraterpenes. These compounds are composed of eight isoprene units. Lycopene is a pigment molecule that imparts color to many fruits and vegetables, including watermelon, grapefruits, apricots, and tomatoes. In the American diet, 85% of lycopene comes from the consumption of tomatoes, specifically in the form of paste or juice. Lycopene has potent antioxidant activities and protects cells from oxidative damage, a major factor in tumor development. It has been shown to prevent CRC progression in mouse xenograft models and induce apoptosis by inhibiting the Akt pathway in CRC cells (Tang et al., 2008, 2011, 2012).

Dietary sources rich in β-carotene are typically in the form of yellow, orange, and green leafy fruits and vegetables. Like lycopene, this carotene has proantioxidant functions and is therefore considered a potential natural therapeutic agent in various cancers. Not much is known about the molecular mechanisms by which this molecule exerts its chemopreventive effects; however, studies in CRC models have found that it can regulate the nuclear factor-κB (NF-κB) pathway (Palozza et al., 2003). β-Carotene interferes with the nuclear translocation of the p65 unit and degrades inhibitor-of-κB protein, preventing DNA binding of the NF-κB complex subunits to DNA and inhibiting gene transcription. Lutein, another tetraterpene found in fruits, vegetables, and egg yolk, exerts chemopreventive effects by modulating KRAS, protein kinase B, and β-catenin levels as was observed in a study on dimethylhydrazine-induced colon cancer in mouse models (Reynoso-Camacho et al., 2011).

Polyphenols

Polyphenols are phytochemicals with phenolic rings that are grouped based on number of rings and the structural characteristics of the functional groups that bind the rings. Flavonoids, phenolic acids, stilbenes, and lignans are subgroups of polyphenols. Currently, there are about 8000 identified polyphenols, which are present as micronutrients in the diet from sources like tea, fruits, vegetables, berries, and plant oils. Polyphenols like quercetin, EGCG, epicatechin, and resveratrol (RES) have been researched extensively for their roles in the treatment of diseases like diabetes, obesity, and cancer. These molecules have been shown to be effective in treating CRCs and will be discussed in the following subsections. Broadly the chemotherapeutic impact of these molecules has been observed in in vitro and in vivo studies, and most of the chemotherapeutic effects of these polyphenols stem from their ability to act as potent antioxidants, preventers of angiogenesis, and anti-inflammation agents (Fig. 10.1).

The molecular mechanisms that govern these effects in CRC are not fully understood and therefore one needs to be cautious about clinical significance and impact until further verified.

Phytosterols

Phytosterols are abundantly found in plant oils and in vegetables, fruits, and nuts. The major phytosterols are beta-sitosterol, campesterol, and stigmasterol. The chemopreventive benefits of phytosterols have been observed in many cancers. However, their chemopreventive role in CRC seems somewhat contradictory. An epidemiological study that followed the intake of phytosterols in a Netherlands cohort showed no association between the intake of dietary phytosterols and a lowered risk of CRC in either males or females. However, when adjusted for energy, there was a positive association between the risk for rectal cancer and the consumption of campesterol and stigmasterol in males (Huang et al., 2017; Normen et al., 2001). Another cohort study in a different ethnic population demonstrated that increased phytosterol consumption lowered the risk of CRC, stigmasterol being the only exception to this finding (Huang et al., 2017). Studies using animal models demonstrated that plant sterols and stanols both promoted tumor and colonic cell proliferation in adenomatous polyposis coli mice (Ramprasath and Awad, 2015). However, in N-desmethyl TAM-induced colonic tumors in rats, beta-sitosterol inhibited tumor development (Baskar et al., 2010, 2012). Absorption and modification of phytochemicals after breakdown of the dietary source in the digestive tract dictates the availability and stability of the compound in circulation, a factor that significantly affects the health benefits of the compound in question. Plant sterols also inhibit the absorption of cholesterol at the small intestine, decreasing plasma cholesterol levels (Jakulj et al., 2005; Lee et al., 2012c). Consequently, a hypercholesteremic environment in the colon can also occur as a result of this inhibition of cholesterol absorption, exposing the colonic epithelium to cholesterol that may promote cancer development. Further studies are necessary to decipher the chemopreventive effects and mechanistic roles of phytosterols in colorectal tissues.

Organosulfur Compounds

Organosulfur compounds are organic molecules that contain sulfur and are associated with the pungent odors characteristic of allium vegetables such as garlic and onions. They are also abundant in cruciferous vegetables such as broccoli and cabbage. Diallyl sulfide (DAS) is a dominant constituent of allyl derivatives found in garlic oils. This molecule, along with other allyl derivatives such as diallyl disulfide (DADS) and diallyl trisulfide (DTS), exhibits pharmacological and biological activities including anti-inflammatory, anticarcinogenic, and antiangiogenic effects. Allyl sulfides have been shown to induce cytotoxic and apoptotic effects in cancer cells. However, the mechanistic action of cancer prevention in CRCs is still unclear. Collectively, these allyl sulfides have been shown to inhibit invasion and metastasis in colon cancer cells, with DTS being more effective than DADS and DAS at concentrations of 10 and 25 μM. These changes were found to be mediated by the downregulation of NF-κB and COX-2, both of which are pathways that affect cell proliferation. Additionally, members of the cell migration and invasion pathway (PI3K, Ras, MEKK3, MKK7, ERK1/2, JNK1/2, p38, and MMPs 2, 7, and 9) were also inhibited (Lai et al., 2013). DADS and its metabolite, allyl mercaptan (AM), have been shown to induce the expression of p21 mediated by a decrease in histone deacetylases (HDACs) and enhanced H3 and H4 hyperacetylation at concentrations of 200 μm in Caco-2 and HT-29 cells. Acetylation of lysine residues at 12 and 16 were preferentially found. AM was found to be more potent as an HDAC inhibitor (92%) compared with

DADS (29%) (Druesne et al., 2004). DADS has also been shown to induce cell cycle arrest in the G2/M phase as a result of an increase in cyclin B1 expression in a p53-independent manner in HCT-116 cells (Jo et al., 2008). Similarly, G2/M arrest was observed in Colo 320 DM colon cancer cells upon treatment with DAS, during which increased caspase-3 expression and ERK2 suppression were observed (Sriram et al., 2008). These and many other studies collectively show the ability of DAS to induce apoptosis, inhibit cell proliferation, and inhibit the migration of colon cancer cells. Nonetheless, future in vivo and clinical studies are required to validate the chemopreventive effects of this allylic compound.

Alkaloids

Alkaloids found abundantly in herbs and higher plants exhibit a spectrum of anticarcinogenic activities and health-promoting cellular benefits. Alkaloids such as vinblastine (antitubulin) and camptothecin (topoisomerase inhibitor) have been extensively used as therapeutic drugs in cancer for years (Lu et al., 2012). Taxols and camptothecins are the dominant alkaloids extracted from plant sources. Paclitaxel is currently the most extensively used taxol against many cancers, including cancers of the breasts and lungs. It is derived from the plant *Taxus brevifolia*. Several other known alkaloids have been proven to be affective in anticancer regimens and other diseases that are well reviewed in a particular study (Isah, 2016). Alkaloids extracted from *Argemone mexicana* have been shown to strongly inhibit cell proliferation in SW480 colon cancer cells. In an investigation of the therapeutic roles of six alkaloid types, at the highest dose of 200 μg/mL, methoxydihydrosanguinarine was found to be the most potent (Singh et al., 2016). Berberine, another alkaloid, has been found to exhibit cytotoxic efficacies at concentrations of 200 μm and to induce G2/M arrest by the induction of p21 in SW480 colon cancer cells. Proteins relevant to the mitochondrial-membrane potential dysfunction-mediated apoptotic pathway were upregulated (Chidambara Murthy et al., 2012b), anti-inflammatory response by the induction of both NF-κB and COX-2 was observed, and tumor necrosis factor-related apoptosis inducing ligand (TRAIL), vascular endothelial growth factor (VEGF), and survivin-affecting angiogenesis were expressed (Chidambara Murthy et al., 2012b). Additionally, the proapoptotic proteins nonsteroidal anti-inflammatory drug-activated gene-1 and activating transcription factor 3 are also induced by berberine treatment (Piyanuch et al., 2007). The spectrum of anticancer molecular benefits promoted by berberine makes it a suitable molecule in the chemoprevention of colon cancer, though in vivo experimentation and clinical studies have yet to determine its clinical significance.

EPIGENETICS AND PHYTOCHEMICALS IN COLORECTAL CANCERS

Of the groups of phytochemicals, polyphenols are prominent epiregulators in the epigenome and most often affect epigenetic processes and changes by the targeting of epigenetic enzymes. Epigenetic processes affected by terpenoids, phytosterols, and alkaloids are beginning to emerge. However, scientific studies and literature are limited in their understanding of the epigenetic role of these groups in CRCs. Certain alkaloids directly affect epi-enzymes such as DNMTs, HDACs, and histone acetyltransferases (HATs), and alter gene expression. However, some of them are indirect epiregulators. Mahanine isolated and purified from *Murraya koenigii* affects DNMT activity and reexpresses silenced Ras-association domain family 1 isoform A, which is often hypermethylated in various tumor types including those originating from prostrate, breast, and lung tissues (Agarwal et al., 2013; Jagadeesh et al., 2007). Additionally, mahanine has been shown to downregulate DNMT1 and DNMT3b in prostate cancer cells, STAT3 in cervical cancer, and Akt in prostate cancer with the upregulation of caspases (Agarwal et al., 2013; Jagadeesh et al., 2007; Sinha et al., 2006). Since the Akt pathway is one among the several pathways deregulated in many cancers, including CRCs, it might be important to look into the pharmacology and bioavailability of the molecule and its relevance as a CRC therapeutic. Sanguinarine, an alkaloid isolated from *Sanguinaria canadensis,* has been used for its many health-promoting benefits. This alkaloid affects the chromatin structure and induces transcriptional repression. It affects p300-and-P300/CBP-associated factor histone acetylation activities and mediates a decrease in the acetylation of histone H3 lysine residues at positions 4 and 9. It also affects the methylation of histone H3 lysine and arginine residues at positions 17 and 4 by inhibiting G9a and protein arginine N-methyltransferase 4/coactivator-associated arginine methyltransferase 1, respectively (Schneider-Stock et al., 2012). The mechanism by which sanguinarine affects cell growth and proliferation has been evaluated in CRCs. Its proapoptotic functions involve an increase in Bax levels, a decrease in Bcl-2 levels, and induction of caspase-3 and caspase-9 (Lee et al., 2012a). Additionally, treatment of CRC cells with this alkaloid induces DNA double-strand breaks resulting in apoptosis independent of p53 activity (Matkar et al., 2008). However, its specific epigenetic function in CRC tumor prevention or treatment is not known. Sanguinarine is also a potent modulator of NF-κB activity (Schneider-Stock et al., 2012). Accounting for this molecule's particular mechanistic function along with its proapoptotic DNA methylation- and acetylation-modulating roles, further studies investigating and confirming its role as a preventive and therapeutic agent in CRCs is warranted.

Terpenoids, as previously discussed, display a broad spectrum of anticarcinogenic activities. Of the many molecules that belong to this group, lycopene, a carotenoid, is prominently studied for its chemopreventive and therapeutic functions in vitro in various cancers. Lycopene is found largely in tomatoes and other red fruits. As a bioactive molecule, it has been shown to have epigenetic effects by modulating DNA methylation. Studies in breast cancer cells have shown that when treated with lycopene, glutathione S-transferase P1, a tumor-suppressor gene, is reexpressed. Other terpenoids, such as triptolide and thymoquinone, inhibit polycomb group proteins that are actively involved in modulating chromatin. Triptolide inhibits histone methyltransferase suppressor of variegation 3–9 homolog 1, enhancer of zeste homolog 2, and HDAC8, and it decreases the methylation and acetylation marks of histones H3 and H4, respectively. Thymoquinone has also been shown to downregulate ubiquitin-like containing PHD and ring finger domains 1 and its partners DNMT1 and HDAC1. However, due to the limited number of studies on this subject, these observations could largely be affiliated with cell type-specific observations and may not be exhibited in all tumor types. Derivatives of allium sulfur compounds have been shown to act as anticancer agents. One such molecule, S-allylmercaptocysteine, modulates HDAC and HAT activities by altering the levels of H2A, H2B, H3, and H4 acetylation. It has been also found to decrease the phosphorylation of H1 and H3, but its epigenetic impact on CRCs is not known. Induction of p21 by the hyperacetylation of histone H3 at lysines 9 and 14 and of H4 at lysines 12 and 16 have been observed in CRC cells mediated by treatment with DADS and AM. Histone H3 hyperacetylation by AM facilitates the binding of Sp3 and p53 to the promotor region of p21 (Nian et al., 2008), inducing an increase in p21 expression in CRC cells.

Resveratrol

RES is a natural phytocompound found in numerous plant products like grape skins, red wine, and peanuts. Many studies indicate promise for RES's future as a chemotherapeutic agent in various cancers, including those of the colon. It has been found that treatment with RES in CRC patients is associated with significant reductions in tumor proliferation (Patel et al., 2010). RES's role as an anticancer agent in CRC has been investigated and its effects on various genes and molecular targets in cancer have been identified. Many of these anticancer effects are associated with various epigenetic markers; hence, the following section will review the existing literature on RES's epigenetically related anticancer effects in colon cancer.

Small noncoding RNA molecules, miRNAs play major roles in gene regulation and DNA silencing. Many miRNAs have been linked to the formation and progression of cancer in the human body. For this reason, investigation into RES's effects on miRNAs in cancer seems promising. For example, in one study, treatment with RES in HCT-116 and HT-29 CRC cells caused reductions in cellular proliferation via upregulation of miR-34c (Yang et al., 2015). Various studies have established that miR-34c plays a well-defined role in tumor suppression. This miRNA is frequently downregulated in many forms of cancer, including colon cancer, and restoration of this miRNA's expression is associated with reductions in malignant growth via repression of genes associated with tumor proliferation (Patel et al., 2010; Yang et al., 2015). It was hypothesized that the increase in the expression of miR-34c following RES treatment was due to promoter demethylation; however, this was disproven. Hence, further investigation is necessary to determine the mechanisms behind miR-34c modulation in colon cancer following RES treatment (Yang et al., 2015). In another study, treatment with RES in SW480 human colon cancer cells was found to decrease the expression levels of 26 miRNAs and to increase the expression levels of 22 miRNAs. Following RES treatment, plausible oncomiRs, including miR-25, miR-21, miR-17, miR-103-1, miR-103-2, and miR-92a-2, were downregulated. Further, treatment with RES has been found to increase the levels of miR-34a in DLD-1 colon cancer cells. MiR-34a, similar to the previously discussed miR-34c, is a tumor-suppressor miRNA that is transcriptionally activated by the tumor-suppressor gene p53 (Fig. 10.1). Increases in the expression of miR-34a lead to increased levels of p53 in HCT116 colon cancer cells (Navarro and Lieberman, 2015). Following RES treatment, it was also found that a target of miR-34a, the E2F transcription factor 3 (E2F3), was downregulated. Furthermore, E2F3's downstream target, Sirt1, a well-known HDAC, was downregulated following RES treatment. These results suggest that RES exhibited anticancer effects via the miR-34a/E2F3/Sirt1 cascade. In another study, it was found that a human-equivalent dose of approximately 210 mg RES/day inhibited tumorigenesis in the colon cancer mouse model by the suppression of the KRAS gene via increases in the expression of miR-96 (Saud et al., 2014). KRAS is protooncogene that promotes tumorigenesis in colon cancer via transcriptional silencing of various tumor-suppressor genes (Armaghany et al., 2012). MiR-96 is known to target KRAS; hence, an increase in miR-96 expression via RES treatment results in various anticancer effects. In another study in which the synergistic effects of natural compound quercetin combined with RES were investigated, a decrease in the levels of oncogenic miR-27a was recorded. MiR-27a is typically overexpressed in CRC stem cells. Increased levels of miR-27a have been found to induce resistance to TRAIL, and knockdown of this miRNA was found to allow CRC cancer cells to regain sensitivity to TRAIL-mediated apoptosis (Zhang et al., 2017). Hence, RES and its combination with other natural compounds, like quercetin, may be of great interest for future experimentation. With all of this information taken into consideration, it can be concluded that RES has much potential as a therapeutic agent in colon cancer via miRNA regulation; however, further experimental models, cohort studies, and clinical investigations are pertinent to confirm and build upon the results of these studies. Since very little comprehensive information exists on the effects of RES as a demethylating agent and an HDAC inhibitor in CRC pathogenesis, expanding the studies in this area may bridge the gap in understanding the interplay of genetic and epigenetic mechanisms in CRC tumor formation and present novel chemopreventive strategies for the same.

Curcumin

Curcumin is a phytocompound found in the *Curcuma longa* plant that is commonly consumed by traditional cultures in the form of turmeric (Wilken et al., 2011). In the Indian subcontinent, turmeric has been lauded for its use as an ancient therapeutic agent for a wide range of diseases. In recent years, interest in the medicinal properties of turmeric have gone beyond the realm of folklore into scientific contexts. Through much investigation, it was found that the medicinal properties of turmeric can be attributed to the natural polyphenol curcumin. Curcumin's therapeutic properties have also found their way into the realm of cancer research, resulting in a plethora of interesting results. Here, we will discuss the possible implications of curcumin in colon cancer via the induction of various epigenetic mechanisms.

Curcumin has been found to exhibit various therapeutic effects via epigenetic mechanisms in the colon cancer model. In one study, it was found that curcumin acts as a demethylating agent in colon cancer (Link et al., 2013). However, alterations in DNA methylation were predominately observed at partially methylated loci; this greatly contrasts with the potent demethylating properties of 5-aza-2′-deoxycytidine that have the ability to demethylate fully methylated CpG sites. Existing Food and Drug Administration-approved DNA demethylating agents, such as 5-azacitidine and decitabine, are known to increase cancer cell sensitivity to chemotherapeutic agents (Reuter et al., 2011). Several studies have indicated that similar results occur following curcumin treatment. A phase 1 clinical trial administered doses of curcumin to 14 patients with advanced metastatic breast cancer every day for several days prior to docetaxel treatment (Bayet-Robert et al., 2010). The study claimed that patients experienced improvements in their condition. Though these results indicate that curcumin may not be the most ideal demethylating agent, these studies form the premise of its demethylating activities, and further investigation is necessary to confirm and build upon these findings. In another study, curcumin was found to inhibit colon cancer cell growth by restoring the expression of tumor-suppressor gene DLEC1 via various epigenetic mechanisms (Guo et al., 2015). Downregulation of the DLEC1 gene is frequently associated with the formation of various human cancers, including those of the colon (Wang et al., 2014). Upregulation in the expressions of β-catenin and glycogen synthase kinase-3β, both of which are involved in the Wnt/β-catenin pathway, was also observed; this could mean that epigenetic modification of DLEC-1 expression following curcumin treatment could result in anticancer effects as a result of Wnt/β-catenin activation. Furthermore, in another study epigenetic restoration of the DLEC-1 gene following curcumin treatment was associated with decreased anchorage-dependent growth in HT29 colon cancer cells (Guo et al., 2015). Moreover, Wnt/β-catenin activation as a result of curcumin treatment in colon cancer has been attributed to downregulation of miR-130a. Typically, high expression of miR-130a is associated with a poor clinical outcome, and some evidence also indicates that this miRNA plays a role in resistance to chemotherapy in colon cancer cells (Dou et al., 2017). In RKO and HCT116 cells, curcumin treatment resulted in significant inhibition of miR-21. In colon cancer, increased miR-21 expression has been found to correlate with carcinogenesis; it plays the role of an oncomiR in colon cancer because it is involved in processes like invasion, proliferation, and metastasis (Yu et al., 2015) (Fig. 10.1). Furthermore, miR-21 has been found to be greatly increased in chemotherapy-resistant CRC patients; hence, miR-21 can be viewed as a possible therapeutic target when treating treatment-resistant colon cancers with curcumin (Yu et al., 2013). Combining curcumin with other natural therapeutics or drugs may prove to be of great use, as such combinations could fill in areas, such as histone deacetylation regulatory functions, that curcumin lacks.

Sulforaphane

Cruciferous vegetables are vegetables derived from the Brassica family. Recently, the cancer-preventative effects of Brassica vegetables have gained great interest in the scientific community, and the results have proven to fruitful. For example, in a clinical trial, subjects consumed a daily intake of 250 g of cruciferous vegetables. The results indicated that the likelihood of colon cancer onset in these patients was possibly reduced, as increased levels of the carcinogen 2-amino-1-methyl-6-phenylimidazo[4,5-*b*]pyridine were found to be expelled in the urine. Many of the chemopreventive effects found in cruciferous vegetable consumption, like the one previously stated, are attributed to various natural phytocompounds found in these vegetables, like indole-3-carbinol, sulforaphane (SFN), and phenethyl isothiocyanate (Higdon et al., 2007). The usage of these natural compounds in chemotherapeutic regimens has been extensively investigated, and these results indicate much promise for the future of cancer therapy; however, their possible effects on the epigenome are less known. Hence, in this section we will be assessing existing information on various cruciferous compounds and the role of the epigenome in their usage as natural cancer-preventive and therapeutic molecules.

One cruciferous compound, SFN, has been found to have numerous anticancer effects in the colon cancer model. Several sources claim that incorporation of SFN-containing foods into the diet could prevent the onset of several human cancers, including cancers of the colon (Ho et al., 2009). Various epigenetic mechanisms have been found to be altered following SFN treatment of colon cancer, making its possible use as a chemotherapeutic regimen even more promising (Ho et al., 2009). SFN has been found to act as an HDAC inhibitor in colon cancer. In HCT116 colon cancer cells, treatment with SFN has been found to increase global histone acetylation as well as selective acetylation of the p21 promoter. P21 is a well-known cyclin-dependent kinase inhibitor that is greatly involved in the process of cell cycle arrest (Abbas and Dutta, 2009). Hence, increased acetylation and therefore transcription of the p21 gene can be viewed as a possible favorable anticancer effect of SFN treatment. Results similar to this were found in vivo as well; mice were fed approximately 6 μmol of SFN, and increased global acetylation and p21 expression were observed, possibly due to acetylation (Myzak et al., 2006). SFN has also been found to exhibit anticancer effects through miRNA regulation. In RKO and HTC116 colon cancer cells, SFN treatment resulted in significant miR-21 downregulation (Martin et al., 2017). As previously stated, miR-21 is considered an oncomiR, an miRNA known to possess oncogenic properties, and it is involved in carcinogenesis of CRCs (Yu et al., 2015). Furthermore, miR-21 expression has been found to be increased in chemotherapy-resistant colon cancer cells; hence, SFN's possible use in treatment resistant colon cancers must be investigated in the future (Yu et al., 2013).

Butyrate

Plant foods are great sources of dietary fiber. The process of fiber fermentation in the colon following consumption results in the formation of various short-chain fatty acids, one of them being butyrate. While consumption of foods like wine and red meat correlates with imbalanced gut microbiota, higher consumption of fibers, possibly as a result of increased butyrate concentrations in the colon, has correlated with favorable and preventative microbial diversity (Tuan and Chen, 2016) (Fig. 10.2). Butyrate has been studied for its use as a therapeutic agent in colon cancers, and the revealed information shows that butyrate has much promise for its future use as a chemotherapeutic regimen. Many studies have linked butyrate's therapeutic properties to epigenetic mechanisms. In this section, we will discuss its chemopreventive role as well as the epigenetic implications of butyrate treatment.

Butyrate has been found to have many effects on the epigenome. In LT97 and HT29 colon cancer cells, butyrate in combination with trichostatin A was found to reduce the expressions of miR-135a, miR-135b, miR-24, miR-106b, and miR-let-7a (Schlormann et al., 2015). Many of these miRNAs are associated with increased cellular proliferation and tumorigenesis; hence, reduced expression of these miRNAs may be a significant sign of butyrate's role as an anticancer agent via modulation of miRNA expression (Valeri et al., 2014; Zhou et al., 2012). Upregulation of these miRNAs induced the expression of

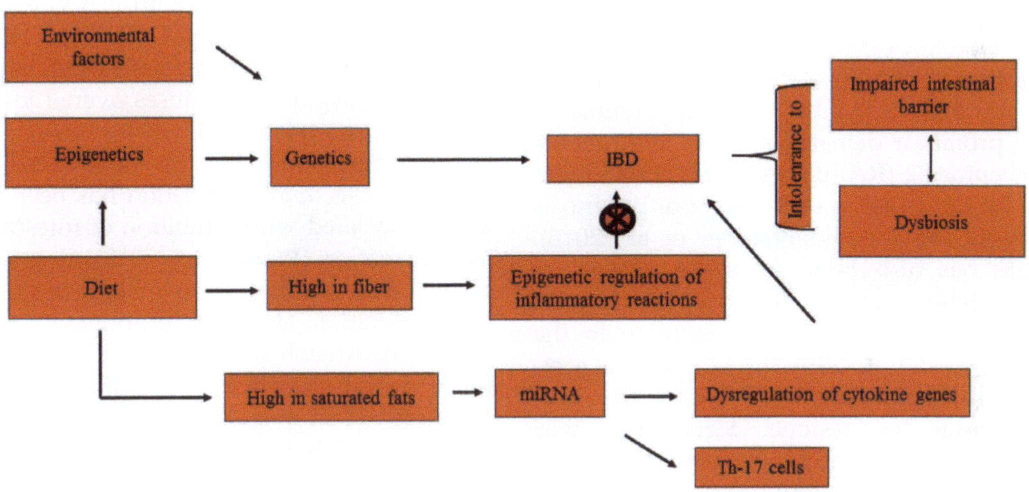

FIGURE 10.2 Diet, epigenetics, and inflammatory bowel disease (IBD). The etiology of IBD is thought to be governed by the genetic predisposition of the individual along with environmental factors and diet. Diet plays a crucial role, as it alters the microbiota of the colon. In a normal colon, the dominant phyla include the firmicutes and bacteroidetes, whereas in IBD, firmicutes decrease and proteobacteria increase. Diets high in saturated fats encourage the IBD phenotype by damaging the intestinal epithelial layer, accompanied by the loss of barrier function, allowing bacterial invasion in the mucosal region. Additionally, high-fat diets alter the differential expression of several miRNAs. These miRNAs are known to dysregulate cytokine genes and alter Th-17-mediated inflammation and autophagy. On the other hand, diets high in fiber allow for high production of short-chain fatty acids (SCFAs) that provide energy to the colonocytes, modulate cytokine production of the gut mucosa, and strengthen the immune system. Butyrate, a predominant SCFA, is also a strong epigenetic regulator of histones, and via its histone modulatory properties, affects genes that regulate inflammatory reactions.

tumor-suppressor gene p21. In another study, expression of the oncogenic miR-17-92 cluster was reduced in both HT-29 cells and HCT116 cells following treatment with sodium butyrate, the sodium salt form of butyrate (Humphreys et al., 2013). The miR-17-92 is often significantly overexpressed in colon cancer (Meng et al., 2015; Mogilyansky and Rigoutsos, 2013). With that taken into consideration, it can be easily concluded that reduction of miR-17-92 expression is yet another significant benefit attributed to butyrate treatment in colon cancer. Another study involving both HT-29 cells and HCT116 cells with butyrate found reduction in the overexpression of miR-92a (Hu et al., 2015). Overexpression of this miRNA has been found to promote the growth and migration of colon cancer cells (Lv et al., 2016). In this study, miR-92a was found to target the p57, Drosha, and c-Myc genes, all of which are known to play roles in the onset and progression of colon cancer. In colon cancer cells, butyrate has also been found to act as a potent demethylating agent. In one study, butyrate was found to initiate promotor demethylation of the retinoic acid receptor β2 (RARβ2) gene. Hypermethylation of RARβ2 is associated with poor prognosis in patients (Miladi-Abdennadher et al., 2010). Butyrate has also been found to act as an HDAC inhibitor. In one study, butyrate treatment was associated with an increase in the expression of p21, possibly through histone acetylation. Downregulation of cyclin B1, also possible due to histone acetylation, was observed in this study (Hinnebusch et al., 2002). Though much useful information already exists regarding the chemopreventive and therapeutic potential of butyrate, its role in CRCs via epigenetic modulation can be further strengthened by in vivo models and cohort and epidemiological studies involving sources of dietary fiber, dietary fiber consumption, epigenetic changes, and polyp formation, as this is the first step toward CRC tumor formation.

Epigallocatechin-3-Gallate

EGCG, a phytocompound found in tea, has been extensively researched in cell culture, animal models, and human trials. Many of these studies have found that EGCG possesses extensive anticancer and chemotherapeutic effects (Schramm, 2013). Many of these therapeutic effects have been tied back to epigenetic mechanisms; hence, in this section we will provide existing information as well as possible future directions on this topic.

EGCG treatment has been associated with significant effects on the epigenome. In one study, when treated with EGCG, colon cancer cells exhibited the ability to inhibit promotor methylation. Additionally, the expression of tumor-suppressor genes p15 and p16 increased as a result of promotor methylation inhibition (Berner et al., 2010). In another study, HCT116 colon cancer cells treated with EGCG exhibited degradation of both DNMT3A and HDAC3. HDAC3 is often underexpressed in colon cancer cells, so its degradation causes overall increases in histone acetylation (Godman et al., 2008). Furthermore, DNMT3A is associated with tumorigenesis, and its deletion has been found to be associated with inhibition of intestinal tumor formation (Weis et al., 2015). Degradation of DNMT3A would lead to lower levels of overall methylation and less tumorigenic activity. Hence, treatment with EGCG could possibly cause antiproliferative and anticancer effects via increases in overall histone acetylation and decreases in DNA methylation. In a combinatorial study of EGCG and butyrate treatment in RKO, HCT-116, and HT-29 cell lines, overall reduction in HDAC activity was observed (Saldanha et al., 2014). The resulting histone hyperacetylation is known to inhibit cell growth. In RKO cells, the percentage of CpG methylation significantly decreased following treatment with both butyrate and ECGC. Though ECGC is known to be a potent DNA demethylating and

HDAC inhibitor, its effects on miRNA are less known. In one study, human hepatocellular carcinoma cells were treated with ECGC. The treatment resulted in an increase in the expression of miR-16, which is considered a contributor to colon cancer progression when downregulated (Qian et al., 2013). We believe that a similar study on the effects of ECGC in colon cancer cell lines may be important in order to analyze its miRNA regulatory effects. In another study involving hepatocellular carcinoma, EGCG was found to bind to both miR-122 and miR-33a, modulating their transcription levels (Baselga-Escudero et al., 2014). In colon cancer cells, miR-33a has been found to inhibit the progression of colon cancer; hence, if ECGC exhibits similar effects in colon cancer cells, cell growth may halt as a result (Tsang and Kwok, 2010). EGCG is therefore an important phytochemical with the potential for strong preventive and therapeutic roles in CRC pathogenesis.

Genistein

The subject of soy consumption often brings on mixed results and opinions. Some studies claim that soy consumption could induce and increase cancer risk, while others present conflicting evidence. Soy consumption has been found to increase the risk of breast cancer diagnosis by some studies. Conversely, some studies also claim that soy consumption throughout childhood and adolescence is associated with a decreased risk of breast cancer onset (Korde et al., 2009; Thanos et al., 2006). Genistein, a soy-derived compound, has been shown to have anticancer effects. Genistein is a phytoestrogen that has been extensively investigated in various cancers, including CRCs.

Some soy products are rich sources of phytoestrogens. One of these phytoestrogens, genistein, has been extensively investigated for its therapeutic properties in various cancer models. Furthermore, some studies have found that numerous therapeutic effects attributed to genistein treatment could result from epigenetic changes. In one study, genistein treatment in SW1116 colon cancer cells was found to reactivate the WNT5a gene, an antagonist of the Wnt signaling pathway (Wang and Chen, 2010). Overactivation of the Wnt signaling pathway is known to be a hallmark of colon cancer (Basu et al., 2016). Furthermore, genistein has been found to act as a DNMT inhibitor in DLD-1 colon cancer cells, causing reexpression of secreted frizzled related protein 2, another antagonist of the Wnt signaling pathway. Genistein has also been found to inhibit colon cancer growth by suppressing miR-95, an miRNA known to promote tumorgenicity (Schetter et al., 2012; Qin et al., 2015). A large majority of studies performed on genistein as an epigenetic modulator in cancers have focused on gynecological cancers and prostate cancer. In the future, more studies must be conducted with regard to colon cancer to substantiate the results of existing studies and further build upon existing information.

Other Phytochemicals

In the plethora of studies undertaken to understand the cancer-preventive and therapeutic roles of pathways affected and altered by bioactive molecules, a few phytochemicals are better studied than others and have been discussed in this chapter. Popularity of the food source and the geographical location in which these plants or fruits are found not only determine the frequency at which the dietary substance/molecule is made available in the diet, but also help in determining the efficacy of the dietary molecule when used. Gingerol from ginger (Aggarwal et al., 2013), silibinin obtained from extract of milk thistle seeds (Raina et al., 2016), thymoquinone found in *Nigella sativa* (Khalife et al., 2016), and ursolic acid found in the epicuticular waxes of apples are some of the other phytochemicals that have been shown to be effective against CRCs, albeit that only a few studies have been conducted (Aggarwal et al., 2013;

Raina et al., 2016). Gingerol and its modified forms have been shown to regulate molecules such as NF-κB, p53, VEGF, and COX-2 that affect cell proliferation and apoptosis in addition to reducing inflammation. Ginger, like turmeric, exhibits its chemopreventive properties via upregulation of the MAP kinase phosphatase-5 pathway. A pilot study of oral silibinin in CRC patients showed that silibinin accumulated in the colonic mucosa at concentrations high enough to exhibit its chemopreventive properties (Hoh et al., 2006). Silibinin and ursolic acid have been shown to target molecules such as cMyc (proliferation), VEGF (metastasis), and survivin (antiapoptotic) in in vitro studies in SW480 and HCT116 CRC cell lines, respectively (Raina et al., 2013; Hoh et al., 2006; Singh et al., 2008; Wang et al., 2013a,b; Xavier et al., 2013; Lin et al., 2013a,b). In vivo studies in murine colon cancer models have shown thymoquinone to be effective in reducing tumor cell invasion and to inhibit tumor growth. Additionally, in silico studies in HT29 cells show that HDAC2 is a direct target of thymoquinone, contributing to the epigenetic properties that this molecule may have in its proapoptotic functions via inhibition of the Akt pathway (Attoub et al., 2013). Further studies are warranted to determine the efficacies of these molecules in CRCs before they can reach clinical trials.

CONCLUSIONS

From the wide spectrum of in vitro studies conducted using phytochemicals in various cancers, it can be easily concluded that all the natural phytochemicals discussed in this chapter have great potential for the future of cancer prevention and therapy. The promise exhibited by the future usage of these phytochemicals is further substantiated by ever-increasing evidence that links the anticancer effects of these compounds with epigenetics. Moreover, there is an urgent need for further investigations into the effects of these compounds on the epigenome, specifically with respect to colon cancer. Though information exists on the epigenetic influences of these compounds among other cancer types, far less is known specifically about colon cancer. For example, the effects of EGCG on DNA methylation and miRNA modulation in prostate cancer is known; however, little information exists for the same with regard to colon cancer. Interestingly, gut microbiota has also been found to be affected by several of these compounds; however, more investigations must be conducted before conclusions are drawn. As the third most common cancer among both men and women and the second-leading cause of cancer-related deaths in the United States, colon cancer is a pressing issue, so it is imperative to further our research and fill in existing gaps in our knowledge. Understanding the epigenetic influences behind the anticancer effects we observe with these compounds would allow us to reach our goal of possibly turning these natural compounds into viable preventive and therapeutic options in the future.

References

Abbas, T., Dutta, A., 2009. p21 in cancer: intricate networks and multiple activities. Nat. Rev. Cancer 9, 400–414.

Aggarwal, B., Prasad, S., Sung, B., Krishnan, S., Guha, S., 2013. Prevention and treatment of colorectal cancer by natural agents from mother nature. Curr. Colorectal Cancer Rep. 9, 37–56.

Agarwal, S., Amin, K.S., Jagadeesh, S., Baishay, G., Rao, P.G., Barua, N.C., Bhattacharya, S., Banerjee, P., 2013. Mahanine restores RASSF1A expression by down-regulating DNMT1 and DNMT3B in prostate cancer cells. Mol. Cancer 12, 1–12.

Al Dhaheri, Y., Attoub, S., Ramadan, G., Arafat, K., Bajbouj, K., Karuvantevida, N., AbuQamar, S., Eid, A., Iratni, R., 2014. Carnosol induces ROS-mediated beclin1-independent autophagy and apoptosis in triple negative breast cancer. PLoS One 9, 1–14.

Alexander, D.D., Weed, D.L., Cushing, C.A., Lowe, K.A., 2011. Meta-analysis of prospective studies of red meat consumption and colorectal cancer. Eur. J. Cancer Prev. 20, 293–307.

Ananthakrishnan, A.N., Du, M., Berndt, S.I., Brenner, H., Caan, B.J., Casey, G., Chang-Claude, J., Duggan, D., Fuchs, C.S., Gallinger, S., Giovannucci, E.L., Harrison, T.A., Hayes, R.B., Hoffmeister, M., Hopper, J.L., Hou, L., Hsu, L., Jenkins, M.A., Kraft, P., Ma, J., Nan, H., Newcomb, P.A., Ogino, S., Potter, J.D., Seminara, D., Slattery, M.L., Thornquist, M., White, E., Wu, K., Peters, U., Chan, A.T., 2015. Red meat intake, NAT2, and risk of colorectal cancer: a pooled analysis of 11 studies. Cancer Epidemiol. Biomarkers Prev. 24, 198–205.

Armaghany, T., Wilson, J.D., Chu, Q., Mills, G., 2012. Genetic alterations in colorectal cancer. Gastrointest. Cancer Res. 5, 19–27.

Assumpcao, C.B., Calcagno, D.Q., Araujo, T.M., Santos, S.E., Santos, A.K., Riggins, G.J., Burbano, R.R., Assumpcao, P.P., et al., 2015. The role of piRNA and its potential clinical implications in cancer. Epigenomics 7, 975–984.

Attoub, S., Sperandio, O., Raza, H., Arafat, K., Al-Salam, S., Al Sultan, M.A., Al Safi, M., Takahashi, T., Adem, A., 2013. Thymoquinone as an anticancer agent: evidence from inhibition of cancer cells viability and invasion in vitro and tumor growth in vivo. Fundam. Clin. Pharmacol. 27, 557–569.

Bardon, S., Foussard, V., Fournel, S., Loubat, A., 2002. Monoterpenes inhibit proliferation of human colon cancer cells by modulating cell cycle-related protein expression. Cancer Lett. 181, 187–194.

Baselga-Escudero, L., Blade, C., Ribas-Latre, A., Casanova, E., Suarez, M., Torres, J.L., Salvadó, M.J., Arola, L., Arola-Arnal, A., 2014. Resveratrol and EGCG bind directly and distinctively to miR-33a and miR-122 and modulate/-divergently their levels in hepatic cells. Nucleic Acids Res. 42, 882–892.

Baskar, A.A., Al Numair, K.S., Gabriel Paulraj, M., Alsaif, M.A., Muamar, M.A., Ignacimuthu, S., 2012. beta-sitosterol prevents lipid peroxidation and improves antioxidant status and histoarchitecture in rats with 1,2-dimethylhydrazine-induced colon cancer. J. Med. Food 15, 335–343.

Baskar, A.A., Ignacimuthu, S., Paulraj, G.M., Al Numair, K.S., 2010. Chemopreventive potential of beta-Sitosterol in experimental colon cancer model-an in vitro and in vivo study. BMC Complement. Altern. Med. 10, 10–24.

Basu, S., Haase, G., Ben-Ze'ev, A., 2016. Wnt signaling in cancer stem cells and colon cancer metastasis. F1000Res. 5, 1–10.

Bayet-Robert, M., Kwiatkowski, F., Leheurteur, M., Gachon, F., Planchat, E., Abrial, C., Mouret-Reynier, M.A., Durando, X., Barthomeuf, C., Chollet, P., 2010. Phase I dose escalation trial of docetaxel plus curcumin in patients with advanced and metastatic breast cancer. Cancer Biol. Ther. 9, 8–14.

Beaver, L.M., Kuintzle, R., Buchanan, A., Wiley, M.W., Glasser, S.T., Wong, C.P., Johnson, G.S., Chang, J.H., Löhr, C.V., Williams, D.E., Dashwood, R.H., Hendrix, D.A., Ho, E., 2017. Long noncoding RNAs and sulforaphane: a target for chemoprevention and suppression of prostate cancer. J. Nutr. Biochem. 42, 72–83.

Belanger, J.T., 1998. Perillyl alcohol: applications in oncology. Altern. Med. Rev. 3, 448–457.

Berner, C., Aumuller, E., Gnauck, A., Nestelberger, M., Just, A., Haslberger, A.G., 2010. Epigenetic control of estrogen receptor expression and tumor suppressor genes is modulated by bioactive food compounds. Ann. Nutr. Metab. 57, 183–189.

Cavin, C., Holzhaeuser, D., Scharf, G., Constable, A., Huber, W.W., Schilter, B., 2002. Cafestol and kahweol, two coffee specific diterpenes with anticarcinogenic activity. Food Chem. Toxicol. 40, 1155–1163.

Chandler, L.A., Jones, P.A., 1988. Hypomethylation of DNA in the regulation of gene expression. Dev. Biol. 5, 335–349.

Chen, T., Yang, P., Wang, H., He, Z.Y., 2017. Silence of long noncoding RNA PANDAR switches low-dose curcumin-induced senescence to apoptosis in colorectal cancer cells. OncoTargets Ther. 10, 483–491.

Chidambara Murthy, K.N., Jayaprakasha, G.K., Patil, B.S., 2012a. D-limonene rich volatile oil from blood oranges inhibits angiogenesis, metastasis and cell death in human colon cancer cells. Life Sci. 91, 429–439.

Chidambara Murthy, K.N., Jayaprakasha, G.K., Patil, B.S., 2012b. The natural alkaloid berberine targets multiple pathways to induce cell death in cultured human colon cancer cells. Eur. J. Pharmacol. 688, 14–21.

Chiyomaru, T., Fukuhara, S., Saini, S., Majid, S., Deng, G., Shahryari, V., Chang, I., Tanaka, Y., Enokida, H., Nakagawa, M., Dahiya, R., Yamamura, S., 2014. Long noncoding RNA HOTAIR is targeted and regulated by miR-141 in human cancer cells. J. Biol. Chem. 289, 12550–12565.

Chun, K.S., Kundu, J., Chae, I.G., Kundu, J.K., 2014. Carnosol: a phenolic diterpene with cancer chemopreventive potential. J. Cancer Prev. 19, 103–110.

Dong, X., Weng, Z., 2013. The correlation between histone modifications and gene expression. Epigenomics 5, 113–116.

Dou, H., Shen, R., Tao, J., Huang, L., Shi, H., Chen, H., Wang, Y., Wang, T., 2017. Curcumin suppresses the colon cancer proliferation by inhibiting Wnt/beta-catenin pathways via miR-130a. Front. Pharmacol. 8, 1–9.

Druesne, N., Pagniez, A., Mayeur, C., Thomas, M., Cherbuy, C., Duee, P.H., Martel, P., Chaumontet, C., 2004. Diallyl disulfide (DADS) increases histone acetylation and p21(waf1/cip1) expression in human colon tumor cell lines. Carcinogenesis 25, 1227–1236.

Ferracin, M., Lupini, L., Mangolini, A., Negrini, M., 2016. Circulating non-coding RNA as biomarkers in colorectal cancer. Adv. Exp. Med. Biol. 937, 171–181.

Gao, X., Wen, J., Gao, P., Zhang, G., Zhang, G., 2017. Overexpression of the long non-coding RNA, linc-UBC1, is associated with poor prognosis and facilitates cell proliferation, migration, and invasion in colorectal cancer. Onco Targets Ther. 10, 1017–1026.

Godman, C.A., Joshi, R., Tierney, B.R., Greenspan, E., Rasmussen, T.P., Wang, H.W., Shin, D.G., Rosenberg, D.W., Giardina, C., 2008. HDAC3 impacts multiple oncogenic pathways in colon cancer cells with effects on Wnt and vitamin D signaling. Cancer Biol. Ther. 7, 1570–1580.

Guilleret, I., Yan, P., Grange, F., Braunschweig, R., Bosman, F.T., Benhattar, J., 2002. Hypermethylation of the human telomerase catalytic subunit (hTERT) gene correlates with telomerase activity. Int. J. Cancer 101, 335–341.

Guo, Y., Shu, L., Zhang, C., Su, Z.Y., Kong, A.N., 2015. Curcumin inhibits anchorageindependent growth of HT29 human colon cancer cells by targeting epigenetic restoration of the tumor suppressor gene DLEC1. Biochem. Pharmacol. 94, 69–78.

Han, Y., Yang, Y.N., Yuan, H.H., Zhang, T.T., Sui, H., Wei, X.L., Liu, L., Huang, P., Zhang, W.J., Bai, Y.X., 2014. UCA1, a long noncoding RNA up-regulated in colorectal cancer influences cell proliferation, apoptosis and cell cycle distribution. Pathology 46, 396–401.

Higdon, J.V., Delage, B., Williams, D.E., Dashwood, R.H., 2007. Cruciferous vegetables and human cancer risk: epidemiologic evidence and mechanistic basis. Pharmacol. Res. 55, 224–236.

Hinnebusch, B.F., Meng, S., Wu, J.T., Archer, S.Y., Hodin, R.A., 2002. The effects of short-chain fatty acids on human colon cancer cell phenotype are associated with histone hyperacetylation. J. Nutr. 132, 1012–1017.

Ho, E., Clarke, J.D., Dashwood, R.H., 2009. Dietary sulforaphane, a histone deacetylase inhibitor for cancer prevention. J. Nutr. 139, 2393–2396.

Hoh, C., Boocock, D., Marczylo, T., Singh, R., Berry, D.P., Dennison, A.R., Hemingway, D., Miller, A., West, K., Euden, S., Garcea, G., Farmer, P.B., Steward, W.P., Gescher, A.J., 2006. Pilot study of oral silibinin, a putative chemopreventive agent, in colorectal cancer patients: silibinin levels in plasma, colorectum, and liver and their pharmacodynamic consequences. Clin. Cancer Res 12, 2944–2950.

Hu, S., Liu, L., Chang, E.B., Wang, J.Y., Raufman, J.P., 2015. Butyrate inhibits pro-proliferative miR-92a by diminishing c-Myc-induced miR-17-92a cluster transcription in human colon cancer cells. Mol. Cancer 14, 1–15.

Huang, J., Xu, M., Fang, Y.J., Lu, M.S., Pan, Z.Z., Huang, W.Q., Chen, Y.M., Zhang, C.X., 2017. Association between phytosterol intake and colorectal cancer risk: a case-control study. Br. J. Nutr. 117, 839–850.

Huang, M., Lu, J.J., Huang, M.Q., Bao, J.L., Chen, X.P., Wang, Y.T., 2012. Terpenoids: natural products for cancer therapy. Expert Opin. Investig. Drugs 21, 1801–1818.

Huber, W.W., Scharf, G., Nagel, G., Prustomersky, S., Schulte-Hermann, R., Kaina, B., 2003. Coffee and its chemopreventive components Kahweol and Cafestol increase the activity of O6-methylguanine-DNA methyltransferase in rat liver—comparison with phase II xenobiotic metabolism. Mutat. Res. 522, 57–68.

Hughes, L.A.E., Simons, C., van den Brandt, P.A., van Engeland, M., Weijenberg, M.P., 2017. Lifestyle, diet, and colorectal cancer risk according to (Epi)genetic Instability: current evidence and future directions of molecular pathological epidemiology. Curr. Colorectal Cancer Rep. 13, 455–469.

Humphreys, K.J., Cobiac, L., Le Leu, R.K., Van Der Hoek, M.B., Michael, M.Z., 2013. Histone deacetylase inhibition in colorectal cancer cells reveals competing roles for members of the oncogenic miR-17-92 cluster. Mol. Carcinog. 52, 459–474.

Isah, T., 2016. Anticancer alkaloids from trees: development into drugs. Pharmacogn. Rev. 10, 90–99.

Jagadeesh, S., Sinha, S., Pal, B.C., Bhattacharya, S., Banerjee, P.P., 2007. Mahanine reverses an epigenetically silenced tumor suppressor gene RASSF1A in human prostate cancer cells. Biochem. Biophys. Res. Commun. 362, 212–217.

Jakulj, L., Trip, M.D., Sudhop, T., Von Bergmann, K., Kastelein, J.J., Vissers, M.N., 2005. Inhibition of cholesterol absorption by the combination of dietary plant sterols and ezetimibe: effects on plasma lipid levels. J. Lipid Res. 46, 2692–2698.

Jedinak, A., Thyagarajan-Sahu, A., Jiang, J., Sliva, D., 2011. Ganodermanontriol, a lanostanoid triterpene from Ganoderma lucidum, suppresses growth of colon cancer cells through sscatenin signaling. Int. J. Oncol. 38, 761–767.

Ji, J., Zhang, L., Wu, Y.Y., Zhu, X.Y., Lv, S.Q., Sun, X.Z., 2006. Induction of apoptosis by dlimonene is mediated by a caspase-dependent mitochondrial death pathway in human leukemia cells. Leuk. Lymphoma 47, 2617–2624.

Ji, Q., Liu, X., Fu, X., Zhang, L., Sui, H., Zhou, L., Sun, J., Cai, J., Qin, J., Ren, J., Li, Q., 2013. Resveratrol inhibits invasion and metastasis of colorectal cancer cells via MALAT1 mediated Wnt/beta-catenin signal pathway. PLoS One 8, 1–12.

Jia, S.S., Xi, G.P., Zhang, M., Chen, Y.B., Lei, B., Dong, X.S., Yang, Y.M., 2013. Induction of apoptosis by D-limonene is mediated by inactivation of Akt in LS174T human colon cancer cells. Oncol. Rep. 29, 349–354.

Jiang, P., Wu, X., Wang, X., Huang, W., Feng, Q., 2016. NEAT1 upregulates EGCG-induced CTR1 to enhance cisplatin sensitivity in lung cancer cells. Oncotarget 7, 1–21.

Jo, H.J., Song, J.D., Kim, K.M., Cho, Y.H., Kim, K.H., Park, Y.C., 2008. Diallyl disulfide induces reversible G2/M phase arrest on a p53-independent mechanism in human colon cancer HCT-116 cells. Oncol. Rep. 19, 275–280.

Johnson, J.J., 2011. Carnosol: a promising anti-cancer and anti-inflammatory agent. Cancer Lett. 305, 1–7.

Johnson, J.J., Syed, D.N., Heren, C.R., Suh, Y., Adhami, V.M., Mukhtar, H., 2008. Carnosol, a dietary diterpene, displays growth inhibitory effects in human prostate cancer PC3 cells leading to G2-phase cell cycle arrest and targets the 5′-AMP-activated protein kinase (AMPK) pathway. Pharm. Res. (N.Y.) 25, 2125–2134.

Juan, M.E., Planas, J.M., Ruiz-Gutierrez, V., Daniel, H., Wenzel, U., 2008. Antiproliferative and apoptosis-inducing effects of maslinic and oleanolic acids, two pentacyclic triterpenes from olives, on HT-29 colon cancer cells. Br. J. Nutr. 100, 36–43.

Kato, K., Maesawa, C., Itabashi, T., Fujisawa, K., Otsuka, K., Kanno, S., Tada, H., Tatemichi, Y., Kotani, K., Oikawa, H., Sugai, T., Wakabayashi, G., Masuda, T., 2009. DNA hypomethylation at the CpG island is involved in aberrant expression of the L1 cell adhesion molecule gene in colorectal cancer. Int. J. Oncol. 35, 467–476.

Khalife, R., Hodroj, M.H., Fakhoury, R., Rizk, S., 2016. Thymoquinone from Nigella sativa seeds promotes the antitumor activity of noncytotoxic doses of Topotecan in human colorectal cancer cells in vitro. Planta Med. 82, 312–321.

Kim, J.H., Kim, Y.H., Song, G.Y., Kim, D.E., Jeong, Y.J., Liu, K.H., Chung, Y.H., Oh, S., 2014. Ursolic acid and its natural derivative corosolic acid suppress the proliferation of APC-mutated colon cancer cells through promotion of beta-catenin degradation. Food Chem. Toxicol. 67, 87–95.

Kim, J.Y., Kim, D.H., Jeong, H.G., 2006. Inhibitory effect of the coffee diterpene kahweol on carrageenan-induced inflammation in rats. Biofactors 26, 17–28.

Korde, L.A., Wu, A.H., Fears, T., Nomura, A.M., West, D.W., Kolonel, L.N., Pike, M.C., Hoover, R.N., Ziegler, R.G., 2009. Childhood soy intake and breast cancer risk in Asian American women. Cancer Epidemiol. Biomark. Prev. 18, 1050–1059.

Kury, S., Buecher, B., Robiou-Du-Pont, S., Scoul, C., Sebille, V., Colman, H., Le Houerou, C., Le Neel, T., Bourdon, J., Faroux, R., Ollivry, J., Lafraise, B., Chupin, L.D., Bezieau, S., 2007. Combinations of cytochrome P450 gene polymorphisms enhancing the risk for sporadic colorectal cancer related to red meat consumption. Cancer Epidemiol. Biomark. Prev. 16, 1460–1467.

Lai, K.C., Hsu, S.C., Kuo, C.L., Yang, J.S., Ma, C.Y., Lu, H.F., Tang, N.Y., Hsia, T.C., Ho, H.C., Chung, J.G., 2013. Diallyl sulfide, diallyl disulfide, and diallyl trisulfide inhibit migration and invasion in human colon cancer colo 205 cells through the inhibition of matrix metalloproteinase-2, -7, and -9 expressions. Environ. Toxicol. 28, 479–488.

Lee, J.H., Khor, T.O., Shu, L., Su, Z.Y., Fuentes, F., Kong, A.N., 2013. Dietary phytochemicals and cancer prevention: Nrf2 signaling, epigenetics, and cell death mechanisms in blocking cancer initiation and progression. Pharmacol. Ther. 137, 153–171.

Lee, J.S., Jung, W.K., Jeong, M.H., Yoon, T.R., Kim, H.K., 2012a. Sanguinarine induces apoptosis of HT-29 human colon cancer cells via the regulation of Bax/Bcl-2 ratio and caspase-9dependent pathway. Int. J. Toxicol. 31, 70–77.

Lee, K.A., Chae, J.I., Shim, J.H., 2012b. Natural diterpenes from coffee, cafestol and kahweol induce apoptosis through regulation of specificity protein 1 expression in human malignant pleural mesothelioma. J. Biomed. Sci. 19, 2–10.

Lee, K.H., Lee, M.S., Cha, E.Y., Sul, J.Y., Lee, J.S., Kim, J.S., Park, J.B., Kim, J.Y., 2017. Inhibitory effect of emodin on fatty acid synthase, colon cancer proliferation and apoptosis. Mol. Med. Rep. 15, 2163–2173.

Lee, S.D., Gershkovich, P., Darlington, J.W., Wasan, K.M., 2012c. Inhibition of cholesterol absorption: targeting the intestine. Pharm. Res. (N.Y.) 29, 3235–3250.

Li, L., Lin, J., Sun, G., Wei, L., Shen, A., Zhang, M., Peng, J., 2016a. Oleanolic acid inhibits colorectal cancer angiogenesis in vivo and in vitro via suppression of STAT3 and Hedgehog pathways. Mol. Med. Rep. 13, 5276–5282.

Li, L., Wei, L., Shen, A., Chu, J., Lin, J., Peng, J., 2015. Oleanolic acid modulates multiple intracellular targets to inhibit colorectal cancer growth. Int. J. Oncol. 47, 2247–2254.

Li, W., Guo, Y., Zhang, C., Wu, R., Yang, A.Y., Gaspar, J., Kong, A.N., 2016b. Dietary phytochemicals and cancer chemoprevention: a perspective on oxidative stress, inflammation, and epigenetics. Chem. Res. Toxicol. 29, 2071–2095.

Lin, J., Chen, Y., Wei, L., Hong, Z., Sferra, T.J., Peng, J., 2013a. Ursolic acid inhibits colorectal cancer angiogenesis through suppression of multiple signaling pathways. Int. J. Oncol. 43, 1666–1674.

Lin, J., Chen, Y., Wei, L., Shen, A., Sferra, T.J., Hong, Z., Peng, J., 2013b. Ursolic acid promotes colorectal cancer cell apoptosis and inhibits cell proliferation via modulation of multiple signaling pathways. Int. J. Oncol. 43, 1235–1243.

Link, A., Balaguer, F., Shen, Y., Lozano, J.J., Leung, H.C., Boland, C.R., Goel, A., 2013. Curcumin modulates DNA methylation in colorectal cancer cells. PLoS One 8, 1–13.

Liu, C., Dai, L., Liu, Y., Rong, L., Dou, D., Sun, Y., Ma, L., 2016. Antiproliferative activity of triterpene glycoside nutrient from monk fruit in colorectal cancer and throat cancer. Nutrients 8, 1–11.

Liu, W.H., Tao, K.S., You, N., Liu, Z.C., Zhang, H.T., Dou, K.F., 2011. Differences in the properties and mirna expression profiles between side populations from hepatic cancer cells and normal liver cells. PLoS One 6, 1–14.

Lu, J.J., Bao, J.L., Chen, X.P., Huang, M., Wang, Y.T., 2012. Alkaloids isolated from natural herbs as the anticancer agents. Evid. Based Complement. Alternat. Med. 2012, 1–12.

Lv, H., Zhang, Z., Wang, Y., Li, C., Gong, W., Wang, X., 2016. MicroRNA-92a promotes colorectal cancer cell growth and migration by inhibiting KLF4. Oncol. Res. 23, 283–290.

Martin, S.L., Kala, R., Tollefsbol, T.O., 2017. Mechanisms for inhibition of colon cancer cells by sulforaphane through epigenetic modulation of microRNA-21 and human telomerase reverse transcriptase (hTERT) down-regulation. Curr. Cancer Drug Targets 18, 97–106.

Matkar, S.S., Wrischnik, L.A., Hellmann-Blumberg, U., 2008. Sanguinarine causes DNA damage and p53-independent cell death in human colon cancer cell lines. Chem. Biol. Interact. 172, 63–71.

Meadows, S.M., Mulkerin, D., Berlin, J., Bailey, H., Kolesar, J., Warren, D., Thomas, J.P., 2002. Phase II trial of perillyl alcohol in patients with metastatic colorectal cancer. Int. J. Gastrointest. Cancer 32, 125–128.

Meng, W.J., Yang, L., Ma, Q., Zhang, H., Adell, G., Arbman, G., Wang, Z.Q., Li, Y., Zhou, Z.G., Sun, X.F., 2015. MicroRNA expression profile reveals miR-17-92 and miR-143-145 cluster in synchronous colorectal cancer. Medicine (Baltim.) 94, 1–12.

Miladi-Abdennadher, I., Abdelmaksoud-Damak, R., Ayadi, L., Khabir, A., Frikha, F., Kallel, L., Amouri, A., Frikha, M., Sellami-Boudawara, T., Gargouri, A., Mokdad-Gargouri, R., 2010. Hypermethylation of RAR-beta2 correlates with high COX-2 expression and poor prognosis in patients with colorectal carcinoma. Tumour Biol. 31, 503–511.

Mogilyansky, E., Rigoutsos, I., 2013. The miR-17/92 cluster: a comprehensive update on its genomics, genetics, functions and increasingly important and numerous roles in health and disease. Cell Death Differ. 20, 1603–1614.

Motawi, T.K., Shaker, O.G., Ismail, M.F., Sayed, N.H., 2017. Peroxisome proliferator-activated receptor gamma in obesity and colorectal cancer: the role of epigenetics. Sci. Rep. 7, 1–8.

Muller, S., Raulefs, S., Bruns, P., Afonso-Grunz, F., Plotner, A., Thermann, R., Jäger, C., Schlitter, A.M., Kong, B., Regel, I., Roth, W.K., Rotter, B., Hoffmeier, K., Kahl, G., Koch, I., Theis, F.J., Kleeff, J., Winter, P., Michalski, C.W., 2015. Erratum to: next-generation sequencing reveals novel differentially regulated mRNAs, lncRNAs, miRNAs, sdRNAs and a piRNA in pancreatic cancer. Mol. Cancer 14, 1–18.

Myzak, M.C., Dashwood, W.M., Orner, G.A., Ho, E., Dashwood, R.H., 2006. Sulforaphane inhibits histone deacetylase in vivo and suppresses tumorigenesis in Apc-minus mice. FASEB J. 20, 506–508.

Navarro, F., Lieberman, J., 2015. miR-34 and p53: new insights into a complex functional relationship. PLoS One 10, 1–23.

Nian, H., Delage, B., Pinto, J.T., Dashwood, R.H., 2008. Allyl mercaptan, a garlic-derived organosulfur compound, inhibits histone deacetylase and enhances Sp3 binding on the P21WAF1 promoter. Carcinogenesis 29, 1816–1824.

Normen, A.L., Brants, H.A., Voorrips, L.E., Andersson, H.A., van den Brandt, P.A., Goldbohm, R.A., 2001. Plant sterol intakes and colorectal cancer risk in The Netherlands Cohort Study on Diet and Cancer. Am. J. Clin. Nutr. 74, 141–148.

Palozza, P., Serini, S., Torsello, A., Di Nicuolo, F., Piccioni, E., Ubaldi, V., Pioli, C., Wolf, F.I., Calviello, G., 2003. Beta carotene regulates NF-kappaB DNA-binding activity by a redox mechanism in human leukemia and colon adenocarcinoma cells. J. Nutr. 133, 381–388.

Park, J.I., Kwak, J.Y., 2012. The role of peroxisome proliferator-activated receptors in colorectal cancer. PPAR Res. 2012, 1–13.

Park, K.W., Kundu, J., Chae, I.G., Kim, D.H., Yu, M.H., Kundu, J.K., Chun, K.S., 2014. Carnosol induces apoptosis through generation of ROS and inactivation of STAT3 signaling in human colon cancer HCT116 cells. Int. J. Oncol. 44, 1309–1315.

Patel, K.R., Brown, V.A., Jones, D.J., Britton, R.G., Hemingway, D., Miller, A.S., West, K.P., Booth, T.D., Perloff, M., Crowell, J.A., Brenner, D.E., Steward, W.P., Gescher, A.J., Brown, K., 2010. Clinical pharmacology of resveratrol and its metabolites in colorectal cancer patients. Cancer Res. 70, 7392–7399.

Piyanuch, R., Sukhthankar, M., Wandee, G., Baek, S.J., 2007. Berberine, a natural isoquinoline alkaloid, induces NAG-1 and ATF3 expression in human colorectal cancer cells. Cancer Lett. 258, 230–240.

Pozo, O.J., Pujadas, M., Gleeson, S.B., Mesa-Garcia, M.D., Pastor, A., Kotronoulas, A., Fitó, M., Covas, M.I., Navarro, J.R.F., Espejo, J.A., Sanchez-Rodriguez, E., Marchal, R., Calleja, M.A., de la Torre, R., 2017. Liquid chromatography tandem mass spectrometric determination of triterpenes in human fluids: evaluation of markers of dietary intake of olive oil and metabolic disposition of oleanolic acid and maslinic acid in humans. Anal. Chim. Acta 990, 84–95.

Qian, J., Jiang, B., Li, M., Chen, J., Fang, M., 2013. Prognostic significance of microRNA-16 expression in human colorectal cancer. World J. Surg. 37, 2944–2949.

Qin, J., Chen, J.X., Zhu, Z., Teng, J.A., 2015. Genistein inhibits human colorectal cancer growth and suppresses miR-95, Akt and SGK1. Cell. Physiol. Biochem. 35, 2069–2077.

Rabi, T., Bishayee, A., 2009. D-Limonene sensitizes docetaxel-induced cytotoxicity in human prostate cancer cells: generation of reactive oxygen species and induction of apoptosis. J. Carcinog. 8, 1–15.

Raina, K., Agarwal, C., Agarwal, R., 2013. Effect of silibinin in human colorectal cancer cells: targeting the activation of NF-kappaB signaling. Mol. Carcinog. 52, 195–206.

Raina, K., Kumar, S., Dhar, D., Agarwal, R., 2016. Silibinin and colorectal cancer chemoprevention: a comprehensive review on mechanisms and efficacy. J. Biomed. Res. 30, 452–465.

Ramprasath, V.R., Awad, A.B., 2015. Role of phytosterols in cancer prevention and treatment. J. AOAC Int. 98, 735–738.

Reuter, S., Gupta, S.C., Park, B., Goel, A., Aggarwal, B.B., 2011. Epigenetic changes induced by curcumin and other natural compounds. Genes Nutr. 6, 93–108.

Reyes-Zurita, F.J., Rufino-Palomares, E.E., Lupianez, J.A., Cascante, M., 2009. Maslinic acid, a natural triterpene from Olea europaea L., induces apoptosis in HT29 human colon-cancer cells via the mitochondrial apoptotic pathway. Cancer Lett. 273, 44–54.

Reyes-Zurita, F.J., Rufino-Palomares, E.E., Medina, P.P., Leticia Garcia-Salguero, E., Peragon, J., Cascante, M., Lupiáñez, J.A., 2013. Antitumour activity on extrinsic apoptotic targets of the triterpenoid maslinic acid in p53-deficient Caco-2 adenocarcinoma cells. Biochimie 95, 2157–2167.

Reynoso-Camacho, R., Gonzalez-Jasso, E., Ferriz-Martinez, R., Villalon-Corona, B., Loarca-Pina, G.F., Salgado, L.M., Ramos-Gomez, M., 2011. Dietary supplementation of lutein reduces colon carcinogenesis in DMH-treated rats by modulating K-ras, PKB, and beta-catenin proteins. Nutr. Cancer 63, 39–45.

Sahnane, N., Magnoli, F., Bernasconi, B., Tibiletti, M.G., Romualdi, C., Pedroni, M., Ponz de Leon, M., Magnani, G., Reggiani-Bonetti, L., Bertario, L., Signoroni, S., Capella, C., Sessa, F., Furlan, D., 2015. Aberrant DNA methylation profiles of inherited and sporadic colorectal cancer. Clin. Epigenet. 7, 1–12.

Sakai, E., Nakajima, A., Kaneda, A., 2014. Accumulation of aberrant DNA methylation during colorectal cancer development. World J. Gastroenterol. 20, 978–987.

Saldanha, S.N., Kala, R., Tollefsbol, T.O., 2014. Molecular mechanisms for inhibition of colon cancer cells by combined epigenetic-modulating epigallocatechin gallate and sodium butyrate. Exp. Cell Res. 324, 40–53.

Saldanha, S.N., Tollefsbol, T.O., 2012. The role of nutraceuticals in chemoprevention and chemotherapy and their clinical outcomes. J. Oncol. 2012, 1–24.

Sanchez-Camargo, A.D., Garcia-Canas, V., Herrero, M., Cifuentes, A., Ibanez, E., 2016. Comparative study of green sub- and supercritical processes to obtain carnosic acid and carnosol-enriched rosemary extracts with in vitro anti-proliferative activity on colon cancer cells. Int. J. Mol. Sci. 17, 1–18.

Saud, S.M., Li, W., Morris, N.L., Matter, M.S., Colburn, N.H., Kim, Y.S., Young, M.R., 2014. Resveratrol prevents tumorigenesis in mouse model of Kras activated sporadic colorectal cancer by suppressing oncogenic Kras expression. Carcinogenesis 35, 2778–2786.

Schetter, A.J., Okayama, H., Harris, C.C., 2012. The role of microRNAs in colorectal cancer. Cancer J. 18, 244–252.

Schlormann, W., Naumann, S., Renner, C., Glei, M., 2015. Influence of miRNA-106b and miRNA-135a on butyrate-regulated expression of p21 and Cyclin D2 in human colon adenoma cells. Genes Nutr. 10, 1–21.

Schneider-Stock, R., Ghantous, A., Bajbouj, K., Saikali, M., Darwiche, N., 2012. Epigenetic mechanisms of plant-derived anticancer drugs. Front. Biosci. (Landmark Ed.) 17, 129–173.

Schramm, L., 2013. Going green: the role of the green tea component EGCG in chemoprevention. J. Carcinog. Mutagen. 4, 1–18.

Shan, J.Z., Xuan, Y.Y., Zheng, S., Dong, Q., Zhang, S.Z., 2009. Ursolic acid inhibits proliferation and induces apoptosis of HT-29 colon cancer cells by inhibiting the EGFR/MAPK pathway. J. Zhejiang Univ. Sci. B 10, 668–674.

Sharma, S.H., Thulasingam, S., Nagarajan, S., 2017. Terpenoids as anti-colon cancer agents - a comprehensive review on its mechanistic perspectives. Eur. J. Pharmacol. 795, 169–178.

Singh, R.P., Gu, M., Agarwal, R., 2008. Silibinin inhibits colorectal cancer growth by inhibiting tumor cell proliferation and angiogenesis. Cancer Res. 68, 2043–2050.

Singh, S., Verma, M., Malhotra, M., Prakash, S., Singh, T.D., 2016. Cytotoxicity of alkaloids isolated from Argemone mexicana on SW480 human colon cancer cell line. Pharm. Biol. 54, 740–745.

Sinha, S., Pal, B.C., Jagadeesh, S., Banerjee, P.P., Bandyopadhaya, A., Bhattacharya, S., 2006. Mahanine inhibits growth and induces apoptosis in prostate cancer cells through the deactivation of Akt and activation of caspases. Prostate 66, 1257–1265.

Slattery, M.L., Potter, J.D., Ma, K.N., Caan, B.J., Leppert, M., Samowitz, W., 2000. Western diet, family history of colorectal cancer, NAT2, GSTM-1 and risk of colon cancer. Cancer Causes Control 11, 1–8.

Sobral, M.V., Xavier, A.L., Lima, T.C., De Sousa, D.P., 2014. Antitumor activity of monoterpenes found in essential oils. Sci. World J. 2014, 1–36.

Sriram, N., Kalayarasan, S., Ashokkumar, P., Sureshkumar, A., Sudhandiran, G., 2008. Diallyl sulfide induces apoptosis in Colo 320 DM human colon cancer cells: involvement of caspase-3, NF-kappaB, and ERK-2. Mol. Cell. Biochem. 311, 157–165.

Sun, J., 2007. D-Limonene: safety and clinical applications. Altern. Med. Re.v 12, 259–264.

Takasawa, K., Arai, Y., Yamazaki-Inoue, M., Toyoda, M., Akutsu, H., Umezawa, A., Nishino, K., 2018. DNA hypermethylation enhanced telomerase reverse transcriptase expression in humaninduced pluripotent stem cells. Hum. Cell 31, 78–86.

Tang, F.Y., Pai, M.H., Kuo, Y.H., Wang, X.D., 2012. Concomitant consumption of lycopene and fish oil inhibits tumor growth and progression in a mouse xenograft model of colon cancer. Mol. Nutr. Food Res. 56, 1520–1531.

Tang, F.Y., Pai, M.H., Wang, X.D., 2011. Consumption of lycopene inhibits the growth and progression of colon cancer in a mouse xenograft model. J. Agric. Food Chem. 59, 9011–9021.

Tang, F.Y., Shih, C.J., Cheng, L.H., Ho, H.J., Chen, H.J., 2008. Lycopene inhibits growth of human colon cancer cells via suppression of the Akt signaling pathway. Mol. Nutr. Food Res. 52, 646–654.

Thanos, J., Cotterchio, M., Boucher, B.A., Kreiger, N., Thompson, L.U., 2006. Adolescent dietary phytoestrogen intake and breast cancer risk (Canada). Cancer Causes Control 17, 1253–1261.

Thorenoor, N., Faltejskova-Vychytilova, P., Hombach, S., Mlcochova, J., Kretz, M., Svoboda, M., Slaby, O., 2016. Long non-coding RNA ZFAS1 interacts with CDK1 and is involved in p53dependent cell cycle control and apoptosis in colorectal cancer. Oncotarget 7, 622–637.

Tuan, J., Chen, Y.X., 2016. Dietary and lifestyle factors associated with colorectal cancer risk and interactions with microbiota: fiber, red or processed meat and alcoholic drinks. Gastrointest. Tumors 3, 17–24.

Tsang, W.P., Kwok, T.T., 2010. Epigallocatechin gallate upregulation of miR-16 and induction of apoptosis in human cancer cells. J. Nutr. Biochem. 21, 140–146.

Tse, J.W.T., Jenkins, L.J., Chionh, F., Mariadason, J.M., 2017. Aberrant DNA methylation in colorectal cancer: what should we target? Trends Cancer 3, 698–712.

Upadhyay, S., Dixit, M., 2015. Role of polyphenols and other phytochemicals on molecular signaling. Oxid. Med. Cell Longev. 2015, 1–33.

Valeri, N., Braconi, C., Gasparini, P., Murgia, C., Lampis, A., Paulus-Hock, V., Hart, J.R., Ueno, L., Grivennikov, S.I., Lovat, F., Paone, A., Cascione, L., Sumani, K.M., Veronese, A., Fabbri, M., Carasi, S., Alder, H., Lanza, G., Gafa, R., Moyer, M.P., Ridgway, R.A., Cordero, J., Nuovo, G.J., Frankel, W.L., Rugge, M., Fassan, M., Groden, J., Vogt, P.K., Karin, M., Sansom, O.J., Croce, C.M., 2014. MicroRNA-135b promotes cancer progression by acting as a downstream effector of oncogenic pathways in colon cancer. Cancer Cell 25, 469–483.

Vergara, D., Simeone, P., Bettini, S., Tinelli, A., Valli, L., Storelli, C., Leo, S., Santino, A., Maffia, M., 2014. Antitumor activity of the dietary diterpene carnosol against a panel of human cancer cell lines. Food Funct. 5, 1261–1269.

Wang, C., Wang, J., Liu, H., Fu, Z., 2014. Tumor suppressor DLC-1 induces apoptosis and inhibits the growth and invasion of colon cancer cells through the Wnt/beta-catenin signaling pathway. Oncol. Rep. 31, 2270–2278.

Wang, H., Khor, T.O., Shu, L., Su, Z.Y., Fuentes, F., Lee, J.H., Kong, A.N., 2012. Plants vs. cancer: a review on natural phytochemicals in preventing and treating cancers and their druggability. Anticancer Agents Med. Chem. 12, 1281–1305.

Wang, J., Liu, L., Qiu, H., Zhang, X., Guo, W., Chen, W., Tian, Y., Fu, L., Shi, D., Cheng, J., Huang, W., Deng, W., 2013a. Ursolic acid simultaneously targets multiple signaling pathways to suppress proliferation and induce apoptosis in colon cancer cells. PLoS One 8, 1–19.

Wang, W., Zhao, C., Jou, D., Lu, J., Zhang, C., Lin, L., Lin, J., 2013b. Ursolic acid inhibits the growth of colon cancer-initiating cells by targeting STAT3. Anticancer Res. 33, 4279–4284.

Wang, L., Steele, I., Kumar, J.D., Dimaline, R., Jithesh, P.V., Tiszlavicz, L., Reisz, Z., Dockray, G.J., Varro, A., 2016. Distinct miRNA profiles in normal and gastric cancer myofibroblasts and significance in Wnt signaling. Am. J. Physiol. Gastrointest. Liver Physiol. 310, 1–17.

Wang, Z., Chen, H., 2010. Genistein increases gene expression by demethylation of WNT5a promoter in colon cancer cell line SW1116. Anticancer Res. 30, 4537–4545.

Weis, B., Schmidt, J., Maamar, H., Raj, A., Lin, H., Toth, C., Riedmann, K., Raddatz, G., Seitz, H.K., Ho, A.D., Lyko, F., Linhart, H.G., 2015. Inhibition of intestinal tumor formation by deletion of the DNA methyltransferase 3a. Oncogene 34, 1822–1830.

Wilken, R., Veena, M.S., Wang, M.B., Srivatsan, E.S., 2011. Curcumin: a review of anti-cancer properties and therapeutic activity in head and neck squamous cell carcinoma. Mol. Cancer 10, 1–45.

Xavier, C.P., Lima, C.F., Pedro, D.F., Wilson, J.M., Kristiansen, K., Pereira-Wilson, C., 2013. Ursolic acid induces cell death and modulates autophagy through JNK pathway in apoptosis-resistant colorectal cancer cells. J. Nutr. Biochem. 24, 706–712.

Yadav, N., Yadav, R., Goyal, A., 2014. Chemistry of terpenoids. Int. J. Pharm. Sci. Rev. Res. 27, 272–278.

Yang, S., Li, W., Sun, H., Wu, B., Ji, F., Sun, T., Chang, H., Shen, P., Wang, Y., Zhou, D., 2015. Resveratrol elicits anti-colorectal cancer effect by activating miR-34c-KITLG in vitro and in vivo. BMC Cancer 15, 1–20.

Yeh, C.T., Rao, Y.K., Yao, C.J., Yeh, C.F., Li, C.H., Chuang, S.E., Luong, J.H., Lai, G.M., Tzeng, Y.M., 2009. Cytotoxic triterpenes from Antrodia camphorata and their mode of action in HT-29 human colon cancer cells. Cancer Lett. 285, 73–79.

Yeruva, L., Hall, C., Elegbede, J.A., Carper, S.W., 2010. Perillyl alcohol and methyl jasmonate sensitize cancer cells to cisplatin. Anti Cancer Drugs 21, 1–9.

Yu, Y., Nangia-Makker, P., Farhana, L., G, R.S., Levi, E., Majumdar, A.P., 2015. miR-21 and miR-145 cooperation in regulation of colon cancer stem cells. Mol. Cancer 14, 1–20.

Yu, Y., Sarkar, F.H., Majumdar, A.P., 2013. Down-regulation of miR-21 induces differentiation of chemoresistant colon cancer cells and enhances susceptibility to therapeutic regimens. Transl. Oncol. 6, 180–186.

Zhang, R., Xu, J., Zhao, J., Bai, J., 2017. Knockdown of miR-27a sensitizes colorectal cancer stem cells to TRAIL by promoting the formation of Apaf-1-caspase-9 complex. Oncotarget 8, 45213–45223.

Zhang, Y.H., Wang, Y.L., Wei, Q.Y., Cai, Y.J., Wang, Q., Liu, Z.L., 2005. Diterpenoids from the Chinese herb Caryopteris terniflora and their antibacterial and antitumor activity. Pharmazie 60, 551–553.

Zhao, L., Sun, Y., Hou, Y., Peng, Q., Wang, L., Luo, H., Tang, X., Zeng, Z., Liu, M., 2012. MiRNA expression analysis of cancer-associated fibroblasts and normal fibroblasts in breast cancer. Int. J. Biochem. Cell Biol. 44, 2051–2059.

Zhou, W., Li, X., Liu, F., Xiao, Z., He, M., Shen, S., Liu, S., 2012. MiR-135a promotes growth and invasion of colorectal cancer via metastasis suppressor 1 in vitro. Acta Biochim. Biophys. Sin. 44, 838–846.

CHAPTER

11

Epigenetic Aspects of Hepatocellular Carcinoma Chemoprevention

Renato Heidor, Ernesto Vargas-Mendez, Fernando Salvador Moreno

Department of Food and Experimental Nutrition, Faculty of Pharmaceutical Sciences of University of São Paulo, São Paulo, Brazil

Abstract

During the development of hepatocellular carcinoma (HCC), there is deregulation of cell systems involved in DNA repair, cell cycle control, apoptosis, HMG-CoA reductase regulation, and membrane fluidity, among others. It is thought that epigenetic alterations may be linked to hepatocarcinogenesis. Because these modifications, including epimutations, are potentially reversible, they are important targets for HCC chemoprevention. Bioactive food compounds (BFCs) and several drugs have the potential to reverse epigenetic events associated with the development of HCC, such as DNA methylation, histone modifications, and microRNA expression. According to their pleiotropic properties and possible selectivity to preneoplastic and neoplastic cells, BFCs can also act on autophagy and in signaling pathways from stem cells. Thus, since chemopreventive agents can have multiple roles in hepatocarcinogenesis, strategies for HCC chemoprevention can be designed, including the use of these agents either intermittently or continuously as adjuvant therapies to obtain a synergistic effect with or without synthetic lethality against cancer.

Keywords

Bioactive food compounds; Cancer; Chemopreventive agents; Epigenetic modulation; Hepatocarcinogenesis; Synthetic agents

INTRODUCTION

Hepatocellular carcinoma (HCC) is the sixth most common cancer in the world and accounts for 90% of all primary hepatic neoplasia (Bray et al., 2013). More than 80% of reported cases of HCC occur in East Asia and sub-Saharan Africa, where its incidence is higher than 20 patients per 100,000 individuals (El-Serag, 2012). Although the incidence of HCC in East Asia is stable, with a predicted reduction as a consequence of immunization against hepatitis B, in Western countries its incidence is increasing when compared with other types of cancers (Jemal et al., 2010; Simard et al., 2012; Singh et al., 2014). Among the etiological factors in the development of HCC, excessive alcohol consumption and ingestion of food contaminated by aflatoxins, hepatitis B and C infections, and nonalcoholic liver steatosis due to increasing obesity should be highlighted (Blonski et al., 2010; Hoshida et al., 2010; He et al., 2011). HCC is often diagnosed at advanced stages, limiting therapeutic options such as percutaneous and surgical

Epigenetics of Cancer Prevention
https://doi.org/10.1016/B978-0-12-812494-9.00011-1

ablation or liver transplant. Approximately 70% of patients subjected to these therapies show relapse within 5 years, reaching stages more aggressive than the primary neoplasia (Llovet et al., 2008; Bruix et al., 2012). The most promising pharmacological therapy so far uses the multikinase inhibitor sorafenib, which increases patient survival from 7.9 to 10.7 months on average (Llovet et al., 2012; Di Marco et al., 2013).

Carcinogenesis is traditionally considered a multistep process in which the accumulation of alterations in genes involved in the regulation of important cellular events, such as DNA repair, cell cycle control, apoptosis, cell adhesion, and metabolism occurs (Vogelstein et al., 1993; Liu et al., 2014). Gene alterations in HCC can be detected precociously in preneoplastic lesions and accumulate with progression of the disease (Thorgeirsson et al., 2002; Hussain et al., 2007; Pitot, 2007) (Fig. 11.1). For instance, it has been suggested that upregulation of the oncogene Myc could be an early event in hepatocarcinogenesis (Lin et al., 2010), while the tumor-suppressor gene TP53 is commonly observed downregulated in HCC (Meng et al., 2014). Recent evidence points to the role of members

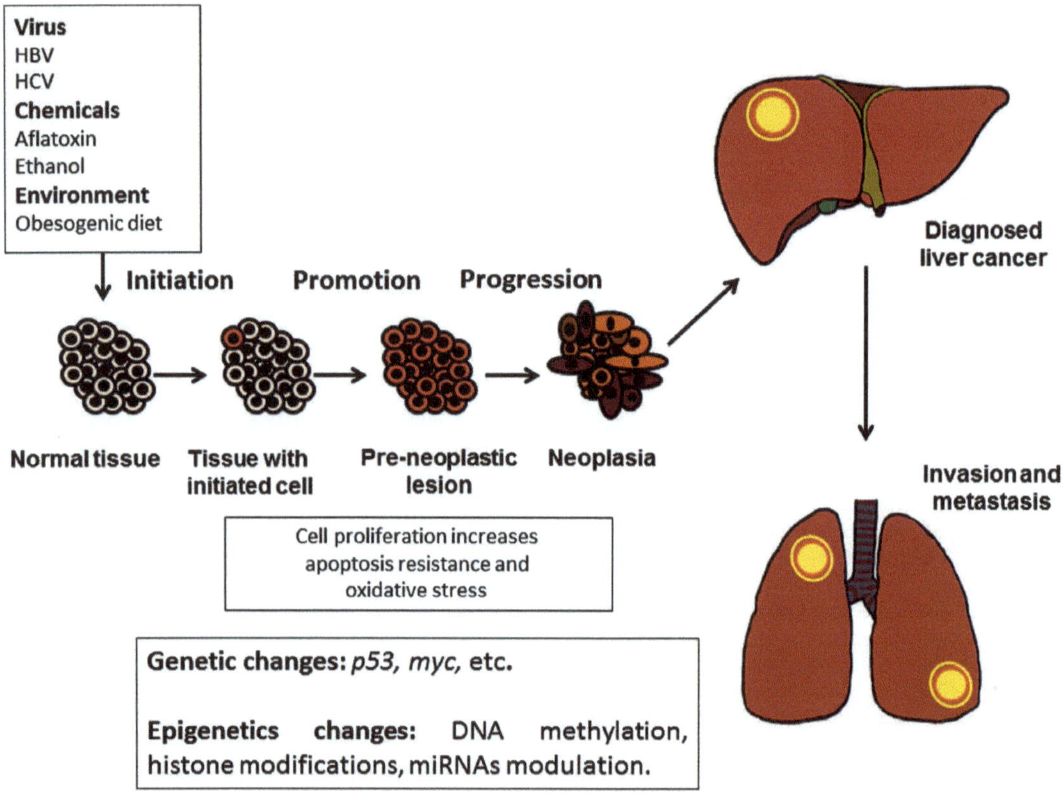

FIGURE 11.1 Evolution of hepatocarcinogenesis and its molecular aspects. Simple mutations and epigenetic alterations occasioned by viruses, chemicals, and nutritional factors may result in the irreversible stage of initiation. The stage of promotion is characterized by the enhancement of cell proliferation, oxidative stress, and apoptosis inhibition of initiated cells with the formation of preneoplastic lesions. The irreversible stage of progression results in karyotypic alterations that evolve into a higher degree of genomic instability, leading from preneoplastic lesions to cancer and metastasis. Epigenetic and genetic mechanisms can drive all stages of hepatocarcinogenesis. *Adapted from Hussain et al., 2007 and Pitot, 2007.*

of the flotillin protein family (i.e., FLOT1) as lipid raft markers that are expressed in HCC patients with poorer prognoses (Zhang et al., 2013). The role of lipid rafts in the initiation and progression of human cancers is well established. These lipid rafts are cellular membrane regions containing the molecular complexes involved in vesicular trafficking and signal transduction. FLOT1 members are also regulators of the oncogene ErbB2, and their inhibition has been associated with a decrease in breast cancer cell proliferation (Paris et al., 2010).

HCC development is a long process that involves dysregulation of epigenetic events (Pogribny et al., 2008). These alterations in genetic expression can be hereditary at the cellular level, also known as epimutations, and occur independently of changes in the nucleotide sequence within the DNA (Jiang et al., 2004; Peltomäki et al., 2012). Because epigenetic events are present at early stages of the carcinogenic process and are potentially reversible, they represent targets for cancer prevention (Fig. 11.2) (Heidor et al., 2012; Vanden Berghe, 2012;

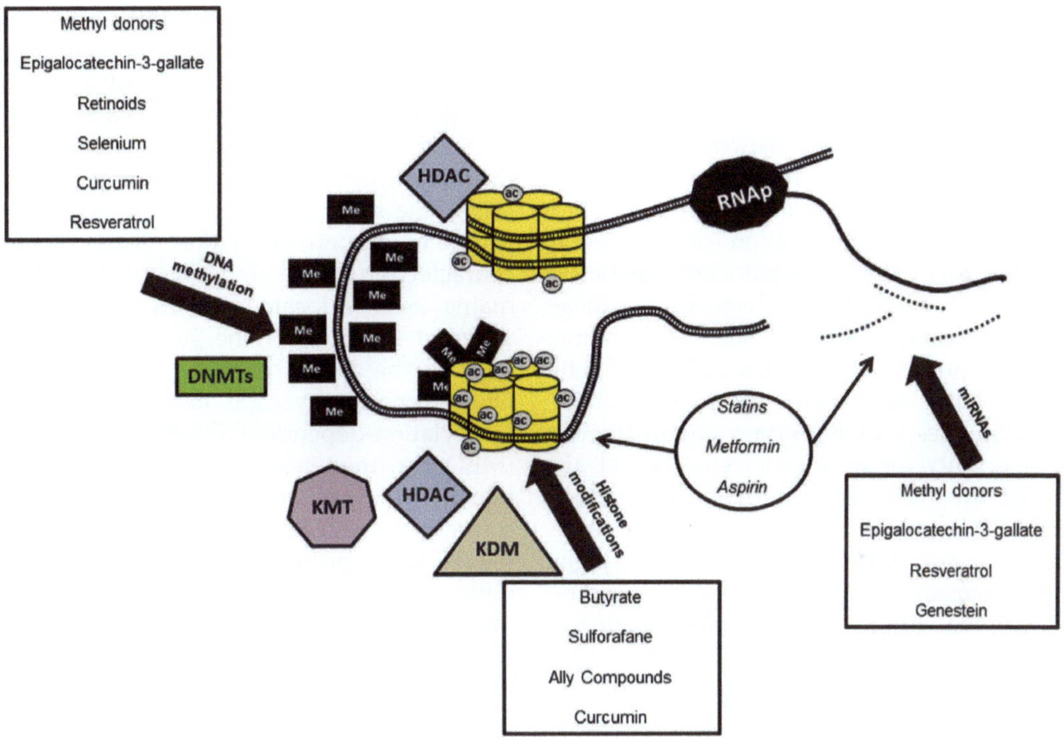

FIGURE 11.2 Role of chemopreventive agents in the modulation of epigenetic events during hepatocarcinogenesis. Chemopreventive agents may act on DNA methylation through the inhibition of DNMTs and reexpression of tumor-suppressor genes that were previously silenced by the hypermethylation of their promoter regions. Further, chemopreventive agents can modulate trimethylation of histones, particularly H3K9me3, which is regulated by histone lysine methyltransferase and histone lysine demethylase enzymes. Chemopreventive agents may also inhibit HDACs, promoting histone hyperacetylation such as in H3K9ac. Removal of methyl groups from gene-promoter regions, reduction in trimethylation of histones, and increased histone acetylation facilitate the access of transcription factors to DNA. Finally, chemopreventive agents can also regulate the expression of miRNAs through translational control of target genes.

Gerhauser, 2013). A possible strategy for prevention of neoplasia is chemoprevention in which the early stages of carcinogenesis are targeted with synthetic or natural compounds, such as bioactive food compounds (BFCs), before progression is established (Ong et al., 2012). This chapter focuses on HCC chemoprevention using BFCs and synthetic compounds, such as statins, metformin, and aspirin, with an emphasis on epigenetic effects.

CHEMOPREVENTION OF HEPATOCELLULAR CARCINOMA BY BIOACTIVE FOOD COMPOUND–INDUCED EPIGENETIC MODULATION

Various studies suggest that consumption of BFCs may be linked to a reduction in the risk of cancer. BFCs have anticancer properties that may be related to modulation of epigenetic events that occur during hepatocarcinogenesis, such as DNA methylation, histone modifications, and the expression of small noncoding microRNAs (miRNAs) (Ong et al., 2011). In general, BFCs have low toxicity and present pleiotropic properties that in certain circumstances show selectivity for neoplastic cells (Wu et al., 2011). In the next sections, we will explore different BFCs and their cancer chemopreventive effects that are linked with one or more epigenetic mechanism. Thus, a comprehensive summary of the BFCs that will be explored throughout this chapter is presented along with their epigenetic mechanisms of action in chemoprevention (Table 11.1).

DNA methylation is the best studied epigenetic modification (Shukla et al., 2014). DNA methylation occurs due to the activity of a family of DNA methyltransferases (DNMTs) that transfer a methyl group from *S*-adenosylmethionine (SAM) into the carbon 5 of cytosine. This nitrosylated base is found mainly in regions rich in cytosine and guanine called CpG islands that are frequently associated with transcription activation sites (Bird, 1986). A common feature in several neoplasms is global DNA hypomethylation that occurs as a consequence of the reduction in the number of methylated cytosines in comparison with nonneoplastic tissue. Global DNA hypomethylation could be a direct consequence of reduced availability of SAM, compromised genomic integrity due to unrepaired DNA lesions, or altered expression of DNMTs (Subramaniam et al., 2014). It can contribute to the carcinogenic process through different mechanisms, including chromosome instability (Eden et al., 2003), derepression of imprinted genes (Feinberg et al., 2002), activation of retrotransposons (Slotkin et al., 2007), and induction of oncogenes, such as c-Myc (Calvisi et al., 2007).

Another important epigenetic event involving the transfer of methyl groups during carcinogenesis is the silencing of specific genes following promoter hypermethylation. This process is characterized by methylation of the DNA domains usually located within CpG islands in different locations of the genome, but particularly in the 5′-end of gene promoter regions. DNA hypermethylation favors recruitment of methylation-dependent DNA binding proteins (MBDs) to the sites of hypermethylated DNA, which in turn recruit proteins that form transcription-inhibitory complexes. These complexes have been observed to be associated with promoters of tumor-suppressor genes in several types of cancer (Shukla et al., 2014), including HCC (Pogribny et al., 2014). Studies show that HCC has different patterns of methylation when compared with normal surrounding tissue. *PROX1* Many genes were found to be silenced due to hypermethylation of their promoter regions, including the tumor suppressors RASSF1A, $p16^{INK4a}$, $p15^{INK4b}$, RB1, SOCS1, SOCS3, SYC, GSTP1, NQO1, PROX1, NORE1B, RIZ1, RELN, FBLN1, and PAX5 (36–38). These changes should not be considered isolated events, as they are also involved in alterations

TABLE 11.1 Epigenetic Mechanisms Induced by Natural and Synthetic Chemopreventive Agents in Cancer

Epigenetic Mechanism	Chemopreventive Agent	Biological Effects	References
DNA methylation	Folic acid and other methyl donors	Inhibition of c-Myc and angiogenesis (in vivo)	Morishita et al. (2015) and Supic et al. (2013)
		Global hypermethylation of suppressor genes (in vivo)	James et al. (2003)
	Epigallocathechin-3-gallate	Antioxidant, prooxidant and induction of apoptosis (in vitro and in vivo)	Wainfan et al. (1992)
		Reexpression of silenced tumor-suppressor genes (in vitro)	Ong et al. (2011) and Tryndyak et al. (2011)
	Retinoids	Chemoprevention independent of c-Myc and HMG-CoA methylation status (in vivo)	Tamura et al. (1997)
		Upregulation of glycine-N-methyltransferase (in vivo)	Sampaio et al. (2007)
	Selenium (selenite)	DNMT1 inhibition, global DNA hypomethylation, and reexpression of GSTP1 (in vitro)	Gerhauser (2013)
	Curcumin	Inhibition of DNMT1 and global DNA hypomethylation (in vitro)	Ghosh et al. (2012)
	Resveratrol	Incidence reduction of liver hyperplastic nodules in HBx transgenic mice (in vivo)	Kang et al. (2005)
		Moderate DNMT inhibitory activity (in vitro)	Lin et al. (2012)
Histone modification	Butyrate and tributyrin	Cell differentiation and apoptosis induction by inhibition of HDAC classes 1 and 2 (in vitro)	Astbury et al. (2012) and Davie (2003)
		Downregulation of HDAC4 and upregulation of H3 and H3K9ac (in vitro)	Davie (2003)
		Incidence reduction of preneoplastic lesions, apoptosis induction, and upregulation of tumor-suppressor genes (in vivo)	Wang et al., (2013) and de Conti et al. (2012)

Continued

TABLE 11.1 Epigenetic Mechanisms Induced by Natural and Synthetic Chemopreventive Agents in Cancer—cont'd

Epigenetic Mechanism	Chemopreventive Agent	Biological Effects	References
	Sulforaphane	Inhibition of HDAC and upregulation of p21 and Ras (in vitro)	Myzak et al. (2004)
	Allyl compounds	HDAC inhibition (in vitro)	Chu et al. (2013)
	Curcumin	Global histone hypomethylation by HAT and HDAC modulation (in vitro and in vivo)	Chen et al. (2013) and Teiten et al. (2013)
	Statins	HDAC inhibition and downregulation of DNA methyltransferases (in vitro)	Karlic et al. (2015)
	Metformin	Downregulation of SIRT1 and increased acetylation of p53 (in vitro and in vivo)	Yu et al. (2017) and Zhang et al. (2015)
		Inhibition of H3 acetylation (in vivo)	Zhang et al. (2015)
	Aspirin	Downregulation of HDAC genetic expression associated with increased levels of H3K27ac (in vitro)	Guo et al. (2016)
miRNA modulation	Methyl donors	Downregulation of miR-122 and miR-29 and upregulation of miR-34a, miR-155, and miR-221 (in vivo)	Chen et al. (2015)
	Epigallocathechin-3-gallate	Modulation of miR-16 (in vitro)	Darvesh et al. (2013)
	β-ionone	Modulation of miRNAs within preneoplastic lesions of the liver (in vivo) and in neoplastic cells (in vitro).	Furtado et al. (2017)
	Statins	Downregulation of miR-17/92 cluster members and upregulation of miR-124a with potential effects in creatine metabolism (in vitro)	Zambrano et al. (2015) and Phulukdaree et al. (2015)
	Metformin	Modulation of 51 miRNAs including upregulation of let-7 family members (in vitro and in vivo)	Miyoshi et al. (2014)

of the epigenome, which include histone modifications (Pogribny et al., 2014).

Histones modifications include acetylation, methylation, phosphorylation, ubiquitination, biotinylation, sumoylation, deamination, proline-isomerization, propionylation, and ADP-ribosylation of amino acid residues (Pogribny et al., 2014; Ma et al., 2014; Chen et al., 2015). Among these modifications, acetylation and methylation are the most studied in HCC (Pogribny et al., 2014). Histone acetylation promotes chromatin relaxation, which is normally found tightly wrapped. This process allows MBDs and transcription factors to access promoter sequences of genes. Modulation of histone acetylation is carried out by histone acetyltransferases (HATs), which add acetyl groups to lysine or arginine residues, as well as histone deacetylases (HDACs), which remove the acetyl radicals from those amino acids. Dysregulation of HDACs such as HDAC1, HDAC2, HDAC3, and SIRT1 occurs in HCC (Pogribny et al., 2014; Ma et al., 2014; Chen et al., 2015).

Methylation of histones also occurs in lysine and arginine residues and can activate or suppress transcription factors according to the number of methyl radicals involved. Histone lysine methyltransferases (KMTs), such as SMYD3, RIZ1, and EZH2, catalyze the transfer of up to three methyl groups from the metabolite SAM to specific residues in histones. On the other hand, removal of methyl groups from histones is carried out by histone lysine demethylases (Pogribny et al., 2012). The silencing of tumor-suppressor genes during hepatocarcinogenesis is linked not only to DNA methylation but also to histone modification in the promoter sequence of genes (Pogribny et al., 2014). In fact, silencing of the genes that encode for the zinc-finger protein RIZ1 and tumor suppressors $p16^{INK4a}$ and RASSF1A in HCC has been related to an increased level of trimethylated histone H3, with methylation occurring at lysines 9 (H3K9me3) and 27 (H3K27me3). Further, hepatocarcinogenesis has a characteristic reduction in the levels of trimethylated histone H4 at lysine residue 20 (H4K20me3) as well as increased levels of H3K27me3 (Pogribny et al., 2014).

Several studies suggest that miRNAs may be responsible for an alternative epigenetic event regulating gene expression (Song et al., 2015a,b). MiRNAs are small noncoding RNA sequences of 16–29 nucleotides that act as negative regulators of genes at the posttranscriptional level. These molecules are synthesized by RNA polymerase II as longer polyadenylated primary miRNAs in a hairpin loop structure. After being transported from the nucleus to the cytoplasm by the transporter protein exportin 5, these macromolecules are processed by the enzyme Dicer, generating miRNA:miRNA hybrids. After unpairing, one strand is degraded and the other originates a mature miRNA. The miRNA then pairs with homologous nontranslated sequences of mRNAs, thus regulating posttranscriptional expression of target genes (Pogribny et al., 2014). Changes in the expression patterns of miRNAs are linked to diverse neoplasia (Romero-Cordoba et al., 2014), including HCC (Song et al., 2015a,b), because many miRNAs were identified to potentially target the genes involved in cell proliferation, differentiation, and apoptosis (Song et al., 2015a,b). In fact, it was recently described that the levels of miR-221, miR-181b-1, miR-155-5p, miR-25, and miR-17-5p were upregulated in murine HCC as well as in human liver tumors (Morishita et al., 2015).

Evidence suggests that environmental factors, particularly ingested BFCs, may modulate changes in DNA methylation, histone methylation, and histone acetylation as well as the expression of miRNAs (Supic et al., 2013). Therefore, BFCs may be considered promissory agents for the chemoprevention of hepatocarcinogenesis. Some of these BFC-induced epigenetic modifications will be explored in the following sections.

Methyl Donors

Methyl donors, such as folic acid, which is found in many fruits and vegetables, are key elements in the synthesis and methylation of DNA. Studies with rodents showed that deficiency in compounds involved in SAM metabolism may lead to hepatocarcinogenesis (Pogribny et al., 2012). Treatment of rats subjected to the "resistant hepatocyte" (RH) model of hepatocarcinogenesis with folic acid showed a reduction in the number of hepatic foci stained for placental glutathione S-transferase (GST-P). In this study, it was also observed that folic acid inhibited c-Myc expression specifically within these GST-P-positive foci (Chagas et al., 2011). In another study, it was demonstrated that the administration of a lower dose of folic acid in rats subjected to the same model of hepatocarcinogenesis resulted in downregulation of the expression of genes associated with angiogenesis (Guariento et al., 2014). Studies in humans suggested that low plasma folate concentrations are associated with an increased risk of developing HCC, and that patients with low plasma folate have a higher risk of tumor progression (Areekul et al., 1977; Welzel et al., 2007). One of the most studied hypotheses regarding the role of folate in hepatocarcinogenesis is that deficiency of this vitamin is linked to dysregulation of epigenetic events (41). Deficiency in methyl donors promotes demethylation of genomic DNA in the liver of rodents (Pogribny et al., 2004, 2012; James et al., 2003) with subsequent activation of oncogenes, including c-Myc, c-Fos, and H-Ras (Wainfan et al., 1992). Animals treated with diets poor in methyl donors showed hypermethylation of tumor-suppressor genes, such as p53, $p16^{INK4a}$, PTPRO, CDH1, and Cx26 (Pogribny et al., 2012). This event appears to occur before the establishment of preneoplastic lesions, as observed in mice treated with a diet low in folic acid (Tryndyak et al., 2011). These deficient diets in methyl donors have also been associated with decreased levels of hepatic H3K9me3, histone H3 at lysine residue 9 (H3K9ac), and H4K20me3. The increase in H3K9 and H4K20 trimethylation in the liver of mice fed diets deficient in methyl donors may be explained by the reduced KMT activity that resulted from insufficient SAM. Another possible factor that could explain this epigenetic modification is altered expression of enzymes related to the methylation status of the histones. Deficiency of methyl donors is involved in the downregulation of SUV39H1, PRDM2/RIZ1, and SUV420H2 KMTs and may result in increased levels of H3K9me3 and H4K20me3 in the hepatic tissue. These alterations are associated with initiation and promotion of HCC (Pogribny et al., 2012).

Regarding the modulation of miRNA expression, experiments in rodents treated with diets deficient in methyl donors showed a reduction in the expressions of miR-122 and miR-29. The former is responsible for liver stem cell differentiation, and its expression is lost in HCC, while the latter is downregulated in liver cancer. Besides, the increased expressions of miR-34a, miR-155, and miR-221 were associated with HCC induced by diets deficient in methyl donors (Pogribny et al., 2012).

Epigallocatechin-3-Gallate

Among polyphenols, emphasis has been given in the literature to (−)-epigallocatechin-3-gallate (EGCG), which is found in green tea and has shown anticancer activity in vitro and in vivo (Ong et al., 2011; Gerhauser, 2013), including against HCC (Darvesh et al., 2013). The effects of EGCG against cancer can be attributed to its biphasic redox activity, acting as an antioxidant by chelating metals or as a prooxidant, inducing expression of antioxidant enzymes and apoptosis. Another mechanism of EGCG anticancer activity is its ability to inhibit the nuclear factor κB signaling pathway (Darvesh et al., 2013).

The chemopreventive activity of many polyphenols found in food can also be attributed to modulation of DNA hypomethylation (Shukla et al., 2014). EGCG-induced DNA hypomethylation may occur directly by the generation of hydrogen bonds between this BFC and active sites within DNMTs or indirectly by reducing the bioavailability of SAM and consequently of DNMTs (Shukla et al., 2014). Inhibition of DNMTs can lead to reexpression of tumor-suppressor genes previously silenced by hypermethylation of promoter sequences. In support of this hypothesis, cells from esophageal cancer treated with EGCG showed reexpression of many tumor-suppressor genes, including $p16^{INK4a}$, MGMT, hMLH1, and RARβ (Fang et al., 2003; Shukla et al., 2014).

EGCG can also modify the expression of many miRNAs in human HCC cells. In this regard, it was observed that this polyphenol modulated the expression of members of the let-7 family, such as miR-16, miR-20a, and miR-221, mediating the induction of apoptosis that followed a Bcl-2 inhibition (Tsang et al., 2010).

Retinoids

The liver is a target organ for retinoids—for instance, retinol, retinaldehyde, and retinoic acid. It is possible that this class of natural and synthetic compounds exerts some activity during the hepatocarcinogenic process (Ong et al., 2012). In humans, there is a correlation between serum levels of retinol and reduced risk of HCC development (Yuan et al., 2006). Retinoids have chemopreventive activity against hepatocarcinogenesis as demonstrated in several studies (Tamura et al., 1997; Bishayee et al., 2000; Moreno et al., 2002; de Almeida et al., 2005; Sampaio et al., 2007). Vitamin A (retinol) showed chemoprevention against hepatocarcinogenesis, specifically during the progression phase. The mechanism by which retinol interfered with the progression of these tumors was independent of changes in the methylation of c-Myc and 3-hydroxy-3-methylglutaryl-coenzyme A (HMG-CoA) reductase (Moreno et al., 2002). There is evidence that the promoter sequence of CRBP1 and RARβ, genes involved in the intracellular transport of retinol and retinoic acid signaling, respectively, are hypermethylated in many cancer cells (Esteller et al., 2002). Further, in vivo treatment with retinyl palmitate, 13-cis, and all-trans retinoic acid promoted upregulation of glycine-N-methyltransferase, an enzyme necessary for the ideal supply of methyl groups. Therefore, these retinoids may be acting as epigenetic modulators by controlling the availability of methyl groups for DNA and histone methylation (Rowling et al., 2002).

Selenium

Selenium is a constituent of selenoproteins, which are enzymatic complexes involved in antioxidant metabolism and detoxification (Darvesh et al., 2010). Many studies have shown that this mineral has anticancer activity (Darvesh et al., 2010; de Miranda et al., 2014; Fernandes et al., 2015; Shen et al., 2015). There is an inverse correlation between plasma levels of selenium and the risk of developing HCC in humans (Yu et al., 1997, 1999). Evidence from in vitro and in vivo studies suggests that the chemopreventive activity of selenium is linked to alterations in DNA methylation (Davis et al., 2000; Ong et al., 2011). This effect may be a consequence of DNMT1 inhibition by selenium as demonstrated in vitro (Ong et al., 2011; Barrera et al., 2012). The modulation of epigenetic events by selenium depends on its chemical structure. Treatment of LNCaP prostate cancer cells with selenite, a mineral form of selenium, restored expression of GSTP1 that was silenced by methylation of the promoter sequence. This effect was associated with downregulation of DNMT1 that led to general DNA hypomethylation. However, treatment of the same cells with selenomethionine, an organic form of selenium, did not reproduce such effects (Ong et al., 2011).

Curcumin

Curcumin, a polyphenol widely used as a food seasoning and food coloring, has shown chemopreventive activity against different types of cancer (31) as well as in vivo models of hepatocarcinogenesis (Darvesh et al., 2012). Given curcumin's promising chemopreventive activity in hepatocarcinogenesis, there has been interest in developing pharmacological delivery systems such as microcapsules (Ghosh et al., 2012) and nanoparticles of this polyphenol in association with sorafenib (Cao et al., 2015).

Curcumin and EGCG act by inhibiting DNMT1. Evidence suggests that curcumin can also induce DNA hypomethylation (Fu et al., 2010), histone methylation, and histone acetylation (Chen et al., 2013). Curcumin can induce hypoacetylation of histones by a mechanism that may involve both the inhibition of HAT enzymatic activity and the modulation of gene expression (Teiten et al., 2013). Studies based on protein–protein interactions have associated curcumin to specific inhibition of p300/CBP's activity, whereas in other studies it was associated with a reduction in HDAC1, HDAC3, and HDAC8 levels (Kang et al., 2005). In the context of liver cancer, Hep3B cells treated with curcumin showed global histone hypoacetylation related to HAT modulation without alterations in HDAC levels (Kang et al., 2005). Thus, curcumin may modulate in vitro histone acetylation by multiple mechanisms involving inhibition of HATs and downregulation of HDACs. However, further studies are needed to confirm whether both mechanisms coexist in vivo.

Resveratrol

Resveratrol, a polyphenol found mainly in the skin of grapes and in red wine, also showed antineoplastic properties. In vitro studies suggest that this BFC regulates intracellular pathways involved in apoptosis, inflammation, angiogenesis, and metastasis, among others. In vivo, it was shown that resveratrol is a chemopreventive agent of hepatocarcinogenesis caused by chronic infection with the hepatitis B virus. In this study, HBx-transgenic mice were treated with resveratrol (30 mg/kg/day) for 7 and 14 days. The treatment with resveratrol decreased intracellular reactive oxygen species and stimulated hepatocyte proliferation in order to replace damaged hepatocytes. It was also observed that resveratrol inhibited LXRα and downregulated the expression of PPARγ and SREBP-1c genes involved in the lipogenic process. In later stages of HBx-mediated hepatocarcinogenesis, resveratrol treatment for 4 months reduced HCC incidence (Lin et al., 2012). Resveratrol has reduced DNMT inhibitory activity and is less efficacious in reversing DNA hypomethylation than EGCG is (Singh et al., 2013).

Butyrate

Butyrate is the main by-product of the fermentation of edible fibers and was the first compound to be identified that showed inhibition of HDACs. It is because of this HDAC-inhibitory activity that it is considered a potential agent for cancer chemoprevention and chemotherapy (Astbury et al., 2012; Heidor et al., 2012). Butyrate is a short-chain fatty acid that acts as a weak ligand of HDAC, inhibiting HDACs from classes 1 and 2 (Davie, 2003) and thus promoting the histone acetylation and subsequent expression of genes involved in cell differentiation and apoptosis (Supic et al., 2013). In SMMC-7721 and HepG2 cells, butyrate inhibits the expression of HDAC4 in vitro and increases the acetylation of H3K9ac. Butyrate also inhibited the cell migration and invasion of these cell lines (Wang et al., 2013).

Despite its chemoprevention potential as an HDAC inhibitor, there are pharmacokinetic limitations for the oral administration of butyric acid. Prodrugs of butyric acid, such as tributyrin, are found in milk fat. This triacylglycerol showed chemoprevention activity in rats subjected to hepatocarcinogenesis (Kuroiwa-Trzmielina et al., 2009; de Conti et al., 2012, 2013; Guariento et al., 2014). Tributyrin induced apoptosis, increased the expression of p21 and H3K9ac, and elevated the levels of H3K18ac and H4K16ac (Kuroiwa-Trzmielina et al., 2009). Further, it inhibited deacetylation of p53, a key event in maintaining the function of this protein in controlling the balance between apoptosis and cell proliferation (de Conti et al., 2013).

Autophagy, a catabolic process that degrades damaged components of the cell, is essential for differentiation, survival, and cell homeostasis (Yang et al., 2015). HDCA6 plays an important role in the removal of proteins deformed as a consequence of autophagy. It was shown that HCC patients have reduced expression of HDAC6, which is associated with a poorer prognosis (Jung et al., 2012). In addition, ectopic overexpression of HDAC6 inhibited the proliferation and growth of neoplastic cells and increased the conversion of LC3B-II and the formation of autophagy vacuoles without inducing apoptosis or cell cycle arrest. HDAC6-dependent inhibition of liver cancer is due to the activation of autophagy-induced cell death via JNK/Beclin 1 independently of caspases (Jung et al., 2012). Because BFCs can act as HDACs inhibitors, induction of autophagy via HDAC6 may be an important target for HCC chemoprevention. Treatment of rats subjected to hepatocarcinogenesis with tributyrin showed reactivation of the p53 signaling pathway with induction of proapoptotic genes and consequently increased apoptosis. This was lined by an induction in autophagy in the liver that was evidenced by an increase in the hepatic levels of proteins LC3B and BECN1 (de Conti et al., 2013).

Sulforaphane

Among isothiocyanates, sulforaphane and its metabolites have shown antineoplastic activity (Li et al., 2013; Kensler et al., 2013; Lenzi et al., 2014). Erucin, a metabolite of sulforaphane, has chemopreventive activity against HCC in vivo (Lamy et al., 2013). One mechanism responsible for sulforaphane's antineoplastic activity involves the inhibition of HDACs (Shukla et al., 2014). Molecular modeling studies suggest that there is interaction between the active site of HDAC with the carboxylate group within a lysine residue of sulforaphane that is positioned as a bidentate ligand for zinc (Myzak et al., 2004). It was shown in vitro that this BFC inhibited HDAC activity and increased the expression of tumor-suppressor gene $p21^{WAF1}$ and proapoptotic protein Bax (Myzak et al., 2006). In humans, consumption of bagels containing 68 g of broccoli, rich in sulforaphane, inhibited HDAC activity in mononuclear blood cells after 3 h of ingestion, an effect that was not observed 24 h postingestion (Myzak et al., 2007).

Allyl Compounds

Allyl compounds are found in garlic and other vegetables of the genus *Allium*, such as alliin, allicin, S-allylcysteine, S-allylmercaptocysteine, diallyl sulfide, diallyl disulfide, and diallyl trisulfide (Ong et al., 2011). These compounds show anticancer activity against many types of cancers, including HCC (Zhang et al., 2007; Ng et al., 2012; Chu et al., 2013). One metabolite of allyl compounds, allyl mercaptan, was identified as a potent HDAC inhibitor, suggesting that this molecule is a potential candidate for the antineoplastic properties of garlic (Nian et al., 2008).

β-Ionone

β-ionone is a sesquiterpene with a structure analogous to the cyclical structure of β-carotene.

This BFC occurring in grapes has antineoplastic properties, according to several studies (de Moura Espíndola et al., 2005; Cardozo et al., 2011; Scolastici et al., 2014). Rats treated with β-ionone and submitted to the RH model of hepatocarcinogenesis showed a decrease in the number of preneoplastic lesions when this sesquiterpene was administrated in different phases of hepatocarcinogenesis (de Moura Espíndola et al., 2005; Cardozo et al., 2011). Data from our laboratory suggest that treatment of rats with β-ionone modulates the expression of miRNAs in microdissected preneoplastic lesions of the liver. We observed that the expression of miR-122 and miR-34a was increased by β-ionone treatment, whereas miR-181b and miR-708 were downregulated (Furtado et al., 2017). It has been suggested that miR-122 and miR-34a are hepatic tumor suppressors, and that the inhibition of miR-181b and miR-708 is associated with a reduction in their targets, TIMP3 and MTSS-1 (Furtado et al., 2017).

CHEMOPREVENTION AND EPIGENETIC MODULATION OF HEPATOCARCINOGENESIS WITH SYNTHETIC COMPOUNDS

Statins

Considering that cholesterol concentration is high in neoplastic cells, the role of compounds showing the potential to inhibit the mevalonate pathway for cancer prevention is of particular concern. This pathway produces essential metabolites for sterol synthesis and other derivatives required for cell metabolism. Its rate-limiting reaction is the reduction of HMG-CoA in mevalonate catalyzed by HMG-CoA reductase. This enzyme can be inhibited at several levels by many BFCs (Ong et al., 2012) with chemopreventive activity in hepatocarcinogenesis, such as β-carotene (Moreno et al., 1995) and farnesol (Ong et al., 2006), as well as by the blood cholesterol-lowering drugs known as statins (Karlic et al., 2015).

Statins can also modulate epigenetic modifications. Lovastatin, simvastatin, pravastatin, fluvastatin, and atorvastatin demonstrated HDAC inhibition and increased levels of acetylated histone H3 in A529 lung carcinoma cells treated with 10 μmol/L each of statin (Lin et al., 2008). In another study, treatment of breast cancer cells MDA-MB-231, prostate cancer cells MG-63, and osteosarcoma cells U2-OS with simvastatin at 32 μM resulted in downregulation of DNMTs and HDACs (Karlic et al., 2015).

The epigenetic actions of statins seem also to involve miRNA modulation. Following treatment with 10 μM atorvastatin, HepG2 human hepatoma cells showed reduced expressions of miR-17-5p and miR-20a-5p, which are often overexpressed in HCC (Zambrano et al., 2015). The treatment of HepG2 cells with 20 μM atorvastatin also increased the expression of miRNA-124a. This miRNA has guanidinoacetate N-methyltransferase as one of its targets, and thus it may mediate a reduction in creatine synthesis (Phulukdaree et al., 2015). Given that upregulated creatine metabolism associated with creatine kinase activity has been observed in HCC and that increased energetic metabolism is a general hallmark of cancer, atorvastatin may be showing chemopreventive potential by interfering with creatine metabolic pathway (Soroida et al., 2012).

Metformin

Among synthetic compounds related to HCC chemoprevention, recent attention has been given to metformin (Ma et al., 2016). This drug showed chemopreventive activity against liver neoplasia in several preclinical studies (Sui et al., 2015). Based on epidemiological data and clinical trials, metformin may present HCC chemopreventive potential in diabetic patients (Ma et al., 2016). Its mechanism of action involves uncoupling of the respiratory chain at mitochondrial complex I, leading to a reduction in ATP synthesis and an increased AMP/ATP ratio. The latter is responsible for the activation of AMP-activated protein kinase (AMPK), which is the underlying mechanism of the glucose-lowering effect of metformin in diabetic patients. Activation of AMPK also inhibits mammalian target of rapamycin, which constitutes a postulated antineoplastic mechanism by which metformin may act, given the association found between AMPK activation and p53-induced apoptosis (Yu et al., 2017).

Metformin can also modulate epigenetic mechanisms, mainly those related to histone modifications. In this regard, HepG2 cells treated with 1 mM metformin reduced the expression of SIRT1 and increased the acetylation levels of p53 in cells undergoing senescence (Yi et al., 2013). In PCL5 human hepatoma cells, silencing of SIRT1 (PCL5-shSIRT1) did not interfere with the reduction in cellular proliferation mediated by 100 mM metformin. When those cells were implanted in mice, the treatment with 150 mg/kg/day metformin for 6 weeks resulted in growth arrest of tumors derived from cells transfected with shSIRT1 as well as PCL5 control cells, but with a more accentuated effect in the case of PCL5-shSIRT1 (Zhang et al., 2015). Epigenetic alterations such as the inhibition of histone H3 acetylation have been observed in animal models of diabetes. This alteration can be reverted with the treatment of metformin as demonstrated in Sprague–Dawley rats treated with metformin at 150 mg/kg twice/day for 10 consecutive weeks (Khan et al., 2016).

Expression of miRNAs can be also modulated by metformin. Human hepatoma HCC cells treated with 10 mM metformin altered the expression of 51 miRNAs from which members of the let-7 family (let-7a, let-7b, and let-7e) were upregulated (Miyoshi et al., 2014). It has been observed that let-7 members are downregulated in several types of cancer, including HCC. Thus, these miRNAs showed tumor-suppressor properties given not only their expression pattern but also their involvement in oncogene downregulation as in the case of RAS (Masood et al., 2017).

Aspirin

The inflammatory microenvironment is distinctive of the evolution of several types of cancer, including HCC, related to viral hepatitis and alcohol abuse. Still, in the context of inflammation, the relationship between nonalcoholic steatohepatitis and HCC should be stressed. In this regard, anti-inflammatory agents may present chemopreventive activity in neoplasia having an inflammatory background. Among drugs with anti-inflammatory action, aspirin, also known as acetylsalicylic acid, has been used for the prevention of cardiovascular diseases. Aspirin has shown chemopreventive activity against several types of cancers including HCC (Singh et al., 2014), acting mainly as an inhibitor of cyclooxygenases 1 and 2. Other mechanisms of chemoprevention are modulation of immune response, inhibition of pro- and preinflammatory transcription factors, and modulation of glucose oxidation in neoplastic cells (Todoric et al., 2016).

Epigenetic mechanisms can also be considered targets of aspirin for cancer chemoprevention. Treatment of CF-1 mice with chow containing 0.02% aspirin and subjected to colon carcinogenesis showed, besides a reduction in the incidence and multiplicity of tumors, reduced expression of several HDACs. This event was associated with higher H3K27ac immunohistochemical staining and reduced expression of genes related to the inflammatory process, such as inducible nitric oxide synthase, tumor necrosis factor-α, and interleukin 6 (Guo et al., 2016). It has been observed that treatment of HepG2 cells with aspirin at 50 μM upregulates SIRT1, an HDAC involved in mitochondrial function (Kamble et al., 2013). Therefore, it is possible that aspirin exerts chemopreventive activities in hepatocarcinogenesis that modulate epigenetic mechanisms along with its canonical anti-inflammatory properties.

CONCLUSION AND PERSPECTIVES

Many BFCs show chemopreventive activity against diverse types of cancer, including HCC (Singh et al., 2014). Chemoprevention involves mechanisms related to the inhibition of cell proliferation and induction of apoptosis and cell differentiation. The molecular pathways linked to these processes can be genetically controlled or even show epigenetic regulation. However, inhibition of a particular gene or molecular pathway alone may not be enough to significantly reduce the viability of neoplastic cells. Thus, concomitant inhibition of multiple targets linked to carcinogenesis may be a fruitful therapeutic intervention against cancer. Such a strategy, called synthetic lethality (Wu et al., 2011), has been gaining interest, and therapeutic approaches combining chemopreventive agents with different mechanisms of action that act simultaneously in multiple pathways involved in carcinogenesis have been proposed (Wu et al., 2011) (Fig. 11.3). For instance, BFCs that alter DNA methylation patterns could be administered along with other BFCs that affect chromatin structure. Because the restoration of tumor-suppressor gene expression is an important antineoplastic mechanism of action of EGCG, it has been suggested that BFC could be combined with other potential chemopreventive agents—for example, butyric acid (Fang et al., 2003). It has been proposed in turn that butyric acid could be administered together with vitamin A analogs—for example, all-trans retinoic acid. This association inhibited the growth of breast cancer cells MCF-7 and restored the expression of RARβ (Andrade et al., 2012). Tributyrin was also used in association with in vivo models of hepatocarcinogenesis. Animals treated with a combination of vitamin A and tributyrin and subjected to hepatocarcinogenesis showed inhibited growth of preneoplastic lesions, increased levels of H3K9ac, and increased expression of $p21^{cip1/waf1}$ (de Conti et al., 2012). Further, lipid modification, such as synthesis of structured lipids, may present a promising tool for HCC chemoprevention. Data from our group suggest that structured lipids obtained after interesterification of tributyrin with flaxseed oil show antineoplastic activity in rats subjected to

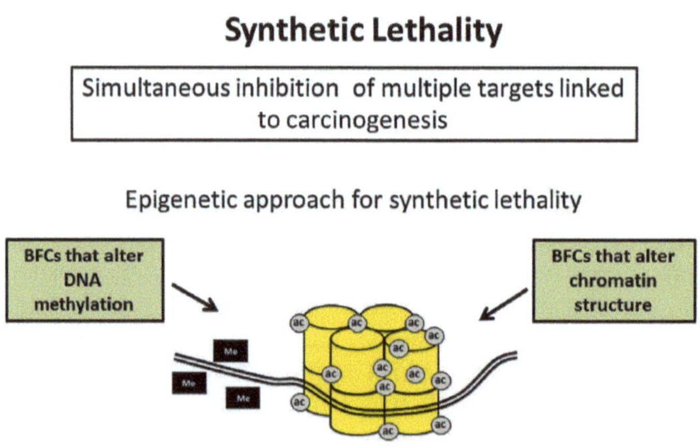

FIGURE 11.3 Synthetic lethality: an approach for cancer control combining different chemopreventive agents.

hepatocarcinogenesis, which is associated with inhibition of HCC-related genes such as c-Myc and CCND1. This inhibition occurred as a consequence of the increased levels of H3K9me3 and H3K27me3 in their promoter regions (Heidor et al., 2016).

Some chemopreventive agents, such as selenium and retinoids, show toxicity at therapeutic doses. An alternative for reducing the possible toxic effects in individuals treated for extensive periods is the intermittent administration of these BFCs. Such a strategy focuses on the elimination of preneoplastic lesions by chemopreventive agents that act selectively on preneoplastic cells by modulating pathways linked to carcinogenesis, particularly epigenetic events (Wu et al., 2011). The intermittent use of chemopreventive agents has been applied in carcinogenesis models using mice with mutations in APC and KRAS genes (Wu et al., 2011). However, this approach is not effective for prevention of all cancers because molecular targets are not fully understood. Thus, studies assessing genetic and epigenetic mechanisms of action for potential chemopreventive agents against HCC are needed.

Besides toxicity, knowledge concerning the bioavailability of BFCs is essential for better understanding the underlying mechanisms of their chemopreventive activity in HCC. Because pharmacological therapies targeting hepatic neoplasia are limited, the concomitant use of BFCs with drugs, such as the administration of sorafenib with vitamin K (Wei et al., 2010) or metformin (Ling et al., 2017), may be an important therapeutic strategy for treating HCC. Further, elucidation of the mechanisms of action of chemopreventive agents may lead to the future development of personalized chemoprevention. Individuals at risk of developing HCC may use chemopreventive agents in an association, either continuously or intermittently, considering their toxicity and bioavailability in such a way that they would exert synthetic lethality specifically in preneoplastic cells.

References

Andrade, F.O., et al., 2012. Efficacy of the dietary histone deacetylase inhibitor butyrate alone or in combination with vitamin A against proliferation of MCF-7 human breast cancer cells. Braz. J. Med. Biol. Res. 45 (9), 841–850.

Areekul, S., et al., 1977. Serum vitamin B12 and folate levels, vitamin B12 and folic acid binding proteins in patients with primary carcinoma of the liver. Southeast Asian J. Trop. Med. Public Health 8 (4), 519–524.

Astbury, S.M., et al., 2012. Uptake and metabolism of the short-chain fatty acid butyrate, a critical review of the literature. Current Drug Metab. 13 (6), 815–821.

Barrera, L.N., et al., 2012. Epigenetic and antioxidant effects of dietary isothiocyanates and selenium: potential implications for cancer chemoprevention. Proc. Nutr. Soc. 71 (2), 237–245.

Bird, A.P., 1986. CpG-rich islands and the function of DNA methylation. Nature 321 (6067), 209–213.

Bishayee, A., et al., 2000. Further evidence for chemopreventive potential of beta-carotene against experimental carcinogenesis: diethylnitrosamine-initiated and phenobarbital-promoted hepatocarcinogenesis is prevented more effectively by beta-carotene than by retinoic acid. Nutr. Cancer 37 (1), 89–98.

Blonski, W., et al., 2010. Non-viral causes of hepatocellular carcinoma. World J. Gastroenterol. 16 (29), 3603–3615.

Bray, F., et al., 2013. Global estimates of cancer prevalence for 27 sites in the adult population in 2008. Int. J. Cancer 132 (5), 1133–1145.

Bruix, J., et al., 2012. Efficacy and safety of sorafenib in patients with advanced hepatocellular carcinoma: subanalyses of a phase III trial. J. Hepatol. 57 (4), 821–829.

Calvisi, D.F., et al., 2007. Altered methionine metabolism and global DNA methylation in liver cancer: relationship with genomic instability and prognosis. Int. J. Cancer 121 (11), 2410–2420.

Cao, H., et al., 2015. Codelivery of sorafenib and curcumin by directed self-assembled nanoparticles enhances therapeutic effect on hepatocellular carcinoma. Mol. Pharm. 12 (3), 922–931.

Cardozo, M.T., et al., 2011. Chemopreventive effects of β-ionone and geraniol during rat hepatocarcinogenesis promotion: distinct actions on cell proliferation, apoptosis, HMGCoA reductase, and RhoA. J. Nutr. Biochem. 22 (2), 130–135.

Chagas, C.E., et al., 2011. Folic acid supplementation during early hepatocarcinogenesis: cellular and molecular effects. Int. J. Cancer 129 (9), 2073–2082.

Chen, C.Q., et al., 2013. Pure curcumin increases the expression of SOCS1 and SOCS3 in myeloproliferative neoplasms through suppressing class I histone deacetylases. Carcinogenesis 34 (7), 1442–1449.

Chen, H.P., et al., 2015. Histone deacetylases and mechanisms of regulation of gene expression. Crit. Rev. Oncog. 20 (1–2), 35–47.

Chu, Y.L., et al., 2013. Allicin induces anti-human liver cancer cells through the p53 gene modulating apoptosis and autophagy. J. Agric. Food Chem. 61 (41), 9839–9848.

Darvesh, A.S., et al., 2010. Selenium in the prevention and treatment of hepatocellular carcinoma. Anticancer Agents Med. Chem. 10 (4), 338–345.

Darvesh, A.S., et al., 2012. Curcumin and liver cancer: a review. Curr. Pharm. Biotechnol. 13 (1), 218–228.

Darvesh, A.S., et al., 2013. Chemopreventive and therapeutic potential of tea polyphenols in hepatocellular cancer. Nutr. Cancer 65 (3), 329–344.

Davie, J.R., 2003. Inhibition of histone deacetylase activity by butyrate. J. Nutr. 133 (7 Suppl.), 2485S–2493S.

Davis, C.D., et al., 2000. Dietary selenium and arsenic affect DNA methylation in vitro in Caco-2 cells and in vivo in rat liver and colon. J. Nutr. 130 (12), 2903–2909.

de Almeida, V.F.E.M., et al., 2005. All-trans and 9-cis retinoic acids, retinol and beta-carotene chemopreventive activities during the initial phases of hepatocarcinogenesis involve distinct actions on glutathione S-transferase positive preneoplastic lesions remodeling and DNA damage. Carcinogenesis 26 (11), 1940–1946.

de Conti, A., 2012. Chemopreventive effects of the dietary histone deacetylase inhibitor tributyrin alone or in combination with vitamin A during the promotion phase of rat hepatocarcinogenesis. J. Nutr. Biochem. 23 (8), 860–866.

de Conti, A., et al., 2013. The chemopreventive activity of the butyric acid prodrug tributyrin in experimental rat hepatocarcinogenesis is associated with p53 acetylation and activation of the p53 apoptotic signaling pathway. Carcinogenesis 34 (8), 1900–1906.

de Miranda, J.X., et al., 2014. Effects of selenium compounds on proliferation and epigenetic marks of breast cancer cells. J. Trace Elem. Med. Biol. 28 (4), 486–491.

de Moura Espíndola, R., et al., 2005. Geranylgeraniol and beta-ionone inhibit hepatic preneoplastic lesions, cell proliferation, total plasma cholesterol and DNA damage during the initial phases of hepatocarcinogenesis, but only the former inhibits NF-kappaB activation. Carcinogenesis 26 (6), 1091–1099.

Di Marco, V., et al., 2013. Sorafenib: from literature to clinical practice. Ann. Oncol. 24 (Suppl. 2), 30–37 ii.

Eden, A., et al., 2003. Chromosomal instability and tumors promoted by DNA hypomethylation. Science 300 (5618), 455.

El-Serag, H.B., 2012. Epidemiology of viral hepatitis and hepatocellular carcinoma. Gastroenterology 142 (6), 1264–1273.

Esteller, M., et al., 2002. Hypermethylation-associated inactivation of the cellular retinol-binding-protein 1 gene in human cancer. Cancer Res. 62 (20), 5902–5905.

Fang, M.Z., et al., 2003. Tea polyphenol (-)-epigallocatechin-3-gallate inhibits DNA methyltransferase and reactivates methylation-silenced genes in cancer cell lines. Cancer Res. 63 (22), 7563–7570.

Feinberg, A.P., et al., 2002. DNA methylation and genomic imprinting: insights from cancer into epigenetic mechanisms. Semin. Cancer Biol. 12 (5), 389–398.

Fernandes, A.P., et al., 2015. Selenium compounds as therapeutic agents in cancer. Biochim. Biophys. Acta 1850 (8), 1642–1660.

Fu, S., et al., 2010. Development of curcumin as an epigenetic agent. Cancer 116 (20), 4670–4676.

Furtado, K.S., et al., 2017. β-ionone modulates the expression of miRNAs and genes involved in the metastatic phenotype of microdissected persistent preneoplastic lesions in rats submitted to hepatocarcinogenesis. Mol. Carcinog. 56 (1), 184–196.

Gerhauser, C., 2013. Cancer chemoprevention and nutriepigenetics: state of the art and future challenges. Top. Curr. Chem. 329, 73–132.

Ghosh, D., et al., 2012. Nanocapsulated curcumin: oral chemopreventive formulation against diethylnitrosamine induced hepatocellular carcinoma in rat. Chem. Biol. Interact. 195 (3), 206–214.

Guariento, A.H., et al., 2014. Transcriptomic responses provide a new mechanistic basis for the chemopreventive effects of folic acid and tributyrin in rat liver carcinogenesis. Int. J. Cancer 135 (1), 7–18.

Guo, Y., et al., 2016. The epigenetic effects of aspirin: the modification of histone H3 lysine 27 acetylation in the prevention of colon carcinogenesis in azoxymethane- and dextran sulfate sodium-treated CF-1 mice. Carcinogenesis 37 (6), 616–624.

He, B., et al., 2011. A comprehensive analysis of the dynamic biological networks in HCV induced hepatocarcinogenesis. PLoS One 6, e18516.

Heidor, R., et al., 2012. Anticarcinogenic actions of tributyrin, a butyric acid prodrug. Curr. Drug Targets 13 (14), 1720–1729.

Heidor, R., et al., 2016. The chemopreventive activity of butyrate-containing structured lipids in experimental rat hepatocarcinogenesis. Mol. Nutr. Food Res. 60 (2), 420–429.

Hoshida, Y., et al., 2010. Molecular classification and novel targets in hepatocellular carcinoma: recent advancements. Semin. Liver Dis. 30 (1), 35–51.

Hussain, S.P., et al., 2007. TP53 mutations and hepatocellular carcinoma: insights into the etiology and pathogenesis of liver cancer. Oncogene 26, 2166–2176.

James, S.J., et al., 2003. Mechanisms of DNA damage, DNA hypomethylation, and tumor progression in the folate/methyl-deficient rat model of hepatocarcinogenesis. J. Nutr. 133 (11 Suppl. 1), 3740S–3747S.

Jemal, A., et al., 2010. Cancer statistics. CA Cancer J. Clin. 60 (5), 277–300.

Jiang, Y.H., et al., 2004. Epigenetics and human disease. Annu. Rev. Genomics Hum. Genet. 5, 479–510.

Jung, K.H., et al., 2012. Histone deacetylase 6 functions as a tumor suppressor by activating c-Jun NH2-terminal kinase-mediated beclin 1-dependent autophagic cell death in liver cancer. Hepatology 56 (2), 644–657.

Kamble, P., et al., 2013. Aspirin may promote mitochondrial biogenesis via the production of hydrogen peroxide and the induction of Sirtuin1/PGC-1α genes. Eur. J. Pharmacol. 699 (1–3), 55–61.

Kang, J., et al., 2005. Curcumin-induced histone hypoacetylation: the role of reactive oxygen species. Biochem. Pharmacol. 69, 1205–1213.

Karlic, H., et al., 2015. Inhibition of the mevalonate pathway affects epigenetic regulation in cancer cells. Cancer Genet. 208 (5), 241–252.

Kensler, T.W., et al., 2013. Keap1-nrf2 signaling: a target for cancer prevention by sulforaphane. Top. Curr. Chem. 329, 163–177.

Khan, S., et al., 2016. Sodium butyrate reduces insulin-resistance, fat accumulation and dyslipidemia in type-2 diabetic rat: a comparative study with metformin. Chem. Biol. Interact. 25 (254), 124–134.

Kuroiwa-Trzmielina, J., et al., 2009. Chemoprevention of rat hepatocarcinogenesis with histone deacetylase inhibitors: efficacy of tributyrin, a butyric acid prodrug. Int. J. Cancer 124 (11), 2520–2527.

Lamy, E., et al., 2013. Preclinical evaluation of 4-methylthiobutyl isothiocyanate on liver cancer and cancer stem cells with different p53 status. PLoS One 8 (8), e70846.

Lenzi, M., et al., 2014. Sulforaphane as a promising molecule for fighting cancer. Cancer Treat. Res. 159, 207–223.

Li, Y., et al., 2013. Targeting cancer stem cells with sulforaphane, a dietary component from broccoli and broccoli sprouts. Future Oncol. 9 (8), 1097–1103.

Lin, C.P., et al., 2010. Targeting c-Myc as a novel approach for hepatocellular carcinoma. World J. Hepatol. 2 (1), 16–20.

Lin, H.C., et al., 2012. Resveratrol helps recovery from fatty liver and protects against hepatocellular carcinoma induced by hepatitis B virus X protein in a mouse model. Cancer Prev. Res. 5 (7), 952–962.

Lin, Y.C., et al., 2008. Statins increase p21 through inhibition of histone deacetylase activity and release of promoter-associated HDAC1/2. Cancer Res. 68 (7), 2375–2383.

Ling, S., et al., 2017. Combination of metformin and sorafenib suppresses proliferation and induces autophagy of hepatocellular carcinoma via targeting the mTOR pathway. Int. J. Oncol. 50 (1), 297–309.

Liu, M., et al., 2014. The genetic and epigenetic alterations in human hepatocellular carcinoma: a recent update. Protein Cell 5 (9), 673–691.

Llovet, J.M., et al., 2008. Novel advancements in the management of hepatocellular carcinoma in 2008. Bruix J J. Hepatol. 48 (Suppl. 1), S20–S37.

Llovet, J.M., et al., 2012. Plasma biomarkers as predictors of outcome in patients with advanced hepatocellular carcinoma. Clin. Cancer Res. 18 (8), 2290–2300.

Ma, L., et al., 2014. Epigenetics in hepatocellular carcinoma: an update and future therapy perspectives. World J. Gastroenterol. 20 (2), 333–345.

Ma, S.J., et al., 2016. Metformin use improves survival of diabetic liver cancer patients: systematic review and meta-analysis. Oncotarget 7 (40), 66202–66211.

Masood, N., et al., 2017. Entangling relation of micro RNA-let7, miRNA-200 and miRNA-125 with various cancers. Pathol. Oncol. Res. https://doi.org/10.1007/s12253-016-0184-0.

Meng, X., et al., 2014. MDM2-p53 pathway in hepatocellular carcinoma. Cancer Res. 74 (24), 7161–7167.

Miyoshi, H., et al., 2014. Effect of the anti-diabetic drug metformin in hepatocellular carcinoma in vitro and in vivo. Int. J. Oncol. 45 (1), 322–332.

Moreno, F.S., et al., 1995. Effect of beta-carotene on the expression of 3-hydroxy-3-methylglutaryl coenzyme A reductase in rat liver. Cancer Lett. 96 (2), 201–208.

Moreno, F.S., et al., 2002. Inhibitory effects of beta-carotene and vitamin a during the progression phase of hepatocarcinogenesis involve inhibition of cell proliferation but not alterations in DNA methylation. Nutr. Cancer 44 (1), 80–88.

Morishita, A., et al., 2015. miRNA in hepatocellular carcinoma. Hepatol. Res. 45 (2), 128–141.

Myzak, M.C., et al., 2004. A novel mechanism of chemoprotection by sulforaphane: inhibition of histone deacetylase. Cancer Res. 64 (16), 5767–5774.

Myzak, M.C., et al., 2006. Sulforaphane inhibits histone deacetylase activity in BPH-1, LnCaP and PC-3 prostate epithelial cells. Carcinogenesis 27 (4), 811–819.

Myzak, M.C., et al., 2007. Sulforaphane retards the growth of human PC-3 xenografts and inhibits HDAC activity in human subjects. Exp. Biol. Med. 232 (2), 227–234.

Ng, K.T., et al., 2012. A garlic derivative, S-allylcysteine (SAC), suppresses proliferation and metastasis of hepatocellular carcinoma. PLoS One 7 (2), e31655.

Nian, H., et al., 2008. Allyl mercaptan, a garlic-derived organosulfur compound, inhibits histone deacetylase and enhances Sp3 binding on the P21WAF1 promoter. Carcinogenesis 29 (9), 1816–1824.

Ong, T.P., et al., 2006. Farnesol and geraniol chemopreventive activities during the initial phases of hepatocarcinogenesis involve similar actions on cell proliferation and DNA damage, but distinct actions on apoptosis, plasma cholesterol and HMGCoA reductase. Carcinogenesis 27 (6), 1194–1203.

Ong, T.P., et al., 2011. Moreno FS, Ross SA. Targeting the epigenome with bioactive food components for cancer prevention. J. Nutrigenet. Nutrigenomics 4 (5), 275–292.

Ong, T.P., et al., 2012. Chemoprevention of hepatocarcinogenesis with dietary isoprenic derivatives: cellular and molecular aspects. Curr. Cancer Drug Targets 12 (9), 1173–1190.

Paris, L., et al., 2010. Inhibition of phosphatidylcholine-specific phospholipase C downregulates HER2 overexpression on plasma membrane of breast cancer cells. Breast Cancer Res. 12 (3), R27.

Peltomäki, P., 2012. Mutations and epimutations in the origin of cancer. Exp. Cell Res. 318 (4), 299–310.

Phulukdaree, A., et al., 2015. Atorvastatin increases miR-124a expression: a mechanism of Gamt modulation in liver cells. J. Cell. Biochem. 116 (11), 2620–2627.

Pitot, H.C., 2007. Adventures in hepatocarcinogenesis. Annu. Rev. Pathol. 2, 1–29.

Pogribny, I.P., et al., 2004. Genomic hypomethylation is specific for preneoplastic liver in folate/methyl deficient rats and does not occur in non-target tissues. Mutat. Res. 548 (1–2), 53–59.

Pogribny, I.P., et al., 2008. Epigenetic aspects of genotoxic and non-genotoxic hepatocarcinogenesis: studies in rodents. Environ. Mol. Mutagen. 49 (1), 9–15.

Pogribny, I.P., et al., 2012. Molecular alterations in hepatocarcinogenesis induced by dietary methyl deficiency. Mol. Nutr. Food Res. 56 (1), 116–125.

Pogribny, I.P., et al., 2014. Role of epigenetic aberrations in the development and progression of human hepatocellular carcinoma. Cancer Lett. 342 (2), 223–230.

Romero-Cordoba, S.L., et al., 2014. miRNA biogenesis: biological impact in the development of cancer. Cancer Biol. Ther. 15 (11), 1444–1455.

Rowling, M.J., et al., 2002. Vitamin A and its derivatives induce hepatic glycine N-methyltransferase and hypomethylation of DNA in rats. J. Nutr. 132 (3), 365–369.

Sampaio, A.R., et al., 2007. Vitamin A and beta-carotene inhibitory effect during 1,2-dimethylhydrazine induced hepatocarcinogenesis potentiated by 5- azacytidine. Food Chem. Toxicol. 45 (4), 563–567.

Scolastici, C., et al., 2014. β-ionone inhibits persistent preneoplastic lesions during the early promotion phase of rat hepatocarcinogenesis: TGF-α, NF-κB, and p53 as cellular targets. Nutr. Cancer 66 (2), 234–241.

Shen, F., et al., 2015. The association between serum levels of selenium, copper, and magnesium with thyroid cancer: a meta-analysis. Biol. Trace Elem. Res. 167 (2), 225–235.

Shukla, S., et al., 2014. Epigenetic regulation by selected dietary phytochemicals in cancer chemoprevention. Cancer Lett. 355 (1), 9–17.

Simard, E.P., et al., 2012. Cancers with increasing incidence trends in the United States: 1999 through 2008. CA Cancer J. Clin. 62 (2), 118–128.

Singh, S., et al., 2014. Chemopreventive strategies in hepatocellular carcinoma. Nat. Rev. Gastroenterol. Hepatol. 11 (1), 45–54.

Singh, V., et al., 2013. DNA methyltransferase-1 inhibitors as epigenetic therapy for cancer. Curr. Cancer Drug Targets 13 (4), 379–399.

Slotkin, R.K., et al., 2007. Transposable elements and the epigenetic regulation of the genome. Nat. Rev. Genet. 8 (4), 272–285.

Song, J.L., et al., 2015a. microRNA regulation of Wnt signaling pathways in development and disease. Cell. Signal. 27 (7), 1380–1391.

Song, Y., et al., 2015b. MicroRNAs contribute to hepatocellular carcinoma. Mini Rev. Med. Chem. 15 (6), 459–466.

Soroida, Y., et al., 2012. Increased activity of serum mitochondrial isoenzyme of creatine kinase in hepatocellular carcinoma patients predominantly with recurrence. J. Hepatol. 57 (2), 330–336.

Subramaniam, D., et al., 2014. DNA methyltransferases: a novel target for prevention and therapy. Front. Oncol. https://doi.org/10.3389/fonc.2014.00080.

Sui, X., et al., 2015. Metformin: a novel but controversial drug in cancer prevention and treatment. Mol. Pharm. 12 (11), 3783–3791.

Supic, G., et al., 2013. Epigenetics: a new link between nutrition and cancer. Nutr. Cancer 65 (6), 781–792.

Tamura, K., et al., 1997. Inhibition by N-(4-hydroxyphenyl) retinamide and all-trans-retinoic acid of exogenous and endogenous development of putative preneoplastic, glutathione S-transferase placental form-positive lesions in the livers of rats. Carcinogenesis 18 (11), 2133–2141.

Teiten, M.H., et al., 2013. Curcumin as a regulator of epigenetic events. Mol. Nutr. Food Res. 57, 1619–1629.

Thorgeirsson, S.S., et al., 2002. Molecular pathogenesis of human hepatocellular carcinoma. Nat. Genet. 31 (4), 339–346.

Todoric, J., et al., 2016. Targeting inflammation in cancer prevention and therapy. Cancer Prev. Res. 9 (12), 895–905.

Tryndyak, V.P., et al., 2011. Coupling global methylation and gene expression profiles reveal key pathophysiological events in liver injury induced by a methyl-deficient diet. Mol. Nutr. Food Res. 55 (3), 411–418.

Tsang, W.P., et al., 2010. Epigallocatechin gallate up-regulation of miR-16 and induction of apoptosis in human cancer cells. J. Nutr. Biochem. 21 (2), 140–146.

Vanden Berghe, W., 2012. Epigenetic impact of dietary polyphenols in cancer chemoprevention: lifelong remodeling of our epigenomes. Pharmacol. Res. 65 (6), 565–576.

Vogelstein, B., et al., 1993. The multistep nature of cancer. Trends Genet. 9 (4), 138–141.

Wainfan, E., et al., 1992. Methyl groups in carcinogenesis: effects on DNA methylation and gene expression. Cancer Res. 52 (7 Suppl.), 2071s–2077s.

Wang, H.G., et al., 2013. Anticancer effects of sodium butyrate on hepatocellular carcinoma cells in vitro. Int. J. Mol. Med. 31 (4), 967–974.

Wei, G., et al., 2010. Vitamin K enhancement of sorafenib-mediated HCC cell growth inhibition in vitro and in vivo. Int. J. Cancer 127 (12), 2949–2958.

Welzel, T.M., et al., 2007. Blood folate levels and risk of liver damage and hepatocellular carcinoma in a prospective high-risk cohort. Cancer Epidemiol. Biomark. Prev. 16 (6), 1279–1282.

Wu, X., et al., 2011. An intermittent approach for cancer chemoprevention. Nat. Rev. Cancer 11 (12), 879–885.

Yang, X., et al., 2015. The role of autophagy induced by tumor microenvironment in different cells and stages of cancer. Cell Biosci. 5, 14.

Yi, G., et al., 2013. Low concentration of metformin induces a p53-dependent senescence in hepatoma cells via activation of the AMPK pathway. Int. J. Oncol. 43 (5), 1503–1510.

Yu, M.W., et al., 1999. Plasma selenium levels and risk of hepatocellular carcinoma among men with chronic hepatitis virus infection. Am. J. Epidemiol. 150 (4), 367–374.

Yu, S.Y., et al., 1997. Protective role of selenium against hepatitis B virus and primary liver cancer in Qidong. Biol. Trace Elem. Res. 56 (1), 117–124.

Yu, X., et al., 2017. Anti-tumor activity of metformin: from metabolic and epigenetic perspectives. Oncotarget 8 (3), 5619–5628.

Yuan, J.M., et al., 2006. Prediagnostic level of serum retinol in relation to reduced risk of hepatocellular carcinoma. J. Natl. Cancer Inst. 98 (7), 482–490.

Zambrano, T., et al., 2015. Altered microRNome profiling in statin-induced HepG2 cells: a pilot study identifying potential new biomarkers involved in lipid-lowering treatment. Cardiovasc. Drugs Ther. 29 (6), 509–518.

Zhang, S.H., et al., 2013. High expression of FLOT1 is associated with progression and poor prognosis in hepatocellular carcinoma. PLoS One 8 (6), e64709.

Zhang, Z.M., et al., 2007. Anti-tumor effects of polybutylcyanoacrylate nanoparticles of diallyl trisulfide on orthotopic transplantation tumor model of hepatocellular carcinoma in BALB/c nude mice. Chin. Med. J. 120 (15), 1336–1342.

Zhang, Z.Y., et al., 2015. SIRT1 regulates oncogenesis via a mutant p53-dependent pathway in hepatocellular carcinoma. J. Hepatol. 62 (1), 121–130.

CHAPTER

12

Epigenetic Control of Pancreatic Carcinogenesis and Its Regulation by Natural Products

Haseeb Zubair[1], Shafquat Azim[1], Mohammad Aslam Khan[1], Girijesh Kumar Patel[1], Aamir Ahmad[1], Sachin Pai[2], Seema Singh[1,3], Ajay P. Singh[1,3]

[1]Department of Oncologic Sciences, Mitchell Cancer Institute, University of South Alabama, Mobile, AL, United States; [2]Department of Interdisciplinary clinical Oncology, Mitchell Cancer Institute, University of South Alabama, Mobile, AL, United States; [3]Department of Biochemistry and Molecular Biology, College of Medicine, University of South Alabama, Mobile, AL, United States

Abstract

Pathogenesis of pancreatic cancer (PC) involves both genetic and epigenetic alterations. While genetic aberrations have been the focus of a majority of studies, recent research has highlighted the benefit of therapeutically targeting epigenetic changes because of their reversible nature. Differential promoter DNA methylation of tumor-suppressor genes/oncogenes and chromatin remodeling through histone methylation/acetylation represent some of the more well-characterized epigenetic changes that have been associated with PC onset and progression. A number of key enzymes, such as DNA methyltransferases, histone deacetylases, and enhancer of zeste homolog 2 that play important roles in epigenetic modifications in PC, are targets of anticancer agents obtained from natural products. Prominent among these natural products are 3-3′-diindolylmethane, epigallocatechin-3-gallate, and resveratrol as well as derivatives of curcumin. This chapter provides a comprehensive overview of epigenetic events that are relevant to PC pathogenesis, along with promising emerging data on the regulation of these epigenetic events by natural products.

Keywords

DNA methylation; Epigenetics; Histone modification; MiRNA; Pancreatic cancer; Pancreatic carcinogenesis; Tumor suppression.

INTRODUCTION

The term epigenetics, first used by Waddington in 1942, refers to any heritable trait, not involving DNA sequence, that influences the phenotype of a

Epigenetics of Cancer Prevention
https://doi.org/10.1016/B978-0-12-812494-9.00012-3

developing organism (Waddington, 2012). Epigenetics encompasses the idea that all cells within a species have the same genotype but have different phenotypes that persist through many generations (Henikoff and Greally, 2016). It is inheritance based on traits passed to the next generation via DNA and histone modifications (Fig. 12.1). Epigenetic modifications affect the genome by either inducing or suppressing gene expression, thus resulting in altered phenotypes. Most importantly, the epigenetic regulation of gene expression not only is applicable to normal cellular development and maintenance but can also be responsible for the deregulation of gene expression associated with disease states. A noteworthy characteristic of epigenetic-based inheritance is its reversibility, in contrast to the stable nature of DNA sequence–based alterations.

Deregulation of epigenetic mechanisms can contribute to cancer development (Piunti and Pasini, 2011; Ahmad, 2016). Key tumor suppressors that play a role in carcinogenesis may be altered through hypermethylation, and oncogenes can be upregulated secondary to permissive histone modifications. Further, factors involved in tumor growth and invasiveness can be aberrantly expressed through regulation by microRNAs (miRNAs). While genetic drivers of pancreatic ductal adenocarcinoma (PDAC) have been well studied, they do not account for all phenotypic and molecular alterations in PDAC cells (Sato and Goggins, 2006; Khan et al., 2017). Some tumor-suppressor genes and oncogenes involved in PDAC pathogenesis have aberrant expression and function not due to genetic causes, but rather because of epigenetic mechanisms regulating their expression. Thus, pancreatic cancer (PC) initiation and progression is the result of a heterogeneous and dynamic combination of both genetic and epigenetic mechanisms. The identification of epigenetic alterations in early pancreatic intraepithelial neoplasia (PanIN) lesions through to PDAC suggests epigenetic changes that occur in parallel to genetic changes (Mccleary-Wheeler et al., 2013).

Recent years have witnessed a renewed interest in the anticancer activity of natural products. This is in part due to emerging evidence documenting the ability of these anticancer agents to modulate epigenetic events (Ahmad et al., 2014). In this chapter, we focus on DNA- and histone-based epigenetic modifications that

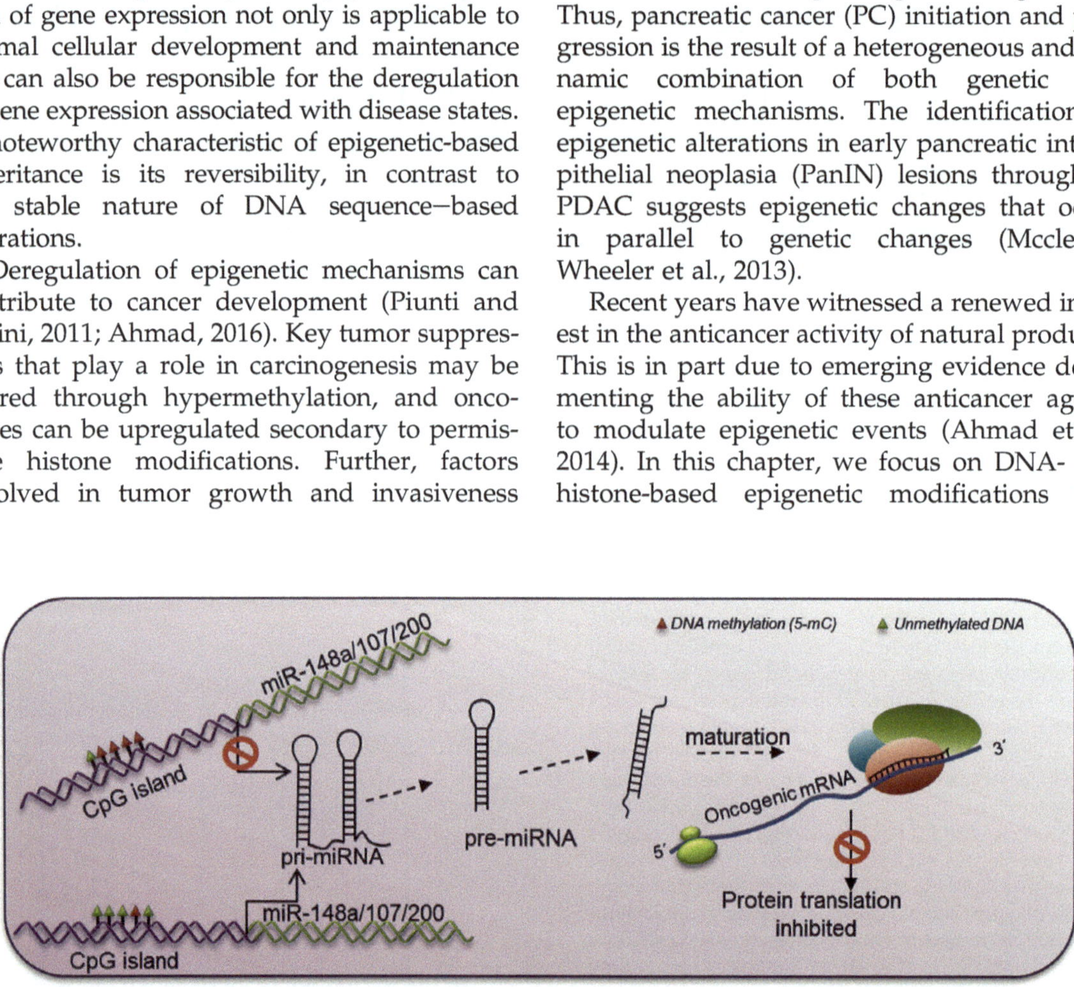

FIGURE 12.1 microRNAs can be subjected to epigenetic control as observed in the case of miRNAs miR-148a, miR-107, and miR-200 in pancreatic cancer. The promoter region of these miRNAs becomes methylated, resulting in reduced binding of transcription factors and leading to inhibition of their expression.

have been observed to regulate PC pathogenesis and the ability of natural products to target the epigenetic modifications as a basis of their therapeutic mechanism of action.

PANCREATIC CANCER

The American Cancer Society estimates that about 55,440 people (26,240 women and 29,200 men) will be diagnosed with PC, and 44,330 individuals will succumb to the disease in the United States alone in 2018. PC will potentially be the third-leading cause of cancer-related deaths (Siegel et al., 2018) in 2018. Overall survival of PC patients is a dismal 2–8 months with a 5-year survival of only 8% (Siegel et al., 2018). The cryptic progression of PC causes late-stage diagnosis, and it is often diagnosed at a stage when it has already metastasized or advanced locally, limiting therapeutic options and interventions (Kleeff et al., 2016; Garrido-Laguna and Hidalgo, 2015). Thus, there is an urgent need for the identification of biomarkers that could help in early detection and a therapeutic outcome of PC. Although CA19-9 (sialyl-Lewis carbohydrate antigen 19-9) is a widely used noninvasive biomarker for PC detection, it has limitations of its own with sensitivity and selectivity. One, it is expressed only in individuals with the Lewis a^+/b^- or Lewis a^+/b^+ genotypes and two, CA19-9 can also be detected in other cancers and some noncancerous conditions such as pancreatitis (Ballehaninna and Chamberlain, 2013).

While certain environmental and genetic factors have been identified as potential risk factors, their precise roles are not well understood. Active smoking is a well-documented risk factor associated with the onset of PC (Piciucchi et al., 2015; Iodice et al., 2008; Lynch et al., 2009). Interestingly, while passive smoking has not been associated with PC, a positive association of PC with maternal smoking and exposure in utero or early in life has been observed as a risk factor for PC (Ding et al., 2015). Furthermore, heavy alcohol consumption has also been associated with an increased risk of PC (Talamini et al., 2010), and heavy alcohol consumption along with heavy smoking has an additive effect leading to increased risk (Talamini et al., 2010). Inconsistent reports about the association of dietary habits are present in the literature. For example, the consumption of red and processed meats has been reported as a risk factor for PC in a number of articles (Zheng and Lee, 2009; Paluszkiewicz et al., 2012; Taunk et al., 2016); however, a study by Jansen et al. (2013), reports no association of meat mutagens with PC. Similarly, research on the health benefit of consuming fruits, vegetables, and certain beverages (such as green tea) against PC is also at odds (Koushik et al., 2012; Shigihara et al., 2014; Liu et al., 2014; Paluszkiewicz et al., 2012; Jansen et al., 2011; Bae et al., 2009; Chen et al., 2014). Most studies do indicate a reduced risk of PC upon consumption of fruits, vegetables, and beverages. Interestingly, carbonated beverages and juices laden with sugar that increase the glycemic load in an individual have been reported to increase the risk of PC (Larsson et al., 2006; Mueller et al., 2010). This has been attributed to the presence of different phytochemicals and the nutrients present in them. Assessments of the studies should be taken with caution as they usually involve self-assessment questionnaires and may not be reflective of the true scenario.

DNA METHYLATION-MEDIATED EPIGENETIC MODIFICATIONS IN PANCREATIC CANCER

One of the better understood epigenetic modifications of transcriptional regulation is DNA methylation, which is most often associated with the repression of gene expression. Methylation of DNA is the addition of a methyl ($-CH_3$) group to cytosine (C) residue preceding a guanosine (G) residue (i.e., CpG, wherein *p* denotes phosphate) by the DNA methyltransferase (DNMT) family of enzymes, yielding a 5-methylcytosine in the CpG. The CpG regions generally occur as extended stretches of CpGs called CpG islands and are present mostly in the promoter regions of genes. As a general observation, hypermethylation of CpG islands is often linked to downregulation of the gene's transcription, and demethylation or hypomethylation of the promoter's CpG island is associated with overexpression of the gene. Gene bodies may also contain CpG repeats that may be methylated; however, the functions associated with gene body methylation remain elusive. Global hypomethylation is one of the hallmarks of cancer, which increases instability of the genome through increased mutational events.

Methylation interferes with the proper binding of transcription factors to their target sequence, partially through the initiation of chromatin compaction, resulting in gene silencing. Physiologically in embryonic development, X-chromosome inactivation, prevention of genomic instability, and genomic imprinting, all employ various degrees of DNA methylation in normal cellular functioning (Rozek et al., 2014). DNMT1 is considered as maintaining methylated patterns of the parent DNA strand and copying it to the synthesized daughter strand (Sen et al., 2010). However, DNMT3a and DNMT3b mainly function as mediators of de novo DNA methylation and are frequently observed as deregulated in cancers (Chedin, 2011). The upregulation of different oncogenic signaling networks has been proposed as a mechanism of DNMT overexpression in PDAC (Azizi et al., 2014). Moreover, nuclear protein 1, frequently upregulated in pancreatic tumors and pancreatitis, has also been observed to bind to the promoter region of DNMT1 to enhance its expression (Grasso et al., 2014, 2015). Additionally, the upregulation of DNMT proteins could itself result from hypomethylation of their promoter, as was observed for DNMT3b in PC tissues (Elliott et al., 2016). Alternatively, it is possible that the expression of DNMTs is not amplified; rather, the activity is enhanced due to increased interaction between DNMTs and their binding proteins.

Regulation of Oncogenes and Tumor-Suppressor Genes by DNA Methylation

PDAC and other tumors have been observed to exhibit aberrant DNA methylation patterns as compared with normal cells or matched normal neighboring tissues (Delpu et al., 2011). This includes a global reduction in methylation along with regional hypermethylation of CpG islands of different gene promoters. The altered methylated epigenome is observed to emerge early in the development of PDAC. It is now well established that inactivation of the CDKN2A tumor suppressor by methylation of the gene promoter in PDAC is an essential alteration during pancreatic tumorigenesis (Khan et al., 2017). About 95% of all PDAC tumors have been observed to demonstrate a loss of p16 (encoded by CDKN2A) expression. The p16 protein functions as an inhibitor of cyclin-dependent kinases such as CDK4 and CDK6 that can phosphorylate retinoblastoma protein, resulting in progression from the G1 phase to the S phase and enhanced growth. Aberrant promoter hypomethylation-induced expression of proto-oncogene guanine nucleotide-exchange factor VAV1 has also been reported in PDAC (Huang et al., 2017). Normally, expression of VAV1 is limited to hematopoietic cells; however, its expression has been observed in human PDAC samples as a result of DNA hypomethylation. Hypomethylation of the VAV1 promoter sequence has also been correlated with poor PDAC patient outcomes and was identified as resulting from transforming growth factor-β (TGF-β)-induced dissociation of DNMT1 from the VAV1 promoter (Huang et al., 2017). Further, VAV1 protein was reportedly expressed in nearly half of all PDAC cases tested, wherein it promoted the activation of Rac and Cdc42 proteins, enhancing the invasion, migration, and survival of pancreatic tumor cells (Razidlo et al., 2015).

In earlier studies, researchers would analyze the methylation status of the promoter region of select genes that were identified to be differentially expressed between normal and cancer cells. For example, Sato et al. observed hypomethylation of genes MSLN (mesothelin) and CLDN4 (claudin 4) in about 92% and 89% of PDAC cases, respectively, and demonstrated an elevated expression compared with that of a normal pancreas (Sato et al., 2003b). Mesothelin has been identified as a potential biomarker for PC and is co-overexpressed with cancer antigen 125/mucin 16 at the invading edges of PC (Sato et al., 2003b). Claudin 4, a tight junction protein required for cell adhesion, has been demonstrated as overexpressed in PC (Kojima et al., 2012) and proposed as a promising biomarker and therapeutic target for various epithelial malignancies (Neesse et al., 2012). Similarly, in a different study when the promoter methylation status of select genes (viz. APC, BRCA1, CDH1, CDKN2A, DAPK1, GSTP1, and TIMP3) was analyzed in multistage carcinogenesis of the pancreas (from normal tissue to inflamed ducts to PanIN lesions through to PDAC), a progressive increase in the methylation of these genes was observed. Interestingly, the increase in the methylation status of genes was related to poor differentiation and high DNMT1 expression (Peng et al., 2006).

Current advances in sequencing technology have provided impetus for the number of studies employing genome-wide approaches for the identification of differential methylation in PDAC samples compared with normal pancreas. For example, when gene expression was analyzed after performing methylated CpG island amplification in PDAC cell lines and compared with control pancreatic tissue samples, several genes were found to be modulated by DNA methylation. The most significant category of genes among these included oncogenes (MYB, JUNB, and FOS) and chromatin modifiers (SET8, KDM6A, and EP400) (Vincent et al., 2011). We have recently identified MYB as an important regulator of PC pathogenesis that regulates a number of vital pathogenic properties such as cell growth, invasion, metastasis, and RNA modification and splicing (Srivastava et al., 2015; Azim et al., 2016). Fos and Jun are also widely recognized as oncogenic transcription factors that can regulate the development and progression of cancers (Curran, 1992). Common genes silenced across all PC cells were found to be involved in a variety of different functions such as cell adhesion (PCDH1, PCDH10, CDH2, and CDH4), WNT signaling (SOX1, APC2, and WNT5A), and stem cell pluripotency (BMP3, FOXD3, and BMI1) (Vincent et al., 2011). A larger study, with 167 resected PDAC samples and 29 samples from untransformed regions of the pancreas, has identified ~3500 aberrantly methylated genes in PDAC. Pathways associated with these modulated genes indicated the hypermethylation of the ROBO1, ROBO3, SLIT3, and SLIT2 genes associated with the axon guidance signaling pathway and have been shown to be altered in a variety of cancers (Nones et al., 2014). Frequent mutations in ROBO and SLIT genes have been reported in PDAC tissues, and while their precise function in PDAC is still under investigation, studies suggest that they might possess tumor-suppressive properties. This is further supported by the observed downregulation of these genes in PDAC. In another study that analyzed alterations in the progression of PDAC, a cytoplasmic serine/threonine kinase WNK2 was identified as significantly hypermethylated in both chronic pancreatitis and PDAC with a concomitant decrease in its mRNA levels. Because an ectopic expression of WNK2 managed to suppress tumor growth and decrease pERK1/2 levels, it was concluded that the hypermethylation occurs in early PDAC pathogenesis to support tumor growth through the ERK/MAPK pathway

(Nones et al., 2014). Omura and Goggins, using high-throughput technologies, compared whole methylome of PDAC with that of normal tissues and observed the most frequent methylation in miR-9-1, MDF1, CNTNAP2, ZNF415, and EVOLV-4 (Omura and Goggins, 2009). Another study analyzed aberrant methylation patterns between PanIN and normal tissue types and found eight genes (ST14, SPARC, NPTX2, CLDN5, CDH3, SARP2, and reprimo) to be differentially methylated (Sato et al., 2008). Also, in intraductal papillary mucinous neoplasm, another study observed a differential methylation pattern of a different subset of genes—i.e., p14, p15, p16, p17, human mutL homolog 1, E-cadherin, and O-6-methylguanine-DNA methyltransferase (Asokan et al., 2014). These studies provide further evidence that deregulation of DNA methylation occurs as an early event in the development and progression of PC.

Regulation of microRNAs Through DNA Methylation

MiRNAs are a family of small noncoding RNAs that control diverse biological processes through posttranscriptional regulation of gene expression. Therefore, dysregulation of miRNAs is observed in a number of diseases including cancer. A number of miRNAs have been observed to be aberrantly expressed in PC with key functions in its development, progression, and chemoresistance. While miRNAs are considered a separate class of modifiers that influence cellular events, these miRNAs may exert another level of epigenetic control through promoter DNA methylation modification events similar to the case of genes expressing mRNAs (Fig. 12.1). MiRNAs are well known to control the initiation, development, and progression of cancers (Khan et al., 2015b). In PC, myriad data are present on the differential expression profile of miRNAs between pancreatic tumor tissues and the normal pancreas; however, not much information is available on the early players of PC progression. MiR-148a is observed to be downregulated by DNA hypermethylation during PanIN-1 and PanIN-2 (Hanoun et al., 2010). The expression of miR-148a has been observed to be downregulated in a number of cancers such as gastric, colorectal, pancreatic, hepatocellular, esophageal, breast, and non-small-cell lung as well as urogenital system tumors, glioma, and osteosarcoma. In almost all these cancers, loss of expression was due to hypermethylation of the promoter region (Chen et al., 2013). Mir-148a is a tumor-suppressor miRNA in PC that can suppress epithelial-to-mesenchymal transition (EMT) and the invasion of PC cells through the inhibition of Wnt/β-catenin signaling pathways (Peng et al., 2017). Other miRNAs linked to cancer initiation and development and observed to be hypermethylated are miR-9-3, miR-34a, miR-107, and miR-200 (Kunej et al., 2011). Unlike miR-148a, promoter hypermethylation of miR-34a and miR-200 is a late event in PC development (Srivastava et al., 2014). MiR-34a is a downstream target of the tumor suppressor p53, and its reexpression in cancer cells can induce cell cycle arrest and cellular senescence (Srivastava et al., 2014). Thus, these observations imply the requirement that PDAC models study impaired miRNA expression during the development of PC.

CHROMATIN-REMODELING-BASED REGULATION OF GENE EXPRESSION

The macromolecular structure in the nucleus composed of DNA spooled around a complex of histones and other proteins is called chromatin. Although the precise structure of chromatin is poorly understood, it has been an area of extensive investigation. The histone protein is an octamer of four core proteins—histones 2A, 2B, 3 (H3), and 4 (H4)—and upon this histone protein complex are wrapped about 146 base pairs of genomic DNA and a linker, histone 1, that helps stabilize DNA around the histone core. Remodeling of the chromatin is a dynamic process, with modifications of the chromatin architecture allowing or inhibiting access of regulatory transcription machinery proteins to genomic DNA and thus controlling expression of the gene. Covalent histone modifications are carried out by histone acetyltransferases (HATs), histone deacetylases (HDACs), kinases, and methyltransferases, which are considered the principal writers and erasers of the chromatin remodeling system. The modifications result from either addition or removal of acetylation, methylation, ubiquitination, phosphorylation, and sumoylation on histone proteins with potentially different degrees of modification, generally termed "*histone marks*" (Esteller, 2008; Strahl and Allis, 2000). The types, positions, and combinations of marks collectively determine whether a given gene may be expressed or silenced, encompassing the "*histone code*" (Strahl and Allis, 2000; Cosgrove et al., 2004). In addition, nucleosome remodeling, wherein DNA regions are either spooled on or off of nucleosomes, is an important chromatin remodeling mechanism and ATP-dependent mechanism (Vignali et al., 2000).

Acetylation of the histone tails usually indicates an increase in transcriptional activation of the genes; however, the location of methyl residue within the histone tail and its number determine the functional consequence of methylation of the histone tail. For example, monomethylation of H3 at lysine 4 (H3K4me1) and acetylation of H3 at lysine 27 (H3K27ac) can influence gene expression even if present in the enhancer elements at a large distance from the genes. It has been observed that H3K27ac marks are enriched at active enhancer regions, and the regions bearing H3K4me1 are activated in response to a stimulus (Creyghton et al., 2010). In addition, di- and trimethylation of histone 3 at lysine 4 and monomethylation of histone 3 at lysine 9 can lead to the open chromatin structure mediating active gene expression, while di- and trimethylation of histone 3 at lysine 9 mediate the repression of gene expression.

PDAC cells have been identified as having a number of chromatin modifiers that are aberrantly expressed and differentially activated. The enhancer of zeste homolog 2 (EZH2) protein, a component of polycomb repressive complex 2 (PRC2) and an H3K27-methylating protein, has been observed to be overexpressed in PDAC as compared with normal pancreatic tissues (Crea et al., 2012). EZH2 overexpression is associated with the poor prognosis of different types of cancer, including PDAC (Crea et al., 2012). Moreover, elevated nuclear accumulation suggestive of increased activity has been observed in PC (Ougolkov et al., 2008). EZH2 is a well-known modulator of DNA-damage repair pathways and regulates cell-cycle checkpoints (Sauvageau and Sauvageau, 2010). It has been observed to result in the repression of different tumor-suppressor genes such as E-cadherin (Cao et al., 2008), BRCA1 (Gonzalez et al., 2009), CDKN1C (Yang et al., 2009), and CDKN1A (Ougolkov et al., 2008). Inhibition of EZH2 expression by RNA interference–mediated technology in cancer cells in vitro can inhibit the growth of PC cells, increase chemosensitivity, and inhibit the growth of PC stem cells (Chen et al., 2010). Oncogenic Ras signaling in PC has been attributed to upregulation of the expression of EZH2 in PC cell lines. Interestingly, in a spontaneous PC mouse model with pancreas-specific expression of mutant KRAS ($KRAS^{G12D}$) and $EZH2^{-/-}$, enhanced-stage neoplasia was observed in $KRAS^{G12D}/EZH2^{-/-}$ when compared with mice having the $KRAS^{G12D}$ mutation alone (Mallen-St Clair et al., 2012). The authors also observed that EZH2 is required for tissue repair following chemically induced injury, thereby potentially inhibiting carcinogenesis. Furthermore, the loss of trimethylation at H3K27, a hallmark property of EZH2, has been associated with poor prognosis for PC patients (Wei et al., 2008). Thus, these observations present conflicting properties for EZH2 in PC pathogenesis. A potential cause of the observed anomalies could be that in the study by Ougolkov et al. (2008), the authors employed MIA PaCa-2 and Panc 04.03 PC cell lines, which harbor mutations in TP53, for their mechanistic studies (Deer et al., 2010; Forbes et al., 2015); however, in the genetic mouse model *KRAS*^{G12D}, *EZH2*$^{-/-}$ had a wild-type p53. This relationship becomes particularly important because not only can activated wild-type p53 inhibit the expression of EZH2 (Tang et al., 2004); mutant-p53 protein, due to its gain-of-function property, can also enhance the expression of EZH2 and accelerate tumor progression and metastasis in an EZH2-dependent manner (Jiang et al., 2015). However, this does not explain the discrepancy in the observed upregulation of EZH2 in PC while at low H3K27me3 levels and warrants further investigation. This could be the result of alterations in the expression of other members of PRC2. Moreover, the potential targeting of EZH2 by 3-deazaneplanocin has been observed to significantly hamper TGF-β1-induced EMT in PC cells. DZNep treatment of PC cells reprogrammed the noncoding miRNA expression patterns, inducing reexpression of miR-663a and miR-4787-5p and leading to reduced tumor burden and decreased metastasis in an orthotropic mouse model of PC (Mody et al., 2016).

The MLL (mixed lineage leukemia) family, responsible for activating H3K4me marks, has been observed to be mutated in a number of different cancers, including PDAC (Rao and Dou, 2015). The MLL family consists of KMT2A (MLL1), KMT2B (MLL2), KMT2C (MLL3), and KMT2D (MLL4), which are involved in the formation of larger complexes that confer to them the ability to perform di- or trimethylation of H3K4; however, the precise functions of the proteins in the complex have not been clearly delineated. Mutations in MLL proteins and decreased expression of MLL3 and MLL4 in PDAC patients have been associated with better survival of PDAC patients. Interestingly, in addition to the presence of mutation, truncation, and deletion in the genes of MLL family proteins, in 2%–7% of ~100 samples tested, amplification of MLL genes has been reported. Knockdown of MLL4 in PDAC cells in vitro has been observed to reduce growth through cell cycle inhibition and increased apoptosis. Furthermore, MLL1 and MLL2 complexes have been observed to bind to the menin protein (encoded by the MEN1 gene). Loss of the MEN1 gene has been demonstrated to lower H3K4me3 levels and result in the downregulation of gene expression, and correlates with increased H3K27me3 levels that result in the development of mouse pancreatic neuroendocrine tumor development (Lin et al., 2015). Interestingly, these alterations in large organized chromatin H3K9-modifications (LOCKs) have been observed to become substantially lost in cancer cell lines in order to provide the cancer cells with phenotypic plasticity in development and disease (Wen et al., 2009). Based on the observation that a majority of all pancreatic tumors subclone and that metastatic tumors share principal driver mutations and genetic anomalies with those of the primary tumor, McDonald and colleagues proposed that epigenomic regulation of individual subclones might provide the prometastatic mechanism for this anomaly (Mcdonald et al., 2017). Widespread global reprogramming of histone H3K9 methylation within LOCKs in the absence of metastasis-specific driver mutations leads to regional changes in gene expression. Interestingly, the metastatic subclones were identified as having a heightened degree of oxidative pentose phosphate pathway (oxPPP), contrary to the observation that mutKRAS induces the non-oxPPP to drive primary pancreatic tumor growth (Mcdonald et al., 2017). In another study, Raju et al. (2016) observed that expression of KLK7 in MIA PaCa-2 and PK-1 PDAC cells corresponded to active H3K9 histone marks. In the PANC-1 cell line, which lacks KLK7 expression, treatment with trichostatin A induced its expression. Upregulation of KLK7 in PC cells leads to reduced cell adhesion to the surrounding matrix, enhancing the ability of tumor cells to disseminate from the primary tumor site (Ramani and Haun, 2008). Thus, these studies suggest a context-dependent activity and expression of H3K9 methylations in PC.

The other most common epigenetic modification of histones observed is acetylation of the lysines on H3. Acetylation of histone lysines leads to neutralization of the positive charges on the tails and results in relaxation of chromatin compaction, providing access for gene transcription, and is thus associated with gene activation (Bedi et al., 2014; Simon et al., 2016). It has been observed that the acetylation marks at H3K18 and H4K12 in PDAC samples are associated with lower overall survival (Juliano et al., 2016). The transcriptional cofactor p300 has HAT activity and is associated with chromatin remodeling and DNA damage (Vo and Goodman, 2001; Chan and La Thangue, 2001). It has been observed to be present in PDAC cell lines and patient tumors; Ono and colleagues observed that p300 can protect PC cells from gemcitabine-induced DNA damage and apoptosis, as the inhibition of expression or activity of p300 in PDAC cells enhances the cytotoxic efficacy of gemcitabine (Ono et al., 2016). In addition, the overexpression of NAD-dependent HDAC sirtuin (SIRT) 1 has been observed in PDAC and correlated with increased viability of PC cells and reduced patient survival (Stenzinger et al., 2013). Furthermore, Kugel et al. observed that SIRT6, a nutrient sensor and HDAC, is downregulated in PDAC relative to normal tissue, and downregulation of SIRT6 is directly associated with mutations in KRAS. They demonstrated that the loss of SIRT6 in PDAC results in hyperacetylation of H3K9 and H3K56 at the promoter of the oncofetal protein LIN28B, increasing its expression and promoting tumor growth (Kugel et al., 2016). Additionally, a number of HDACs have been observed to be associated with PDAC/PNET growth and patient survival. Klieser et al. observed HDAC5 to be a predictor of poor clinical outcome for PNET patients (Klieser et al., 2017). Similarly, HDAC7 is overexpressed in PDAC samples and associated with low patient survival as well as disease recurrence (Ouaissi et al., 2014). Molecular experiments determining the alterations in gene expression that are caused by aberrant acetylation in PDAC need to be obtained.

Histone acetylation has also been observed as responsible for recruiting the bromodomain-and-extraterminal domain (BET) protein family to chromatin (Marmorstein and Zhou, 2014). BET family members are epigenetic domain readers that can recognize acetylated lysine residues on the tails of histone proteins (Muller et al., 2011). The bromodomain (BRD) BRD4 in PC has been observed to be significantly upregulated, supporting growth and tumorigenicity. Gemcitabine treatment of PDAC cells enhanced the expression of BRD4, and its silencing improves the gemcitabine sensitivity of PC cells. Moreover, BRD2, BRD3, and BRD4 can increase the expression of sonic hedgehog pathway members in a ligand-independent manner, thus promoting tumor–stromal cross talk (Wang et al., 2015b; Khan et al., 2017). In fact, BRD2 and BRD3 contribute to PDAC cell growth through direct interaction with the GLI family of transcription factors and modulate their activity (Huang et al., 2016). Furthermore, BET proteins have been suggested to increase desmoplastic stroma by the stimulation of cancer-associated fibroblasts (CAFs) through the recruitment of transcriptional machinery responsible for the activation of TGF-β-dependent gene expression in CAFs (Yamamoto et al., 2016). Inversely, stroma-activated PDAC cells demonstrate a widespread increase in histone acetylation at transcriptionally enhanced genes, implicating a requirement for the BET protein family in the interpretation of the PDAC epigenome and thus providing a point of convergence between transcriptional upregulation and enhanced histone acetylation. Interestingly, inhibition of BRD2 blocked stroma-inducible transcriptional regulation in vitro and tumor progression in vivo. Thus, histone modification has a direct influence on the growth and development of PDAC, and further investigations are imperative to precisely understand the role and implication of these alterations.

EPIGENETICS-BASED DIAGNOSTIC BIOMARKERS FOR PANCREATIC CANCER

The identification of novel biomarkers for PC is an urgent requirement, and rapidly expanding epigenetics-based research shows significant promise. For effective biomarkers in PC diagnosis, noninvasive or minimally invasive techniques should be employed for the identification of these epigenetic markers, and they should outperform CA19-9 in selectivity and specificity. In a pilot study, whole blood was analyzed for the presence of promoter CpG sites of tumor-associated genes TNFRSF10 C, RARβ, APC, DAPK1 3OST2, p16, CD44, BCL2, and 3OST2 from 30 PDAC patients and their 49 matched controls (Dauksa et al., 2012). A slightly higher degree of methylation was observed in tumor-suppressor genes at CpG sites compared with that of healthy donors. More important, it was observed that higher methylation levels at TNFRSCF10C correlate with the positive perineural spread of tumor cells, and significantly shorter patient survival was associated with higher methylation levels of TNFRSF10C and ACIN1 (Dauksa et al., 2012). In the search for another biomarker, NPTX2 protein has been investigated as a potential serum methylation marker for PC (Park et al., 2012). It is reported as a tumor suppressor that can inhibit the proliferation and invasion of PC (Zhang et al., 2011) and is known to be downregulated in PC by hypermethylation (Sato et al., 2003a).

In another approach, Pedersen et al. performed DNA methylation analysis of leukocyte DNA by comparing the global profile of 60 healthy controls and 132 PDAC patients (Pedersen et al., 2011), and a number of CpG islands were found to be differentially methylated between the two samples. In a more targeted approach, they evaluated a panel of five CpG sites—viz. IL10_P348, ZAP70_P220, TAL1_P817, LCN2_P86, and AIM2_P624—with the most differential expression between the two samples. The validation study provided a sensitivity and specificity of 72% and 70%, respectively (Pedersen et al., 2011). Although the exact mechanism of aberrant promoter methylation changes in leukocytes is not known, it is highly likely that the cancer promotes alterations in immune cells to potentially dampen immune response. These studies have provided evidence that epigenetic-based modifications can serve as potential diagnostic biomarkers with minimally invasive techniques. Moreover, since epigenetic changes are an early event in tumorigenesis, there is potential for exploring these markers as biomarkers.

MODULATION OF EPIGENETIC ALTERATIONS BY PHYTOCHEMICALS

Dietary phytochemicals have attracted reasonable attention in the past few decades for their potent biological properties. It has been observed that these phytochemicals can have an effective impact on nearly all epigenetic modifications and thus elicit effective chemopreventive and therapeutic properties. As mentioned above, although heritable, epigenetic modifications are readily reversible, making them potential therapeutic targets in many different diseases. The targeting of epigenetic modifiers has gained much interest in recent years.

Methylation-based silencing can be rendered inactive by demethylation, and histone complexes be made transcriptionally active through modification of histone methylation and acetylation. HDAC inhibitors (HDACIs) have been observed to demonstrate in vitro and preclinical antitumor activity in a number of cancer types including PC (Mishra et al., 2017; Zagorac et al., 2016). Many HDACIs are in clinical trials for different cancer types, with some approved for therapy of cutaneous and peripheral T-cell lymphoma (Mottamal et al., 2015). Moreover, a growing body of evidence suggests the possibility of supplementing HDACIs with standard line-of-care for a more efficient and tumor-specific activity (Mottamal et al., 2015; Li et al., 2010). Thus, epigenetics seems to offer a potentially effective therapeutic opportunity from a clinical standpoint, as opposed to genetic changes.

A number of studies have provided evidence that phytochemicals can work through a range of overlapping molecular mechanisms through their antioxidant and prooxidant effects, activation of detoxification system, modulation of carcinogenic signaling, etc. In addition, about a decade of research has identified that phytochemicals can target various epigenetic modifiers, such as DNMTs and HDACs, and potentially function as chemopreventive and therapeutic agents by targeting more than just the signaling pathways in different cancers. In PC, only a small number of studies have analyzed the effects of natural products on epigenetic modifiers; therefore, in subsequent paragraphs we will also discuss phytochemicals that have been demonstrated to alter epigenetic modifiers, relevant to PC, in other cancer types.

In a recent study, Xie et al. (2017) demonstrated that dietary 3,3′-diindolylmethane (DIM) has potent anticancer activity against pancreatic tumor cells. DIM is a phytonutrient present in cruciferous vegetables such as cabbage, broccoli, brussels sprouts, and cauliflower (Ahmad et al., 2010). Xie et al. observed that DIM treatment of PC cells leads to downregulation of DNMT1 that decreases DNA methylation at the KLF4 promoter site. Upregulation of KLF4 in these cells was observed to cause differentiation and significant growth inhibition of PC cells (Xie et al., 2017). Furthermore, treatment of PC cells with DIM upregulated E-cadherin, an epithelial marker gene commonly downregulated due to promoter methylation in PC and an important regulator of EMT (Shankar et al., 2013). DIM treatment has also been demonstrated to induce significant inhibition of HDACs in both androgen-sensitive and androgen-insensitive prostate cancer cell lines. DIM specifically downregulated the expression of HDAC2 and significantly altered HDAC activity, resulting in increased expression of p21, a known target of HDACIs (Beaver et al., 2012). A novel formulation of DIM (abbreviated BR-DIM, 3,3′-diindolylmethane by BioResponse Nutrients, Boulder, CO) was demonstrated to regulate the expression of EZH2 in castration-resistant prostate cancer cells. BR-DIM upregulated the miRNA let-7 that downregulated the expression of EZH2 in these cells, thereby decreasing cancer aggressiveness (Kong et al., 2012).

Epigallocatechin-3-gallate (EGCG), a major tea polyphenol, has been demonstrated to have significant PC-suppressive properties (Kumazoe et al., 2017; Lu et al., 2015). EGCG treatment of lung cancer cells could resensitize NSCLC to cisplatin (DDP) treatment (Zhang et al., 2015). In these cells, EGCG was observed to cause inhibition of DNMT and HDAC activity leading to increased expression of GAS1, TIMP4, ICAM1 and WISP2 genes in A549/DDP cells (Zhang et al., 2015). Similarly, two common coffee polyphenols, namely caffeic acid and chlorogenic acid, have been observed to reduce DNMT1 activity in MCF7 and MDA-MB-231 breast cancer cells, inhibiting their growth (Lee and Zhu, 2006). EGCG has also been demonstrated to downregulate the expression of HDACs when acting in synergism with Am80, a synthetic retinoid with a structure from all-trans-retinoic acid.

Although the authors did not observe a change in H3/H4 acetylation, an increase in acetylated p53 and acetylated-α-tubulin in the cytosolic fraction indicated decreased HDAC activity. Moreover, an increase in p21 expression was observed that corresponded with increased apoptosis upon treatment (Oya et al., 2017). EGCG has been determined to bind directly to DNMT3B and HDAC1 (through molecular modeling) and decrease their activity in HeLa cells, upregulating tumor-suppressor genes and decreasing cancer cell growth (Khan et al., 2015a). Moreover, EGCG has been demonstrated to inhibit proliferation and induce apoptosis in H9C2 cardiomyocytes in a SIRT1-dependent manner (Cai et al., 2015).

Curcumin, the bioactive phytochemical present in turmeric, has been demonstrated in several in vitro and preclinical studies to possess anticancer properties (Nagaraju et al., 2013; Ahmad et al., 2016). However, the clinical efficacy of curcumin is marred due to its low bioavailability. To overcome this, several semisynthetic derivatives of curcumin have been synthesized with higher efficacy and better bioavailability than those of the parent compound. Two curcumin analogues, EF31 and UBS109, have potent DNA-hypomethylating properties against PC, a mechanism observed to be partly dependent on the inhibition of NF-κB signaling pathway−mediated upregulation of DNMT1 expression (Nagaraju et al., 2013). Similarly, a curcumin analog, dimethoxycurcumin, in combination with a DNMT inhibitor, has been demonstrated to potentiate the reexpression of promoter-methylated tumor-suppressor genes in leukemic cell lines, thereby inhibiting the growth of cancer cells (Hassan et al., 2016). Inhibition of histone acetylation upon curcumin treatment has also been observed. For example, Muta et al. demonstrated that the kidneys of high-salt- and curcumin-treated rats exhibited fewer H3K9ac cells compared with those from the kidneys of only high-salt-fed rats. Thus, they concluded that suppression of histone acetylation by curcumin ameliorates nephrosclerosis (Muta et al., 2016).

Curcumin-based chromatin remodeling has also been identified in human hepatoma HepG2 and Huh7 cell lines and HaCaT keratinocytes (Mohammadi-Bardbori et al., 2016). At higher concentrations, curcumin was observed to decrease histone acetylation in these cells to induce CYP1A1 gene expression (Mohammadi-Bardbori et al., 2016). Although some literature might suggest that CYP1A1 activates environmental carcinogens (Perepechaeva et al., 2017; Shimada, 2017), significant reports have demonstrated that CYP1A1 is required for the metabolic activation of dietary compounds with cancer-preventative activity (Androutsopoulos et al., 2009; Perepechaeva et al., 2017). This suggests that the contribution of CYP1A1 to cancer progression or prevention may depend on the extent of detoxification and activation of procarcinogens and dietary natural agents (Androutsopoulos et al., 2009). Another analog of curcumin, difluorinated curcumin (CDF), can inhibit the expression of EZH2 in PC cells (Bao et al., 2012b). Mechanistically, CDF induced the expression of tumor-suppressor miRNAs that target EZH2 and are typically lost in PC (Bao et al., 2012b). This effect was also recapitulated under hypoxia, suggesting the potential for development of CDF as an important anticancer agent (Bao et al., 2012a). Another curcumin analog, C66, has been demonstrated to reduce HAT activity and reduce p300/CBP HAT expression, thus decreasing the H3-lysine9/14-acetylation (H3K9/14Ac) level and the p300/CBP occupancy at the CTGF, PAI-1, and FN-1 gene promoters (Wang et al., 2015a).

Resveratrol, a grape polyphenol, has been the subject of intense research over the past couple of years for its varied biological activities, including anticancer properties against multiple cancers (Jiang et al., 2016; Xu et al., 2015; Carter et al., 2014; Zubair et al., 2017). Resveratrol has been demonstrated to inhibit the expression of DNMT enzymes and reactivate tumor-suppressor genes to constrain cancer growth. Similar to the case with curcumin, resveratrol derivatives having better potency and bioavailability are being synthesized to achieve better anticancer activity in vivo. Aldawsari et al. generated salicylate derivatives of resveratrol as selective DNMT3 inhibitors that were confirmed using an in vitro DNMT inhibition assay (Aldawsari et al., 2016). Furthermore, in breast cancer cells, resveratrol was shown to reduce methylation of the promoter region of tumor suppressor BRCA1 by inhibiting DNMTs (Papoutsis et al., 2012). Treatment of hepatocellular carcinoma cell line HepG2 yielded specific inhibition of HDACs with a significant increase in hyperacetylated histones. Moreover, high doses of resveratrol were well tolerated by normal hepatocytes, suggesting an important pan-HDACI activity of resveratrol (Venturelli et al., 2013).

Inhibition of telomerase activity is considered a viable antitumor mechanism, as activation of telomerase is considered an integral part of carcinogenesis. Methyl-2-cyano-3,12-dioxooleana-1,9(11)-dien-28-oate (CDDO-Me), a synthetic oleanane triterpenoid, has been observed to reduce the expressions of DNMT1 and DNMT3a in PC, resulting in hypomethylation of the promoter region of human telomerase reverse transcriptase (Deeb et al., 2014). However, there was an overall reduction in the 'transcriptionally active' chromatin markers in PC cells treated with CDDO-Me, along with decreased histone deacetylation and histone demethylation at hTERT promoter. This suggested that CDDO-Me-mediated epigenetic downregulation of telomerase is sufficient to induce apoptosis in PC cells. Apart from dietary phytochemicals, different agents from other sources are also being tested for their anticancer activities. The cyclic tetrapeptide 1-alaninechlamydocin from *Tolypocladium* sp., a Great Lakes-derived fungus, has been observed to inhibit MIA PaCa-2 PC cells in vitro by the induction of G2/M cell cycle arrest and apoptosis in an HADC inhibition-based mechanism (Du et al., 2014). Based on these observations, it appears that the targeting of epigenetic modifiers by natural products is emerging as a mechanism of their anticancer activity.

CONCLUSIONS AND PERSPECTIVES

Recent studies have demonstrated the importance of epigenetic modifications in PC pathogenesis, as DNA methylation and chromatin-based epigenetic events are now believed to be significantly altered in PDAC. The growing importance of epigenetics in cancer therapy is evidenced by several clinical trials of HDACI that are currently underway. Most inhibitors of epigenetic events are synthetic chemicals with associated toxicity, therefore creating renewed interest in natural products (phytochemicals/nutraceuticals) that have long been evaluated for their anticancer potential against a variety of cancers, including PC, and are usually well tolerated with minimal to no toxicity. The ability of natural compounds to favorably alter epigenetic events has been demonstrated in several other cancers, with such reports in PC just beginning to emerge. Although only a few studies have explored the modulation of epigenetic modifiers by natural compound phytochemicals in PC, a convincing body of evidence from multiple cancer models suggests that these compounds modulate epigenetic modifiers relevant to PC, thus implying a hypothetical mechanistic link between the two (Fig. 12.2). It is important to point out that while natural compounds are tolerated well, they have their own associated drawbacks, with bioavailability a major point of concern. This is partly being overcome through synthesis of novel derivatives with significantly greater bioavailability, but often these derivatives have increased toxicity. Thus, a major challenge now is to fine-tune the balance between bioavailability and toxicity. Further, it has been suggested that natural compounds can sensitize tumor cells to conventional

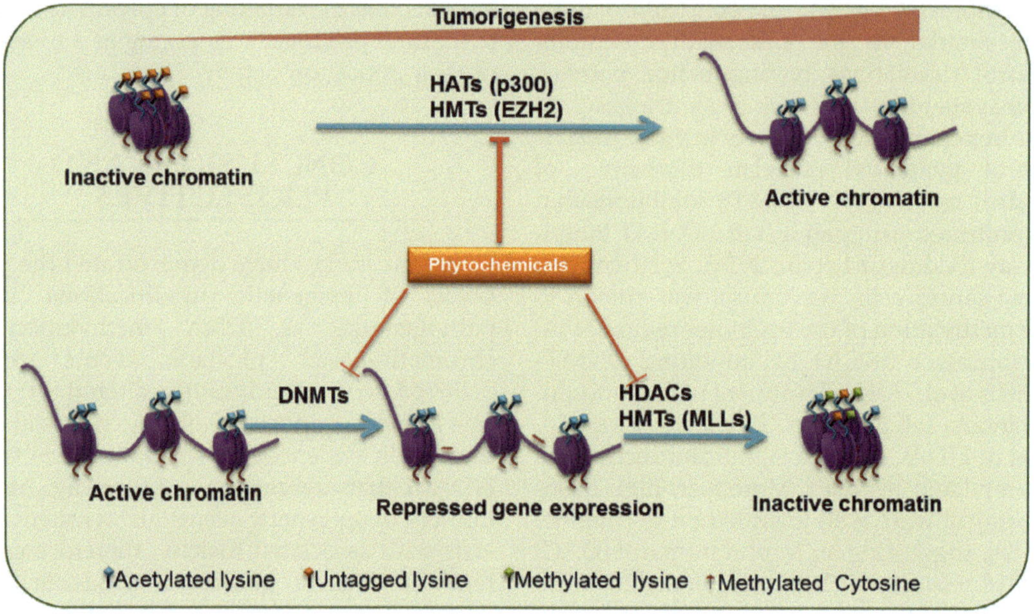

FIGURE 12.2 During tumorigenesis, oncogenic histone writers such as p300 and EZH2 can lead to enhanced histone acetylation and methylation, respectively, generating an active chromatin status that has been observed to result in the upregulation of oncogene expression. At the same time, tumor-suppressor genes can be inactivated through DNMT activity that increases the methylation of cytosine residues in CpG islands, repressing mRNA expression. Furthermore, HDACs and HMTs can deacetylate histones and introduce repressive histone methylation, generating inactive chromatin. It has been demonstrated that several phytochemicals can inhibit pancreatic cancer growth and progression. Different mechanisms of action have been identified, and phytochemicals and their synthetic derivatives can inhibit different stages of epigenetic modulation that affect pancreatic pathogenesis.

chemotherapies. Future studies should address the multitude of queries pertaining to the feasibility of using phytochemicals alone or in combination with other cytotoxic drugs for better patient outcomes, what dosage to use, the sequence of administration, etc. Furthermore, since most literature discussed in this chapter is based on in vitro evidence, detailed in vivo and preclinical studies are needed in order to verify the promising results from in vitro studies. Also, given the important role that the tumor microenvironment and its various components (such as immune and stromal cells) plays in PC progression, the therapeutic potential of phytochemicals needs to be evaluated with appropriate model systems that are reminiscent of the complex tumor microenvironment.

In conclusion, preliminary evidence suggests that epigenetic alterations in PDAC can be modulated by several phytochemicals, but more mechanistic studies need to be performed in appropriate PC models before phytochemicals/nutraceuticals can find a place in clinics.

Acknowledgments

The authors would like to thankfully acknowledge the editors of this book who provided us the opportunity to write this chapter. This work is financially supported by NIH/NCI [R01CA175772, and U01CA185490 (to APS)] and USA-MCI.

References

Ahmad, A., 2016. Epigenetics in personalized management of lung cancer. Adv. Exp. Med. Biol. 890, 111–122.

Ahmad, A., Li, Y., Bao, B., Kong, D., Sarkar, F.H., 2014. Epigenetic regulation of miRNA-cancer stem cells nexus by nutraceuticals. Mol. Nutr. Food Res. 58, 79–86.

Ahmad, A., Li, Y., Sarkar, F.H., 2016. The bounty of nature for changing the cancer landscape. Mol. Nutr. Food Res. 60, 1251–1263.

Ahmad, A., Sakr, W.A., Rahman, K.M., 2010. Anticancer properties of indole compounds: mechanism of apoptosis induction and role in chemotherapy. Curr. Drug Targets 11, 652–666.

Aldawsari, F.S., Aguayo-Ortiz, R., Kapilashrami, K., Yoo, J., Luo, M., Medina-Franco, J.L., Velazquez-Martinez, C.A., 2016. Resveratrol-salicylate derivatives as selective DNMT3 inhibitors and anticancer agents. J. Enzyme Inhib. Med. Chem. 31, 695–703.

Androutsopoulos, V.P., Tsatsakis, A.M., Spandidos, D.A., 2009. Cytochrome P450 CYP1A1: wider roles in cancer progression and prevention. BMC Cancer 9, 187.

Asokan, G.S., Jeelani, S., Gnanasundaram, N., 2014. Promoter hypermethylation profile of tumour suppressor genes in oral leukoplakia and oral squamous cell carcinoma. J. Clin. Diagn. Res. 8, ZC09–12.

Azim, S., Zubair, H., Srivastava, S.K., Bhardwaj, A., Zubair, A., Ahmad, A., Singh, S., Khushman, M., Singh, A.P., 2016. Deep sequencing and in silico analyses identify MYB-regulated gene networks and signaling pathways in pancreatic cancer. Sci. Rep. 6, 28446.

Azizi, M., Teimoori-Toolabi, L., Arzanani, M.K., Azadmanesh, K., Fard-Esfahani, P., Zeinali, S., 2014. MicroRNA-148b and microRNA-152 reactivate tumor suppressor genes through suppression of DNA methyltransferase-1 gene in pancreatic cancer cell lines. Cancer Biol. Ther. 15, 419–427.

Bae, J.M., Lee, E.J., Guyatt, G., 2009. Citrus fruit intake and pancreatic cancer risk: a quantitative systematic review. Pancreas 38, 168–174.

Ballehaninna, U.K., Chamberlain, R.S., 2013. Biomarkers for pancreatic cancer: promising new markers and options beyond CA 19-9. Tumour Biol. 34, 3279–3292.

Bao, B., Ali, S., Ahmad, A., Azmi, A.S., Li, Y., Banerjee, S., Kong, D., Sethi, S., Aboukameel, A., Padhye, S.B., Sarkar, F.H., 2012a. Hypoxia-induced aggressiveness of pancreatic cancer cells is due to increased expression of VEGF, IL-6 and miR-21, which can be attenuated by CDF treatment. PLoS One 7, e50165.

Bao, B., Ali, S., Banerjee, S., Wang, Z., Logna, F., Azmi, A.S., Kong, D., Ahmad, A., Li, Y., Padhye, S., Sarkar, F.H., 2012b. Curcumin analogue CDF inhibits pancreatic tumor growth by switching on suppressor microRNAs and attenuating EZH2 expression. Cancer Res. 72, 335–345.

Beaver, L.M., Yu, T.W., Sokolowski, E.I., Williams, D.E., Dashwood, R.H., Ho, E., 2012. 3,3′-Diindolylmethane, but not indole-3-carbinol, inhibits histone deacetylase activity in prostate cancer cells. Toxicol. Appl. Pharmacol. 263, 345–351.

Bedi, U., Mishra, V.K., Wasilewski, D., Scheel, C., Johnsen, S.A., 2014. Epigenetic plasticity: a central regulator of epithelial-to-mesenchymal transition in cancer. Oncotarget 5, 2016–2029.

Cai, Y., Zhao, L., Qin, Y., He, Y., 2015. High dose of epigallocatechin-3-gallate inhibits proliferation and induces apoptosis of H9C2 cardiomyocytes through downregulation of SIRT1. Pharmazie 70, 12–16.

Cao, Q., Yu, J., Dhanasekaran, S.M., Kim, J.H., Mani, R.S., Tomlins, S.A., Mehra, R., Laxman, B., Cao, X., Yu, J., Kleer, C.G., Varambally, S., Chinnaiyan, A.M., 2008. Repression of E-cadherin by the polycomb group protein EZH2 in cancer. Oncogene 27, 7274–7284.

Carter, L.G., D'orazio, J.A., Pearson, K.J., 2014. Resveratrol and cancer: focus on in vivo evidence. Endocr. Relat. Cancer 21, R209–R225.

Chan, H.M., La Thangue, N.B., 2001. p300/CBP proteins: HATs for transcriptional bridges and scaffolds. J. Cell Sci. 114, 2363–2373.

Chedin, F., 2011. The DNMT3 family of mammalian de novo DNA methyltransferases. Prog. Mol. Biol. Transl. Sci. 101, 255–285.

Chen, K., Zhang, Q., Peng, M., Shen, Y., Wan, P., Xie, G., 2014. Relationship between tea consumption and pancreatic cancer risk: a meta-analysis based on prospective cohort studies and case-control studies. Eur. J. Cancer Prev. 23, 353–360.

Chen, Y., Song, Y.X., Wang, Z.N., 2013. The microRNA-148/152 family: multi-faceted players. Mol. Cancer 12, 43.

Chen, Y., Xie, D., Yin Li, W., Man Cheung, C., Yao, H., Chan, C.Y., Chan, C.Y., Xu, F.P., Liu, Y.H., Sung, J.J., Kung, H.F., 2010. RNAi targeting EZH2 inhibits tumor growth and liver metastasis of pancreatic cancer in vivo. Cancer Lett. 297, 109–116.

Cosgrove, M.S., Boeke, J.D., Wolberger, C., 2004. Regulated nucleosome mobility and the histone code. Nat. Struct. Mol. Biol. 11, 1037–1043.

Crea, F., Paolicchi, E., Marquez, V.E., Danesi, R., 2012. Polycomb genes and cancer: time for clinical application? Crit. Rev. Oncol. Hematol. 83, 184–193.

Creyghton, M.P., Cheng, A.W., Welstead, G.G., Kooistra, T., Carey, B.W., Steine, E.J., Hanna, J., Lodato, M.A., Frampton, G.M., Sharp, P.A., Boyer, L.A., Young, R.A., Jaenisch, R., 2010. Histone H3K27ac separates active from poised enhancers and predicts developmental state. Proc. Natl. Acad. Sci. U.S.A. 107, 21931–21936.

Curran, T., 1992. Fos and Jun: oncogenic transcription factors. Tohoku J. Exp. Med. 168, 169–174.

Dauksa, A., Gulbinas, A., Barauskas, G., Pundzius, J., Oldenburg, J., El-Maarri, O., 2012. Whole blood DNA aberrant methylation in pancreatic adenocarcinoma shows association with the course of the disease: a pilot study. PLoS One 7, e37509.

Deeb, D., Brigolin, C., Gao, X., Liu, Y., Pindolia, K.R., Gautam, S.C., 2014. Induction of apoptosis in pancreatic cancer cells by CDDO-me involves repression of telomerase through epigenetic pathways. J. Carcinog. Mutagen. 5, 177.

Deer, E.L., Gonzalez-Hernandez, J., Coursen, J.D., Shea, J.E., Ngatia, J., Scaife, C.L., Firpo, M.A., Mulvihill, S.J., 2010. Phenotype and genotype of pancreatic cancer cell lines. Pancreas 39, 425–435.

Delpu, Y., Hanoun, N., Lulka, H., Sicard, F., Selves, J., Buscail, L., Torrisani, J., Cordelier, P., 2011. Genetic and epigenetic alterations in pancreatic carcinogenesis. Curr. Genomics 12, 15–24.

Ding, Y., Yu, C., Han, Z., Xu, S., Li, D., Meng, X., Chen, D., 2015. Environmental tobacco smoke and pancreatic cancer: a case-control study. Int. J. Clin. Exp. Med. 8, 16729–16732.

Du, L., Risinger, A.L., King, J.B., Powell, D.R., Cichewicz, R.H., 2014. A potent HDAC inhibitor, 1-alaninechlamydocin, from a Tolypocladium sp. induces G2/M cell cycle arrest and apoptosis in MIA PaCa-2 cells. J. Nat. Prod. 77, 1753–1757.

Elliott, E.N., Sheaffer, K.L., Kaestner, K.H., 2016. The 'de novo' DNA methyltransferase Dnmt3b compensates the Dnmt1-deficient intestinal epithelium. Elife 5.

Esteller, M., 2008. Epigenetics in cancer. N. Engl. J. Med. 358, 1148–1159.

Forbes, S.A., Beare, D., Gunasekaran, P., Leung, K., Bindal, N., Boutselakis, H., Ding, M., Bamford, S., Cole, C., Ward, S., Kok, C.Y., Jia, M., De, T., Teague, J.W., Stratton, M.R., Mcdermott, U., Campbell, P.J., 2015. COSMIC: exploring the world's knowledge of somatic mutations in human cancer. Nucleic Acids Res. 43, D805–D811.

Garrido-Laguna, I., Hidalgo, M., 2015. Pancreatic cancer: from state-of-the-art treatments to promising novel therapies. Nat. Rev. Clin. Oncol. 12, 319–334.

Gonzalez, M.E., Li, X., Toy, K., Duprie, M., Ventura, A.C., Banerjee, M., Ljungman, M., Merajver, S.D., Kleer, C.G., 2009. Downregulation of EZH2 decreases growth of estrogen receptor-negative invasive breast carcinoma and requires BRCA1. Oncogene 28, 843–853.

Grasso, D., Bintz, J., Lomberk, G., Molejon, M.I., Loncle, C., Garcia, M.N., Lopez, M.B., Urrutia, R., Iovanna, J.L., 2015. Pivotal role of the chromatin protein Nupr1 in Kras-induced senescence and transformation. Sci. Rep. 5, 17549.

Grasso, D., Garcia, M.N., Hamidi, T., Cano, C., Calvo, E., Lomberk, G., Urrutia, R., Iovanna, J.L., 2014. Pancreatitis promotes oncogenic Kras(G12D)-induced pancreatic transformation through activation of Nupr1. Mol. Cell. Oncol. 1, e29913.

Hanoun, N., Delpu, Y., Suriawinata, A.A., Bournet, B., Bureau, C., Selves, J., Tsongalis, G.J., Dufresne, M., Buscail, L., Cordelier, P., Torrisani, J., 2010. The silencing of microRNA 148a production by DNA hypermethylation is an early event in pancreatic carcinogenesis. Clin. Chem. 56, 1107–1118.

Hassan, H.E., Keita, J.A., Narayan, L., Brady, S.M., Frederick, R., Carlson, S., Glass, K.C., Natesan, S., Buttolph, T., Fandy, T.E., 2016. The combination of dimethoxycurcumin with DNA methylation inhibitor enhances gene re-expression of promoter-methylated genes and antagonizes their cytotoxic effect. Epigenetics 1–10.

Henikoff, S., Greally, J.M., 2016. Epigenetics, cellular memory and gene regulation. Curr. Biol. 26, R644–R648.

Huang, P.H., Lu, P.J., Ding, L.Y., Chu, P.C., Hsu, W.Y., Chen, C.S., Tsao, C.C., Chen, B.H., Lee, C.T., Shan, Y.S., Chen, C.S., 2017. TGFbeta promotes mesenchymal phenotype of pancreatic cancer cells, in part, through epigenetic activation of VAV1. Oncogene 36, 2202–2214.

Huang, Y., Nahar, S., Nakagawa, A., Fernandez-Barrena, M.G., Mertz, J.A., Bryant, B.M., Adams, C.E., Mino-Kenudson, M., Von Alt, K.N., Chang, K., Conery, A.R., Hatton, C., Sims 3rd, R.J., Fernandez-Zapico, M.E., Wang, X., Lillemoe, K.D., Fernandez-Del Castillo, C., Warshaw, A.L., Thayer, S.P., Liss, A.S., 2016. Regulation of GLI underlies a role for BET bromodomains in pancreatic cancer growth and the tumor microenvironment. Clin. Cancer Res. 22, 4259–4270.

Iodice, S., Gandini, S., Maisonneuve, P., Lowenfels, A.B., 2008. Tobacco and the risk of pancreatic cancer: a review and meta-analysis. Langenbeck's Arch. Surg. 393, 535–545.

Jansen, R.J., Robinson, D.P., Frank, R.D., Stolzenberg-Solomon, R.Z., Bamlet, W.R., Oberg, A.L., Rabe, K.G., Olson, J.E., Petersen, G.M., Sinha, R., Anderson, K.E., 2013. Meat-related mutagens and pancreatic cancer: null results from a clinic-based case-control study. Cancer Epidemiol. Biomark. Prev. 22, 1336–1339.

Jansen, R.J., Robinson, D.P., Stolzenberg-Solomon, R.Z., Bamlet, W.R., De Andrade, M., Oberg, A.L., Hammer, T.J., Rabe, K.G., Anderson, K.E., Olson, J.E., Sinha, R., Petersen, G.M., 2011. Fruit and vegetable consumption is inversely associated with having pancreatic cancer. Cancer Causes Control 22, 1613–1625.

Jiang, F.Z., He, Y.Y., Wang, H.H., Zhang, H.L., Zhang, J., Yan, X.F., Wang, X.J., Che, Q., Ke, J.Q., Chen, Z., Tong, H., Zhang, Y.L., Wang, F.Y., Li, Y.R., Wan, X.P.,

2015. Mutant p53 induces EZH2 expression and promotes epithelial-mesenchymal transition by disrupting p68-Drosha complex assembly and attenuating miR-26a processing. Oncotarget 6, 44660–44674.

Jiang, Z., Chen, X., Chen, K., Sun, L., Gao, L., Zhou, C., Lei, M., Duan, W., Wang, Z., Ma, Q., Ma, J., 2016. YAP inhibition by resveratrol via activation of AMPK enhances the sensitivity of pancreatic cancer cells to gemcitabine. Nutrients 8.

Juliano, C.N., Izetti, P., Pereira, M.P., Dos Santos, A.P., Bravosi, C.P., Abujamra, A.L., Prolla, P.A., Osvaldt, A.B., Edelweiss, M.I., 2016. H4K12 and H3K18 acetylation associates with poor prognosis in pancreatic cancer. Appl. Immunohistochem. Mol. Morphol. 24, 337–344.

Khan, M.A., Azim, S., Zubair, H., Bhardwaj, A., Patel, G.K., Khushman, M., Singh, S., Singh, A.P., 2017. Molecular drivers of pancreatic cancer pathogenesis: looking inward to move forward. Int. J. Mol. Sci. 18.

Khan, M.A., Hussain, A., Sundaram, M.K., Alalami, U., Gunasekera, D., Ramesh, L., Hamza, A., Quraishi, U., 2015a. (-)-Epigallocatechin-3-gallate reverses the expression of various tumor-suppressor genes by inhibiting DNA methyltransferases and histone deacetylases in human cervical cancer cells. Oncol. Rep. 33, 1976–1984.

Khan, M.A., Zubair, H., Srivastava, S.K., Singh, S., Singh, A.P., 2015b. Insights into the role of microRNAs in pancreatic cancer pathogenesis: potential for diagnosis, prognosis, and therapy. Adv. Exp. Med. Biol. 889, 71–87.

Kleeff, J., Korc, M., Apte, M., La Vecchia, C., Johnson, C.D., Biankin, A.V., Neale, R.E., Tempero, M., Tuveson, D.A., Hruban, R.H., Neoptolemos, J.P., 2016. Pancreatic cancer. Nat. Rev. Dis. Primers 2, 16022.

Klieser, E., Urbas, R., Stattner, S., Primavesi, F., Jager, T., Dinnewitzer, A., Mayr, C., Kiesslich, T., Holzmann, K., Di Fazio, P., Neureiter, D., Swierczynski, S., 2017. Comprehensive immunohistochemical analysis of histone deacetylases in pancreatic neuroendocrine tumors: HDAC5 as a predictor of poor clinical outcome. Hum. Pathol. 65, 41–52.

Kojima, T., Kyuno, D., Sawada, N., 2012. Targeting claudin-4 in human pancreatic cancer. Expert Opin. Ther. Targets 16, 881–887.

Kong, D., Heath, E., Chen, W., Cher, M.L., Powell, I., Heilbrun, L., Li, Y., Ali, S., Sethi, S., Hassan, O., Hwang, C., Gupta, N., Chitale, D., Sakr, W.A., Menon, M., Sarkar, F.H., 2012. Loss of let-7 up-regulates EZH2 in prostate cancer consistent with the acquisition of cancer stem cell signatures that are attenuated by BR-DIM. PLoS One 7, e33729.

Koushik, A., Spiegelman, D., Albanes, D., Anderson, K.E., Bernstein, L., Van Den Brandt, P.A., Bergkvist, L., English, D.R., Freudenheim, J.L., Fuchs, C.S., Genkinger, J.M., Giles, G.G., Goldbohm, R.A., Horn-Ross, P.L., Mannisto, S., Mccullough, M.L., Millen, A.E., Miller, A.B., Robien, K., Rohan, T.E., Schatzkin, A., Shikany, J.M., Stolzenberg-Solomon, R.Z., Willett, W.C., Wolk, A., Ziegler, R.G., Smith-Warner, S.A., 2012. Intake of fruits and vegetables and risk of pancreatic cancer in a pooled analysis of 14 cohort studies. Am. J. Epidemiol. 176, 373–386.

Kugel, S., Sebastian, C., Fitamant, J., Ross, K.N., Saha, S.K., Jain, E., Gladden, A., Arora, K.S., Kato, Y., Rivera, M.N., Ramaswamy, S., Sadreyev, R.I., Goren, A., Deshpande, V., Bardeesy, N., Mostoslavsky, R., 2016. SIRT6 suppresses pancreatic cancer through control of Lin28b. Cell 165, 1401–1415.

Kumazoe, M., Takai, M., Hiroi, S., Takeuchi, C., Yamanouchi, M., Nojiri, T., Onda, H., Bae, J., Huang, Y., Takamatsu, K., Yamashita, S., Yamada, S., Kangawa, K., Takahashi, T., Tanaka, H., Tachibana, H., 2017. PDE3 inhibitor and EGCG combination treatment suppress cancer stem cell properties in pancreatic ductal adenocarcinoma. Sci. Rep. 7, 1917.

Kunej, T., Godnic, I., Ferdin, J., Horvat, S., Dovc, P., Calin, G.A., 2011. Epigenetic regulation of microRNAs in cancer: an integrated review of literature. Mutat. Res. 717, 77–84.

Larsson, S.C., Bergkvist, L., Wolk, A., 2006. Consumption of sugar and sugar-sweetened foods and the risk of pancreatic cancer in a prospective study. Am. J. Clin. Nutr. 84, 1171–1176.

Lee, W.J., Zhu, B.T., 2006. Inhibition of DNA methylation by caffeic acid and chlorogenic acid, two common catechol-containing coffee polyphenols. Carcinogenesis 27, 269–277.

Li, A., Omura, N., Hong, S.M., Goggins, M., 2010. Pancreatic cancer DNMT1 expression and sensitivity to DNMT1 inhibitors. Cancer Biol. Ther. 9, 321–329.

Lin, W., Watanabe, H., Peng, S., Francis, J.M., Kaplan, N., Pedamallu, C.S., Ramachandran, A., Agoston, A., Bass, A.J., Meyerson, M., 2015. Dynamic epigenetic regulation by menin during pancreatic islet tumor formation. Mol. Cancer Res. 13, 689–698.

Liu, S.Z., Chen, W.Q., Wang, N., Yin, M.M., Sun, X.B., He, Y.T., 2014. Dietary factors and risk of pancreatic cancer: a multi-centre case-control study in China. Asian Pac. J. Cancer Prev. 15, 7947–7950.

Lu, Q.Y., Zhang, L., Yee, J.K., Go, V.W., Lee, W.N., 2015. Metabolic consequences of LDHA inhibition by epigallocatechin gallate and Oxamate in MIA PaCa-2 pancreatic cancer cells. Metabolomics 11, 71–80.

Lynch, S.M., Vrieling, A., Lubin, J.H., Kraft, P., Mendelsohn, J.B., Hartge, P., Canzian, F., Steplowski, E., Arslan, A.A., Gross, M., Helzlsouer, K., Jacobs, E.J., Lacroix, A., Petersen, G., Zheng, W., Albanes, D.,

Amundadottir, L., Bingham, S.A., Boffetta, P., Boutron-Ruault, M.C., Chanock, S.J., Clipp, S., Hoover, R.N., Jacobs, K., Johnson, K.C., Kooperberg, C., Luo, J., Messina, C., Palli, D., Patel, A.V., Riboli, E., Shu, X.O., Rodriguez Suarez, L., Thomas, G., Tjonneland, A., Tobias, G.S., Tong, E., Trichopoulos, D., Virtamo, J., Ye, W., Yu, K., Zeleniuch-Jacquette, A., Bueno-De-Mesquita, H.B., Stolzenberg-Solomon, R.Z., 2009. Cigarette smoking and pancreatic cancer: a pooled analysis from the pancreatic cancer cohort consortium. Am. J. Epidemiol. 170, 403–413.

Mallen-St Clair, J., Soydaner-Azeloglu, R., Lee, K.E., Taylor, L., Livanos, A., Pylayeva-Gupta, Y., Miller, G., Margueron, R., Reinberg, D., Bar-Sagi, D., 2012. EZH2 couples pancreatic regeneration to neoplastic progression. Genes Dev. 26, 439–444.

Marmorstein, R., Zhou, M.M., 2014. Writers and readers of histone acetylation: structure, mechanism, and inhibition. Cold Spring Harb. Perspect. Biol. 6, a018762.

Mccleary-Wheeler, A.L., Lomberk, G.A., Weiss, F.U., Schneider, G., Fabbri, M., Poshusta, T.L., Dusetti, N.J., Baumgart, S., Iovanna, J.L., Ellenrieder, V., Urrutia, R., Fernandez-Zapico, M.E., 2013. Insights into the epigenetic mechanisms controlling pancreatic carcinogenesis. Cancer Lett. 328, 212–221.

Mcdonald, O.G., Li, X., Saunders, T., Tryggvadottir, R., Mentch, S.J., Warmoes, M.O., Word, A.E., Carrer, A., Salz, T.H., Natsume, S., Stauffer, K.M., Makohon-Moore, A., Zhong, Y., Wu, H., Wellen, K.E., Locasale, J.W., Iacobuzio-Donahue, C.A., Feinberg, A.P., 2017. Epigenomic reprogramming during pancreatic cancer progression links anabolic glucose metabolism to distant metastasis. Nat. Genet. 49, 367–376.

Mishra, V.K., Wegwitz, F., Kosinsky, R.L., Sen, M., Baumgartner, R., Wulff, T., Siveke, J.T., Schildhaus, H.U., Najafova, Z., Kari, V., Kohlhof, H., Hessmann, E., Johnsen, S.A., 2017. Histone deacetylase class-I inhibition promotes epithelial gene expression in pancreatic cancer cells in a BRD4- and MYC-dependent manner. Nucleic Acids Res. 45, 6334–6349.

Mody, H.R., Hung, S.W., Alsaggar, M., Griffin, J., Govindarajan, R., 2016. Inhibition of S-Adenosy lmethionine-Dependent methyltransferase attenuates TGFbeta1-induced EMT and metastasis in pancreatic cancer: putative roles of miR-663a and miR-4787-5p. Mol. Cancer Res. 14, 1124–1135.

Mohammadi-Bardbori, A., Akbarizadeh, A.R., Delju, F., Rannug, A., 2016. Chromatin remodeling by curcumin alters endogenous aryl hydrocarbon receptor signaling. Chem. Biol. Interact. 252, 19–27.

Mottamal, M., Zheng, S., Huang, T.L., Wang, G., 2015. Histone deacetylase inhibitors in clinical studies as templates for new anticancer agents. Molecules 20, 3898–3941.

Mueller, N.T., Odegaard, A., Anderson, K., Yuan, J.M., Gross, M., Koh, W.P., Pereira, M.A., 2010. Soft drink and juice consumption and risk of pancreatic cancer: the Singapore Chinese Health Study. Cancer Epidemiol. Biomark. Prev. 19, 447–455.

Muller, S., Filippakopoulos, P., Knapp, S., 2011. Bromodomains as therapeutic targets. Expert Rev. Mol. Med. 13, e29.

Muta, K., Obata, Y., Oka, S., Abe, S., Minami, K., Kitamura, M., Endo, D., Koji, T., Nishino, T., 2016. Curcumin ameliorates nephrosclerosis via suppression of histone acetylation independent of hypertension. Nephrol. Dial. Transplant. 31, 1615–1623.

Nagaraju, G.P., Zhu, S., Wen, J., Farris, A.B., Adsay, V.N., Diaz, R., Snyder, J.P., Mamoru, S., El-Rayes, B.F., 2013. Novel synthetic curcumin analogues EF31 and UBS109 are potent DNA hypomethylating agents in pancreatic cancer. Cancer Lett. 341, 195–203.

Neesse, A., Griesmann, H., Gress, T.M., Michl, P., 2012. Claudin-4 as therapeutic target in cancer. Arch. Biochem. Biophys. 524, 64–70.

Nones, K., Waddell, N., Song, S., Patch, A.M., Miller, D., Johns, A., Wu, J., Kassahn, K.S., Wood, D., Bailey, P., Fink, L., Manning, S., Christ, A.N., Nourse, C., Kazakoff, S., Taylor, D., Leonard, C., Chang, D.K., Jones, M.D., Thomas, M., Watson, C., Pinese, M., Cowley, M., Rooman, I., Pajic, M., Butturini, G., Malpaga, A., Corbo, V., Crippa, S., Falconi, M., Zamboni, G., Castelli, P., Lawlor, R.T., Gill, A.J., Scarpa, A., Pearson, J.V., Biankin, A.V., Grimmond, S.M., 2014. Genome-wide DNA methylation patterns in pancreatic ductal adenocarcinoma reveal epigenetic deregulation of SLIT-ROBO, ITGA2 and MET signaling. Int. J. Cancer 135, 1110–1118.

Omura, N., Goggins, M., 2009. Epigenetics and epigenetic alterations in pancreatic cancer. Int. J. Clin. Exp. Pathol. 2, 310–326.

Ono, H., Basson, M.D., Ito, H., 2016. P300 inhibition enhances gemcitabine-induced apoptosis of pancreatic cancer. Oncotarget 7, 51301–51310.

Ouaissi, M., Silvy, F., Loncle, C., Ferraz Da Silva, D., Martins Abreu, C., Martinez, E., Berthezene, P., Cadra, S., Le Treut, Y.P., Hardwigsen, J., Sastre, B., Sielezneff, I., Benkoel, L., Delgrande, J., Ouaissi, A., Iovanna, J., Lombardo, D., Mas, E., 2014. Further characterization of HDAC and SIRT gene expression patterns in pancreatic cancer and their relation to disease outcome. PLoS One 9, e108520.

Ougolkov, A.V., Bilim, V.N., Billadeau, D.D., 2008. Regulation of pancreatic tumor cell proliferation and chemoresistance by the histone methyltransferase enhancer of zeste homologue 2. Clin. Cancer Res. 14, 6790–6796.

Oya, Y., Mondal, A., Rawangkan, A., Umsumarng, S., Iida, K., Watanabe, T., Kanno, M., Suzuki, K., Li, Z., Kagechika, H., Shudo, K., Fujiki, H., Suganuma, M., 2017. Down-regulation of histone deacetylase 4, -5 and -6 as a mechanism of synergistic enhancement of apoptosis in human lung cancer cells treated with the combination of a synthetic retinoid, Am80 and green tea catechin. J. Nutr. Biochem. 42, 7–16.

Paluszkiewicz, P., Smolinska, K., Debinska, I., Turski, W.A., 2012. Main dietary compounds and pancreatic cancer risk. The quantitative analysis of case-control and cohort studies. Cancer Epidemiol. 36, 60–67.

Papoutsis, A.J., Borg, J.L., Selmin, O.I., Romagnolo, D.F., 2012. BRCA-1 promoter hypermethylation and silencing induced by the aromatic hydrocarbon receptor-ligand TCDD are prevented by resveratrol in MCF-7 cells. J. Nutr. Biochem. 23, 1324–1332.

Park, J.K., Ryu, J.K., Yoon, W.J., Lee, S.H., Lee, G.Y., Jeong, K.S., Kim, Y.T., Yoon, Y.B., 2012. The role of quantitative NPTX2 hypermethylation as a novel serum diagnostic marker in pancreatic cancer. Pancreas 41, 95–101.

Pedersen, K.S., Bamlet, W.R., Oberg, A.L., De Andrade, M., Matsumoto, M.E., Tang, H., Thibodeau, S.N., Petersen, G.M., Wang, L., 2011. Leukocyte DNA methylation signature differentiates pancreatic cancer patients from healthy controls. PLoS One 6, e18223.

Peng, D.F., Kanai, Y., Sawada, M., Ushijima, S., Hiraoka, N., Kitazawa, S., Hirohashi, S., 2006. DNA methylation of multiple tumor-related genes in association with overexpression of DNA methyltransferase 1 (DNMT1) during multistage carcinogenesis of the pancreas. Carcinogenesis 27, 1160–1168.

Peng, L., Liu, Z., Xiao, J., Tu, Y., Wan, Z., Xiong, H., Li, Y., Xiao, W., 2017. MicroRNA-148a suppresses epithelial-mesenchymal transition and invasion of pancreatic cancer cells by targeting Wnt10b and inhibiting the Wnt/beta-catenin signaling pathway. Oncol. Rep. 38, 301–308.

Perepechaeva, M.L., Seredina, T.A., Sidorova, Y.A., Pivovarova, E.N., Markel, A.L., Lyakhovich, V.V., Grishanova, A.Y., 2017. Quercetin attenuates benzo(alpha)pyrene-induced CYP1A expression. Biomed. Environ. Sci. 30, 308–313.

Piciucchi, M., Capurso, G., Valente, R., Larghi, A., Archibugi, L., Signoretti, M., Stigliano, S., Zerboni, G., Barucca, V., La Torre, M., Cavallini, M., Costamagna, G., Marchetti, P., Ziparo, V., Delle Fave, G., 2015. Early onset pancreatic cancer: risk factors, presentation and outcome. Pancreatology 15, 151–155.

Piunti, A., Pasini, D., 2011. Epigenetic factors in cancer development: polycomb group proteins. Future Oncol. 7, 57–75.

Raju, I., Kaushal, G.P., Haun, R.S., 2016. Epigenetic regulation of KLK7 gene expression in pancreatic and cervical cancer cells. Biol. Chem. 397, 1135–1146.

Ramani, V.C., Haun, R.S., 2008. Expression of kallikrein 7 diminishes pancreatic cancer cell adhesion to vitronectin and enhances urokinase-type plasminogen activator receptor shedding. Pancreas 37, 399–404.

Rao, R.C., Dou, Y., 2015. Hijacked in cancer: the KMT2 (MLL) family of methyltransferases. Nat. Rev. Cancer 15, 334–346.

Razidlo, G.L., Magnine, C., Sletten, A.C., Hurley, R.M., Almada, L.L., Fernandez-Zapico, M.E., Ji, B., Mcniven, M.A., 2015. Targeting pancreatic cancer metastasis by inhibition of Vav1, a driver of tumor cell invasion. Cancer Res. 75, 2907–2915.

Rozek, L.S., Dolinoy, D.C., Sartor, M.A., Omenn, G.S., 2014. Epigenetics: relevance and implications for public health. Annu. Rev. Public Health 35, 105–122.

Sato, N., Fukushima, N., Hruban, R.H., Goggins, M., 2008. CpG island methylation profile of pancreatic intraepithelial neoplasia. Mod. Pathol. 21, 238–244.

Sato, N., Fukushima, N., Maitra, A., Matsubayashi, H., Yeo, C.J., Cameron, J.L., Hruban, R.H., Goggins, M., 2003a. Discovery of novel targets for aberrant methylation in pancreatic carcinoma using high-throughput microarrays. Cancer Res. 63, 3735–3742.

Sato, N., Goggins, M., 2006. The role of epigenetic alterations in pancreatic cancer. J. Hepatobiliary Pancreat. Surg. 13, 286–295.

Sato, N., Maitra, A., Fukushima, N., Van Heek, N.T., Matsubayashi, H., Iacobuzio-Donahue, C.A., Rosty, C., Goggins, M., 2003b. Frequent hypomethylation of multiple genes overexpressed in pancreatic ductal adenocarcinoma. Cancer Res. 63, 4158–4166.

Sauvageau, M., Sauvageau, G., 2010. Polycomb group proteins: multi-faceted regulators of somatic stem cells and cancer. Cell Stem Cell 7, 299–313.

Sen, G.L., Reuter, J.A., Webster, D.E., Zhu, L., Khavari, P.A., 2010. DNMT1 maintains progenitor function in self-renewing somatic tissue. Nature 463, 563–567.

Shankar, S., Kumar, D., Srivastava, R.K., 2013. Epigenetic modifications by dietary phytochemicals: implications for personalized nutrition. Pharmacol. Ther. 138, 1–17.

Shigihara, M., Obara, T., Nagai, M., Sugawara, Y., Watanabe, T., Kakizaki, M., Nishino, Y., Kuriyama, S., Tsuji, I., 2014. Consumption of fruits, vegetables, and seaweeds (sea vegetables) and pancreatic cancer risk: the Ohsaki Cohort Study. Cancer Epidemiol. 38, 129–136.

Shimada, T., 2017. Inhibition of carcinogen-activating cytochrome P450 enzymes by xenobiotic chemicals in relation to antimutagenicity and anticarcinogenicity. Toxicol Res. 33, 79–96.

Siegel, R.L., Miller, K.D., Jemal, A., 2018. Cancer statistics, 2018. CA Cancer J. Clin. 68, 7–30.

Simon, R.P., Robaa, D., Alhalabi, Z., Sippl, W., Jung, M., 2016. Katching-up on small molecule modulators of lysine acetyltransferases. J. Med. Chem. 59, 1249–1270.

Srivastava, S.K., Arora, S., Singh, S., Bhardwaj, A., Averett, C., Singh, A.P., 2014. MicroRNAs in pancreatic malignancy: progress and promises. Cancer Lett. 347, 167–174.

Srivastava, S.K., Bhardwaj, A., Arora, S., Singh, S., Azim, S., Tyagi, N., Carter, J.E., Wang, B., Singh, A.P., 2015. MYB is a novel regulator of pancreatic tumour growth and metastasis. Br. J. Cancer 113, 1694–1703.

Stenzinger, A., Endris, V., Klauschen, F., Sinn, B., Lorenz, K., Warth, A., Goeppert, B., Ehemann, V., Muckenhuber, A., Kamphues, C., Bahra, M., Neuhaus, P., Weichert, W., 2013. High SIRT1 expression is a negative prognosticator in pancreatic ductal adenocarcinoma. BMC Cancer 13, 450.

Strahl, B.D., Allis, C.D., 2000. The language of covalent histone modifications. Nature 403, 41–45.

Talamini, R., Polesel, J., Gallus, S., Dal Maso, L., Zucchetto, A., Negri, E., Bosetti, C., Lucenteforte, E., Boz, G., Franceschi, S., Serraino, D., La Vecchia, C., 2010. Tobacco smoking, alcohol consumption and pancreatic cancer risk: a case-control study in Italy. Eur. J. Cancer 46, 370–376.

Tang, X., Milyavsky, M., Shats, I., Erez, N., Goldfinger, N., Rotter, V., 2004. Activated p53 suppresses the histone methyltransferase EZH2 gene. Oncogene 23, 5759–5769.

Taunk, P., Hecht, E., Stolzenberg-Solomon, R., 2016. Are meat and heme iron intake associated with pancreatic cancer? Results from the NIH-AARP diet and health cohort. Int. J. Cancer 138, 2172–2189.

Venturelli, S., Berger, A., Bocker, A., Busch, C., Weiland, T., Noor, S., Leischner, C., Schleicher, S., Mayer, M., Weiss, T.S., Bischoff, S.C., Lauer, U.M., Bitzer, M., 2013. Resveratrol as a pan-HDAC inhibitor alters the acetylation status of histone [corrected] proteins in human-derived hepatoblastoma cells. PLoS One 8, e73097.

Vignali, M., Hassan, A.H., Neely, K.E., Workman, J.L., 2000. ATP-dependent chromatin-remodeling complexes. Mol. Cell Biol. 20, 1899–1910.

Vincent, A., Omura, N., Hong, S.M., Jaffe, A., Eshleman, J., Goggins, M., 2011. Genome-wide analysis of promoter methylation associated with gene expression profile in pancreatic adenocarcinoma. Clin. Cancer Res. 17, 4341–4354.

Vo, N., Goodman, R.H., 2001. CREB-binding protein and p300 in transcriptional regulation. J. Biol. Chem. 276, 13505–13508.

Waddington, C.H., 2012. The epigenotype. 1942. Int. J. Epidemiol. 41, 10–13.

Wang, Y., Wang, Y., Luo, M., Wu, H., Kong, L., Xin, Y., Cui, W., Zhao, Y., Wang, J., Liang, G., Miao, L., Cai, L., 2015a. Novel curcumin analog C66 prevents diabetic nephropathy via JNK pathway with the involvement of p300/CBP-mediated histone acetylation. Biochim. Biophys. Acta 1852, 34–46.

Wang, Y.H., Sui, Y.N., Yan, K., Wang, L.S., Wang, F., Zhou, J.H., 2015b. BRD4 promotes pancreatic ductal adenocarcinoma cell proliferation and enhances gemcitabine resistance. Oncol. Rep. 33, 1699–1706.

Wei, Y., Xia, W., Zhang, Z., Liu, J., Wang, H., Adsay, N.V., Albarracin, C., Yu, D., Abbruzzese, J.L., Mills, G.B., Bast Jr., R.C., Hortobagyi, G.N., Hung, M.C., 2008. Loss of trimethylation at lysine 27 of histone H3 is a predictor of poor outcome in breast, ovarian, and pancreatic cancers. Mol. Carcinog. 47, 701–706.

Wen, B., Wu, H., Shinkai, Y., Irizarry, R.A., Feinberg, A.P., 2009. Large histone H3 lysine 9 dimethylated chromatin blocks distinguish differentiated from embryonic stem cells. Nat. Genet. 41, 246–250.

Xie, V.K., Li, Z., Yan, Y., Jia, Z., Zuo, X., Zhenjin, J., Wang, J., Du, J., Xie, D., Wei, D., Xie, K., 2017. DNA-methyltransferase 1 induces dedifferentiation of pancreatic cancer cells through silencing of Kruppel-like factor 4 expression. Clin. Cancer Res. 23, 5585–5597.

Xu, Q., Zong, L., Chen, X., Jiang, Z., Nan, L., Li, J., Duan, W., Lei, J., Zhang, L., Ma, J., Li, X., Wang, Z., Wu, Z., Ma, Q., Ma, Z., 2015. Resveratrol in the treatment of pancreatic cancer. Ann. N.Y. Acad. Sci. 1348, 10–19.

Yamamoto, K., Tateishi, K., Kudo, Y., Hoshikawa, M., Tanaka, M., Nakatsuka, T., Fujiwara, H., Miyabayashi, K., Takahashi, R., Tanaka, Y., Ijichi, H., Nakai, Y., Isayama, H., Morishita, Y., Aoki, T., Sakamoto, Y., Hasegawa, K., Kokudo, N., Fukayama, M., Koike, K., 2016. Stromal remodeling by the BET bromodomain inhibitor JQ1 suppresses the progression of human pancreatic cancer. Oncotarget 7, 61469–61484.

Yang, X., Karuturi, R.K., Sun, F., Aau, M., Yu, K., Shao, R., Miller, L.D., Tan, P.B., Yu, Q., 2009. CDKN1C (p57) is a direct target of EZH2 and suppressed by multiple epigenetic mechanisms in breast cancer cells. PLoS One 4, e5011.

Zagorac, S., Alcala, S., Fernandez Bayon, G., Bou Kheir, T., Schoenhals, M., Gonzalez-Neira, A., Fernandez Fraga, M., Aicher, A., Heeschen, C., Sainz Jr., B., 2016. DNMT1 inhibition reprograms pancreatic cancer stem cells via upregulation of the miR-17-92 cluster. Cancer Res. 76, 4546–4558.

Zhang, L., Gao, J., Li, L., Li, Z., Du, Y., Gong, Y., 2011. The neuronal pentraxin II gene (NPTX2) inhibit proliferation and invasion of pancreatic cancer cells in vitro. Mol. Biol. Rep. 38, 4903–4911.

Zhang, Y., Wang, X., Han, L., Zhou, Y., Sun, S., 2015. Green tea polyphenol EGCG reverse cisplatin resistance of A549/DDP cell line through candidate genes demethylation. Biomed. Pharmacother. 69, 285–290.

Zheng, W., Lee, S.A., 2009. Well-done meat intake, heterocyclic amine exposure, and cancer risk. Nutr. Cancer 61, 437–446.

Zubair, H., Azim, S., Ahmad, A., Khan, M.A., Patel, G.K., Singh, S., Singh, A.P., 2017. Cancer chemoprevention by phytochemicals: nature's healing touch. Molecules 22.

CHAPTER

13

Phytochemicals as Epigenetic Modifiers for Cancer Management With Special Reference to Lung Cancer

Tamilselvam Rajavel, Kasi Pandima Devi

Department of Biotechnology, Alagappa University (Science Campus), Karaikudi, India

Abstract

Lung cancer is one of the deadliest types of cancers and a major health burden around the globe. It has a low survival rate due to late diagnosis of the disease. In addition, poor understanding and the multifactorial nature of the disease lead to low treatment outcomes and the failure of chemotherapeutic drugs. Recent findings suggest that epigenetic alteration contributes to the genesis of lung cancer development and chemoresistance. Interestingly, different kinds of epigenetic alterations have occurred in the different stages of cancer development, such as initiation, promotion and metastasis. The term "epigenetics" refers to reversible, heritable genetic changes without causing any alterations in the genetic code of DNA. Epigenetic modifications include DNA methylation, histone modification, and noncoding microRNA (miRNA) expression. Numerous phytochemicals identified from dietary foods and medicinal plants have been found to act against a vast number of cancer models. Interestingly, molecular studies have concluded that phytochemicals abrogate cancer growth by modulating several epigenetic mechanisms, such as DNA hypermethylation, miRNA overexpression, and histone modification. In this chapter, we focus on common epigenetic modifications in lung cancer and other carcinogenesis, highlighting the role of phytochemicals in modulating therapeutically important epigenetic modifications in human lung cancer and other cancer cell models.

Keywords

DNA methylation; Epigenetics; Histone modification; Lung cancer; MiRNA; Phytochemicals

INTRODUCTION

Lung cancer cases have become more common around the world, and this aggressive malignant tumor results in increased death rates every year. Global statistical reports have revealed that lung cancer prevalence is common in both developed and developing countries. For instance, the International Agency for Research on Cancer (IARC) estimated nearly 1.8 million new lung cancer cases in 2012, of which 58% occurred in less-developed countries (Jemal et al., 2010). Based on histology, lung cancer has been classified into small-cell lung cancer (SCLC) and non-small-cell lung cancer (NSCLC),

Epigenetics of Cancer Prevention
https://doi.org/10.1016/B978-0-12-812494-9.00013-5

of which the NSCLC subtype is quite common and has an extremely low 5-year survival rate (15%) during late diagnosis. Traditionally, cancer development occurs through the accumulation of genetic defects, such as mutations, amplifications, translocation, and deletions. These genetic changes alter the normal cell machinery and contribute to cancer cell survival and migration. Further, interactions between tumor cells and the microenvironment also support the growth of cancer. More recently, intriguing evidence has suggested that cancer epigenetic modifications are ideal alternative targets, rather than targeting genetic mutations, in human cancer treatment (Dawson and Kouzarides., 2012; Mai, 2014; Jones et al., 2016). Epigenetic modification refers to studying the reversible genetic changes that occur during the process of carcinogenesis (Inbar-Feigenberg et al., 2013). DNA methylation, chromatin condensation, and noncoding microRNA (miRNA) expression are commonly associated with epigenetic modification and play a key role in all stages of cancer development. Importantly, epigenetic modification affects the functions of macromolecules, such as DNA, RNA, and proteins. For instance, DNA methylation suppresses the transcriptional activity of particular gene functions through the gene-silencing process without causing any changes in genetic code. Hence, reactivation of gene function would be possible if drugs could target epigenetic processes in cancer cells (Wee et al., 2014). Due to their reversible nature, epigenetic modifications have gained significant attention from the scientific community in recent years. Studies also suggest that epigenetic-targeted therapies will be more potent than current chemotherapeutical treatments in terms of clinical outcome. Moreover, several small molecules recently have been approved by the United States Food and Drug Administration (FDA) for chemotherapeutical treatments. For example, the DNA methyltransferase (DNMT) inhibitors azacitidine and decitabine have been used in human leukemic cancer treatment (Dhanak and Jackson, 2014). Similarly, a number of epigenetic-targeting drugs are underway in clinical trials against different kinds of solid tumors (Nervi et al., 2015). In addition, synergistic combinations of conventional chemotherapeutical drugs and epigenetic-targeting drugs are relatively promising in cancer patients (Qin et al., 2015). Recently, numerous phytochemicals have been reported as inhibiting cancer growth by modulating the cancer epigenetic mechanism. The present review highlights potential targets of phytochemicals on cancer epigenetic pathways in both *in vitro* and *in vivo* lung cancer and other cancer models.

EPIGENETIC MODIFICATION IN LUNG CANCER DNA METHYLATION

DNA methylation–mediated gene silencing is the most extensively studied pathogenic hallmark of lung cancer tissues (Zochbauer-Muller, 2002). In addition, epigenetic inactivation of certain tumor-suppressor genes has been associated with aberrant hypermethylation of the promoter region. DNA methylation is the most extensively studied epigenetic mechanism in the mammalian system that results in suppressed gene expression or gene silencing (Fig. 13.1A). This process is catalyzed by a DNMT enzyme that transfers a methyl group from S-adenosyl-L-methionine (SAM) to cytosine. Four different DNMT enzymes have been identified in mammals: DNMT1, DNMT2, DNMT3A, and DNMT3B (Jin and Robertson, 2013). Among these, DNMT3A and DNMT3B play crucial roles in lung cancer carcinogenesis (Jia et al., 2016). Moreover, the addition of a methyl group to the cytosine nucleotide, especially next to guanosine in the CpG island region, could alter transcription efficacy and chromatin-associated proteins of particular genes in the genome. Methylated DNA also activates methyl-binding proteins

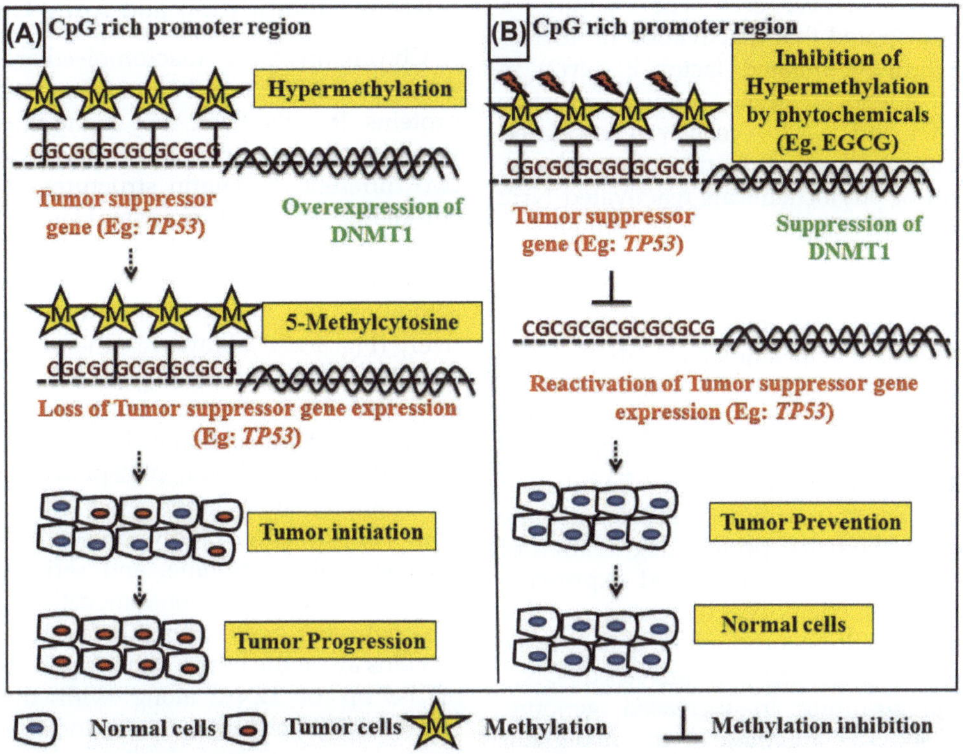

FIGURE 13.1 **Schematic representation showing DNA methylation on tumor development and the role of phytochemicals in cancer epigenetic therapy.** (A) Overview of malignant transformation of normal cells by aberrant DNA methylation and inactivation of the tumor-suppressor gene function. (B) Blocking the overexpression of DNMT1 by phytochemicals and reactivating the tumor-suppressor gene function.

that further suppress gene function through the histone deacetylation mechanism (Jones and Baylin, 2002). CpG islands are conserved sequences clustered within the entire human genome, especially toward the 5′ end of the gene, with an average length of 0.5–4 kb. Additionally, 60% of the human genes in the genome have CpG-rich repeats on their promoter regions. Methylation of CpG islands controls the DNA-replication process in proliferating cells, and the aberrant promoter hypermethylation pattern contributes the driving event in lung tumorigenesis. Studies also suggest that methylated CpG islands are promising biomarkers for early detection of lung cancer (Rauch et al., 2012).

During the past few years, numerous promoter-hypermethylated genes have been reported in clinical lung tumor samples. A pilot study report on 766 genes found differential DNA methylation patterns in human lung adenocarcinoma samples. Among those, 57 genes gained function by hypomethylation and 164 genes were down-regulated by hypermethylation with respect to nonmalignant samples. This study also highlighted certain novel hyperactivated genes for epigenetic therapy, such as ABCA3 (involved in lung development), SOX17 (tumor-suppressor protein), and TMEM204 (cell adhesion) (Selamat et al., 2012). However, blocking promoter methylation restored the expression of genes associated with cancer cells. For example, a study demonstrated that inactivation of nuclear factor erythroid 2–related factor 2 (Nrf2) in transgenic adenocarcinoma mouse prostate (TRAMP) C1 and JB6 P+ cells was associated with promoter hypermethylation. Surprisingly, cotreatment of phytochemicals reactivated Nrf2 expression by blocking promoter hypermethylation, which suggested an alternative approach in cancer epigenetic therapy (Zhang et al., 2013; Yu et al., 2010). DNA methylation occurs at a higher frequency in smoking-related cancer cases (Liu et al., 2006). Overexpression of the enzyme DNA (cytosine-5)-methyltransferase 1 (DNMT1) was found in patients having smoking habits. Interestingly, studies found that tobacco-specific nitrosamines prevented the ubiquitination-mediated protein degradation of DNMT1 that resulted in increased expression (Lin et al., 2010). Moreover, increased reactive oxygen species (ROS) production and chronic inflammation contributed to increased DNA methylation resulting in increased genomic instability and the initiation of the malignant transformation of cells (O'Hagan et al., 2011). Two cigarette smoke carcinogens, namely methylnitrosourea (MNU) and benzo(a)pyrene-diolepoxide 1 (BPDE), were exposed to normal human bronchial epithelial cells (HBECs) for 12 weeks to demonstrate the role of DNMT1 in smoking-mediated lung epithelial cancer. Stable knockdown of DNMT1 was observed in HBECs, which prevented the malignant transformation of cells, whereas wild DNMT1 containing HBECs showed transformation of the fibroblast-like mesenchymal appearance from epithelial–mesenchymal transition (EMT), and downregulation of EMT-negative regulators such as miR-200b and miR-200c, in the fourth week of exposure to MNU and BPDE (Damiani et al., 2008). This study confirms that DNA methylation is an initial epigenetic modification in the lung carcinogenic pathway.

Histone Modification and Lung Cancer

Chromatin is a macromolecular complex made up of DNA and histone and nonhistone proteins. It is the basic unit of the nucleosomes that play a crucial role in gene expression. The two different chromatin structures are heterochromatin and euchromatin. Heterochromatin and euchromatin are tightly and loosely packed forms of chromatin that control the transcription activation and inactivation processes, respectively (Fig. 13.2A). However, the transcriptional activity of any particular gene is controlled by epigenetic modification of histone proteins through post-translational changes that include acetylation, methylation, phosphorylation, ubiquitinylation, sumoylation, and ADP ribosylation. Additionally, highly conserved residues, such as lysine, arginine, and serine, undergo extensive post-translational modification within the controlled gene expression system. Histone proteins are at the heart of chromatin and consist of 147 bp of DNA along with the histone octamer. This histone octamer includes H3/H4 tetramer and two H2A/H2B dimers.

Histone acetylation is the most dynamic and well-studied post-translational modification in the gene expression system. For example, acetylation of lysine residues in histone proteins alters nucleosomal conformational changes by neutralizing the positive charges that result in the increased accessibility of transcriptional factors. However, gene expression is negatively regulated by the histone deacetylation mechanism. Both histone acetylation and deacetylation are catalyzed by two enzymes, histone acetyltransferase (HAT) and histone deacetylase (HDAC), respectively (Struhl, 1998). Indeed, studies also suggest that gene expression is suppressed by the combined interaction of DNA methylation and histone deacetylation (Jones et al., 1998). Surprisingly, a synergistic effect was demonstrated by combining DNMT inhibitors and

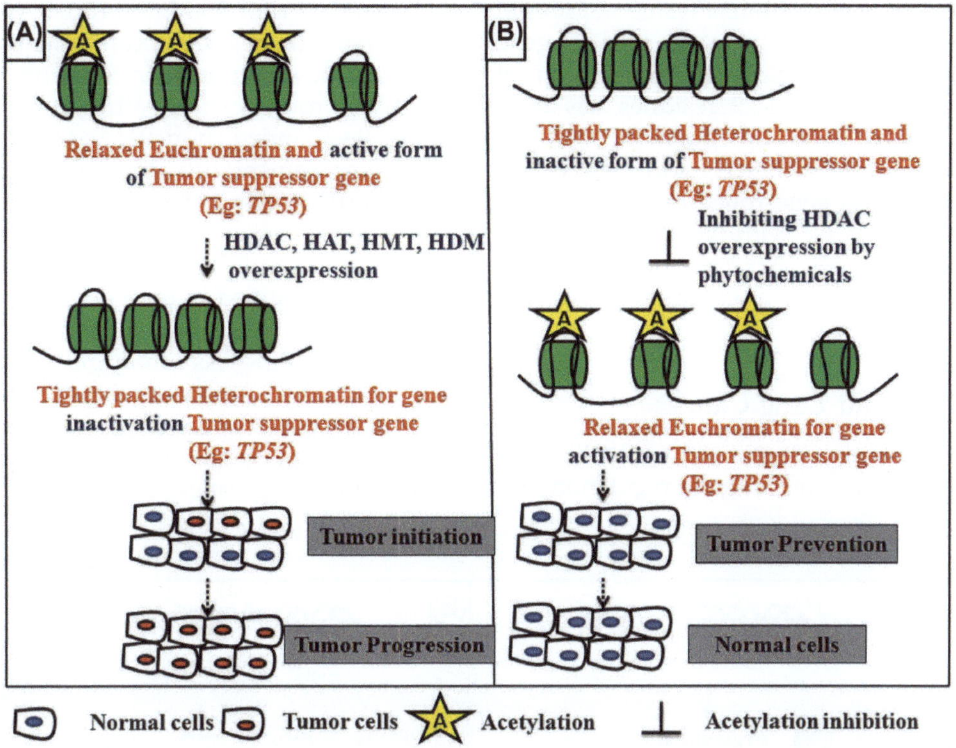

FIGURE 13.2 **Effect of bioactive phytochemicals on cancer-associated histone modification.** (A) Overview of malignant transformation of normal cells by aberrant histone modification and inactivation of tumor-suppressor gene function. (B) Blocking the aberrant expression of histone deacetylase (HDAC) proteins by phytochemicals and reactivating the tumor-suppressor gene function.

HDAC inhibitors, which reactivated certain tumor-suppressor genes (Steele et al., 2009). In clinical cases, hyperactivation of HDAC expression was found in numerous lung cancer cases (Kim et al., 2006). Indeed, knockdown or inhibition of HDAC was found to have a significant anticancer effect in cancer cell lines and animal models. Moreover, the FDA has approved several HDAC inhibitors, such as vorinostat, trichostatin A, and entinostat, for treating numerous types of solid cancers. Carcinogenic elements found in cigarette smoke, including nickel, chromate, and arsenite, also induce histone deacetylation and methylation in lung epithelial cells (Stojanović et al., 2004; Cantone et al., 2011). In addition, acrolein, the major component of cigarette smoke, forms an adduct with free histone proteins and significantly modulates its nucleosome assembly through the deacetylation mechanism. As a result of this interaction, it influences genome instability and initiates lung carcinogenesis (Chen et al., 2013). Recent studies have demonstrated that some dietary phytochemicals act significantly against HDAC expression.

Histone methylation and demethylation also contribute to lung carcinogenesis. Histone methyltransferases and histone demethylases dynamically control gene expression, and aberrant expression was found in human cancer tissues.

By contrast, the methylation of lysine residues in histone protein can activate or repress gene expression. For example, the methylation of H3K4, H3K36, and H3K79 residues has been associated with gene upregulation, whereas methylation of H3K9, H3K27, and H4K20 was found to suppress gene function (Vakoc et al., 2006). Unlike the case with the histone acetylation process, the role of histone methylation in lung carcinogenesis is a complex process to understand.

MicroRNAs and Lung Cancer

As short, endogenous, noncoding RNAs, miRNAs play a crucial role in gene regulation and malignant transformation. In recent years, miRNAs have attracted significant attention within the scientific community in an effort to understand their role in cancer pathogenesis. miRNAs are single-stranded short RNAs (20–22 bp) that bind complementarily to target messenger RNAs (mRNAs) and block their translation process by repression and mRNA degradation. Large numbers of differentially regulated miRNAs were identified in both normal and malignant cells (www.mirbase.org). However, deregulated miRNAs contributed to cancer development by repressing the mRNA expression of tumor-suppressor genes (Fig. 13.3A). Oncogenic expressions of miRNAs are also associated with lung cancer development and are

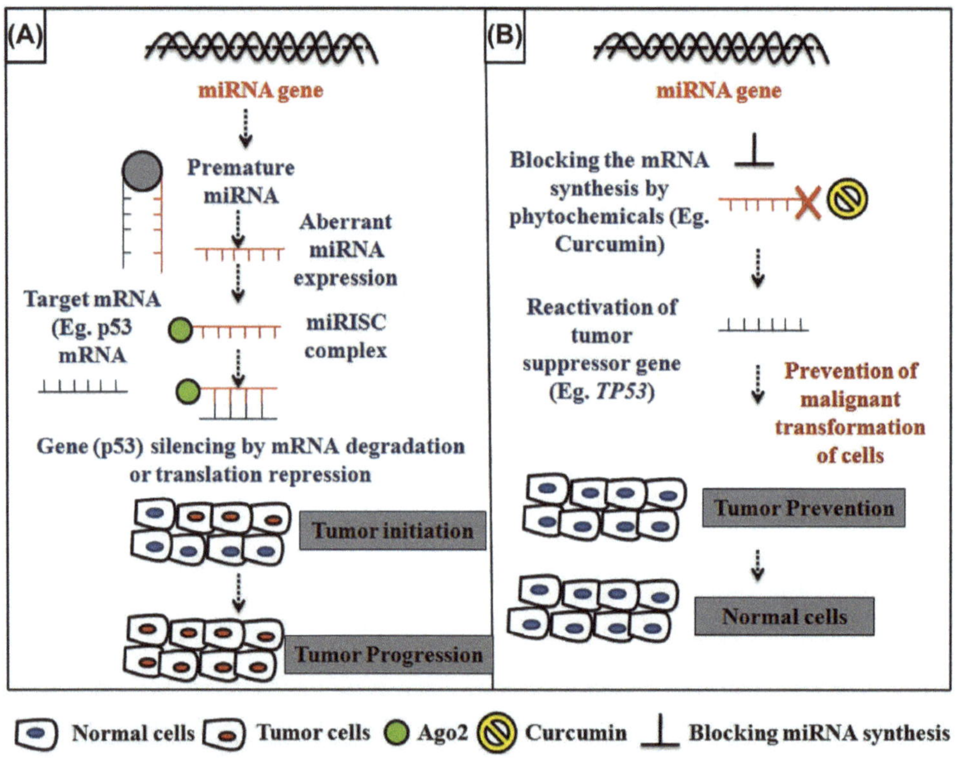

FIGURE 13.3 **Effect of bioactive phytochemicals on cancer-associated microRNAs (miRNAs).** (A) Overview of malignant transformation of normal cells by hyperactivation of miRNAs and inactivation of the tumor-suppressor gene function. (B) Blocking hyperactive miRNAs by phytochemicals and regaining the expression of the tumor-suppressor gene function.

called oncogenic miRNAs. Indeed, suppressed expression of let-7 miRNA was found in human lung cancers. Interestingly, studies have demonstrated that miRNAs can activate or cause overexpression of DNMT1 in lung cancer tissues. They also revealed a complex network of epigenetic processes that are involved in cancer development (Fabbri et al., 2007).

Tumor-suppressor miRNAs negatively regulate gene expression by downregulating the mRNA expression of oncogenic genes, such as RAS, CDK1, WNT, and epidermal growth factor receptor (EGFR) (Johnson et al., 2005). For example, let-7 is the first identified miRNA in a mammalian system later recognized to play a crucial role in tumor suppression and the DNA repair process (Johnson et al., 2007). However, the suppressed effect of the let-7 family of miRNAs was associated with human cancer cases. Indeed, the tumor-suppressor mechanism of let-7 and its family of miRNAs is still unknown. The miR-34 family of tumor-suppressor miRNAs is activated by the TP3 gene during DNA damage, cell cycle arrest, and apoptosis. However, in recent years, much scientific literature has reported tumor-suppressor miRNAs in the mammalian system.

Oncogenic miRNAs or hyperactive miRNAs are commonly observed in many pathogenic conditions, including human cancer. For instance, overexpression of miR-21 has been largely observed in various kinds of human cancers. Studies have found that overexpression of miR-21 acts as a driving event in the carcinogenesis pathway. Surprisingly, miR-21 targets the tumor-suppressor gene phosphatase and tensin homolog (PTEN) that leads to suppression of apoptosis in NSCLC cancer cells (Liu et al., 2013). Apoptosis resistance is a known hallmark in human cancer, and miR-21 indirectly favors the growth and survival of cancer cells. Additionally, the miR-17–92 polycistronic cluster (polycistronic miRNA), miR-221, and miR-222 were also found in hyperactive forms in cancer tissues (Hayashita et al., 2005; Garofalo et al., 2008). In lung cancer cases, increased expression of miR-21 was associated with reduced survival rate and apoptosis resistance (Nasser et al., 2008). In addition, aberrant expression of miR-21 inhibits the expression of many tumor-suppressor proteins such as programmed cell death protein 4 (PDCD4) (Chan et al., 2005). The miR-17–92 cluster was found in an aberrant-expression form, particularly in SCLC cells, and exposure of the miR-17–92 cluster increases the cell proliferation rate in in vitro conditions (Hayashita et al., 2005). Indeed, blocking the expression of miR-17–92 clusters in lung cancer cells through antisense miRNAs selectively induced apoptotic cell death (Matsubara et al., 2007). Recent studies suggest that miRNAs also regulate the expression of growth factors like EGFR. EGFR overexpression was found in numerous clinical lung cancer cases and has become a major therapeutic target in chemotherapy (Pao and Miller, 2005). Deletion of miRNA-128b has frequently been found in lung cancer cases that are correlated with the increased expression of EGFR, suggesting the tumor-suppressor potential of miRNA-128b. Another study demonstrated that let-7 regulates EGFR expression, and inactivation of let-7 contributes to aberrant expression of EGFR (Webster et al., 2009). Indeed, global miRNA expression studies have revealed that several signature miRNAs (miR-137, miR-182, miR-372, miR-155, 17-3p, and let-7a-2) were predicted in the prognosis of patients with lung cancer (Lewis et al., 2005; Yanaihara et al., 2006). Hence, targeting oncogenic and tumor-suppressor miRNAs in chemotherapeutical treatment could prevent the oncogenic transformation of cells.

TARGETING CANCER EPIGENETIC MODIFICATION BY PHYTOCHEMICALS

Over the last 2 decades, phytochemical application has contributed remarkably in cancer chemotherapeutical treatment. Naturally occurring plant secondary metabolites, especially from fruits, vegetables, and medicinal plants, provide health benefits against human cancer through various pharmacological actions, including reducing oxidative stress, promoting apoptosis, blocking the cell cycle, and abrogating cell signaling. Accumulating evidence also has demonstrated that phytochemicals block the malignant transformation of cells by altering epigenetic modification. Indeed, epigenetic therapy has produced significant clinical outcomes because they reverse epigenetic changes (reactivation of silenced tumor-suppressor protein) in cancer cells and eliminate the cells via apoptosis. In the future, the combination of epigenetic therapy and conventional chemotherapy will be a novel approach in human lung cancer management. The following sections summarize phytochemical-based epigenetic therapy in the treatment of lung cancer.

Epigallocatechin-3-Gallate

Downregulated expression of tumor-suppressor genes has been associated with methylation in CpG promoter regions. Moreover, the presence of normal copies of tumor-suppressor genes provides a large platform for cancer treatment. However, conventional chemotherapy kills cancer cells without recovering the function of tumor-suppressor genes. Surprisingly, DNA methylation inhibitors efficiently recover the silenced function of tumor-suppressor genes without causing toxic effects in normal cells. Hence, reducing DNA methylation in cancer cells using a phytochemical-based approach is a more promising approach in lung cancer management (Fig. 13.1A). Tea leaves are rich in antioxidant compounds that are consumed in different forms around the world. Among the different phytochemicals present in tea, epigallocatechin-3-gallate (EGCG) is an abundant catechin that exhibits numerous pharmacological properties. EGCG targets cancer cells in multiple ways including cell cycle arrest, the triggering of apoptosis, and the targeting of cancer angiogenesis (Du et al., 2012). Recent studies have also demonstrated that EGCG targets DNA methylation in various solid tumors. For example, hyperactivation of the Wnt signaling pathway is commonly observed in lung cancer cases. Overexpression of the Wnt protein associated with the methylated Wnt inhibitory factor 1 (WIF-1) gene results in aberrant cell proliferation and survival. Interestingly, treatment with EGCG in lung cancer cell lines H460 and A549 inhibits their proliferation rate by blocking hyperactivation of the Wnt protein and restores WIF-1 gene expression by the demethylating process (Gao et al., 2009). Similarly, numerous studies have reported that EGCG significantly reactivates various tumor-suppressor proteins by decreasing promoter CpG DNA methylation. EGCG was demonstrated to effectively recover silenced function of tumor-suppressor genes in breast cancer cells. EGCG treatment inhibits the function of DNMT1 in both mRNAs and the protein level, resulting in restoration of estrogen receptor (ER) α, progesterone receptor B, TMS1, cyclin D2, and O6-methylguanine-DNMT (MGMT) (Mirza et al., 2013). Moreover, EGCG treatment also recovers the expression of p16, retinoic acid receptor beta (RARβ), MGMT, and human mutL homologue 1 (hMLH1) genes. Western blot analysis also confirmed the upregulation of MGMT and hMLH1in EGCG-treated colon and prostate cancer cells (Fang et al., 2003). Moreover, the results were similar to those of the clinically approved drugs 5-aza-2′-deoxycytidine and zebularine. In addition, molecular modeling studies confirmed that EGCG effectively binds and forms hydrogen bonding in the active site of the DNMT1 substrate pocket that results in the loss of DNMT1 activity. SAM is the methyl donor for DNMT1 methylation activity, and studies have reported that EGCG reduces the level of SAM in breast cancer cells. Similarly, EGCG treatment promotes the DNA methylation inhibitor

S-adenosyl-L-homocysteine. Interestingly, it was found that EGCG decreases telomerase activity in breast cancer cells by downregulating the hTERT gene. Surprisingly, a time-dependent decrease in the methylation pattern was found in the hTERT gene, leading to cell growth inhibition. Overall, these studies confirm that EGCG exerts a chemopreventive effect through direct and indirect inhibition of DNMT1 methylation activity (Shukla and Meeran, 2013).

Apart from its DNA methylation inhibition efficacy, EGCG also inhibits the aberrant expression of miRNAs associated with human cancers. In our previous review article, we reported the significant potential of EGCG against oncogenic miRNAs and the activation of tumor-suppressor miRNAs (Devi et al., 2016). Moreover, numerous anticancer studies have also concluded that EGCG exerts anticancer properties by targeting aberrant expression of miRNAs. For example, studies have demonstrated that miR-16 overexpression is associated with constant activation of antiapoptotic Bcl-2 protein in various human cancers. Interestingly, EGCG treatment induces apoptosis in HepG2 cells by downregulating both miR-16 and bcl-2 expression. Surprisingly, the apoptotic effect of EGCG was abrogated when the cells were transfected with antimiR-16 inhibitor. Overall, this study concluded that EGCG exerts its apoptotic effect by targeting the hyperactivation of miR-16 (Tsang and Kwok, 2010). EGCG-treated NSCLC cells showed inhibition of cell proliferation and cell cycle arrest at the G2/M phase. Moreover, molecular studies have found that EGCG treatment upregulates the expression levels of let-7a-1 and let-7g, which leads to downregulation of oncogenic protein c-Myc. Downregulation of c-Myc contributed to the growth-inhibitory effect of EGCG on HepG2 cells (Zhong et al., 2012). EGCG also restored the tumor-suppressor property of let-7 miRNA in many human cancer cell lines. EGCG exposure in skin melanoma cells upregulated the let-7b tumor-suppressor miRNA. Moreover, the upregulation of let-7b was demonstrated to be associated with the activation of novel signaling pathway cAMP/protein kinase A/protein phosphatase 2A (Yamada et al., 2016). Similarly, EGCG treatment significantly recovered the expression of the let-7 miRNA by blocking overexpression of oncogenic protein K-Ras. Overall, these studies have confirmed that EGCG will be a potent and safe natural anticancer agent for reactivation of tumor-suppressor proteins by modulating cancer-associated epigenetic modifications.

Curcumin

Curcumin is a golden pharmacological agent that has a pleiotropic effect on numerous cancer models. In recent decades, curcumin has received special attention in the scientific community because of its multitargeted effects on various human cancer cells. Curcumin exerts its chemopreventive effect by modulating cancer signaling pathways, triggering apoptosis, blocking cell cycle progression, and inhibiting cell invasion and metastasis (Kasi et al., 2016). Recent studies have demonstrated that curcumin inhibits cancer growth by targeting epigenetic modifications such as DNA methylation. RARβ is the RAR receptor family considered to play a role in the tumor-suppressor effect. Reduced expression of RARβ has been reported in lung cancer and other solid tumors, mainly due to DNA hypermethylation of the promoter region of RARβ. Curcumin treatment in NSCLC cells (A549 and H460) decreased the methylation frequency of RARβ and reactivated its expression. Moreover, the study also demonstrated that curcumin treatment significantly downregulates the mRNA expression level of the DNMT3b gene (Jiang et al., 2015). In addition, curcumin recovers the suppressed antioxidant defense mechanism, thereby inhibiting cancer cell growth. The silenced function of Nrf2 (a master regulator in antioxidant expression) commonly associated with cancer cells results in the loss of the antioxidant defense mechanism. TRAMP C1 prostate cancer cells in mice lost their Nrf2 expression by hypermethylation in the CpG promoter region. Interestingly, curcumin treatment decreased the Nrf2 CpG promoter methylation level and reactivated the expression of Nrf2 in both mRNA and protein levels. In addition, enhanced expression of NAD(P)H quinone dehydrogenase 1 (a downstream target gene of Nrf2) also supported the reactivation of Nrf2 expression during curcumin treatment (Khor et al., 2011).

Curcumin has been demonstrated to dynamically modulate HDAC and HAT activity through the targeting of epigenetic modifications (Fig. 13.2B). Curcumin exposure in K562 and HEL cell lines restores the cytokine signaling negative regulator suppressor of cytokine signaling 1 (SOCS1) and 3 (SOCS3) via inhibiting the overexpression of HDAC activity. Moreover, it recovers the expression of SOCS1 and SOCS3 by histone acetylation in promoter regions (Chen et al., 2013). Curcumin treatment in Burkitt lymphoma model Raji cells significantly inhibited the activity of HDACs 1, 3, and 8 as well as that of HAT p300. Moreover, downregulation of nuclear factor-κB and the Notch 1 signaling pathway was associated with the inhibition of HDAC enzymes. A similar study found that curcumin treatment in Raji cells blocked the expression of class I HDACs while upregulating the expression of Ac-histone H4 (Liu et al., 2005). In the case of breast cancer MCF-7 cells, curcumin inhibits the activity of HAT with increased levels of acetylated H3K18 and H4K16, leading to cell-proliferation inhibition (Collins et al., 2013).

Curcumin also modulates the expression of numerous miRNAs for their anticancer properties (Fig. 13.3B). For example, curcumin inhibits oncogenic miR-21 expression in the colon cancer cell lines RKO and HCT116, which leads to inhibition of invasion and metastatic properties (Mudduluru et al., 2011). Curcumin also restores the expression of tumor-suppressor proteins. A study reported that curcumin exposure in phorbol-12-myristate-13-acetate-induced colon cancer cells recovers the silenced the expression of PDCD4. In addition, curcumin blocks the binding efficacy of activator protein 1 (AP-1) in the miR-21 promoter region, leading to downregulation of miR-21 expression (Mudduluru et al., 2011). The miRNA profile of curcumin-treated human pancreatic carcinoma BxPC-3 cells showed altered expression for 29 miRNAs. Among these, upregulation of miR-22 contributed to the growth inhibition property of curcumin against of BxPC-3 cells. This study also concluded that miR-22 upregulation leads to downregulation of its target genes, Sp1 and ESR1 (Sun et al., 2008). Curcumin inhibits breast cancer growth and the pro-inflammatory cytokines CXCL1 and CXCL2 through upregulation of miR-181b (Kronski et al., 2014). The curcumin analogue difluorinated-curcumin inhibits pancreatic tumor growth by reducing enhancer of zeste homolog 2 expression. Moreover, the study also demonstrated the significant upregulation of tumor-suppressor let-7 miRNA family members, such as miR-26a, miR-146a, miR-101, and miR-200b/c (Bao et al., 2012). Overall, studies on curcumin have revealed that it targets numerous miRNAs to exert its anticancer activity. In the future, many unexplored miRNAs will need to be identified in curcumin-treated human cancer cells and animal models.

Genistein

Genistein is a polyphenolic isoflavone that belongs to the flavonoid group and is commonly found in various dietary vegetables, such as soy beans and fava beans. Epidemiological studies have shown that the intake of dietary-rich isoflavones reduces the risk of various human cancers. Genistein is a well-studied anticancer agent that affects cancer growth through multiple mechanisms. Interestingly, recent studies have found that genistein treatment altered epigenetic characteristics such as DNA methylation in cancer cells and activated the silenced function of tumor-suppressor proteins

(Kikuno et al., 2008). For example, genistein inhibited the growth of NSCLC A549 cells without harming the normal human lung cell line MRC-5. Moreover, genistein generated ROS-mediated oxidative stress in A549 cells and blocked the antioxidant-defense mechanism by downregulating Nrf2 expression. Surprisingly, molecular studies have found that genistein treatment decreases methylation in the Kelch-like ECH-associated protein 1 (KEAP1) promoter region. KEAP1 is the negative regulator for Nrf2 protein, and its activation suppresses Nrf2 expression. Activation of KEAP1 in genistein treatment sustained oxidative stress in A549 cells, resulting in apoptotic cell death. Moreover, the study also demonstrated that combination with radiation therapy further enhances the anticancer effect of genistein (Liu et al., 2016). Genistein also exerts its anticancer effect by strongly inhibiting DNMT activity. Genistein treatment also reactivates the silenced function of tumor-suppressor proteins. In breast cancer cells, genistein treatment recovered the expression of tumor-suppressor proteins, such as $p16^{INK4a}$, RARβ, and MGMT, resulting in tumor growth suppression and the activation of apoptosis (Fang et al., 2005). In addition, a combination of genistein and daidzein treatment reactivates glutathione S-transferase P1 and ephrin receptor expression by decreasing the DNA methylation in PC-3, DU-145, and LNCaP prostate cancer cell lines (Vardi et al., 2010). Overall, the studies clearly confirmed that genistein could be a potent DNA methylation inhibitor in cancer cells that is devoid of toxic effects to normal cells.

The role of genistein on histone-mediated epigenetic modification is poorly studied. Very few reports have demonstrated the chemopreventive effect of genistein on histone epigenetic modification. For instance, it was found that genistein treatment reactivates the tumor-suppressor function of secreted frizzled-related protein 1 (SFRP1) and Smad4 through demethylation of their promoter regions and histone proteins in prostate cancer cells (Hirata et al., 2014). Previous studies demonstrated that genistein inhibits the prostate cancer LNCaP cell line by targeting the androgen receptor (AR), and the mechanism was shown to be mediated through histone epigenetic modifications. Downregulated expression of AR was mediated through ubiquitin-mediated protein degradation of the AR protein. Genistein treatment also decreased Hsp90 chaperone (AR stabilizer) activity by the formation of an acetylated form of Hsp90, which further decreased expression of the HDAC6 protein. The results revealed that genistein significantly modulated the HDAC6-Hsp90 chaperone function to exert its chemopreventive effect (Basak et al., 2008).

Studies also have reported that genistein targets various cancer-associated miRNAs for its chemopreventive potential. Genistein restored the expression of sFRP1 and Smad4 in prostate cancer cells by inhibiting oncogenic miR-1260b, thereby inhibiting cell proliferation and invasion (Hirata et al., 2014). In addition, in clinically developed ovarian cancer model UL-3A and UL-3B cells treated with genistein, nearly 53 differentially regulated miRNAs were observed. Moreover, upregulation of ERα and ERβ expression (mRNA and protein level), and cell growth and migration inhibition, were noted during genistein exposure. However, this study did not emphasize the role of miRNAs that activated ERα and ERβ protein expression (Parker et al., 2009). In another study, genistein treatment upregulated the miR-1296 level and decreased the mRNA and protein levels of the minichromosome maintenance gene in prostate cancer cells (Majid et al., 2010).

Sulforaphane

Sulforaphane is an isothiocyanate compound commonly found in many cruciferous vegetables, such as broccoli, brussels sprouts, kale, and cabbages. Similar to curcumin, sulforaphane has gained significant attention in anticancer research. It has been demonstrated to possess various anticancer mechanisms in cancer cells. Studies have demonstrated that sulforaphane is also a potent competitive inhibitor against DNMT enzymes. For example, sulforaphane induced demethylation at cyclin D2 promoter regions that resulted in growth inhibition of prostate cancer cells and enhanced the expression of cyclin D2 (Hsu et al., 2011). In addition, sulforaphane significantly inhibited the activity of DNMT1 expression in colon cancer CaCo2 cells. Importantly, the role of sulforaphane on Nrf2 activation is well documented. Treatment of sulforaphane ameliorates the growth of prostate cancer model TRAMP C1 cells. In addition, the growth-inhibition mechanism revealed decreased methylation and enhanced expression of the Nrf2 protein. Reactivation of Nrf2 enhances the antioxidant defense mechanism and thereby inhibits the growth of prostate cancer cells (Zhang et al., 2013). Sulforaphane also induces translocation of Nrf2 and its target proteins in a TPA-induced mouse skin cancer model. In addition, it blocks the expressions of HDAC1, HDAC2, HDAC3, and HDAC4 to reactivate Nrf2 expression (Su et al., 2014). Preclinical experiments concluded that an oral dose of sulforaphane in wild mice causes HDAC inhibition while increasing the acetylation of H3 and H4. APC (Min/+) mice (a mouse model for studying human colon cancer) administered sulforaphane showed increased acetylation of p21 and BAX gene promoters that resulted in tumor growth inhibition and the induction of apoptosis (Myzak et al., 2007, 2008). A significant decrease in HDAC activity, increase in global histone acetylation, and increase in the expression of BAX protein was also demonstrated in the prostate cancer PC3 xenograft mice model. Interestingly, a pilot study reported the inhibition of HDAC activity and induced acetylation of histones H3 and H4, after 6 h consumption of 68 g of broccoli sprouts in a single dose. However, the effect of sulforaphane on histone epigenetic modification and miRNA regulation was poorly studied. Also, there is a lack of study for understanding the chemopreventive effects of sulforaphane, especially in the area of miRNA regulation.

Other Phytochemicals

Apart from the aforementioned compounds, other phytochemicals also act as epigenetic modifiers. Resveratrol (3,5,4′-trihydroxy-*trans*-stilbene) is a distinct type of polyphenol that belongs to the stilbenoid group. Grape skins, blueberries, peanuts, and mulberries are common dietary sources of resveratrol. Apart from its direct antioxidant property, resveratrol has significant anticancer effect against various human cancers cells. Preclinical study has demonstrated that administration of resveratrol downregulates the overexpression of DNMT3b in estradiol-induced mammary tumor tissue in female ACI rat strains (Qin et al., 2014). Resveratrol reactivates the epigenetically silenced expression of BRCA-1 in breast cancer MCF-7 cells by modulating the acetylation of H3K9, H4, and DNMT1 in the BRCA-1 gene promoter region (Papoutsis et al., 2010). In addition, reactivation of BRAC-1 was associated with increased expression of sirtuin 1 (SIRT1). Moreover, SIRT1 activation leads downregulation of the survivin protein and acetylation of histone H3K9 (Tili et al., 2010). Resveratrol also modulated miRNA expression for mediation of its anticancer effect. For example, resveratrol treatment inhibited several oncogenic miRNAs and reactivated certain tumor-suppressor proteins, such as Dicer 1, PDCD4, and PTEN, in SW480 colon cancer cells (Tili et al., 2010).

Diallyl disulfide (DADS) belongs to the group of organosulfur compounds and is abundantly present in garlic species. Studies have demonstrated that allyl mercaptan, a metabolic product of DADS, is a potent HDAC inhibitor in many cancer cells. Allyl mercaptan treatment in colon cancer cells effectively inhibited histone deacetylase activity by enhancing the binding effect of

SP3 on the p21 promoter region. Moreover, recruitment of p53 and p21 was found at 3–72 h of treatment with allyl mercaptan in colon cancer cells (Nian et al., 2008). Quercetin, a major dietary polyphenol commonly found in red wine, blueberries, apples, red onions, and green tea, has been demonstrated to decrease the hypermethylation of p16INK4a and thereby inhibit the growth of colon cancer RKO cells (Tan et al., 2008). Quercetin also inhibited interferon-γ-inducible protein 10 and macrophage inflammatory protein 2 expression by blocking CBP/p300 activity and acetylating H3 on promoter regions in a TNF-induced condition (Ruiz et al., 2007). In human leukemia HL-60 cells, quercetin treatment induced extrinsic mediated apoptosis through c-Jun/AP-1 pathway upregulation and activation of histone H3 acetylation (Lee et al., 2011). In addition, a preclinical study demonstrated that quercetin posttreatment delayed tumor growth, arrested the cell cycle, and promoted apoptosis in 7,12-dimethyl benz[a]anthracene-induced breast cancer in animal models. The chemopreventive effect of quercetin was associated with the inhibition of HDAC1 and DNMT1 (Priyadarsini et al., 2011).

CONCLUSION AND FUTURE PERSPECTIVES

Over the last 3 decades, research reports have provided convincing evidence about the ability of natural compounds, such as phytochemicals, to reactivate many tumor-suppressor proteins that are silenced by epigenetic modifications. Epigenetic therapy differs slightly from conventional treatment because it recovers silenced gene function by attenuating epigenetic modifications. In this chapter, we have discussed the role of phytochemicals in mediating epigenetic modification in cancer cells. Among the phytochemicals, EGCG is more potent in terms of its many pharmacological actions against cancer, as well as its safety. Importantly, it alters all three epigenetic hallmarks, namely DNA methylation, histone modification, and miRNA expression. Curcumin and genistein are also effective in modulating epigenetic histone modification in many human cancer cells. However, we did not come across any combinational synergistic studies of phytochemical effects on cancer epigenetic modifications. Similarly, a lack of studies was found for the role of quercetin on DNA methylation and miRNA expression. We also found few anticancer studies on lung cancer models rather than other cancers. Only two reports were available on DNA methylation inhibition that suggest efficacy for phytochemicals on epigenetic modifications in lung cancer cell lines. Because there is a complete lack of studies of lung cancer–associated epigenetic management using phytochemicals, we suggest the following future directions in this regard, which includes

1. Understanding early lung cancer–associated epigenetic modifications in both smokers and nonsmokers
2. Developing a lung cancer model to understand lung cancer–associated epigenetic modifications
3. Screening phytochemicals against lung cancer–associated epigenetic modifications in both *in vitro* and *in vivo* studies
4. Assessing the clinical efficacy of conventional anticancer agents in combination with epigenetic modifiers such as EGCG and curcumin

Acknowledgments

The authors gratefully acknowledge the Bioinformatics Infrastructure Facility provided by the Alagappa University (funded by Department of Biotechnology, Government of India; Grant No. BT/BI/25/015/2012). Tamilselvam Rajavel wishes to thank University Grants Commission, New Delhi, India for offering financial support in the form of a fellowship [UGC-BSR (F.25-1/2013-14 (BSR)/7–326/2011 (BSR) dated 30.05.2014].

References

Bao, B., et al., 2012. Curcumin analogue CDF inhibits pancreatic tumor growth by switching on suppressor microRNAs and attenuating EZH2 expression. Canc. Res. 72 (1), 335–345.

Basak, S., et al., 2008. Genistein down-regulates androgen receptor by modulating HDAC6-Hsp90 chaperone function. Mol. Canc. Therapeut. 7 (10), 3195–3202.

Cantone, L., et al., 2011. Inhalable metal-rich air particles and histone H3K4 dimethylation and H3K9 acetylation in a cross-sectional study of steel workers. Environ. Health Perspect. 119 (7), 964–969.

Chan, J.A., Krichevsky, A.M., Kosik, K.S., 2005. MicroRNA-21 is an antiapoptotic factor in human glioblastoma cells. Canc. Res. 65 (14), 6029–6033.

Chen, C.Q., et al., 2013. Pure curcumin increases the expression of SOCS1 and SOCS3 in myeloproliferative neoplasms through suppressing class I histone deacetylases. Carcinogenesis 34 (7), 1442–1449.

Chen, D., et al., 2013. Cigarette smoke component acrolein modulates chromatin assembly by inhibiting histone acetylation. J. Biol. Chem. 288 (30), 21678–21687.

Collins, H.M., et al., 2013. Differential effects of garcinol and curcumin on histone and p53 modifications in tumour cells. BMC Canc. 13 (1), 37.

Damiani, L.A., et al., 2008. Carcinogen-induced gene promoter hypermethylation is mediated by DNMT1 and causal for transformation of immortalized bronchial epithelial cells. Canc. Res. 68 (21), 9005–9014.

Dawson, M.A., Kouzarides, T., 2012. Cancer epigenetics: from mechanism to therapy. Cell 150 (1), 12–27.

Devi, K.P., et al., 2016. Targeting miRNAs by polyphenols: novel therapeutic strategy for cancer. Semin. Canc. Biol.

Dhanak, D., Jackson, P., 2014. Development and classes of epigenetic drugs for cancer. Biochem. Biophys. Res. Commun. 455 (1–2), 58–69.

Du, G.J., et al., 2012. Epigallocatechin gallate (EGCG) is the most effective cancer chemopreventive polyphenol in green tea. Nutrients 4 (11), 1679–1691.

Fabbri, M., et al., 2007. MicroRNA-29 family reverts aberrant methylation in lung cancer by targeting DNA methyltransferases 3A and 3B. Proc. Natl. Acad. Sci. Unit. States Am. 104 (40), 15805–15810.

Fang, M.Z., et al., 2003. Epigallocatechin-3-Gallate inhibits DNA methyltransferase and reactivates methylation-silenced genes in cancer cell lines. Canc. Res. 63 (22), 7563–7570.

Fang, M.Z., et al., 2005. Reversal of hypermethylation and reactivation of p16INK4a, RARβ, and MGMT genes by genistein and other isoflavones from soy. Clin. Canc. Res. 11 (19), 7033–7041.

Gao, Z., et al., 2009. Promoter demethylation of WIF-1 by epigallocatechin-3-gallate in lung cancer cells. Anticancer Res. 29 (6), 2025–2030.

Garofalo, M., et al., 2008. MicroRNA signatures of TRAIL resistance in human non-small cell lung cancer. Oncogene 27 (27), 3845–3855.

Hayashita, Y., et al., 2005. A polycistronic MicroRNA cluster, miR-17-92, is overexpressed in human lung cancers and enhances cell proliferation. Canc. Res. 65 (21), 9628–9632.

Hirata, H., et al., 2014. Genistein downregulates onco-miR-1260b and upregulates sFRP1 and Smad4 via demethylation and histone modification in prostate cancer cells. Br. J. Canc. 110 (6), 1645–1654.

Hsu, A., et al., 2011. Promoter de-methylation of cyclin D2 by sulforaphane in prostate cancer cells. Clin. Epigenet. 3 (1), 3.

Inbar-Feigenberg, M., et al., 2013. Basic concepts of epigenetics. Fertil. Steril. 99 (3), 607–615.

Jemal, A., et al., 2010. Cancer statistics, 2010. CA Cancer J. Clin. 60 (5), 277–300.

Jia, Y., et al., 2016. Negative regulation of DNMT3A de novo DNA methylation by frequently overexpressed UHRF family proteins as a mechanism for widespread DNA hypomethylation in cancer. Cell Dis. 2, 16007.

Jiang, A., et al., 2015. Curcumin reactivates silenced tumor suppressor gene RARβ by reducing DNA methylation. Phytother Res. 29 (8), 1237–1245.

Jin, B., Robertson, K.D., 2013. DNA methyltransferases, DNA damage repair, and cancer. Adv. Exp. Med. Biol. 754, 3–29.

Johnson, C.D., et al., 2007. The let-7 microRNA represses cell proliferation pathways in human cells. Canc. Res. 67 (16), 7713–7722.

Johnson, S.M., et al., 2005. RAS is regulated by the let-7 microRNA family. Cell 120 (5), 635–647.

Jones, P.A., Issa, J.P., Baylin, S., 2016. Targeting the cancer epigenome for therapy. Nat. Rev. Genet. 17 (10), 630–641.

Jones, P.A., Baylin, S.B., 2002. The fundamental role of epigenetic events in cancer. Nat. Rev. Genet. 3 (6), 415–428.

Jones, P.L., et al., 1998. 2–1 Methylated DNA and MeCP2 recruit histone deacetylase to repress transcription. Nat. Genet. 19 (2), 187–191.

Kasi, P.D., et al., 2016. Molecular targets of curcumin for cancer therapy: an updated review. Tumor Biol. 37 (10), 13017–13028.

Khor, T.O., et al., 2011. Pharmacodynamics of curcumin as DNA hypomethylation agent in restoring the expression of Nrf2 via promoter CpGs demethylation. Biochem. Pharmacol. 82 (9), 1073–1078.

Kikuno, N., et al., 2008. Genistein mediated histone acetylation and demethylation activates tumor suppressor genes in prostate cancer cells. Int. J. Canc. 123 (3), 552–560.

Kim, H., et al., 2006. Elevated mRNA levels of DNA methyltransferase-1 as an independent prognostic factor in primary nonsmall cell lung cancer. Cancer 107 (5), 1042–1049.

Kronski, E., et al., 2014. MiR181b is induced by the chemopreventive polyphenol curcumin and inhibits breast cancer metastasis via down-regulation of the inflammatory cytokines CXCL1 and -2. Mol. Oncol. 8 (3), 581–595.

Lee, D.E., et al., 2011. 7,3′,4′-trihydroxyisoflavone, a metabolite of the soy isoflavone Daidzein, suppresses ultraviolet B-induced skin cancer by targeting Cot and MKK4. J. Biol. Chem. 286 (16), 14246–14256.

Lewis, B.P., Burge, C.B., Bartel, D.P., 2005. Conserved seed pairing, often flanked by adenosines, indicates that thousands of human genes are microRNA targets. Cell 120 (1), 15–20.

Lin, R., et al., 2010. The tobacco-specific carcinogen NNK induces DNA methyltransferase 1 accumulation and tumor suppressor gene hypermethylation in mice and lung cancer patients. J. Clin. Invest. 120 (2).

Liu, H.L., et al., 2005. Curcumin, a potent anti-tumor reagent, is a novel histone deacetylase inhibitor regulating B-NHL cell line Raji proliferation. Acta Pharmacol. Sin. 26 (5), 603–609.

Liu, X., et al., 2016. Genistein mediates the selective radiosensitizing effect in NSCLC A549 cells via inhibiting methylation of the keap1 gene promoter region. Oncotarget 7 (19), 27267–27279.

Liu, Y., et al., 2006. Aberrant promoter methylation of p16 and MGMT genes in lung tumors from smoking and never-smoking lung cancer patients. Neoplasia 8 (1), 46–51.

Liu, Z.-L., et al., 2013. MicroRNA-21 (miR-21) expression promotes growth, metastasis, and chemo- or radioresistance in non-small cell lung cancer cells by targeting PTEN. Mol. Cell. Biochem. 372 (1–2), 35–45.

Mai, A., 2014. Targeting epigenetics in drug discovery. ChemMedChem 9 (3), 415–417.

Majid, S., et al., 2010. Regulation of minichromosome maintenance gene family by MicroRNA-1296 and genistein in prostate cancer. Canc. Res. 70 (7), 2809–2818.

Matsubara, H., et al., 2007. Apoptosis induction by antisense oligonucleotides against miR-17-5p and miR-20a in lung cancers overexpressing miR-17-92. Oncogene 26 (41), 6099–6105.

Mirza, S., et al., 2013. Expression of DNA methyltransferases in breast cancer patients and to analyze the effect of natural compounds on DNA methyltransferases and associated proteins. J. Breast Cancer 16 (1), 23–31.

Mudduluru, G., et al., 2011. Curcumin regulates miR-21 expression and inhibits invasion and metastasis in colorectal cancer. Biosci. Rep. 31 (3), 185–197.

Myzak, M.C., et al., 2007. Sulforaphane retards the growth of human PC-3 xenografts and inhibits HDAC activity in human subjects. Exp. Biol. Med. 232 (2), 227–234.

Myzak, M.C., Dashwood, W.M., Orner, G.A., Ho, E., D, R., 2008. Sulforaphane inhibits histone deacetylase in vivo and suppresses tumorigenesis in Apcmin mice. FASEB J. 23 (1), 1–7.

Nasser, M.W., et al., 2008. Down-regulation of micro-RNA-1 (miR-1) in lung cancer: suppression of tumorigenic property of lung cancer cells and their sensitization to doxorubicin-induced apoptosis by miR-1. J. Biol. Chem. 283 (48), 33394–33405.

Nervi, C., De Marinis, E., Codacci-Pisanelli, G., 2015. Epigenetic treatment of solid tumours: a review of clinical trials. Clin. Epigenet. 7 (1), 127.

Nian, H., et al., 2008. Allyl mercaptan, a garlic-derived organosulfur compound, inhibits histone deacetylase and enhances Sp3 binding on the P21WAF1 promoter. Carcinogenesis 29 (9), 1816–1824.

O'Hagan, H.M., et al., 2011. Oxidative damage targets complexes containing DNA methyltransferases, SIRT1, and polycomb members to promoter CpG Islands. Canc. Cell 20 (5), 606–619.

Pao, W., Miller, V.A., 2005. Epidermal growth factor receptor mutations, small-molecule kinase inhibitors, and non-small-cell lung cancer: current knowledge and future directions. J. Clin. Oncol. 23 (11), 2556–2568.

Papoutsis, A.J., et al., 2010. Resveratrol prevents epigenetic silencing of BRCA-1 by the aromatic hydrocarbon receptor in human breast cancer cells. J. Nutr. 140 (9), 1607–1614.

Parker, L.P., et al., 2009. Modulation of microRNA associated with ovarian cancer cells by genistein. Eur. J. Gynaecol. Oncol. 30 (6), 616–621.

Priyadarsini, R.V., et al., 2011. The flavonoid quercetin modulates the hallmark capabilities of hamster buccal pouch tumors. Nutr. Canc. 63 (2), 218–226.

Qin, T., et al., 2015. Epigenetic synergy between decitabine and platinum derivatives. Clin. Epigenet. 7 (1), 97.

Qin, W., et al., 2014. Methylation and miRNA effects of resveratrol on mammary tumors vs. Normal tissue. Nutr. Canc. 66 (2), 270–277.

Rauch, T.A., et al., 2012. DNA methylation biomarkers for lung cancer. Tumor Biol. 33 (2), 287–296.

Ruiz, P.A., et al., 2007. Quercetin inhibits TNF-induced NF-kappaB transcription factor recruitment to proinflammatory gene promoters in murine intestinal epithelial cells. J. Nutr. 137 (5), 1208–1215.

Selamat, S.A., et al., 2012. Genome-scale analysis of DNA methylation in lung adenocarcinoma and integration with mRNA expression. Genome Res. 22 (7), 1197–1211.

Shukla, S., Meeran, S.M., 2013. Epigenetic factors in breast cancer progression. Breast Cancer Meta. Drug Res. 341–365.

Steele, N., et al., 2009. Combined inhibition of DNA methylation and histone acetylation enhances gene re-expression and drug sensitivity in vivo. Br. J. Canc. 100 (5), 758–763.

Stojanović, D., Nikić, D., Lazarević, K., 2004. The level of nickel in smoker's blood and urine. Cent. Eur. J. Publ. Health 12 (4), 187–189.

Struhl, K., 1998. Histone acetylation and transcriptional regulatory mechanisms. Genes Dev. 12 (5), 599–606.

Su, Z.Y., et al., 2014. Requirement and epigenetics reprogramming of Nrf2 in suppression of tumor promoter TPA-induced mouse skin cell transformation by sulforaphane. Canc. Prev. Res. 7 (3), 319–329.

Sun, M., et al., 2008. Curcumin (diferuloylmethane) alters the expression profiles of microRNAs in human pancreatic cancer cells. Mol. Canc. Therapeut. 7 (3), 464–473.

Tan, S., et al., 2008. Quercetin is able to demethylate the p16INK4a gene promoter. Chemotherapy 55 (1), 6–10.

Tili, E., et al., 2010. Resveratrol modulates the levels of microRNAs targeting genes encoding tumor-suppressors and effectors of TGFβ signaling pathway in SW480 cells. Biochem. Pharmacol. 80 (12), 2057–2065.

Tsang, W.P., Kwok, T.T., 2010. Epigallocatechin gallate upregulation of miR-16 and induction of apoptosis in human cancer cells. J. Nutr. Biochem. 21 (2), 140–146.

Vakoc, C.R., et al., 2006. Profile of histone lysine methylation across transcribed mammalian chromatin. Mol. Cell Biol. 26 (24), 9185–9195.

Vardi, A., et al., 2010. Soy phytoestrogens modify DNA methylation of GSTP1, RASSF1A, EPH2 and BRCA1 promoter in prostate cancer cells. In Vivo 24 (4), 393–400.

Webster, R.J., et al., 2009. Regulation of epidermal growth factor receptor signaling in human cancer cells by microRNA-7. J. Biol. Chem. 284 (9), 5731–5741.

Wee, S., et al., 2014. Targeting epigenetic regulators for cancer therapy. Ann. N. Y. Acad. Sci. 1309 (1), 30–36.

Yamada, S., et al., 2016. Epigallocatechin-3-O-gallate upregulates microRNA-let-7b expression by activating 67-kDa laminin receptor signaling in melanoma cells. Sci. Rep. 6, 19225.

Yanaihara, N., et al., 2006. Unique microRNA molecular profiles in lung cancer diagnosis and prognosis. Canc. Cell 9 (3), 189–198.

Yu, S., et al., 2010. Nrf2 expression is regulated by epigenetic mechanisms in prostate cancer of TRAMP Mice. PLoS One 5 (1), e8579.

Zhang, C., et al., 2013. Sulforaphane enhances Nrf2 expression in prostate cancer TRAMP C1 cells through epigenetic regulation. Biochem. Pharmacol. 85 (9), 1398–1404.

Zhong, Z., et al., 2012. Inhibition of proliferation of human lung cancer cells by green tea catechins is mediated by upregulation of let-7. Exp. Ther. Med. 4 (2), 267–272.

Zochbauer-Muller, S., 2002. Aberrant DNA methylation in lung cancer: biological and clinical implications. Oncol. 7 (5), 451–457.

CHAPTER

14

Epigenetics of Prostate Cancer and Novel Chemopreventive and Therapeutic Approaches

Divya Bhagirath, Thao L Yang, Rajvir Dahiya, Shahana Majid, Sharanjot Saini

Department of Urology, Veterans Affairs Medical Center, San Francisco and University of California, San Francisco, CA, United States

Abstract

Prostate cancer is a commonly diagnosed cancer and a leading cause of male cancer-related mortality. Risk factors associated with prostate cancer include age, race, family history, and dietary factors. Although several etiological factors govern prostate cancer occurrence, epigenetic mechanisms are an important means of gene regulation in prostate cancer. In this chapter, we will briefly review the causal genomic factors involved in prostate tumorigenesis and will then elaborate on the roles of different epigenetic components, including DNA methylation, histone modifications, microRNAs, and their alterations in prostate tumor cell lines and in patient-derived tumors. Frequent epigenetic aberrations, such as DNA hypermethylation and hypomethylation as well as altered histone acetylation and methylation, have been observed in prostate cancer leading to tumorigenesis, tumor progression, and metastasis. In addition, small noncoding RNAs such as microRNAs are involved in regulating the prostate cancer epigenome. Recent studies have highlighted the important cross talk between microRNAs and cellular epigenetic mechanisms and will be discussed here. Further, dietary factors/nutraceuticals play an important role in regulation of the epigenome and the chemoprevention of prostate cancer. In this chapter, we examine the current literature regarding diet-mediated epigenetic alterations in prostate cancer with a particular focus on microRNAs. We will review how different dietary components regulate DNA methylation, histone modifications, and microRNA expression and can be utilized as chemopreventive agents in prostate cancer.

Keywords

Chemoprevention; Epigenetics; microRNAs; Nutraceuticals; Prostate cancer

INTRODUCTION

Prostate cancer (PCa) is the most commonly occurring cancer and the third-leading cause of cancer-related deaths among men in the United States. About 164,690 cases of PCa are expected to be diagnosed, and approximately 29,430

https://doi.org/10.1016/B978-0-12-812494-9.00014-7

deaths are expected from the disease in 2018 (Siegel et al., 2018). Age remains the most significant risk factor for the occurrence of PCa, as most cases are diagnosed at an average age of 65 years and above. PCa occurrence is less than 1% for men below 50 years of age (Gronberg, 2003). Ethnicity is another important risk factor for PCa occurrence, with an overall higher incidence of the disease in African American men than in those from Asia (Patel and Klein, 2009). In addition, dietary, hormonal, and hereditary factors have been shown to influence the risk of PCa occurrence, although these factors lack a strong association with the disease (Witte, 2009).

Evaluation of serum prostate-specific antigen (PSA), a noninvasive marker secreted by the prostate, has been used as a biomarker for PCa detection, as it is highly elevated under tumorigenic conditions. However, elevated PSA levels are also seen under other conditions, such as inflammation, infection, and prostate enlargement, reducing the efficacy of PSA as a specific biomarker for PCa diagnosis or tumor aggressiveness. Considering these observations, many researchers have characterized the cancer genome from large cohorts of patient-derived tumor tissues using high-resolution next-generation sequencing tools to stratify the disease into different clinical subtypes and to identify more specific and sensitive biomarkers that can accurately predict disease prognosis, recurrence, and treatment response (Cancer Genome Atlas Research, 2015; Fraser et al., 2017; Robinson et al., 2015; Taylor et al., 2010; Witte, 2009). These studies have led to the identification of several gene alterations, including gene fusions, chromosomal rearrangements and aberrations, such as translocations/insertions/deletions, and somatic mutations, in novel and known cancer-causing genes (Baca et al., 2013; Barbieri et al., 2012; Berger et al., 2011). Some frequently observed chromosomal alterations include: loss on chromosomes 1q, 6q, 8p, 10q, 13q, and 16q, gain on chromosomes 1q, 3q, 7q, 8q, 17q, and Xq (Alers et al., 2000; Boyd et al., 2012; Clark et al., 2003; Joos et al., 1995), gene fusions of AR-responsive gene *TMPRSS2* with ETS transcription factor (Boyd et al., 2012; Kumar-Sinha et al., 2008; Tomlins et al., 2005), loss of tumor suppressor *PTEN*, and amplification of oncogene *Myc* (Fraser et al., 2017). Together, these gene alterations have been shown to be good predictors for disease prognosis (Boyd et al., 2012; Fraser et al., 2017; Squire, 2009; Trock et al., 2016). In addition, mutations in several known and putative oncogenes and tumor suppressors such as *RAS, AR, PTEN, TP53, FOXA1, SPOP*, and *PIK3CA* (Barbieri et al., 2012; Boyd et al., 2012; Taylor et al., 2010) have been reported. However, the relative occurrence of some somatic mutations is quite rare and does not significantly correlate with PCa severity. Copy number alterations resulting from chromosomal abnormalities are frequently observed in most PCa cases and show a stronger association with disease prognosis (Boyd et al., 2012; Fraser et al., 2017; Taylor et al., 2010). These frequently deleted/amplified regions of chromosomes harbor several important regulatory genes (Calin and Croce, 2007). Expression of *NKX3.1*, a prostate-specific homeobox tumor-suppressor gene involved in prostate differentiation (Bhatia-Gaur et al., 1999), has been shown to decrease with the progression of disease (Bowen et al., 2000; He et al., 1997). This gene is located on chromosome 8p21, which is well known to be frequently deleted in most advanced prostate tumors, suggesting tumor cell mechanisms that evade tumor-suppressor responses (Bhatia-Gaur et al., 1999; Bowen et al., 2000; Boyd et al., 2012). Similarly, *PTEN*, another tumor-suppressor gene deleted in prostate tumors, is mapped to chromosome10q23 (Yoshimoto et al., 2012). Recent investigations from our laboratory have shown important biological implications of these chromosomal losses in regulating microRNA (miRNA) genes (Bucay et al., 2016; Bucay et al., 2017; Saini et al., 2014).

In addition to gene alterations, epigenetic mechanisms, such as altered patterns of

DNA methylation and histone modifications, contribute to PCa initiation and progression (Chinaranagari et al., 2015). The effect of various risk factors including age, race, and environment on PCa incidence is primarily mediated through epigenetic mechanisms (Chinaranagari et al., 2015). PCa stratification based on tumor epigenetic signatures has also been performed (Borno et al., 2012; Cancer Genome Atlas Research, 2015; Kim et al., 2011). Tumors harboring *ERG* gene fusions and *SPOP/FOXA1* mutations have a distinct epigenetic profile with respect to DNA methylation (Cancer Genome Atlas Research, 2015). A recent whole-genome sequencing study using tumors from localized nonindolent PCa patients showed that hypermethylation of transcription elongation regulator *TCERG1L* strongly correlated with poor prognosis and disease relapse (Fraser et al., 2017). These studies signify the importance of PCa epigenetic regulation in predicting disease outcomes and serving as an important biomarker for disease progression.

PROSTATE CANCER EPIGENOME AND ITS REGULATION BY DIFFERENT EPIGENETIC MODULATORS

The chromatin structure comprises DNA strands wrapped with an octamer of basic proteins known as histones. The term "epigenetics" refers to heritable changes within the genome, primarily through modification of DNA and histones, that regulate gene expression without altering underlying DNA sequences (Yegnasubramanian, 2016). In addition to various genetic aberrations that contribute to PCa pathogenesis, alterations in gene expression by epigenetic mechanisms are equally essential in maintaining the tumor genome and its characteristics for survival and propagation. This is done primarily by DNA-modifying enzymes, such as DNA methyltransferases (DNMTs), that directly add methyl groups to DNA or do so indirectly by modification of histone proteins that form the core of the chromatin structure. Histones undergo reversible posttranslational modifications including acetylation, methylation, and phosphorylation of their N-terminal tails to regulate gene expression (Chinaranagari et al., 2015; Kouzarides, 2007). Histone acetylation regulated by histone deacetylases (HDACs) and histone acetyltransferases (HATs), and histone methylation by histone methyltransferases (HMTs), are well studied in PCa and will be discussed in this chapter (Table 14.1). In addition, small noncoding RNAs, such as miRNAs, are involved in regulating the PCa epigenome and will be discussed here.

TABLE 14.1 Common Epigenetic Gene Targets Altered by DNA Methylation and Histone Modifications in Prostate Cancer

PCa Disease Stage	Genes Altered by DNA Methylation	Genes Altered by Histone Modifications
Indolent or nonaggressive	*GSTP1, RASSF1A, PTGS2*	
Aggressive	*APC, PLAU, WNT5A, HPSE, S100P, EZH2, RASSF1A, GSTP1, RUNX3, CDH1*	*AR* and *AR* gene targets, *EZH2, HDAC1, HDAC2, HDAC3*

DNA Methylation in Prostate Cancer

The 5′-carbon of cytosine within the DNA backbone undergoes covalent addition of a methyl group in a reaction catalyzed by DNMTs using S-adenosyl methyl (SAM) as a methyl donor. Mammalian cells have specialized regions in the DNA known as cytosine-phosphate-guanine (CpG) islands that are sites for these DNMT-mediated methyl additions. Methylation of CpG islands functions as a means of gene regulation, as most CpG sequences fall within the promoter region of key genes involved in both normal and tumor development (Chinaranagari et al., 2015; Wu et al., 2015). Methylation at CpG sites blocks transcription either by directly preventing the binding of transcription factors at the start site or by binding methyl-CpG binding proteins that in association with other proteins (such as Mi2-nucleosome remodeling deacetylase complex) form a chromatin-repressing complex (Chin et al., 2011; Yegnasubramanian, 2016). Thus, covalent methylation of the DNA sequence functions as a gene-expression regulator, where hypermethylation of CpG sites inhibits transcription, and hypomethylation leads to gene expression. Alterations in the methylation of various CpG islands have been found during the initiation and progression of PCa. Hypermethylation of many tumor-suppressor genes involved in DNA damage, cell cycle regulation, and oxidative damage is seen in most prostate tumors (Chin et al., 2011; Kang et al., 2004) (Table 14.1). Hypermethylation of glutathione *S*-transferase pi 1 (*GSTP1*), a gene that functions to protect the genome from oxidative damage, is a frequently observed key epigenetic event that occurs in more than 90% of prostate tumors (Brooks et al., 1998; Chin et al., 2011; Graca et al., 2016; Jeronimo et al., 2011; Lee et al., 1994). In addition, DNA hypermethylation of other important genes, such as adenomatous polyposis coli (*APC*), Ras-associated domain family 1A (*RASSF1A*), cyclin D2, prostaglandin endoperoxide synthase, methylguanine-DNMT, and runt-related transcription factor 3 (*RUNX3*), occur in prostate tumors and lead to their altered expression (Chin et al., 2011; Henrique et al., 2007; Kang et al., 2004; Liu et al., 2002; Rosenbaum et al., 2005; Wu et al., 2015; Yegnasubramanian, 2016). Hypermethylation frequencies of *APC*, *RASSF1A*, and *RUNX3* were reported to be higher in high-PSA and high-Gleason score tumors than they were in low-PSA and low-Gleason tumors (Kang et al., 2004). Similarly, E-cadherin (*CDH1*) promoter has been reported to be hypermethylated in PCa, with hypermethylation increasing in higher-grade tumors and leading to the loss of *CDH1* expression in addition to tumor progression (Graff et al., 1995; Li et al., 2001).

Hypomethylation of the long-interspersed element-1 repetitive DNA regions is frequently observed in PCa and has been shown to have prognostic and diagnostic significance (Chin et al., 2011; Kim et al., 2011). Furthermore, hypomethylation of various genes involved in tumor invasion such as *PLAU* (plasminogen activator, urokinase), *WNT5A* (Wnt Family member 5A), *S100P* (S100 calcium-binding protein P), and *HPSE* (Heparanase-1) is frequently observed in epigenetic alterations in PCa (Jeronimo et al., 2011) (Table 14.1). More recent studies on whole-genome DNA methylation have identified several important signature genes that are altered in indolent PCa patients and are predictive of aggressive disease (Bhasin et al., 2015; Fraser et al., 2017; Mundbjerg et al., 2017). In addition to aberrant methylation in promoter regions, recent investigations have revealed the presence of altered DNA methylation patterns within many intra- and intergenic DNA regions in prostate tumors (Hulf et al., 2011) that are thought to be transcriptional start sites for noncoding RNAs (Yegnasubramanian, 2016). Additionally, several miRNAs, including miR-34a, miR-126, miR-145, and miR-615, are known to be regulated by DNA methylation (Jeronimo et al., 2011; Suh et al., 2011), suggesting cross talk between different epigenetic mechanisms in regulating the tumor epigenome.

Histone Modifications in Prostate Cancer

Histones comprise different subunits that have peptide tails amenable to various posttranslational modifications, such as acetylation, methylation, phosphorylation, ubiquitination, and sumoylation, by specific enzymes. Various combinations of posttranslational histone modifications make up the "histone-code" that is responsible for opening and compressing chromatin, thereby regulating gene expression. Histone acetylation and methylation are the most common types of chromatin modifications. These modifications are regulated by different enzymes; for example, histone acetylation and deacetylation are regulated by HATs and HDACs, respectively, and histone methylation and demethylation by HMTs and histone demethylases (HDMs), respectively (Bannister and Kouzarides, 2011; Chin et al., 2011). Acetylation of the H3 lysine at positions 4, 36, and 79 is associated with activation of gene expression, and di- and trimethylation of H3 lysines 3 and 9 are associated with gene repression. In prostate tumorigenesis, global alterations in the pattern of histone modifications are observed. Immunohistochemical staining of prostate tumor tissues with antibodies against acetylated and methylated histone residues revealed a distinct modification pattern that can predict the risk of recurrence in patients with low-grade tumors from those with high-grade tumors, thereby emphasizing their importance as clinical markers for disease outcome (Seligson et al., 2005, 2009). These altered histone codes can further be linked to the altered expression of various chromatin-modifying enzymes that is frequently observed in PCa. The HDM lysine-specific demethylase 1A (LSD1) that relieves chromatin repression by removing a methyl group from H3K9 has been shown to be involved in ligand-dependent AR-mediated transcription by interacting with AR and functioning as a coactivator that removes repressing methyl groups from AR target promoters (Metzger et al., 2005). However, it was also shown to function as a corepressor of AR and its target genes by repressing AR expression at its intronic promoter (Cai et al., 2011). HDAC1 expression is found to be elevated in PCa and has been linked to disease recurrence as well as increased genomic instability (Burdelski et al., 2015). Furthermore, a set of transcriptional corepressors including HDACs 1, 2, and 3 in association with polycomb group gene enhancer of zeste homolog 2 (EZH2) is shown to be involved in AR-mediated transcription in *ERG*-overexpressing tumors and is associated with PCa progression (Chng et al., 2012) (Table 14.1). The HMT EZH2 is the most commonly overexpressed chromatin-modifying enzyme in prostate tumors and is associated with aggressive PCa (Chin et al., 2011; Jeronimo et al., 2011). Overexpression of EZH2 is associated with increased cellular proliferation in prostate tumors and is elevated in metastatic disease, therefore serving as a marker for disease progression (Bachmann et al., 2006; Bracken et al., 2003; Varambally et al., 2002). It is an important HMT that has been linked to more global alterations seen in the PCa epigenome. It is also involved in DNA methylation and has been shown to interact with DNMTs via PRC2/3 complexes (Kobayashi et al., 2011; Vire et al., 2006). Interestingly, EZH2 expression is also directly regulated by miR-101 (Friedman et al., 2009; Varambally et al., 2008). DNA methylation studies in non-*TMPRSS2-ERG* fusion tumors showed hypermethylation of miR-26a that correlated with increased EZH2 levels in these tumors. Overexpression of miR-26a in PCa cell lines leads to reduction of EZH2 expression (Borno et al., 2012). Likewise, knockdown of EZH2 in PCa cell line DU145 leads to upregulation of miRNAs miR-26a, miR-31, and miR-203 (Asangani et al., 2013), emphasizing the importance of miRNAs in regulating the epigenome. Taken together, these studies indicate that different epigenetic mechanisms function in concert, precisely regulating the epigenome. The function and role of miRNAs in epigenetic regulation will be discussed in more detail below.

microRNAs: Important Regulators of the Prostate Cancer Epigenome

miRNAs are small noncoding single-stranded RNAs, 22–23nt in length, that play a very important role in the regulation of gene expression by controlling the mRNA levels of several genes involved in cell differentiation, apoptosis, cell cycle, and cell growth (Di Leva et al., 2014). Since their discovery in *Caenorhabditis elegans* in 1993 (Lee et al., 1993), their biogenesis and functions have been well studied. miRNA genes are in intergenic, exonic, and intronic regions of the chromosome. Most miRNA genes cluster together and are transcribed as a single polycistronic transcript that is processed by different ribonucleases such as DROSHA and DICER in nuclear and cytoplasmic compartments, respectively, resulting in a single-stranded 22–23nt long mature RNA molecule. Together with Argonaute, these RNA molecules form the RISC complex, and by partial or complete complementarity associate with the 3′-UTRs of cytoplasmic mRNAs, leading to their degradation by either exo- or endonucleases (Di Leva et al., 2014). About 60% of human gene expression is regulated by miRNA-mediated decay. In cancer, miRNA levels are often deregulated via various mechanisms, including genomic loss or deletion, promoter methylation, or other tumor-promoting factors (Calin and Croce, 2006). Depending upon their loss or gain of expression in tumors, they are functionally classified as tumor suppressors (loss of expression) or oncomiRs (overexpressed). However, these functions are context dependent and may vary from one cell type to another (Di Leva et al., 2014).

Alterations in miRNA expression are frequently observed in prostate tumors in comparison with normal/benign prostate tissues. miRNA expression in tumors has been shown to stratify PCa tumors based on their androgen dependence, implicating their usefulness as PCa diagnostic and prognostic markers (Ozen et al., 2008; Porkka et al., 2007). Gene expression analysis of miRNAs along with their respective mRNA targets in prostate tumors shows significant alterations in their expression with respect to normal tissues. Several important oncomiRs, including miR-106b-25 cluster, miR-1, and miR-32, were found to be responsible for targeting proteins p21/WAF, E2F1 (cell cycle inhibitors), and Bim (antiapoptotic), thus contributing to prostate tumorigenesis (Ambs et al., 2008; Bracken et al., 2003). Research studies from our laboratory and other scientific groups have provided important insights into the role of several miRNAs in PCa pathogenesis. We found that overall there is loss in expression of miRNAs that are actively involved in controlling cell cycle, proliferation, and motility under normal conditions. Tumor-suppressor miRNAs including miR-34a, miR-34b, miR-23b, miR-4534, and miR-4723 were found to be downregulated in most prostate tumors compared with normal tissues and play an important role in negatively regulating transcription factors, receptor tyrosine kinases, and other signaling pathways such as C-myc, Akt, Src kinase, PTEN, and c-Abl (Arora et al., 2013; Liu et al., 2011; Majid et al., 2012, 2013; Nip et al., 2016; Yamamura et al., 2012). In addition, miRNAs are involved in inhibiting tumor progression by controlling epithelial-to-mesenchymal transition and other metastatic pathways (Sekhon et al., 2016). miR-203 and mir-195 have been shown to regulate the progression, invasion, and migration of prostate tumor cells by regulating metastatic genes such as CDH1, Bmi, and ZEB2 (Bonci et al., 2016; Saini et al., 2011). More recent investigations from our laboratory have identified many tumor-suppressor miRNAs (miR-3607, miR-383, miR-3622a and miR-3622b) that are lost due to chromosomal deletions at 5q and 8p regions in PCa and are important mediators of prostate tumorigenesis (Bucay et al., 2016; Bucay et al., 2017; Saini et al., 2014). We have shown that miR-3607 levels are attenuated in prostate tumors and that it functions as a tumor suppressor that regulates the activity of oncogene SRC kinase (Saini et al., 2014). miR-3622a/b and miR-383, located at chromosome 8p, were found to be downregulated in prostate tumors and play important roles in disease progression and metastasis (Bucay et al., 2016; Bucay et al., 2017). Tumor-suppressor miR-466 directly controls transcription factor RUNX2 and its associated targets osteopontin, osteocalcin, angiopoietin, and the matrix metalloproteinase (MMP) MMP11, thereby inhibiting PCa bone metastasis (Colden et al., 2017). Tumor-initiating stem cells are well-known metastatic disease precursors (Oskarsson et al., 2014), and miRNAs are involved in regulating these cell populations in PCa. miR-708 and miR-320 alter the expression of Akt and β-catenin, which in turn regulate the stem-like properties of these cells (Hsieh et al., 2013; Saini et al., 2012).

As described previously, various epigenetic mechanisms are tightly regulated, cross talk with one another, and function to maintain the epigenome. Recent studies have highlighted the important cross talk between miRNAs and cellular epigenetic mechanisms. In addition, expression of miRNA genes is controlled by DNA methylation and histone modification patterns (discussed in Epigenetic Regulation of microRNA Expression in Prostate Cancer section). While there is evidence about how aberrant miRNA gene expression is mediated in tumors by the tumor epigenome, the reverse regulation—i.e., how miRNAs control the epigenome—is not very well known. Identification of miR-101 as a regulator of histone methylation enzyme EZH2 is an important example of the regulation of epigenetic machinery by miRNAs (Varambally et al., 2008). Genomic loss of miR-101 expression controls EZH2 expression in prostate tumors (Varambally et al., 2008). In addition to miR-101, EZH2 expression is regulated in *TMPRSS2-ERG* fusion-positive and fusion-negative prostate tumors via miR-26a (Borno et al., 2012). Polycomb group proteins are involved in regulating the self-renewal of normal and tumor stem cells and are elevated in aggressive prostate tumors. A reciprocal regulatory control of polycomb repressive complex proteins PRC1 and PRC2 by miR-181, miR-200a, miR-200b, and miR-203 has been demonstrated (Cao et al., 2011). These miRNAs were shown to be downregulated in metastatic prostate tumors with a concomitant upregulation of EZH2 and PRC1 protein BMI1 and RING2 (Cao et al., 2011; Saini et al., 2011). miR-449a is downregulated in PCa and has been shown to be a direct regulator of HDAC1 expression via direct binding to the 3′-UTR of *HDAC1* mRNA. Overexpression of miR449a in prostate carcinoma cell line PC3 leads to cell-cycle arrest owing to increased p27 expression that results from HDAC1 repression in tumor cells (Noonan et al., 2009). Recently, miR-449a has also been shown to target SIRT1, an NAD^+-dependent HDAC in *TMPRSS2-ERG* fusion-positive prostate tumors (Kumar et al., 2016). Our group has shown a direct epigenetic regulation by miR-34b. miR-34b is downregulated in prostate tumors owing to CpG methylation of its promoter region. Forced re-expression of this tumor-suppressor miRNA led to downregulation of DNMTs (DNMT1 and DNMT3b) and HDACs (HDACs 1, 2, 3, and 4) and caused partial demethylation and alteration in chromatin modifications (Majid et al., 2013). Another study done by Xue et al. (2015) showed similar downregulation of DNMT3b via miR-145. Overexpression of miR-145 in PCa cells not only abrogated DNMT3b expression but also sensitized the cells to X-ray irradiation (Xue et al., 2015). miR-152 has been shown to be decreased in aggressive prostate tumors. Overexpression of miR-152 directly targets DNMT1, leading to a decrease in the migration, invasion, and growth of tumor cells, all of which are responsible for the aggressive phenotype of prostate tumor cells (Theodore et al., 2014). miR-137 has been shown to repress expression of transcriptional coregulators with the lysine-specific HDMs KDMA1, KDMA2, KDMA4A, KDMA5B, and KDMA7A in different complexes in an androgen-dependent manner (Nilsson et al., 2015). Taken together, these studies indicate an important feedback mechanism for miRNAs and other components of the epigenetic system (DNMTs/HDACs) in regulating the PCa epigenome. Aberrant changes during tumorigenesis influence this cross talk to facilitate tumor growth.

EPIGENETIC REGULATION OF MICRORNA EXPRESSION IN PROSTATE CANCER

Due to the important role of miRNAs in controlling vital cellular functions under normal conditions, their expression is strictly regulated via multiple mechanisms. However, there is a loss of miRNA regulation in tumors. Epigenetic components including DNA methylation and histone modifications are important players that regulate miRNA expression. These essential components of the epigenetic system are themselves significantly altered in tumors, thereby altering not only the associated coding genes but the noncoding RNAs as well. We will discuss each of these regulatory mechanisms below.

Regulation of microRNA Expression by DNA Methylation in Prostate Cancer

It is well known that miRNA gene expression is controlled by DNMTs in tumor cells (Iorio et al., 2010; Lujambio et al., 2007). Due to their intergenic location, most miRNA genes are silenced in tumors by aberrant DNA hypermethylation of CpG islands, leading to downregulation of several tumor-suppressor miRNAs. Methylation of miRNA gene promoters is as an important tumor cell mechanism that represses expression of several tumor-suppressor miRNAs, including miR-124, miR-9, miR-145, miR-200c, miR-141, miR-375, and miR-205 (Hulf et al., 2013; Lynch et al., 2016; Suh et al., 2011; Suzuki et al., 2012) in prostate and other cancers. CpG methylation of the miR-34a promoter is a commonly observed phenomenon in most tumors, including PCa, and forced re-expression of miR-34a has been shown to induce cell cycle arrest via miR-34a-mediated targeting of CDK6 in prostate carcinoma cell lines, confirming its tumor-suppressive role (Lodygin et al., 2008). Recently, it was shown that the promoter of miR130b-301b cluster, which is involved in inducing a senescence-associated phenotype upon oncogene activation and DNA damage response, undergoes methylation leading to its reduced expression in prostate tumors (Ramalho-Carvalho et al., 2017). In a recent study, Torres-Ferreira et al. (2017) showed the clinical significance of miR-193b, miR-129-2, and miR-34b/c promoter methylation and its use as a PCa diagnostic and prognostic biomarker. Similarly, the methylation status of GABRE/miR-452/miR-224 in prostate tumors is significantly associated with recurrence of the disease after prostatectomy (Kristensen et al., 2014). Certain miRNA gene promoters also undergo hypomethylation that results in their increased expression in tumors (Suzuki et al., 2012). A member of the let-7 family, namely let-7a-3, is an oncomiR whose gene locus is methylated under normal conditions. However, the onset of tumorigenesis induces hypomethylation of let-7a-3 locus, leading to epigenetic reactivation of its expression in tumors. This oncomiR is involved in regulating cell proliferation genes *CDK6*, *PCNA*, and the well-known *RAS* oncogene (Brueckner et al., 2007). Studies from our laboratory have demonstrated that miR-23b is epigenetically silenced in prostate tumors via methylation of its promotor. It functions as a tumor-suppressor miRNA and directly targets proto-oncogene Src kinase and Akt (Majid et al., 2012). Another tumor-suppressor miRNA, miR-31, is methylated in prostate tumors, and its expression is involved in regulating AR signaling and cell-cycle-related genes (Lin et al., 2013). Hulf et al. (2011) performed an integrative analysis on DNA methylation status, transcriptional alterations and histone H3K9 acetylation in normal and prostate tumor cell lines to identify miRNAs that are altered in tumors as a result of epigenetics.

They identified miR-615, miR-196b, miR-21, and miR-205 as altered in tumor cells as compared with normal cells. They also reported that the miR-205 gene locus is transcriptionally repressed in tumors, and its repression is enforced by a combination of DNA methylation, gain of H3K9 methylation, and loss of H3K4 trimethylation (Hulf et al., 2011). This suggests that the tumor epigenome enforces repression of crucial tumor-suppressor miRNAs via several means to control their expression.

Regulation of microRNA Expression by Histone Modifications in Prostate Cancer

Histone-modifying enzymes modulate miRNA expression as an epigenetic regulatory mechanism to control miRNA levels in tumors. The *Myc* oncogene, which is well known to be overexpressed in many tumors, is involved in the epigenetic regulation of miRNAs by controlling histone deacetylation and trimethylation patterns of promoters via HDACs and HMTs (Liu et al., 2013). miR-29, let-7 family members, the miR-15a/16 cluster, and miR-23a are deregulated in tumors, and changes in their expression have been attributed to a combined effect of DNA methylation and *MYC*-induced histone modifications through HDACs. EZH2, a well-known HMT that is overexpressed in prostate tumors, has been shown to regulate miR-181a, miR-181b, miR-203, let-7, and miR-200 family members by repressing their expression in prostate tumors (Liu et al., 2013). EZH2 is involved in H3K27 trimethylation in miR-31 promoters, thereby reducing its expression in prostate tumor cells and conferring resistance to apoptosis in response to docetaxel treatment (Zhang et al., 2014). Recently, it has been shown that treatment of prostate carcinoma cell lines with HDAC inhibitor leads to upregulation of miR-320a via H3K9 acetylation of its promoter. Epigenetic reexpression of miR-320a is involved in posttranscriptional regulation of AR in these cells (Sato et al., 2016).

PROSTATE CANCER EPIGENOME AND THERAPEUTIC APPROACHES

Given the interdependence of the various components of epigenetic pathways, they are important clinical biomarkers as well as therapeutic targets. Epigenetic modifications are reversible as opposed to genetic aberrations, which are irreversible. Since epigenetic modifications target key genes involved in PCa initiation, progression, and metastasis (Table 14.1 and Fig. 14.1), epigenetic modulators hold great potential as therapeutic agents. Several DNMT- and histone-modifying enzyme inhibitors have been tested and used as anticancer therapeutics (Table 14.2). 5-azacytidine and 5-aza-2′-deoxycytidine are nucleoside analogs that inhibit DNMTs. These compounds function by being incorporated into the growing DNA molecule during active replication and binding to DNMTs covalently, inhibiting them on the DNA strand and eventually leading to the loss of DNMT activity (Graca et al., 2016). These demethylating agents have been approved by the Food and Drug Administration (FDA) as clinical drugs for the treatment of myelodysplastic syndrome and acute myeloid leukemia (AML) (Liu et al., 2013). Both drugs inhibit cell growth and proliferation by reversing methylation at CpG islands of tumor-suppressor genes. Some of their tumor-suppressive effects are believed to be due to loss of methylation of tumor-suppressor miRNAs. Procainamide, another DNMT inhibitor, has been shown to reverse *GSTP1* hypermethylation in PCa cell lines (Lin et al., 2001). Despite their potential as therapeutic drugs in fast-growing tumors such as MDS and AML, their efficacy is limited in slow-growing solid tumors (Graca et al., 2016). Moreover, they are believed to have cytotoxic and mutagenic behaviors in normal cells (Chin et al., 2011).

HDAC inhibitors fall into different chemical classes, namely benzamides, aliphatic chain, cyclic peptides, and hydroxamic acids.

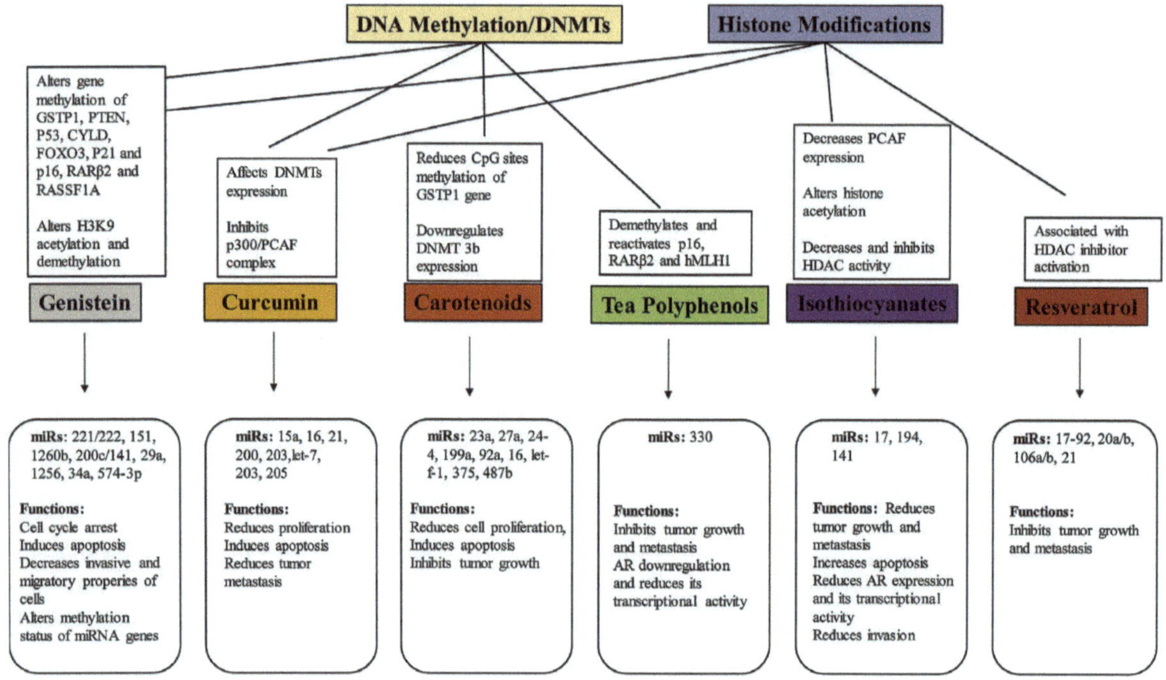

FIGURE 14.1 Schematic summary depicting the roles of different dietary supplements in regulating prostate cancer epigenetic components. Figure summarizes the role of dietary factors in the regulation of the prostate cancer epigenome via their effects on various epigenetic components including DNA methylation, histone modifications, and microRNA expression patterns. microRNAs reported to be altered by various components are listed below each component along with the functional effects.

Suberoylanilide hydroxamic acid (SAHA) has been used in clinical trials for advanced PCa and has also been approved by the FDA for the treatment of cutaneous T-cell lymphoma (Garcia-Manero et al., 2008; Mann et al., 2007). SAHA has been shown to reduce cell proliferation of PCa cell lines LNCaP, PC3, and TSU-Pr1, as well as xenografted prostate tumors (Butler et al., 2000). However, use of this drug in phase II clinical trials showed limited efficacy and substantial associated toxicity (Graca et al., 2016). HAT inhibitor C646 has been shown to

TABLE 14.2 Inhibitors for DNMTs and Histone-Modifying Enzymes and Epigenetic Gene Targets in Prostate Cancer

Genes Altered by DNMT Inhibitors	Genes Altered by Inhibitors for Histone-Modifying Enzymes
Procainamide—reverses *GSTP1* hypermethylation (Lin et al., 2001)	C646—inhibits p300/CBP (Santer et al., 2011) 3-Deazaneplanocin A—alters EZH2 expression (Crea et al., 2011) Pargyline—inhibits AR target genes (Graca et al., 2016) Suberoylanilide hydroxamic acid (Butler et al., 2000)

inhibit p300/CBP and induce apoptosis in prostate carcinoma cell lines PC3, DU145, LNCaP, and LAPC-4 (Santer et al., 2011). 3-deazaneplanocin A is an HMT inhibitor that affects EZH2 and polycomb repressor protein complex and has been shown to cause cell cycle arrest and apoptosis in PCa cell lines (Crea et al., 2011). Lastly, monoamine oxidase inhibitors (such as pargyline) are HDM inhibitors that have been shown to reduce LSD1-mediated H3K9 demethylation and inhibition of AR target genes in PCa cell lines (Graca et al., 2016) (Table 14.2).

Despite the efficacy of chemical drugs in shown in promising results in PCa cell lines and some PCa patients, their positive effects are countered by their associated toxic side effects and bystander effects on normal cells, making their use as clinical drugs less efficient. Conversely, natural derivatives from dietary nutrients have minimal known side effects. In addition to acting selectively on a target/effector, they are harmless to normal cells near tumors, thereby adding to their efficacy as potential tumor therapeutics.

ROLE OF DIETARY SUPPLEMENTS/ NUTRACEUTICALS IN REGULATING THE PROSTATE CANCER EPIGENOME

A growing amount of evidence suggests the importance of nutraceutical components in regulating tumor growth in cancer. The occurrence and risk of PCa is variable among different population groups. For example, it is the most commonly diagnosed cancer among men in the northern hemisphere. However, the incidence of PCa is significantly less among Asian men. Furthermore, Asian men who have migrated to or are born in the United States have a relatively higher risk than their Asian counterparts, suggesting that lifestyle and dietary intake, known variables among the two groups, could play an important role in regulating PCa incidence (Labbe et al., 2015). Diets include various bioactive components, such as resveratrol (red grapes), sulforaphane, and isothiocyanates (cruciferous vegetables), catechin and epigallocatechins (tea polyphenols), genistein (soy isoflavonoid), phenethyl isothiocyanate (PEITC), and curcumin. These dietary nutraceuticals are known to have antitumor effects *in vitro* and *in vivo* and have been demonstrated as chemopreventive agents in several cancers. Given that most of these dietary components regulate DNA methylation via DNMTs and histone modifications, their consumption is sometimes referred to as the "epigenetic diet." We will briefly discuss how different nutraceuticals function in regulating the epigenome, particularly DNA methylation and histone modifications in PCa. Fig. 14.1 summarizes the roles of different dietary supplements in regulating prostate epigenetic components.

Dietary Modulators of DNA Methylation/DNA Methyltransferases in Prostate Cancer

Genistein is a natural isoflavonoid that is found in soy plants, such as lupin and beans, and has been shown to function as a DNMT inhibitor. It allows partial demethylation of several tumor-suppressor genes, including *GSTP1, PTEN, TP53, CYLD, FOXO3, CDKN1A, CDKN2A, RARβ2,* and *RASSF1A,* in prostate and other tumor cell lines, thereby leading to their re-expression followed by a reduction in tumor cell growth (Graca et al., 2016; Labbe et al., 2015; Majid et al., 2008) (Fig. 14.1). The tea polyphenol epigallocatechin-3-gallate (EGCG) is a natural product found in green tea that inhibits DNMT1 activity in the PC3 PCa cell line (Wu et al., 2015). Its inhibitory effect on DNMT activity is associated with demethylation and reactivation of *CDKN2A, RARβ2,* and *hMLH1* genes in PCa (Wu et al., 2015). SGI1027, a small quinoline-based compound, has shown inhibitory effects on DNMT1, DNMT3A, and DNMT3B (Datta et al., 2009). Mahanine is a plant-based carbazole alkaloid that induces degradation of DNMT1 and DNMT3B via proteosomal ubiquitination in PCa cells, causing reexpression of *RASSF1A* (Agarwal et al., 2013). Curcumin and selenium also affect the activity and expression of DNMTs (Wu et al., 2015). Other dietary components, such as choline, methionine, and folate, function as methyl donors to SAM during the one-carbon metabolic cycle and are also known to influence PCa risk (Ho et al., 2011). Recently, the red dietary carotenoid astaxanthin was shown to reduce methylation of CpG sites of *GSTP1* gene and cause downregulation of DNMT3b expression in LNCaP cells (Fig. 14.1) (Yang et al., 2017).

Dietary Modulators of Histone Modifications in Prostate Cancer

In addition to inhibiting DNMT activity, the isoflavonoid genistein is also involved in H3K9 acetylation and demethylation (Fig. 14.1). Together, both functional effects of genistein on epigenetic modulators lead to reactivation of tumor-suppressor genes (Labbe et al., 2015). Apigenin, a plant-derived flavone, has been shown to inhibit HDAC1 and HDAC3 and causes growth arrest in PCa cells by mediating histone H3 acetylation on the p21/*WAF1* promoter (Pandey et al., 2012). Organosulfur compounds, such as sulforaphane and PEITC found in cruciferous vegetables, are also shown to have HDAC-inhibitory properties (Fig. 14.1) (Graca et al., 2016). Sulforaphane induces cell cycle arrest and apoptosis in PCa cells and alters histone acetylation by decreasing HDAC activity (Clarke et al., 2011). In addition, it has been shown to affect the gene expression of nuclear factor erythroid-2-related factor-2 in PCa TRAMP C1 cells by demethylation of its CpG island (Zhang et al., 2013). Curcumin, anacardic acid, and garcinol are some other dietary factors that function as inhibitors of the p300/PCAF complex (Fig. 14.1). Anacardic acid has a growth-inhibitory effect in LNCaP cells, where it induces G1/S cell cycle arrest and apoptosis (Graca et al., 2016). PEITC has been shown to upregulate expression of miR-17, leading to a decrease in PCAF expression in PCa cells (Fig. 14.1) (Bosutti et al., 2016). This suggests that dietary components can also modulate miRNA expression by targeting cellular methylation and histone modification components. We will discuss how dietary nutraceuticals affect miRNA expression in PCa in more detail in the following section.

ROLE OF NUTRACEUTICALS AS THERAPEUTICS FOR REGULATION OF EPIGENOME AND CHEMOPREVENTION OF PROSTATE CANCER VIA MICRORNAS

MiRNA targeting involves suppressing oncomiRs and inducing expression of tumor-suppressor miRNAs. Several dietary components have associated demethylation activity, which in addition to targeting various tumor-suppressor genes can also unmask the promoter regions of tumor-suppressor miRNAs by demethylation. A wide range of nutraceuticals are now shown to have growth-inhibitory effects on tumor growth via modulation of miRNA expression in tumor cells. Fig. 14.1 summarizes the roles of different dietary supplements in regulating the PCa epigenome through the modulation of miRNA expression. We discuss here the major dietary components involved in miRNA alterations in PCa.

Genistein

It has previously been demonstrated that genistein functions by inhibiting nuclear factor-κB

(NF- κB) and Akt signaling pathways (Bosutti et al., 2016; Saini et al., 2010). They are very selective antitumor compounds and have demonstrated no associated cytotoxicity in normal prostate cells (Bosutti et al., 2016). These soy isoflavonoids, apart from inducing re-expression of many tumor-suppressor genes (*GSTP1, PTEN, TP53, CDKN1A, CDKN2A, RASSF1A*), are also involved in modulating miRNA expression in PCa cells. In PCa cells, downregulation of oncogenic miR-221/222 is seen in response to treatment with genistein followed by induction of *ARH1*, a tumor-suppressor gene that is normally methylated in prostate tumors (Chen et al., 2011). These isoflavones are involved in inhibiting and promoting the expression of miRNAs in PCa. While treatment with genistein leads to downregulation of oncomiRs miR-151 and miR-1260b, it induces reexpression of miR-29a, miR-1256, miR-34a, and miR574-3p by promoter demethylation in PCa cells (Chiyomaru et al., 2012, 2013a; Hirata et al., 2014; Li et al., 2012). Upregulated miR-29a and miR-1256 cause suppression of *TRIM68* and *PGK1* and cell growth arrest (Li et al., 2012). High miR-151 expression is associated with poor survival in PCa patients. Genistein causes downregulation of miR-151, leading to suppression of its oncogenic targets *SOX17* and *ARHGDIA*, and is therefore a suggested chemopreventive therapy for prostate tumors (Chiyomaru et al., 2012). Upregulation of miR574-3p in DU145 PCa cells by genistein causes decreased proliferative, migratory, and invasive properties and induces apoptosis (Chiyomaru et al., 2013a). HOTAIR, a noncoding RNA that is significantly expressed in PCa, interacts with PRC2 complex to suppress expression of HOX genes in tumors. Induction of miR-34a expression by genistein alters the HOTAIR expression in prostate tumor cell lines and causes changes in cell proliferation, invasion, and migration *in vitro* and *in vivo* (Chiyomaru et al., 2013b). Genistein also has an inhibitory effect on miR-1260b that is upregulated in prostate tumors. Reduction in miR-1260b expression by genistein leads to upregulated expression of *SFRP1* and *SMAD4,* which are known tumor suppressors and direct targets of miR-1260b. In addition to altering expression via miR-1260b, genistein is also involved in altering the DNA and histone methylation patterns of *SFRP1* and *SMAD4* gene promoters (Hirata et al., 2014). Recently, genistein has been shown to influence the methylation status of the miR-200c/miR-141 loci, which leads to upregulated expression of miR-200c in PCa cell lines, inducing apoptosis and downregulation of DNMT3a and *TET1/TET3* genes (Lynch et al., 2016).

Curcumin

Diferuloylmethane (curcumin) is a polyphenolic compound present in the roots of *Curcuma longa*. Several research studies have highlighted the antitumoral, anti-inflammatory, and antimicrobial effects of curcumin. Its antitumor activity is mainly due to its targeted effect on NF-κB, Akt, and MAPK signaling pathways (Momtazi et al., 2016). Apart from regulating histone acetylation via p300 HAT components (Graca et al., 2016), curcumin functions as an miRNA modulator. Several miRNAs, including miR-15a, miR-16, miR-21, miR-101, miR-200, miR-203, and let-7, are known to be altered by curcumin (Teiten et al., 2013). Recently, curcumin was shown to inhibit the growth of PCa cells by upregulation of cell cycle inhibitor *CDKN1A* and inhibition of the miR-208 expression, a *CDKN1A*-targeting miRNA (Guo et al., 2015). PC3 cells treated with CDF show downregulation in the expression of miR-21 under hypoxic conditions. miR-21 is involved in regulating the expression of VEGF and IL6 and confers cancer stem cell-like ability to PC3 cells under hypoxia. However, these effects of miR-21 are attenuated by CDF treatment in PCa cell lines (Bao et al., 2012). CXCL1 and CXCL2 are important proinflammatory cytokines that are shown to promote prostate tumor metastasis. However, curcumin downregulates the expression of these two cytokines via inhibition of the NF-κB signaling pathway, thereby reducing tumor metastasis from these cells (Killian et al., 2012). Curcumin-tagged nanoparticles have been shown to specifically target PCa cells, causing increased apoptosis and tumor regression *in vivo* due in part to alterations in the levels of miR-21 and miR-205 (Yallapu et al., 2014).

Resveratrol

3,5,4′-trihydroxy-*trans*-stilbene is a natural phenol found in the skin of red grapes, mulberries, blueberries, raspberries, and peanuts. It has been reported to induce cell cycle arrest and apoptosis in PCa cell lines PC3 and LNCaP, causing accumulation of cells in G0/G1 phase and alteration in the activity and expression of cell cycle modulators, including cyclins B, D, and E, and Cdk 1 and 4 (Benitez et al., 2007). A similar effect of cell cycle inhibition and induction of apoptosis was observed in PC3 xenografts, where resveratrol in conjunction with tumor necrosis factor–related apoptosis-inducing ligand (TRAIL) functions to inhibit tumor growth and metastasis (Ganapathy et al., 2010). Importantly, treatment of PCa cell line LNCaP with resveratrol has been shown to alter expression of several oncogenic and tumor-suppressor miRNAs. Oncogenic miRNAs, including miR-17-92, miR-20a-b, and miR-106a-b, were demonstrated to be downregulated after resveratrol treatment. Furthermore, some miRNAs known to regulate the PTEN pathway were altered in DU145 cells, leading to upregulation in expression of PTEN (Dhar et al., 2011). The tumor growth-inhibitory and antimetastatic effects of resveratrol in PCa cell lines was further shown to be exerted by downregulation of oncomiR miR-21. Resveratrol treatment in aggressive PCa cell line PC3-3M-MM2 leads to reduction in both miR-21 and phospho-Akt that results in decreased tumor growth and a reduction in lung metastasis in tumor xenografts in SCID mice (Sheth et al., 2012). Resveratrol and its analogue pterostilbene have been shown to mediate their anticancer effects by targeting metastasis-associated protein 1, a component of nucleosomal remodeling and deacetylase repressor complex, HDACs, and miRNAs in PCa cell lines (Dhar et al., 2015). It has been shown that both resveratrol and pterostilbene inhibit the tumor-promoting effects of miR-17-5p and miR-106-5p in DU145 cells. The oncomiRs miR-17 and miR-106 directly target PTEN and promote tumor growth. Treatment with either dietary stilbene leads to downregulation of miR-17 and miR-106 (Dhar et al., 2015).

Isothiocyanates

These are sulfur-containing organic compounds naturally found in vegetables belonging to the Cruciferae family, such as mustard, brussels sprouts, cabbage, cauliflower, broccoli, and others. PEITCs and sulforaphane are important derivatives of isothiocyanates that have been shown to have anticancer properties (Ho et al., 2011). PEITC treatment in PC3 and LNCaP cells has been demonstrated to increase apoptosis *in vitro* and *in vivo* and to downregulate apoptotic proteins XIAP and survivin (Sakao et al., 2012). Administration of PEITC and curcumin in combination or alone has been shown to reduce tumor growth in PC3 xenografts (Khor et al., 2006). A similar effect on tumor reduction was also observed in PC3 xenografts upon sulforaphane treatment. It induces expression of pro-apoptotic protein Bax and downregulates Bcl2 to induce apoptosis in PCa cells (Singh et al., 2004). It has been shown to modulate immune natural killer T-cell and reduces tumorigenesis and lung metastasis in the TRAMP mice model (Singh et al., 2009). An important functional effect of sulforaphane administration is a reduction in AR and its target gene expression due to destabilization resulting from reduced HDAC6 deacetylase activity conferred by sulforaphane treatment (Gibbs et al., 2009). PEITC has been also demonstrated to inhibit AR transcriptional activity by upregulating miR-17, which represses PCAF, a coregulator for AR (Yu et al., 2013). Furthermore, in LNCaP cells it shows a similar reduction in AR transcriptional activity by modulating the expression of AR corepressor SHP (small heterodimer partner) via upregulation of miR-141 (Xiao et al., 2012). PEITC treatment in PC3 cells upregulates miR-194 and has been shown to target BMP1, causing a reduction in invasion in tumor cells via reduction in proinvasion proteins, such as MMP2 and MMP9 (Zhang et al., 2016).

Carotenoids

Carotenoids are large group of bright yellow-, orange-, or red-colored pigments found in fruits and vegetables, including α-carotene, β-carotene, lycopene, lutein, astaxanthin, and many more. Several studies have reported the benefits of dietary carotenoid intake for reducing PCa risk (Chang et al., 2005; Giovannucci et al., 1995; Norrish et al., 2000). Lycopene, found in tomatoes, is an effective dietary source for reducing the risk of PCa occurrence (Chang et al., 2005; Giovannucci et al., 1995). Treatment of primary prostate epithelial cells with lycopene has been shown to upregulate proteins involved in apoptosis, GSTP1, and represses the Akt/mTOR pathway (Qiu et al., 2013). It reduces cell growth in PCa cell line LNCaP by inducing expression of cell cycle inhibitors p27, p21, and p53 and causes reduction in phospho-Akt levels in cells (Palozza et al., 2010). It has also been shown to target androgen pathway genes in the TRAMP PCa model (Wan et al., 2014). In PC3 cells, treatment with lycopene has been shown to upregulate let-7f-1, which targets Akt2 and inhibits cancer cell proliferation (Li et al., 2016). However, it was demonstrated recently that inclusion of higher doses of lycopene exerts an opposite effect in patients with prostatic intraepithelial neoplasia and atypical small acinar proliferation. Supplementing lycopenes in the diet was associated with upregulation of tumor-suppressor miRNAs (miR-23a, −27a, -24-2) and oncomiRs (miR-199a, −92a, −16) (Gontero et al., 2015).

Astaxanthin, a xanthophyll carotenoid found in seafood including salmon, crab, and shrimps, has also demonstrated antitumor effects that target cell proliferation, apoptosis, and signaling pathways such as PI3K-Akt, MAPK, and PPARγ (Zhang and Wang, 2015). Administration of astaxanthin in xenograft mice with PC3-derived tumors causes reduction in Ki67 and PCNA expression in tumors along with upregulated expression of tumor-suppressor miRNAs, including miR-375 and miR-487b, thus inhibiting tumor growth (Ni et al., 2017).

Tea Polyphenols

EGCG is a polyphenol found abundantly in green tea. EGCG has shown some anticancer effects in PCa cells *in vitro* as well as in *in vivo* models (Johnson et al., 2010). However, these effects are not as substantial as those observed for the other nutraceuticals described above. EGCG has been shown to increase the sensitivity of LNCaP cells for TRAIL, thereby showing an inhibition of angiogenic and metastatic ability of these cells (Siddiqui et al., 2008). Furthermore, treatment of LNCaP cells with EGCG causes downregulation in AR expression (Ren et al., 2000). Importantly, in addition to altering AR-mediated transcriptional activity in PCa cell lines, it reduces AR nuclear translocation in PCa xenografts and has been shown to promote expression of tumor suppressor miR-330 (Siddiqui et al., 2011).

CONCLUSIONS

Epigenetic factors, in addition to genetic mechanisms, play an important role in PCa pathogenesis. Recent evidence highlights an important role of miRNAs in mediating prostate tumorigenesis. More significantly, it is now becoming apparent that miRNAs mediate cross talk with epigenetic components to regulate the prostate tumor epigenome. Increasing evidence indicates the role of specific dietary components in determining PCa risk through regulation of the PCa epigenome. Naturally derived products, such as genistein, curcumin, tea polyphenols, isothiocyanates, and carotenoids, are well-known anticancer agents and have shown promising results in reducing tumor cell proliferation, apoptosis, and metastasis in many PCa cancer cell lines, xenograft models, and some clinical trials. Chemoprevention using natural analogs or in general by epigenetic gene modulators holds promise for PCa therapy.

Recent evidence shows that these nutraceuticals employ miRNA-based regulation in controlling tumor growth. These dietary modulators can hence be used as chemopreventive and therapeutic agents to alter the epigenome via regulation of miRNA expression. Despite emerging data, our current understanding in this area is still very limited, and more rigorous studies in this direction are warranted. The causality between diet and the PCa epigenome needs to be experimentally validated more rigorously. Overall, a better understanding of dietary-regulated epigenetic pathways in PCa will yield critical information and potentially aid in developing novel chemopreventive and therapeutic strategies for the management of PCa.

Acknowledgments

We thank Dr. Roger Erickson for his support with preparation of the manuscript. Research in authors' laboratory is supported by the National Cancer Institute at the NIH (Grant Number RO1CA177984 to SS; RO1CA160079 to RD).

References

Agarwal, S., Amin, K.S., Jagadeesh, S., Baishay, G., Rao, P.G., Barua, N.C., et al., 2013. Mahanine restores RASSF1A expression by down-regulating DNMT1 and DNMT3B in prostate cancer cells. Mol. Cancer 12 (1), 99.

Alers, J.C., Rochat, J., Krijtenburg, P.J., Hop, W.C., Kranse, R., Rosenberg, C., et al., 2000. Identification of genetic markers for prostatic cancer progression. Lab. Investig. 80 (6), 931–942.

Ambs, S., Prueitt, R.L., Yi, M., Hudson, R.S., Howe, T.M., Petrocca, F., et al., 2008. Genomic profiling of microRNA and messenger RNA reveals deregulated microRNA expression in prostate cancer. Cancer Res. 68 (15), 6162–6170.

Arora, S., Saini, S., Fukuhara, S., Majid, S., Shahryari, V., Yamamura, S., et al., 2013. MicroRNA-4723 inhibits prostate cancer growth through inactivation of the Abelson family of nonreceptor protein tyrosine kinases. PLoS One 8 (11), e78023.

Asangani, I.A., Ateeq, B., Cao, Q., Dodson, L., Pandhi, M., Kunju, L.P., et al., 2013. Characterization of the EZH2-MMSET histone methyltransferase regulatory axis in cancer. Mol. Cell 49 (1), 80–93.

Baca, S.C., Prandi, D., Lawrence, M.S., Mosquera, J.M., Romanel, A., Drier, Y., et al., 2013. Punctuated evolution of prostate cancer genomes. Cell 153 (3), 666–677.

Bachmann, I.M., Halvorsen, O.J., Collett, K., Stefansson, I.M., Straume, O., Haukaas, S.A., et al., 2006. EZH2 expression is associated with high proliferation rate and aggressive tumor subgroups in cutaneous melanoma and cancers of the endometrium, prostate, and breast. J. Clin. Oncol. 24 (2), 268–273.

Bannister, A.J., Kouzarides, T., 2011. Regulation of chromatin by histone modifications. Cell Res. 21 (3), 381–395.

Bao, B., Ahmad, A., Kong, D., Ali, S., Azmi, A.S., Li, Y., et al., 2012. Hypoxia induced aggressiveness of prostate cancer cells is linked with deregulated expression of VEGF, IL-6 and miRNAs that are attenuated by CDF. PLoS One 7 (8), e43726.

Barbieri, C.E., Baca, S.C., Lawrence, M.S., Demichelis, F., Blattner, M., Theurillat, J.P., et al., 2012. Exome sequencing identifies recurrent SPOP, FOXA1 and MED12 mutations in prostate cancer. Nat. Genet. 44 (6), 685–689.

Benitez, D.A., Pozo-Guisado, E., Alvarez-Barrientos, A., Fernandez-Salguero, P.M., Castellon, E.A., 2007. Mechanisms involved in resveratrol-induced apoptosis and cell cycle arrest in prostate cancer-derived cell lines. J. Androl. 28 (2), 282–293.

Berger, M.F., Lawrence, M.S., Demichelis, F., Drier, Y., Cibulskis, K., Sivachenko, A.Y., et al., 2011. The genomic complexity of primary human prostate cancer. Nature 470 (7333), 214–220.

Bhasin, J.M., Lee, B.H., Matkin, L., Taylor, M.G., Hu, B., Xu, Y., et al., 2015. Methylome-wide sequencing detects DNA hypermethylation distinguishing indolent from aggressive prostate cancer. Cell Rep. 13 (10), 2135–2146.

Bhatia-Gaur, R., Donjacour, A.A., Sciavolino, P.J., Kim, M., Desai, N., Young, P., et al., 1999. Roles for Nkx3.1 in prostate development and cancer. Genes Dev. 13 (8), 966–977.

Bonci, D., Coppola, V., Patrizii, M., Addario, A., Cannistraci, A., Francescangeli, F., et al., 2016. A microRNA code for prostate cancer metastasis. Oncogene 35 (9), 1180–1192.

Borno, S.T., Fischer, A., Kerick, M., Falth, M., Laible, M., Brase, J.C., et al., 2012. Genome-wide DNA methylation events in TMPRSS2-ERG fusion-negative prostate cancers implicate an EZH2-dependent mechanism with miR-26a hypermethylation. Cancer Discov. 2 (11), 1024–1035.

Bosutti, A., Zanconati, F., Grassi, G., Dapas, B., Passamonti, S., Scaggiante, B., 2016. Epigenetic and miRNAs dysregulation in prostate cancer: the role of nutraceuticals. Anticancer Agents Med. Chem. 16 (11), 1385–1402.

Bowen, C., Bubendorf, L., Voeller, H.J., Slack, R., Willi, N., Sauter, G., et al., 2000. Loss of NKX3.1 expression in human prostate cancers correlates with tumor progression. Cancer Res. 60 (21), 6111–6115.

Boyd, L.K., Mao, X., Lu, Y.J., 2012. The complexity of prostate cancer: genomic alterations and heterogeneity. Nat. Rev. Urol. 9 (11), 652–664.

Bracken, A.P., Pasini, D., Capra, M., Prosperini, E., Colli, E., Helin, K., 2003. EZH2 is downstream of the pRB-E2F pathway, essential for proliferation and amplified in cancer. EMBO J. 22 (20), 5323–5335.

Brooks, J.D., Weinstein, M., Lin, X., Sun, Y., Pin, S.S., Bova, G.S., et al., 1998. CG island methylation changes near the GSTP1 gene in prostatic intraepithelial neoplasia. Cancer Epidemiol. Biomark. Prev. 7 (6), 531–536.

Brueckner, B., Stresemann, C., Kuner, R., Mund, C., Musch, T., Meister, M., et al., 2007. The human let-7a-3 locus contains an epigenetically regulated microRNA gene with oncogenic function. Cancer Res. 67 (4), 1419–1423.

Bucay, N., Sekhon, K., Majid, S., Yamamura, S., Shahryari, V., Tabatabai, Z.L., et al., 2016. Novel tumor suppressor microRNA at frequently deleted chromosomal region 8p21 regulates epidermal growth factor receptor in prostate cancer. Oncotarget 7 (43), 70388–70403.

Bucay, N., Sekhon, K., Yang, T., Majid, S., Shahryari, V., Hsieh, C., et al., 2017. MicroRNA-383 located in frequently deleted chromosomal locus 8p22 regulates CD44 in prostate cancer. Oncogene 39 (19), 2667–2679.

Burdelski, C., Ruge, O.M., Melling, N., Koop, C., Simon, R., Steurer, S., et al., 2015. HDAC1 overexpression independently predicts biochemical recurrence and is associated with rapid tumor cell proliferation and genomic instability in prostate cancer. Exp. Mol. Pathol. 98 (3), 419–426.

Butler, L.M., Agus, D.B., Scher, H.I., Higgins, B., Rose, A., Cordon-Cardo, C., et al., 2000. Suberoylanilide hydroxamic acid, an inhibitor of histone deacetylase, suppresses the growth of prostate cancer cells in vitro and in vivo. Cancer Res. 60 (18), 5165–5170.

Cai, C., He, H.H., Chen, S., Coleman, I., Wang, H., Fang, Z., et al., 2011. Androgen receptor gene expression in prostate cancer is directly suppressed by the androgen receptor through recruitment of lysine-specific demethylase 1. Cancer Cell 20 (4), 457–471.

Calin, G.A., Croce, C.M., 2006. MicroRNA signatures in human cancers. Nat. Rev. Cancer 6 (11), 857–866.

Calin, G.A., Croce, C.M., 2007. Chromosomal rearrangements and microRNAs: a new cancer link with clinical implications. J. Clin. Investig. 117 (8), 2059–2066.

Cancer Genome Atlas Research, N, 2015. The molecular taxonomy of primary prostate cancer. Cell 163 (4), 1011–1025.

Cao, Q., Mani, R.S., Ateeq, B., Dhanasekaran, S.M., Asangani, I.A., Prensner, J.R., et al., 2011. Coordinated regulation of polycomb group complexes through microRNAs in cancer. Cancer Cell 20 (2), 187–199.

Chang, S., Erdman Jr., J.W., Clinton, S.K., Vadiveloo, M., Strom, S.S., Yamamura, Y., et al., 2005. Relationship between plasma carotenoids and prostate cancer. Nutr. Cancer 53 (2), 127–134.

Chen, Y., Zaman, M.S., Deng, G., Majid, S., Saini, S., Liu, J., et al., 2011. MicroRNAs 221/222 and genistein-mediated regulation of ARHI tumor suppressor gene in prostate cancer. Cancer Prev. Res. (Phila.) 4 (1), 76–86.

Chin, S.P., Dickinson, J.L., Holloway, A.F., 2011. Epigenetic regulation of prostate cancer. Clin. Epigenet. 2 (2), 151–169.

Chinaranagari, S., Sharma, P., Bowen, N.J., Chaudhary, J., 2015. Prostate cancer epigenome. Methods Mol. Biol. 1238, 125–140.

Chiyomaru, T., Yamamura, S., Fukuhara, S., Hidaka, H., Majid, S., Saini, S., et al., 2013a. Genistein up-regulates tumor suppressor microRNA-574-3p in prostate cancer. PLoS One 8 (3), e58929.

Chiyomaru, T., Yamamura, S., Fukuhara, S., Yoshino, H., Kinoshita, T., Majid, S., et al., 2013b. Genistein inhibits prostate cancer cell growth by targeting miR-34a and oncogenic HOTAIR. PLoS One 8 (8), e70372.

Chiyomaru, T., Yamamura, S., Zaman, M.S., Majid, S., Deng, G., Shahryari, V., et al., 2012. Genistein suppresses prostate cancer growth through inhibition of oncogenic microRNA-151. PLoS One 7 (8), e43812.

Chng, K.R., Chang, C.W., Tan, S.K., Yang, C., Hong, S.Z., Sng, N.Y., et al., 2012. A transcriptional repressor coregulatory network governing androgen response in prostate cancers. EMBO J. 31 (12), 2810–2823.

Clark, J., Edwards, S., Feber, A., Flohr, P., John, M., Giddings, I., et al., 2003. Genome-wide screening for complete genetic loss in prostate cancer by comparative hybridization onto cDNA microarrays. Oncogene 22 (8), 1247–1252.

Clarke, J.D., Hsu, A., Yu, Z., Dashwood, R.H., Ho, E., 2011. Differential effects of sulforaphane on histone deacetylases, cell cycle arrest and apoptosis in normal prostate cells versus hyperplastic and cancerous prostate cells. Mol. Nutr. Food Res. 55 (7), 999–1009.

Colden, M., Dar, A.A., Saini, S., Dahiya, P.V., Shahryari, V., Yamamura, S., et al., 2017. MicroRNA-466 inhibits tumor growth and bone metastasis in prostate cancer by direct regulation of osteogenic transcription factor RUNX2. Cell Death Dis. 8 (1), e2572.

Crea, F., Hurt, E.M., Mathews, L.A., Cabarcas, S.M., Sun, L., Marquez, V.E., et al., 2011. Pharmacologic disruption of polycomb repressive complex 2 inhibits tumorigenicity and tumor progression in prostate cancer. Mol. Cancer 10, 40.

Datta, J., Ghoshal, K., Denny, W.A., Gamage, S.A., Brooke, D.G., Phiasivongsa, P., et al., 2009. A new class of quinoline-based DNA hypomethylating agents reactivates tumor suppressor genes by blocking DNA methyltransferase 1 activity and inducing its degradation. Cancer Res. 69 (10), 4277–4285.

Dhar, S., Hicks, C., Levenson, A.S., 2011. Resveratrol and prostate cancer: promising role for microRNAs. Mol. Nutr. Food Res. 55 (8), 1219–1229.

Dhar, S., Kumar, A., Rimando, A.M., Zhang, X., Levenson, A.S., 2015. Resveratrol and pterostilbene epigenetically restore PTEN expression by targeting oncomiRs of the miR-17 family in prostate cancer. Oncotarget 6 (29), 27214–27226.

Di Leva, G., Garofalo, M., Croce, C.M., 2014. MicroRNAs in cancer. Annu. Rev. Pathol. 9, 287–314.

Fraser, M., Sabelnykova, V.Y., Yamaguchi, T.N., Heisler, L.E., Livingstone, J., Huang, V., et al., 2017. Genomic hallmarks of localized, non-indolent prostate cancer. Nature 541 (7637), 359–364.

Friedman, J.M., Liang, G., Liu, C.C., Wolff, E.M., Tsai, Y.C., Ye, W., et al., 2009. The putative tumor suppressor microRNA-101 modulates the cancer epigenome by repressing the polycomb group protein EZH2. Cancer Res. 69 (6), 2623–2629.

Ganapathy, S., Chen, Q., Singh, K.P., Shankar, S., Srivastava, R.K., 2010. Resveratrol enhances antitumor activity of TRAIL in prostate cancer xenografts through activation of FOXO transcription factor. PLoS One 5 (12), e15627.

Garcia-Manero, G., Yang, H., Bueso-Ramos, C., Ferrajoli, A., Cortes, J., Wierda, W.G., et al., 2008. Phase 1 study of the histone deacetylase inhibitor vorinostat (suberoylanilide hydroxamic acid [SAHA]) in patients with advanced leukemias and myelodysplastic syndromes. Blood 111 (3), 1060–1066.

Gibbs, A., Schwartzman, J., Deng, V., Alumkal, J., 2009. Sulforaphane destabilizes the androgen receptor in prostate cancer cells by inactivating histone deacetylase 6. Proc. Natl. Acad. Sci. U.S.A. 106 (39), 16663–16668.

Giovannucci, E., Ascherio, A., Rimm, E.B., Stampfer, M.J., Colditz, G.A., Willett, W.C., 1995. Intake of carotenoids and retinol in relation to risk of prostate cancer. J. Natl. Cancer Inst. 87 (23), 1767–1776.

Gontero, P., Marra, G., Soria, F., Oderda, M., Zitella, A., Baratta, F., et al., 2015. A randomized double-blind placebo controlled phase I-II study on clinical and molecular effects of dietary supplements in men with precancerous prostatic lesions. Chemoprevention or "chemopromotion"? Prostate 75 (11), 1177–1186.

Graca, I., Pereira-Silva, E., Henrique, R., Packham, G., Crabb, S.J., Jeronimo, C., 2016. Epigenetic modulators as therapeutic targets in prostate cancer. Clin. Epigenet. 8, 98.

Graff, J.R., Herman, J.G., Lapidus, R.G., Chopra, H., Xu, R., Jarrard, D.F., et al., 1995. E-cadherin expression is silenced by DNA hypermethylation in human breast and prostate carcinomas. Cancer Res. 55 (22), 5195–5199.

Gronberg, H., 2003. Prostate cancer epidemiology. Lancet 361 (9360), 859–864.

Guo, H., Xu, Y., Fu, Q., 2015. Curcumin inhibits growth of prostate carcinoma via miR-208-mediated CDKN1A activation. Tumour Biol. 36 (11), 8511–8517.

He, W.W., Sciavolino, P.J., Wing, J., Augustus, M., Hudson, P., Meissner, P.S., et al., 1997. A novel human prostate-specific, androgen-regulated homeobox gene (NKX3.1) that maps to 8p21, a region frequently deleted in prostate cancer. Genomics 43 (1), 69–77.

Henrique, R., Ribeiro, F.R., Fonseca, D., Hoque, M.O., Carvalho, A.L., Costa, V.L., et al., 2007. High promoter methylation levels of APC predict poor prognosis in sextant biopsies from prostate cancer patients. Clin. Cancer Res. 13 (20), 6122–6129.

Hirata, H., Hinoda, Y., Shahryari, V., Deng, G., Tanaka, Y., Tabatabai, Z.L., et al., 2014. Genistein downregulates onco-miR-1260b and upregulates sFRP1 and Smad4 via demethylation and histone modification in prostate cancer cells. Br. J. Cancer 110 (6), 1645–1654.

Ho, E., Beaver, L.M., Williams, D.E., Dashwood, R.H., 2011. Dietary factors and epigenetic regulation for prostate cancer prevention. Adv. Nutr. 2 (6), 497–510.

Hsieh, I.S., Chang, K.C., Tsai, Y.T., Ke, J.Y., Lu, P.J., Lee, K.H., et al., 2013. MicroRNA-320 suppresses the stem cell-like characteristics of prostate cancer cells by downregulating the Wnt/beta-catenin signaling pathway. Carcinogenesis 34 (3), 530–538.

Hulf, T., Sibbritt, T., Wiklund, E.D., Bert, S., Strbenac, D., Statham, A.L., et al., 2011. Discovery pipeline for epigenetically deregulated miRNAs in cancer: integration of primary miRNA transcription. BMC Genom. 12, 54.

Hulf, T., Sibbritt, T., Wiklund, E.D., Patterson, K., Song, J.Z., Stirzaker, C., et al., 2013. Epigenetic-induced repression of microRNA-205 is associated with MED1 activation and a poorer prognosis in localized prostate cancer. Oncogene 32 (23), 2891–2899.

Iorio, M.V., Piovan, C., Croce, C.M., 2010. Interplay between microRNAs and the epigenetic machinery: an intricate network. Biochim. Biophys. Acta 1799 (10–12), 694–701.

Jeronimo, C., Bastian, P.J., Bjartell, A., Carbone, G.M., Catto, J.W., Clark, S.J., et al., 2011. Epigenetics in prostate cancer: biologic and clinical relevance. Eur. Urol. 60 (4), 753–766.

Johnson, J.J., Bailey, H.H., Mukhtar, H., 2010. Green tea polyphenols for prostate cancer chemoprevention: a translational perspective. Phytomedicine 17 (1), 3–13.

Joos, S., Bergerheim, U.S., Pan, Y., Matsuyama, H., Bentz, M., du Manoir, S., et al., 1995. Mapping of chromosomal gains and losses in prostate cancer by comparative genomic hybridization. Genes Chromosomes Cancer 14 (4), 267–276.

Kang, G.H., Lee, S., Lee, H.J., Hwang, K.S., 2004. Aberrant CpG island hypermethylation of multiple genes in prostate cancer and prostatic intraepithelial neoplasia. J. Pathol. 202 (2), 233–240.

Khor, T.O., Keum, Y.S., Lin, W., Kim, J.H., Hu, R., Shen, G., et al., 2006. Combined inhibitory effects of curcumin and phenethyl isothiocyanate on the growth of human PC-3 prostate xenografts in immunodeficient mice. Cancer Res. 66 (2), 613–621.

Killian, P.H., Kronski, E., Michalik, K.M., Barbieri, O., Astigiano, S., Sommerhoff, C.P., et al., 2012. Curcumin inhibits prostate cancer metastasis in vivo by targeting the inflammatory cytokines CXCL1 and -2. Carcinogenesis 33 (12), 2507–2519.

Kim, J.H., Dhanasekaran, S.M., Prensner, J.R., Cao, X., Robinson, D., Kalyana-Sundaram, S., et al., 2011. Deep sequencing reveals distinct patterns of DNA methylation in prostate cancer. Genome Res. 21 (7), 1028–1041.

Kobayashi, Y., Absher, D.M., Gulzar, Z.G., Young, S.R., McKenney, J.K., Peehl, D.M., et al., 2011. DNA methylation profiling reveals novel biomarkers and important roles for DNA methyltransferases in prostate cancer. Genome Res. 21 (7), 1017–1027.

Kouzarides, T., 2007. Chromatin modifications and their function. Cell 128 (4), 693–705.

Kristensen, H., Haldrup, C., Strand, S., Mundbjerg, K., Mortensen, M.M., Thorsen, K., et al., 2014. Hypermethylation of the GABRE~miR-452~miR-224 promoter in prostate cancer predicts biochemical recurrence after radical prostatectomy. Clin. Cancer Res. 20 (8), 2169–2181.

Kumar-Sinha, C., Tomlins, S.A., Chinnaiyan, A.M., 2008. Recurrent gene fusions in prostate cancer. Nat. Rev. Cancer 8 (7), 497–511.

Kumar, P., Sharad, S., Petrovics, G., Mohamed, A., Dobi, A., Sreenath, T.L., et al., 2016. Loss of miR-449a in ERG-associated prostate cancer promotes the invasive phenotype by inducing SIRT1. Oncotarget 7 (16), 22791–22806.

Labbe, D.P., Zadra, G., Ebot, E.M., Mucci, L.A., Kantoff, P.W., Loda, M., et al., 2015. Role of diet in prostate cancer: the epigenetic link. Oncogene 34 (36), 4683–4691.

Lee, R.C., Feinbaum, R.L., Ambros, V., 1993. The *C. elegans* heterochronic gene lin-4 encodes small RNAs with antisense complementarity to lin-14. Cell 75 (5), 843–854.

Lee, W.H., Morton, R.A., Epstein, J.I., Brooks, J.D., Campbell, P.A., Bova, G.S., et al., 1994. Cytidine methylation of regulatory sequences near the pi-class glutathione S-transferase gene accompanies human prostatic carcinogenesis. Proc. Natl. Acad. Sci. U.S.A. 91 (24), 11733–11737.

Li, D., Chen, L., Zhao, W., Hao, J., An, R., 2016. MicroRNA-let-7f-1 is induced by lycopene and inhibits cell proliferation and triggers apoptosis in prostate cancer. Mol. Med. Rep. 13 (3), 2708–2714.

Li, L.C., Zhao, H., Nakajima, K., Oh, B.R., Ribeiro Filho, L.A., Carroll, P., et al., 2001. Methylation of the E-cadherin gene promoter correlates with progression of prostate cancer. J. Urol. 166 (2), 705–709.

Li, Y., Kong, D., Ahmad, A., Bao, B., Dyson, G., Sarkar, F.H., 2012. Epigenetic deregulation of miR-29a and miR-1256 by isoflavone contributes to the inhibition of prostate cancer cell growth and invasion. Epigenetics 7 (8), 940–949.

Lin, P.C., Chiu, Y.L., Banerjee, S., Park, K., Mosquera, J.M., Giannopoulou, E., et al., 2013. Epigenetic repression of miR-31 disrupts androgen receptor homeostasis and contributes to prostate cancer progression. Cancer Res. 73 (3), 1232–1244.

Lin, X., Asgari, K., Putzi, M.J., Gage, W.R., Yu, X., Cornblatt, B.S., et al., 2001. Reversal of GSTP1 CpG island hypermethylation and reactivation of pi-class glutathione S-transferase (GSTP1) expression in human prostate cancer cells by treatment with procainamide. Cancer Res. 61 (24), 8611–8616.

Liu, C., Kelnar, K., Liu, B., Chen, X., Calhoun-Davis, T., Li, H., et al., 2011. The microRNA miR-34a inhibits prostate cancer stem cells and metastasis by directly repressing CD44. Nat. Med. 17 (2), 211–215.

Liu, L., Yoon, J.H., Dammann, R., Pfeifer, G.P., 2002. Frequent hypermethylation of the RASSF1A gene in prostate cancer. Oncogene 21 (44), 6835–6840.

Liu, X., Chen, X., Yu, X., Tao, Y., Bode, A.M., Dong, Z., et al., 2013. Regulation of microRNAs by epigenetics and their interplay involved in cancer. J. Exp. Clin. Cancer Res. 32, 96.

Lodygin, D., Tarasov, V., Epanchintsev, A., Berking, C., Knyazeva, T., Korner, H., et al., 2008. Inactivation of miR-34a by aberrant CpG methylation in multiple types of cancer. Cell Cycle 7 (16), 2591–2600.

Lujambio, A., Ropero, S., Ballestar, E., Fraga, M.F., Cerrato, C., Setien, F., et al., 2007. Genetic unmasking of an epigenetically silenced microRNA in human cancer cells. Cancer Res. 67 (4), 1424–1429.

Lynch, S.M., O'Neill, K.M., McKenna, M.M., Walsh, C.P., McKenna, D.J., 2016. Regulation of miR-200c and miR-141 by methylation in prostate cancer. Prostate 76 (13), 1146–1159.

Majid, S., Dar, A.A., Saini, S., Arora, S., Shahryari, V., Zaman, M.S., et al., 2012. miR-23b represses proto-oncogene Src kinase and functions as methylation-silenced tumor suppressor with diagnostic and prognostic significance in prostate cancer. Cancer Res. 72 (24), 6435–6446.

Majid, S., Dar, A.A., Saini, S., Shahryari, V., Arora, S., Zaman, M.S., et al., 2013. miRNA-34b inhibits prostate cancer through demethylation, active chromatin modifications, and AKT pathways. Clin. Cancer Res. 19 (1), 73–84.

Majid, S., Kikuno, N., Nelles, J., Noonan, E., Tanaka, Y., Kawamoto, K., et al., 2008. Genistein induces the p21WAF1/CIP1 and p16INK4a tumor suppressor genes in prostate cancer cells by epigenetic mechanisms involving active chromatin modification. Cancer Res. 68 (8), 2736–2744.

Mann, B.S., Johnson, J.R., Cohen, M.H., Justice, R., Pazdur, R., 2007. FDA approval summary: vorinostat for treatment of advanced primary cutaneous T-cell lymphoma. Oncologist 12 (10), 1247–1252.

Metzger, E., Wissmann, M., Yin, N., Muller, J.M., Schneider, R., Peters, A.H., et al., 2005. LSD1 demethylates repressive histone marks to promote androgen-receptor-dependent transcription. Nature 437 (7057), 436–439.

Momtazi, A.A., Shahabipour, F., Khatibi, S., Johnston, T.P., Pirro, M., Sahebkar, A., 2016. Curcumin as a MicroRNA regulator in cancer: a review. Rev. Physiol. Biochem. Pharmacol. 171, 1–38.

Mundbjerg, K., Chopra, S., Alemozaffar, M., Duymich, C., Lakshminarasimhan, R., Nichols, P.W., et al., 2017. Identifying aggressive prostate cancer foci using a DNA methylation classifier. Genome Biol. 18 (1), 3.

Ni, X., Yu, H., Wang, S., Zhang, C., Shen, S., 2017. Astaxanthin inhibits PC-3 xenograft prostate tumor growth in nude mice. Mar. Drugs 15 (3).

Nilsson, E.M., Laursen, K.B., Whitchurch, J., McWilliam, A., Odum, N., Persson, J.L., et al., 2015. MiR137 is an androgen regulated repressor of an extended network of transcriptional coregulators. Oncotarget 6 (34), 35710–35725.

Nip, H., Dar, A.A., Saini, S., Colden, M., Varahram, S., Chowdhary, H., et al., 2016. Oncogenic microRNA-4534 regulates PTEN pathway in prostate cancer. Oncotarget 7 (42), 68371–68384.

Noonan, E.J., Place, R.F., Pookot, D., Basak, S., Whitson, J.M., Hirata, H., et al., 2009. miR-449a targets HDAC-1 and induces growth arrest in prostate cancer. Oncogene 28 (14), 1714–1724.

Norrish, A.E., Jackson, R.T., Sharpe, S.J., Skeaff, C.M., 2000. Prostate cancer and dietary carotenoids. Am. J. Epidemiol. 151 (2), 119–123.

Oskarsson, T., Batlle, E., Massague, J., 2014. Metastatic stem cells: sources, niches, and vital pathways. Cell Stem Cell 14 (3), 306–321.

Ozen, M., Creighton, C.J., Ozdemir, M., Ittmann, M., 2008. Widespread deregulation of microRNA expression in human prostate cancer. Oncogene 27 (12), 1788–1793.

Palozza, P., Colangelo, M., Simone, R., Catalano, A., Boninsegna, A., Lanza, P., et al., 2010. Lycopene induces cell growth inhibition by altering mevalonate pathway and Ras signaling in cancer cell lines. Carcinogenesis 31 (10), 1813–1821.

Pandey, M., Kaur, P., Shukla, S., Abbas, A., Fu, P., Gupta, S., 2012. Plant flavone apigenin inhibits HDAC and remodels chromatin to induce growth arrest and apoptosis in human prostate cancer cells: in vitro and in vivo study. Mol. Carcinog. 51 (12), 952–962.

Patel, A.R., Klein, E.A., 2009. Risk factors for prostate cancer. Nat. Clin. Pract. Urol. 6 (2), 87–95.

Porkka, K.P., Pfeiffer, M.J., Waltering, K.K., Vessella, R.L., Tammela, T.L., Visakorpi, T., 2007. MicroRNA expression profiling in prostate cancer. Cancer Res. 67 (13), 6130–6135.

Qiu, X., Yuan, Y., Vaishnav, A., Tessel, M.A., Nonn, L., van Breemen, R.B., 2013. Effects of lycopene on protein expression in human primary prostatic epithelial cells. Cancer Prev. Res. (Phila.) 6 (5), 419–427.

Ramalho-Carvalho, J., Graca, I., Gomez, A., Oliveira, J., Henrique, R., Esteller, M., et al., 2017. Downregulation of miR-130b ~ 301b cluster is mediated by aberrant promoter methylation and impairs cellular senescence in prostate cancer. J. Hematol. Oncol. 10 (1), 43.

Ren, F., Zhang, S., Mitchell, S.H., Butler, R., Young, C.Y., 2000. Tea polyphenols down-regulate the expression of the androgen receptor in LNCaP prostate cancer cells. Oncogene 19 (15), 1924–1932.

Robinson, D., Van Allen, E.M., Wu, Y.M., Schultz, N., Lonigro, R.J., Mosquera, J.M., et al., 2015. Integrative clinical genomics of advanced prostate cancer. Cell 161 (5), 1215–1228.

Rosenbaum, E., Hoque, M.O., Cohen, Y., Zahurak, M., Eisenberger, M.A., Epstein, J.I., et al., 2005. Promoter hypermethylation as an independent prognostic factor for relapse in patients with prostate cancer following radical prostatectomy. Clin. Cancer Res. 11 (23), 8321–8325.

Saini, S., Majid, S., Dahiya, R., 2010. Diet, microRNAs and prostate cancer. Pharm. Res. (N.Y.) 27 (6), 1014–1026.

Saini, S., Majid, S., Shahryari, V., Arora, S., Yamamura, S., Chang, I., et al., 2012. miRNA-708 control of CD44(+) prostate cancer-initiating cells. Cancer Res. 72 (14), 3618–3630.

Saini, S., Majid, S., Shahryari, V., Tabatabai, Z.L., Arora, S., Yamamura, S., et al., 2014. Regulation of SRC kinases by microRNA-3607 located in a frequently deleted locus in prostate cancer. Mol. Cancer Ther. 13 (7), 1952–1963.

Saini, S., Majid, S., Yamamura, S., Tabatabai, L., Suh, S.O., Shahryari, V., et al., 2011. Regulatory role of mir-203 in prostate cancer progression and metastasis. Clin. Cancer Res. 17 (16), 5287–5298.

Sakao, K., Desineni, S., Hahm, E.R., Singh, S.V., 2012. Phenethyl isothiocyanate suppresses inhibitor of apoptosis family protein expression in prostate cancer cells in culture and in vivo. Prostate 72 (10), 1104–1116.

Santer, F.R., Hoschele, P.P., Oh, S.J., Erb, H.H., Bouchal, J., Cavarretta, I.T., et al., 2011. Inhibition of the acetyltransferases p300 and CBP reveals a targetable function for p300 in the survival and invasion pathways of prostate cancer cell lines. Mol. Cancer Ther. 10 (9), 1644–1655.

Sato, S., Katsushima, K., Shinjo, K., Hatanaka, A., Ohka, F., Suzuki, S., et al., 2016. Histone deacetylase inhibition in prostate cancer triggers mir-320-mediated suppression of the androgen receptor. Cancer Res. 76 (14), 4192–4204.

Sekhon, K., Bucay, N., Majid, S., Dahiya, R., Saini, S., 2016. MicroRNAs and epithelial-mesenchymal transition in prostate cancer. Oncotarget 7 (41), 67597–67611.

Seligson, D.B., Horvath, S., McBrian, M.A., Mah, V., Yu, H., Tze, S., et al., 2009. Global levels of histone modifications predict prognosis in different cancers. Am. J. Pathol. 174 (5), 1619–1628.

Seligson, D.B., Horvath, S., Shi, T., Yu, H., Tze, S., Grunstein, M., et al., 2005. Global histone modification patterns predict risk of prostate cancer recurrence. Nature 435 (7046), 1262–1266.

Sheth, S., Jajoo, S., Kaur, T., Mukherjea, D., Sheehan, K., Rybak, L.P., et al., 2012. Resveratrol reduces prostate cancer growth and metastasis by inhibiting the Akt/MicroRNA-21 pathway. PLoS One 7 (12), e51655.

Siddiqui, I.A., Asim, M., Hafeez, B.B., Adhami, V.M., Tarapore, R.S., Mukhtar, H., 2011. Green tea polyphenol EGCG blunts androgen receptor function in prostate cancer. FASEB J. 25 (4), 1198–1207.

Siddiqui, I.A., Malik, A., Adhami, V.M., Asim, M., Hafeez, B.B., Sarfaraz, S., et al., 2008. Green tea polyphenol EGCG sensitizes human prostate carcinoma LNCaP cells to TRAIL-mediated apoptosis and synergistically inhibits biomarkers associated with angiogenesis and metastasis. Oncogene 27 (14), 2055–2063.

Siegel, R.L., Miller, K.D., Jemal, A., 2018. Cancer statistics. CA Cancer J. Clin. 68 (1), 7–30.

Singh, A.V., Xiao, D., Lew, K.L., Dhir, R., Singh, S.V., 2004. Sulforaphane induces caspase-mediated apoptosis in cultured PC-3 human prostate cancer cells and retards growth of PC-3 xenografts in vivo. Carcinogenesis 25 (1), 83–90.

Singh, S.V., Warin, R., Xiao, D., Powolny, A.A., Stan, S.D., Arlotti, J.A., et al., 2009. Sulforaphane inhibits prostate carcinogenesis and pulmonary metastasis in TRAMP mice in association with increased cytotoxicity of natural killer cells. Cancer Res. 69 (5), 2117–2125.

Squire, J.A., 2009. TMPRSS2-ERG and PTEN loss in prostate cancer. Nat. Genet. 41 (5), 509–510.

Suh, S.O., Chen, Y., Zaman, M.S., Hirata, H., Yamamura, S., Shahryari, V., et al., 2011. MicroRNA-145 is regulated by DNA methylation and p53 gene mutation in prostate cancer. Carcinogenesis 32 (5), 772–778.

Suzuki, H., Maruyama, R., Yamamoto, E., Kai, M., 2012. DNA methylation and microRNA dysregulation in cancer. Mol. Oncol. 6 (6), 567–578.

Taylor, B.S., Schultz, N., Hieronymus, H., Gopalan, A., Xiao, Y., Carver, B.S., et al., 2010. Integrative genomic profiling of human prostate cancer. Cancer Cell 18 (1), 11–22.

Teiten, M.H., Dicato, M., Diederich, M., 2013. Curcumin as a regulator of epigenetic events. Mol. Nutr. Food Res. 57 (9), 1619–1629.

Theodore, S.C., Davis, M., Zhao, F., Wang, H., Chen, D., Rhim, J., et al., 2014. MicroRNA profiling of novel African American and Caucasian Prostate Cancer cell lines reveals a reciprocal regulatory relationship of miR-152 and DNA methyltranferase 1. Oncotarget 5 (11), 3512–3525.

Tomlins, S.A., Rhodes, D.R., Perner, S., Dhanasekaran, S.M., Mehra, R., Sun, X.W., et al., 2005. Recurrent fusion of TMPRSS2 and ETS transcription factor genes in prostate cancer. Science 310 (5748), 644–648.

Torres-Ferreira, J., Ramalho-Carvalho, J., Gomez, A., Menezes, F.D., Freitas, R., Oliveira, J., et al., 2017. MiR-193b promoter methylation accurately detects prostate cancer in urine sediments and miR-34b/c or miR-129-2 promoter methylation define subsets of clinically aggressive tumors. Mol. Cancer 16 (1), 26.

Trock, B.J., Fedor, H., Gurel, B., Jenkins, R.B., Knudsen, B.S., Fine, S.W., et al., 2016. PTEN loss and chromosome 8 alterations in Gleason grade 3 prostate cancer cores predicts the presence of un-sampled grade 4 tumor: implications for active surveillance. Mod. Pathol. 29 (7), 764–771.

Varambally, S., Cao, Q., Mani, R.S., Shankar, S., Wang, X., Ateeq, B., et al., 2008. Genomic loss of microRNA-101 leads to overexpression of histone methyltransferase EZH2 in cancer. Science 322 (5908), 1695–1699.

Varambally, S., Dhanasekaran, S.M., Zhou, M., Barrette, T.R., Kumar-Sinha, C., Sanda, M.G., et al., 2002. The polycomb group protein EZH2 is involved in progression of prostate cancer. Nature 419 (6907), 624–629.

Vire, E., Brenner, C., Deplus, R., Blanchon, L., Fraga, M., Didelot, C., et al., 2006. The Polycomb group protein EZH2 directly controls DNA methylation. Nature 439 (7078), 871–874.

Wan, L., Tan, H.L., Thomas-Ahner, J.M., Pearl, D.K., Erdman Jr., J.W., Moran, N.E., et al., 2014. Dietary tomato and lycopene impact androgen signaling- and carcinogenesis-related gene expression during early TRAMP prostate carcinogenesis. Cancer Prev. Res. (Phila.) 7 (12), 1228–1239.

Witte, J.S., 2009. Prostate cancer genomics: towards a new understanding. Nat. Rev. Genet. 10 (2), 77–82.

Wu, Y., Sarkissyan, M., Vadgama, J.V., 2015. Epigenetics in breast and prostate cancer. Methods Mol. Biol. 1238, 425–466.

Xiao, J., Gong, A.Y., Eischeid, A.N., Chen, D., Deng, C., Young, C.Y., et al., 2012. miR-141 modulates androgen receptor transcriptional activity in human prostate cancer cells through targeting the small heterodimer partner protein. Prostate 72 (14), 1514–1522.

Xue, G., Ren, Z., Chen, Y., Zhu, J., Du, Y., Pan, D., et al., 2015. A feedback regulation between miR-145 and DNA methyltransferase 3b in prostate cancer cell and their responses to irradiation. Cancer Lett. 361 (1), 121–127.

Yallapu, M.M., Khan, S., Maher, D.M., Ebeling, M.C., Sundram, V., Chauhan, N., et al., 2014. Anti-cancer activity of curcumin loaded nanoparticles in prostate cancer. Biomaterials 35 (30), 8635–8648.

Yamamura, S., Saini, S., Majid, S., Hirata, H., Ueno, K., Deng, G., et al., 2012. MicroRNA-34a modulates c-Myc transcriptional complexes to suppress malignancy in human prostate cancer cells. PLoS One 7 (1), e29722.

Yang, Y., Fuentes, F., Shu, L., Wang, C., Pung, D., Li, W., et al., 2017. Epigenetic CpG methylation of the promoter and reactivation of the expression of GSTP1 by astaxanthin in human prostate LNCaP cells. AAPS J. 19 (2), 421–430.

Yegnasubramanian, S., 2016. Prostate cancer epigenetics and its clinical implications. Asian J. Androl. 18 (4), 549–558.

Yoshimoto, M., Ludkovski, O., DeGrace, D., Williams, J.L., Evans, A., Sircar, K., et al., 2012. PTEN genomic deletions that characterize aggressive prostate cancer originate close to segmental duplications. Genes Chromosomes Cancer 51 (2), 149–160.

Yu, C., Gong, A.Y., Chen, D., Solelo Leon, D., Young, C.Y., Chen, X.M., 2013. Phenethyl isothiocyanate inhibits androgen receptor-regulated transcriptional activity in prostate cancer cells through suppressing PCAF. Mol. Nutr. Food Res. 57 (10), 1825–1833.

Zhang, C., Shu, L., Kim, H., Khor, T.O., Wu, R., Li, W., et al., 2016. Phenethyl isothiocyanate (PEITC) suppresses prostate cancer cell invasion epigenetically through regulating microRNA-194. Mol. Nutr. Food Res. 60 (6), 1427–1436.

Zhang, C., Su, Z.Y., Khor, T.O., Shu, L., Kong, A.N., 2013. Sulforaphane enhances Nrf2 expression in prostate cancer TRAMP C1 cells through epigenetic regulation. Biochem. Pharmacol. 85 (9), 1398–1404.

Zhang, L., Wang, H., 2015. Multiple mechanisms of anti-cancer effects exerted by astaxanthin. Mar. Drugs 13 (7), 4310–4330.

Zhang, Q., Padi, S.K., Tindall, D.J., Guo, B., 2014. Polycomb protein EZH2 suppresses apoptosis by silencing the pro-apoptotic miR-31. Cell Death Dis. 5, e1486.

CHAPTER

15

Prevention of Breast Cancer by Food Bioactives in Relation to Cancer Subtypes: Epigenetic Mechanisms

Micah G. Donovan[1,2], *Ornella I. Selmin*[1,2], *Donato F. Romagnolo*[1,2]

[1]**Department of Nutritional Sciences, The University of Arizona, Tucson, AZ, United States;** [2]**The University of Arizona Cancer Center, The University of Arizona, Tucson, AZ, United States**

Abstract

Breast cancer is the most common type of cancer and the leading cause of mortality among women worldwide. Aberrations in epigenetic regulation of tumor-suppressor genes contribute to hereditary and sporadic breast cancers. Accumulating evidence suggests a role for dietary compounds as epigenetic modifiers of breast cancer risk. However, given that breast cancer is a heterogeneous disease, it is important to consider variations in the epigenetic profile and potential response to food compounds across breast cancer subtypes. In this chapter, we first summarize research evidence regarding differences in epigenotype across breast tumor subtypes with emphasis on changes in DNA methylation, histone modifications, and the expression of noncoding RNA. Second, we discuss the role of epigenetic mechanisms in the silencing of breast cancer 1 (*BRCA1*), selected as a prototype early-onset breast cancer gene and epigenetic target by environmental and endogenous compounds that activate the aryl hydrocarbon receptor (AhR). Third, we highlight research evidence about food components (e.g., resveratrol and genistein) as dietary preventatives against epigenetic repression of *BRCA1* by AhR agonists and the development of estrogen receptor-negative and triple-negative breast cancer.

Keywords

Aryl hydrocarbon receptor; BRCA1; Breast cancer; Diet; DNA methylation; Epigenetics; Histone posttranslational modifications; Noncoding RNA.

INTRODUCTION

Estimates by the American Cancer Society indicate that approximately one in eight women will develop breast cancer during her lifetime (Nomura et al., 2016). In 2017 alone, the number of new breast cancer cases in the United States is expected to be ~255,180, whereas the number of deaths due to breast cancer is estimated at ~41,070 (American Cancer Society, 2017). Breast

Epigenetics of Cancer Prevention
https://doi.org/10.1016/B978-0-12-812494-9.00015-9

cancer is the leading cause of cancer mortality among women worldwide (Torre et al., 2017). Despite advances in screening and diagnostic and therapeutic technologies, breast cancer incidence and mortality increased by approximately 18 percent from 2008 to 2012. Recent projections suggested that global incidence will reach ~3.2 million new breast cancer cases per year by 2050 (Tao et al., 2015b). Yet, large differences exist in the development rates between Western countries (e.g., the United States and Western Europe) and other regions such as Eastern Asia and Central Africa. These variations are likely influenced by differences in access to early screening and detection technologies as well as lifestyle, diet, and environmental exposures (Torre et al., 2015). Given that most breast cancer cases (~90%) are sporadic—i.e., not linked to hereditary mutations—a better understanding of the mechanisms through which diet and environmental exposures affect the expression of breast cancer susceptibility genes is needed to improve the prevention and treatment of this malignancy.

Combined mutations in breast cancer 1 and 2 (*BRCA1* and *BRCA2*) susceptibility genes account for ~20% of familial cases (Curtit et al., 2015). Other mutated genes in hereditary tumors include tumor protein 53; phosphatase and tensin homolog (*PTEN*); cadherin 1; checkpoint kinase 2; serine/threonine kinase 11; ATM serine/threonine kinase (*ATM*); and partner and localizer of *BRCA2* (Sheikh et al., 2015; Apostolou and Fostira, 2013). The identification of these tumor-suppressor genes has provided new insight into genetic aberrations that drive mammary carcinogenesis. However, unlike other tumor types (e.g., colon cancer), defining a stepwise etiology of sporadic breast cancer has proven difficult (Tsubura et al., 2007; Malhotra et al., 2010). A traditional multistep model of breast carcinogenesis suggests that invasive carcinomas (ICs) arise from carcinoma in situ (CIS) that evolve from atypical hyperplasia in ductal or lobular regions of the mammary tree (Beckmann et al., 1997). However, recent evidence suggests that although precursor lesions (e.g., atypical hyperplasia and CIS) are high index risk factors for invasive tumors, lesions that transform likely generate breast tumor subtypes that are genetically and epigenetically distinct, requiring their further molecular characterization for therapy optimization (Makki, 2015).

Breast Cancer Subtypes

A recent study (Makki, 2015) proposed that there may be at least 29 distinct histological subtypes of breast carcinomas, which are usually classified as either CIS or IC. CIS are subdivided into ductal (DCIS) or lobular (LCIS). DCIS can be further subdivided into five distinct architectural subtypes. Unlike DCIS, subtyping of LCIS has not been implemented due to low histological variation. Invasive carcinomas are classified as either invasive ductal (IDC) or lobular (ILC). Fifteen subtypes of IDC have been described based on architectural features; cell type; amount, type, and location of secretion; and immunohistochemical profile. Five histological variants contribute to the ILC classification. The majority of ICs (~80%) are categorized as IDCs, whereas ILCs account for ~10–15% of breast tumor cases (Russo et al., 2005).

Current molecular classification of breast tumors is based on hormone receptor (HR) status comprising estrogen receptor α (ERα), progesterone receptor (PR), and human epidermal growth factor receptor 2 (HER2), a receptor tyrosine kinase overexpressed in some breast cancers by amplification of the *ERbB-2* gene. This receptor classification has led to the definition of the luminal A (ER+, PR+ and HER2-), luminal B (ER+, PR ± and HER2+/−), HER2-enriched (ER/PR- and HER2+), and basal-like and non-basal-like triple-negative breast cancer (TNBC, ER/PR/HER2) subtypes (Table 15.1) (Turner et al., 2004). These tumor subgroups complement other classifications and account for

TABLE 15.1 Intrinsic Molecular Subtypes of Breast Cancer

Subtype	Immunoprofile	Prevalence	Available Therapy
Luminal A	"Luminal A-like" ER & PR positive HER2 negative Ki-67 low (<14%)	30%–70%	α-Endocrine therapy[a]
Luminal B	"Luminal B-like" (HER2-) ER positive HER2 negative Ki-67 high (>20%) or PR negative	10%–20%	α-Endocrine therapy + chemotherapy (e.g., anthracycline-cyclophosphamide ± docetaxel)
	"Luminal B-like" (HER2+) ER positive HER2 amplified/overexpressed Any Ki-67 Any PR		α-Endocrine therapy + α-HER2[b] therapy + Chemotherapy (e.g., epirubicin and cyclophosphamide)
HER2-Enriched	"HER2 positive" (nonluminal) ER & PR negative HER2 amplified/overexpressed	5%–15%	α-HER2 therapy + chemotherapy (e.g., epirubicin and cyclophosphamide)
TNBC	"Basal like" ER, PR, and HER2 negative CK5/6 and/or EGFR positive EGFR positive	15%–20%	Chemotherapy (e.g., anthracycline-cyclophosphamide ± docetaxel)
	"Non-basal-like" ER, PR, and HER2 negative CK5/6 and EGFR negative		

[a] *SERMS (e.g., tamoxifen) and aromatase inhibitors.*
[b] *Trastuzumab (Herceptin).*
CK, cytokeratin; *EGFR*, epidermal growth factor receptor; *ER*, estrogen receptor; *HER2*, human epidermal growth factor receptor 2; *PR*, progesterone receptor; *TNBC*, triple-negative breast cancer.
Adapted from Toss, A., Cristofanilli, M. 2015. Molecular characterization and targeted therapeutic approaches in breast cancer. Breast Cancer Res., 17, 6; Rapoport, B.L., Demetriou, G.S., Moodley, S.D., Benn, C.A. 2014. When and how do I use neoadjuvant chemotherapy for breast cancer? Curr. Treat. Options Oncol., 15, 86–98.

heterogeneity regarding tumor behavior, response to targeted therapy, and prognosis (Ahn et al., 2016; Prat et al., 2015; Prat and Perou, 2011; Makki, 2015). Recently, microarray and RNA-Seq analyses of patient-derived breast samples identified transcriptional variations and unique molecular profiles within TNBC cases (Lehmann et al., 2011; Jovanović et al., 2017). Thus, TNBC could be regarded as a heterogeneous cancer comprising various subtypes including basal-like 1, basal-like 2, mesenchymal, mesenchymal stem-like, luminal androgen receptor, and immunomodulatory. Although TNBC represents only ~15–20% of breast tumor cases compared with other molecular subtypes, they are more aggressive in terms of metastatic potential, and as a result patients with TNBC have worse breast cancer prognosis overall.

Breast Cancer Subtypes and Therapies

Aromatase inhibitors (AIs) and selective estrogen receptor modulators (SERMs) have been shown to antagonize the proliferative effects of ERα signaling in HR-positive luminal A and luminal B breast tumors (Toss and Cristofanilli, 2015). AIs prevent the synthesis of estrogen (E2) from androgenic precursors by the enzyme aromatase, a cytochrome P450 (CYP19A1) (Ahmad and Shagufta, 2015). SERMs such as tamoxifen exert partial agonist/antagonist effects on ERα signaling, depending on cell and tissue type, and pure antagonistic effects on ERβ (Barkhem et al., 1998). Historically, agonists of ER were thought to mimic the actions of E2, whereas ER antagonists were believed to act through competitive inhibition of E2 binding (Wardell et al., 2014). However, molecular studies have demonstrated that SERMs may induce conformational changes in ERα and alter its ability to interact with transcriptional coactivators.

Antiendocrine therapies (SERMs and AIs) have been shown to significantly reduce HR-positive breast cancer recurrence and mortality (Dowsett et al., 2015). Similar success has been observed with targeted therapies in patients with tumors overexpressing *ERbB-2*. For example, trastuzumab is a humanized monoclonal antibody utilized for the treatment of HR-positive/luminal B HER2-positive and HR-negative/HER2-positive tumors. Compared with results from control patients, trastuzumab was shown to improve survival by 7% and 9.4% and overall survival by 3.8% and 8.8% in patients with HR-positive and -negative breast tumors, respectively (O'Sullivan et al., 2015). Although the arsenal of targeted therapies has expanded in parallel with increased molecular profiling, there unfortunately are no current specific therapies for TNBC (Toss and Cristofanilli, 2015), for which the prospects for treatment remain poor (Nass and Kalinski, 2015).

Epigenetics of Breast Cancer

Epigenetic events contribute to determining cell fate and phenotype in response to developmental and environmental cues (Felsenfeld, 2014; Boland et al., 2014). An epigenetic trait is defined as a stably heritable phenotype resulting from changes in gene expression without alterations in the DNA sequence and comprises changes in histone posttranslational modifications, DNA methylation, expression of DNA-binding proteins, and noncoding RNAs (ncRNAs) (Berger et al., 2009). Epigenetic dysregulation such as silencing of tumor-suppressor genes often occurs in the early stages of carcinogenesis and has been linked to neoplastic transformation and the development of breast cancer (Kanwal et al., 2015). Given that epigenetic changes are potentially reversible, epigenetic monitoring may improve the efficacy of breast cancer surveillance, and possibly treatment in refractory or otherwise nontargetable tumors (e.g., TNBC) (Esteller, 2007; Romagnolo et al., 2016).

DNA Methylation

DNA methylation is an epigenetic modification that refers to covalent addition of a methyl group (CH_3) to the C5 position of cytosine residues in cytosine-phosphate-guanine (CpG) dinucleotides (Marzese and Hoon, 2015; Dean, 2014). Methylation-sensitive gene promoters typically harbor CpG islands—i.e., a CpG-rich ($\geq$50%) region of DNA $\geq$200 bp in length with an observed/expected Cp:G ratio $\geq$60% (Saxonov et al., 2006; Gardiner-Garden and Frommer, 1987). DNA hypermethylation near gene promoters is generally associated with downregulation or gene silencing (Egger et al., 2004). In contrast, actively transcribed genes are typically CpG hypomethylated (Lister et al., 2009; Laurent et al., 2010; Zhang et al., 2009). However, exceptions to this model have been documented. For example, hypermethylation of the human telomerase reverse transcriptase gene has been classified as transcriptionally permissive (Won et al., 2002; Meeran et al., 2010; Berletch et al., 2008). Hyper- and hypomethylation of DNA have been observed in breast cancers, generally with silencing effects on tumor-suppressor genes and activating effects on proto-oncogenes. In breast tumors, genes silenced by DNA methylation include those encoding for factors involved in cell cycle regulation, DNA repair, growth inhibition, cell adhesion, angiogenesis, detoxification, hormonal regulation, and apoptosis (Tyagi et al., 2015).

DNA methylation is catalyzed by DNA methyltransferases (DNMTs), which transfer CH_3 from S-adenosyl-L-methionine (SAM) to cytosine substrates in CpG dinucleotides (Jurkowska et al., 2011). SAM is then converted to S-adenosyl-L-homocysteine (SAH), which in turn can act as a feedback inhibitor of DNMT (Saavedra et al., 2009; Isakovic et al., 2009). Three isoforms of DNMT have been identified in humans (DNMT1, DNMT3a, and DNMT3b) (Lee et al., 2005). The general structure of DNMT consists of an N-terminal regulatory and a C-terminal catalytic region tethered by a lysine- and glycine-rich KG linker domain (Yoo et al., 2013). Methylation reactions are catalyzed by nucleophilic attack at the C6 position of a CpG dinucleotide by a cysteine residue in motif IV of the DNMT catalytic domain (Jurkowska et al., 2011). Formation of a covalent bond between the enzyme and the nucleotide base imparts a partial negative charge on the C5 atom that is resolved through β-elimination. This allows the transfer of CH_3 from SAM to the C5 and release of the DNMT enzyme. Methylation at the C5 position of a CpG situates CH_3 within the major groove of DNA and is efficiently recognized by DNA-interacting proteins.

DNMT3a and DNMT3b are regarded as de novo DNMTs with similar affinity for both unmethylated and hemimethylated CpG sites (Okano et al., 1999). De novo DNMTs are responsible for adding new methylation CpG marks (Jurkowska et al., 2011) that are then carried through cell division by DNMT1, the "maintenance" DNMT, which has higher affinity for hemimethylated DNA (Lee et al., 2005). After DNA replication, parental strands harbor prereplication methylation patterns, and newly synthesized daughter strands are devoid of methylation marks. The affinity of DNMT1 for hemimethylated DNA allows for methylation of daughter strands and maintenance of the prereplication pattern. Therefore, inhibition of DNMT1 is a mechanism of passive DNA demethylation that can reduce global methylation by approximately 50% for every round of DNA replication.

DNMT inhibitors have been grouped into nucleoside and nonnucleoside analogues (Shilpi et al., 2015) (Table 15.2). Nucleoside analogues, such as 5-azacytidine (5-aza), function as antimetabolites and have gained United States Food and Drug Administration approval for the treatment of myelodysplastic syndrome (DeZern, 2015). Following cellular uptake, azacytidines are metabolically activated to 5-aza-2′-deoxycytidine-triphosphate nucleotides (Stresemann and Lyko, 2008). Then, 5-aza triphosphates enter the pool of available nucleotides and are substituted for cytosines during DNA replication. DNMT1 recognizes 5-aza-guanine dinucleotides as methylation substrates (Santi et al., 1984) with subsequent DNMT/5-aza adduct formation. Nitrogen substitution at the C5 atom in the 5-aza prevents β-elimination of the covalent bond and stabilizes the DNMT/5-aza adduct. This results

TABLE 15.2 Mechanisms of DNMT Inhibition

Mechanism	Compound
Antimetabolite; forms irreversible covalent bonds with DNMT1	5-Azacytidine
Competitive binding at cofactor binding site	SAH
Direct binding at catalytic site	EGCG, genistein
COMT-dependent SAH accumulation	Catechin, epicatechin, quercetin[a], myricetin[a], fisetin

[a] *May also have COMT-independent mechanism;* COMT, *catechol-O-methyltransferase;* EGCG, *epigallocatechin gallate;* SAH, *S-adenosyl-L-methionine.*

in DNMT1 degradation and passive DNA demethylation (Stresemann and Lyko, 2008).

Examples of nonnucleoside DNMT inhibitors include SAH, the synthetic analogues RG108 and NSC14778, and several dietary phytochemicals including epigallocatechin gallate (EGCG), genistein, curcumin, myricetin, and quercetin (Shilpi et al., 2015; King-Batoon et al., 2008). Nonnucleoside analogues suppress DNMT activity by binding to the active site or cofactor binding pockets (Shilpi et al., 2015). For example, genistein has been shown to inhibit DNMT1 activity through direct binding to the catalytic site (Xie et al., 2014). Catechol-containing polyphenols exert inhibitory effects on DNMT via a catechol-*O*-methyltransferase (COMT)-dependent mechanism (Lee et al., 2005). COMT-mediated reactions utilize SAM as a cofactor. The *O*-methylation of catechol-containing polyphenolic substrates leads to accumulation of SAH, which acts as a potent inhibitor of DNMT. EGCG found in tea is a potent DNMT inhibitor with an IC_{50} of $\sim$0.47 μM. Molecular modeling studies have demonstrated that the gallic acid moiety of EGCG can bind directly to the catalytic site of DNMT1 (Lee et al., 2005). In addition to EGCG, genistein and resveratrol have been shown to modulate DNA methylation in preclinical models of breast cancer (Xie et al., 2014; Mirza et al., 2013; King-Batoon et al., 2008; Berletch et al., 2008; Stefanska et al., 2012; Papoutsis et al., 2012) (Table 15.3).

Characterization of the "epigenotype" of breast tumors has shown that several genes are

TABLE 15.3 Effects of Selected Bioactive Compounds on DNA Methylation Based on Preclinical Breast Cancer Subtype

Class	Compound	Gene(s)	Subtype	Model	Dose (μM)	Time	Specific Effect	Ref
Catechins	EGCG	*hTERT*	Luminal A	MCF-7 cells	100	6/12d	Demethylation	Berletch et al. (2008)
		RARβ2, CCND2, MGMT	Luminal A	MCF-7 cells	10	96h	Demethylation	Mirza et al. (2013)
			TNBC	MDA-MB-231 cells	15			
Isoflavones	Genistein	*GSTP1*	Basal-like	MDA-MB-468 cells	3.125	6d	Demethylation	King-Batoon et al. (2008)
		PTEN, ATM, APC, SERPINB5	Luminal A	MCF-7 cells	60/100	48/72h	Demethylation	Xie et al. (2014)
			TNBC	MDA-MB-231 cells				
Stilbenoids	Resveratrol	*BRCA1*	Luminal A	MCF-7 cells	1	6d	Demethylation	Papoutsis et al. (2012)
		PTEN	Luminal A	MCF-7 cells	14	72h	Demethylation	Stefanska et al. (2012)
			TNBC	MDA-MB-231 cells	30			

APC, adenomatous polyposis coli; *ATM*, ataxia telangiectasia mutated; *BRCA1*, breast cancer 1; *CCND2*, cyclin D2; *EGCG*, epigallocatechin gallate; *GSTP1*, glutathione *S*-transferase; *hTERT*, human telomerase; *MGMT*, O6-alkylguanine DNA alkyltransferase; *PTEN*, phosphatase and tensin homolog; *RARβ2*, retinoic acid receptor β2; *TNBC*, triple-negative breast cancer.

differentially methylated across subtypes (Mathe et al., 2015) (Table 15.4). In TNBC, genes shown to be uniquely hypermethylated include *ESR1* (ERα), *PTEN*, and *BRCA1*, among others (Stefansson et al., 2011; Stefanska et al., 2012; Judes et al., 2016; Holm et al., 2010; Hafez et al., 2015; Coyle et al., 2016; Conway et al., 2014; Bediaga et al., 2010). Breast cancers overexpressing HER2 (HER2-enriched and luminal B HER2-positive) had hypermethylated *NPY, DBC1, FGF2, RASSF1, HS3ST2*, Let-7a, Sry-related HMG box member 17 (*SOX17*), and *TNFRS10D* (Bediaga et al., 2010). In contrast, HER2-enriched tumors harbored hypomethylated *Alu* and long interspersed nucleotide element-1 noncoding elements (Park et al., 2014). Luminal B tumors were found to harbor hypermethylated *GSTP1* and chitinase 3 like 2 (Holm et al., 2010). Luminal A tumors tended to have hypomethylated *SOX1* and *SOX17*, but harbored hypermethylated *MMP7, RASSF1, RASSF2, HDAC9, HIN-1*, and *17βHSD-1* (Perez-Janices et al., 2015; Hosny et al., 2016; Holm et al., 2010; Conway et al., 2014; Bediaga et al., 2010).

Histone Posttranslational Modifications

Histone posttranslational modifications (e.g., methylation, acetylation, ubiquitination, and phosphorylation) impart a transcriptionally permissive or repressive chromatin state (Egger et al., 2004). For example, dimethylation of histone 3 (H3) at lysine 4, trimethylation of H3 at lysine 4 (H3K4me3), and acetylation of H3 at lysine 4 (H3K4ac) associate with a transcriptionally permissive state (Messier et al., 2016; Baylin and Jones, 2011). On the other hand, trimethylation of H3 at lysine 9 (H3K9me3) and trimethylation of H3 at lysine 27 (H3K27me3) are generally considered transcriptionally repressive marks. Histone acetylation and deacetylation are carried out by histone acetyltransferases (HATs) and deacetylases (HDACs), respectively.

As summarized in Table 15.5, various dietary compounds influence histone posttranslational modifications in breast cancer models (Zhang et al., 2009; Tseng et al., 2017; Meeran et al., 2010; Markaverich et al., 2011; Li et al., 2010, 2013; Degner et al., 2009; Deb et al., 2015; Berletch et al., 2008). From these studies, it appears that EGCG may represent a promising

TABLE 15.4 Methylation Status of Selected Genes Based on breast Cancer Subtype

Subtype	Hypermethylated Genes	Hypomethylated Genes
Luminal A	*ABP3, FGF2, FZD9, GAS7, HDAC9, HIN-1, HOXA11, MME, MMP7, PAX6, PEG10, POMC, PTG2, RASSF1, RASSF2, RBP, VAMP8, 17βHSD-1*	*SOX1, SOX17*
Luminal B	*CHI3L2, DBC1*[a], *FGF2*[a], *GSTP1, HS3ST2*[a], *Let-7a*[a], *NPY*[a], *RASSF1*[a], *SOX17*[a], *TNFRS10D*[a]	N/A
HER2++	*DBC1, FGF2, HS3ST2, Let-7a, NPY, RASSF1, SOX17, TNFRS10D*	*Alu, LINE*
TNBC	*ARHGDIB, BCR, BRCA1,C4B, DAB2IP, DAPK1, ESR1, GRB7, MEST, PTEN, RARα, Rb, RNF8, SEMA3B, SEPT5, SERPINA5, TFF1, THY1, TWIST, VAMP8*	*DBC, EZH2, FGF2, GSTP1, HS3ST2, JAK3, NPY, PRKCDBP, RARRES1, RASSF1, SPARC, TNFRS10D*

[a] *Indicates HER2-positive luminal B cases specifically.*

TABLE 15.5 Effects of Selected Bioactive Compounds on Histone Modifications in Preclinical Breast Cancer Subtypes

Class	Compound	Gene	Subtype	Model	Dose	Time	Epigenetic Effect	Ref
FLAVONES								
	Luteolin	*PLK1*	Luminal A	MCF-7 cells	17.5 μM	24h	Deacetylation of H4	Markaverich et al. (2011)
	Apigenin	*p21*$^{WAF1/CIP1}$	TNBC	MDA-MB-231 cells MDA-MB-231 xenografts	40 μM 25 mg/kg/d	48h 9w	HDAC inhibition, Increased acetyl-H3	Tseng et al. (2017)
CATECHINS								
	EGCG	*hTERT*	Luminal A	MCF-7 cells	100 μM	6/12d	Increased H3K9Ac	Berletch et al. (2008)
		TIMP3	Luminal A TNBC	MCF-7 cells MDA-MB-231 cells	20 μM	3/7d	Increased H3K9Ac & H3K18Ac; Decreased H3K27me3; HDAC8 inhibition; HDAC1 & 3 inhibition	Deb et al. (2015)
		ERα	TNBC	MDA-MB-231 cells	10 μM	72h	Increased acetyl-H3, acetyl-H4, H3K9Ac, & H3K4me2; Decreased H3K9me3; HAT activation; HDAC inhibition	Li et al. (2010)
ISOFLAVONES								
	Genistein	*ERα*	TNBC	MDA-MB-231 cells MDA-MB-157 cells	25 μM	72h	Increased acetyl-H3, acetyl-H4, H3K9Ac, H3K4me2; HDAC1 inhibition	Li et al. (2013)
ISOTHIOCYANATES								
	DIM	*COX-2*	Luminal A	MCF-7 cells	10 μM	1.5h	Decreased acetyl-H4	Degner et al. (2009)
	Sulforaphane	*hTERT*	Luminal A TNBC	MCF-7 cells MDA-MB-231 cells	10 μM	6d	Increased acetyl-H3, acetyl-H4; Decreased H3K9me3, H3K27me3.	Meeran et al. (2010)

DIM, diindolylmethane; *EGCG*, epigallocatechin gallate.

bioactive-based HDAC inhibitor compared with other compounds, based on the relatively lower concentration (10 μM) needed to impart an effect. However, further studies, particularly in animal models, are necessary in order to investigate the impact of EGCG on breast tumor development and progression.

Luminal A tumors (MCF-7 cells) and TNBC (MDA-MB-231 cells) likely differ in regard to their H3K4me3 and H3K4ac profiles. For example, enrichment of H3K4me3 was seen at chromosome 19 in MCF-7 cells, in contrast to MDA-MB-231 cells, which had reduced H3K4me3 at chromosome 22. In regard to histone acetylation, H3K4ac was markedly higher in MCF-7 than it was in MDA-MB-231 cells and was associated with genes involved in E2 and EMT pathways. An increase in global H3K4ac has been associated with both early- and late-stage cell phenotypes, whereas accumulation of H3K4me3 has been linked to later tumor stages (Messier et al., 2016).

Studies with breast tumor samples have sought to identify subtype-specific differences in the enrichment of acetyl-H3K4, acetyl-H3K9, and H3K27me3 across a panel of genes implicated in tumor initiation and progression (*BRCA1, ERS2, PGR, ESR1, EZH2, P300,* and *SRC3*) (Judes et al., 2016). Tumors overexpressing HER2 (HER2-enriched and luminal B HER2-positive) displayed enrichment of acetyl-H3K4 and −H3K9, and decreased H3K27me3, associated with the *BRCA1* promoter, compared with their presence in other tumor subtypes. Conversely, in HER2-negative tumors (i.e., luminal A, luminal B and HER2-negative, and basal-like), there were no significant differences in H3K4ac, H3K9ac, and H3K27me3 marks at the *BRCA1* promoter. Luminal A and B/HER2-negative tumors had decreased acetylated histone marks and increased H3K27me3 associated with *ESR1, PGR,* and *ESR2* compared with levels observed in luminal B HER2-positive, HER2-enriched, and basal-like tumors. Finally, luminal B/HER2-positive, HER2-enriched, and basal-like tumors had increased acetylated histone marks and decreased H3K27me3 associated with *EZH2, p300,* and *SRC3* (Judes et al., 2016). Overall, these observations suggest that systematic cataloging of the differences in histone modifications (i.e., acetylation, methylation, and phosphorylation) at specific promoters could be useful for profiling tumor subtypes and developing targeted therapies.

Noncoding RNA

ncRNAs are transcripts that are not translated but that regulate the expression of target genes (Wang et al., 2017). Small ncRNAs are less than 200 base pairs long and include tRNA, rRNA, microRNA (miRNA), and Piwi-interacting RNA. Long ncRNAs (lncRNAs) are >200 base pairs long and are classified based on genetic location (template or nontemplate strand) and transcriptional direction (sense or antisense) (Wang et al., 2009). Aberrant expression of miRNA and lncRNA has been reported in all subtypes of breast cancer (Xu et al., 2016). MicroRNAs target mRNA transcripts and block their translation (Ambros, 2004). Signal, guide, and scaffold lncRNAs positively regulate gene expression, whereas decoy lncRNAs have a negative influence on expression.

As demonstrated in Table 15.6, dietary bioactives including luteolin, quercetin, genistein, glabridin, indol-3-carbinol (I3C), diindolylmethane (DIM), and resveratrol exert modifying effects on miRNA expression (Tao et al., 2015a; Sun et al., 2015; Jiang et al., 2016; Hargraves et al., 2016; Avci et al., 2015; Vislovukh et al., 2013; Venkatadri et al., 2016; Singh et al., 2014; Jin, 2011). Despite this accumulating evidence, in vitro studies have demonstrated that supraphysiological levels (>100 μM) are needed in order to significantly impact miRNA expression, which may represent a clinical limitation. On the other hand, studies using xenograft models have shown that modulation of miRNA expression is achievable through dietary exposure to certain compounds (e.g., DIM and quercetin).

TABLE 15.6 Effects of Selected Bioactive Compounds on miRNA Expression in Breast Cancer Subtypes

Class	Compound	miRNA (Up- or Downregulation)	Subtype	Model	Concentration/ Dose	Time	Gene/ Pathway	Ref
FLAVONES								
	Luteolin	miR-181a, miR-139-5p, miR-224, and miR-246 (up); miR-155 (down)	Luminal A	MCF-7	100 μM	24 h	Notch	Sun et al. (2015)
			TNBC	MDA-MB-231				
FLAVONOLS								
	Quercetin	miR-146a (up)	Luminal A	MCF-7	25–100 μM	48 h	EGFR	Tao et al. (2015a)
			TNBC	MDA-MB-231				
				MDA-MB-231 xenografts	10 mg/kg/d	8 weeks		
ISOFLAVONES								
	Genistein	miR-23b (up)	Luminal A	MCF-7	175 μM	48 h	*PAK2*	Avci et al. (2015)
ISOFLAVANES								
	Glabridin	miR-148a (up)	TNBC	MDA-MB-231 Hs-578T	10 μM	72 h	TGFβ/ SMAD	Jiang et al. (2016)
				MDA-MB-231 xenografts	20 mg/kg/d	6 weeks		
ISOTHIOCYANATES								
	I3C	miR-34a (up)	Luminal A	MCF-7	200 μM	48 h	*CDK4*	Hargraves et al. (2016)
	DIM	miR-21 (up)	Luminal A	MCF-7 cells MCF-7 xenografts MDA-MB-469 cells	30 or 60 μM 5 mg/kg/d 30 or 60 μM	24–96 h 7 weeks 24–96 h	*CDC25 A*	Jin (2011)
			TNBC	MDA-MB-468 xenografts	5 mg/kg/d	7 weeks		
Stilbenoids	Resveratrol	miR-663, miR-744 (up) miR-125b-5p, miR-200c-3p, miR-409-3p, miR-122-5p and miR-542-3p (up) miR-93	Luminal A	MCF-7 cells	100 μM	24 h	eEF1A2	Vislovukh et al. (2013) Venkatadri et al. (2016)
			Luminal A	MCF-7 cells	200 μM	12 h	Apoptosis	
			TNBC	MDA-MB-231 cells	50 mg SubQ pellet each month	8 weeks	Cell cycle Akt signaling	
			HR-positive	ACI rats				Singh et al. (2014)

Sequencing studies revealed that 511 miRNAs and 3580 lncRNAs were differentially expressed in breast tumors compared with their expression in normal tissue (Xu et al., 2016). Further examination of breast tumor subtypes informed that 79 and 157 miRNAs were aberrantly expressed in luminal and basal-like tumors, respectively. Differential expression was observed for 1224 lncRNAs in luminal and 511 lncRNAs in basal-like breast cancers. The majority of deregulated ncRNAs (~70%) were noted in luminal and basal subtypes, although the mechanisms leading to their deregulation remains largely unknown. miR-200b-3p was overexpressed in both luminal and basal tumors; however, this was associated with increased enrichment of H3K4me1, H3K4me3, and H3K27ac, as well as decreased H3K27me3, in the luminal subtypes, whereas reduced H3K27ac and H3K27me3 were observed in the basal subtype. Overexpression of miR-21 and downregulation of miR-125b-1 have been documented in luminal and basal subtypes. The expression of lncRNA HOTAIR, which has been linked to induction of proliferation, metastasis, and angiogenesis, was higher in luminal and basal-like subtypes and accompanied by enrichment of H3K4me1, H3K4me3, and H3K27ac, and a reduction in H3K27me3 (Xu et al., 2016; Wang et al., 2017). In contrast, the expression of lncRNA GAS5, which promotes growth arrest and apoptosis and prevents metastasis, was downregulated in parallel with decreased H3K4me and H3K27me3, and increased H3K27ac, in luminal subtypes. Overexpression of lncRNA GAS5 in basal tumors was associated with enrichment of H3K36me3.

Studies of 823 breast tumor subtypes revealed deregulation of 42 lncRNAs in luminal A; 9 in luminal B; 14 in HER2-enriched; and 74 in basal-like tumors (Van Grembergen et al., 2016). In luminal A tumor samples, deregulation of lncRNA was associated with activation of the PI3K pathway and EMT. Signatures associated with luminal B tumors included activation of RAF and Myc, which are strong proliferative factors. Aberrant lncRNA signatures observed in HER2-enriched tumors were related to activation of mTOR and PI3K pathways. In basal-like tumors, lncRNA coexpressed with E2F1 target genes, interleukin-6, and epidermal growth factor receptor (EGFR).

To identify breast tumor subtype-specific differences in miRNA profiles, a study analyzed the expression level of 725 miRNAs in 51 breast cancer cell lines (Riaz et al., 2013). Results indicated that 40 miRNAs were differentially expressed between basal-like and non-basal-like TNBC cells. In regard to luminal tumors, expression of 39 miRNAs was associated with HER2 overexpression, and 24 miRNAs were associated with luminal tumors harboring mutated E-cadherin. In nonluminal breast tumor cells, which often have hypermethylated E-cadherin and overexpress HER2, expressions of hsa-let-7b, miR-640, miR-200c, miR-378, miR-141, miR-196a, miR-29c, and miR-18a were higher than in other breast tumor subtypes. Moreover, overexpression of miR-501-5p, miR-202, miR-760, and miR-62 was associated with HER2-negative luminal cells. In claudin-low cells, expression of miR-22, miR-532-3p, miR-125b, miR-501-5p, and miR-155 was elevated, whereas miR-492, miR-26b, miR-617, and miR-155 were highly expressed in basal-like breast cancer subtypes. Furthermore, studies showed opposing expression patterns for oncogenic hsa-miR-155 versus hsa-miR-155* when comparing basal-like and claudin-low breast cancer cell subtypes (Riaz et al., 2013). Finally, studies with formalin-fixed paraffin-embedded breast tumors reported that TNBC ($n = 24$) had higher expressions of miR-17-5p, miR-18a-5p, and miR-20a-5 than those of luminal A ($n = 20$) tumors (Calvano Filho et al., 2014). Overall, these cumulative data suggest that cataloging differential miRNA expression may be useful in profiling breast tumor subtypes and developing targeted therapies.

Breast Cancer Susceptibility Genes and Dietary Prevention

Breast cancer susceptibility genes (e.g., *BRCA1*) are clearly major targets for epigenetic therapy, particularly in ERα-negative breast cancer and TNBC. For example, estimates indicate an approximately fourfold increase in the lifetime risk of breast cancer development in women who inherit loss-of-function mutations in *BRCA1* (Pan et al., 2014). The tumor-suppressor activities of the BRCA1 protein are attributed to its role in the regulation of DNA repair, cell cycle progression, transcription, apoptosis, and mRNA splicing (Savage and Harkin, 2015). Breast cells that lack BRCA1 functions have inefficient repair of DNA double-strand breaks via homologous recombination and utilize nonconservative, potentially mutagenic, pathways such as nonhomologous end joining and single-strand annealing (Turner et al., 2004; Savage and Harkin, 2015). BRCA1-deficient breast cells evade the cell cycle arrest and apoptosis that are normally induced by the unrepaired DNA damage, also resulting from p53 mutations, that has been detected in approximately 67 percent of familial *BRCA1* tumors (Greenblatt et al., 2001; Crook et al., 1998). Amplification of proliferative c-MYC has been observed in familial *BRCA1*-mutated (~53 percent) and sporadic *BRCA1*-hypermethylated breast cancers (Palacios et al., 2003; Turner et al., 2004; Grushko et al., 2004). *BRCA1*-mutation tumors also showed overexpression of *EGFR* (Turner et al., 2004) linked to the development of basal-like TNBC (Lehmann et al., 2011).

Epigenetic Silencing of Breast Cancer 1

BRCA1-mutations are associated with poor tumor prognosis (Antoniou et al., 2006). The vast majority of familial *BRCA1* tumors (greater than 80 percent) are TNBC (Lakhani et al., 2002; Turner et al., 2007; Joosse et al., 2009; Silver et al., 2010) and share gene expression profiles with the basal-like sporadic subtype (Sorlie et al., 2003). This overlap between hereditary and sporadic breast tumors associated with reduced BRCA1 has led to the definition of a "BRCAness" phenotype (Turner et al., 2004). Importantly, tumors arising from germline mutations contribute only ~5%–10% of cases (Duncan et al., 1998; Apostolou and Fostira, 2013), whereas somatic *BRCA1* mutations are not found in sporadic cases (Futreal et al., 1994). Thus, the loss of BRCA1 in sporadic breast tumors could result from epigenetic silencing (Turner et al., 2004) that is particularly prevalent in ERα-negative breast cancer and TNBC. Bisulfite sequencing of genomic DNA obtained from ERα-negative cells (UACC-3199) revealed that the proximal *BRCA1* promoter was methylated at all 36 CpG sites, whereas ERα-positive breast cancer cells had little or no *BRCA1* CpG methylation (Rice et al., 1998). Similarly, patient-derived TNBC showed a ~6.5-fold higher incidence of *BRCA1* CpG methylation than that of healthy control tissue and tumor samples from other molecular subtypes (Romagnolo et al., 2015). Methylation-specific analysis of DNA from 24 formalin-fixed paraffin-embedded TNBCs indicated that ~40% of samples had hypermethylated *BRCA1* (Foedermayr et al., 2014). A meta-analysis (Zhang and Long, 2015) of the association between *BRCA1* methylation and pathological features of breast cancer subtypes reported strong associations between *BRCA1* methylation and ER-negative (OR = 2.36), PR-negative (OR = 2.14), and triple-negative (OR = 2.79) phenotypes. Additionally, hypermethylation of *BRCA1* was significantly associated with increased risk of breast cancer overall (OR = 3.15), lymph node metastasis (OR = 1.25), and higher histological grade (OR = 2.29). Thus, unraveling the mechanisms contributing to hypermethylation of *BRCA1* may hold clinical relevance for the prevention of hereditary and sporadic ERα-negative breast cancer and TNBC.

The *BRCA1* gene spans ~80 kb at the chromosomal 17q21.31 location and consists of 22 coding and 2 noncoding exons (Xu et al., 1995; Karami and Mehdipour, 2013). Two distinct transcriptional start sites have been identified in exon 1a and exon 1b (Xu et al., 1995; Papoutsis et al., 2012). Activator protein-1 binding sites, which interact with Jun and Fos transcription factors, have been located upstream of exon 1b (Jeffy et al., 2005). *BRCA1* also harbors consensus xenobiotic response elements (XRE = GCGTG) for the aryl hydrocarbon receptor (AhR) in exon 1a, the intervening sequence between exon 1a and 1b, and exon 1b (Hockings et al., 2006). Ligands of the AhR include exogenous (environmental and dietary) and endogenous compounds (Table 15.7). The 5′ region of the *BRCA1* promoter (~1.6 Kb) harbors a CpG island based on its 56% CpG content (Rice et al., 1998). Sodium bisulfite sequencing of genomic DNA from normal human mammary epithelial cells, peripheral blood lymphocytes, and six sporadic breast cancer cell lines confirmed that 36 CpG sites harbored in the 5′ promoter region (−567 to +255) of *BRCA1* were targets for methylation (Rice et al., 1998).

TABLE 15.7 Exogenous and Endogenous AhR Ligands

Source	Examples
EXOGENOUS	
Xenobiotic	2,3,7,8-Tetrachlorodibenzo-p-dioxin 7,12-Dimethyl-benz[a]anthracene Benzo[a]pyrene Biphenyls Polycyclic aromatic hydrocarbons Halogenated aromatic hydrocarbons 3,3′,4,4′,5-Pentachlorobiphenyl
Dietary	Quercetin Galangin Indole 3-carbinol 3-3′-Diindoylmethane Resveratrol Curcumin
ENDOGENOUS	
Tryptophan metabolites	6-Formylindolo[3,2b]carbazole Kynurenic acid
Arachidonic acid metabolites	Lipoxin 4a
Heme degradation	Bilirubin Biliverdin

In ERα-positive breast epithelial cells, E2 was found to induce *BRCA1* transcription through a noncanonical pathway (Powell et al., 2012; Marks et al., 1997; Lipovka and Konhilas, 2016) involving the recruitment of a heterocomplex comprising ERα and the coactivators p300, SRC-1 (Jeffy et al., 2005), Sp-1, and Sp-4 (Hockings et al., 2008). The activation of *BRCA1* by ERα is believed to be a mechanism that controls DNA repair as cells progress through the cell cycle under the proliferative pressure of E2. Thus, exposure to agents that disrupt *BRCA1* expression is expected to increase the susceptibility to tumor development, in particular in E2-responsive tissues (i.e., breast and ovary). For example, work by our group highlighted the potential role of AhR ligands as an epigenetic disruptor of *BRCA1*. We showed that recruitment of activated AhR to XRE sequences harbored in *BRCA1* antagonized the stimulatory effect of E2 on *BRCA1* transcription (Papoutsis et al., 2010, 2012). Epigenetic signatures contributing to transcriptional repression of *BRCA1* by the AhR included CpG hypermethylation, enrichment of H3K9me3, and loss of acetyl-H4, acetyl-H3, and AcH3K9. Other repressive changes noted at the *BRCA1* gene were decreased recruitment of p300 (an HAT) and SRC-1 as well as increased association with DNMT (−1, -3a, and -3b), methyl-binding proteins (i.e., MBD2), and HDACs (i.e., HDAC1) (Papoutsis et al., 2012).

Prevention of Breast Cancer 1–Silencing With Dietary AhR Antagonists

Many dietary compounds exert antagonistic effects toward the AhR (Table 15.8). For example, we reported that the grape phytoalexin resveratrol prevented recruitment of the AhR to XRE sequences in the *BRCA1* promoter and its epigenetic repression (Papoutsis et al., 2010). This work established a proof-of-principle for

TABLE 15.8 Dietary AhR Antagonists

Dietary Bioactive	Mechanism of AhR Inhibition
FLAVONES	
Flavone Apigenin Luteolin	Competitive inhibitors of TCDD binding, prevent nuclear translocation and dissociation from chaperone complex
FLAVONOLS	
Galangin Kaempferol Quercetin	Competitive inhibitors of TCDD binding, prevent nuclear translocation and dissociation from chaperone complex
FLAVANONES	
Naringenin Eriodictyol	Prevention of ARNT/AhR phosphorylation through MEK/ERK inhibition
CATECHINS	
Epigallocatechin EGCG	Prevention of ARNT/AhR phosphorylation through MEK/ERK inhibition
ISOFLAVONES	
Genistein	Decrease AhR nuclear localization
ISOTHIOCYANATES	
DIM	Induce conformational change in AhR that allows DNA binding but prevents recruitment of RNA Pol II and CBP (rat hepatoma cells)
STILBENOIDS	
Resveratrol	Decrease AhR binding at DNA XRE sequences

dietary compounds with antagonism toward the AhR as tools against epigenetic disruption. In support of this idea, flavones from celery, parsley, pepper (e.g., flavones, such as apigenin and luteolin) and flavonols from onions, kale, and cruciferous vegetables (galangin and kaempferol) were shown to act as competitive inhibitors of the AhR and prevent its nuclear translocation to target genes (Mukai et al., 2010). In MCF-7 breast cancer cells, DIM, a condensation product from I3C found in cruciferous vegetables, was shown to antagonize AhR-dependent activation of *CYP1A1*, a putative AhR target gene (Hestermann and Brown, 2003). Results from our laboratory also documented that in ERα-positive MCF-7 cells, treatment with DIM abrogated recruitment of the AhR to XRE sequences harbored in the proinflammatory *COX-2* gene (Degner et al., 2009). Flavanones commonly found in *Citrus* fruits (e.g., naringenin and eriodictyol) and catechins present in tea (epigallocatechins such as EGCG) may be effective breast cancer preventive agents because they inhibit binding of AhR and ARNT to XRE. Therefore, bioactive compounds that exert inhibitory effects on the AhR (Beedanagari et al., 2009) may offer a new line of prevention against epigenetic repression of tumor suppressors (e.g., *BRCA1*), and activation of tumor promoters (e.g., *COX-2* and *CYP1B1*) in breast tissues.

The effect of dietary compounds on the AhR may be dose-dependent and influenced by interactions with other bioactive compounds and carcinogens. For example, biochanin A, an *O*-methylated metabolite of genistein, was shown to antagonize 7,12-dimethylbenz[a]anthracene (DMBA)-induced recruitment of the AhR to *CYP1A1*. DMBA is a mammary carcinogen and strong AhR agonist. However, when biochanin A was used alone or at high doses (50 μM), it actually enhanced AhR recruitment (Han et al., 2006). These apparently contrasting dose-dependent agonist/antagonist effects on the AhR have been well characterized using α-naphthoflavone (NF, 7,8-benzoflavone), a synthetic compound and prototypical AhR modulator. Depending on concentration, NF may act as an agonist ($\geq$10 μM) or antagonist ($\leq$10 μM) of AhR activity (Cui and Li, 2013). In human MCF-7 cells, the antagonistic effects of NF have been attributed to direct binding to the AhR and production of an AhR complex with lower binding affinity for XRE (Merchant et al., 1993; Gasiewicz and Rucci, 1991). These examples with biochanin A and NF clearly suggest the need for testing the dose-dependent effects of dietary compounds against epigenetic regulation by the AhR in normal breast tissue and breast tumors.

Soy Isoflavones

A diet high in soy is thought to underlie the lower rates of breast cancer incidence among Asian women compared with those of women residing in Western countries (Tham and Haskell, 1998). For Asian women who have migrated to Western countries, the risk of breast cancer has been shown to increase over several generations to levels comparable to those of Western white women. Interestingly, migrant women living in the United States for ≥10 years appeared to have ~80% greater risk of breast cancer compared with that of women who migrated recently (Ziegler et al., 1993). These increases in breast cancer risk associated with migration have been attributed at least in part to changes in dietary pattern (Russo et al., 2016; Yang et al., 2012). In support of this hypothesis, estimates from meta-analyses found a 25%–39% decrease in the risk of breast cancer that was associated with the higher consumption of soy and soy foods (Wu et al., 2015; Woo et al., 2014; Qin et al., 2006; Liu et al., 2014).

Genistein and daidzein are major soy isoflavones. Studies reported an ~11–40% decrease in the risk of breast cancer incidence among Asian women who had higher intakes of soy isoflavones (Dong and Qin, 2011; Chen et al., 2014). Intake of soy and soy isoflavones has been associated with lower risk of breast cancer in pre- and postmenopausal women (Trock et al., 2006; Chen et al., 2014). Food sources of isoflavones include fava and garbanzo beans, coffee, and various botanicals (Kaufman et al., 1997; Alves et al., 2010). In Asian populations overall, consumption estimates suggest intakes of up to ~80 mg of total isoflavones/day (Barnes et al., 2011), whereas the average daily intake for Japanese people ranged from 39.4 to 47.4 mg/d (Chen et al., 1999; Arai et al., 2000). In a multiethnic cohort that consisted of Japanese American, White, Latino, African American, and Native Hawaiian subjects, quintiles for the median daily intakes of isoflavones were lower (1.7, 4.8, 9.1, 16.0, and 29.6 mg/d) (Morimoto et al., 2014), and in US adults, consumption of soy isoflavones averaged only ~1.0 mg/day (Chun et al., 2009).

In cells, concentrations of genistein at <10 μM tend to have proliferative effects, whereas levels >10 μM of genistein are usually antiproliferative (Taylor et al., 2009). Genistein is considered a phytoestrogen as it binds to ER, and two isoforms, ERα and ERβ, have been identified (Russo et al., 2016). Importantly, activation of ERβ counteracts the proproliferative signaling of ERα in breast cancer cells (Wisinski et al., 2016; Paruthiyil et al., 2004). Therefore, the greater affinity (6–10-fold) of genistein for ERβ over ERα is believed to account for some of its antiproliferative actions (Margeat et al., 2003; Kuhl, 2005). Additionally, genistein preferentially stimulates ERβ binding at ERE (Kostelac et al., 2003) and is more efficient at recruiting coactivators (e.g., SRC-1) to ERβ than to ERα (Margeat et al., 2003).

Genistein may exert anticancer effects through repression of DNA methylation. For example, in MDA-MB-468 cells, which are a model of TNBC, a 6-day treatment with noncytotoxic levels of genistein (3.125 μM) decreased CpG methylation and increased the expression of *GSTP1*. However, in ERα- and PR-positive but HER2-negative MCF-7 cells, equimolar genistein failed to elicit an epigenetic response (King-Batoon et al., 2008). At higher doses (60 and 100 μM), genistein decreased global DNA methylation irrespective of ERα status (Xie et al., 2014). Genistein was also shown to CpG-demethylate various tumor-suppressor genes including *PTEN*, *ATM*, *APC*, and *SERPINB5*. These changes were associated with decreased DNMT activity and DNMT1 protein levels in MCF-7 and MDA-MB-231 cell lines, whereas the levels of DNMT3a and DNMT3b were not affected. Modeling studies revealed that genistein bound to the catalytic domain of

DNMT1 and interfered with its enzymatic activity (Xie et al., 2014).

Changes in histone modifications have also been reported for genistein in cell culture models of TNBC (MDA-MB-231 and MDA-MB-157). A significant increase in ERα transcription was observed after exposure to genistein (25 μM) for 3 days. This was associated with enrichment of acetyl-H3, acetyl-H3K9, acetyl-H4, and dimethyl-H3K4 at the *ESR1* promoter; overall inhibition of HDAC and DNMT; and decreased binding of HDAC1 and DNMT1 at the *ESR1* promoter (Li et al., 2013). Additionally, genistein in combination with trichostatin A (TSA), an HDAC inhibitor, reactivated ERα expression, which sensitized cells to the cytotoxic effects of tamoxifen. In ERα-positive MCF-7 cells, cytotoxicity induced by genistein ($IC_{50} = 175$ μM within 48 h) has been linked to reactivation of miR-23b expression (~57-fold) (Avci et al., 2015), which is usually silenced via DNA methylation in cancers. miR-23b exerts a putative tumor-suppressive role through the inhibition of EMT, invasion, and metastasis (Zhang et al., 2011; Pellegrino et al., 2013; Majid et al., 2012). These cumulative data support the notion that regimens based on genistein alone or in combination with epigenetic drugs (TSA, 5-aza) may offer novel opportunities for epigenetic therapy of breast tumors irrespective of ER status.

With respect to the epigenetic effects of genistein on AhR-induced breast tumorigenesis, results have been somewhat inconclusive. Some studies have shown that genistein can synergize with ERα and AhR-carcinogens to induce mammary tumorigenesis in rodent models (Hsieh et al., 1998; Day et al., 2001). Conversely, other studies have found that prepubertal exposure to genistein protected against adult AhR-induced mammary tumorigenesis (Murrill et al., 1996; de Assis et al., 2011), and that physiological concentrations (0.5–1.0 μM) of genistein stimulated *BRCA1* expression in cell culture models (Fan et al., 2006). Recently, our laboratory reported that genistein reduced the expression of DNMT1 and antagonized AhR-induced *BRCA1* CpG methylation in MCF-7 breast cancer cells (Romagnolo et al., 2017). We also documented that genistein demethylated *BRCA1* and *ESR1* in ERα-negative UACC-3199 cells harboring constitutively high levels of AhR. Similar protective effects have been reported for the flavonol galangin (3,5,7-trihydroxyflavone), an AhR antagonist(IC_{50} ~0.2 μM) (Ashida et al., 2000) that shares structural similarities with genistein. Galangin was shown to inhibit the growth of ERα-negative breast cancer cells that expressed high levels of AhR (Murray et al., 2006). Therefore, future studies should investigate the epigenetic responses to compounds such as genistein and galangin in AhR-rich tumors and interactions with ERα status.

CONCLUSIONS

Breast cancer subtypes harbor distinct epigenetic signatures that may influence responsiveness to therapy (Miozzo et al., 2015). Given that epigenetic changes are reversible and often occur early in neoplastic evolution, food components may represent attractive breast cancer preventatives. Dietary compounds, such as flavones, flavonols, catechins, isoflavones, isothiocyanates, and stilbenoids, influence DNA methylation, histone modification, and miRNA expression in breast cancer models. Epigenetic silencing of *BRCA1* increases breast cancer risk and is associated with ERα-negative breast cancer and the TNBC phenotype. Work by our group demonstrated a role for the activated AhR in the induction of CpG methylation and epigenetic repression of *BRCA1*, and epigenetic protection by dietary compounds (e.g., resveratrol, genistein) (Fig. 15.1). More research is needed to elucidate the potential for the epigenetic prevention of hereditary and sporadic breast tumors using food bioactives.

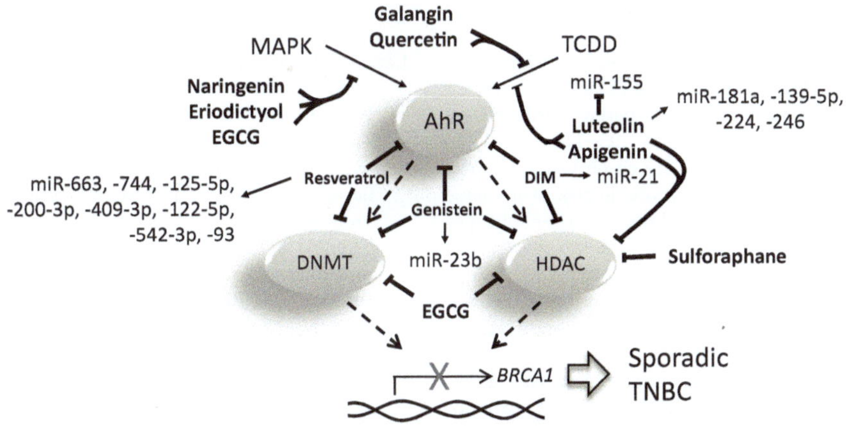

FIGURE 15.1 Potential opportunities for dietary bioactive epigenetic therapy in triple-negative breast cancer (TNBC). TNBC etiology is associated with loss of breast cancer 1 gene (BRCA1) activity. BRCA1 is transcriptionally repressed by epigenetic mechanisms (e.g., DNA methylation and histone deacetylation), which are coordinated by the activated aryl hydrocarbon receptor (AhR). AhR is activated by exposure to exogenous (2,3,7,8-Tetrachlorodibenzo-p-dioxin) and endogenous ligands and is stimulated by the mitogen-activated protein kinase (MAPK) pathway. Dietary bioactives may play a preventative role at several levels in this pathway. At the level of the AhR, flavones (e.g., luteolin and apigenin) and flavonols (e.g., galangin and quercetin) inhibit AhR activation through competitive inhibition of ligand binding; flavanones (e.g., naringenin and eriodictyol) and catechins [e.g., epigallocatechin and epigallocatechin gallate (EGCG)] suppress AhR activation through inhibition of the MAPK pathway; and genistein (isoflavone), resveratrol (stilbenoid), and DIM (isothiocyanate) have all demonstrated antagonistic effects on AhR transcriptional activity. At the level of epigenetic machinery, dietary compounds have been shown to inhibit DNA methyltransferases (DNMTs) (resveratrol), histone deacetylases (HDACs) (sulforaphane, DIM, luteolin, and apigenin), and both DNMTs and HDACs (genistein and EGCG). Several of these compounds also modulate the expression of various miRNAs involved in breast cancer tumorigenesis. Dashed lines with arrows show the flow of the model pathway. Filled arrows represent stimulatory effects. Blunted arrows represent inhibition.

Acknowledgments

This work was supported by grants from the Arizona Biomedical Research Commission (QSR14082995); the Soy Health Research Program; and the United States Department of Defense Breast Cancer Program (DAMD-14-1-0470 and DAMD-15- 1–0387).

References

Ahmad, I., Shagufta, 2015. Recent developments in steroidal and nonsteroidal aromatase inhibitors for the chemoprevention of estrogen-dependent breast cancer. Eur. J. Med. Chem. 102, 375–386.

Ahn, S.G., Kim, S.J., Kim, C., Jeong, J., 2016. Molecular classification of triple-negative breast cancer. J. Breast Cancer 19, 223–230.

Alves, R.C., Almeida, I.M., Casal, S., Oliveira, M.B., 2010. Isoflavones in coffee: influence of species, roast degree, and brewing method. J. Agric. Food Chem. 58, 3002–3007.

Ambros, V., 2004. The functions of animal microRNAs. Nature 431, 350–355.

Antoniou, A.C., Durocher, F., Smith, P., Simard, J., Easton, D.F., 2006. BRCA1 and BRCA2 mutation predictions using the BOADICEA and BRCAPRO models and penetrance estimation in high-risk French-Canadian families. Breast Cancer Res. 8, R3.

Apostolou, P., Fostira, F., 2013. Hereditary breast cancer: the era of new susceptibility genes. BioMed Res. Int. 2013, 747318.

Arai, Y., Uehara, M., Sato, Y., Kimira, M., Eboshida, A., Adlercreutz, H., Watanabe, S., 2000. Comparison of isoflavones among dietary intake, plasma concentration and urinary excretion for accurate estimation of phytoestrogen intake. J. Epidemiol. 10, 127–135.

Ashida, H., Fukuda, I., Yamashita, T., Kanazawa, K., 2000. Flavones and flavonols at dietary levels inhibit a transformation of aryl hydrocarbon receptor induced by dioxin. FEBS Lett. 476, 213–217.

Avci, C.B., Susluer, S.Y., Caglar, H.O., Balci, T., Aygunes, D., Dodurga, Y., Gunduz, C., 2015. Genistein-induced mir-23b expression inhibits the growth of breast cancer cells. Contemp. Oncol. 19, 32–35.

Barkhem, T., Carlsson, B., Nilsson, Y., Enmark, E., Gustafsson, J., Nilsson, S., 1998. Differential response of estrogen receptor alpha and estrogen receptor beta to partial estrogen agonists/antagonists. Mol. Pharmacol. 54, 105–112.

Barnes, B.B., Steindorf, K., Hein, R., Flesch-Janys, D., Chang-Claude, J., 2011. Population attributable risk of invasive postmenopausal breast cancer and breast cancer subtypes for modifiable and non-modifiable risk factors. Cancer Epidemiol. 35, 345–352.

Baylin, S.B., Jones, P.A., 2011. A decade of exploring the cancer epigenome - biological and translational implications. Nat. Rev. Cancer 11, 726–734.

Beckmann, M.W., Niederacher, D., Schnurch, H.G., Gusterson, B.A., Bender, H.G., 1997. Multistep carcinogenesis of breast cancer and tumour heterogeneity. J. Mol. Med. (Berl.) 75, 429–439.

Bediaga, N.G., Acha-Sagredo, A., Guerra, I., Viguri, A., Albaina, C., Ruiz Diaz, I., Rezola, R., Alberdi, M.J., Dopazo, J., Montaner, D., Renobales, M., Fernandez, A.F., Field, J.K., Fraga, M.F., Liloglou, T., De Pancorbo, M.M., 2010. DNA methylation epigenotypes in breast cancer molecular subtypes. Breast Cancer Res. 12, R77.

Beedanagari, S.R., Bebenek, I., Bui, P., Hankinson, O., 2009. Resveratrol inhibits dioxin-induced expression of human CYP1A1 and CYP1B1 by inhibiting recruitment of the aryl hydrocarbon receptor complex and RNA Polymerase II to the regulatory regions of the Corresponding genes. Toxicol. Sci. 110, 61–67.

Berger, S.L., Kouzarides, T., Shiekhattar, R., Shilatifard, A., 2009. An operational definition of epigenetics. Genes Dev. 23, 781–783.

Berletch, J.B., Liu, C., Love, W.K., Andrews, L.G., Katiyar, S.K., Tollefsbol, T.O., 2008. Epigenetic and genetic mechanisms contribute to telomerase inhibition by EGCG. J. Cell. Biochem. 103, 509–519.

Boland, M.J., Nazor, K.L., Loring, J.F., 2014. Epigenetic regulation of pluripotency and differentiation. Circ. Res. 115, 311–324.

Calvano Filho, C.M., Calvano-Mendes, D.C., Carvalho, K.C., Maciel, G.A., Ricci, M.D., Torres, A.P., Filassi, J.R., Baracat, E.C., 2014. Triple-negative and luminal A breast tumors: differential expression of miR-18a-5p, miR-17-5p, and miR-20a-5p. Tumour Biol. 35, 7733–7741.

Chen, M., Rao, Y., Zheng, Y., Wei, S., Li, Y., Guo, T., Yin, P., 2014. Association between soy isoflavone intake and breast cancer risk for pre- and post-menopausal women: a meta-analysis of epidemiological studies. PLoS One 9, e89288.

Chen, Z., Zheng, W., Custer, L.J., Dai, Q., Shu, X.O., Jin, F., Franke, A.A., 1999. Usual dietary consumption of soy foods and its correlation with the excretion rate of isoflavonoids in overnight urine samples among Chinese women in Shanghai. Nutr. Cancer 33, 82–87.

Chun, O.K., Chung, S.J., Song, W.O., 2009. Urinary isoflavones and their metabolites validate the dietary isoflavone intakes in US adults. J. Am. Diet Assoc. 109, 245–254.

Conway, K., Edmiston, S.N., May, R., Kuan, P.F., Chu, H., Bryant, C., Tse, C.K., Swift-Scanlan, T., Geradts, J., Troester, M.A., Millikan, R.C., 2014. DNA methylation profiling in the Carolina Breast Cancer Study defines cancer subclasses differing in clinicopathologic characteristics and survival. Breast Cancer Res. 16, 450.

Coyle, K.M., Murphy, J.P., Vidovic, D., Vaghar-Kashani, A., Dean, C.A., Sultan, M., Clements, D., Wallace, M., Thomas, M.L., Hundert, A., Giacomantonio, C.A., Helyer, L., Gujar, S.A., Lee, P.W., Weaver, I.C., Marcato, P., 2016. Breast cancer subtype dictates DNA methylation and ALDH1A3-mediated expression of tumor suppressor RARRES1. Oncotarget 7, 44096–44112.

Crook, T., Brooks, L.A., Crossland, S., Osin, P., Barker, K.T., Waller, J., Philp, E., Smith, P.D., Yulug, I., Peto, J., Parker, G., Allday, M.J., Crompton, M.R., Gusterson, B.A., 1998. p53 mutation with frequent novel condons but not a mutator phenotype in BRCA1- and BRCA2-associated breast tumours. Oncogene 17, 1681–1689.

Cui, J., Li, S., 2013. Structure, chemistry and pharmacology of naphthoflavones. Mini Rev. Med. Chem. 13, 1357–1368.

Curtit, E., Benhamo, V., Gruel, N., Popova, T., Manie, E., Cottu, P., Mariani, O., Stoppa-Lyonnet, D., Pivot, X., Stern, M.H., Vincent-Salomon, A., 2015. First description of a sporadic breast cancer in a woman with BRCA1 germline mutation. Oncotarget 6, 35616–35624.

Day, J.K., Besch-Williford, C., Mcmann, T.R., Hufford, M.G., Lubahn, D.B., Macdonald, R.S., 2001. Dietary genistein increased DMBA-induced mammary adenocarcinoma in wild-type, but not ER alpha KO, mice. Nutr. Cancer 39, 226–232.

De Assis, S., Warri, A., Benitez, C., Helferich, W., Hilakivi-Clarke, L., 2011. Protective effects of prepubertal genistein exposure on mammary tumorigenesis are dependent on BRCA1 expression. Cancer Prev. Res. (Phila) 4, 1436–1448.

Dean, W., 2014. DNA methylation and demethylation: a pathway to gametogenesis and development. Mol. Reprod. Dev. 81, 113–125.

Deb, G., Thakur, V.S., Limaye, A.M., Gupta, S., 2015. Epigenetic induction of tissue inhibitor of matrix metalloproteinase-3 by green tea polyphenols in breast cancer cells. Mol. Carcinog. 54, 485–499.

Degner, S.C., Papoutsis, A.J., Selmin, O., Romagnolo, D.F., 2009. Targeting of aryl hydrocarbon receptor-mediated activation of cyclooxygenase-2 expression by the indole-

3-carbinol metabolite 3,3′-diindolylmethane in breast cancer cells. J. Nutr. 139, 26–32.

Dezern, A.E., 2015. Nine years without a new FDA-approved therapy for MDS: how can we break through the impasse? Hematol. Am. Soc. Hematol. Educ. Program 2015, 308–316.

Dong, J.Y., Qin, L.Q., 2011. Soy isoflavones consumption and risk of breast cancer incidence or recurrence: a meta-analysis of prospective studies. Breast Cancer Res. Treat. 125, 315–323.

Dowsett, M., Forbes, J.F., Bradley, R., Ingle, J., Aihara, T., Bliss, J., Boccardo, F., Coates, A., Coombes, R.C., Cuzick, J., Dubsky, P., Gnant, M., Kaufmann, M., Kilburn, L., Perrone, F., Rea, D., Thurlimann, B., Van De Velde, C., Pan, H., Peto, R., Davies, C., Gray, R., 2015. Aromatase inhibitors versus tamoxifen in early breast cancer: patient-level meta-analysis of the randomised trials. Lancet 386, 1341–1352.

Duncan, J.A., Reeves, J.R., Cooke, T.G., 1998. BRCA1 and BRCA2 proteins: roles in health and disease. Mol. Pathol. 51, 237–247.

Egger, G., Liang, G., Aparicio, A., Jones, P.A., 2004. Epigenetics in human disease and prospects for epigenetic therapy. Nature 429, 457–463.

Esteller, M., 2007. Cancer epigenomics: DNA methylomes and histone-modification maps. Nat. Rev. Genet. 8, 286–298.

Fan, S., Meng, Q., Auborn, K., Carter, T., Rosen, E.M., 2006. BRCA1 and BRCA2 as molecular targets for phytochemicals indole-3-carbinol and genistein in breast and prostate cancer cells. Br. J. Cancer 94, 407–426.

Felsenfeld, G., 2014. A Brief history of epigenetics. Cold Spring Harb. Perspect. Biol. 6.

Foedermayr, M., Sebesta, M., Rudas, M., Berghoff, A.S., Promberger, R., Preusser, M., Dubsky, P., Fitzal, F., Gnant, M., Steger, G.G., Weltermann, A., Zielinski, C.C., Zach, O., Bartsch, R., 2014. BRCA-1 methylation and TP53 mutation in triple-negative breast cancer patients without pathological complete response to taxane-based neoadjuvant chemotherapy. Cancer Chemother. Pharmacol. 73, 771–778.

Futreal, P.A., Liu, Q., Shattuck-Eidens, D., Cochran, C., Harshman, K., Tavtigian, S., Bennett, L.M., Haugen-Strano, A., Swensen, J., Miki, Y., et al., 1994. BRCA1 mutations in primary breast and ovarian carcinomas. Science 266, 120–122.

Gardiner-Garden, M., Frommer, M., 1987. CpG islands in vertebrate genomes. J. Mol. Biol. 196, 261–282.

Gasiewicz, T.A., Rucci, G., 1991. Alpha-naphthoflavone acts as an antagonist of 2,3,7, 8-tetrachlorodibenzo-p-dioxin by forming an inactive complex with the Ah receptor. Mol. Pharmacol. 40, 607–612.

Greenblatt, M.S., Chappuis, P.O., Bond, J.P., Hamel, N., Foulkes, W.D., 2001. TP53 mutations in breast cancer associated with BRCA1 or BRCA2 germ-line mutations: distinctive spectrum and structural distribution. Cancer Res. 61, 4092–4097.

Grushko, T.A., Dignam, J.J., Das, S., Blackwood, A.M., Perou, C.M., Ridderstrale, K.K., Anderson, K.N., Wei, M.J., Adams, A.J., Hagos, F.G., Sveen, L., Lynch, H.T., Weber, B.L., Olopade, O.I., 2004. MYC is amplified in BRCA1-associated breast cancers. Clin. Cancer Res. 10, 499–507.

Hafez, M.M., Al-Shabanah, O.A., Al-Rejaie, S.S., Al-Harbi, N.O., Hassan, Z.K., Alsheikh, A., Al Theyab, A.I., Aldelemy, M.L., Sayed-Ahmed, M.M., 2015. Increased hypermethylation of glutathione S-transferase P1, DNA-binding protein inhibitor, death associated protein kinase and paired box protein-5 genes in triple-negative breast cancer Saudi females. Asian Pac. J. Cancer Prev. 16, 541–549.

Han, E.H., Kim, J.Y., Jeong, H.G., 2006. Effect of biochanin A on the aryl hydrocarbon receptor and cytochrome P450 1A1 in MCF-7 human breast carcinoma cells. Arch Pharm. Res. (Seoul) 29, 570–576.

Hargraves, K.G., He, L., Firestone, G.L., 2016. Phytochemical regulation of the tumor suppressive microRNA, miR-34a, by p53-dependent and independent responses in human breast cancer cells. Mol. Carcinog. 55, 486–498.

Hestermann, E.V., Brown, M., 2003. Agonist and chemopreventative ligands induce differential transcriptional cofactor recruitment by aryl hydrocarbon receptor. Mol. Cell Biol. 23, 7920–7925.

Hockings, J.K., Degner, S.C., Morgan, S.S., Kemp, M.Q., Romagnolo, D.F., 2008. Involvement of a specificity proteins-binding element in regulation of basal and estrogen-induced transcription activity of the BRCA1 gene. Breast Cancer Res. 10, R29.

Hockings, J.K., Thorne, P.A., Kemp, M.Q., Morgan, S.S., Selmin, O., Romagnolo, D.F., 2006. The ligand status of the aromatic hydrocarbon receptor modulates transcriptional activation of BRCA-1 promoter by estrogen. Cancer Res. 66, 2224–2232.

Holm, K., Hegardt, C., Staaf, J., Vallon-Christersson, J., Jonsson, G., Olsson, H., Borg, A., Ringner, M., 2010. Molecular subtypes of breast cancer are associated with characteristic DNA methylation patterns. Breast Cancer Res. 12, R36.

Hosny, M.M., Sabek, N.A., El-Abaseri, T.B., Hassan, F.M., Farrag, S.H., 2016. Promoter methylation status of breast cancer susceptibility gene 1 and 17 beta Hydroxysteroid Dehydrogenase type 1 gene in sporadic breast cancer patients. Int. J. Breast Cancer 2016, 9545241.

Hsieh, C.Y., Santell, R.C., Haslam, S.Z., Helferich, W.G., 1998. Estrogenic effects of genistein on the growth of estrogen receptor-positive human breast cancer (MCF-7) cells in vitro and in vivo. Cancer Res. 58, 3833–3838.

Isakovic, L., Saavedra, O.M., Llewellyn, D.B., Claridge, S., Zhan, L., Bernstein, N., Vaisburg, A., Elowe, N., Petschner, A.J., Rahil, J., Beaulieu, N., Gauthier, F., Macleod, A.R., Delorme, D., Besterman, J.M., Wahhab, A., 2009. Constrained (l-)-S-adenosyl-l-homocysteine (SAH) analogues as DNA methyltransferase inhibitors. Bioorg. Med. Chem. Lett 19, 2742–2746.

Jeffy, B.D., Hockings, J.K., Kemp, M.Q., Morgan, S.S., Hager, J.A., Beliakoff, J., Whitesell, L.J., Bowden, G.T., Romagnolo, D.F., 2005. An estrogen receptor-alpha/p300 complex activates the BRCA-1 promoter at an AP-1 site that binds Jun/Fos transcription factors: repressive effects of p53 on BRCA-1 transcription. Neoplasia 7, 873–882.

Jiang, F., Li, Y., Mu, J., Hu, C., Zhou, M., Wang, X., Si, L., Ning, S., Li, Z., 2016. Glabridin inhibits cancer stem cell-like properties of human breast cancer cells: an epigenetic regulation of miR-148a/SMAd2 signaling. Mol. Carcinog. 55, 929–940.

Jin, Y., 2011. 3,3′-Diindolylmethane inhibits breast cancer cell growth via miR-21-mediated Cdc25A degradation. Mol. Cell. Biochem. 358, 345–354.

Joosse, S.A., Van Beers, E.H., Tielen, I.H., Horlings, H., Peterse, J.L., Hoogerbrugge, N., Ligtenberg, M.J., Wessels, L.F., Axwijk, P., Verhoef, S., Hogervorst, F.B., Nederlof, P.M., 2009. Prediction of BRCA1-association in hereditary non-BRCA1/2 breast carcinomas with array-CGH. Breast Cancer Res. Treat. 116, 479–489.

Jovanović, B., Sheng, Q., Seitz, R.S., Lawrence, K.D., Morris, S.W., Thomas, L.R., Hout, D.R., Schweitzer, B.L., Guo, Y., Pietenpol, J.A., Lehmann, B.D., 2017. Comparison of triple-negative breast cancer molecular subtyping using RNA from matched fresh-frozen versus formalin-fixed paraffin-embedded tissue. BMC Canc. 17.

Judes, G., Dagdemir, A., Karsli-Ceppioglu, S., Lebert, A., Echegut, M., Ngollo, M., Bignon, Y.J., Penault-Llorca, F., Bernard-Gallon, D., 2016. H3K4 acetylation, H3K9 acetylation and H3K27 methylation in breast tumor molecular subtypes. Epigenomics 8, 909–924.

Jurkowska, R.Z., Jurkowski, T.P., Jeltsch, A., 2011. Structure and function of mammalian DNA methyltransferases. Chembiochem 12, 206–222.

Kanwal, R., Gupta, K., Gupta, S., 2015. Cancer epigenetics: an introduction. Methods Mol Biol 1238, 3–25.

Karami, F., Mehdipour, P., 2013. A comprehensive focus on global spectrum of BRCA1 and BRCA2 mutations in breast cancer. BioMed Res. Int. 2013, 928562.

Kaufman, P.B., Duke, J.A., Brielmann, H., Boik, J., Hoyt, J.E., 1997. A comparative survey of leguminous plants as sources of the isoflavones, genistein and daidzein: implications for human nutrition and health. J. Altern. Complement Med. 3, 7–12.

King-Batoon, A., Leszczynska, J.M., Klein, C.B., 2008. Modulation of gene methylation by genistein or lycopene in breast cancer cells. Environ. Mol. Mutagen. 49, 36–45.

Kostelac, D., Rechkemmer, G., Briviba, K., 2003. Phytoestrogens modulate binding response of estrogen receptors alpha and beta to the estrogen response element. J. Agric. Food Chem. 51, 7632–7635.

Kuhl, H., 2005. Pharmacology of estrogens and progestogens: influence of different routes of administration. Climacteric 8 (Suppl 1), 3–63.

Lakhani, S.R., Van De Vijver, M.J., Jacquemier, J., Anderson, T.J., Osin, P.P., Mcguffog, L., Easton, D.F., 2002. The pathology of familial breast cancer: predictive value of immunohistochemical markers estrogen receptor, progesterone receptor, HER-2, and p53 in patients with mutations in BRCA1 and BRCA2. J. Clin. Oncol. 20, 2310–2318.

Laurent, L., Wong, E., Li, G., Huynh, T., Tsirigos, A., Ong, C.T., Low, H.M., Kin Sung, K.W., Rigoutsos, I., Loring, J., Wei, C.L., 2010. Dynamic changes in the human methylome during differentiation. Genome Res. 20, 320–331.

Lee, W.J., Shim, J.Y., Zhu, B.T., 2005. Mechanisms for the inhibition of DNA methyltransferases by tea catechins and bioflavonoids. Mol. Pharmacol. 68, 1018–1030.

Lehmann, B.D., Bauer, J.A., Chen, X., Sanders, M.E., Chakravarthy, A.B., Shyr, Y., Pietenpol, J.A., 2011. Identification of human triple-negative breast cancer subtypes and preclinical models for selection of targeted therapies. J. Clin. Invest. 121, 2750–2767.

Li, Y., Meeran, S.M., Patel, S.N., Chen, H., Hardy, T.M., Tollefsbol, T.O., 2013. Epigenetic reactivation of estrogen receptor-alpha (ERalpha) by genistein enhances hormonal therapy sensitivity in ERalpha-negative breast cancer. Mol. Cancer 12, 9.

Li, Y., Yuan, Y.Y., Meeran, S.M., Tollefsbol, T.O., 2010. Synergistic epigenetic reactivation of estrogen receptor-alpha (ERalpha) by combined green tea polyphenol and histone deacetylase inhibitor in ERalpha-negative breast cancer cells. Mol. Cancer 9, 274.

Lipovka, Y., Konhilas, J.P., 2016. The complex nature of oestrogen signalling in breast cancer: enemy or ally? Biosci. Rep. 36.

Lister, R., Pelizzola, M., Dowen, R.H., Hawkins, R.D., Hon, G., Tonti-Filippini, J., Nery, J.R., Lee, L., Ye, Z., Ngo, Q.M., Edsall, L., Antosiewicz-Bourget, J., Stewart, R., Ruotti, V., Millar, A.H., Thomson, J.A., Ren, B., Ecker, J.R., 2009. Human DNA methylomes at base resolution show widespread epigenomic differences. Nature 462, 315–322.

Liu, X.O., Huang, Y.B., Gao, Y., Chen, C., Yan, Y., Dai, H.J., Song, F.J., Wang, Y.G., Wang, P.S., Chen, K.X., 2014. Association between dietary factors and breast cancer risk among Chinese females: systematic review and meta-analysis. Asian Pac. J. Cancer Prev. 15, 1291–1298.

Majid, S., Dar, A.A., Saini, S., Arora, S., Shahryari, V., Zaman, M.S., Chang, I., Yamamura, S., Tanaka, Y., Deng, G., Dahiya, R., 2012. miR-23b represses proto-oncogene Src kinase and functions as methylation-silenced tumor suppressor with diagnostic and prognostic significance in prostate cancer. Cancer Res. 72, 6435–6446.

Makki, J., 2015. Diversity of breast carcinoma: histological subtypes and clinical relevance. Clin. Med. Insights Pathol. 8, 23–31.

Malhotra, G.K., Zhao, X., Band, H., Band, V., 2010. Histological, molecular and functional subtypes of breast cancers. Cancer Biol. Ther. 10, 955–960.

Margeat, E., Bourdoncle, A., Margueron, R., Poujol, N., Cavailles, V., Royer, C., 2003. Ligands differentially modulate the protein interactions of the human estrogen receptors alpha and beta. J. Mol. Biol. 326, 77–92.

Markaverich, B.M., Shoulars, K., Rodriguez, M.A., 2011. Luteolin regulation of estrogen signaling and cell cycle pathway genes in MCF-7 human breast cancer cells. Int. J. Biomed. Sci. 7, 101–111.

Marks, J.R., Huper, G., Vaughn, J.P., Davis, P.L., Norris, J., Mcdonnell, D.P., Wiseman, R.W., Futreal, P.A., Iglehart, J.D., 1997. BRCA1 expression is not directly responsive to estrogen. Oncogene 14, 115–121.

Marzese, D.M., Hoon, D.S., 2015. Emerging technologies for studying DNA methylation for the molecular diagnosis of cancer. Expert Rev. Mol. Diagn 15, 647–664.

Mathe, A., Scott, R.J., Avery-Kiejda, K.A., 2015. MiRNAs and other epigenetic changes as biomarkers in triple negative breast cancer. Int. J. Mol. Sci. 16, 28347–28376.

Meeran, S.M., Patel, S.N., Tollefsbol, T.O., 2010. Sulforaphane causes epigenetic repression of hTERT expression in human breast cancer cell lines. PLoS One 5, e11457.

Merchant, M., Krishnan, V., Safe, S., 1993. Mechanism of action of alpha-naphthoflavone as an Ah receptor antagonist in MCF-7 human breast cancer cells. Toxicol. Appl. Pharmacol. 120, 179–185.

Messier, T.L., Gordon, J.A., Boyd, J.R., Tye, C.E., Browne, G., Stein, J.L., Lian, J.B., Stein, G.S., 2016. Histone H3 lysine 4 acetylation and methylation dynamics define breast cancer subtypes. Oncotarget 7, 5094–5109.

Miozzo, M., Vaira, V., Sirchia, S.M., 2015. Epigenetic alterations in cancer and personalized cancer treatment. Future Oncol. 11, 333–348.

Mirza, S., Sharma, G., Parshad, R., Gupta, S.D., Pandya, P., Ralhan, R., 2013. Expression of DNA methyltransferases in breast cancer patients and to analyze the effect of natural compounds on DNA methyltransferases and associated proteins. J. Breast Cancer 16, 23–31.

Morimoto, Y., Maskarinec, G., Park, S.Y., Ettienne, R., Matsuno, R.K., Long, C., Steffen, A.D., Henderson, B.E., Kolonel, L.N., Le Marchand, L., Wilkens, L.R., 2014. Dietary isoflavone intake is not statistically significantly associated with breast cancer risk in the multiethnic cohort. Br. J. Nutr. 112, 976–983.

Mukai, R., Shirai, Y., Saito, N., Fukuda, I., Nishiumi, S., Yoshida, K., Ashida, H., 2010. Suppression mechanisms of flavonoids on aryl hydrocarbon receptor-mediated signal transduction. Arch. Biochem. Biophys. 501, 134–141.

Murray, T.J., Yang, X., Sherr, D.H., 2006. Growth of a human mammary tumor cell line is blocked by galangin, a naturally occurring bioflavonoid, and is accompanied by down-regulation of cyclins D3, E, and A. Breast Cancer Res. 8, R17.

Murrill, W.B., Brown, N.M., Zhang, J.X., Manzolillo, P.A., Barnes, S., Lamartiniere, C.A., 1996. Prepubertal genistein exposure suppresses mammary cancer and enhances gland differentiation in rats. Carcinogenesis 17, 1451–1457.

Nass, N., Kalinski, T., 2015. Tamoxifen resistance: from cell culture experiments towards novel biomarkers. Pathol. Res. Pract. 211, 189–197.

Nomura, S.J., Inoue-Choi, M., Lazovich, D., Robien, K., 2016. WCRF/AICR recommendation adherence and breast cancer incidence among postmenopausal women with and without non-modifiable risk factors. Int. J. Cancer 138, 2602–2615.

O'sullivan, C.C., Bradbury, I., Campbell, C., Spielmann, M., Perez, E.A., Joensuu, H., Costantino, J.P., Delaloge, S., Rastogi, P., Zardavas, D., Ballman, K.V., Holmes, E., De Azambuja, E., Piccart-Gebhart, M., Zujewski, J.A., Gelber, R.D., 2015. Efficacy of adjuvant trastuzumab for patients with human epidermal growth factor receptor 2–positive early breast cancer and tumors ≤ 2 cm: a meta-analysis of the randomized trastuzumab trials. J. Clin. Oncol. 33, 2600–2608.

Okano, M., Bell, D.W., Haber, D.A., Li, E., 1999. DNA methyltransferases Dnmt3a and Dnmt3b are essential for de novo methylation and mammalian development. Cell 99, 247–257.

Palacios, J., Honrado, E., Osorio, A., Cazorla, A., Sarrio, D., Barroso, A., Rodriguez, S., Cigudosa, J.C., Diez, O., Alonso, C., Lerma, E., Sanchez, L., Rivas, C., Benitez, J., 2003. Immunohistochemical characteristics defined by tissue microarray of hereditary breast cancer not attributable to BRCA1 or BRCA2 mutations: differences from breast carcinomas arising in BRCA1 and BRCA2 mutation carriers. Clin. Cancer Res. 9, 3606–3614.

Pan, H., He, Z., Ling, L., Ding, Q., Chen, L., Zha, X., Zhou, W., Liu, X., Wang, S., 2014. Reproductive factors and breast cancer risk among BRCA1 or BRCA2 mutation carriers: results from ten studies. Cancer Epidemiol. 38, 1–8.

Papoutsis, A.J., Borg, J.L., Selmin, O.I., Romagnolo, D.F., 2012. BRCA-1 promoter hypermethylation and silencing induced by the aromatic hydrocarbon receptor-ligand TCDD are prevented by resveratrol in MCF-7 cells. J. Nutr. Biochem. 23, 1324–1332.

Papoutsis, A.J., Lamore, S.D., Wondrak, G.T., Selmin, O.I., Romagnolo, D.F., 2010. Resveratrol prevents epigenetic silencing of BRCA-1 by the aromatic hydrocarbon receptor in human breast cancer cells. J. Nutr. 140, 1607–1614.

Park, S.Y., Seo, A.N., Jung, H.Y., Gwak, J.M., Jung, N., Cho, N.Y., Kang, G.H., 2014. Alu and LINE-1 hypomethylation is associated with HER2 enriched subtype of breast cancer. PLoS One 9, e100429.

Paruthiyil, S., Parmar, H., Kerekatte, V., Cunha, G.R., Firestone, G.L., Leitman, D.C., 2004. Estrogen receptor beta inhibits human breast cancer cell proliferation and tumor formation by causing a G2 cell cycle arrest. Cancer Res. 64, 423–428.

Pellegrino, L., Stebbing, J., Braga, V.M., Frampton, A.E., Jacob, J., Buluwela, L., Jiao, L.R., Periyasamy, M., Madsen, C.D., Caley, M.P., Ottaviani, S., Roca-Alonso, L., El-Bahrawy, M., Coombes, R.C., Krell, J., Castellano, L., 2013. miR-23b regulates cytoskeletal remodeling, motility and metastasis by directly targeting multiple transcripts. Nucleic Acids Res. 41, 5400–5412.

Perez-Janices, N., Blanco-Luquin, I., Torrea, N., Liechtenstein, T., Escors, D., Cordoba, A., Vicente-Garcia, F., Jauregui, I., De La Cruz, S., Illarramendi, J.J., Coca, V., Berdasco, M., Kochan, G., Ibanez, B., Lera, J.M., Guerrero-Setas, D., 2015. Differential involvement of RASSF2 hypermethylation in breast cancer subtypes and their prognosis. Oncotarget 6, 23944–23958.

Powell, E., Shanle, E., Brinkman, A., Li, J., Keles, S., Wisinski, K.B., Huang, W., Xu, W., 2012. Identification of estrogen receptor dimer selective ligands reveals growth-inhibitory effects on cells that Co-Express ERα and ERβ. PLoS One 7.

Prat, A., Perou, C.M., 2011. Deconstructing the molecular portraits of breast cancer. Mol. Oncol. 5, 5–23.

Prat, A., Pineda, E., Adamo, B., Galvan, P., Fernandez, A., Gaba, L., Diez, M., Viladot, M., Arance, A., Munoz, M., 2015. Clinical implications of the intrinsic molecular subtypes of breast cancer. Breast 24 (Suppl. 2), S26–S35.

Qin, L.Q., Xu, J.Y., Wang, P.Y., Hoshi, K., 2006. Soyfood intake in the prevention of breast cancer risk in women: a meta-analysis of observational epidemiological studies. J. Nutr. Sci. Vitaminol. 52, 428–436.

Rapoport, B.L., Demetriou, G.S., Moodley, S.D., Benn, C.A., 2014. When and how do I use neoadjuvant chemotherapy for breast cancer? Curr. Treat. Options Oncol. 15, 86–98.

Riaz, M., Van Jaarsveld, M.T., Hollestelle, A., Prager-Van Der Smissen, W.J., Heine, A.A., Boersma, A.W., Liu, J., Helmijr, J., Ozturk, B., Smid, M., Wiemer, E.A., Foekens, J.A., Martens, J.W., 2013. miRNA expression profiling of 51 human breast cancer cell lines reveals subtype and driver mutation-specific miRNAs. Breast Cancer Res. 15, R33.

Rice, J.C., Massey-Brown, K.S., Futscher, B.W., 1998. Aberrant methylation of the BRCA1 CpG island promoter is associated with decreased BRCA1 mRNA in sporadic breast cancer cells. Oncogene 17, 1807–1812.

Romagnolo, D.F., Daniels, K.D., Grunwald, J.T., Ramos, S.A., Propper, C.R., Selmin, O.I., 2016. Epigenetics of breast cancer: modifying role of environmental and bioactive food compounds. Mol. Nutr. Food Res. 60, 1310–1329.

Romagnolo, D.F., Donovan, M.G., Papoutsis, A.J., Doetschman, T.C., Selmin, O.I., 2017. Genistein prevents *BRCA1* CpG methylation and proliferation in human breast cancer cells with activated aromatic hydrocarbon receptor. Curr. Dev. Nutr. 1.

Romagnolo, D.F., Papoutsis, A.J., Ladukaitis, C., Selmin, O.I., 2015. Constitutive expression of AhR and BRCA-1 promoter CpG hypermethylation as biomarkers of ERα-negative breast tumorigenesis. BMC Canc. 15.

Russo, J., Mailo, D., Hu, Y.F., Balogh, G., Sheriff, F., Russo, I.H., 2005. Breast differentiation and its implication in cancer prevention. Clin. Cancer Res. 11, 931s-6s.

Russo, M., Russo, G.L., Daglia, M., Kasi, P.D., Ravi, S., Nabavi, S.F., Nabavi, S.M., 2016. Understanding genistein in cancer: the "good" and the "bad" effects: a review. Food Chem. 196, 589–600.

Saavedra, O.M., Isakovic, L., Llewellyn, D.B., Zhan, L., Bernstein, N., Claridge, S., Raeppel, F., Vaisburg, A., Elowe, N., Petschner, A.J., Rahil, J., Beaulieu, N., Macleod, A.R., Delorme, D., Besterman, J.M., Wahhab, A., 2009. SAR around (l)-S-adenosyl-l-homocysteine, an inhibitor of human DNA methyltransferase (DNMT) enzymes. Bioorg. Med. Chem. Lett 19, 2747–2751.

Santi, D.V., Norment, A., Garrett, C.E., 1984. Covalent bond formation between a DNA-cytosine methyltransferase and DNA containing 5-azacytosine. Proc. Natl. Acad. Sci. USA 81, 6993–6997.

Savage, K.I., Harkin, D.P., 2015. BRCA1, a 'complex' protein involved in the maintenance of genomic stability. FEBS J. 282, 630–646.

Saxonov, S., Berg, P., Brutlag, D.L., 2006. A genome-wide analysis of CpG dinucleotides in the human genome distinguishes two distinct classes of promoters. Proc. Natl. Acad. Sci. USA 103, 1412–1417.

Sheikh, A., Hussain, S.A., Ghori, Q., Naeem, N., Fazil, A., Giri, S., Sathian, B., Mainali, P., Al Tamimi, D.M., 2015. The spectrum of genetic mutations in breast cancer. Asian Pac. J. Cancer Prev. 16, 2177–2185.

Shilpi, A., Parbin, S., Sengupta, D., Kar, S., Deb, M., Rath, S.K., Pradhan, N., Rakshit, M., Patra, S.K., 2015.

Mechanisms of DNA methyltransferase-inhibitor interactions: procyanidin B2 shows new promise for therapeutic intervention of cancer. Chem. Biol. Interact. 233, 122–138.

Silver, D.P., Richardson, A.L., Eklund, A.C., Wang, Z.C., Szallasi, Z., Li, Q., Juul, N., Leong, C.O., Calogrias, D., Buraimoh, A., Fatima, A., Gelman, R.S., Ryan, P.D., Tung, N.M., De Nicolo, A., Ganesan, S., Miron, A., Colin, C., Sgroi, D.C., Ellisen, L.W., Winer, E.P., Garber, J.E., 2010. Efficacy of neoadjuvant Cisplatin in triple-negative breast cancer. J. Clin. Oncol. 28, 1145–1153.

Singh, B., Shoulson, R., Chatterjee, A., Ronghe, A., Bhat, N.K., Dim, D.C., Bhat, H.K., 2014. Resveratrol inhibits estrogen-induced breast carcinogenesis through induction of NRF2-mediated protective pathways. Carcinogenesis 35, 1872–1880.

Society, A.C., 2017. Cancer Facts & Figures 2017. American Cancer Society, Atlanta.

Sorlie, T., Tibshirani, R., Parker, J., Hastie, T., Marron, J.S., Nobel, A., Deng, S., Johnsen, H., Pesich, R., Geisler, S., Demeter, J., Perou, C.M., Lonning, P.E., Brown, P.O., Borresen-Dale, A.L., Botstein, D., 2003. Repeated observation of breast tumor subtypes in independent gene expression data sets. Proc. Natl. Acad. Sci. USA 100, 8418–8423.

Stefanska, B., Salame, P., Bednarek, A., Fabianowska-Majewska, K., 2012. Comparative effects of retinoic acid, vitamin D and resveratrol alone and in combination with adenosine analogues on methylation and expression of phosphatase and tensin homologue tumour suppressor gene in breast cancer cells. Br. J. Nutr. 107, 781–790.

Stefansson, O.A., Jonasson, J.G., Olafsdottir, K., Hilmarsdottir, H., Olafsdottir, G., Esteller, M., Johannsson, O.T., Eyfjord, J.E., 2011. CpG island hypermethylation of BRCA1 and loss of pRb as co-occurring events in basal/triple-negative breast cancer. Epigenetics 6, 638–649.

Stresemann, C., Lyko, F., 2008. Modes of action of the DNA methyltransferase inhibitors azacytidine and decitabine. Int. J. Cancer 123, 8–13.

Sun, D.W., Zhang, H.D., Mao, L., Mao, C.F., Chen, W., Cui, M., Ma, R., Cao, H.X., Jing, C.W., Wang, Z., Wu, J.Z., Tang, J.H., 2015. Luteolin inhibits breast cancer development and progression in vitro and in vivo by suppressing notch signaling and regulating MiRNAs. Cell. Physiol. Biochem. 37, 1693–1711.

Tao, S.F., He, H.F., Chen, Q., 2015a. Quercetin inhibits proliferation and invasion acts by up-regulating miR-146a in human breast cancer cells. Mol. Cell. Biochem. 402, 93–100.

Tao, Z., Shi, A., Lu, C., Song, T., Zhang, Z., Zhao, J., 2015b. Breast cancer: epidemiology and etiology. Cell Biochem. Biophys. 72, 333–338.

Taylor, C.K., Levy, R.M., Elliott, J.C., Burnett, B.P., 2009. The effect of genistein aglycone on cancer and cancer risk: a review of in vitro, preclinical, and clinical studies. Nutr. Rev. 67, 398–415.

Tham, D.G.C., Haskell, W., 1998. Clinical review 97: potential health benefits of dietary phytoestrogens: a review of the clinical, epidemiological, and mechanistic evidence. J. Clin. Endocrinol. Metab 83, 2223–2235.

Torre, L.A., Bray, F., Siegel, R.L., Ferlay, J., Lortet-Tieulent, J., Jemal, A., 2015. Global cancer statistics, 2012. CA Cancer J. Clin. 65, 87–108.

Torre, L.A., Islami, F., Siegel, R.L., Ward, E.M., Jemal, A., 2017. Global cancer in women: burden and trends. Cancer Epidemiol. Biomarkers Prev.

Toss, A., Cristofanilli, M., 2015. Molecular characterization and targeted therapeutic approaches in breast cancer. Breast Cancer Res. 17, 60.

Trock, B.J., Hilakivi-Clarke, L., Clarke, R., 2006. Meta-analysis of soy intake and breast cancer risk. J. Natl. Cancer Inst. 98, 459–471.

Tseng, T.H., Chien, M.H., Lin, W.L., Wen, Y.C., Chow, J.M., Chen, C.K., Kuo, T.C., Lee, W.J., 2017. Inhibition of MDA-MB-231 breast cancer cell proliferation and tumor growth by apigenin through induction of G2/M arrest and histone H3 acetylation-mediated p21WAF1/CIP1 expression. Environ. Toxicol. 32, 434–444.

Tsubura, A., Yoshizawa, K., Uehara, N., Yuri, T., Matsuoka, Y., 2007. Multistep mouse mammary tumorigenesis through pre-neoplasia to neoplasia and acquisition of metastatic potential. Med. Mol. Morphol. 40, 9–17.

Turner, N., Tutt, A., Ashworth, A., 2004. Hallmarks of 'BRCAness' in sporadic cancers. Nat. Rev. Cancer 4, 814–819.

Turner, N.C., Reis-Filho, J.S., Russell, A.M., Springall, R.J., Ryder, K., Steele, D., Savage, K., Gillett, C.E., Schmitt, F.C., Ashworth, A., Tutt, A.N., 2007. BRCA1 dysfunction in sporadic basal-like breast cancer. Oncogene 26, 2126–2132.

Tyagi, T., Treas, J.N., Mahalingaiah, P.K., Singh, K.P., 2015. Potentiation of growth inhibition and epigenetic modulation by combination of green tea polyphenol and 5-aza-2'-deoxycytidine in human breast cancer cells. Breast Cancer Res. Treat. 149, 655–668.

Van Grembergen, O., Bizet, M., De Bony, E.J., Calonne, E., Putmans, P., Brohee, S., Olsen, C., Guo, M., Bontempi, G., Sotiriou, C., Defrance, M., Fuks, F., 2016. Portraying breast cancers with long noncoding RNAs. Sci. Adv. 2, e1600220.

Venkatadri, R., Muni, T., Iyer, A.K., Yakisich, J.S., Azad, N., 2016. Role of apoptosis-related miRNAs in resveratrol-induced breast cancer cell death. Cell Death Dis. 7, e2104.

Vislovukh, A., Kratassiouk, G., Porto, E., Gralievska, N., Beldiman, C., Pinna, G., El'skaya, A., Harel-Bellan, A., Negrutskii, B., Groisman, I., 2013. Proto-oncogenic isoform A2 of eukaryotic translation elongation factor eEF1 is a target of miR-663 and miR-744. Br. J. Cancer 108, 2304–2311.

Wang, J., Ye, C., Xiong, H., Shen, Y., Lu, Y., Zhou, J., Wang, L., 2017. Dysregulation of long non-coding RNA in breast cancer: an overview of mechanism and clinical implication. Oncotarget 8, 5508–5522.

Wang, P., Zou, F., Zhang, X., Li, H., Dulak, A., Tomko Jr., R.J., Lazo, J.S., Wang, Z., Zhang, L., Yu, J., 2009. microRNA-21 negatively regulates Cdc25A and cell cycle progression in colon cancer cells. Cancer Res. 69, 8157–8165.

Wardell, S.E., Nelson, E.R., Mcdonnell, D.P., 2014. From empirical to mechanism-based discovery of clinically useful Selective Estrogen Receptor Modulators (SERMs). Steroids 90, 30–38.

Wisinski, K.B., Xu, W., Tevaarwerk, A.J., Saha, S., Kim, K., Traynor, A., Dietrich, L., Hegeman, R., Patel, D., Blank, J., Harter, J., Burkard, M.E., 2016. Targeting estrogen receptor beta in a phase 2 study of high-dose estradiol in metastatic triple-negative breast cancer: a wisconsin oncology network study. Clin. Breast Cancer 16, 256–261.

Won, J., Yim, J., Kim, T.K., 2002. Opposing regulatory roles of E2F in human telomerase reverse transcriptase (hTERT) gene expression in human tumor and normal somatic cells. FASEB J. 16, 1943–1945.

Woo, H.D., Park, S., Oh, K., Kim, H.J., Shin, H.R., Moon, H.K., Kim, J., 2014. Diet and cancer risk in the Korean population: a meta- analysis. Asian Pac. J. Cancer Prev. 15, 8509–8519.

Wu, Y.C., Zheng, D., Sun, J.J., Zou, Z.K., Ma, Z.L., 2015. Meta-analysis of studies on breast cancer risk and diet in Chinese women. Int. J. Clin. Exp. Med. 8, 73–85.

Xie, Q., Bai, Q., Zou, L.Y., Zhang, Q.Y., Zhou, Y., Chang, H., Yi, L., Zhu, J.D., Mi, M.T., 2014. Genistein inhibits DNA methylation and increases expression of tumor suppressor genes in human breast cancer cells. Genes Chromosomes Cancer 53, 422–431.

Xu, C.F., Brown, M.A., Chambers, J.A., Griffiths, B., Nicolai, H., Solomon, E., 1995. Distinct transcription start sites generate two forms of BRCA1 mRNA. Hum. Mol. Genet. 4, 2259–2264.

Xu, J., Wang, Z., Li, S., Chen, J., Zhang, J., Jiang, C., Zhao, Z., Li, J., Li, Y., Li, X., 2016. Combinatorial epigenetic regulation of non-coding RNAs has profound effects on oncogenic pathways in breast cancer subtypes. Brief Bioinform.

Yang, Z., Kulkarni, K., Zhu, W., Hu, M., 2012. Bioavailability and Pharmacokinetics of genistein: mechanistic studies on its ADME. Anticancer Agents Med. Chem. 12, 1264–1280.

Yoo, J., Choi, S., Medina-Franco Jé, L., 2013. Molecular modeling studies of the novel inhibitors of DNA methyltransferases SGI-1027 and CBC12: implications for the mechanism of inhibition of DNMTs. PLoS One 8.

Zhang, H., Hao, Y., Yang, J., Zhou, Y., Li, J., Yin, S., Sun, C., Ma, M., Huang, Y., Xi, J.J., 2011. Genome-wide functional screening of miR-23b as a pleiotropic modulator suppressing cancer metastasis. Nat. Commun. 2, 554.

Zhang, L., Long, X., 2015. Association of BRCA1 promoter methylation with sporadic breast cancers: evidence from 40 studies. Sci. Rep. 5.

Zhang, Y., Rohde, C., Tierling, S., Jurkowski, T.P., Bock, C., Santacruz, D., Ragozin, S., Reinhardt, R., Groth, M., Walter, J., Jeltsch, A., 2009. DNA methylation analysis of chromosome 21 gene promoters at single base pair and single allele resolution. PLoS Genet. 5, e1000438.

Ziegler, R.G., Hoover, R.N., Pike, M.C., Hildesheim, A., Nomura, A.M., West, D.W., Wu-Williams, A.H., Kolonel, L.N., Horn-Ross, P.L., Rosenthal, J.F., Hyer, M.B., 1993. Migration patterns and breast cancer risk in Asian-American women. J. Natl. Cancer Inst. 85, 1819–1827.

CHAPTER

16

Epigenomics of Ovarian Cancer and Its Chemoprevention

Ourania Koukoura[1], *Stavros Sifakis*[2], *Nikolaos Goutsias*[1], *Irontianta Gkorezi-Ntavela*[1], *Jiannis Hajiioannou*[3]

[1]Department of Obstetrics and Gynecology, University Hospital of Larissa, Larissa, Greece; [2]Department of Obstetrics and Gynecology, "Mitera" Hospital, Herakleion, Greece; [3]Department of ENT, Head and Neck Surgery, University Hospital of Larissa, Larissa, Greece

Abstract

Ovarian cancer is the most lethal gynecologic malignancy. Lack of early detection markers and the development of drug resistance following chemotherapy are the main obstacles to effective treatment strategies. Despite advances in cancer research and treatment, survival statistics have remained largely unchanged for many years. Genetic alterations alone cannot account for the complexity of ovarian cancer. The potential reversibility of epigenetic mechanisms makes them attractive candidates for the prevention and treatment of ovarian carcinoma. Detection of the epigenetic signature of each cancer cell may be useful in the identification of candidate biomarkers for disease detection, classification, and monitoring and may also facilitate personalized cancer treatment. Herein we summarize the current developments in epigenetic research in ovarian cancer and evaluate the potential role of epigenetic markers as diagnostic and prognostic tools. Furthermore, novel epigenetic therapies that target cancer cells, reactivate tumor-suppressor genes, and resensitize tumor cells to chemotherapy, will be discussed.

Keywords

Biomarkers; DNA methylation; Epigenetics; Histone acetylation; MiRNA; Ovarian cancer

INTRODUCTION

Ovarian cancer is the leading cause of gynecologic cancer death worldwide while constituting only 3% of all female cancers (Hennessy et al., 2009). In the United States, ovarian cancer is the fifth-leading cause of cancer death in women with an incidence of 14,180 deaths in 2015 (Siegel et al., 2015). Due to the lack of specific symptoms in the early stage, 70% of cases are not diagnosed until the cancer has reached an advanced stage, FIGO Stages IIB to IV (spread of tumor within the pelvis or elsewhere in the abdomen) (Jemal et al., 2009). Early detection of ovarian cancer reportedly increases the 5-year survival rate by up to 92%; however, the actual overall 5-year survival rate is only 15%–45%. Despite advances in cancer research and treatment, survival statistics have remained largely unchanged for many years. A better understanding of the molecular pathogenesis of ovarian cancer is needed in order to develop new drug therapies or diagnostic

Epigenetics of Cancer Prevention
https://doi.org/10.1016/B978-0-12-812494-9.00016-0

biomarkers and elucidate the role of environmental exposures to individual predisposition to the disease.

Ovarian epithelial carcinoma (OEC) is the most common ovarian malignancy, with substantial histopathological heterogeneity. According to the 2003 World Health Organization classification scheme, the most common histologic subtype is serous ovarian carcinoma (~60%), while other subtypes include endometrioid (10%–20%), clear cell (10%), transitional (6%), mucinous (<5%), and undifferentiated (<1%). The underlying genetic basis of ovarian cancer contributes to this heterogeneity. Most OECs (90%) are sporadic, with the remaining few being inherited. Inherited ovarian cancers account for 5%–10% of all ovarian cancers and are characterized by the development of highly aggressive neoplasms at an earlier age of onset than their sporadic counterparts.

Several studies (Tian et al., 2012) have reported the use of the candidate gene approach in the search for common risk variants associated with ovarian cancer. However, the genes that participate in the development of ovarian cancer represent only a small portion of ovarian cancer-associated genes, as many of them are merely associated with ovarian cancer development but do not contribute to its initiation and progression. Moreover, molecular pathways in different ovarian tumors may vary significantly. Thus, genetic modifications alone cannot account for the complexity of the disease. Because genetic factors are almost impossible to reverse, the potential reversibility of epigenetic mechanisms makes them attractive candidates for the prevention and treatment of ovarian carcinoma (Baylin and Ohm, 2006; Yoo and Jones, 2006).

Epigenetics involve the interplay between DNA methylation, histone modifications, and expression of noncoding RNAs in the regulation of gene transcription. Increasing evidence has shown that epigenetic alterations play a significant role in cancer, from the silencing of tumor suppressors to activation of oncogenes, promotion of metastasis, and resistance to chemotherapy (Wilting and Dannenberg, 2012). DNA methylation is a key element in tissue differentiation; the diversion of normal cell cycles to those of a less differentiated status is the initial step in tumorigenesis (Koukoura et al., 2012). Aberrant DNA methylation is now recognized as one of the most common molecular abnormalities in cancer and it is frequently associated with drug resistance (Wilting and Dannenberg, 2012). Posttranscriptional histone modifications represent an important epigenetic mechanism associated with ovarian cancer progression and moreover, a significant research field for personalized treatment in these cases. They are characterized by modifications of the N-terminal tails of histones (H2A, H2B, H3, and H4) including acetylation, methylation, phosphorylation, biotinylation, and ubiquitination (Waldmann and Schneider, 2013). Accumulating evidence has revealed that histone acetylation plays an essential role in a variety of biological processes including inflammation, cell proliferation, apoptosis, and carcinogenesis. MicroRNAs (miRNAs) represent a group of small noncoding RNAs involved in the posttranscriptional regulation of gene expression. They have been recognized as an essential component of biological processes such as proliferation, differentiation, immune response, and angiogenesis (Mahdian-Shakib et al., 2016). Dysregulation of numerous miRNAs in cancer cells has been associated with different steps in tumorigenesis, highlighting the potential role of miRNAs as diagnostic and prognostic markers.

Understanding the epigenetic mechanisms associated with lethal malignancies such as ovarian cancer is an ongoing effort targeted on elucidating the exact pathophysiology of the disease, yielding potential diagnostic markers, and determining the etiology of resistance in several chemotherapeutic agents. Recent research advances that have unraveled the whole epigenotype of a tumor cell will guide further studies on mechanisms in specific sites of the epigenome. The optimistic result of this effort will be personalization of the diagnostic process and treatment based on individual epigenetic biomarkers, which will improve prognosis and outcome (Ivanov et al., 2012).

DNA METHYLATION AND OVARIAN CANCER

The DNA methylation profile of a tumor cell reflects its somatic lineage, environmental exposure, and genetic predisposition. The DNA methylation profile is therefore distinct for each histological subtype, suggesting different tumorigenic mechanisms. The detection of the epigenetic signature of each cancer cell may be useful in the identification of candidate biomarkers for disease detection, classification, and monitoring, as well as facilitating cancer treatment (Laird, 2003).

In ovarian cancer as well as nongynecologic cancers, two opposite epigenetic phenomena occur: (1) an overall global decrease in DNA methylation of heterochromatin leading to demethylation of several oncogenes and chromatin instability; and (2) specific CpG island hypermethylation associated with the promoters of tumor-suppressor genes (TSGs) (Baylin and Ohm, 2006; Esteller et al., 2000; Esteller, 2008) (Fig. 16.1). The aberrant methylation of CpG islands in gene promoters has been correlated with the loss of a gene's expression, and thus DNA methylation provides an alternative pathway to deletion and mutation of TSGs (Baylin and Chen, 2005). The epigenetic silencing of TSGs induces mechanisms such as

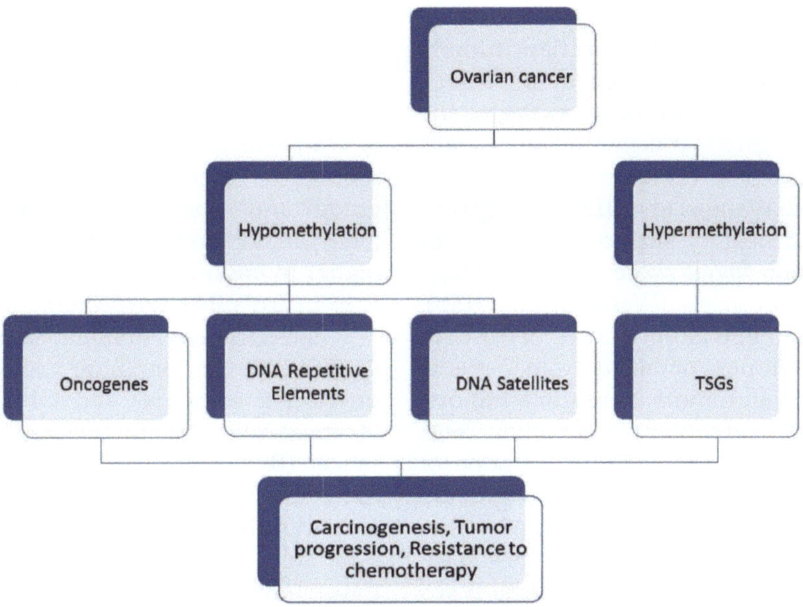

FIGURE 16.1 Methylation mechanisms implicated in the initiation and progression of ovarian cancer. Two opposite epigenetic phenomena have been observed in ovarian carcinogenesis: (1) an overall global hypomethylation of heterochromatin leading to demethylation of several oncogenes and chromatin instability, and (2) specific CpG island hypermethylation associated with the promoters of tumor-suppressor genes. Both hypo- and hypermethylation have been associated with cancer initiation, progression, and response to treatment.

uncontrolled cell division, the ability to infiltrate surrounding tissues, metastasis, and avoiding apoptosis or sustaining angiogenesis, all of which are responsible for promoting tumor development. In ovarian cancer, many TSGs have been found to undergo hypermethylation (Petrocca et al., 2006; Ozdemir et al., 2012; Yu et al., 2003).

One of the most studied genes in ovarian cancer is breast cancer early-onset gene 1 (BRCA1) due to its role in inherited and sporadic forms of the disease (Baldwin et al., 2000; Hilton et al., 2002). BRCA1 is important in maintaining genomic stability (Li and Greenberg, 2012) and interacts with numerous proteins, forming complexes that are involved in recognizing and subsequently repairing DNA. Evidence suggests that in cases of sporadic ovarian cancer, promoter hypermethylation is responsible for BRCA1 inactivation (McCoy et al., 2003). Complete or partial inactivation of BRCA1 through hypermethylation of its promoters has been reported in 15% of sporadic ovarian tumors (Baldwin et al., 2000; Strathdee et al., 2001). Hypermethylation leads to the silencing of the gene in ovarian tumors and methylation levels correlate to those of BRCA1 gene's expression (Chan et al., 2002; Wang et al., 2013b). Compared with stage I and healthy subjects, BRCA1 promoter methylation levels were higher in stage II and III ovarian cancers (Wang et al., 2013b). Comparing the methylation status of BRCA1 among tumor samples obtained from patients with benign ovarian tumors, borderline tumors, and carcinomas, promoter methylation was detected in 31% of carcinomas but in none of the benign and borderline tumors (Wang et al., 2004). Hypermethylation of BRCA1 was detected at a significantly higher frequency in serous carcinomas than in other histological types (Yang et al., 2006). BRCA2 does not exhibit a similar methylation profile in ovarian cancer. Findings of previous studies have shown that levels of methylated CpGs at the BRCA2 promoter were either nondetectable or very low in tumor DNA samples compared with those of normal tissues (Chan et al., 2002).

TSGs involved in DNA mismatch repair (MMR) possess a distinct carcinogenic effect in ovarian tumors. DNA MMR is an endogenous molecular mechanism that reverses the replication errors that escape correction by replicative DNA polymerases. In MMR-defective cells, both base-to-base mismatches and insertion–deletion loops are left uncorrected (Blasi et al., 2006). This results in increased spontaneous somatic mutations. Approximately 10% of ovarian cancers are related to this molecular pathway (Murphy and Wentzensen, 2011). Defective MMR is often a consequence of germline mutations of hMLH1, hMSH2, MGMT, and occasionally MSH6 and PMS2 genes. Hypermethylation of the MLH1 gene, accompanied by loss of the gene's expression, has been reported in 10%–30% of ovarian malignancies, whereas in cases with acquired resistance to platinum-based chemotherapy, hMLH1 promoter hypermethylation has been identified in 56% of cases (Watanabe et al., 2007). The methylation frequency of hMSH2 promoters has been reported as high as 57% in ovarian cancers. Methylation levels of hMSH2 have been correlated with histological grade and lymphatic spread. Additionally, methylation levels of hMSH2 were significantly higher in endometrioid adenocarcinoma tissues than in any other type (Zhang et al., 2008).

Ras-association domain family protein 1a (RASSF1A)—an inhibitor of the anaphase-promoting complex—and OPCML are among the most frequently methylated genes in ovarian cancer (Barton et al., 2008). Genes involved in cell cycle pathways such as p16 and p15 have also been affected by altered methylation of their promoters (Moselhy et al., 2013). E-cadherin is a transmembrane glycoprotein that mediates calcium-dependent interactions between adjacent epithelial cells. It has been found that the rate of E-cadherin hypermethylation was 1.347-fold higher among patients with ovarian cancer than in patients with benign ovarian tumors

(Moselhy et al., 2013). Other genes involved in cell adherence, such as H-cadherin and CDH1, have shown similar results (Dhillon et al., 2004). hSulf-1, which encodes an arylsulfatase that acts on cell surface heparin sulfate proteoglycans and inhibits growth factor signaling, has been found to be methylated in more than 50% of ovarian tumors and ovarian cancer cell lines (Staub et al., 2007).

Methylation profiles of several genes belonging to the homeobox (HOX) gene family have also been investigated in cases of ovarian carcinomas. HOX genes constitute a family of transcriptional factors that, during embryonic development, control pattern formation, differentiation, and proliferation (Samuel and Naora, 2005). Findings from previous studies suggest that abnormal expression of specific HOX genes is associated with ovarian carcinomas (Kelly et al., 2011). Methylation of the HOXA9 gene has been observed in 95% of patients with high-grade serous ovarian carcinoma (Montavon et al., 2012). It has been suggested that the methylation statuses of HOXA9 and HOXAD11 genes may serve as potential diagnostic and prognostic biomarkers of the disease (Widschwendter et al., 2009). Hypermethylation in ovarian cancer, however, has been found to be associated with the inactivation of almost every pathway involved in ovarian cancer development, including DNA repair, cell cycle regulation, apoptosis, cell adherence, and detoxification pathways (Ibanez De Caceres et al., 2004; Makarla et al., 2005; Rathi et al., 2002; Strathdee et al., 2001; Tam et al., 2007).

In addition to the hypermethylation of promoter-associated CpG islands, global hypomethylation and specific hypomethylation of protein-expressed genes that subsequently become overexpressed, play significant role in ovarian cancer. Hypomethylation in the centromere and subtelomeric regions is involved in the induction of genomic instability (GI) that leads to chromosomal translocations and gene disruption through the reactivation of transposable elements (Esteller, 2008). Decreased methylation of LINE-1 elements is correlated with high grade, advanced stage, and poor prognosis in ovarian cancer patients (Pattamadilok et al., 2008). Satellite DNA hypomethylation is an independent marker of poor prognosis. Hypomethylation gradually increases from non-neoplastic tissue to ovarian cancer, and it also advances alongside the grade and stage of the disease (Widschwendter et al., 2004).

Several oncogenes have been reported as having an increased epigenetically induced expression. Oncogenes such as CLDN4 (encoding an integral component of tight junctions), myelin and lymphocyte protein (MAL, a T-cell differentiation protein), and BORIS (brother of the regulator of imprinted sites) are among several oncogenes that contribute to drug resistance and are associated with the overall prognosis of ovarian cancer patients (Honda et al., 2006; Lee et al., 2010; Woloszynska-Read et al., 2007). Upregulation, together with hypomethylation of the ABCG2 multidrug transporter and class III β-tubulin (TUBB3) genes has been observed in cases of advanced-stage drug-acquired chemoresistance-associated ovarian carcinoma (Izutsu et al., 2008). Other cancer-associated genes, including methylation-controlled DnaJ (MCJ) (Strathdee et al., 2004, 2005) and synuclein-γ, encoding an activator of the MAPK and Elk-1 signaling cascades (Gupta et al., 2003; Woloszynska-Read et al., 2007), are upregulated in ovarian cancer in association with DNA hypomethylation. Table 16.1 demonstrates the methylation patterns of selected genes and their clinical significance.

HISTONE MODIFICATIONS AND OVARIAN CANCER

The fundamental building blocks of the chromatin, the nucleosomes, are formed by an octamer of histones. The nucleosome contains 147 base pairs that are wrapped around the

TABLE 16.1 Methylation Patterns of Selected Genes and Their Clinical Significance in Ovarian Cancer

Gene	Methylation Status	Clinical Correlation	References
RASSF1A	Hypermethylation	Detection of OC. Methylation in bodily fluids in early stages. Paclitaxel-like activity	Ibanez de Caceres et al. (2004), Makarla et al. (2005), Liggett et al. (2011); Liu et al. (2003)
BRCA1	Hypermethylation	Detection of OC. Poor prognosis. Improved chemotherapy response. Methylation in bodily fluids in early stages	Teodoridis et al. (2005); Ibanez de Caceres et al. (2004), Chiang et al. 2006)
APC	Hypermethylation	Serum/ascites diagnosis of OC	Ibanez de Caceres et al. (2004), Makarla et al. (2005)
MGMT	Hypermethylation	Detection of OC. Improved response to chemotherapy in late stage disease	Makarla et al. (2005), Teodoridis et al. (2005)
hMLH1	Hypermethylation	Poor prognosis. Improved chemotherapy response. Reversal of drug resistance	Gifford et al. (2004), Plumb et al. (2000), Strathdee et al. (1999)
HOXA9	Hypermethylation	Detection of OC	Montavon et al. (2012)
HOXA11	Hypermethylation	Residual disease after surgery and poor prognosis	Fiegl et al. (2008)
OPCML	Hypermethylation	Detection of OC	Zhou et al. (2014)
SFRP-1,-2,-3,-4,-5	Hypermethylation	Detection of OC. Poor prognosis. Recurrence	Su et al. (2009)
FZD4, DVL1, NFATC3, ROCK1, LRP5, AXIN1, NKD1	Hypermethylation	Poor prognosis	Dai W et al. (2011)
FBXO32	Hypermethylation	Poor prognosis	Chou et al. (2010)
FANCF	Hypermethylation	Cisplatin resistance	Taniguchi et al. (2003)
TMS1	Hypermethylation	More frequent in clear cell histology	Terasawa et al., 2004
WT1	Hypermethylation	More frequent in clear cell histology	Kaneuchi et al. (2005)
GSTP1	Hypermethylation	Detection of OC. Improved response to chemotherapy in late-stage disease	Makarla et al. (2005), Teodoridis et al. (2005)
DAPK	Hypermethylation	Methylation in bodily fluids (serum, plasma) in early-stage disease	Ibanez De Caceres et al. (2004), Collins et al. (2006)
CDKN2A	Hypermethylation	Methylation in bodily fluids (serum, plasma) in early-stage disease	Ibanez De Caceres et al. (2004)
IGFBP-3	Hypermethylation	Associated with disease progression in early-stage disease	Wiley A et al. (2006)

TABLE 16.1 Methylation Patterns of Selected Genes and Their Clinical Significance in Ovarian Cancer—cont'd

Gene	Methylation Status	Clinical Correlation	References
SOX1, LMX1A	Hypermethylation	Diagnosis and recurrence	Su et al. (2009)
PAX1	Hypermethylation	Diagnosis	Su et al. (2009)
SNCG	Hypomethylation	Tumor metastasis	Gupta et al. (2003)
hSulf-1	Hypomethylation	Sensitization to chemotherapy	Staub et al. (2007), Lai et al. (2003)
MCJ	Hypomethylation	Upregulation induces sensitivity to cisplatin, topotecan and paclitaxel	Shridhar et al. (2001)
TUBB3	Hypermethylation	Contributes to taxane resistance	Izutsu et al. (2008)
BORIS	Hypomethylation	Advanced stage and decreased survival	Woloszynska-Read et al. (2011)

histone octamer with two each of histones H2A, H2B, H3, and H4. The lysine-rich tails of the histones provide sites for reversible modifications that alter DNA structure and thus the expression of genes. Histones can become methylated or acetylated on specific amino acid residues located on their N- and C-terminal tails that correlate with active or repressive transcription (Esteller, 2007; Kouzarides, 2007). In normal cells, genes available for transcription are those with open nucleosome spacing around the transcription start site. By contrast, gene expression is silenced by condensing chromatin, methylated DNA, and deacetylating histones. Histones store epigenetic information through posttranslational modifications such as lysine acetylation, methylation, and ubiquitination, arginine methylation, and serine and theanine phosphorylation. Based on a new terminology, the main epigenetic regulators can be categorized as writers, erasers, and readers of posttranscriptional modifications (Dawson et al., 2012). Epigenetic writers catalyze the deposition of the posttranscriptional changes on proteins and introduce dynamic modifications that respond to environmental influences. Epigenetic erasers remove epigenetic modifications, and epigenetic readers are proteins with specific domains that are involved in the interpretation of signals that may influence subsequent changes.

Histone acetylation neutralizes the positive charge of the core histones and "unties" nucleosome-associated DNA, consequently allowing binding of transcription factors and polymerases. Histones acetylation is a dynamic process controlled by histone acetyltransferases and histone deacetylases (HDACs). Hypoacetylated histones correlate with gene silencing. Methylation at specific histone tail residues is associated with transcriptional activation as well as repression. Histone methylation occurs at both arginine and lysine residues on the tails of histone proteins H3 and H4 (Chervona and Costa, 2012).

Loss of the trimethylated histone 3 lysine is associated with overexpression of claudin-3 and claudin-4 oncogenes in ovarian cancer (Kwon et al., 2010). Transforming growth factor-beta (TGF-β)/SMAD signaling is a key growth regulatory pathway that may alter the expression of downstream targets and promote ovarian carcinogenesis. RunX1T1, a SMAD4 target, has epigenetically been silenced in a panel of ovarian cancer cells by promoter hypermethylation and histone methylation at H3 lysine 9. SMAD4 depletion increased repressive histone modifications of RunX1T1 promoter without affecting promoter methylation in ovarian cancer cells (Yeh et al., 2011). GATA6 is lost or excluded from the nucleus in 85% of ovarian tumors, and GATA4 expression is absent in most ovarian cancer cell lines. GATA4 and GATA6 gene silencing was found to correlate with hypoacetylation of histones H3 and H4 and loss of histone H3/lysine K4 trimethylation at their promoters in ovarian cancer cell lines (Caslini et al., 2006). Silencing of the osteoprotegerin (Lu et al., 2009) gene, which controls a protein with growth-inhibiting properties in cancer cells, has been shown to be associated with reduced histone 3 lysine 4 trimethylation and increased histone 3 lysine 27 trimethylation. Histone deacetylation of the regulator of the G-protein signaling 10 gene that regulates ovarian cancer cell growth and survival has been reported to be associated with ovarian cancer chemoresistance. Histone modification also has been implicated with other genes involved in cell cycle regulation such as p21 (Richon et al., 2000), cyclinB1 (Valls et al., 2005) and ADAM19 (Chan et al., 2008).

MICRORNAS AND OVARIAN CANCER

miRNAs are small (21–25 nucleotides) evolutionarily conserved noncoding RNAs that regulate the expression of about 30% of genes (Hata and Lieberman, 2015). They exhibit posttranscriptional regulation by binding to the 3′-untranslated region of their target genes. Half of human miRNAs have been identified as being in cancer-associated genomic regions. Overexpressed miRNAs may function as oncogenes by downregulating TSGs, whereas downregulated miRNAs may act as tumor suppressors by negatively regulating oncogenes (Ruan et al., 2009). Recent evidence suggests that the base pairing between an miRNA and its target gene not only results in repressed expression but also has an impact on the levels of the miRNA. This interplay between miRNAs and mRNAs results in a complex network of interactions that control the apoptosis, proliferation, and differentiation that lead to carcinogenesis (Zhang and Ng, 2016).

The oncogenic involvement of miRNAs is not limited to their direct impact on specific genes and their regulatory role as promoters or inhibitors of ovarian carcinogenesis. Indirect influences of miRNAs that involve angiogenesis, inflammation, and oxidative cell damage may also contribute to ovarian carcinogenesis, progression, and poor response to therapy (Hill et al., 2014). Conversely, the microenvironmental status of ovarian cancer cells may trigger miRNA dysregulation that in turn promotes further cancer progression. In experimental models where members of the miR-200 family were delivered into the tumor endothelium, inhibition of angiogenesis and subsequent normalization of vascular patterns was achieved by targeting interleukin-8 and CXCL1, resulting in ovarian cancer regression (Pecot et al., 2013). Similarly, indirect downregulation of hypoxia-inducible factor 1 and vascular endothelial growth factor (VEGF) by miR-145, miR-125b, and miR-199 resulted in inhibition of tumor angiogenesis (Xu et al., 2012; He et al., 2013). mir-141 and miR-200a also act on ovarian tumorigenesis by controlling oxidative stress response (Mateescu et al., 2011). Overexpression of miR-184 could suppress EOC cell proliferation and inflammation, and induce apoptosis in vitro (Qin et al., 2015). IKKβ, which is a subunit of a complex that directly activates NF-κB and promotes a proinflammatory environment, has been found to be regulated by miR-199 (Chen et al., 2008).

So far, most studies have demonstrated different miRNA expression in different histologic types and stages of ovarian carcinoma.

miR-200a and miR-200c were upregulated in all subtypes (mucinous, endometrioid, and clear cells), miR-200b and miR-141 were upregulated in serous as well as endometrioid carcinomas, and miR-21, miR-203, and miR-205 were upregulated only in endometrioid carcinomas (Iorio et al., 2007). miR-200a and miR-141 were highly upregulated in cancer, whereas miR-199a, miR-140, miR-145, and miR-125b1 were downregulated. In a genome-wide analysis of 498 high-grade serous epithelial ovarian carcinomas performed by the Cancer Genome Atlas (TCGA) group (2011), 89% of the genes found to be affected, were regulated by eight miRNAs. Transcriptomic analyses proposed a classification of four different molecular subtypes of ovarian cancer, not significantly associated with survival, named immunoreactive, differentiated, proliferative, and mesenchymal, based on the enrichment of specific genes in each subtype. The miRNAs' expression, on the other hand, differentiated the cancer samples into three main subtypes. Subtype 1 overlapped with the mRNA proliferative subtype, and miRNA subtypes 2–3 overlapped with the mRNA mesenchymal subtype. Subtype 1 was associated with worse patient survival. A more recent analysis of the TCGA data performed by Yang et al. (2013) suggested that an integrated analysis of miRNAs and the transcriptome can group the transcriptional subtypes into two clinically relevant subtypes, one mesenchymal and one epithelial. The authors highlighted the important role of an miRNA regulatory network consisting of eight key miRNAs for the mesenchymal subtype including miR-141, miR-200, miR-29c, miR-101, miR-506, and miR-128.

Most members of the miR-200 family, which consists of miR-200a, 200b, 200c, 141, and 429, have been extensively studied in ovarian cancer samples. They share sequence homology and possibly the same target genes (Pecot et al., 2013). miR-200c was identified to be associated with progression-free survival (PFS), overall survival, or both in stage I ovarian cancers (Marchini et al., 2011). Chen et al. (2013), after reviewing eight different studies, concluded that not all members of the miR-200 family are upregulated in ovarian cancer. This aforementioned inconsistency may be due to differences in the study design and the selection of control groups. Disease stage, histologic type, and the presence of metastasis may alter miRNA functions in ovarian cancer, and therefore a careful selection of specimens is important. An elevated expression of most members of the miR-200 family, however, is generally reported (Cao et al., 2014).

Forty-four miRNAs demonstrated different expression profiles between early- and late-stage ovarian carcinomas, with complete downregulation of all miRNAs in late-stage tumors (Zhang et al., 2008). Nam et al. (2008) identified 23 aberrantly expressed miRNAs in at least 60% of ovarian cancer samples, with miR-21 being the most upregulated (85% of samples) and miR-125b the most downregulated (95% of samples). miR-200a, miR-34a, and miR-449b were identified as mostly downregulated in advanced-stage ovarian tumors (Eitan et al., 2009). In a cohort of 55 advanced ovarian carcinomas, miR200a was identified as predictor of favorable outcome (Hu et al., 2009). miRNAs with tumor-suppressor properties, such as let-7d and miR-127, have been reported as downregulated in ovarian cancer samples (Johnson et al., 2005; Saito et al., 2006). Similarly, significantly decreased miR-215 expression in ovarian cancer tissues and cell lines was associated with the stage of the disease and lymph node metastasis (Lin et al., 2017). Calura et al. (2013) investigated the miRNA profiles of each histologic type and found that clear cell ovarian carcinoma is characterized by higher expression of miR-30a-5p and miR-30a-3p, whereas the mucinous type displays higher levels of miR-192 and miR-194 (Fig. 16.2).

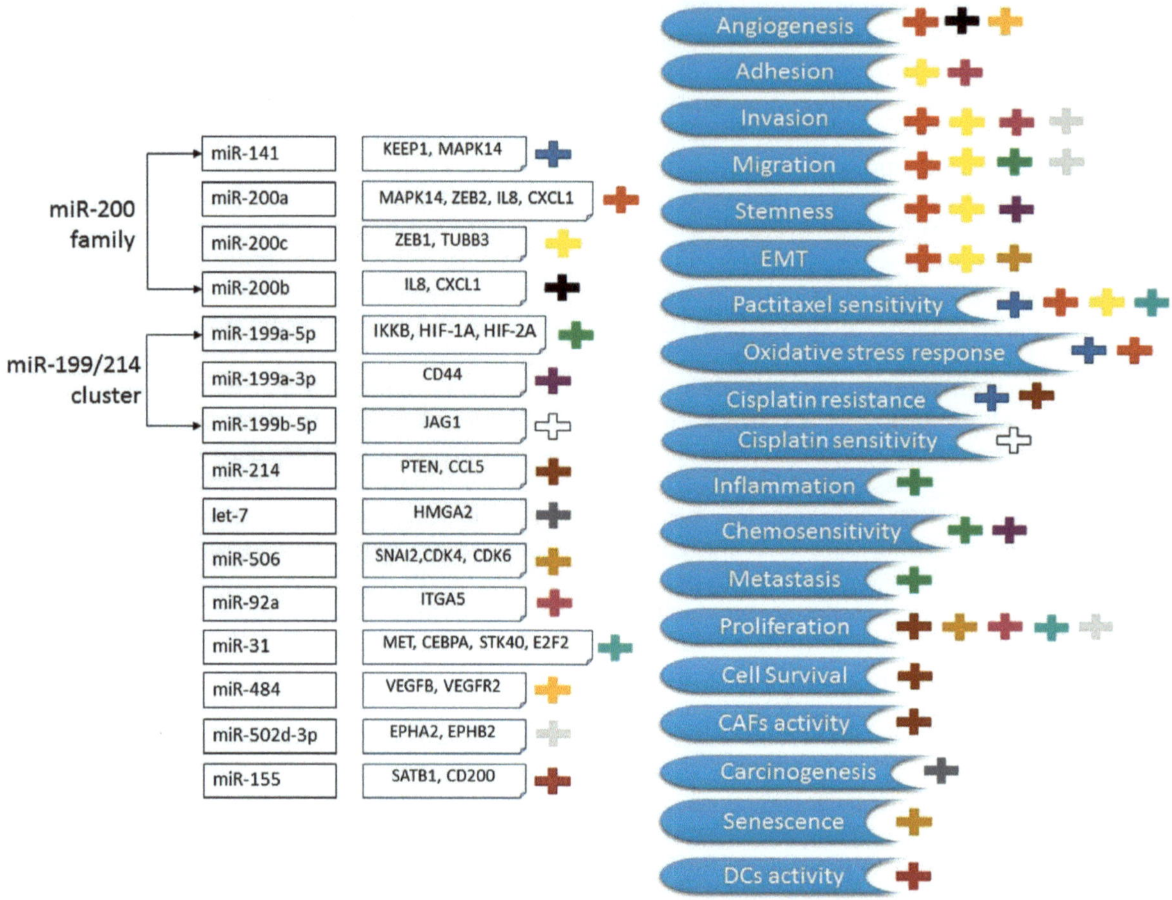

FIGURE 16.2 Potential therapeutic effects of miRNAs in ovarian cancer.

EPIGENETIC BIOMARKERS OF OVARIAN CANCER DETECTION

Because aberrant methylation is one of the earliest molecular alterations during tumorigenesis, it has been proposed as a promising strategy for early ovarian cancer detection. However, methylation of single genes may have limited clinical application. At present, no single epigenetic biomarker can accurately detect early ovarian cancer in either tissues or body fluids. Analysis of the methylation status of multiple genes simultaneously in a blood-based assay may provide a more sensitive and specific method for the molecular classification and prognosis of ovarian cancer.

A genome-wide DNAm profiling in a large ovarian cancer case control cohort study demonstrated that active ovarian cancer has a significant impact on the DNAm pattern in peripheral blood (Teschendorff et al., 2009). A microarray-based analysis of ovarian tumors identified 112 methylated loci prognostic for PFS in advanced ovarian cancer patients

(Wei et al., 2006). The data suggested that a higher degree of CpG island methylation is associated with disease recurrence following chemotherapy (Wei et al., 2002). Promoter hypermethylation of at least one of six genes (BRCA1, RASSF1A, APC, p14ARF, p16INK4A, and DAPK) was observed in 41 of 50 ovarian cancer serum specimens. Thus, hypermethylation of certain genes may present as an early event in ovarian tumorigenesis that can be detected in the serum of patients with ovary-confined (stage IA or B) tumors and cytologically negative peritoneal fluid (Ibanez et al., 2004). A recent study that used multiplex methylation-specific PCR to analyze the methylation status of cell-free serum DNA of seven candidate genes (APC, RASSF1A, CDH1, RUNX3, TFPI2, SFRP5, and OPCML) achieved sensitivity and specificity of 85.3% and 90.5%, respectively, for stage I OEC. The detection rates were markedly higher than those of CA125, which produced a sensitivity of 56.1% and a specificity of 64.15% (Zhang et al., 2013).

The role of DNA methylation biomarkers in ovarian cancer is promising. However, progression toward clinical practice is hampered by a lack of detection techniques that combine high accuracy with low cost. The main obstacles that have to be overcome are the standardization of analysis techniques and establishment of reliable reference values.

Most promising results seem to emerge from miRNAs used as markers for early detection of ovarian carcinoma. Circulating miRNAs in patients' blood may reflect the disease's status and therefore aid in early detection, accurate prediction of disease, progression, and monitoring. Circulating miRNAs are resistant to degradation by RNase, and they also exert control to distant target genes. The stability of miRNAs is attributed to their association with protein complexes such as Ago2 and the presence of miRNAs in circulating microvesicles such as exosomes.

A realistic diagnostic approach to ovarian cancer detection at an early stage relies mainly on a potential diagnostic test that can be easily applied and it is cheap, specific, and sensitive. To date such a biomarker is not available. miRNAs, although stable, have short sequences, and therefore interference of the cross-hybridization of sequences between homologous miRNAs is possible. In a study where whole blood samples from 24 patients with recurrent ovarian carcinoma and 15 healthy subjects were analyzed with microarray screening, 147 miRNAs were deregulated in ovarian cancer patients. In the subgroup of patients with serous carcinoma compared with an extended control group, diagnostic accuracy, specificity, and sensitivity increased by up to 85% (Häusler et al., 2010). A daring assumption has been made by Kan et al. in a study that demonstrated that miR-182, miR-200a, miR-200b, and miR-200c were highly overexpressed in serous epithelial ovarian cancer cell lines relative to normal human ovarian surface epithelial cells. The authors concluded that miRNAs in peripheral blood alone or in combination with other testing regimens can discriminate patients with high-grade serous epithelial ovarian carcinoma (Kan et al., 2012).

Eight miRNAs (miR-21, miR-141, miR-200a, miR-200b, miR-200c, miR-203, miR-205, and miR-214) with established overexpression in ovarian cancer tissues have been studied with respect to their expression in the peripheral blood of the corresponding patients. The exosomal miRNA profiles of the ovarian cancer patients were elevated, whereas the exosomal miRNAs could not be detected in healthy controls (Taylor and Gercel-Taylor, 2008). In a similar study by Resnick et al. (2009), miR-21, miR-29a, miR-92, miR-93, and miR-126 were significantly overexpressed in the serum of ovarian cancer patients compared with controls, while miR-99b, miR-127, and miR-155 were significantly underexpressed.

Serum miR-26a, miR-132, miR-145, and let-7b were all downregulated in the serum of ovarian cancer patients compared with controls (Chung et al., 2013). Zhang et al. evaluated 360 plasma samples from patients with EOC and compared them with those of 200 healthy subjects. They found higher plasma levels of miR-205 and lower levels of let-7f in the EOC cases than in the controls. Combined miR-205 and let-7f provided high diagnostic accuracy for EOC, especially for early-stage disease. Addition of CA-125 to the test improved detection accuracy (Zheng et al., 2013). In a meta-analysis of all published results from studies that evaluated the relative expression of miR-145 in the serum of patients with ovarian malignancy, miR-145 demonstrated good performance as a biomarker for ovarian cancer detection (Liang et al., 2015).

Overall, miRNAs have a potential clinical utility in the diagnosis of ovarian cancer. Which one will perform best remains to be proven by future large-scale studies. What we already know suggests that specific miRNAs or a panel of several ovarian cancer–associated miRNAs in peripheral blood could become valuable methods for early disease detection.

EPIGENETICS AND RESISTANCE TO CHEMOTHERAPY

While many ovarian cancer patients respond completely to chemotherapy, the majority of initial responders eventually develop chemoresistance. Ovarian cancer drug resistance can be de novo (intrinsic) or acquired (extrinsic). De novo chemoresistance is attributed to the existence of cancer stem cells, whereas genetic and epigenetic alterations of oncogenes and TSGs contribute to the acquired type.

In addition to mutations, DNA methylation–induced silencing of various drug-response genes and pathways also facilitates the development of ovarian tumor cell drug resistance. It was shown that the silencing of SFRP5, which is a Wnt antagonist, by DNA hypermethylation was associated with platinum resistance in ovarian cancer (Su et al., 2010). Similarly, hypermethylation of several genes such as hMLH1, the arginine biosynthesis-related gene ASS1, and ESR2 (encoding the ER-b) are involved in platinum resistance (Nicholson et al., 2009; Strathdee et al., 1999; Yap et al., 2009). Platinum resistance has also been correlated with stage-progressive of the MCJ gene hypermethylation associated with poor response to chemotherapy (Strathdee et al., 2005). DAPK, which is a gene involved in apoptosis, has also been shown to be silenced in drug-resistant cancer due to methylation (Lehmann et al., 2002).

In addition to the loss of expression due to DNA methylation, it was shown that hypomethylation along with an increase in the expression of the MAL gene is associated with platinum resistance (Lee et al., 2010). Hypomethylation and upregulation of the ABCG2 multidrug transporter gene was also shown to occur during chemoresistance in two ovarian carcinoma cell lines (Curley et al., 2009). Based on the association of DNA methylation of specific genes with platinum sensitivity, it was shown that the hypomethylation-mediated activation of the cell growth–promoting pathways, PI3K/Akt and TGF-β, and cell cycle progression, may contribute to cisplatin resistance in ovarian cancer cells (Li et al., 2009).

Teodoridis et al. (2005) demonstrated that methylation of at least one of three genes involved in DNA repair and drug detoxification, BRCA1, GSTP1, and MGMT, is associated with improved response to chemotherapy in patients with late-stage EOC. Patients with advanced-stage EOC that overexpress hSulf-1 demonstrated an increased response rate to chemotherapy compared with that of patients whose tumors express low or moderate levels of hSulf-1 (Staub et al., 2007). Methylation-associated silencing of hSulf-1, which leads to its downregulation, results in the reduction of cisplatin-induced cytotoxicity (Lai et al., 2003).

miRNAs have already demonstrated their ability to predict clinical response to chemotherapy. In a study where miRNA expression of patients with no recurrence within 6 months of treatment's end was compared with that of patients who progressed rapidly after treatment, seven miRNAs were significantly differentially expressed, including hsa-miR-27a, miR-23a, and miR-378 (Eitan et al., 2009). The differential expression of these three miRNAs between sensitive and resistant tumors was also observed in a subset of stage III patients treated with combined paclitaxel/carboplatin. miRNAs miR-199a, −200a, and −214 were found to be upregulated in ovarian tumors, and miR-214 was demonstrated to target the tumor suppressor PTEN and to have an association with resistance to platinum therapy (Iorio et al., 2007). Likewise, women lacking complete response to paclitaxel-carboplatin therapy had tumors with significantly lower miR-200c levels than the levels of those with complete response to treatment (Leskelä et al., 2010). The miRNA let-7i was significantly downregulated in platinum-resistant ovarian tumors, and when its function was restored, it led to chemosensitivity of ovarian cancer cells (Yang et al., 2008). Two other studies showed that miR-182 and miR-125b were associated with resistance to cisplatin due to the repression of two TSGs, PDCD4 and Bcl-2 antagonist killer 1, respectively (Kong et al., 2011; Wang et al., 2013a).

Based on the known association between chemotherapy response and miRNA expression, many studies have attempted to disturb the mechanisms responsible for chemoresistance by altering the levels of miRNAs. Prislei and colleagues (2013) showed that repression of TUBB3 genes by overexpression of miR-200c increased the sensitivity to paclitaxel and cisplatin. Similar results were demonstrated in another study where restoration of miR-200c increased sensitivity to microtubule-targeting agents and moderated invasiveness by 85% (Cochrane et al., 2009). A chimera composed of an MUC1 aptamer and let-7i delivered let-7i into paclitaxel/cisplatin-resistant ovarian tumor cells and resensitized the tumor to the apoptotic properties of the drug (Liu et al., 2012).

CHEMOPREVENTION

Cancer chemoprevention is a major challenge in cancer-preventive strategies that utilize naturally occurring dietary compounds or therapeutic drugs with relatively low toxicity. So far, the cost of any cancer treatment is the sacrifice of many normal noncancerous cells in the human body. In an effort to identify and develop anticancer agents that cause minimal harm to normal cells while effectively attacking cancer cells, a number of naturally occurring phytochemicals in food and medicinal plants have been investigated. In the past two decades, several studies have shown that dietary supplements and natural compounds are potent epigenetic regulators that could target or revert abnormal epigenetic alterations in various human diseases including cancer (Huang et al., 2011; Hauser and Jung, 2008; Davis and Ross, 2008). These dietary phytochemicals have shown potential as epigenetic modulators such as DNA methyltransferase (DNMT) inhibitors, histone modifiers, and the varying of miRNA levels, thus altering gene expression and restoring the expression of TSGs.

Curcumin is the active component of the popular South Asian spice turmeric (*Curcuma longa*). It has been proven to possess antimicrobial, antioxidant, anti-inflammatory, antiallergic, and anticarcinogenic effects (Prasad et al., 2014; Bar-Sela et al., 2010). Several studies have reported that curcumin is a potential modifier of DNMTs, HDACs, and miRNAs. Curcumin reactivates TSGs such as RARB and RASSF1A by reducing DNA methylation (Jiang et al., 2015; Du et al., 2012). Both TSGs were found hypermethylated in ovarian carcinoma. Treatment with curcumin resulted in an increase in miR-9 expression in SKOV3 ovarian cancer cells (Zhao et al., 2014). Formulated curcumin-loaded nanostructured lipid carriers have been shown to be more effective in suppressing the proliferation of ovarian cancer cell lines than is the administration of free curcumin (Bondi et al., 2017).

Resveratrol is a phytoalexin produced by plants that is present in grapes, wine, and root extracts of the weed *Polygonum cuspidatum*. It has been identified as a potent anti-inflammatory, antiaging, and chemopreventive agent. Resveratrol elicits its antitumor effect in ovarian cancer through epigenetic upregulation of ARH-I and downregulation of STAT3 expression that induces autophagy of the migrated cells (Ferraresi et al., 2017). Resveratrol inhibits the growth of BRCA1 mutant tumors both in vitro and in vivo by upregulating the activity of SIRT1, a well-known deacetylase, followed by reduction of survivin expression (Wang et al., 2008). Berberine is an isoquinoline quaternary alkaloid that has been used in traditional Chinese and Indian medicine. It has anti-inflammatory, antimicrobial, and antidiabetic properties. Berberine has been shown to augment the effects of cisplatin in A2780 ovarian carcinoma cells by inhibiting miR-93 expression and increasing the expression of tumor suppressor PTEN, which is the target of miR-93 (Chen et al., 2015).

Genistein is a soy phytoestrogen that has been shown to possess anticancer properties through modulation of the genes controlling cell-cycle progression. Parker et al. (2009) studied the impact of genistein treatment on miRNA expression in ovarian cell lines from one patient before and after cancer recurrence post treatment. Several miRNAs were upregulated by genistein, while miR-135 and miR-765 exhibited higher expression in both cell lines (Parker et al., 2009). In another study, treatment of ovarian cancer cells with genistein reduced the oncogenic miR-27a and consequently repressed cell proliferation and migration (Xu et al., 2013). The tumor suppressor SPRY2, which is a target of miR-27a and known to inhibit ERK signaling pathways, was significantly increased, probably through miR-27a downregulation induced by genistein.

EPIGENETIC THERAPY

Unlike cancer-associated gene mutations, DNA methylation and other epigenetic modifications are potentially reversible. This makes epigenetic agents attractive candidates for disease prevention and resensitization to chemotherapeutic agents (Chen et al., 2011). The merit of developing pharmacologic inhibitors of DNA methylation and deacetylation is to reverse the epigenetic silencing of key genes, leading to their reexpression in cancer cells and the reactivation of tumor-suppression mechanisms.

There are two types of DNA methylation inhibitors, nucleoside and nonnucleoside analogues. Nucleoside analogues inhibit methylation when they are integrated into DNA and block the release of DNMTs by forming a covalent complex with these enzymes (Santi et al., 1984). They have been found to have clinical activities, especially on hematopoietic malignancies (Issa et al., 2005). These inhibitors have been used to induce the reexpression of TSGs that have been silenced by hypermethylation. Although aberrant promoter methylation is corrected by DNA methylation inhibitors, when the drug is stopped, aberrant methylation and gene silencing are reestablished (Herman and Baylin, 2003). Nonnucleoside analogues are small molecular inhibitors that bind to the catalytic region of DNMTs and suppress translation.

Azacytidine and decitabine are the first two DNMT inhibitors (DNMTIs) approved for therapy of myelodysplastic syndromes. The demethylating effect of decitabine is stronger than that of azacytidine because the former only binds DNA, unlike 5-azacytidine, which incorporates into both DNA and RNA (Hellebrekers et al., 2007). Decitabine has been shown to cause demethylation in numerous ovarian cell lines, reversing the silencing of several TSGs such as hMLH1 (Takai et al., 2005). Decitabine has also been reported to decrease cisplatin resistance in both ovarian cancer cells and a mouse xenograft through demethylation of the hMLH1 promoter (Plumb et al., 2000). Moreover, reversal of lysine methylation at K9 of histone H3 is induced by decitabine (Nguyen et al., 2002). Decitabine has also been reported to upregulate a group of six miRNAs clustered on chromosome 19 and seven clustered on chromosome 14 (Zhang et al., 2008). Two clinical trials have provided evidence that azacytidine and decitabine are capable of reversing platinum resistance in ovarian cancer patients (Fang et al., 2010; Fu et al., 2011). However, DNMTIs may simultaneously cause widespread genomic hypomethylation that potentially leads to GI (Chen et al., 1998).

Zebularine is a recently developed cytosine analog-based DNMTI with marked stable chemical properties for oral administration, decreased toxicity and high selectivity for tumor cells (Cheng et al., 2004). Zebularine forms a complex with DNMTs in order to deplete DNMT1 or partially deplete DNMT3a and Dnmt3b (Zhou et al., 2002). Zebularine proved to be able to demethylate and reactivate the silenced p16 gene in vitro and in vivo (Cheng et al., 2003). High levels of the drug are required to achieve efficacy, however, and its potential application in clinical settings is therefore currently limited. Other cytidine analogs, such as arabinosyl-5-aza-cytosine (fazarabine) and dihydro-5-aza-cytidine, demonstrated disappointing results in clinical trials (Goffin and Eisenhauer, 2002).

Nonnucleotide DNMTIs, which share the same mechanism of action by binding to cytosines and interfering with DNMTs, have recently been developed to avoid the inherent toxicity of nucleotide analogs. These include an antiarrhythmic (procainamide), an anesthetic (procaine), an antihypertensive (hydralazine), epigallocathechin-3-gallate, and the novel compound RG108 (Brueckner et al., 2005; Fang et al., 2003; Segura-Pacheco et al., 2003). MG98 represents a newly launched therapeutic agent that specifically inhibits DNMT1 function (Yan et al., 2003).

A rapidly expanding field of epigenetic therapy is the development of drugs that target histone modifications. HDAC inhibitors (HDACIs) have been shown to decrease cancer cell growth, induce apoptosis, and promote cell differentiation. A wide variety of agents that function as HDACs have been developed, including organic hydroxamic acids, short-chain fatty acids, benzamides, cyclic tetrapeptides, and sulfonamides (Takai and Narahara, 2010). HDACIs that have been studied in ovarian cancer include suberoylanilide hydroxamic acid (SAHA), valproic acid (VPA), and romidepsin, vorinostat, and belinostat. Other HDACIs not currently approved by the FDA have shown potential benefit in ovarian cancer treatment. Trichostatin A, which specifically inhibits class I and II mammalian HDAC families, was highly cytotoxic with limited clinical value (Vigushin et al., 2001).

SAHA is an HDACI that can promote cell cycle arrest, apoptosis, and decreased cell viability in ovarian cancer cells (Takai et al., 2004). SAHA exhibits synergistic action when combined with paclitaxel, while SAHA in conjunction with decitabine inhibits ovarian cancer cell growth by increasing the expression of tumor-suppressor imprinted genes and thus increasing apoptosis, cell cycle arrest, and autophagy (Chen MY et al., 2011). The antiproliferative properties of SAHA have been attributed to the accumulation of acetylated proteins and core histones. SAHA is FDA approved for the treatment of cutaneous T-cell lymphoma and is generally well tolerated. VPA, which has been used previously as an anticonvulsant, is a short-chain fatty acid drug approved by the FDA. VPA promotes cell cycle arrest and apoptosis in ovarian cancer cell lines and also sensitizes and resensitizes ovarian cancer to cisplatin treatment (Lin et al., 2008; Takai et al., 2004). VPA was effective in ovarian cancer cell lines, and treatment upregulated E-cadherin expression while it downregulated MMP9 and VEGF (Shan et al., 2012). Monti et al. provided an extensive review of VPA function (Monti et al., 2009). Romidepsin has received FDA approval for cutaneous T-cell lymphoma. Romidepsin inhibited cell viability and induced apoptosis in ovarian cell lines in an earlier report, and moreover showed promising results in combination with cisplatin by increasing cell apoptosis in in vivo and in vitro models (Wilson et al., 2012). Table 16.2 demonstrates several epigenetic agents and their use in cancer treatment.

While HDACIs and DNMTIs have demonstrated clinical activity as single-agent therapies for hematopoietic malignancies, DNA methylation and histone deacetylation often coordinate to inhibit gene transcription, and restoration of both silencing mechanisms may be necessary for maximal gene derepression. Treatment with a DNMTI/HDACI combination in ovarian cancer cases was synergistic for upregulation of the proapoptotic gene TMS1/ASC, in contrast to either agent alone (Terasawa et al., 2004). An earlier integrated microarray analysis demonstrated that DNMTI/HDACI-combined treatment of ovarian cancer cells affects more genes than either agent individually (Shi et al., 2003).

While the general concept of the reversal of oncogenic-induced epigenetic changes in ovarian cancer patients by administration of epigenetic agents seems promising, the actual efficacy of such therapies in actual patients is scarce. Most data derive from small or ongoing clinical trials, and therefore no single agents have been proven superior or equal to standard therapy. Most trials include patients with aggressive disease or those who demonstrate resistance to standard cytotoxic chemotherapy. Fazarabine, a nucleoside analogue that consists of the arabinoside ring of 1-beta-D-arabinofuranosylcytosine and the pyrimidine base of 5-azacytidine, was administered to 19 patients with recurrent disease. Although 44% of patients had stable disease, no complete or partial responses were observed. The major toxicity of this agent was severe neutropenia, which was more prominent in 4 patients (Manetta et al., 1995). In a phase II study with 27 patients with recurrent platinum-sensitive ovarian cancer where vorinostat was administered, only 2 patients had PFS longer than 6 months (Modesitt et al., 2008). In a cohort of 32 patients with platinum-resistant recurrent disease that consisted of 18 epithelial ovarian cancer patients and 14 with low malignancy potential, belinostat as a single agent achieved PFS of 2.3 months, while only 9 patients had the best response to treatment (Mackay et al., 2010). Given the lack of promising results, attention has been shifted toward the combination of several epigenetic agents with standard chemotherapy.

Restoration of chemotherapeutic sensitivity has been assessed in patients with platinum-resistant ovarian cancer. In a phase 1b-a clinical trial study with sequential combination of azacytidine and carboplatin in 30 patients with either platinum-resistant or platinum refractory epithelial ovarian cancer, objective response rate (ORR) in patients with platinum-resistant disease was 22% with a median PFS of 5.6 months (Fu et al., 2011). In a similar study where low-dose decitabine was administered before carboplatin in 17 platinum-resistant ovarian cancer patients, ORR was 35% and PFS 10.2 months, with 9 patients free of progression for 6 months (Matei et al., 2012). HDACIs, however, have not demonstrated similar results in restoring chemosensitivity. In a phase II study by Dizon et al. that included 27 patients with platinum-resistant disease, a combination of belinostat with carboplatin resulted in an ORR of only 7.4%. The study was terminated early due to the disappointing preliminary results (Dizon et al., 2012a). In another study where two epigenetic agents (an HDACI and a DNMTI) were combined in 10 platinum-refractory ovarian cancer patients, only 3 patients exhibited stable disease while 80% of patients experienced severe adverse effects.

Clinical trials have also investigated the potential benefit of combining epigenetic agents with chemotherapy in disease progression. Low-dose decitabine combined with paclitaxel

TABLE 16.2 Known Epigenetic Agents and Their Uses in Cancer Treatment

Class	Drug	Advantages	Disadvantages	References
DNA METHYLTRANSFERASE INHIBITORS				
Nucleoside analog DNMT inhibitor	Azacitidine (5-azacytidine)	US FDA approved Reverse platinum resistance	Less inhibitory effect than that of other DNMTIs Chemically unstable in water Induces global hypomethylation Short half-life, thus repeated daily doses are needed	Tam et al. (2007), Graham et al. (2009)
	Decitabine (5-aza-2′-deoxycytidine)	FDA approved Stronger DNMTI effect than azacytidine Selective for cancer cells Enhances the effects of HDACIs	Chemically unstable in water Induces global hypomethylation Mutagenic in vivo Enhances the apoptotic effects of HDACIs Short half-life, thus repeated daily doses are needed	Tam et al. (2007), Graham et al. (2009), Chuang et al. (2005), Kattan et al. (2003), Chen et al. (1998)
	Zebularine	More stable in water. Suitable for oral preparations Less toxic Selective for cancer cells	High doses of drug are required to achieve efficacy. Short half-life, thus repeated daily doses are needed	Cheng et al., 2004; Graham et al., 2009.
Nonnucleoside DNMT inhibitor	Hydralazine	Reexpresses the functional products of TSGs in cancer cell lines. No cytotoxicity	Induces a lupus-like syndrome in high doses	Yung et al. (1997), Segura-Pacheco et al. (2003)
	RG108	Less toxic Specific for hypermethylated TSGs	Hydrophobic Limited value as an anticancer drug	Mai and Altucci (2009), Brueckner et al. (2005)
HISTONE DEACETYLASE INHIBITORS				
Hydroxamic acids	Suberoylanilide hydroxamic acid	FDA approved Well tolerated and acceptable safety Synergistic action with paclitaxel and decitabine	Side effects Limited ability to be a selective HDAC isoform inhibitor	Butler and Kozikowski (2008)
Short-chain fatty acid	Valproic acid	Reduces tumor growth and metastasis Resensitizes ovarian cancer to cisplatin	Side effects	Gottlicher et al. (2001)

and platinum chemotherapy administered to 17 patients with platinum-resistant or platinum-refractory disease produced partial response or no progression in 70% of patients with no significant adverse effects (Fu et al., 2015). A combination of decitabine with carboplatin in a randomized phase II trial, versus carboplatin alone, in patients with recurrent partially platinum-sensitive ovarian cancer resulted in reduction of the efficacy of carboplatin. Neutropenia and carboplatin hypersensitivity were also more frequent in patients when the combined regimen was used (Glasspool et al., 2014). High toxicity also has been noted in patients where vorinostat was combined with chemotherapy. Vorinostat in combination with carboplatin and gemcitabine in platinum-sensitive patients who have been diagnosed with recurrent disease resulted in an unacceptably high rate of hematologic toxicity and termination of the study (Matulonis et al., 2015). Similarly, in a study where vorinostat was combined with paclitaxel and carboplatin, gastrointestinal perforation occurred in 3 out of 11 patients (Mendivil et al., 2013). Belinostat, on the other hand, seems to be better tolerated in combination with chemotherapy. In a phase 1b-2 trial that included 19 patients with recurrent platinum-sensitive disease and 16 patients with platinum-refractory disease, belinostat with paclitaxel and carboplatin resulted in an ORR of 43%. Toxicity was not significant, with grade 4 neutropenia in 14% of patients (Dizon et al., 2012b).

CONCLUSIONS

Epigenetic alterations are clearly involved in ovarian cancer initiation and progression. The complexity of ovarian cancer and its aggressive nature stress the need for novel diagnostic biomarkers and effective management strategies. Epigenetic mechanisms have been implicated in ovarian cancer progression, and moreover can distinguish different subtypes and stages of the disease. Promising results have been demonstrated from studies that assess circulating epigenetic markers in patient blood that can predict which cancer types are susceptible to chemotherapy. Clinical trials on epigenetic agents such as DNA demethylating agents and HDACIs have so far exhibited limited clinical application, and therefore their exact therapeutic roles are yet to be discovered. Future studies should focus on the detection of specific epigenetic markers that could discriminate ovarian cancer patients who will be susceptible to different chemotherapeutic agents. Epigenetic profiling of different histologic types and stages of the disease might provide an individualized approach to therapy, monitoring of the disease, metastasis, and response to therapy. Chemoprevention in ovarian cancer is a most promising field of research and a substantial step in cancer prevention/treatment. Dietary supplements and phytochemicals exhibit their antiproliferative anticancer properties through epigenetic pathways. Because these agents have minimal adverse effects in normal cells, lowering treatment toxicity in aggressive neoplasms such as ovarian cancer should direct our future studies. Altering epigenetic mechanisms and inhibiting cancer progression, or even preventing carcinogenesis, is in accordance with the foundation of Hippocratic medicine: "first do no harm."

References

Baldwin, R.L., Nemeth, E., Tran, H., Shvartsman, H., Cass, I., Narod, S., Karlan, B.Y., 2000. BRCA1 promoter region hypermethylation in ovarian carcinoma: a population-based study. Cancer. Res. 60, 5329–5333.

Bar-Sela, G., Epelbaum, R., Schaffer, M., 2010. Curcumin as an anti-cancer agent: review of the gap between basic and clinical applications. Curr. Med. Chem. 17, 190–197.

Barton, C.A., Hacker, N.F., Clark, S.J., O'Brien, P.M., 2008. DNA methylation changes in ovarian cancer: implications for early diagnosis, prognosis and treatment. Gynecol. Oncol. 109, 129–139.

Baylin, S.B., Chen, W.Y., 2005. Aberrant gene silencing in tumor progression: implications for control of cancer. Cold. Spring. Harb. Symp. Quant. Biol. 70, 427–433.

Baylin, S.B., Ohm, J.E., 2006. Epigenetic gene silencing in cancer - a mechanism for early oncogenic pathway addiction? Nat. Rev. Cancer 6, 107–116.

Blasi, M.F., Ventura, I., Aquilina, G., Degan, P., Bertario, L., Bassi, C., Radice, P., Bignami, M., 2006. A human cell-based assay to evaluate the effects of alterations in the MLH1 mismatch repair gene. Cancer. Res. 66, 9036–9044.

Bondì, M.L., Emma, M.R., Botto, C., Augello, G., Azzolina, A., Di Gaudio, F., Craparo, E.F., Cavallaro, G., Bachvarov, D., Cervello, M., 2017. Biocompatible lipid nanoparticles as carriers to improve curcumin efficacy in ovarian cancer treatment. J. Agric. Food Chem. 65, 1342–1352.

Brueckner, B., Boy, R.G., Siedlecki, P., Musch, T., Kliem, H.C., Zielenkiewicz, P., Suhai, S., Wiessler, M., Lyko, F., 2005. Epigenetic reactivation of tumor suppressor genes by a novel small-molecule inhibitor of human DNA methyltransferases. Cancer. Res. 65, 6305–6311.

Butler, K.V., Kozikowski, A.P., 2008. Chemical origins of isoform selectivity in histone deacetylase inhibitors. Curr. Pharm. Des. 14, 505–528.

Calura, E., Fruscio, R., Paracchini, L., Bignotti, E., Ravaggi, A., Martini, P., Sales, G., Beltrame, L., Bignotti, E., Tognon, G., Milani, R., Clivio, L., Dell'Anna, T., Cattoretti, G., Katsaros, D., Sartori, E., Mangioni, C., Ardighieri, L., D'Incalci, M., Marchini, S., Romualdi, C., 2013. miRNA landscape in stage I epithelial ovarian cancer defines the histotype specificities. Clin. Cancer. Res. 19, 4114–4123.

Cancer Genome Atlas Research Network, 2011. Integrated genomic analyses of ovarian carcinoma. Nature 474, 609–615.

Cao, Q., Lu, K., Dai, S., Hu, Y., Fan, W., 2014. Clinicopathological and prognostic implications of the miR-200 family in patients with epithelial ovarian cancer. Int. J. Clin. Exp. Pathol. 7, 2392–2401.

Caslini, C., Capo-chichi, C.D., Roland, I.H., Nicolas, E., Yeung, A.T., Xu, X.X., 2006. Histone modifications silence the GATA transcription factor genes in ovarian cancer. Oncogene 25, 5446–5461.

Chan, M.W., Huang, Y.W., Hartman-Frey, C., Kuo, C.T., Deatherage, D., Qin, H., Cheng, A.S., Yan, P.S., Davuluri, R.V., Huang, T.H., Nephew, K.P., Lin, H.J., 2008. Aberrant transforming growth factor beta1 signaling and SMAD4 nuclear translocation confer epigenetic repression of ADAM19 in ovarian cancer. Neoplasia 10, 908–919.

Chan, K.Y., Ozcelik, H., Cheung, A.N., Ngan, H.Y., Khoo, U.S., 2002. Epigenetic factors controlling the BRCA1 and BRCA2 genes in sporadic ovarian cancer. Cancer. Res. 62, 4151–4156.

Chen, R., Alvero, A.B., Silasi, D.A., Kelly, M.G., Fest, S., Visintin, I., Leiser, A., Schwartz, P.E., Rutherford, T., Mor, G., 2008. Regulation of IKKbeta by miR-199a affects NF-kappaB activity in ovarian cancer cells. Oncogene 27, 4712–4723.

Chen, H., Hardy, T.M., Tollefsbol, T.O., 2011. Epigenomics of ovarian cancer and its chemoprevention. Front. Genet. 2, 67.

Chen, M.Y., Liao, W.S., Lu, Z., Bornmann, W.G., Hennessey, V., Washington, M.N., Rosner, G.L., Yu, Y., Ahmed, A.A., Bast Jr., R.C., 2011. Decitabine and suberoylanilide hydroxamic acid (SAHA) inhibit growth of ovarian cancer cell lines and xenografts while inducing expression of imprinted tumor suppressor genes, apoptosis, G2/M arrest, and autophagy. Cancer 117, 4424–4438.

Chen, R.Z., Pettersson, U., Beard, C., Jackson-Grusby, L., Jaenisch, R., 1998. DNA hypomethylation leads to elevated mutation rates. Nature 395, 89–93.

Chen, Q., Qin, R., Fang, Y., Li, H., 2015. Berberine sensitizes human ovarian cancer cells to cisplatin through miR-93/PTEN/Akt signaling pathway. Cell. Physiol. Biochem. 36, 956–965.

Chen, Y., Zhang, L., Hao, Q., 2013. Candidate microRNA biomarkers in human epithelial ovarian cancer: systematic review profiling studies and experimental validation. Cancer. Cell. Int. 13, 86.

Cheng, J.C., Matsen, C.B., Gonzales, F.A., Ye, W., Greer, S., Marquez, V.E., Jones, P.A., Selker, E.U., 2003. Inhibition of DNA methylation and reactivation of silenced genes by zebularine. J. Natl. Cancer. Inst. 95, 399–409.

Cheng, J.C., Yoo, C.B., Weisenberger, D.J., Chuang, J., Wozniak, C., Liang, G., Marquez, V.E., Greer, S., Orntoft, T.F., Thykjaer, T., Jones, P.A., 2004. Preferential response of cancer cells to zebularine. Canc. Cell 6, 151–158.

Chervona, Y., Costa, M., 2012. Histone modifications and cancer: biomarkers of prognosis? Am. J. Cancer. Res. 2, 589–597.

Chiang, J.W., Karlan, B.Y., Cass, L., Baldwin, R.L., 2006. BRCA1 promoter methylation predicts adverse ovarian cancer prognosis. Gynecol. Oncol. 101, 403–410.

Chou, J.L., Su, H.Y., Chen, L.Y., Liao, Y.P., Hartman-Frey, C., Lai, Y.H., Yang, H.W., Deatherage, D.E., Kuo, C.T., Huang, Y.W., Yan, P.S., Hsiao, S.H., Tai, C.K., Lin, H.J., Davuluri, R.V., Chao, T.K., Nephew, K.P., Huang, T.H., Lai, H.C., Chan, M.W., 2010. Promoter hypermethylation of FBXO32, a novel TGF-beta/SMAD4 target gene and tumor suppressor, is associated with poor prognosis in human ovarian cancer. Lab. Invest. 90, 414–425.

Chuang, J.C., Yoo, C.B., Kwan, J.M., Li, T.W., Liang, G., Yang, A.S., Jones, P.A., 2005. Comparison of biological effects of non-nucleoside DNA methylation inhibitors versus 5-aza-2′-deoxycytidine. Mol. Cancer. Ther. 4, 1515–1520.

Chung, Y.W., Bae, H.S., Song, J.Y., Lee, J.K., Lee, N.W., Kim, T., Lee, K.W., 2013. Detection of microRNA as novel biomarkers of epithelial ovarian cancer from the serum of ovarian cancer patients. Int. J. Gynecol. Cancer 23, 673–679.

Cochrane, D.R., Spoelstra, N.S., Howe, E.N., Nordeen, S.K., Richer, J.K., 2009. MicroRNA-200c mitigates Invasiveness and restores sensitivity to microtubule-targeting chemotherapeutic agents. Mol. Cancer. Ther. 8, 1055–1066.

Collins, Y., Dicioccio, R., Keitz, B., et al., 2006. Methylation of death-associated protein kinase in ovarian carcinomas. Int. J. Gynecol. Cancer 16 (Suppl. 1), 195–199.

Curley, M.D., Therrien, V.A., Cummings, C.L., Sergent, P.A., Koulouris, C.R., Friel, A.M., Roberts, D.J., Seiden, M.V., Scadden, D.T., Rueda, B.R., Foster, R., 2009. CD133 expression defines a tumor initiating cell population in primary human ovarian cancer. Stem Cell. 27, 2875–2883.

Dai, W., Teodoridis, J.M., Zeller, C., Graham, J., Hersey, J., Flanagan, J.M., Stronach, E., Millan, D.W., Siddiqui, N., Paul, J., Brown, R., 2011. Systematic CpG islands methylation profiling of genes in the wnt pathway in epithelial ovarian cancer identifies biomarkers of progression-free survival. Clin. Cancer. Res. 17, 4052–4062.

Davis, C.D., Ross, S.A., 2008. Evidence for dietary regulation of microRNA expression in cancer cells. Nutr. Rev. 66, 477–482.

Dawson, M.A., Kouzarides, T., Huntly, B.J., 2012. Targeting epigenetic readers in cancer. N. Engl. J. Med. 367, 647–657.

Dhillon, V.S., Young, A.R., Husain, S.A., Aslam, M., 2004. Promoter hypermethylation of MGMT, CDH1, RAR-beta and SYK tumour suppressor genes in granulosa cell tumours (GCTs) of ovarian origin. Br. J. Cancer 90, 874–881.

Dizon, D.S., Blessing, J.A., Penson, R.T., Drake, R.D., Walker, J.L., Johnston, C.M., Disilvestro, P.A., Fader, A.N., 2012a. A phase II evaluation of belinostat and carboplatin in the treatment of recurrent or persistent platinum-resistant ovarian, fallopian tube, or primary peritoneal carcinoma: a Gynecologic Oncology Group study. Gynecol. Oncol. 125, 367–371.

Dizon, D.S., Damstrup, L., Finkler, N.J., Lassen, U., Celano, P., Glasspool, R., Crowley, E., Lichenstein, H.S., Knoblach, P., Penson, R.T., 2012b. Phase II activity of belinostat (PXD-101), carboplatin, and paclitaxel in women with previously treated ovarian cancer. Int. J. Gynecol. Cancer 22, 979–986.

Du, L., Xie, Z., Wu, L.C., Chiu, M., Lin, J., Chan, K.K., Liu, S., Liu, Z., 2012. Reactivation of RASSF1A in breast cancer cells by curcumin. Nutr. Cancer 64, 1228–1235.

Eitan, R., Kushnir, M., Lithwick-Yanai, G., David, M.B., Hoshen, M., Glezerman, M., Hod, M., Sabah, G., Rosenwald, S., Levavi, H., 2009. Tumor microRNA expression patterns associated with resistance to platinum-based chemotherapy and survival in ovarian cancer patients. Gynecol. Oncol. 114, 253–259.

Esteller, M., 2007. Cancer epigenomics: DNA methylomes and histone-modification maps. Nat. Rev. Genet. 8, 286–298.

Esteller, M., 2008. Epigenetics in cancer. N. Engl. J. Med. 358, 1148–1159.

Esteller, M., Silva, J.M., Dominguez, G., Bonilla, F., Matias-Guiu, X., Lerma, E., Bussaglia, E., Prat, J., Harkes, I.C., Repasky, E.A., Gabrielson, E., Schutte, M., Baylin, S.B., Herman, J.G., 2000. Promoter hypermethylation and BRCA1 inactivation in sporadic breast and ovarian tumors. J. Natl. Cancer. Inst. 92, 564–569.

Fang, F., Balch, C., Schilder, J., Breen, T., Zhang, S., Shen, C., Li, L., Kulesavage, C., Snyder, A.J., Nephew, K.P., Matei, D.E., 2010. A phase 1 and pharmacodynamic study of decitabine in combination with carboplatin in patients with recurrent, platinum-resistant, epithelial ovarian cancer. Cancer 116, 4043–4053.

Fang, M.Z., Wang, Y., Ai, N., Hou, Z., Sun, Y., Lu, H., Welsh, W., Yang, C.S., 2003. Tea polyphenol (−)-epigallocatechin-3-gallate inhibits DNA methyltransferase and reactivates methylation-silenced genes in cancer cell lines. Cancer. Res. 63, 7563–7570.

Ferraresi, A., Phadngam, S., Morani, F., et al., 2017. Resveratrol inhibits IL-6-induced ovarian cancer cell migration through epigenetic up-regulation of autophagy. Mol. Carcinog. 56, 1164–1181.

Fiegl, H., Windbichler, G., Mueller-Holzner, E., Goebel, G., Lechner, M., Jacobs, I.J., Widschwendter, M., 2008. HOXA11 DNA methylation − a novel prognostic biomarker in ovarian cancer. Int. J. Cancer 123, 725–729.

Fu, S., Hu, W., Iyer, R., Kavanagh, J.J., Coleman, R.L., Levenback, C.F., Sood, A.K., Wolf, J.K., Gershenson, D.M., Markman, M., Hennessy, B.T., Kurzrock, R., Bast Jr., R.C., 2011. Phase 1b-2a study to reverse platinum resistance through use of a hypomethylating agent, azacitidine, in patients with platinum-resistant or platinum-refractory epithelial ovarian cancer. Cancer 117, 1661–1669.

Fu, X., Zhang, Y., Wang, X., Chen, M., Wang, Y., Nie, J., Meng, Y., Han, W., 2015. Low dose decitabine combined with taxol and platinum chemotherapy to treat refractory/recurrent ovarian cancer: an open-label, single-arm, phase I/II study. Curr. Protein Pept. Sci. 16, 329–336.

Gifford, G., Paul, J., Vasey, P.A., Kaye, S.B., Brown, R., 2004. The acquisition of hMLH1 methylation in plasma DNA after chemotherapy predicts poor survival for ovarian cancer patients. Clin. Cancer. Res. 10, 4420–4426.

Glasspool, R.M., Brown, R., Gore, M.E., Rustin, G.J., McNeish, I.A., Wilson, R.H., Pledge, S., Paul, J., Mackean, M., Hall, G.D., Gabra, H., Halford, S.E., Walker, J., Appleton, K., Ullah, R., Kaye, S., Scottish Gynaecological Trials Group, 2014. A randomised, phase II trial of the DNA-hypomethylating agent 5-aza-2′-deoxycytidine (decitabine) in combination with carboplatin vs carboplatin alone in patients with recurrent, partially platinum-sensitive ovarian cancer. Br. J. Cancer 110, 1923–1929.

Goffin, J., Eisenhauer, E., 2002. DNA methyltransferase inhibitors-state of the art. Ann. Oncol. 13, 1699–1716.

Gottlicher, M., Minucci, S., Zhu, P., Krämer, O.H., Schimpf, A., Giavara, S., Sleeman, J.P., Lo Coco, F., Nervi, C., Pelicci, P.G., Heinzel, T., 2001. Valproic acid defines a novel class of HDAC inhibitors inducing differentiation of transformed cells. EMBO J. 20, 6969–6978.

Graham, J.S., Kaye, S.B., Brown, R., 2009. The promises and pitfalls of epigenetic therapies in solid tumours. Eur. J. Cancer 45, 1129–1136.

Gupta, A., Godwin, A.K., Vanderveer, L., Lu, A., Liu, J., 2003. Hypomethylation of the synuclein gamma gene CpG island promotes its aberrant expression in breast carcinoma and ovarian carcinoma. Cancer. Res. 63, 664–673.

Hata, A., Lieberman, J., 2015. Dysregulation of microRNA biogenesis and gene silencing in cancer. Sci. Signal. 8, re3.

Hauser, A.T., Jung, M., 2008. Targeting epigenetic mechanisms: potential of natural products in cancer chemoprevention. Planta Med. 74, 1593–1601.

Häusler, S.F.M., Keller, A., Chandran, P.A., Ziegler, K., Zipp, K., Heuer, S., Krockenberger, M., Engel, J.B., Hönig, A., Scheffler, M., Dietl, J., Wischhusen, J., 2010. Whole blood-derived miRNA profiles as potential new tools for ovarian cancer screening. Br. J. Cancer 103, 693–700.

He, J., Jing, Y., Li, W., Qian, X., Xu, Q., Li, F.S., Liu, L.Z., Jiang, B.H., Jiang, Y., 2013. Roles and mechanism of miR-199a and miR-125b in tumor angiogenesis. PLoS One 8, e56647.

Hellebrekers, D.M., Griffioen, A.W., Van Engeland, M., 2007. Dual targeting of epigenetic therapy in cancer. Biochim. Biophys. Acta 1775, 76–91.

Hennessy, B.T., Coleman, R.L., Markman, M., 2009. Ovarian cancer. Lancet 374, 1371–1382.

Herman, J.G., Baylin, S.B., 2003. Gene silencing in cancer in association with promoter hypermethylation. N. Engl. J. Med. 349, 2042–2054.

Hill, C.G., Matyunina, L.V., Walker, D., Benigno, B.B., McDonald, J.F., 2014. Transcriptional override: a regulatory network model of indirect responses to modulations in microRNA expression. BMC Syst. Biol. 25, 36.

Hilton, J.L., Geisler, J.P., Rathe, J.A., Hattermann-Zogg, M.A., DeYoung, B., Buller, R.E., 2002. Inactivation of BRCA1 and BRCA2 in ovarian cancer. J. Natl. Cancer. Inst. 94, 1396–1406.

Honda, H., Pazin, M.J., Ji, H., Wernyj, R.P., Morin, P.J., 2006. Crucial roles of Sp1 and epigenetic modifications in the regulation of the CLDN4 promoter in ovarian cancer cells. J. Biol. Chem. 281, 21433–21444.

Hu, X., Macdonald, D.M., Huettner, P.C., Feng, Z., El Naqa, I.M., Schwarz, J.K., Mutch, D.G., Grigsby, P.W., Powell, S.N., Wang, X., 2009. A miR-200 microRNA cluster as prognostic marker in advanced ovarian cancer. Gynecol. Oncol. 114, 457–464.

Huang, J., Plass, C., Gerhauser, C., 2011. Cancer chemoprevention by targeting the epigenome. Curr. Drug Targets 12, 1925–1956.

Ibanez De Caceres, I., Battagli, C., Esteller, M., Herman, J.G., Dulaimi, E., Edelson, M.I., Bergman, C., Ehya, H., Eisenberg, B.L., Cairns, P., 2004. Tumor cell-specific brca1 and RASSF1A hypermethylation in serum, plasma, and peritoneal fluid from ovarian cancer patients. Cancer. Res. 64, 6476–6481.

Iorio, M.V., Visone, R., Di Leva, G., Donati, V., Petrocca, F., Casalini, P., Taccioli, C., Volinia, S., Liu, C.G., Alder, H., Calin, G.A., Ménard, S., Croce, C.M., 2007. MicroRNA signatures in human ovarian cancer. Cancer. Res. 67, 8699–8707.

Issa, J.P., Gharibyan, V., Cortes, J., Jelinek, J., Morris, G., Verstovsek, S., Talpaz, M., Garcia-Manero, G., Kantarjian, H.M., et al., 2005. Phase II study of low-dose decitabine in patients with chronic myelogenous leukemia resistant to imatinib mesylate. J. Clin. Oncol. 23, 3948–3956.

Ivanov, M., Kacevska, M., Ingelman-Sundberg, M., 2012. Epigenomics and interindividual differences in drug response. Clin. Pharmacol. Ther. 92, 727–736.

Izutsu, N., Maesawa, C., Shibazaki, M., Oikawa, H., Shoji, T., Sugiyama, T., Masuda, T., 2008. Epigenetic modification is involved in aberrant expression of class III β-tubulin, TUBB3, in ovarian cancer cells. Int. J. Oncol. 32, 1227–1235.

Jemal, A., Siegel, R., Ward, E., Hao, Y., Xu, J., Thun, M.J., 2009. Cancer statistics 2009. CA Cancer J. Clin. 5, 225–249.

Jiang, A., Wang, X., Shan, X., Li, Y., Wang, P., Jiang, P., Feng, Q., 2015. Curcumin reactivates silenced tumor suppressor gene RARbeta by reducing DNA methylation. Phytother Res. 29, 1237–1245.

Johnson, S., Grosshans, H., Shingara, J., Byrom, M., Jarvis, R., Cheng, A., Labourier, E., Reinert, K.L., Brown, D., Slack, F.J., 2005. RAS is regulated by the let-7 microRNA family. Cell 120, 635–647.

Kan, C.W.S., Hahn, M.A., Gard, G.B., Maidens, J., Huh, J.Y., Marsh, D.J., Howell, V.M., 2012. Elevated levels of circulating microRNA-200 family members correlate with serous epithelial ovarian cancer. BMC Canc. 12, 627.

Kaneuchi, M., Sasaki, M., Tanaka, Y., Shiina, H., Yamada, H., Yamamoto, R., Sakuragi, N., Enokida, H., Verma, M., Dahiya, R., 2005. WT1 and WT1-AS genes are inactivated by promoter methylation in ovarian clear cell adenocarcinoma. Cancer 104, 1924–1930.

Kattan, M.W., Shariat, S.F., Andrews, B., Zhu, K., Canto, E., Matsumoto, K., Muramoto, M., Scardino, P.T., Ohori, M., Wheeler, T.M., Slawin, K.M., 2003. The addition of interleukin-6 soluble receptor and transforming growth factor-β1 improves a preoperative nomogram for predicting biochemical progression in patients with clinically localized prostate cancer. J. Clin. Oncol. 21, 3573–3579.

Kelly, Z.L., Michael, A., Butler-Manuel, S., Pandha, H.S., Morgan, R.G., 2011. HOX genes in ovarian cancer. J. Ovarian. Res. 4, 16.

Kong, F., Sun, C., Wang, Z., Han, L., Weng, D., Lu, Y., Chen, G., 2011. miR-125b confers resistance of ovarian cancer cells to cisplatin by targeting pro-apoptotic Bcl-2 antagonist killer 1. J. Huazhong Univ. Sci. Technol. - Med. Sci. 31, 543–549.

Koukoura, O., Sifakis, S., Spandidos, D.A., 2012. DNA methylation in the human placenta and fetal growth (Review). Mol. Med. Rep. 5, 883–889.

Kouzarides, T., 2007. Chromatin modifications and their function. Cell 128, 693–705.

Kwon, M.J., Kim, S.S., Choi, Y.L., Jung, H.S., Balch, C., Kim, S.H., Song, Y.S., Marquez, V.E., Nephew, K.P., Shin, Y.K., 2010. Derepression of CLDN3 and CLDN4 during ovarian tumori genesis is associated with loss of repressive histone modifications. Carcinogenesis 31, 974–983.

Lai, J., Chien, J., Staub, J., Avula, R., Greene, E.L., Matthews, T.A., Smith, D.I., Kaufmann, S.H., Roberts, L.R., Shridhar, V., 2003. Loss of HSulf-1 up-regulates heparin-binding growth factor signaling in cancer. J. Biol. Chem. 278, 23107–23117.

Laird, P.W., 2003. The power and the promise of DNA methylation markers. Nat. Rev. Cancer 3, 253–266.

Lee, P.S., Teaberry, V.S., Bland, A.E., Huang, Z., Whitaker, R.S., Baba, T., Fujii, S., Secord, A.A., Berchuck, A., Murphy, S.K., 2010. Elevated MAL expression is accompanied by promoter hypomethylation and platinum resistance in epithelial ovarian cancer. Int. J. Cancer 126, 1378–1389.

Lehmann, U., Celikkaya, G., Hasemeier, B., Länger, F., Kreipe, H., 2002. Promoter hypermethylation of the death-associated protein kinase gene in breast cancer is associated with the invasive lobular subtype. Cancer Res. 62, 6634–6638.

Leskelä, S., Leandro-García, L.J., Mendiola, M., Barriuso, J., Inglada-Pérez, L., Muñoz, I., Martínez-Delgado, B., Redondo, A., de Santiago, J., Robledo, M., Hardisson, D., Rodríguez-Antona, C., 2010. The miR-200 family controls beta-tubulin III expression and is associated with paclitaxel-based treatment response and progression-free survival in ovarian cancer patients. Endocr. Relat. Cancer 18, 85–95.

Li, M., Balch, C., Montgomery, J.S., Jeong, M., Chung, J.H., Yan, P., Huang, T.H., Kim, S., Nephew, K.P., 2009. Integrated analysis of DNA methylation and gene expression reveals specific signaling pathways associated with platinum resistance in ovarian cancer. BMC. Med. Genomics 2, 34.

Li, M.L., Greenberg, R.A., 2012. Links between genome integrity and BRCA1 tumor suppression. Trends Biochem. Sci. 37, 418–424.

Liang, H., Jiang, Z., Xie, G., Lu, Y., 2015. Serum microRNA-145 as a novel biomarker in human ovarian cancer. Tumour. Biol. 36, 5305–5313.

Liggett, T.E., Melnikov, A., Yi, Q., Replogle, C., Hu, W., Rotmensch, J., Kamat, A., Sood, A.K., Levenson, V., 2011. Distinctive DNA methylation patterns of cell-free plasma DNA in women with malignant ovarian tumors. Gynecol. Oncol. 120, 113–120.

Lin, Y., Jin, Y., Xu, T., Zhou, S., Cui, M., 2017. MicroRNA-215 targets NOB1 and inhibits growth and invasion of epithelial ovarian cancer. Am. J. Transl. Res. 9, 466–477.

Lin, C.T., Lai, H.C., Lee, H.Y., Lin, W.H., Chang, C.C., Chu, T.Y., Lin, Y.W., Lee, K.D., Yu, M.H., 2008. Valproic acid resensitizes cisplatin-resistant ovarian cancer cells. Cancer. Sci. 99, 1218–1226.

Liu, L., Tommasi, S., Lee, D.H., Dammann, R., Pfeifer, G.P., 2003. Control of microtubule stability by the RASSF1A tumor suppressor. Oncogene 22, 8125–8136.

Liu, N., Zhou, C., Zhao, J., Chen, Y., 2012. Reversal of paclitaxel resistance in epithelial ovarian carcinoma cells by a MUC1 aptamer-let-7i chimera. Cancer. Invest. 30, 577–582.

Lu, T.Y., Kao, C.F., Lin, C.T., Huang, D.Y., Chiu, C.Y., Huang, Y.S., Wu, H.C., 2009. DNA methylation and histone modification regulate silencing of OPG during tumor progression. J. Cell. Biochem. 108, 315–325.

Mackay, H.J., Hirte, H., Colgan, T., Covens, A., MacAlpine, K., Grenci, P., Wang, L., Mason, J., Pham, P.A., Tsao, M.S., Pan, J., Zwiebel, J., Oza, A.M., 2010. Phase II trial of the histone deacetylase inhibitor belinostat in women with platinum resistant epithelial ovarian cancer and micropapillary (LMP) ovarian tumours. Eur. J. Cancer 46, 1573–1579.

Mahdian-Shakib, A., Dorostkar, R., Tat, M., Hashemzadeh, M.S., Saidi, N., 2016. Differential role of microRNAs in prognosis, diagnosis, and therapy of ovarian cancer. Biomed. Pharmacother. 84, 592–600.

Mai, A., Altucci, L., 2009. Epi-drugs to fight cancer: from chemistry to cancer treatment, the road ahead. Int. J. Biochem. Cell Biol. 41, 199–213.

Makarla, P.B., Saboorian, M.H., Ashfaq, R., Toyooka, K.O., Toyooka, S., Minna, J.D., Gazdar, A.F., Schorge, J.O., 2005. Promoter hypermethylation profile of ovarian epithelial neoplasms. Clin. Cancer. Res. 11, 5365–5369.

Manetta, A., Blessing, J.A., Look, K.Y., 1995. A phase II study of fazarabine in patients with advanced ovarian cancer. A Gynecologic Oncology Group study. Am. J. Clin. Oncol. 18, 156–157.

Marchini, S., Cavalieri, D., Fruscio, R., Calura, E., Garavaglia, D., Fuso Nerini, I., Mangioni, C., Cattoretti, G., Clivio, L., Beltrame, L., Katsaros, D., Scarampi, L., Menato, G., Perego, P., Chiorino, G., Buda, A., Romualdi, C., D'Incalci, M., 2011. Association between miR-200c and the survival of patients with stage I epithelial ovarian cancer: a retrospective study of two independent tumour tissue collections. Lancet Oncol. 12, 273–285.

Mateescu, B., Batista, L., Cardon, M., Gruosso, T., de Feraudy, Y., Mariani, O., Nicolas, A., Meyniel, J.P., Cottu, P., Sastre-Garau, X., Mechta-Grigoriou, F., 2011. miR-141 and miR-200a act on ovarian tumorigenesis by controlling oxidative stress response. Nat. Med. 17, 1627–1635.

Matei, D., Fang, F., Shen, C., Schilder, J., Arnold, A., Zeng, Y., Berry, W.A., Huang, T., Nephew, K.P., 2012. Epigenetic resensitization to platinumin ovarian cancer. Cancer Res. 72, 2197–2205.

Matulonis, U., Berlin, S., Lee, H., Whalen, C., Obermayer, E., Penson, R., Liu, J., Campos, S., Krasner, C., Horowitz, N., 2015. Phase I study of combination of vorinostat, carboplatin, and gemcitabine in women with recurrent, platinum-sensitive epithelial ovarian, fallopian tube, or peritoneal cancer. Cancer. Chemother. Pharmacol. 76, 417–423.

McCoy, M.L., Mueller, C.R., Roskelley, C.D., 2003. The role of the breast cancer susceptibility gene 1 (BRCA1) in sporadic epithelial ovarian cancer. Reprod. Biol. Endocrinol. 1, 72.

Mendivil, A.A., Micha, J.P., Brwon 3rd, J.V., Rettenmaier, M.A., Abaid, L.N., Lopez, K.L., Goldstein, B.H., 2013. Increased incidence of severe gastrointestinal events with first-line paclitaxel, carboplatin, and vorinostat chemotherapy for advanced-stage epithelial ovarian, primary peritoneal, and fallopian tube cancer. Int. J. Gynecol. Cancer 23, 533–539.

Modesitt, S.C., Sill, M., Hoffman, J.S., Bender, D.P., Gynecologic Oncology Grou, 2008. A phase II study of vorinostat in the treatment of persistent or recurrent epithelial ovarian or primary peritoneal carcinoma: a Gynecologic Oncology Group study. Gynecol. Oncol. 109, 182–186.

Montavon, C., Gloss, B.S., Warton, K., Barton, C.A., Statham, A.L., Scurry, J.P., Tabor, B., Nguyen, T.V., Qu, W., Samimi, G., Hacker, N.F., Sutherland, R.L., Clark, S.J., O'Brien, P.M., 2012. Prognostic and diagnostic significance of DNA methylation patterns in high grade serous ovarian cancer. Gynecol. Oncol. 124, 582–588.

Monti, B., Polazzi, E., Contestabile, A., 2009. Biochemical molecular and epigenetic mechanisms of valproic acid neuroprotection. Curr. Mol. Pharmacol. 2, 95–109.

Moselhy, S.S., Kumosani, T.A., Kamal, I.H., Jalal, J.A., Jabaar, H.S., Dalol, A., 2013. Hypermethylation of P15, P16, and E-cadherin genes in ovarian cancer. Toxicol. Ind. Health. Apr 9, (Epub ahead of print).

Murphy, M.A., Wentzensen, N., 2011. Frequency of mismatch repair deficiency in ovarian cancer: a systematic review. This article is a US Government work and, as such, is in the public domain of the United States of America. Int. J. Cancer 129, 1914–1922.

Nam, E.J., Yoon, H., Kim, S.W., Kim, H., Kim, Y.T., Kim, J.H., Kim, J.W., Kim, S., 2008. MicroRNA expression profiles in serous ovarian carcinoma. Clin. Cancer. Res. 14, 2690–2695.

Nguyen, C.T., Weisenberger, D.J., Velicescu, M., Gonzales, F.A., Lin, J.C., Liang, G., Jones, P.A., 2002. Histone H3-lysine 9 methylation is associated with aberrant gene silencing in cancer cells and is rapidly reversed by aza-2′-deoxycytidine. Cancer. Res. 62, 6456–6461.

Nicholson, L.J., Smith, P.R., Hiller, L., Szlosarek, P.W., Kimberley, C., Sehouli, J., Koensgen, D., Mustea, A., Schmid, P., Crook, T., 2009. Epigenetic silencing of argininosuccinate synthetase confers resistance to platinum-induced cell death but collateral sensitivity to arginine auxotrophy in ovarian cancer. Int. J. Cancer 125, 1454–1463.

Ozdemir, F., Altinisik, J., Karateke, A., Coksuer, H., Buyru, N., 2012. Methylation of tumor suppressor genes in ovarian cancer. Exp. Ther. Med. 4, 1092–1096.

Parker, L.P., Taylor, D.D., Kesterson, J., , et al.Metzinger, D.S., Gercel-Taylor, C., 2009. Modulation of microRNA associated with ovarian cancer cells by genistein. Eur. J. Gynaecol. Oncol. 30, 616–621.

Pattamadilok, J., Huapai, N., Rattanatanyong, P., Vasurattana, A., Triratanachat, S., Tresukosol, D., Mutirangura, A., 2008. LINE-1 hypomethylation level as a potential prognostic factor for epithelial ovarian cancer. Int. J. Gynecol. Cancer 18, 711–717.

Pecot, C.V., Rupaimoole, R., Yang, D., Akbani, R., Ivan, C., Lu, C., Wu, S., Han, H.D., Shah, M.Y., Rodriguez-Aguayo, C., Bottsford-Miller, J., Liu, Y., Kim, S.B., et al., 2013. Tumour angiogenesis regulation by the miR-200 family. Nat. Commun. 4, 2427.

Petrocca, F., Iliopoulos, D., Qin, H.R., Nicoloso, M.S., Yendamuri, S., Wojcik, S.E., Shimizu, M., Di Leva, G., Vecchione, A., Trapasso, F., Godwin, A.K., Negrini, M., Calin, G.A., Croce, C.M., 2006. Alterations of the tumor suppressor gene ARLTS1 in ovarian cancer. Cancer. Res. 66, 10287–10291.

Plumb, J.A., Strathdee, G., Sludden, J., Kaye, S.B., Brown, R., 2000. Reversal of drug resistance in human tumor xenografts by 2′-deoxy-5-azacytidine-induced demethylation of the hMLH1 gene promoter. Cancer. Res. 60, 6039–6044.

Prasad, S., Gupta, S.C., Tyagi, A.K., Aggarwal, B.B., 2014. Curcumin, a component of golden spice: from bedside to bench and back. Biotechnol. Adv. 32, 1053–1064.

Prislei, S., Martinelli, E., Mariani, M., Raspaglio, G., Sieber, S., Ferrandina, G., Shahabi, S., Scambia, G., Ferlini, C., 2013. miR-200c and HuR in ovarian cancer. BMC Canc. 13, 72.

Qin, C.Z., Lou, X.Y., Lv, Q.L., Cheng, L., Wu, N.Y., Hu, L., Zhou, H.H., 2015. MicroRNA-184 acts as a potential diagnostic and prognostic marker in epithelial ovarian cancer and regulates cell proliferation, apoptosis and inflammation. Pharmazie 70, 668–673.

Rathi, A., Virmani, A.K., Schorge, J.O., Elias, K.J., Maruyama, R., Minna, J.D., Mok, S.C., Girard, L., Fishman, D.A., Gazdar, A.F., 2002. Methylation profiles of sporadic ovarian tumors and nonmalignant ovaries from high-risk women. Clin. Cancer. Res. 8, 3324–3331.

Resnick, K.E., Alder, H., Hagan, J.P., Richardson, D.L., Croce, C.M., Cohn, D.E., 2009. The detection of differentially expressed microRNAs from the serum of ovarian cancer patients using a novel real-time PCR platform. Gynecol. Oncol. 112, 55–59.

Richon, V.M., Sandhoff, T.W., Rifkind, R.A., Marks, P.A., 2000. Histone deacetylase inhibitor selectively induces p21WAF1 expression and gene-associated histone acetylation. Proc. Natl. Acad. Sci. USA 97, 10014–10019.

Ruan, K., Fang, X., Ouyang, G., 2009. MicroRNAs: novel regulators in the hallmarks of human cancer. Canc. Lett. 285, 116–126.

Saito, Y., Liang, G., Egger, G., Friedman, J.M., Chuang, J.C., Coetzee, G.A., Jones, P.A., 2006. Specific activation of microRNA-127 with downregulation of the proto-oncogene BCL6 by chromatin-modifying drugs in human cancer cells. Canc. Cell 9, 435–443.

Samuel, S., Naora, H., 2005. Homeobox gene expression in cancer: insights from developmental regulation and deregulation. Eur. J. Cancer 41, 2428–2437.

Santi, D.V., Norment, A., Garrett, C.E., 1984. Covalent bond formation between a DNA-cytosine methyltransferase and DNA containing 5-azacytosine. Proc. Natl. Acad. Sci. USA 81, 6993–6997.

Segura-Pacheco, B., Trejo-Becerril, C., Perez-Cardenas, E., Taja-Chayeb, L., Mariscal, I., Chavez, A., Acuña, C., Salazar, A.M., Lizano, M., Dueñas-Gonzalez, A., 2003. Reactivation of tumor suppressor genes by the cardiovascular drugs hydralazine and procainamide and their potential use in cancer therapy. Clin. Cancer. Res. 9, 1596–1603.

Shan, Z., Feng-Nian, R., Jie, G., Ting, Z., 2012. Effects of valproic acid on proliferation, apoptosis, angiogenesis and metastasis of ovarian cancer in vitro and in vivo. Asian. Pac. J. Cancer. Prev. 13, 3977–3982.

Shi, H., Wei, S.H., Leu, Y.W., Rahmatpanah, F., Liu, J.C., Yan, P.S., Nephew, K.P., Huang, T.H., 2003. Triple analysis of the cancer epigenome: an integrated microarray system for assessing gene expression, DNA methylation, and histone acetylation. Cancer. Res. 63, 2164–2171.

Shridhar, V., Bible, K.C., Staub, J., Avula, R., Lee, Y.K., Kalli, K., Huang, H., Hartmann, L.C., Kaufmann, S.H., Smith, D.I., 2001. Loss of expression of a new member of the DNAJ protein family confers resistance to chemotherapeutic agents used in the treatment of ovarian cancer. Cancer. Res. 61, 4258–4265.

Siegel, R.L., Miller, K.D., Jemal, A., 2015. Cancer statistics. CA. Cancer. J. Clin 65, 5–29.

Staub, J., Chien, J., Pan, Y., Qian, X., Narita, K., Aletti, G., Scheerer, M., Roberts, L.R., Molina, J., Shridhar, V., 2007. Epigenetic silencing of HSulf-1 in ovarian cancer: implications in chemoresistance. Oncogene 26, 4969–4978.

Strathdee, G., Appleton, K., Illand, M., Millan, D.W., Sargent, J., Paul, J., Brown, R., 2001. Primary ovarian carcinomas display multiple methylator phenotypes involving known tumor suppressor genes. Am. J. Pathol. 158, 1121–1127.

Strathdee, G., Davies, B.R., Vass, J.K., Siddiqui, N., Brown, R., 2004. Cell type-specific methylation of an intronic CpG island controls expression of the MCJ gene. Carcinogenesis 25, 693–701.

Strathdee, G., MacKean, M.J., Illand, M., Brown, R., 1999. A role for methylation of the hMLH1 promoter in loss of hMLH1 expression and drug resistance in ovarian cancer. Oncogene 18, 2335–2341.

Strathdee, G., Vass, J.K., Oien, K.A., Siddiqui, N., Curto-Garcia, J., Brown, R., 2005. Demethylation of the MCJ gene in stage III/IV epithelial ovarian cancer and response to chemotherapy. Gynecol. Oncol. 97, 898–903.

Su, H.Y., Lai, H.C., Lin, Y.W., Chou, Y.C., Liu, C.Y., Yu, M.H., 2009. An epigenetic marker panel for screening and prognostic prediction of ovarian cancer. Int. J. Cancer 124, 387–393.

Su, H.Y., Lai, H.C., Lin, Y.W., Liu, C.Y., Chen, C.K., Chou, Y.C., Lin, S.P., Lin, W.C., Lee, H.Y., Yu, M.H., 2010. Epigenetic silencing of SFRP5 is related to malignant phenotype and chemoresistance of ovarian cancer through Wnt signaling pathway. Int. J. Cancer 127, 555–567.

Takai, N., Kawamata, N., Gui, D., Said, J.W., Miyakawa, I., Koeffler, H.P., 2004. Human ovarian carcinoma cells: histone deacetylase inhibitors exhibit antiproliferative activity and potently induce apoptosis. Cancer 101, 2760–2770.

Takai, N., Kawamata, N., Walsh, C.S., Gery, S., Desmond, J.C., Whittaker, S., Said, J.W., Popoviciu, L.M., Jones, P.A., Miyakawa, I., Koeffler, H.P., 2005. Discovery of epigenetically masked tumor suppressor genes in endometrial cancer. Mol. Cancer. Res. 3, 261–269.

Takai, N., Narahara, H., 2010. Histone deacetylase inhibitor therapy in epithelial ovarian cancer. J. Oncol 2010, 458431.

Tam, K.F., Liu, V.W., Liu, S.S., Tsang, P.C., Cheung, A.N., Yip, A.M., Ngan, H.Y., 2007. Methylation profile in benign, borderline and malignant ovarian tumors. J. Cancer. Res. Clin. Oncol. 133, 331–341.

Taniguchi, T., Tischkowitz, M., Ameziane, N., Hodgson, S.V., Mathew, C.G., Joenje, H., Mok, S.C., D'Andrea, A.D., 2003. Disruption of the fanconi anemia–BRCA pathway in cisplatin-sensitive ovarian tumors. Nat. Med. 9, 5668–5674.

Taylor, D.D., Gercel-Taylor, C., 2008. MicroRNA signatures of tumor-derived exosomes as diagnostic biomarkers of ovarian cancer. Gynecol. Oncol. 110, 13–21.

Teodoridis, J.M., Hall, J., Marsh, S., Kannall, H.D., Smyth, C., Curto, J., Siddiqui, N., Gabra, H., McLeod, H.L., Strathdee, G., Brown, R., 2005. CpG island methylation of DNA damage response genes in advanced ovarian cancer. Cancer. Res. 65, 8961–8967.

Terasawa, K., Sagae, S., Toyota, M., Tsukada, K., Ogi, K., Satoh, A., Mita, H., Imai, K., Tokino, T., Kudo, R., 2004. Epigenetic inactivation of TMS1/ASC in ovarian cancer. Clin. Cancer. Res. 10, 2000–2006.

Teschendorff, A.E., Menon, U., Gentry-Maharaj, A., Ramus, S.J., Gayther, S.A., Apostolidou, S., Jones, A., Lechner, M., Beck, S., Jacobs, I.J., Widschwendter, M., 2009. An epigenetic signature in peripheral blood predicts active ovarian cancer. PLoS One 4, e8274.

Tian, C., Ambrosone, C.B., Darcy, K.M., Krivak, T.C., Armstrong, D.K., Bookman, M.A., Davis, W., Zhao, H., Moysich, K., Gallion, H., DeLoia, J.A., 2012. Common variants in ABCB1, ABCC2 and ABCG2 genes and clinical outcomes among women with advanced stage ovarian cancer treated with platinum and taxane-based chemotherapy: a Gynecologic Oncology Group study. Gynecol. Oncol. 124, 575–581.

Valls, E., Sánchez-Molina, S., Martínez-Balbás, M.A., 2005. Role of histone modifications in marking and activating genes through mitosis. J. Biol. Chem. 280, 42592–42600.

Vigushin, D.M., Ali, S., Pace, P.E., Mirsaidi, N., Ito, K., Adcock, I., Coombes, R.C., 2001. Trichostatin A is a histone deacetylase inhibitor with potent antitumor activity against breast cancer in vivo. Clin. Cancer. Res. 7, 971–976.

Waldmann, T., Schneider, R., 2013. Targeting histone modifications–epigenetics in cancer. Curr. Opin. Cell Biol. 25, 184–189, 2013.

Wang, Y.Q., Guo, R.D., Guo, R.M., Sheng, W., Yin, L.R., 2013a. MicroRNA-182 promotes cell growth, invasion, and chemoresistance by targeting programmed cell death 4 (PDCD4) in human ovarian carcinomas. J. Cell. Biochem. 114, 1464–1473.

Wang, C., Horiuchi, A., Imai, T., Ohira, S., Itoh, K., Nikaido, T., Katsuyama, Y., Konishi, I., 2004. Expression of BRCA1 protein in benign, borderline, and malignant epithelial ovarian neoplasms and its relationship to methylation and allelic loss of the BRCA1 gene. J. Pathol. 202, 215–223.

Wang, Y.Q., Yan, Q., Zhang, J.R., Li, S.D., Yang, Y.X., Wan, X.P., 2013b. Epigenetic inactivation of BRCA1 through promoter hypermethylation in ovarian cancer progression. J. Obstet. Gynaecol. Res. 39, 549–554.

Wang, R.H., Zheng, Y., Kim, H.S., Xu, X., Cao, L., Luhasen, T., Lee, M.H., Xiao, C., Vassilopoulos, A., Chen, W., Gardner, K., Man, Y.G., Hung, M.C., Finkel, T., Deng, C.X., 2008. Interplay among BRCA1, SIRT1, and Survivin during BRCA1-associated tumorigenesis. Mol. Cell 32, 11–20.

Watanabe, Y., Ueda, H., Etoh, T., Koike, E., Fujinami, N., Mitsuhashi, A., Hoshiai, H., 2007. A change in promoter methylation of hMLH1 is a cause of acquired resistance to platinum-based chemotherapy in epithelial ovarian cancer. Anticancer Res. 27, 1449–1452.

Wei, S.H., Balch, C., Paik, H.H., Kim, Y.S., Baldwin, R.L., Liyanarachchi, S., Li, L., Wang, Z., Wan, J.C., Davuluri, R.V., Karlan, B.Y., Gifford, G., Brown, R., Kim, S., Huang, T.H., Nephew, K.P., 2006. Prognostic DNA methylation biomarkers in ovarian cancer. Clin. Cancer. Res. 12, 2788–2794.

Wei, S.H., Chen, C.M., Strathdee, G., Harnsomburana, J., Shyu, C.R., Rahmatpanah, F., Shi, H., Ng, S.W., Yan, P.S., Nephew, K.P., Brown, R., Huang, T.H., 2002. Methylation microarray analysis of late-stage ovarian carcinomas distinguishes progression-free survival in patients and identifies candidate epigenetic markers. Clin. Cancer. Res. 8, 2246–2252.

Widschwendter, M., Apostolidou, S., Jones, A.A., Fourkala, E.O., Arora, R., Pearce, C.L., Frasco, M.A., Ayhan, A., Zikan, M., Cibula, D., Iyibozkurt, C.A., Yavuz, E., Hauser-Kronberger, C., Dubeau, L., Menon, U., Jacobs, I.J., 2009. HOXA methylation in normal endometrium from premenopausal women is associated with the presence of ovarian cancer: a proof of principle study. Int. J. Cancer 125, 2214–2218.

Widschwendter, M., Jiang, G., Woods, C., Müller, H.M., Fiegl, H., Goebel, G., Marth, C., Müller-Holzner, E., 2004. DNA hypomethylation and ovarian cancer biology. Cancer. Res. 64, 4472–4480.

Wiley, A., Katsaros, D., Fracchioli, S., Yu, H., 2006. Methylation of the insulin-like growth factor binding protein-3 gene and prognosis of epithelial ovarian cancer. Int. J. Gynecol. Cancer 16, 210–218.

Wilson, A.J., Lalani, A.S., Wass, E., Saskowski, J., Khabele, D., 2012. Romidepsin (FK228) combined with cisplatin stimulates DNA damage-induced cell death in ovarian cancer. Gynecol. Oncol. 127, 579–586.

Wilting, R.H., Dannenberg, J.H., 2012. Epigenetic mechanisms in tumorigenesis, tumor cell heterogeneity and drug resistance. Drug. Resist. Updat. 15, 21–38.

Woloszynska-Read, A., James, S.R., Link, P.A., Yu, J., Odunsi, K., Karpf, A.R., 2007. DNA methylation-dependent regulation of BORIS/CTCFL expression in ovarian cancer. Cancer. Immun. 7, 21.

Woloszynska-Read, A., Zhang, W., Yu, J., Mhawech-Fauceglia, P., Collamat, G., Akers, S.N., Ostler, K.R., Godley, L.A., Odunsi, K., Karpf, A.R., et al., 2011. Coordinated cancer germline antigen promoter and global DNA hypomethylation in ovarian cancer: association with the BORIS/CTCF expression ratio and advanced stage. Clin. Cancer. Res. 17, 2170–2180.

Xu, Q., Liu, L., Qian, X., et al., 2012. MiR-145 directly targets p70S6K1 in cancer cells to inhibit tumor growth and angiogenesis. Nucleic Acids Res. 40, 761–774.

Xu, L., Xiang, J., Shen, J., Zou, X., Zhai, S., Yin, Y., Li, P., Wang, X., Sun, Q., 2013. Oncogenic MicroRNA-27a is a target for genistein in ovarian cancer cells. Anticancer. Agents. Med. Chem. 13, 1126–1132.

Yan, L., Nass, S.J., Smith, D., Nelson, W.G., Herman, J.G., Davidson, N.E., 2003. Specific inhibition of DNMT1 by antisense oligonucleotides induces re-expression of estrogen receptor-α (ER) in ER-negative human breast cancer cell lines. Cancer Biol. Ther. 2, 552–556.

Yang, N., Kaur, S., Volinia, S., et al., 2008. MicroRNA microarray identifies Let-7i as a novel biomarker and therapeutic target in human epithelial ovarian cancer. Cancer. Res. 68, 10307–10314.

Yang, H.J., Liu, V.W., Wang, Y., Tsang, P., Ngan, H., 2006. Differential DNA methylation profiles in gynecological cancers and correlation with clinico-pathological data. BMC Canc. 6, 212.

Yang, D., Sun, Y., Hu, L., Zheng, H., Ji, P., Pecot, C.V., Zhao, Y., Reynolds, S., Cheng, H., Rupaimoole, R., Cogdell, D., Nykter, M., Broaddus, R., Rodriguez-Aguayo, C., Lopez-Berestein, G., Liu, J., Shmulevich, I., Sood, A.K., Chen, K., Zhang, W., 2013. Integrated analyses identify a master microRNA regulatory network for the mesenchymal subtype in serous ovarian cancer. Canc. Cell 23, 186–199.

Yap, O.W., Bhat, G., Liu, L., 2009. Epigenetic modifications of the estrogen receptor beta gene in epithelial ovarian cancer cells. Anticancer Res. 29, 139–144.

Yeh, K.T., Chen, T.H., Yang, H.W., Chou, J.L., Chen, L.Y., Yeh, C.M., Chen, Y.H., Lin, R.I., Su, H.Y., Chen, G.C., Deatherage, D.E., Huang, Y.W., Yan, P.S., Lin, H.J., Nephew, K.P., Huang, T.H., Lai, H.C., Chan, M.W., 2011. Aberrant TGFβ/SMAD4 signaling contributes to epigenetic silencing of a putative tumor suppressor, RunX1T1 in ovarian cancer. Epigenetics 6, 727–739.

Yoo, C.B., Jones, P.A., 2006. Epigenetic therapy of cancer: past, present and future. Nat. Rev. Drug Discov. 5, 37–50.

Yu, Y., Fujii, S., Yuan, J., Luo, R.Z., Wang, L., Bao, J., Kadota, M., Oshimura, M., Dent, S.R., Issa, J.P., Bast Jr., R.C., 2003. Epigenetic regulation of ARHI in breast and ovarian cancer cells. Ann. NY. Acad. Sci. 983, 268–277.

Yung, R., Chang, S., Hemati, N., Johnson, K., Richardson, B., 1997. Mechanisms of drug-induced lupus. IV. Comparison of procainamide and hydralazine with analogs in vitro and in vivo. Arthritis Rheum. 40, 1436–1443.

Zhang, Q., Hu, G., Yang, Q., Dong, R., Xie, X., Ma, D., Shen, K., Kong, B., 2013. A multiplex methylation-specific PCR assay for the detection of early-stage ovarian cancer using cell-free serum DNA. Gynecol. Oncol. 130, 132–139.

Zhang, S., Ng, M.K., 2016. Gene-microRNA network module analysis for ovarian cancer. BMC Syst. Biol. 10 (Suppl. 4), 117.

Zhang, L., Volinia, S., Bonome, T., Calin, G., Greshock, J., Yang, N., Liu, C., Giannakakis, A., Alexiou, P., Hasegawa, K., Johnstone, C., Megraw, M., Adams, S., Lassus, H., Huang, J., Kaur, S., et al., 2008. Genomic and epigenetic alterations deregulate microRNA expression in human epithelial ovarian cancer. Proc. Natl. Acad. Sci. USA 105, 7004–7009.

Zhang, H., Zhang, S., Cui, J., Zhang, A., Shen, L., Yu, H., 2008. Expression and promoter methylation status of mismatch repair gene hMLH1 and hMSH2 in epithelial ovarian cancer. Aust. N. Z. J. Obstet. Gynaecol. 48, 505–509.

Zhao, S.F., Zhang, X., Zhang, X.J., Shi, X.Q., Yu, Z.J., Kan, Q.C., 2014. Induction of microRNA-9 mediates cytotoxicity of curcumin against SKOV3 ovarian cancer cells. Asian. Pac. J. Cancer. Prev. 15, 3363–3368.

Zheng, H., Zhang, L., Zhao, Y., Yang, D., Song, F., Wen, Y., Hao, Q., Hu, Z., Zhang, W., Chen, K., 2013. Plasma miRNAs as diagnostic and prognostic biomarkers for ovarian cancer. PLoS One 8, e77853.

Zhou, L., Cheng, X., Connolly, B.A., Dickman, M.J., Hurd, P.J., Hornby, D.P., 2002. Zebularine: a novel DNA methylation inhibitor that forms a covalent complex with DNA methyltransferases. J. Mol. Biol. 321, 591–599.

Zhou, F., Ma, M., Tao, G., Chen, X., Xie, W., Wang, Y., Cao, X., 2014. Detection of circulating methylated opioid binding protein/cell adhesion molecule-like gene as a biomarker for ovarian carcinoma. Clin. Lab. 60, 759–765.

PART III

MISCELLANEOUS

CHAPTER

17

Dietary and Lifestyle Factors in Epigenetic Regulation of Cancer

Eswar Shankar[1,2], *Karishma Gupta*[1,2], *Sanjay Gupta*[1,2,3,4,5]

[1]Department of Urology, The James and Eilleen Dicke Research Laboratory, Case Western Reserve University, Cleveland, OH, United States; [2]Department of Urology, The Urology Institute, University Hospitals Cleveland Medical Center, Cleveland, OH, United States; [3]Department of Nutrition, Case Western Reserve University, Cleveland, OH, United States; [4]Division of General Medical Sciences, Case Comprehensive Cancer Center, Cleveland, OH, United States; [5]Department of Urology, Louis Stokes Cleveland Veterans Affairs Medical Center, Cleveland, OH, United States

Abstract

Dietary and lifestyle factors play an important role in human health, as nutritional imbalances are major determinants of various diseases including cancer. Emerging literature suggests that diet and lifestyle factors have a major impact on gene expression through epigenetic mechanisms. Epigenetic modifications are heritable and cause potentially reversible changes in gene expression without altering the DNA sequence. Epigenetic inscriptions include changes in DNA methylation, histone modifications, and small noncoding microRNAs. Atypical epigenetic alterations possibly occur at an early stage during neoplastic development and are widely recognized as essential players in cancer progression. Additionally, environmental cues might lead to epigenetic modifications, thus providing a link between susceptible genes and environmental factors in the etiology of cancer. This chapter primarily highlights the role of various epigenetic mechanisms in the regulation and maintenance of the mammalian genome. This chapter also discusses the impact of endogenous factors including diet, lifestyle, environment, and socioeconomic status on the process of carcinogenesis through various epigenetic modifications. Elucidating the impact of dietary and lifestyle factors on epigenetic mechanisms may serve as a personalized prognostic tool for gauging cancer susceptibility and providing recommendations and guidance for the prevention of cancer.

Keywords

Carcinogens; Chemoprevention; Dietary agents; DNA methylation; Gene–environment interaction; Histone modification; Noncoding RNA

INTRODUCTION

Cancer is a heterogeneous disease driven by the accumulation of genetic and epigenetic alterations. The malignant lesions are inherited by a series of transformations in clonally selected cells that exhibit selective growth advantage, sustain

Epigenetics of Cancer Prevention
https://doi.org/10.1016/B978-0-12-812494-9.00017-2

proliferative signaling, evade growth suppressors, resist cell death, enable replicative immortality, induce angiogenesis, and activate invasion and metastasis (Herceg, 2007; Kanwal and Gupta, 2010). Although genetic lesions drive tumor progression, it is becoming evident that epigenetic modifications are likewise important in cancer development (Kanherkar et al., 2014). The elucidation of the human genome sequence has made it possible to identify genetic and epigenetic alterations in cancers. Epigenetics is referred to as the study of the stable inheritance of gene expression that occurs without changes in DNA sequences (Herceg, 2007; Kanwal and Gupta, 2010; Kanherkar et al., 2014). A majority of cancers result from changes that accumulate throughout life as a result of exposure to dietary, lifestyle, environmental, and socioeconomic/racial factors. Epigenetic disruption of gene expression by these endogenous factors plays an important role in cancer progression. Three major epigenetic mechanisms are known to regulate gene expression. These include DNA methylation, modulation of chromatin structure by posttranslational modification of histone or nonhistone proteins, and small noncoding microRNAs (miRNAs) that alter gene expression by inhibiting translation or causing targeted degradation of specific mRNAs. These mechanisms are critical components in the normal development and growth of cells, and their modifications contribute to the neoplastic phenotype.

MECHANISMS UNDERLYING EPIGENETIC REGULATION

DNA Methylation. Methylation of cytosine residues within the dinucleotide sequence CpG is a unique phenomenon with widely studied epigenetic modifications in mammals (Issa and Kantarjian, 2009). As an essential component of the cellular epigenetic machinery, DNA methylation in association with histone modification regulates gene expression by modulating DNA packaging and chromatin architecture (Dehan et al., 2009). DNA methylation is a chemical modification that comprises the transfer of a methyl (CH_3) moiety from the donor S-adenosyl methionine (SAM) to the 5′ position of cytosine residue that precedes guanine in the CpG dinucleotide sequence, forming 5-methyl cytosine and S-adenosyl-L-homocysteine (SAH) (Issa and Kantarjian, 2009; Gerhauser, 2013a; Bestor, 2000; Santi et al., 1983). The mammalian genome has been reported to harbor 3×10^7 methylated cytosine residues mostly within CpG dinucleotide sequences (Bestor, 2000). Although CpG sequences are unevenly distributed throughout the human genome, they are frequently enriched in gene promoters (often referred to as CpG islands) and large repetitive sequences, such as long interspersed nuclear element (LINE) and Alu retrotransposon elements (Bird, 2002). DNA methylation is catalyzed by a group of enzymes known as DNA methyltransferases (DNMTs) (Issa and Kantarjian, 2009; Bestor, 2000). Three major DNMTs (DNMT1, DNMT3a, and DNMT3b) have been identified in mammals. Evidence from phenotypic analyses of mice with mutant DNMT genes has provided useful mechanistic insights into the role and establishment of DNA methylation patterns during development (Jaenisch and Bird, 2003; Bestor, 2000). The DNMT1 enzyme has been demonstrated to have a 5- to 30-fold greater preference for hemimethylated substrates and therefore is popularly designated as a maintenance methyltransferase. It preserves the existing methylation patterns in daughter DNA strands by adding methyl groups to hemimethylated CpG sequences following replication. However, DNMT1 has also been demonstrated to be involved in de novo methylation activity in embryo lysates, and its sequence specificity was shown to be confined to the 5′-CpG-3′ dinucleotide sequence with little dependence on sequence context or density (Yoder et al., 1997). DNMT3a and DNMT3b enzymes are essential for global de novo methylation, as they preferentially target unmethylated CpG sequences (Okano et al., 1999). Although DNMT3L, the fourth family member, lacks intrinsic DNMT activity by itself, it colocalizes with DNMT3a and DNMT3b to establish genomic imprints in the maternal germ line and facilitate methylation of retroposons (Bourc'his et al., 2001). DNMT2, another member of the DNMT family, was found to lack biochemically detectable DNMT activity, and its deletion in mice had no obvious phenotypic effects on the genomic methylation pattern or methylation of retroviral DNA (Okano et al., 1999). DNMTs have been shown to be highly expressed in developing mouse embryos, and establish methylation patterns postimplantation (Okano et al., 1999). Interestingly, in contrast to the maternal genome, which is only partially demethylated after fertilization, demethylation is a remarkably active and rapid process in the male genome that results in nearly complete removal of methyl groups after fertilization (within hours). Studies indicate that during early embryonic development, promoter methylation is accompanied by specific histone modifications typical of heterochromatin, and that abnormal epigenetic premarking during early embryonic development may predispose certain genes to cancer-promoting DNA methylation events (Issa and Kantarjian, 2009). Although the enzymes catalyzing DNA methylation are well established, recent research has also identified mechanisms involved with the removal of methyl groups. Discovery of the ten-eleven translocation (TET) [which derives its name from the recurrent chromosomal translocation t-(10; 11) (q22; q23)] and activation-induced cytidine deaminase (AID) enzyme families has provided unprecedented information in our understanding of DNA demethylation (Scourzic et al., 2015). DNA demethylation can be achieved by two processes involving passive and active demethylation. Passive demethylation occurs by the failure of the maintenance DNMT enzyme to methylate DNA after replication, whereas active DNA demethylation utilizes TET and AID enzyme families to hydroxylate, oxidize, and deaminate 5-methyl cytosine. Three TET family members have been identified so far (TET1, TET2, and TET3), and each is involved in a distinct cellular process. Hydroxylation of 5-methyl cytosine by TET proteins produces 5-hydroxy methylcytosine and its subsequent conversion into 5-formylcytosine and 5-carboxylcytosine, followed by deamination and entry into the subsequent base excision repair pathway (Zhao and Chen, 2013).

It is well recognized that hypermethylation of CpG islands is typically associated with gene silencing. There are multiple routes through which DNA methylation can suppress gene transcription. The general mechanism involves excluding the binding of proteins that modulate transcription through their DNA-binding domains (Watt and Molloy, 1988). For example, binding of the chromatin boundary element–binding protein CTCF to DNA is blocked by CpG methylation, which allows the enhancer to activate transcription (Bell et al., 1999; Ohlsson et al., 2001). This mechanism has been demonstrated to be essential for imprinting the insulin-like growth factor 2 gene (Hark et al., 2000). In addition, CpG methylation has been shown to block the binding of several other transcription factors; however, the biological consequences of this remain unknown (Tate and Bird, 1993). Another mechanism for DNA methylation-mediated gene repression involves the binding of specific DNA-binding proteins to methylated CpG stretches, which forms repressor complexes with histone deacetylases (HDACs) and causes chromatin compaction (Jones et al., 1998; Nan et al., 1998; Feng and Zhang, 2001). In mammals, six methyl-CpG-binding proteins have been characterized to date, which include MeCp2, the methyl-binding domain proteins (MBDs) 1–4, and Kaiso. Studies demonstrate that all of these (except mammalian MBD3) possess a domain that specifically targets them to methylated

CpG regions in vitro and in vivo (Hendrich and Bird, 1998; Nan et al., 1993). Several detailed reviews are available that discuss DNA methylation and the role of these proteins in cancer development as biomarkers, which is currently beyond the scope of this chapter.

Histone Modifications. In addition to DNA methylation, posttranslational modification of N-terminal histone tails plays a significant role in the epigenetic regulation of gene expression (Luger and Richmond, 1998; Kornberg and Lorch, 1999). A typical nucleosome unit consisting of ~146 bp of DNA wrapped around an octamer of histones (H2A, H2B, H3, and H4) represents the fundamental building unit of eukaryotic chromatin. A diverse array of covalent chemical modifications of the less structured, protruding N-terminal tails of core histones by methylation, acetylation, ubiquitination, phosphorylation, sumoylation, and ADP-ribosylation dictate the dynamics of the chromatin state (Strahl and Allis, 2000). Euchromatin is a lightly packed form of chromatin where DNA is accessible for transcription, whereas heterochromatin represents a tightly packed chromatin state inaccessible to cellular transcriptional machinery. Most chemical modifications occur at lysine (K), arginine (R), and serine residues within the histone tails. These distinct histone modifications on one or more histone tails (often referred to as the "histone code"), which may act sequentially or in combination, are recognized by other proteins that signal further downstream events. A number of enzymes have been implicated in catalyzing (by addition or removal) various histone modifications. These include histone acetyltransferases (HATs), HDACs, histone methyltransferases (HMTs), histone demethylases (HDMs), and histone kinases. In brief, HATs catalyze the addition of an acetyl group on the ε-amino group of K residues in N-terminal histone tails, which neutralizes the positive charge, relaxes the chromatin, and facilitate the binding of transcriptional machinery to the DNA (Struhl, 1998). To date, 25 HATs have been characterized and are divided into four families. Examples include the GNAT (hGCN5 and PCAF), MYST (MYST and Tip60), p300/CREB-binding protein (CBP) (p300/CBP and SRC), and TAFII250 families (Gerhauser, 2013a; Bannister and Kouzarides, 2011). In contrast, HDACs catalyze the removal of acetyl groups from K residues, thus resulting in compaction of the chromatin configuration and repressed transcription (Kuo and Allis, 1998). HDACs are classified into four groups. HDACs 1, 2, 3, and 8 are members of the class I HDAC family, while HDACs 4, 5, 6, 7, 9, and 10 belong to the class II HDAC family. HDAC11 belongs to the Class IV HDAC group. Sirtuins that require NAD + as cofactor for their activity are structurally unrelated to other HDAC classes and constitute the class III HDAC family (Sauve et al., 2006; Mottet and Castronovo, 2008). HMTs catalyze the addition of methyl groups to K or R residues while HDMs act to remove them (Shi et al., 2007; Rice and Allis, 2001; Upadhyay and Cheng, 2011). Examples of histone lysine methyltransferases include enhancer of zeste homolog 2 (EZH2), and examples of histone lysine demethylases include lysine-specific demethylase 1 (Deb et al., 2014; Kouzarides, 2007). Depending on the site of K methylation (K4, K9, K27, etc. in histone H3) and methylation status (mono-, di-, or trimethylation), histone methylation may have an activating or a repressive effect on gene expression (Kouzarides, 2007; Bannister and Kouzarides, 2011). H3K4, H3K36, and H3K79 methylation have activating effects on gene transcription, whereas the methylation of H3K9, H3K27, and H4K20 is generally associated with gene silencing or transcriptional repression (Suzuki and Miyata, 2006; Bannister and Kouzarides, 2011; Upadhyay and Cheng, 2011). Trimethylation of histone H3 at K27 is associated with transcriptional repression and maintenance of silent chromatin through recruitment of polycomb-group protein (PcG) complexes (Deb et al., 2013). Consistent with their important role in transcriptional control, many HMTs are essential for cell proliferation, and their deregulation is implicated in human diseases including cancer (Sutherland and Costa, 2003; Bernstein et al., 2002; Varambally et al., 2008). Several excellent reviews are available on each group of histone-modifying enzymes, their mechanisms of action, and various histone modifications for biomarker and assay development, which are beyond the scope of this chapter.

Noncoding RNAs. Recent evidence indicates that noncoding RNA (ncRNA) transcripts play a fundamental role in epigenetic regulation of gene expression, and they have been implicated in various epigenetic mechanisms such as transposon silencing, X-chromosome inactivation, DNA imprinting, and paramutation (Costa, 2008; Peschansky and Wahlestedt, 2014; Zhou et al., 2010). In humans, ncRNAs include miRNA, small interfering RNA, and piwi-interacting RNA, which combined account for the majority of transcripts in representing approximately 98% of all human transcriptional output (Mattick, 2005; Szymanski et al., 2005). Based on size, ncRNAs can be classified into small ncRNAs that are generally less than 200 nucleotides in length and long noncoding RNA (lncRNA) transcripts that are more than 200 nucleotides in length. They can be divided into further subtypes based on their genomic origins and biogenic processes (Peschansky and Wahlestedt, 2014). Both types of ncRNAs have been shown to be essential "epigenetic modifiers" constituting a hidden layer of complex internal signals that control multiple levels of gene expression associated with the development and physiology of organisms (Ulitsky et al., 2011; Magistri et al., 2012; De Lucia and Dean, 2011; Zhou et al., 2010). lncRNAs have been demonstrated to be involved in gene silencing via mechanisms involving both histone modifications and DNA methylation. For example, the antisense lncRNA located in the p14/p15/INK4 locus, ANRIL, was reported to cause gene silencing via recruitment of PcGs (Pasmant et al., 2007; Kotake et al., 2011). Another well-studied example includes the involvement of a 17 kb lncRNA, XIST, in X-chromosome inactivation that ensures X-linked gene dosage compensation in mammalian females (Mak et al., 2002; Plath et al., 2003; Silva et al., 2003; Zhao et al., 2008). This process involves the recruitment of mammalian PRC2 complex containing the HMT EZH2 to the locus by a short repeat RNA within XIST and the deposition and spread of repressive H3K27me3 marks throughout the X–chromosome. In addition to histone modifications, lncRNAs were reported to mediate gene silencing through DNA methylation. One such example includes Kcnq1ot1, which in addition to interacting with the PRC2 complex and G9a, has been implicated in the recruitment of DNMT1 through a critical 890 bp region to the CpG island of imprinted genes (Mohammad, 2010).

Small ncRNAs, particularly miRNAs, regulate key epigenetic mechanisms. Accumulating evidence indicates that deregulation in miRNAs is linked to several steps of cancer initiation and progression. miRNAs are relatively small noncoding RNAs (usually 20–22 nucleotides long) that are excised from longer (60–110 nucleotides) RNA precursors (Bartel, 2004; Calin and Croce, 2006). miRNAs play essential roles in normal biological processes including development, proliferation, differentiation, and cell death (Bartel, 2004; Pasquinelli et al., 2005). Interestingly, miRNAs appear to be able to act as either tumor suppressors or oncogenes by affecting distinct genes of the gene families involved in critical biological processes such as proliferation and differentiation. Many miRNA genes are located in the genomic loci known as fragile sites and are therefore susceptible to either loss or amplification (Zhang et al., 2006). Short RNAs (50–200 nucleotides) were reported to be transcribed from H3K27me3-enriched

PRC2 target genes and to cause cell-type-specific gene silencing in cis by stabilizing the PRC2 complex near the transcription site through interactions via formation of stem-loop structures (Kanhere et al., 2010).

miRNAs are known to regulate various components of cellular epigenetic machinery, particularly PcGs, and thus affect multiple downstream effects (Deb et al., 2013, 2014; Sander et al., 2008; Varambally et al., 2008). One such example is miR-214, which downregulates EZH2 expression by targeting its 3′-UTR region and accelerating skeletal muscle differentiation and transcription of developmental regulators in embryonic stem cells (Juan et al., 2009). Other miRNAs have been implicated in the repression of B-cell-specific Moloney murine leukemia virus integration site 1 (BMI-1), a component of the PRC1 complex (Liu et al., 2012; Godlewski et al., 2008; Wellner et al., 2009). DNA methylation also has been shown to be modulated by miRNAs. DNMT1 and DNMT3 have been reported as targeted by the miR-29 family in lung cancer and leukemia cells (Fabbri et al., 2007; Garzon et al., 2009). In addition to the role of small ncRNAs as regulators of various epigenetic mechanisms, in many instances they are themselves targets of the same epigenetic processes, which may lead to further downstream alterations. For example, in human breast tumorigenesis and metastasis, decreased expression of a set of miRNAs was attributed to gene hypermethylation (Lehmann et al., 2007; Hsu et al., 2009; Lujambio et al., 2008). In summary, recent evidence suggests that ncRNAs have emerged as key regulators of epigenetic mechanisms, and also that the modulation of these RNA transcripts by the same epigenetic processes may lead to major consequences. Several excellent publications are available on each group of ncRNAs and their mechanisms of action affecting gene regulation, which is beyond the scope of this chapter.

ENDOGENOUS FACTORS AFFECTING THE EPIGENOME

Environmental Factors. Environmental factors including chemical carcinogens, environmental pollutants, dietary contaminants, and physical carcinogens play important roles in the etiology of human cancer. In general, the degree to which environmental factors influence the process of carcinogenesis depends on the presence of specific hazardous entities and the duration of exposure. However, the degree to which hazardous exposures affect cancer largely reflects the variation in susceptibility to a given environmental exposure. Generally, environmental factors capable of initiating tumor development by altering the epigenome include agents capable of inducing changes either directly or indirectly in genomic DNA and agents that affect the critical cellular regulatory processes of gene transcription such as DNA damage and repair, cell cycle control, and the cell death process. Studies demonstrate that the mismatch repair gene MHL1 is frequently hypermethylated in sporadic tumors that exhibit microsatellite instability (MSI) (de Leeuw et al., 2000). Similarly, the silencing of O^6-methylguanine methyltransferase (MGMT), the DNA repair gene—encoding protein responsible for the removal of carcinogen-induced O^6-methylguanine adducts from DNA (which if left unrepaired result in G-to-A transition mutation), appears to increase the mutation rate in critical cellular regulators including tumor suppressors and oncogenes (Khin et al., 2011). These studies provide cues that environmental exposures alter either the expression or the activity of enzymes involved in de novo DNA methylation (DNMT3A and DNMT3B) and/or the maintenance of DNA methylation (DNMT1) that may predispose them to mutational events (Khin et al., 2011; de Leeuw et al., 2000). A study by Baglietto et al. (2017) used an Illumina Infinium HumanMethylation450 array to determine whether methylation in DNA from prediagnostic peripheral blood samples is associated with lung cancer risk. This case-control study, nested within the European Prospective Investigation into Cancer and Nutrition—Italy cohort and a study within the Melbourne Collaborative Cohort Study cohort as discovery sets (a total of 552 case-control pairs), identified six CpGs for which hypomethylation was associated with lung cancer risk. These findings provide convincing evidence that smoking and possibly other factors lead to DNA methylation changes measurable in peripheral blood that may improve the prediction of lung cancer risk. Additionally, different agents in the environment may induce mutational events through preferential binding to hypermethylated DNA. Studies on benzo[a]pyrene diol epoxide (BPDE), a carcinogen from tobacco smoke, exhibits a preference for methylated CpG sites that results in the formation of DNA adducts and G-to-T transversions often found in cancers of the aerodigestive tract in tobacco smokers. It has been shown that certain infectious agents, such as human papillomavirus, induce gene silencing via DNA hypermethylation of the promoters of host genes including CDH1, RB1, INK4a/p16, CDNK2A, MTHFR, PEG3, and others listed in Table 17.1.

Agents categorized in another group have the ability to alter the patterns of chromatin modification (histone code) in a transient manner and are likely to induce changes in key cellular processes including gene transcription, DNA damage response, and DNA repair. Primary epigenetic targets for environmental factors in this group may be the proteins and protein complexes responsible for histone modifications, such as HATs and HDACs, whose activities are often found deregulated in cancer. Recent studies have shown that HATs are involved in the process of DNA repair, suggesting that even moderate and transient inhibition of HAT activity induced by environmental exposures may compromise DNA repair, leading to mutation fixation and genomic instability (Herceg, 2007). Similarly, HDAC was shown to be required for efficient DNA repair, suggesting that the removal of histone acetylation is required for restoration of the normal (default) chromatin structure following the completion of DNA repair. A tight regulation of HAT and HDAC activity is thus essential for the proper regulation of gene transcription and DNA repair. Reduced levels of histone acetylation or enhanced histone deacetylation may result in the compaction of chromatin, blocking access of transcription factors to DNA and/or impeding the progression of RNA polymerases. Therefore, different environmental factors may transiently alter chromatin-modifying/remodeling activities and the patterns of histone modification, thus impeding DNA repair and other chromatin-based processes.

The next possible epigenetic "target" of adverse environmental exposure may be general methyl-C-binding proteins, a group of proteins (including MBD1, MBD2, MBD3, MeCP2, and Kaiso) that bind to methylated CpG sites (Mazzio and Soliman, 2012). Some members of this family, exemplified by MeCP2, were found to bind and recruit HDAC to chromatin. Changes in MeCP2 protein stability and function elicited by hazardous agents in the diet and environment may thus affect normal gene transcription and lead to aberrant cell proliferation and cancer (Liyanage et al., 2014). Given that histone modification and DNA methylation appear to work together to establish a permissive or repressive chromatin state, agents in the environment and diet that affect one of these intimately linked and self-reinforcing mechanisms would inevitably affect the other. Although poorly understood, the molecular mechanisms by which epigenetic carcinogens in the environment and diet may exhibit adverse effects in histone modification are beginning to emerge. Several recent studies have examined the effect of specific environmental carcinogens on histone

TABLE 17.1 Epigenetic Modifications and Targeted Genes Affected by Environmental, Dietary, and Lifestyle Factors

Agents/Exposure	Targeted Genes	Epigenetic Mechanisms	References
ENVIRONMENTAL FACTORS			
Tobacco smoke	CDKN2A, p53, MGMT	Binding of PAHs to methylated CpGs	Belinsky et al. (2004), Chen et al. (1998), Yoon et al. (2001), Osada and Takahashi (2002)
Nickel	CDKN2A, CDKN2B (p15 gene), MLH1	Inhibition of histone acetylation; damage of core histone	Cameron et al. (1999), Sutherland and Costa (2003), Broday et al. (1999), Karaczyn et al. (2005)
AFB1	RASSF1A, MGMT, CDKN2A, SNSG	Preferential binding to methylated lysine or modified histone proteins causing DNA damage	Zhang et al. (2002), Zhang et al. (2002), Zhang et al. (2006)
Cadmium	DNMTs	Inhibition of DNMTs (acute exposure), increased DNMT activity (chronic exposure)	Takiguchi et al. (2003)
Arsenic	c-Myc	Decrease of SAM leading to global hypomethylation and activation of oncogenes	Sutherland and Costa (2003), Zhao et al. (1997), Chen et al. (2001)
Benzene	MAGE-1	Alterations in DNA methylation in LINE-1 repeated elements and gene-specific methylation of the iNOS promoter.	Bollati et al. (2007)
PAHs	GSTM1, HMOX-1	Functional genetic variations in GSTM1 and HMOX-1, both of which are related to defenses against oxidative stress	Chahine et al. (2007)
POPs	miR-222, miR-21	Particulate matter exposure, increased miR-222 and miR-21 expression	Bollati et al. (2010)
Ionizing radiation	p16/INK4A	DNA methylation–mediated silencing of key cellular regulators	Belinsky et al. (200)4
UV radiation	Genes not identified	Global hypomethylation	Van Doorn et al. (2005), Mittal et al. (2003)

Bacteria (H pylori)	Genes not identified	De novo methylation associated with chronic inflammation and cell proliferation	Maekita et al. (2006), Sugiyama et al. (1998), Szaleczky et al. (2000)
HPV, EBV, HBV, HPV, EB	Genes not identified	Methylation of viral genome; binding of viral proteins to host gene promoters; hypermethylation of host gene promoter; changes in chromatin modification; patterns and recruitment of HDACs	De Capoa et al. (2003) Li and Minarovits (2003), Van Tine et al. (2004), Zhang et al. (2006), Kalantari et al. (2004), Wiley et al. (2005), Zhou et al. (2005), Zazula et al. (2006), Jicai et al. (2006), Zhang et al. (2005)
DIETARY AND LIFESTYLE FACTORS			
Alcohol	APC-1A, CDKN2D (p14 gene)	Alcohol metabolites such as acetaldehyde acting as cocarcinogens with HBV and HCV viruses; depleted SAM; aflatoxins; and obesity	Van Engeland et al. (2003), Giovannucci et al. (1995), Marsit et al. (2006)
	NR2B gene	Demethylation of cytosine residues in CpG island	Marutha Ravindran et al. (2009)
Dihydrocoumarin	CDKN2A, hMLH1, MGMT, RASSF1A, SFRP	Disruption of heterochromatic silencing; inhibition of SIRT1	Haigis et al. (2006)
Folate deficiency	Genes not identified	Posttranslational modification of histone and DNA methylation	Waterland et al. (2006)
Low methionine intake	Genes not identified	Modifications in DNA methylation and histone modifications; increased binding of methyl-CpG-binding proteins	Waterland et al. (2006) Giovannucci et al. (1995)
Unknown/Aging	Genes not identified	Reactivation of inactive X-chromosome and silent allele LOI	Richardson (2003), Wilson and Jones (1983), Fraga et al. (2005)
Transgenerational epigenetic inheritance	IGF2, transposable elements	Histone-based germline inheritance	Sakatani et al. (2005), Rakyan et al. (2003), Pembrey et al. (2006), Kaati et al. (2002)

modification and suggest that these agents may affect the patterns of histone modification through different mechanisms (Table 17.1).

Another prospective mechanism by which environmental exposure, including ingestion, affects the epigenome involves transposable elements. Transposons, when activated, may cause genetic mutations and transcriptional noise (Liyanage et al., 2014). For example, the Alu family alone consists of several hundred thousand elements that are shown to be heavily methylated and transcriptionally silent in somatic cells. It is well documented that the activation of transposable element-derived promoters may be a consequence of perturbed DNA methylation; transposable elements were shown to be activated by various kinds of cellular stress. Therefore, stress induced by environmental agents may activate transposable elements, leading to the altered establishment and maintenance of epigenetic states.

Dietary Factors. Studies have demonstrated that maternal nutrition imbalances and metabolic disturbances during embryonic development have a persistent effect on the health of offspring and may be passed down to the next generation (Gallou-Kabani and Junien, 2005). The potential effect of nutritional factors on the phenotype is best demonstrated by studies on the risk of cancer for pregnant women and fetuses. When a mother is exposed to adverse conditions, fetal nutrition may cause alterations in structure, physiology, and metabolism that predispose her to several diseases including cancer. Selected dietary components consumed during early pregnancy may influence the postnatal risk of cancer development, although not all dietary components are harmful. In those cases where adverse effects on fetal development were observed, a proposed mechanism includes the methylation of genes due to dietary food components in the mother's diet. Both hypermethylation and hypomethylation of selected genes were observed. Genes that were overexpressed included Klf6, Klf9, Nid2, Ntn4, Per1, and TXNIP, and genes that were repressed included Bcar3, Cldn12, Csf1, Jag1, Lgals3, Lypd3, Nme1, Ptges2, Ptgs1, and Smarcb1 (Gallou-Kabani and Junien, 2005; Szarc vel et al., 2015). In animal models, deficiencies of macronutrients during placental growth have been shown to affect fetal growth. Most genes that contribute to reduced fetal growth are regulated by imprinting, and the maternal allele is affected in these cases. Functionally, nutrient transport from mother to fetus via the placenta is affected dramatically by hypomethylation of genes in the embryonic trophectoderm (Verma, 2012).

Direct effects of nutritional factors on epigenetic changes are some of the most studied, and the best understood is the relationship between dietary methionine and DNA methylation (Verma, 2012). Methionine, an essential amino acid, plays a central role in epigenetic regulation by serving as a methyl donor for methylation reactions. In the process of cytosine methylation, a DNMT enzyme converts SAM to SAH; therefore, an optimal supply of SAM or the removal of SAH is essential for the normal establishment of genome-wide DNA methylation patterns (Kanwal and Gupta, 2010). CpG methylation patterns are largely erased in early embryos and then reestablished in a tissue-specific manner. Therefore, early embryonic development may represent a sensitive stage, and dietary and environmental factors that affect DNA methylation reaction and the activity of DNMTs may result in permanent fixation of aberrant methylation patterns (Kanwal and Gupta, 2010; Herceg, 2007). In postnatal development and adulthood, established patterns of DNA methylation and histone modification must be maintained through multiple mitotic divisions; therefore, inappropriate quantities of methionine, other food components and environmental agents may affect normal patterns of DNA methylation and histone modification. In this respect, it is interesting to note that in adult men with hyperhomocysteinemia, a disorder occurring in several genetically determined and acquired diseases with uremia, treatment with high doses of folate increase methylation levels at specific genes and restore normal expression (Herceg, 2007). In addition to DNA methylation, the methylation of histones, a distinct epigenetic mechanism dependent on 1-carbon groups, may be affected by the consumption of excessive levels of specific nutritional factors. Therefore, nutrition factors are likely to directly or indirectly (through changes in DNA methylation) affect histone modifications such as histone methylation. In fact, DNA methylation is the only epigenetic mark with strong mechanistic support for both heritability and response to dietary changes (Maddocks et al., 2016; Mentch et al., 2015; Mentch and Locasale, 2016; Rodríguez-Gallego et al., 2015) and represents an epigenetic link that may be translated into clinical investigations.

Epigenetic analysis of colon tumors induced by a dietary carcinogen amino-1-methyl-6-phenylimidazo[4,5-b]pyridine (PhIP) in rats showed altered miRNA signatures linked to dysregulated pluripotency factors, such as c-Myc and Krüppel-like factor 4 (KLF4) (Chen et al., 2017). Downstream targets that corroborated in the rat were investigated in human cancer data sets. The results confirmed that multiple let-7 family members were downregulated in PhIP-induced skin, colon, lung, small intestine, and Zymbal's gland tumors and were associated with c-Myc and high-mobility group AT-hook 2 upregulation. PhIP-signature miRNAs with the profile miR-21-high/miR-126-low/miR-29c-low/miR-215-low/miR-145-low were linked to reduced KLF4 levels in rat tumors and human pancreatic cancer and colorectal cancer (CRC). These data provide a catalog of miRNAs in rats with PhIP signatures that could represent a new family of markers with possible value for human risk assessment. The WCRF and the AICR have found convincing evidence that particular dietary components and cumulative dietary effects, such as obesity, are associated with several cancers. The connection between specific dietary factors and epigenetic alterations is clear in some animal models, but data in human populations are inconsistent. Dietary factors are likely to interact either directly or indirectly with the epigenome to accelerate or decelerate age-related epigenetic changes in cancer-associated genes. Randomized clinical trials have generally failed to show clear and consistent effects of the diet and supplements on cancer risk because of the high phenotypic variability of response (Sapienza and Issa, 2016). Studies have demonstrated that the measurement of metabolic states can be correlated with chromatin states and gene expression. DNA methylation is currently being used to construct models for predicting chronological age at a population level, because regulation of the chromatin landscape can alter life span (,Benayoun et al., 2015). These models have practical implications for studying the role of methylation in age-related diseases and might explain the association of complex metabolic and inflammatory states with the early onset of diseases including cancer (Hannum et al., 2013 and Horvath et al., 2015).

Lifestyle Factors. Lifestyle factors including exercise and diet play an important role in regulating the epigenome and altering gene expression. Exercise can modify the epigenome in order to preserve and prolong life. Exercise has been shown to induce positive changes in DNA methylation within adipose tissue and to regulate metabolism in both healthy and diseased individuals (Rönn et al., 2013). Increased DNA methylation of genes HDAC4 and NCOR2 has been shown to increase lipogenesis following exercise (Rönn et al., 2013). Exercise also leads to beneficial changes in DNA methylation patterns in skeletal muscle (Nitert et al., 2012). Not only is obesity an indicator for diseases such as type 2 diabetes and cardiovascular disease, but it also puts additional stress on the system that can itself negatively impact health (Ronti et al., 2006). Acute exercise is associated with DNA hypomethylation of the entire genome in the skeletal muscle cells of sedentary individuals, and high-intensity exercise tends to cause a reduction in the promoter methylation of certain genes (Ntanasis-Stathopoulos et al., 2013). Exercise is also known to positively influence the expression patterns of miRNAs in leukocyte cells (Radom-Aizik et al., 2010). The health benefits of physical exercise, especially on a long-term and strenuous basis, have a positive effect on epigenetic mechanisms and ultimately may reduce the incidence and severity of cancer (Sanchis-Gomar et al., 2012).

Studies on genomic imprinting have revealed that DNA methylation patterns are influenced by diet and that epigenomic sensitivity to specific diets influences cancer susceptibility. Dietary fat makes up a large part of the Westernized diet, which results in increased adipose tissue via adipocyte hypertrophy and hyperplasia (Kanherkar et al., 2014). Dietary fat influences adipokine release by influencing the epigenome to affect the DNA methylation and posttranslational modification of histone proteins. This represents one method by which dietary fat may influence cancer progression. Overconsumption of well-done meats and saturated fats causes an increase in somatic glutathione S-transferase pi (GSTP1) inactivation by CpG island methylation in the promoter region, thus increasing susceptibility to prostate cancer (Nelson et al., 2014).

Behavioral and lifestyle factors have also been shown to relate to a number of health-related outcomes, and studies examining their relationship to molecular aging rates using aging-related epigenetic biomarkers have previously shown that these factors can predict all-cause mortality, chronic conditions, and age-related functional decline. The studies have observed that extrinsic epigenetic age acceleration (EEAA) exhibited significant associations with fish intake, moderate alcohol consumption, body mass index (BMI) ($P = .01$), and blood carotenoid levels (an indicator of fruit and vegetable consumption), whereas intrinsic epigenetic age acceleration (IEAA) was associated with poultry intake and BMI. Both EEAA and IEAA were found to relate to indicators of metabolic syndrome, which appear to mediate their associations with BMI and suggest that an increase in BMI is associated with increases in both EEAA and IEAA. Age-related epigenetic analysis of blood confirms the conventional wisdom regarding the benefits of eating a high-plant-based diet with lean meats, moderate alcohol consumption, physical activity, and education, as well as the health risks of obesity and metabolic syndrome (Quach et al., 2017).

Several studies have provided evidence that alcohol consumption is associated with different epigenetic changes in human cancer (Ponomarev, 2013). In a large epidemiological study (the Netherlands Cohort Study on diet and cancer), analysis of DNA methylation showed that the prevalence of promoter hypermethylation of several genes including APC-1A, CDKN2D, CDKN2A, human mutL homologue 1 (hMLH1), MGMT, and RASSF1A was higher in CRC patients with high alcohol (and low folate) intake than it was among CRC patients with high folate and low alcohol intakes. In addition, a study of human head and neck squamous cell carcinoma showed that promoter hypermethylation of the MGMT gene and genes known to regulate the WNT pathway occurs more frequently in both heavy and light drinkers than in nondrinkers. The mechanism underlying the epigenetic changes caused by alcohol abuse may also involve SAM. This small metabolite is regenerated from demethylated SAM via the methionine cycle, which involves folate. Therefore, an imbalance in this cycle through alcohol consumption may result in the depletion of SAM and aberrant epigenetic patterns. In addition, it was shown that human class I alcohol dehydrogenase (ADH) genes may be regulated by epigenetic mechanisms. Class I ADH genes were found to be repressed in human hepatoma through epigenetic modification, which suggests that changes associated with alcohol-metabolizing genes may also enhance the other toxic effects of alcohol on different organs, most notably the liver and including hepatic tumorigenesis (Herceg, 2007).

Tobacco smoke is a complex aerosol that contains polycyclic aromatic hydrocarbons (PAHs), mostly benzo[a]pyrene, which is considered the most carcinogenic. Epigenetic targets of PAHs from tobacco smoke induce DNA damage through preferential binding to methylated CpG sites, a phenomenon already demonstrated for BPDE, another carcinogen found in tobacco smoke. Several studies have demonstrated the hypermethylation and silencing of several genes in lung cancer associated with smoking (Steenaard et al., 2015). The genes frequently altered by promoter hypermethylation in lung cancers of smokers are p53, p16, and MGMT. In addition, different components in tobacco smoke induce histone changes and alter histone code. Some potentially novel histone marks including acetylation, monomethylation, and dimethylation in the specific K and R residues of histones H3 and H4 in mouse lungs were elucidated. Cigarette smoking has also been linked to somatic genetic and epigenetic aberrations including CpG island methylator phenotype (CIMP)-high, MSI-high, and *BRAF* mutation. These molecular features have been associated with synchronous primary CRCs. It was observed that when current and former smokers were compared with never-smokers, multivariable HRs for synchronous CRCs were significantly higher than they were for solitary CRCs positive for CIMP-high, MSI-high, or *BRAF* mutation. The evidence suggests that smoking is associated with an elevated risk of synchronous CRCs and contributes to an etiologic field effect that favors these somatic molecular alterations and the development of multiple primary tumors (Drew et al., 2017).

Nutrients extracted from the diet enter metabolic pathways and are transformed into useful molecules. These nutrients are known to have epigenetic targets in cells such that they can be used to modify the epigenome in order to correct abnormally activated or silenced genes and can be combined into an "epigenetic diet" useful as a therapeutic and/or chemopreventive measure. During this transitory phase, methyl groups are formed from key nutrients including folic acid, B vitamins, and SAM, and these methyl groups comprise important epigenetic marks for gene silencing. Diets high in such methyl-rich nutrients may significantly alter gene expression and offer protective health benefits (Kanherkar et al., 2014). Deficiencies in folate and methionine, both of which are involved in cellular processes that supply the methyl groups needed for DNA methylation, can change the expression (imprinting) of growth-factor genes such as insulin-like growth factor 1 to influence cancer progression (Ross and Milner, 2007). In addition, several natural nutrient products have interesting biological properties and structural diversity. These include polyphenols present in fruits, vegetables, and other dietary botanicals. Phenolic acids, flavonoids, stilbenes, and lignans are the most abundantly occurring polyphenols and are an integral part of everyday nutrition in populations worldwide. Certain food components epigenetically increase the levels of DNA repair enzymes such as MGMT and MLH1, and flavonoids such as bilberry anthocyanins, in order to actively decrease DNA damage. Anthocyanin is an effective antioxidant for humans that is found in plants and is easily identified by its potent red or purple pigment. It is found in plants such as eggplant, plums, pomegranate, red onion, cranberries, blueberries, kidney beans, and cherries. This flavonoid serves as a powerful antioxidant that contributes to the scavenging of DNA-damaging free radicals. While the direct fate of anthocyanins in vivo following digestion may be less than 5% (the majority being rapidly excreted), the potent residual antioxidant property remains in blood following the consumption of anthocyanin-rich foods due to the metabolic breakdown of flavonoids and resultant increase in uric acid levels. Some common examples of the most studied and promising cancer preventive polyphenols include epigallocatechin-3-gallate (EGCG) from green tea, curcumin from the curry plant, genistein from soy, resveratrol from grapes and berries, and sulforaphane from broccoli. A large number of dietary agents that affect DNA methylation, histone modification, and the regulation of noncoding miRNA expression in various human cancers are shown in Tables 17.2 and 17.3. Significant gains have been made in understanding the molecular mechanisms that underpin the chemopreventive effects of polyphenols, and consequently a wide range of mechanisms and gene targets have been identified for individual compounds.

Several studies have demonstrated that the green tea polyphenol (GTP) constituent EGCG is a potent demethylating agent that inhibits the enzymes involved in DNA methylation, and is also an effective histone-modifying agent (Pandey et al., 2010; Deb et al., 2014; Li et al., 2010; Nandakumar et al., 2011). It is well known that CpG island hypermethylation at the promoter region leads to epigenetic repression of several critical tumor-suppressor genes during tumorigenesis. A study suggests that EGCG acts as a competitive inhibitor of DNMT ($Ki = 6.89\ \mu M$) that binds to the catalytic pocket and inhibits DNMT activity in a dose-dependent manner (Fang et al., 2003). Furthermore, EGCG treatment (5–50 μM for 12–144 h) was found to effectively reactivate methylation-silenced genes such as p16INK4a, retinoic acid receptor β (RARβ), MGMT, and hMLH1 in human esophageal cancer KYSE 510 cells. EGCG was also reported to inhibit HDACs and increase permissive or active histone modifications such as histone acetylation at target gene promoters. Studies from our laboratory showed that exposure of prostate cancer cells to GTP caused a reexpression of the epigenetically silenced GSTP1 gene that correlated with promoter demethylation due to DNMT1 inhibition and histone modification at the promoter region (Pandey et al., 2010). However, GTP treatment did not show any global hypomethylation effect that could result in genomic instability, because the methylation status of the LINE-1 promoter remained unaffected as demonstrated by methylation-specific PCR. GTP treatment decreased the mRNA and protein levels of MBD1, MBD4, MeCP2, and HDACs 1–3, whereas acetylated histones H3 (LysH9/18) and H4 were found to be elevated. In another study, we demonstrated that GTP treatment caused cell cycle arrest and apoptosis by inducing proteasomal degradation of class I HDACs in human prostate cancer cells (Thakur et al., 2012). Studies by Li et al. (2010) demonstrated that EGCG in combination with trichostatin A (TSA), an organic compound that serves as an antifungal antibiotic and potent inhibitor of HATs, could synergistically reactivate estrogen receptor-α (ER-α) expression in ER-α-negative MDA-MB-231 breast cancer cells by modulating histone methylation and acetylation patterns at the gene promoter. In addition, they also reported that treatment with EGCG and/or TSA contributes to transcriptional activation of ER-α by causing decreased binding of the transcription-repressor complex Rb/p130-E2F4/5-HDAC1-SUV39H1-DNMT1 to the regulatory region of the gene.

EGCG has been reported to modulate PcGs such as BMI-1 and EZH2 (Balasubramanian et al., 2010; Choudhury et al., 2011; Deb et al., 2014). EGCG alone or in combination with DZNep was shown to decrease the levels of PcG proteins including EZH2, embryonic ectoderm development, suppressor of zeste 12, polycomb group ring finger 2, and BMI-1 via a mechanism involving proteasome-associated degradation. The reduction in PcG protein levels correlated with a decrease in repressive chromatin marks such as H3K27me3 and H2AK119ub as well as HDAC1 levels, whereas the accumulation of acetylated H3 levels was found to be elevated. A recent study from our group has shown that exposure of breast cancer cells to GTP or its major constituent EGCG

TABLE 17.2 Epigenetic Modifications Affected by Dietary Agents and Related Compounds in Various Human Cancers

Dietary Agents	Source	Cancer Type	Molecular Targets	Molecular Mechanisms	References
Equol	Soy	Breast	H2A/H2B/H3/H4 Acetylation	HDAC inhibitor	Hong et al. (2004)
Genistein		Esophageal; prostate; breast; renal	H2A/H2B/H3/H4 Acetylation p21, p16, PTEN, p53 FOXA3, SIRT1, hTERT	HAT inhibitor HDAC inhibitor	Fang et al. (2005), Li et al. (2009a,b), Majid et al. (2009), Hong et al. (2004), Majid et al. (2008), Kikuno et al. (2008)
Daidzein		Esophageal; prostate	Histone acetylation	HDAC inhibitor	Fang et al. (2005), Hong et al. (2004)
Biochanin A		Esophageal		DNMT inhibitor	Fang et al., 2005, 2007
Daidzein		Breast	RARβ, MGMT, CDKN2, GSTP1, HMGNS, BTG3	DNMT inhibitor	Fang et al., 2005, 2007
Genistein		Prostate Esophageal	RXRα, CDX2, GSTP1, W1F1	DNMT inhibitor	Fang et al. (2005), Majid et al. (2009)
Fisetin	Poison ivy	Cervix; breast		Unknown	Howitz et al. (2003), Wood et al. (2004)
		Breast, esophageal		DNMT inhibitor	Fang et al. (2007), Lee et al. (2005)
Allyl mercaptan	Garlic	Erythroleukemia Liver, colon	H3/H4 acetylation p21/Waf1	SIRT1 inhibitor	Lea et al. (1999), 2001, Druesne et al. (2004), Nian et al. (2008)
S-allylmercaptocysteine		Erythroleukemia	H3/H4 acetylation	HAT inhibitor	Lea et al. (2001), Nian et al. (2008)
Anacardic acid	Cashew nuts	Leukemia; cervix; breast; prostate	H3K9 and H3K14 deacetylation	HDAC inhibitor	Sbardella et al. (2008), Mai et al. (2006), Balasubramanyam et al. (2003), Sun et al. (2006), Eliseeva et al. (2007), Chandregowda et al. (2009)

Continued

TABLE 17.2 Epigenetic Modifications Affected by Dietary Agents and Related Compounds in Various Human Cancers—cont'd

Dietary Agents	Source	Cancer Type	Molecular Targets	Molecular Mechanisms	References
Curcumin	Turmeric	Cervix; HIV; hepatoma; leukemia; prostate Brain, lymphoma	Histone acetylation GATA4, EOMES, GZMB	HAT & HDAC inhibitor	Singh and Misra (2009), Marcu et al. (2006), Kang et al. (2005), Cui et al. (2007), Sng et al. (2006), Chiu et al. (2009), Tikoo et al. (2008), Li et al. (2008), Morimoto et al. (2008), Liu et al. (2005), Chen et al. (2007), Bora-Tatar et al. (2009)
		Leukemia Esophageal		DNMT inhibitor	Fang et al. (2007), Liu et al. (2009), Kuck et al. (2010)
Catechin	Green tea	Lymphocytes, breast		HAT & DNMT inhibitor	Choi et al. (2009), Lee et al. 2005
Epicatechin		Lymphocytes, esophageal	H3/H4 acetylation	HAT & DNMT inhibitor	Choi et al., 2009, Fang et al. (2003), Lee et al. (2005)
Epicatechin gallate		Colon, prostate, esophageal		HAT & DNMT inhibitor	Choi et al. (2009), Nair et al. (2008), Fang et al. (2003)
Epigallocatechin-3-gallate		Colon, prostate Prostate	H3K27 trimethylation EZH2, SUZ12	HAT inhibitor	Pandey et al. (2010), Choi et al. (2009)
				HMT inhibitor	Balasubramanian et al. (2010), Nair et al. (2008)
			RARβ, MGMT, MLH1, CDKN2A, RECK, TERT, RXRα, CDX2, GSTP1, W1F1	DNMT inhibitor	Gao et al. (2009), Kato et al. (2008), Fang et al. (2003, 2007)
Polyphenon B	Green & black teas	Liver cancer		Increases HDAC1 expression	Murugan et al. (2009)
Isothiocyanates	Broccoli	Prostate, leukemia, erythroleukemia Colon, esophageal	H3/H4 acetylation p21, GSTP1	HDAC inhibitor	Wang et al. (2007), Lea and Randolph (2001), Mai et al. (2006), Wang et al. (2008), Wang et al. (2007), Wilkinson et al. (1995)

Sulforaphane		Prostate, breast, colon, kidney, esophageal	H3/H4 acetylation RARβ, HBD-2, p21, Bax	HDAC inhibitor Decreases DNMT expression	Fang et al. (2005), Nair et al. (2008), Myzak et al. (2004), Schwab et al. (2008), Paluszczak et al. (2010)
Luteolin	Parsley & celery	Cervix, esophageal	TNFα, IL-8, RBP	SIRT1 activator & DNMT inhibitor	Howitz et al. (2003), Fang et al. (2007)
Apigenin		Esophageal		DNMT inhibitor	Fang et al. (2007)
Resveratrol	Grapes, wine & eucalyptus	Colon, lungs, breast		SIRT1 activator & DNMT inhibitor	Yang et al. (2007, 2009), Paluszczak et al. (2010), Stefanska et al. (2010)
Baicalein	Indian trumpet	Breast		DNMT inhibitor	Paluszczak et al. (2010)
Betanin	Beet root	Breast		DNMT inhibitor	Paluszczak et al. (2010)
Lycopene	Tomatoes	Breast		Unknown	King −Batoon et al. (2008)
Hesperidin	Citrus	Esophageal		DNMT inhibitor	Fang et al. (2007)
Quercetin		Breast; colon; esophageal		DNMT inhibitor	Lee et al. (2005), Fang et al. (2007), Paluszczak et al. (2010), Stefanska et al. (2010), Tan et al. (2009)
Ursolic acid	Blueberries, cranberries & apple peels	Skin	DNMT1, DNMT3a, HDAC1, HDAC2, HDAC3, HDAC8, HDAC6, HDAC7; Nrf2, HO-1, NQO1, UGT1A	DNMT inhibitor HDAC inhibitor	Kim et al. (2016) Hwang et al. (2014)
Oleanolic acid	Olive oil, Garlic	Mouse RAW 264.7 cells, lung cancer cells	Nrf2; MafK, p65 acetylation		Zhao et al. (2015)

TABLE 17.3 MicroRNA Changes Affected by Dietary Agents

Dietary agent	Source	Cancer Type	MicroRNA	Reference
EGCG	Green tea	Neuroblastoma	miR-92, miR-93, miR 106b, miR-34a, miR-99a	Chakrabarti et al. (2012)
		Hepatocellular carcinoma	miR-467b, miR-487b, miR-197, miR-805, miR-374, let-7f, miR-350, miR-24-1, miR-137, miR-335-3p, let-7a, miR-222, miR-26b, miR-30c-1, let-7d, miR-98, miR-30c, miR-30b, miR-32, miR-674, miR-532-5p, let-7g,miR-18a, miR-192, miR-302d, miR-30b, miR-802, let-7e, miR-322, miR-720, miR-146b, miR-340-3p, miR-185, miR-425, miR-10a, miR-126-5p, miR-101a, miR-30e, let-7c, miR-141, miR-33, miR-29a, miR-199b,miR-450a-5p,miR-21,miR-23a,miR-101b, miR-148a, miR-193, miR-23b, miR-107, miR-140, miR-551b, miR-466c-5p, miR-106a, miR-590-3p, miR-875-3p, miR-224, miR-292-5p, miR-678, miR-469, let-7b, miR-463, miR-574-3p, miR-201, miR-290-3p, miR-181a, miR-302a, miR-429, miR-133a, miR-190b, miR-710, miR-135b, miR-296-5p, miR-191, miR-188-5p, miR-298, miR-181a-1,miR-466g, miR-26b, miR-466f-3p, miR-29b, miR-1224, miR-291b-5p, miR-324-5p, miR-486, miR-128, miR-450b-3p, miR-135a, miR-294, miR-671-5p, miR-878-3p, miR-801, miR-370, miR-1, miR-494, miR-133b	Tsang and Kwok (2010)
		Lung carcinoma	let-7c, miR-210	Tsang and Kwok (2010)
		Pancreas	let-7c	Tsang and Kwok (2010)
		Colon	miR-27a, miR-20A, miR-17-5P, miR-21	Gandhy et al. (2012)
		Bladder	miR-203	Saini et al. (2011)
Curcumin	Turmeric	Lung	miR-320, miR-26a, let-7i, miR-130a, mir-16, miR-125b, miR-23a, miR-27b, miR-155, miR-625, miR-576-3p, miR-186, miR-9	Zhang et al. (2010)
		Leukemia cells	miR-15a, miR-16-1	Gao et al. (2012)

		Pancreatic cancer cells	miR-103, miR-140, miR-146b, miR-148, miR-15b, miR-181a, miR-181b, miR-181d, miR-195, miR-196a, miR-199a, miR-19a, miR-204, miR-20a, miR-21, miR-22 miR-23a–b, miR-24, miR-25, miR-26a, miR-27a, miR-34a, miR-374, miR-510, miR-7, miR-92, miR-93, miR-98	Sun et al. (2008)
		Prostate cancer cells	miR-151a, miR-151-5p, miR-221, miR222	Chiyomaru et al. (2012)
Genistein	Soy	Ovarian cancer cells	miR-100, miR-122a, miR-125b, miR-126, miR-135, miR-135b, miR-136, miR-137, miR-141, miR-152, miR-190, miR-196a, miR-196b, miR-204, miR-205, miR-206, miR-217, miR-22, miR-296, miR-30a-3p, miR-30a-5p, miR-331, miR-335, miR-342, miR-362, miR-449b, miR-454, miR-497, miR-500, miR-501, miR-503, miR-515, miR-517c, miR-532, miR-565, miR-578, miR-584, miR-585, miR-590, miR-595, miR-625, miR-647, miR-7, miR-765, miR-766	Parker et al. (2009)
		Pancreatic cancer cells	miR-200	Li et al. (2009)
		Prostate cancer cells from lymph node	miR-106b, miR-1224-5p, miR-1228, miR-1231, miR-1246, miR-1260, miR-1267, miR-1268, miR-129, miR-1290, miR-1308, miR-1469, miR-149, miR-150, miR-152, miR-15a, miR-17, miR-1825, miR-185, miR-18b, miR-1908, miR-1915, miR-197, miR-1972, miR-1973, miR-1974, miR-1975, miR-1977, miR-1979, miR-20a, miR-20b, miR-24, miR-296-5p, miR-483-5p, miR-513a-5p, miR-548q, miR-572, miR-575, miR-612, miR-638, miR-654-5p, miR-659, miR-671-5p, miR-7, miR-762, miR-764, miR-874, miR-92b, miR-939	Dhar et al. (2011)
Resveratrol	Grapes, wine & eucalyptus	Prostate cancer cells	miR-101, miR-106a, miR-106b, miR-1274b, miR-136, miR-141, miR-145, miR-17, miR-182, miR-1826, miR-200b, miR-200c, miR-20a, miR-20b, miR-21, miR-214, miR-221, miR-222, miR-302c, miR-375, miR-378, miR-720, miR-768-3, miR-93	Sheth et al. (2012)

Continued

TABLE 17.3 MicroRNA Changes Affected by Dietary Agents—cont'd

Dietary agent	Source	Cancer Type	MicroRNA	Reference
		Colorectal cancer cells	miR-1, miR-100-1/2, miR-102, miR-103-1, miR-103-2, miR-146a, miR-146b-5p, miR-16-0, miR-17, miR-181a2, miR-194-2, miR-196a1, miR-205, miR-206, miR-21, miR-23a,miR-23b, miR-25, miR-26a, miR-29c, miR-30a-3p, miR-30c-1,miR-30d, miR-30e-5p, miR-323, miR-340, miR-363-5p, miR-424, miR-494, miR-497, miR-560, miR-560, miR-565, miR-565, miR-572, miR-574, miR-594, miR-615, miR-622, miR-629, miR-631, miR-638, miR-639, miR-657, miR-659, miR-663, miR-801, miR-92a-2	Tili et al. (2010)
Apigenin	Parsley & celery	Lung cancer cells	miR138, miR-125a-5p	Jiang et al. (2011)
		Neuroblastoma cells	miR-138	Chakrabarti et al. (2013)
Luteolin		Prostate cancer cells	miR-630	Chakrabarti et al. (2013)
		Gastric cancer cells	miR-34a	Wu et al. (2014)
Ellagitannins	Berries	Hepatocellular carcinoma	miR-let-7e, miR-370, let-7a, let-7c, let-7d, let-7a, let-7c, let-7d	Wen et al. (2009)
Ursolic acid	Berries & apple peel	Human glioblastoma cells	miR-21	Saw et al. (2011)
Oleanolic acid	Olive oil & garlic	Human lung cancer cells	miR-122	Zhao et al. (2015)
Diindolylmethane & indole-3-carbinol	Cruciferous vegetables	MCF-7 cells	miR-let-7b, miR-146, miR- let-7e, miR-200b, and miR-200c, miR-21	Jin et al. (2010)

reactivated the tissue inhibitor of the metalloproteinases-3 (TIMP-3) gene that was epigenetically silenced by mechanisms involving EZH2 and class I HDACs independent of promoter DNA methylation (Deb et al., 2015). After GTP or EGCG treatment, the protein levels of class I HDACs and EZH2 showed significant reduction. Interestingly, transcriptional activation of TIMP-3 was associated with decreased EZH2 localization and H3K27me3 at the promoter, with a concomitant elevation in H3K9/18 acetylation levels.

Numerous epidemiological and experimental studies have demonstrated the chemopreventive effects of genistein and other isoflavones on various cancer types (Banerjee et al., 2008). The role of genistein and other soy isoflavones as epigenetic modulators regulating gene expression has been widely reported by several studies. Genistein has been shown to be a more potent DNMT inhibitor than both biochanin A and daidzein. A study reported that genistein (2–20 μM/L) could reactivate methylation-silenced genes such as RARβ, p16INK4a, and MGMT in esophageal squamous carcinoma cells KYSE 510 and prostate cancer cells LNCaP and PC3 (Fang et al., 2005a,b). Another study demonstrated that genistein treatment in benign MCF10AT breast cancer cells as well as MCF-7 cancer cells depletes human telomerase reverse transcriptase (hTERT) activity through epigenetic modulation that involves a genistein-mediated decrease in DNMT1, DNMT3a, and DNMT3b levels (Li et al., 2009a). Furthermore, genistein was shown to repress the hTERT promoter by chromatin remodeling that involved an increase in trimethyl-H3K9 enrichment with a concomitant decrease in dimethyl-H3K4 chromatin marks. A study by King-Batoon et al. (2008) showed that a low nontoxic dose of genistein (3.125 μM, resupplemented every 48 h for 1 week) could partially demethylate GSTP1, a tumor-suppressor gene, in MCF-7, MDA-MB-468, and MCF10 A breast cancer cells. Similar in vitro studies in other cancer types provide evidence that genistein is a potent demethylating and histone-modifying agent that could reverse the silenced state of critical tumor-suppressor genes (Majid et al., 2009; Majid et al., 2010a,b; Day et al., 2002). A study by Basak et al. (2008) demonstrated that AR downregulation in prostate cancer cell line LNCaP by genistein was attributed to the inhibition of the HDAC6-Hsp90 cochaperone function required for AR protein stabilization. Genistein and other soy isoflavones are known to module miRNAs as well (J.S. Parker et al., 2009; Li et al., 2009b; Sun et al., 2009; Majid et al., 2010a). L.P. Parker et al. (2009) performed miRNA profiling of genistein-treated and genistein-untreated UL-3A and UL-3B cell lines and found 53 miRNAs that were differentially expressed. Upregulation of miR-200 and let-7 by isoflavones was shown to downregulate ZEB1, slug, and vimentin and therefore cause reversal of epithelial-to-mesenchymal transition in gemcitabine-resistant pancreatic cancer cells (Li et al., 2009b). In human uveal melanoma cells, genistein treatment was demonstrated to lead to significant growth inhibition by targeting miR-27a and its target ZBTB10 (Sun et al., 2009). However, in vivo clinical studies were inconclusive and did not fall in line with the studies performed in cell line models.

Curcumin (diferuloylmethane) is a biologically active polyphenolic compound found in turmeric, a spice derived from the rhizomes of the plant *Curcuma longa* Linn. Curcumin has been shown to modulate multiple intracellular pathways associated with proliferation, survival, invasion, apoptosis, and inflammation (Teiten et al., 2010). In the context of epigenetic pathways, several studies have reported curcumin as a potent modulator of DNMTs, histone-modifying enzymes such as HDACs and HATs, and miRNAs (Fu and Kurzrock, 2010). In silico molecular docking studies of curcumin with DNMT1 revealed that it can block or inhibit the catalytic thiol group of the C1226-binding site in the enzyme, thus resulting in decreased

DNMT activity (Fu and Kurzrock, 2010; Medina-Franco et al., 2011). This study was further validated by in vitro experimental studies that showed curcumin to be a potent DNA hypomethylating agent (Liu et al., 2009). Curcumin was reported to be an effective HDAC inhibitor. Docking studies performed for curcumin binding to HDAC8 revealed that curcumin was a more potent HDAC inhibitor than known pharmacological inhibitors such as sodium butyrate and valproic acid (Bora-Tatar et al., 2009). Another study reported that curcumin treatment of B-NHL cell line Raji cells could reduce HDAC 1, 3, and 8 protein levels in a dose-dependent manner and increase H4 acetylation levels (Liu et al., 2005). Consistent with earlier findings, studies by Chen et al. (2007) reported a significant reduction in p300/CBP, HDAC1, and HDAC3 levels after exposure of Raji cells to curcumin. Studies revealed curcumin to be a specific inhibitor of p300/CBP HAT, which has emerged as a novel target for cancer treatment (Dekker and Haisma, 2009; Marcu et al., 2006; Balasubramaniam and Eckert, 2004). Exposure to curcumin caused proteasomal degradation of p300/CBP HAT and other closely related CBP proteins with minimal effects on HATs such as GCN5 and PCAF (Marcu et al., 2006). Curcumin has also been closely linked to an ability to modulate miRNAs in cancer cells. A microarray-based study of the effect of curcumin (10 μM) on the miRNA profile of the pancreatic cancer cell line PxBC-3 showed significant changes in the expression of 29 miRNAs (11 upregulated and 18 downregulated) after 72 h treatment (Sun et al., 2008). Further studies confirmed that MiRNA-22, which has tumor-suppressive function, was upregulated after exposure to curcumin, and its downstream target genes SP1 and ESR1 were suppressed in these pancreatic cells. Ali et al. (2012) demonstrated that treatment of pancreatic cancer cells with curcumin and its analogue difluorinated-curcumin could induce gemcitabine sensitivity via induction of miR-200 and inactivation of miR-21 expression.

Epigenetic studies on resveratrol have previously focused on SIRT1 and acetyl transferase p300 (Wang et al., 2008a; Rajendran et al., 2011; Howitz et al., 2003; Wood et al., 2004). Resveratrol (3,4′,5-trihydroxystilbene), a class of polyphenolic compounds called stilbenes, was identified as a potent dietary activator of SIRT1, which lowers Km (Michaelis constant) for both acetylated substrate and NAD+. It was reported to stimulate SIRT1-dependent p53 deacetylation that ultimately contributes to increased cell survival (Howitz et al., 2003). In another study by Wood et al. (2004), resveratrol was shown to activate sirtuins from metazoans, namely *Caenorhabditis elegans* and *Drosophila melanogaster*, and delay aging without any effect on fecundity. The antitumor effect of resveratrol was reported to be mediated partly by SIRT1 (Boily et al., 2009). In addition, resveratrol was shown to have a negative effect on survivin gene expression through histone deacetylation at the gene promoter and to display a more profound inhibitory effect on BRCA-1 mutant cells both in vitro and in vivo (Wang et al., 2008b). In prostate cancer cells, resveratrol was reported to cause downregulation of metastasis-associated protein and destabilize the nucleosome-remodeling deacetylase complex, thus allowing p53 acetylation. Furthermore, activation of p53 was shown to induce proapoptotic pathways and thus cause apoptosis in prostate cancer cells (Kai et al., 2010).

Sulforaphane (SFN) is derived from the principal glucosinolate of broccoli and broccoli sprouts. At a physiological concentration, SFN downregulates DNMT1 gene expression in human colon Caco-2 cells (Traka et al., 2005). Studies by Meeran et al. (2010) demonstrated that in MCF-7 and MDA-MB-231 breast cancer cells, SFN treatment caused dose- and time-dependent inhibition of hTERT via an epigenetic mechanism involving DNA methylation and histone modification. SFN treatment was shown to cause downregulation of DNMT1 and DNMT3a, which induced site-specific demethylation at the hTERT gene first exon and facilitated the binding of CTCF associated with hTERT repression. Furthermore, ChIP analysis of the hTERT promoter revealed that active histone chromatin marks such as acetyl-H3, acetyl-H3K9, and acetyl-H4 were increased, whereas repressive chromatin marks that included trimethyl-H3K9 and trimethyl-H3K27 were reduced after SFN treatment in a dose-dependent manner. SFN-induced hyperacetylation was reported to promote the binding of repressor proteins such as MAD1 and CTCF to the hTERT regulatory region. In another study, Myzak et al. (2004) reported that the SFN metabolites SFN–cysteine and SFN-N-acetylcysteine were more potent HDAC inhibitors in vitro than both SFN and its glutathione conjugate. Furthermore, SFN treatment of HCT116 human CRC cells increased β-catenin-responsive reporter activity in a dose-dependent manner and inhibited HDAC activity. Consequently, there was an induction of acetylated histone levels bound to the p21 (Cip1/Waf1) promoter. In human prostate epithelial cells BPH-1, LNCaP, and PC3, SFN treatment was shown to inhibit HDAC activity, which was accompanied by an increase in acetylated histone levels by 50%–100% and a corresponding induction of p21 and Bax expression that led to downstream events such as cell cycle arrest and apoptosis (Myzak et al., 2006). SFN treatment was shown to inhibit HDAC activity in breast cancer cells, but no change in H3 or H4 acetylation was observed (Pledgie-Tracy et al., 2007). Studies by Myzak et al. (2006) provided the first evidence for the inhibition of in vivo HDAC activity and suppression of tumorigenesis in ApcMin mice.

The effect of apigenin on epigenetic-related enzymes and their mechanisms was not recognized until recently. Apigenin (4′,5,7-trihydroxyflavone) is a natural product belonging to the flavone class and is the aglycone of several naturally occurring glycosides found in many fruits and vegetables. Apigenin treatment has been shown to cause a marked decrease in DNMT activity in vitro (Fang et al., 2007). Studies from our laboratory demonstrated that apigenin-mediated growth arrest and apoptosis in prostate cancer cells was due to the inhibition of class I HDACs (Pandey et al., 2012). In vivo studies using PC-3 xenografts in athymic nude mice further confirmed that oral intake of apigenin (20 and 50 μg/mouse/day over an 8-week period) reduces tumor burden, HDAC activity, and HDAC1/HDAC3 protein levels. HDAC1 and HDAC3 mRNA and protein levels were found to be significantly decreased in apigenin-treated (20–40 μM) PC-3 and 22Rv1 prostate cancer cell lines, which resulted in a global decline in the acetylation levels of histones H3 and H4. A corresponding elevation in p21/Waf1 and Bax levels was observed in both in vitro and in vivo studies, which resulted in the induction of downstream events—i.e., apoptosis and cell cycle arrest. In a recent study by Paredes-Gonzalez et al. (2014), apigenin was shown to activate nuclear factor (erythroid-derived 2)-like 2 factor (Nrf2), which encodes a key transcription factor known for regulating the antioxidative defense system and skin homeostasis, in mouse skin epidermal JB6 P+ cells via epigenetic mechanisms. The hypermethylation of 15 CpG sites in the Nrf2 promoter was demonstrated to be reversed by apigenin treatment in a dose-dependent manner. Furthermore, apigenin treatment resulted in decreased expression of DNMT1, DNMT3A, DNMT3B, and HDAC 1–8 levels. However, the nuclear localization of Nrf2 was shown to be enhanced, and there was increased expression of Nrf2 as well as its target gene NAD(P)H quinone dehydrogenase 1 after apigenin treatment.

Socioeconomic and Racial Factors. Few studies have reported significant epigenetic differences in socioeconomic/racial status that account for differences in cancers and their outcomes (Mohammed et al., 2012). Certain populations are prone to specific types of cancer. For example, African Americans have a 14% higher cancer incidence and a 34% higher cancer death rate than those of Caucasian American men. Although access to quality healthcare, socioeconomic status, and genetic makeup are implicated in this disparity, the fundamental causes of such cancer disparity seem to be a complex phenomenon. Many investigators are trying to address various sociocultural determinants as a major cause of cancer disparity in order to understand and underpin mechanisms for designing better community-specific interventions for different populations. For example, African Americans have been found to have (statistically significant) lower plasma concentrations of certain antioxidants such as vitamin E, alpha-carotene, β-carotene, lutein, and zeaxanthin than those of Caucasian American men (Signorello et al., 2007). This report indicates that low antioxidant levels may affect the epigenome and gene expression, thus leading to higher susceptibility and differential cancer outcomes. More research is needed in order to completely understand how these epigenetic modifications occur and subsequently affect cancer outcomes in diverse populations.

SUMMARY AND CONCLUSIONS

From the studies described herein, it is clear that dietary and lifestyle factors hold great promise for cancer prevention through epigenetic modifications. As the importance of epigenetic modifications in cancer is now well recognized, the precise contributions of epigenetic mechanisms and the cellular targets of epigenetic alterations by various endogenous factors in human cancer need further investigation. In recent years, the development of genome-wide analytical methods, including Next-Gen sequencing, has unlocked the possibility of identifying multiple changes in gene expression at the genetic and epigenetic levels that affect the genomes of cancer cells. These studies determine which alterations, or combinations thereof, lead to neoplastic progression, and their interpretation as reliable biomarkers provides information about the carcinogenesis process. As a result, research on epigenetic changes as potential biomarkers has led to sensitive detection of cancer using epigenetic biomarkers in DNA extracted from plasma/serum and other bodily fluids. It is expected that these universally applicable markers will be made available in a clinical diagnostic setting and in population-based screening.

The intrinsic reversibility of epigenetic alterations represents an exciting opportunity for the development of novel strategies for cancer prevention. Molecular events that occur during early stages of cancer development or in precursor lesions are more likely be influenced by chemoprevention through diet and lifestyle modifications. In addition, recent advances in the field of cancer epigenetics have enhanced our understanding of epigenetic changes in normal cellular processes and the abnormal events leading to tumorigenesis. Moreover, a deeper understanding of the global patterns of epigenetic modifications by dietary compounds and lifestyle factors in cancer will lead to the design of better strategies for preventing and curing cancer. Enough preclinical and clinical data will be required about the epigenetic changes induced by dietary phytochemicals to lead to a better understanding of the epigenetic targets and pathways altered by these agents in eliciting their efficacy with cancer. Additional preclinical and clinical studies are required in order to analyze the safety profile of doses, routes of administration, and organ bioavailability alone and in combination for obtaining maximum beneficial effects. Finally, systematic well-designed and randomized placebo-controlled trials with adequate power and relevant clinical epigenetic endpoints are needed. Despite these challenges, research on diet and nutrition continues to emerge and will offer new epigenetic targets and promising agents with more opportunities for cancer prevention in the near future.

ABBREVIATIONS

AA African American
AID Activation-induced cytidine deaminase
BMI-1 B-cell-specific Moloney murine leukemia virus integration site 1
BPDE Benzo(a)pyrene diol epoxide
CA Caucasian American
DNA Deoxyribonucleic acid
DNMT DNA methyltransferase
EEAA Extrinsic epigenetic age acceleration
EED Embryonic ectoderm development
EGCG Epigallocatechin-3-gallate
EPIC European Prospective Investigation into Cancer and Nutrition
ER Estrogen receptor
EZH2 Enhancer of zeste homolog 2
GSTP1 Glutathione S-transferase pi
GTP Green tea polyphenol
HAT Histone acetyltransferase
HDAC Histone deacetylase
HDM Histone demethylases
HMGA2 High-mobility group AT-hook 2
hMLH1 Human mutL homologue 1
HMT Histone methyltransferase
HPV Human papillomavirus
IEAA Intrinsic epigenetic age acceleration
IGF Insulin-like growth factor
LINE Long interspersed nuclear element
lncRNA Long noncoding RNA
LSD1 Lysine-specific demethylase 1
MBD Methyl-binding domain proteins
MEL18 Polycomb group ring finger 2
MGMT O^6-methylguanine methyltransferase
miRNA MicroRNA
ncRNA Noncoding RNA
PAH Polycyclic aromatic hydrocarbons
PcG Polycomb-group proteins
PhIP Amino-1-methyl-6-phenylimidazo[4,5-b]pyridine
piRNA Piwi-interacting RNA
RARβ Retinoic acid receptor β
RepA Short repeat RNA
SAH S-adenosyl-L-homocysteine
SAM S-adenosyl methionine
SFN Sulforaphane
siRNA Small interfering RNA
SUZ12 Suppressor of zeste 12
TET Ten-eleven translocation
TIMP-3 Tissue inhibitor of metalloproteinases-3
TSA Trichostatin A

Acknowledgments

The original work from the author's laboratory outlined in this chapter was supported by United States Public Health Service Grants R01CA115491, R01CA108512, R21CA193080, and R03CA186179, Department of Veteran Affairs grant 1I01BX002494, and Department of Defense grant W81XWH-15-1-0558 to SG.

Conflict of interest

The authors have no competing interest.

Refernces

Ali, S., et al., 2012. Inactivation of Ink4a/Arf leads to deregulated expression of miRNAs in K-Ras transgenic mouse model of pancreatic cancer. J. Cell. 227, 3373–3380.

Baglietto, L., et al., 2017. DNA methylation changes measured in pre-diagnostic peripheral blood samples are associated with smoking and lung cancer risk. Int. J. Cancer 140, 50–61.

Balasubramanian, S., Adhikary, G., Eckert, R.L., 2010. The Bmi-1 polycomb protein antagonizes the (-)-epigallocatechin-3-gallate-dependent suppression of skin cancer cell survival. Carcinogenesis 31, 496–503.

Balasubramanian, S., Eckert, R.L., 2004. Green tea polyphenol and curcumin inversely regulate human involucrin promoter activity via opposing effects on CCAAT/enhancer-binding protein function. J. Biol. Chem. 279, 24007–24014.

Balasubramanyam, K., et al., 2003. Small molecule modulators of histone acetyltransferase p300. J. Biol. Chem. 278, 19134–19140.

Banerjee, S., et al., 2008. Multi-targeted therapy of cancer by genistein. Cancer Lett. 269, 226–242.

Bannister, A.J., Kouzarides, T., 2011. Regulation of chromatin by histone modifications. Cell Res. 21, 381–395.

Bartel, D.P., 2004. MicroRNAs: genomics, biogenesis, mechanism, and function. Cell 116, 281–297.

Basak, S., et al., 2008. Genistein down-regulates androgen receptor by modulating HDAC6-Hsp90 chaperone function. Mol. Cancer Ther. 7, 3195–3202.

Belinsky, S.A., et al., 2004. Plutonium targets the p16 gene for inactivation by promoter hypermethylation in human lung adenocarcinoma. Carcinogenesis 25, 1063–1067.

Bell, D., et al., 1999. In breast carcinoma tissue, immature dendritic cells reside within the tumor, whereas mature dendritic cells are located in peritumoral areas. J. Exp. Med. 190, 1417–1426.

Benayoun, B.A., Pollina, E.A., Brunet, A., 2015. Epigenetic regulation of ageing: linking environmental inputs to genomic stability. Nat. Rev. Mol. Cell Biol. 16, 593–610.

Bernstein, B.E., et al., 2002. Methylation of histone H3 Lys 4 in coding regions of active genes. Proc. Natl. Acad. Sci. USA 99, 8695–8700.

Bestor, T.H., 2000. The DNA methyltransferases of mammals. Hum. Mol. Genet. 9, 2395–2402.

Bird, A., 2002. DNA methylation patterns and epigenetic memory. Genes Dev. 16, 6–21.

Boily, G., et al., 2009. SirT1-null mice develop tumors at normal rates but are poorly protected by resveratrol. Oncogene 28, 2882–2893.

Bollati, V., et al., 2007. Changes in DNA methylation patterns in subjects exposed to low-dose benzene. Cancer Res. 67, 876–880.

Bollati, V., et al., 2010. Exposure to metal-rich particulate matter modifies the expression of candidate microRNAs in peripheral blood leukocytes. Environ. Health Perspect. 118, 763–768.

Bora-Tatar, G., et al., 2009. Molecular modifications on carboxylic acid derivatives as potent histone deacetylase inhibitors: activity and docking studies. Bioorg. Med. Chem. 17, 5219–5228.

Bourc'his, D., et al., 2001. Dnmt3L and the establishment of maternal genomic imprints. Science 294, 2536–2539.

Broday, L., Cai, J., Costa, M., 1999. Nickel enhances telomeric silencing in *Saccharomyces cerevisiae*. Mutat. Res. 440, 121–130.

Calin, G.A., Croce, C.M., 2006 Nov. MicroRNA signatures in human cancers. Nat. Rev. Cancer 6 (11), 857–866. Review. PubMed PMID: 17060945.

Cameron, E.E., Bachman, K.E., Myohanen, S., 1999. Synergy of demethylation and histone deacetylase inhibition in the re-expression of genes silenced in cancer. Nat. Genet. 21, 103–107.

Chahine, T., et al., 2007. Particulate air pollution, oxidative stress genes, and heart rate variability in an elderly cohort. Environ. Health Perspect. 115, 1617–1622.

Chakrabarti, M., Banik, N.L., Ray, S.K., 2013. miR-138 overexpression is more powerful than hTERT knockdown to potentiate apigenin for apoptosis in neuroblastoma in vitro and in vivo. Exp. Cell Res. 319, 1575–1585.

Chakrabarti, M., et al., 2012. Alterations in expression of specific microRNAs by combination of 4-HPR and EGCG inhibited growth of human malignant neuroblastoma cells. Brain Res. 1454, 1–13.

Chandregowda, V., Kush, A., Reddy, G.C., 2009. Synthesis of benzamide derivatives of anacardic acid and their cytotoxic activity. Eur. J. Med. Chem. 44, 2711–2719.

Chen, H., et al., 2001. Association of c-myc overexpression and hyperproliferation with arsenite-induced malignant transformation. Toxicol. Appl. Pharmacol. 175, 260–268.

Chen, J.X., et al., 1998. Carcinogens preferentially bind at methylated CpG in the p53 mutational hot spots. Cancer Res. 58, 2070–2075.

Chen, Y., et al., 2007. Curcumin, both histone deacetylase and p300/CBP-specific inhibitor, represses the activity of nuclear factor kappa B and Notch 1 in Raji cells. Basic Clin. Pharmacol. Toxicol. 101, 427–433.

Chen, Y.S., et al., 2017. A miRNA signature for an environmental heterocyclic amine defined by a multi-organ carcinogenicity bioassay in the rat. Arch. Toxicol. https://doi.org/10.1007/s00204-017-1945-6 [Epub ahead of print] PMID: 28289824.

Chiu, J., et al., 2009. Curcumin prevents diabetes-associated abnormalities in the kidneys by inhibiting p300 and nuclear factor-kappaB. Nutrition 25, 964–972.

Chiyomaru, T., et al., 2012. Genistein suppresses prostate cancer growth through inhibition of oncogenic microRNA-151. PLoS One 7, e43812.

Choi, K.C., et al., 2009. Epigallocatechin-3-gallate, a histone acetyltransferase inhibitor, inhibits EBV-induced B lymphocyte transformation via suppression of RelA acetylation. Cancer Res. 69, 583–592.

Choudhury, S.R., et al., 2011. (-)-Epigallocatechin-3-gallate and DZNep reduce polycomb protein level via a proteasome-dependent mechanism in skin cancer cells. Carcinogenesis 32, 1525–1532.

Costa, F.F., 2008. Non-coding RNAs, epigenetics and complexity. Gene 410, 9–17.

Cui, L., et al., 2007. PfGCN5-mediated histone H3 acetylation plays a key role in gene expression in Plasmodium falciparum. Eukaryot. Cell 6, 1219–1227.

Day, J.K., et al., 2002. Genistein alters methylation patterns in mice. J. Nutr. 132, 2419S–2423S.

De Capoa, A., et al., 2003. DNA demethylation is directly related to tumour progression: evidence in normal, premalignant and malignant cells from uterine cervix samples. Oncol. Rep. 10, 545–549.

de Leeuw, W.J., et al., 2000. Prediction of a mismatch repair gene defect by microsatellite instability and immunohistochemical analysis in endometrial tumours from HNPCC patients. J. Pathol. 192, 328–335.

De Lucia, F., Dean, C., 2011. Long non-coding RNAs and chromatin regulation. Curr. Opin. Plant Biol. 14, 168–173.

Deb, G., Singh, A.K., Gupta, S., 2014. EZH2: not EZHY (easy) to deal. Mol. Cancer Res. 12, 639–653.

Deb, G., Thakur, V.S., Gupta, S., 2013. Multifaceted role of EZH2 in breast and prostate tumorigenesis: epigenetics and beyond. Epigenetics 8, 464–476.

Deb, G., et al., 2015. Epigenetic induction of tissue inhibitor of matrix metalloproteinase-3 by green tea polyphenols in breast cancer cells. Mol. Carcinog. 54, 485–499.

Dehan, P., et al., 2009. DNA methylation and cancer diagnosis: new methods and applications. Expert Rev. Mol. Diagn. 9, 651–657.

Dekker, F.J., Haisma, H.J., 2009. Histone acetyl transferases as emerging drug targets. Drug Discov. Today 14, 942–948.

Dhar, S., Hicks, C., Levenson, A.S., 2011. Resveratrol and prostate cancer: promising role for microRNAs. Mol. Nutr. Food Res. 55, 1219–1229.

Drew, D.A., et al., 2017. A prospective study of smoking and risk of synchronous colorectal cancers. Am. J. Gastroenterol. 112, 493–501.

Druesne, N., et al., 2004. Diallyl disulfide (DADS) increases histone acetylation and p21(waf1/cip1) expression in human colon tumor cell lines. Carcinogenesis 25, 1227–1236.

Eliseeva, E.D., et al., 2007. Characterization of novel inhibitors of histone acetyltransferases. Mol. Cancer Ther. 6, 2391–2398.

Fabbri, M., et al., 2007. Regulatory mechanisms of microRNAs involvement in cancer. Expert Opin. Biol. Ther. 7, 1009–1019.

Fang, M., Chen, D., Yang, C.S., 2007. Dietary polyphenols may affect DNA methylation. J. Nutr. 137, 223S–228S.

Fang, M.Z., et al., 2005a. Reversal of hypermethylation and reactivation of p16INK4a, RARbeta, and MGMT genes by genistein and other isoflavones from soy. Clin. Cancer Res. 11, 7033–7041.

Fang, M.Z., et al., 2005b. Promoter hypermethylation and inactivation of O (6)-methylguanine-DNA methyltransferase in esophageal squamous cell carcinomas and its reactivation in cell lines. Int. J. Oncol. 26, 615–622.

Fang, M.Z., et al., 2003. Tea polyphenol(-)-epigallocatechin-3-gallate inhibits DNA methyltransferase and reactivates methylation-silenced genes in cancer cell lines. Cancer Res. 63, 7563–7570.

Feng, Q., Zhang, Y., 2001. The MeCP1 complex represses transcription through preferential binding, remodeling, and deacetylating methylated nucleosomes. Genes Dev. 15, 827–832.

Fraga, M.F., et al., 2005. Epigenetic differences arise during the lifetime of monozygotic twins. Proc. Natl. Acad. Sci. USA 102, 10604–10609.

Fu, S., Kurzrock, R., 2010. Development of curcumin as an epigenetic agent. Cancer 116, 4670–4676.

Gallou-Kabani, C., Junien, C., 2005. Nutritional epigenomics of metabolic syndrome: new perspective against the epidemic. Diabetes 54, 1899–1906.

Gandhy, S.U., et al., 2012. Curcumin and synthetic analogs induce reactive oxygen species and decreases specificity protein (Sp) transcription factors by targeting microRNAs. BMC Cancer 12, 564.

Gao, S.M., et al., 2012. Pure curcumin decreases the expression of WT1 by upregulation of miR-15a and miR-16-1 in leukemic cells. J. Exp. Clin. Cancer Res. 31, 2.

Gao, Z., et al., 2009. Promoter demethylation of WIF-1 by epigallocatechin-3-gallate in lung cancer cells. Anticancer Res. 29, 2025–2030.

Garzon, R., Calin, G.A., Croce, C.M., 2009. MicroRNAs in cancer. Annu. Rev. Med. 60, 167–179.

Gerhauser, C., 2013a. Cancer chemoprevention and nutriepigenetics: state of the art and future challenges. Top. Curr. Chem. 329, 73–132.

Giovannucci, E., et al., 1995. Alcohol, low-methionine–low-folate diets, and risk of colon cancer in men. J. Natl. Cancer Inst. 87, 265–273.

Godlewski, J., et al., 2008. Targeting of the Bmi-1 oncogene/stem cell renewal factor by microRNA-128 inhibits glioma proliferation and self-renewal. Cancer Res. 68, 9125–9130.

Hannum, G., et al., 2013. Genome-wide methylation profiles reveal quantitative views of human aging rates. Mol. Cell. 49, 359–367.

Hark, A.T., et al., 2000. CTCF mediates methylation-sensitive enhancer-blocking activity at the H19/Igf2 locus. Nature 405, 486–489.

Hendrich, B., Bird, A., 1998. Identification and characterization of a family of mammalian methyl-CpG binding proteins. Mol. Cell Biol. 18, 6538–6547.

Herceg, Z., 2007. Epigenetics and cancer: towards an evaluation of the impact of environmental and dietary factors. Mutagenesis 22, 91–103.

Haigis, M.C., Guarente, L.P., 2006. Mammalian sirtuins—emerging roles in physiology, aging, and calorie restriction. Genes Dev. 20, 2913–2921.

Hong, T., et al., 2004. Isoflavones stimulate estrogen receptor-mediated core histone acetylation. Biochem. Biophys. Res. Commun. 317, 259–264.

Horvath, S., et al., 2015. Accelerated epigenetic aging in Down syndrome. Aging Cell 14, 491–495.

Howitz, K.T., et al., 2003. Small molecule activators of sirtuins extend *Saccharomyces cerevisiae* lifespan. Nature 425, 191–196.

Hsu, R.J., Yang, H.J., Tsai, H.J., 2009. Labeled microRNA pull-down assay system: an experimental approach for high-throughput identification of microRNA-target mRNAs. Nucl. Acids Res. 37, 77.

Hwang, Y.J., Song, J., Kim, H.R., Hwang, K.A., 2014. Oleanolic acid regulates NF-kappaB signaling by suppressing MafK expression in RAW 264.7 cells. BMB Rep. 47, 524–529.

Issa, J.P., Kantarjian, H.M., 2009. Targeting DNA methylation. Clin. Cancer Res. 15, 3938–3946.

Jaenisch, R., Bird, A., 2003. Epigenetic regulation of gene expression: how the genome integrates intrinsic and environmental signals. Nat. Genet. 33, 245–254.

Jiang, L., et al., 2011. MicroRNA HSA-miR-125a-5p induces apoptosis by activating p53 in lung cancer cells. Exp. Lung Res. 37, 387–398.

Jicai, Z., et al., 2006. Persistent infection of hepatitis B virus is involved in high rate of p16 methylation in hepatocellular carcinoma. Mol. Carcinog. 45, 530–536.

Jin, Y., Zou, X., Feng, X., 2010. Diindolylmethane negatively regulates Cdc25A and induces a G2/M arrest by modulation of microRNA 21 in human breast cancer cells. Anti Canc. Drugs 21, 814–822.

Jones, P.L., et al., 1998. Methylated DNA and MeCP2 recruit histone deacetylase to repress transcription. Nat. Genet. 19, 187–191.

Juan, A.H., et al., 2009. Mir-214-dependent regulation of the polycomb protein Ezh2 in skeletal muscle and embryonic stem cells. Mol. Cell 36, 61–74.

Kaati, G., Bygren, L.O., Edvinsson, S., 2002. Cardiovascular and diabetes mortality determined by nutrition during parents' and grandparents' slow growth period. Eur. J. Hum. Genet. 10, 682–688.

Kai, L., Samuel, S.K., Levenson, A.S., 2010. Resveratrol enhances p53 acetylation an apoptosis in prostate cancer by inhibiting MTA1/NuRD complex. Int. J. Cancer 126, 1538–1548.

Kalantari, M., et al., 2004. Conserved methylation patterns of human papillomavirus type 16 DNA in asymptomatic infection and cervical neoplasia. J. Virol. 78, 12762–12772.

Kang, J., et al., 2005. Curcumin-induced histone hypoacetylation: the role of reactive oxygen species. Biochem. Pharmacol. 69, 1205–1213.

Kanhere, A., et al., 2010. Short RNAs are transcribed from repressed polycomb target genes and interact with polycomb repressive complex-2. Mol. Cell 38, 675–688.

Kanherkar, R.R., Bhatia-Dey, N., Csoka, A.B., 2014. Epigenetics across the human lifespan. Front. Cell Dev. Biol. 2, 49.

Kanwal, R., Gupta, S., 2010. Epigenetics and cancer. J. Appl. Physiol. 109, 598–605.

Karaczyn, A.A., Golebiowski, F., Kasprzak, K.S., 2005. Truncation, deamidation, and oxidation of histone H2B in cells cultured with nickel(II). Chem. Res. Toxicol. 18, 1934–1942.

Kato, K., et al., 2008. Effects of green tea polyphenol on methylation status of RECK gene and cancer cell invasion in oral squamous cell carcinoma cells. Br. J. Cancer 99, 647–654.

Khin, S.S., et al., 2011. Epigenetic alteration by DNA promoter hypermethylation of genes related to transforming growth factor-β (TGF-β) signaling in cancer. Cancers 3, 982–993.

Kikuno, N., et al., 2008. Genistein mediated histone acetylation and demethylation activates tumor suppressor genes in prostate cancer cells. Int. J. Cancer 123, 552–560.

Kim, H., et al., 2016. Epigenetic modifications of triterpenoid ursolic acid in activating Nrf2 and blocking cellular transformation of mouse epidermal cells. J. Nutr. Biochem. 33, 54–62.

King-Batoon, A., Leszczynska, J.M., Klein, C.B., 2008. Modulation of gene methylation by genistein or lycopene in breast cancer cells. Environ. Mol. Mutagen. 49, 36–45.

Kornberg, R.D., Lorch, Y., 1999. Twenty-five years of the nucleosome, fundamental particle of the eukaryote chromosome. Cell 98, 285–294.

Kotake, Y., et al., 2011. Long non-coding RNA ANRIL is required for the PRC2 recruitment to and silencing of p15(INK4B) tumor suppressor gene. Oncogene 30, 1956–1962.

Kouzarides, T., 2007. SnapShot: histone-modifying enzymes. Cell 128, 802.

Kuck, D., Caulfield, T., Lyko, F., Medina-Franco, J.L., 2010 Nov. Nanaomycin A selectively inhibits DNMT3B and reactivates silenced tumor suppressor genes in human cancer cells. Mol. Cancer Ther. 9 (11), 3015–3023. https://doi.org/10.1158/1535-7163.MCT-10-0609. Epub 2010 Sep 10. PubMed PMID: 20833755.

Kuo, M.H., Allis, C.D., 1998. Roles of histone acetyltransferases and deacetylases in gene regulation. Bioessays 20, 615–626.

Lea, M.A., Randolph, V.M., 2001. Induction of histone acetylation in rat liver and hepatoma by organosulfur compounds including diallyl disulfide. Anticancer Res. 21, 2841–2845.

Lea, M.A., et al., 2001. Induction of histone acetylation in mouse erythroleukemia cells by some organosulfur compounds including allyl isothiocyanate. Int. J. Cancer 92, 784–789.

Lea, M.A., Randolph, V.M., Patel, M., 1999. Increased acetylation of histones induced by diallyl disulfide and structurally related molecules. Int. J. Oncol. 15, 347–352.

Lee, W.J., Shim, J.Y., Zhu, B.T., 2005. Mechanisms for the inhibition of DNA methyltransferases by tea catechins and bioflavonoids. Mol. Pharmacol. 68, 1018–1030.

Lehmann, U., et al., 2007. Epigenetic inactivation of microRNA genes in mammary carcinoma. Verh. Dtsch. Ges. Pathol. 91, 214–220.

Li, H., Minarovits, J., 2003. Host cell-dependent expression of latent Epstein-Barr virus genomes: regulation by DNA methylation. Adv. Cancer Res. 89, 133–156.

Li, H.L., et al., 2008. Curcumin prevents and reverses murine cardiac hypertrophy. J. Clin. Invest. 118, 879–893.

Li, Y., et al., 2009a. Genistein depletes telomerase activity through cross-talk between genetic and epigenetic mechanisms. Int. J. Cancer 125, 286–296.

Li, Y., et al., 2009b. Up-regulation of miR-200 and let-7 by natural agents leads to the reversal of epithelial-to-mesenchymal transition in gemcitabine-resistant pancreatic cancer cells. Cancer Res. 69, 6704–6712.

Li, Y., et al., 2010. Synergistic epigenetic reactivation of estrogen receptor-α (ERα) by combined green tea polyphenol and histone deacetylase inhibitor in ERα-negative breast cancer cells. Mol. Cancer 9, 274.

Liu, H.L., et al., 2005. Curcumin, a potent anti-tumor reagent, is a novel histone deacetylase inhibitor regulating B-NHL cell line Raji proliferation. Acta Pharmacol. Sin. 26, 603–609.

Liu, R., et al., 2012. Serum microRNA expression profile as a biomarker in the diagnosis and prognosis of pancreatic cancer. Clin. Chem. 58, 610–618.

Liu, Z., et al., 2009. Curcumin is a potent DNA hypomethylation agent. Bioorg. Med. Chem. Lett 19, 706–709.

Liyanage, V.R., et al., 2014. DNA modifications: function and applications in normal and disease States. Biology 3, 670–723.

Luger, K., Richmond, T.J., 1998. DNA binding within the nucleosome core. Curr. Opin. Struct. Biol. 8, 33–40.

Lujambio, A., et al., 2008. A microRNA DNA methylation signature for human cancer metastasis. Proc. Natl. Acad. Sci. USA 105, 13556–13561.

Maddocks, O.D., et al., 2016. Serine metabolism supports the methionine cycle and DNA/RNA methylation through de novo ATP synthesis in cancer cells. Mol. Cell. 61, 210–221.

Maekita, T., et al., 2006. High levels of aberrant DNA methylation in Helicobacter pylori-infected gastric mucosae and its possible association with gastric cancer risk. Clin. Cancer Res. 12, 989–995.

Magistri, M., et al., 2012. Regulation of chromatin structure by long noncoding RNAs: focus on natural antisense transcripts. Trends Genet. 28, 389–396.

Mai, A., et al., 2006. Small-molecule inhibitors of histone acetyltransferase activity: identification and biological properties. J. Med. Chem. 49, 6897–6907.

Majid, S., et al., 2009. BTG3 tumor suppressor gene promoter demethylation, histone modification and cell cycle arrest by genistein in renal cancer. Carcinogenesis 30, 662–670.

Majid, S., et al., 2010a. Regulation of minichromosome maintenance gene family by microRNA-1296 and genistein in prostate cancer. Cancer Res. 70, 2809–2818.

Majid, S., et al., 2010b. Genistein reverses hypermethylation and induces active histone modifications in tumor suppressor gene B-Cell translocation gene 3 in prostate cancer. Cancer 116, 66–76.

Majid, S., et al., 2008. Genistein induces the p21WAF1/CIP1 and p16INK4a tumor suppressor genes in prostate cancer cells by epigenetic mechanisms involving active chromatin modification. Cancer Res. 68, 2736–2744.

Mak, W., et al., 2002. Mitotically stable association of polycomb group proteins eed and enx1 with the inactive x chromosome in trophoblast stem cells. Curr. Biol. 12, 1016–1020.

Marcu, M.G., et al., 2006. Curcumin is an inhibitor of p300 histone acetylatransferase. Med. Chem. 2, 169–174.

Marsit, C.J., et al., 2006. Epigenetic inactivation of the SFRP genes is associated with drinking, smoking and HPV in head and neck squamous cell carcinoma. Int. J. Cancer 119, 1761–1766.

Marutha Ravindran, C.R., Ticku, M.K., 2009. Effect of 5-azacytidine on the methylation aspects of NMDA receptor NR2B gene in the cultured cortical neurons of mice. Neurochem Res 34 (2), 342–350. https://doi.org/10.1007/s11064-008-9783-9. PMID: 18618247.

Mattick, J.S., 2005. The functional genomics of noncoding RNA. Science 309, 1527–1528.

Mazzio, E.A., Soliman, K.F., 2012. Basic concepts of epigenetics: impact of environmental signals on gene expression. Epigenetics 7, 119–130.

Medina-Franco, J.L., López-Vallejo, F., Kuck, D., Lyko, F., 2011 May. Natural products as DNA methyltransferase inhibitors: a computer-aided discovery approach. Mol. Divers. 15 (2), 293–304. https://doi.org/10.1007/s11030-010-9262-5. Epub 2010 Aug 10. PubMed PMID: 20697809.

Meeran, S.M., Patel, S.N., Tollefsbol, T.O., 2010. Sulforaphane causes epigenetic repression of hTERT expression in human breast cancer cell lines. PLoS One 5, e11457.

Mentch, S.J., Locasale, J.W., 2016. One-carbon metabolism and epigenetics: understanding the specificity. Ann. N. Y. Acad. Sci. 1363, 91–98.

Mentch, S.J., et al., 2015. Histone methylation dynamics and gene regulation occur through the sensing of one-carbon metabolism. Cell Metab. 22, 861–873.

Mittal, A., et al., 2003. Exceptionally high protection of photocarcinogenesis by topical application of (−)-epigallocatechin-3-gallate in hydrophilic cream in SKH-1 hairless mouse model: relationship to inhibition of UVB-induced global DNA hypomethylation. Neoplasia 5, 555–565.

Mohammad, F., 2010. Kcnq1ot1 noncoding RNA mediates transcriptional gene silencing by interacting with Dnmt1. Development 137, 2493–2499.

Mohammed, S.I., Springfield, S., Das, R., 2012. Role of epigenetics in cancer health disparities. Methods Mol. Biol. 863, 395–410.

Morimoto, T., et al., 2008. The dietary compound curcumin inhibits p300 histone acetyltransferase activity and prevents heart failure in rats. J. Clin. Invest. 118, 868–878.

Mottet, D., Castronovo, V., 2008. Histone deacetylases: target enzymes for cancer therapy. Clin. Exp. Metastasis 25, 183–189.

Murugan, R.S., et al., 2009. Black tea polyphenols target matrix metalloproteinases, RECK, proangiogenic molecules and histone deacetylase in a rat hepatocarcinogenesis model. Anticancer Res. 29, 2301–2305.

Myzak, M.C., Dashwood, R.H., 2006. Chemoprotection by sulforaphane: keep one eye beyond Keap1. Cancer Lett. 233, 208–218.

Myzak, M.C., et al., 2006. Sulforaphane inhibits histone deacetylase in vivo and suppresses tumorigenesis in Apc-minus mice. FASEB J. 20, 506–508.

Myzak, M.C., et al., 2004. A novel mechanism of chemoprotection by sulforaphane: inhibition of histone deacetylase. Cancer Res. 64, 5767–5774.

Nair, S., et al., 2008. Synergistic effects of a combination of dietary factors sulforaphane and (-) epigallocatechin-3-gallate in HT-29 AP-1 human colon carcinoma cells. Pharm. Res. 25, 387–399.

Nan, X., Meehan, R.R., Bird, A., 1993. Dissection of the methyl-CpG binding domain from the chromosomal protein MeCP2. Nucl. Acids Res. 21, 4886–4892.

Nan, X., Ng, H.H., Johnson, C.A., 1998. Transcriptional repression by the methyl-CpG-binding protein MeCP2 involves a histone deacetylase complex. Nature 393, 386–389.

Nandakumar, V., Vaid, M., Katiyar, S.K., 2011. (-)-Epigallocatechin-3-gallate reactivates silenced tumor suppressor genes, Cip1/p21 and p16INK4a, by reducing DNA methylation and increasing histones acetylation in human skin cancer cells. Carcinogenesis 32, 537–544.

Nelson, W.G., Demarzo, A.M., Yegnasubramanian, S., 2014. The diet as a cause of human prostate cancer. Cancer Treat. Res. 159, 51–68.

Nian, H., Delage, B., Pinto, J.T., 2008. Allyl mercaptan, a garlic-derived organosulfur compound, inhibits histone deacetylase and enhances Sp3 binding on the P21WAF1 promoter. Carcinogenesis 29, 1816–1824.

Nitert, M.D., et al., 2012. Impact of an exercise intervention on DNA methylation in skeletal muscle from first-degree relatives of patients with type 2 diabetes. Diabetes 61, 3322–3332.

Ntanasis-Stathopoulos, J., et al., 2013. Epigenetic regulation on gene expression induced by physical exercise. J. Musculoskelet. Neuronal Interact. 13, 133–146.

Ohlsson, R., Renkawitz, R., Lobanenkov, V., 2001. CTCF is a uniquely versatile transcription regulator linked to epigenetics and disease. Trends Genet. 17, 520–527.

Okano, M., et al., 1999. DNA methyltransferases Dnmt3a and Dnmt3b are essential for de novo methylation and mammalian development. Cell 99, 247–257.

Osada, H., Takahashi, T., 2002. Genetic alterations of multiple tumor suppressors and oncogenes in the carcinogenesis and progression of lung cancer. Oncogene 21, 7421–7434.

Paluszczak, J., Krajka-Kuźniak, V., Baer-Dubowska, W., 2010. The effect of dietary polyphenols on the epigenetic regulation of gene expression in MCF7 breast cancer cells. Toxicol. Lett. 192, 119–125.

Pandey, M., et al., 2012. Plant flavone apigenin inhibits HDAC and remodels chromatin to induce growth arrest and apoptosis in human prostate cancer cells: in vitro and in vivo study. Mol. Carcinog. 51, 952–962.

Pandey, M., Shukla, S., Gupta, S., 2010. Promoter demethylation and chromatin remodeling by green tea polyphenols leads to re-expression of GSTP1 in human prostate cancer cells. Int. J. Cancer 126, 2520–2533.

Paredes-Gonzalez, X., et al., 2014. Apigenin reactivates Nrf2 anti-oxidative stress signaling in mouse skin epidermal JB6 P + cells through epigenetics modifications. AAPS J. 16, 727–735.

Parker, J.S., et al., 2009. Enhancement of the seed-target recognition step in RNA silencing by a PIWI/MID domain protein. Mol. Cell 33, 204–214.

Parker, L.P., et al., 2009. Modulation of microRNA associated with ovarian cancer cells by genistein. Eur. J. Gynaecol. Oncol. 30, 616–621.

Pasmant, E., et al., 2007. Characterization of a germ-line deletion, including the entire INK4/ARF locus, in a melanoma-neural system tumor family: identification of ANRIL, an antisense noncoding RNA whose expression coclusters with ARF. Cancer Res. 67, 3963–3969.

Pasquinelli, A.E., Hunter, S., Bracht, J., 2005. MicroRNAs: a developing story. Curr. Opin. Genet. Dev. 15, 200–205.

Pembrey, M.E., et al., 2006. Sex-specific, male-line transgenerational responses in humans. Eur. J. Hum. Genet. 14, 159–166.

Peschansky, V.J., Wahlestedt, C., 2014. Non-coding RNAs as direct and indirect modulators of epigenetic regulation. Epigenetics 9, 3–12.

Plath, K., et al., 2003. Role of histone H3 lysine 27 methylation in X inactivation. Science 300, 131–135.

Pledgie-Tracy, A., Sobolewski, M.D., Davidson, N.E., 2007. Sulforaphane induces celltype-specific apoptosis in human breast cancer cell lines. Mol. Cancer Ther. 6, 1013–1021.

Ponomarev, I., 2013. Epigenetic control of gene expression in the alcoholic brain. Alcohol Res. 35, 69–76.

Quach, A., et al., 2017. Epigenetic clock analysis of diet, exercise, education, and lifestyle factors. Aging (Albany NY) 9, 419–446.

Radom-Aizik, S., et al., 2010. Evidence for microRNA involvement in exercise-associated neutrophil gene expression changes. J. Appl. Physiol. 109, 252–261.

Rajendran, P., et al., 2011. Dietary phytochemicals, HDAC inhibition, and DNA damage/repair defects in cancer cells. Clin. Epigenetics 3, 4.

Rakyan, V.K., Chong, S., Champ, M.E., Cuthbert, P.C., Morgan, H.D., Luu, K.V., Whitelaw, E., 2003. Transgenerational inheritance of epigenetic states at the murine Axin(Fu) allele occurs after maternal and paternal transmission. Proc. Natl. Acad. Sci. USA 100 (5), 2538–2543. Epub 2003 Feb 24. PubMed PMID: 12601169; PubMed Central PMCID: PMC151376.

Rice, J.C., Allis, C.D., 2001. Histone methylation versus histone acetylation: new insights into epigenetic regulation. Curr. Opin. Cell Biol. 13, 263–273.

Richardson, B., 2003. Impact of aging on DNA methylation. Ageing Res. Rev. 2, 245–261.

Rodríguez-Gallego, E., et al., 2015. Mapping of the circulating metabolome reveals α-ketoglutarate as a predictor of morbid obesity-associated non-alcoholic fatty liver disease. Int. J. Obes. 39, 279–287.

Rönn, T., Volkov, P., Davegårdh, C., Dayeh, T., Hall, E., Olsson, A.H., Nilsson, E., Tornberg, A., Dekker Nitert, M., Eriksson, K.F., Jones, H.A., Groop, L., Ling, C., 2013 Jun. A six months exercise intervention influences the genome-wide DNA methylation pattern in human adipose tissue. PLoS Genet. 9 (6), e1003572. https://doi.org/10.1371/journal.pgen.1003572. Epub 2013 Jun 27. PMCID: PMC3694844; PMID: 23825961.

Ronti, T., Lupattelli, G., Mannarino, E., 2006. The endocrine function of adipose tissue: an update. Clin. Endocrinol. 64, 355–365.

Ross, S.A., Milner, J.A., 2007. Epigenetic modulation and cancer: effect of metabolic syndrome? Am. J. Clin. Nutr. 86, s872–s877.

Saini, S., et al., 2011. Curcumin modulates microRNA-203-mediated regulation of the Src-Akt axis in bladder cancer. Cancer Prev. Res. 4, 1698–1709.

Sakatani, T., et al., 2005. Loss of imprinting of Igf2 alters intestinal maturation and tumorigenesis in mice. Science 307, 1976–1978.

Sanchis-Gomar, F., et al., 2012. Physical exercise as an epigenetic modulator: eustress, the "positive stress" as an effector of gene expression. J. Strength Cond. Res. 26, 3469–3472.

Sander, S., et al., 2008. MYC stimulates EZH2 expression by repression of its negative regulator miR-26a. Blood 112, 4202–4212.

Santi, D.V., Garrett, C.E., Barr, P.J., 1983. On the mechanism of inhibition of DNA-cytosine methyltransferases by cytosine analogs. Cell 33, 9–10.

Sapienza, C., Issa, J.P., 2016. Diet, nutrition, and cancer epigenetics. Annu. Rev. Nutr. 36, 665–681.

Sauve, A.A., et al., 2006. The biochemistry of sirtuins. Annu. Rev. Biochem. 75, 435–465.

Saw, C.L., et al., 2011. Impact of Nrf2 on UVB-induced skin inflammation/photoprotection and photoprotective effect of sulforaphane Mol. Carcinog 50,479–486.

Sbardella, G., et al., 2008. Identification of long chain alkylidenemalonates as novel small molecule modulators of histone acetyltransferases. Bioorg. Med. Chem. Lett 18, 2788–2792.

Schwab, M., et al., 2008. The dietary histone deacetylase inhibitor sulforaphane induces human beta-defensin-2 in intestinal epithelial cells. Immunology 125, 241–251.

Scourzic, L., Mouly, E., Bernard, O.A., 2015. TET proteins and the control of cytosine demethylation in cancer. Genome Med. 7, 9.

Sheth, S., et al., 2012. Resveratrol reduces prostate cancer growth and metastasis by inhibiting the Akt/MicroRNA-21 pathway. PLoS One 7, e51655.

Shi, X., et al., 2007. Modulation of p53 function by SET8-mediated methylation at lysine 382. Mol. Cell 27, 636–646.

Signorello, L.B., et al., 2007. Comparing diabetes prevalence between African Americans and Whites of similar socioeconomic status. Am. J. Public Health 97, 2260–2267.

Singh, N., Misra, K., 2009. Computational screening of molecular targets in Plasmodium for novel non-resistant antimalarial drugs. Bioinformation 3, 255–262.

Silva, J., et al., 2003. Establishment of histone h3 methylation on the inactive X chromosome requires transient recruitment of Eed-Enx1 polycomb group complexes. Dev. Cell 4, 481–495.

Sng, J.C., Taniura, H., Yoneda, Y., 2006. Histone modifications in kainate-induced status epilepticus. Eur. J. Neurosci. 23, 1269–1282.

Steenaard, R.V., et al., 2015. Tobacco smoking is associated with methylation of genes related to coronary artery disease. Clin. Epigenetics 7, 54.

Stefanska, B., et al., 2010. Hypomethylation and induction of retinoic acid receptor beta 2 by concurrent action of adenosine analogues and natural compounds in breast cancer cells. Eur. J. Pharmacol. 638, 47–53.

Strahl, B.D., Allis, C.D., 2000. The language of covalent histone modifications. Nature 403, 41–45.

Struhl, K., 1998. Histone acetylation and transcriptional regulatory mechanisms. Genes Dev. 12, 599–606.

Sugiyama, A., et al., 1998. *Helicobacter pylori* infection enhances N-methyl-N-nitrosourea-induced stomach carcinogenesis in the Mongolian gerbil. Cancer Res. 58, 2067–2069.

Sun, M., et al., 2008. Curcumin (diferuloylmethane) alters the expression profiles of microRNAs in human pancreatic cancer cells. Mol. Cancer Ther. 7, 464–473.

Sun, Q., et al., 2009. Genistein inhibits growth of human uveal melanoma cells and affects microRNA-27a and target gene expression. Oncol. Rep. 22, 563–567.

Sun, Y., et al., 2006. Inhibition of histone acetyltransferase activity by anacardic acid sensitizes tumor cells to ionizing radiation. FEBS Lett. 580, 4353–4356.

Sutherland, J.E., Costa, M., 2003. Epigenetics and the environment. Ann. N. Y. Acad. Sci. 983, 151–160.

Suzuki, T., Miyata, N., 2006. Rational design of non-hydroxamate histone deacetylase inhibitors. Mini Rev. Med. Chem. 6, 515–526.

Szaleczky, E., Prónai, L., Molnár, B., Berczi, L., Fehér, J., Tulassay, Z., 2000. Increased cell proliferation in chronic Helicobacter pylori positive gastritis and gastric carcinoma–correlation between immuno-histochemistry and Tv image cytometry. Anal Cell Pathol 20 (2-3), 131–139. PubMed PMID: 11153608; PubMed Central PMCID: PMC4615991.

Szymanski, M., et al., 2005. A new frontier for molecular medicine: noncoding RNAs. Biochim. Biophys. Acta 1756, 65–75.

Takiguchi, M., et al., 2003. Effects of cadmium on DNA-(Cytosine-5) methyltransferase activity and DNA methylation status during cadmium-induced cellular transformation. Exp. Cell Res. 286, 355–365.

Tan, S., Wang, C., Lu, C., Zhao, B., Cui, Y., Shi, X., Ma, X., 2009. Quercetin is able to demethylate the p16INK4a gene promoter. Chemotherapy 55 (1), 6–10. https://doi.org/10.1159/000166383. Epub 2008 Oct 31. PubMed PMID: 18974642.

Tate, P.H., Bird, A.P., 1993. Effects of DNA methylation on DNA-binding proteins and gene expression. Curr. Opin. Genet. Dev. 3, 226–231.

Teiten, M.H., et al., 2010. Curcumin-the paradigm of a multi-target natural compound with applications in cancer prevention and treatment. Toxins 2, 128–162. PMID: 22069551.

Thakur, V.S., Gupta, K., Gupta, S., 2012. Green tea polyphenols causes cell cycle arrest and apoptosis in prostate cancer cells by suppressing class I histone deacetylases. Carcinogenesis 33, 377–384.

Tikoo, K., et al., 2008. Change in post-translational modifications of histone H3, heat-shock protein-27 and MAP kinase p38 expression by curcumin in streptozotocin-induced type I diabetic nephropathy. Br. J. Pharmacol. 153, 1225–1231.

Tili, E., et al., 2010. Resveratrol modulates the levels of microRNAs targeting genes encoding tumor-suppressors and effectors of TGFbeta signaling pathway in SW480 cells. Biochem. Pharmacol. 80, 2057–2065.

Traka, M., et al., 2005. Transcriptome analysis of human colon Caco-2 cells exposed to sulforaphane. J. Nutr. 135, 1865–1872.

Tsang, W.P., Kwok, T.T., 2010. Epigallocatechin gallate up-regulation of miR-16 and induction of apoptosis in human cancer cells. J. Nutr. Biochem. 21, 140–146.

Ulitsky, I., et al., 2011. Conserved function of lincRNAs in vertebrate embryonic development despite rapid sequence evolution. Cell 147, 1537–1550.

Upadhyay, A.K., Cheng, X., 2011. Dynamics of histone lysine methylation: structures of methyl writers and erasers. Prog. Drug Res. 67, 107–124.

Van Doorn, R., et al., 2005. Aberrant DNA methylation in cutaneous malignancies. Semin. Oncol. 32, 479–487.

Van Engeland, M., et al., 2003. Effects of dietary folate and alcohol intake on promoter methylation in sporadic colorectal cancer: The Netherlands cohort study on diet and cancer. Cancer Res. 63, 3133–3137.

Van Tine, B.A., et al., 2004. Clonal selection for transcriptionally active viral oncogenes during progression to cancer. J. Virol. 78, 11172–11186.

Varambally, S., et al., 2008. Genomic loss of microRNA-101 leads to overexpression of histone methyltransferase EZH2 in cancer. Science 322, 1695–1699.

Verma, M., 2012. Cancer control and prevention by nutrition and epigenetic approaches. Antioxid. Redox Signal 17, 355–364.

Wang, L.G., et al., 2007. Dual action on promoter demethylation and chromatin by an isothiocyanate restored GSTP1 silenced in prostate cancer. Mol. Carcinog. 46, 24–31.

Wang, L.G., et al., 2008. De-repression of the p21 promoter in prostate cancer cells by an isothiocyanate via inhibition of HDACs and c-Myc. Int. J. Oncol. 33, 375–380.

Wang, R.H., et al., 2008a. Impaired DNA damage response, genome instability, and tumorigenesis in SIRT1 mutant mice. Canc. Cell 14, 312–323.

Wang, R.H., et al., 2008b. Interplay among BRCA1, SIRT1, and Survivin during BRCA1-associated tumorigenesis. Mol. Cell 32, 11–20.

Waterland, R.A., Dolinoy, D.C., Lin, J.R., Smith, C.A., Shi, X., Tahiliani, K.G., 2006. Maternal methyl supplements increase offspring DNA methylation at Axin Fused. Genesis 44 (9), 401–406. PubMed PMID: 16868943.

Watt, F., Molloy, P.L., 1988. Cytosine methylation prevents binding to DNA of a HeLa cell transcription factor required for optimal expression of the adenovirus major late promoter. Genes Dev. 2, 1136–1143.

Wellner, U., et al., 2009. The EMT-activator ZEB1 promotes tumorigenicity by repressing stemness-inhibiting microRNAs. Nat. Cell Biol. 11, 1487–1495.

Wen, X.Y., et al., 2009. Ellagitannin (BJA3121), an anti-proliferative natural polyphenol compound, can regulate the expression of MiRNAs in HepG2 cancer cells. Phytother Res. 23, 778–784.

Wiley, D.J., et al., 2005. Methylation of human papillomavirus genomes in cells of anal epithelia of HIV-infected men. J. Acquir. Immune Defic. Syndr. 39, 143–151.

Wilson, V.L., Jones, P.A., 1983. DNA methylation decreases in aging but not in immortal cells. Science 220, 1055–1057.

Wilkinson, J.T., et al., 1995. Effect of alkyl chain length on inhibition of N nitrosomethylbenzylamine-induced esophageal tumorigenesis and DNA methylation by isothiocyanates. Carcinogenesis 16, 1011–1015.

Wood, J.G., et al., 2004. Sirtuin activators mimic caloric restriction and delay ageing in metazoans. Nature 430, 686–689.

Wu, H., et al., 2014. Luteolin induces apoptosis by upregulating miR-34a in human gastric cancer cells. Technol. Cancer Res. Treat. [Epub ahead of print] PubMed PMID: 24988056.

Yang, J., et al., 2009. Activation of SIRT1 by resveratrol represses transcription of the gene for the cytosolic form of phosphoenolpyruvate carboxykinase (GTP) by deacetylating hepatic nuclear factor 4alpha. J. Biol. Chem. 284, 27042–27053.

Yang, S.R., et al., 2007. Sirtuin regulates cigarette smoke-induced proinflammatory mediator release via RelA/p65 NF-kappaB in macrophages in vitro and in rat lungs in vivo: implications for chronic inflammation and aging. Am. J. Physiol. Lung Cell Mol. Physiol. 292, L567–L576.

Yoder, J.A., Walsh, C.P., Bestor, T.H., 1997. Cytosine methylation and the ecology of intragenomic parasites. Trends Genet. 13, 335–340.

Yoon, J.H., et al., 2001. Methylated CpG dinucleotides are the preferential targets for G-to-T transversion mutations induced by benzo[a]pyrene diol epoxide in mammalian cells: similarities with the p53 mutation spectrum in smoking-associated lung cancers. Cancer Res. 61, 7110–7117.

Zazula, M., et al., 2006. CDH1 gene promoter hypermethylation in gastric cancer: relationship to Goseki grading, microsatellite instability status, and EBV invasion. Diagn. Mol. Pathol. 15, 24–29.

Zhang, J., et al., 2010. Curcumin promotes apoptosis in human lung adenocarcinoma cells through miR-186n signaling pathway. Oncol. Rep. 24, 1217–1223.

Zhang, J., et al., 2005. DNA methylation in anal intraepithelial lesions and anal squamous cell carcinoma. Clin. Cancer Res. 11, 6544–6549.

Zhang, Y.J., et al., 2002. High frequency of promoter hypermethylation of RASSF1A and p16 and its relationship to aflatoxin B1-DNA adduct levels in human hepatocellular carcinoma. Mol. Carcinog. 35, 85–92.

Zhang, Y.J., et al., 2003. Inactivation of the DNA repair gene O6-methylguanine-DNA methyltransferase by promoter hypermethylation and its relationship to aflatoxin B1-DNA adducts and p53 mutation in hepatocellular carcinoma. Int. J. Cancer 103, 440–444.

Zhang, Y.J., et al., 2006. Aflatoxin B1 and polycyclic aromatic hydrocarbon adducts, p53 mutations and p16 methylation in liver tissue and plasma of hepatocellular carcinoma patients. Int. J. Cancer 119, 985–991.

Zhao, C.Q., et al., 1997. Association of arsenic-induced malignant transformation with DNA hypomethylation and aberrant gene expression. Proc. Natl. Acad. Sci. USA 94, 10907–10912.

Zhao, H., Chen, T., 2013. Tet family of 5-methylcytosine dioxygenases in mammalian development. J. Hum. Genet. 58, 421–427.

Zhao, J., et al., 2008. Polycomb proteins targeted by a short repeat RNA to the mouse X chromosome. Science 322, 750–756.

Zhao, X., Liu, M., Li, D., 2015. Oleanolic acid suppresses the proliferation of lung carcinoma cells by miR-122/Cyclin G1/MEF2D axis Mol. Cell. Biochem. 400, 1–7.

Zhou, H., Hu, H., Lai, M., 2010. Non-coding RNAs and their epigenetic regulatory mechanisms. Biol. Cell 102, 645–655.

Zhou, L., et al., 2005. Frequent hypermethylation of RASSF1A and TSLC1, and high viral load of Epstein-Barr Virus DNA in nasopharyngeal carcinoma and matched tumor-adjacent tissues. Neoplasia 7, 809–815.

Further Reading

Belinsky, S.A., 2004. Gene-promoter hypermethylation as a biomarker in lung cancer. Nat. Rev. Cancer 4, 707–717.

Deeb, D., et al., 2014. Induction of apoptosis in pancreatic cancer cells by CDDO-me involves repression of telomerase through epigenetic pathways. J. Carcinog. Mutagen. 5, 177.

Dinkova-Kostova, A.T., et al., 2005. Extremely potent triterpenoid inducers of the phase 2 response: correlations of protection against oxidant and inflammatory stress Proc. Natl. Acad. Sci. USA 102, 4584–4589.

Feng, Q., et al., 2005. Detection of hypermethylated genes in women with and without cervical neoplasia. J. Natl. Cancer Inst. 97, 273–282.

Gerhauser, C., 2013b. Epigenetic impact of dietary isothiocyanates in cancer chemoprevention. Curr. Opin. Clin. Nutr. Metab. Care 16, 405–410.

Issa, J.P.J., et al., 2001. Accelerated age-related CpG island methylation in ulcerative colitis. Cancer Res. 61, 3573–3577.

Kuck, D., et al., 2010. Novel and selective DNA methyltransferase inhibitors: docking-based virtual screening and experimental evaluation. Bioorg. Med. Chem. 18, 822–829.

Marutha Ravindran, C.R., Ticku, M.K., 2004. Changes in methylation pattern of NMDA receptor NR2B gene in cortical neurons after chronic ethanol treatment in mice. Brain Res. Mol. Brain Res. 121, 19–27.

Medina-Franco, J.L., Caulfield, T., 2011. Advances in the computational development of DNA methyltransferase inhibitors. Drug Discov. Today 16, 418–425.

Olaharski, A.J., et al., 2005. The flavoring agent dihydrocoumarin reverses epigenetic silencing and inhibits sirtuin deacetylases. PLoS Genet. 1, e77.

Puri, S.K., et al., 2005. Aberrant promoter hypermethylation of multiple genes in head and neck squamous cell carcinoma. Am. J. Otolaryngol. 26, 12–17.

Rakyan, V., Whitelaw, E., 2003. Transgenerational epigenetic inheritance. Curr. Biol. 13, R6.

Szarc vel Szic, K., et al., 2015. From inflammaging to healthy aging by dietary lifestyle choices: is epigenetics the key to personalized nutrition? Clin. Epigenetics 7, 33.

Tabe, Y., et al., 2007. PPARgamma-active triterpenoid CDDO enhances ATRA-induced differentiation in APL. Cancer Biol. Ther. 6, 1967–1977.

Tarantini, L., et al., 2009. Effects of particulate matter on genomic DNA methylation content and iNOS promoter methylation. Environ. Health Perspect. 117, 217–222.

Waterland, R.A., 2006. Assessing the effects of high methionine intake on DNA methylation. J. Nutr. 136, 1706S–1710S.

Waterland, R.A., Jirtle, R.L., 2003. Transposable elements: targets for early nutritional effects on epigenetic gene regulation. Mol. Cell Biol. 23, 5293–5300.

Zheng, Z.M., Baker, C.C., 2006. Papillomavirus genome structure, expression, and post-transcriptional regulation. Front. Biosci. 11, 2286–2302.

CHAPTER

18

Epigenetics, Public Health, Lifestyle, and Chemoprevention

Massimo Romani[1], *Barbara Banelli*[1,2]

[1]Laboratory of Tumor Epigenetics, IRCCS Ospedale Policlinico San Martino, Genova, Italy; [2]Department of Health Sciences, University of Genova, Genova, Italy

Abstract

Many environmental and lifestyle factors have been linked to cancer development, and it has been hypothesized that their mechanisms of action could be at least partly epigenetic. Epigenetics translates the basic information of the genome, given by the DNA sequence, into a function through several layers of interconnected components. Environment, diet, and lifestyle can act upon each of them and reversibly modify genome functionality without permanently changing the DNA sequence. Thus, molecules acting on the epigenome could in principle be considered potentially useful drugs for therapeutic and chemopreventive purposes. In this chapter, we will discuss the epigenetic action of the environment and environmental, dietary, and lifestyle factors and their possible roles in the development of diseases. We also will summarize the results of several clinical trials that have tested the chemopreventive properties of some of these factors.

Keywords

Chemoprevention; Environment; Epigenetics; Epigenome; Lifestyle; Public health

INTRODUCTION

It is widely recognized that diet, smoking habits, body weight, and in general many additional lifestyle factors contribute to the development of cancer and other diseases, and it has been hypothesized that epigenetics could be the interface between the reversible effects of lifestyle factors and the irreversible changes that occur during carcinogenesis (Choi and Lee, 2013; Kanherkar et al., 2014; Reynolds and Caton, 2012). At the same time, it has been proposed that intervening in the epigenome to revert or counteract the conditions leading to the irreversible alteration of genetic information could be an important chemopreventive action or could increase the efficacy of existing therapies (Lee and Herceg, 2014; Heerboth et al., 2014).

The reversibility and gradual acquisition of epigenetic changes and their involvement in

Epigenetics of Cancer Prevention
https://doi.org/10.1016/B978-0-12-812494-9.00018-4

essentially all aspects of cell physiology and pathology make them the ideal targets for many therapeutic interventions. Moreover, the recognized influence of epigenetics in many complex diseases such hypertension and other cardiovascular diseases (Levy et al., 2017; Bauer and Martin, 2017), diabetes (Zullo et al., 2017; Dunford and Sangster, 2017; Bansal and Pinney, 2017), and neurodegenerative, psychiatric (Banik et al., 2017; Modabbernia et al., 2017; Hwang et al., 2017), and autoimmune disorders (Xiang et al., 2017; Generali et al., 2017; Alvarez-Errico et al., 2017) has expanded the potential application of epigenetics well beyond the boundaries of cancer.

The mechanistic aspects of epigenetics are extraordinarily complex and redundant, involving several deeply interconnected layers, and it is generally agreed that epigenetics can be influenced by the environment (Fig. 18.1).

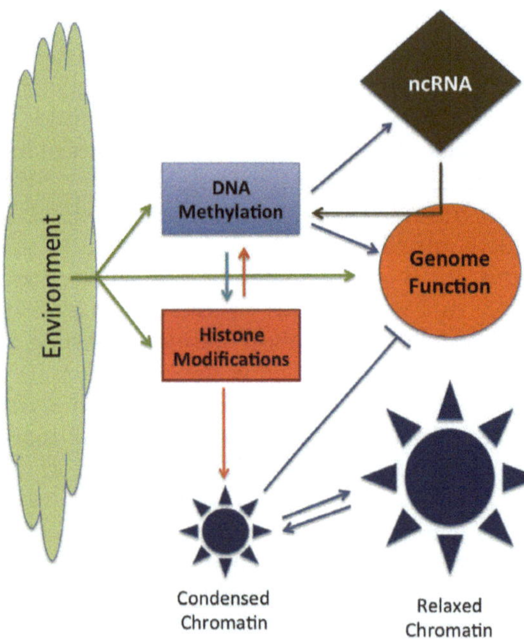

FIGURE 18.1 Schematic representation of the interplay between components of the epigenetic machinery at the interface between environment and genome.

Epigenetic machinery and its components can be envisioned as a "resilient" interface between the environment and the genetic background (Banelli et al., 2015, 2017). This interface responds adaptively and reversibly to external and internal stimuli to fine-tune the functionality of the genome without altering the primary structure of the DNA, thus preserving the original genetic information and ideally allowing for the perfect recovery of cell functionality when the modifying stimulus is withdrawn.

In practice, epigenetics translates the basic layer of information given by the raw sequence of DNA (genotype) into a function (phenotype) through the coordinated and interconnected activities of its components: DNA and histone modifications, chromatin conformation, and noncoding RNA (ncRNA) (Gosden and Feinberg, 2007).

DNA methylation and hydroxymethylation along with some histone modifications (primarily acetylation, methylation, and phosphorylation) are the most known and studied epigenetic alterations and traditionally have been linked to cancer (Riggs and Jones, 1983; Choi and Lee, 2013). Epigenetic modifications are controlled by a set of enzymes that can be functionally defined as "writers" that modify their target by adding residues, "erasers" that modify their target by removing residues, and "readers" that recognize and bind modified targets and act as intermediates or partners in protein–protein interactions. These proteins are collectively denoted "epigenetic modifiers"; their number is steadily increasing, and at the moment this class of proteins includes more than 110 members (Fig. 18.2).

DNA methylation is biologically relevant only when it occurs at cytosine residues within CpG doublets, mainly when clustered in small regions of 200–400 bp with a CpG content of at least 60%, which are known as "CpG islands" (CGIs) (Deaton and Bird, 2011; Herman and Baylin, 2003). In addition, however, methylation of a CpG not included in an island also can have functional consequences (Nile et al., 2008;

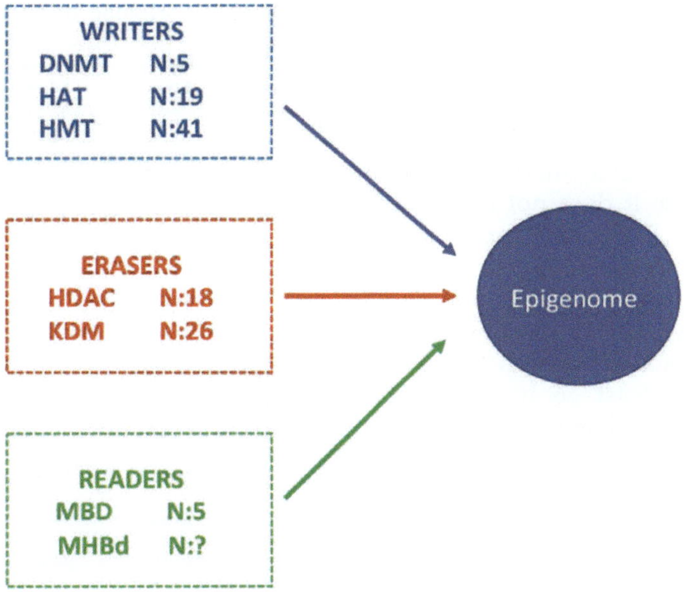

FIGURE 18.2 Different classes of "Epigenetic Modifiers." Writers: add residues to their target; Erasers: remove residues from their target; Readers: recognize and bind the modified substratum. *DNMT*, DNA methyltransferase; *HAT*, histone acetyltransferase; *HDAC*, histone deacetylase; *HMT*, histone methyltransferase; *KDM*, histone lysine demethylase; *MBD*, methyl binding protein; *MHBd*, modified histone binding protein; *N*, number of known different members.

Pogribny et al., 2000). Cytosine methylation is driven by five DNA methyltransferases (DNMTs) that can act predominantly as either de novo methyltransferases (DNMT3a and DNMT3b) or maintenance methyltransferases (DNMT1) (Xu et al., 2010). DNA methylation is inversely correlated with gene expression; historically, only methylation in the promoter regions was considered relevant for transcription control, but it is now clear that methylation outside the canonical CpG-rich regions in the so-called "CGI shores" and "CGI shelves" has functional consequences as well (Irizarry et al., 2009; Kim et al., 2011; Zhang et al., 2015). Inter- and intragenic CpG clusters are normally hypermethylated to avoid transcription from cryptic or internal promoters that could generate spurious products. 5-hydroxymethylcytosine is an oxidation derivative of 5-methylcytosine and acts as a transcriptional activator that displaces "reader" proteins and inhibits their binding (Kraus et al., 2014; Wang et al., 2014).

For many years, histones were considered purely structural proteins necessary for the assembly of chromosomes. In the 1960s, V.G. Allfrey and his collaborators discovered that histones could undergo a variety of posttranslational modifications, and they were able to place them in relation to transcriptional activity (Allfrey et al., 1964; Allfrey and Mirsky, 1964; Vidali et al., 1968). When, in the mid-1970s, the basic "bead-on-a-string" conformation of the chromatin was understood (Olins and Olins, 1974; Thomas and Kornberg, 1975), it was discovered that the tight nucleosome structure could be relaxed during transcription by acetylation of certain lysines of H3 and H4. These modifications weaken protein–DNA interactions, and we now know that the "relaxed" chromatin allows the recruitment of transcription machinery. In practice, HDACs and HATs act as transcriptional repressors and transcriptional activators, respectively.

Histone lysine methylation was discovered along with acetylation by V.G. Allfrey in 1964, but differently from other modifications, it was neglected because it does not change the interactions between DNA and proteins and was considered irreversible. Only in the early 2000s was the first protein with histone lysine demethylase activity, LSD1, discovered and linked to transcriptional repression (Shi et al., 2004). Since then, at least 26 histone lysine demethylases have been identified, and these enzymes and their modulators are now considered one of the most promising avenues to understanding physiological and disease mechanisms and developing innovative therapies (Natoli et al., 2009; Shen et al., 2017; Colvin and Mori, 2017; Jambhekar et al., 2017). Table 18.1 presents a summary of histone modifications and their effects on transcription in mammals.

Although DNA methylation and histone modification can modify the packaging of the chromatin, more profound remodeling is obtained through other mechanisms such as chromatin remodeling complexes (CRCs) and polycomb proteins (PcGs). CRCs include four families of proteins (SWI/SNF, ISWI, IN080, and CHD) that are assembled as a multiunit with an ATPase core (Tyagi et al., 2016). SWI/SNF activates transcription by changing the position and distance between adjacent nucleosomes through the formation of a DNA loop (Dechassa et al., 2010; Havas et al., 2000). SWI/SNF has been strongly implicated in disease, and more than 20% of human cancers harbor mutations in this family (Kadoch et al., 2013).

PcGs are a group of repressive complexes that include three different families (Chittock et al., 2017). PRC1 and PR-DUB have opposite effects on H2A—while PRC1 induces monoubiquitination at Lys119, PR-DUB removes the ubiquitin

TABLE 18.1 Histone Modifications and Their Effect on Transcription in Mammals

Histone	Site	Modification	Effect
H1	Lys26	Methylation	Repression
	Ser27	Phosphorylation	Activation
H2A	Lys5	Acetylation	Activation
	Ser1	Phosphorylation	Repression
H2B	Lys5	Acetylation	Activation
	Lys12	Acetylation	Activation
	Lys15	Acetylation	Activation
	Lys20	Acetylation	Activation
	Lys120	Ubiquitylation	Activation
H3	Lys4	Acetylation	Activation
	Lys9	Acetylation	Activation
	Lys14	Acetylation	Activation
	Lys18	Acetylation	Activation
	Lys23	Acetylation	Activation
	Lys27	Acetylation	Activation
	Lys4	Methylation	Activation
	Arg8	Methylation	Repression
	Lys9	Methylation	Repression
	Arg17	Methylation	Activation
	Lys27	Methylation	Repression
	Lys36	Methylation	Activation
	Lys79	Methylation	Activation
	Ser10	Phosphorylation	Activation
	Ser28	Phosphorylation	Activation
H4	Lys5	Acetylation	Activation
	Lys8	Acetylation	Activation
	Lys12	Acetylation	Activation
	Lys16	Acetylation	Activation
	Lys20	Methylation	Repression
	Lys59	Methylation	Repression

from H2AK119. The PRC2 complex comprises histone methyltransferases that contain an EZH2 catalytic subunit and catalyze the mono- and trimethylation of H3K27 (Aranda et al., 2015).

PcG proteins are potent inhibitors of transcription that compact chromatin by DNA methylation, multiple histone modifications, and the binding of recruited proteins at histone tails to stop mRNA elongation (Aranda et al., 2015). PcG can be targeted by ncRNA, and this interaction is particularly important in X chromosome inactivation (Zhao et al., 2008).

The most recent class of epigenetic modifiers is the ncRNAs; not being transcribed, the exert control of genome functionality by interfering with other RNAs. because their inhibition is sequence-specific and does not involve permanent alteration of the DNA sequence, ncRNAs are considered an integral part of the epigenetic machinery. ncRNAs are divided into different classes, and except for the "long ncRNA" whose length is >200 bp, the other ncRNA (miRNA, siRNA, piRNA, and xiRNA) are only 20–42 bp long.

miRNA is the most abundant and studied class of ncRNA. These short RNAs inhibit mRNAs by the base pairing of nucleotides 2–8 to the 3′UTR of mRNA to form the so-called "silencing complex" (Bartel, 2004; He and Hannon, 2004). miRNAs are involved in a variety of biological processes and can be subjected to epigenetic regulation by DNA methylation and interfere with the epigenetic machinery by targeting the genes coding DNMTs, thus generating a complex "epigenetic interplay" (Fabbri et al., 2007; Parodi et al., 2016; Poddar et al., 2017; Sato et al., 2011).

In this chapter, we will make a summary of recent findings on how the environment can act on the epigenome and how this interaction can influence cell functionality. This survey will include the effects of industrial pollutants and chemicals that are part of our daily life as well as lifestyle habits such as food intake, smoking, and alcohol consumption, which have a clear-cut influence on health. Finally, we will discuss how diet and certain exogenous substances can have a protective effect over the development of common diseases.

ENVIRONMENT, GENOME, AND EPIGENOME ARE THE "BERMUDA TRIANGLE" OF DEVELOPMENTAL PROGRAMMING

One of the most sustained ideas about the risk of developing certain diseases during one's lifespan is that adverse conditions, including nutrient restriction and exposure to pollutants that occur in utero or during early infancy, could have both short- and long-term consequences. This basic concept was originally developed in the 1990s (Bradley, 1991; Barker, 1990) and was later denoted "Developmental Programming" (Reynolds et al., 2010) (Fig. 18.3).

The metabolic syndrome has been considered the paradigm of developmental programming and has been extensively documented, although some inconsistencies have been pinpointed (Armitage et al., 2004). According to NIH guidelines, a metabolic syndrome is a

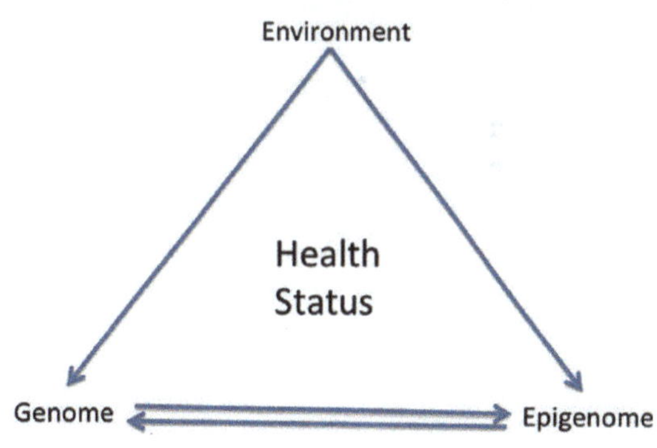

FIGURE 18.3 The "Bermuda Triangle" of developmental programming. The environment interacts with the genetic and epigenetic backgrounds of the individual to determine health status. Genome and epigenome interact with each other to adapt to a variety of stressing conditions.

condition where at least three of five disorders are diagnosed: hypertension, abdominal adiposity, high triglyceride levels, low high-density lipoprotein cholesterol, and fasting hyperglycemia, and can evolve into type 2 diabetes and cardiovascular diseases (https://www.nhlbi.nih.gov/health/health-topics/topics/ms).

Epidemiological studies have supported the idea that adverse conditions such as low birth weight, exceedingly high or low temperature, low pO_2, very low calorie intake, multiple pregnancies, and hormonal exposure could sum with maternal, paternal, and fetal genotypes in determining the risk of developing a variety of major diseases early after birth or at adulthood (Reynolds and Caton, 2012). In experimental animal models and in livestock production, exposure to conditions that favor low fetal or neonatal growth are related to higher morbidity and mortality, metabolic disorders, cardiovascular diseases. and other dysfunctions of specific organs (Wu et al., 2006; Du et al., 2010; Reynolds et al., 2010). Two large retrospective epidemiological studies in human subjects resulted in partially conflicting results. The "Dutch Famine" study conducted on a cohort that suffered extreme starvation in the winter of 1944–45 (Ravelli et al., 1998; Ravelli et al., 1976; Lumey et al., 2007) showed that limited food intake during pregnancy resulted in an increased risk of cardiovascular and metabolic diseases in offspring. Similar studies conducted on other malnourished populations (Levitt et al., 1999; Thame et al., 2000) essentially confirmed the earlier study. Conversely, data taken from children born during the Siege of Leningrad did not confirm the "Dutch Famine" study (Huxley et al., 2002; Stanner et al., 1997). A possible explanation for these conflicting results could reside in the different nutritional status at the end of the famine period in the Netherlands and Leningrad. The post- and perinatal food supply of the Dutch population before and after the starvation period was satisfactory, whereas in Leningrad it was poor. The lack of food could have generated a "predictive adaptive response" (Bateson et al., 2014) in newborns, and when the calorie intake in Dutch children became higher than expected, the disease status made it appear as though the low calorie intake in utero was less harmful than that sustained after birth (Eriksson et al., 1999). Several perinatal insults such as under- and overnutrition, stress, lifestyle exposure to pollutants, and health status of the mother are believed to interfere with developmental programming through mechanisms that are still unclear and certainly much more complex than hypothesized in early studies. Some potential mechanisms have been proposed and fall into two categories, structural defects and functional defects.

The best-known example of structural defects is the reduced number of muscle cells in offspring as a consequence of nutrient restriction during fetal development (Du et al., 2010, 2011). The functional defects in general are a consequence of altered gene expression. Metabolic and steroid hormones are targets of many exogenous factors that could adversely affect many physiological processes (Hiort, 2013; Forhead and Fowden, 2009; Wood, 2014). Oxidative stress induced by obesity, high fat diet, hypertension, and smoking is another proposed mechanism of interference with developmental programming (Valko et al., 2007; Bruce et al., 2009).

The placenta is considered a key organ in developmental programming, because being extremely vascularized, it is where all exchanges between mother and fetus occur. If pregnancy is compromised, placental vascularization is also compromised, with fetal growth reduced as well (Reynolds et al., 2006). Last but not least, it is through the placenta that environmental factors can affect the fetus either by permanently altering the DNA sequence or more often by inducing subtle and less subtle functional changes through epigenetic modifications.

Influence of the Environment on the Epigenome and Epigenetic Reprogramming

DNA, even when packaged into the nucleus, is a large target and is exposed to a variety of effectors that can alter genome functions through genetic and epigenetic mechanisms. An example of the effect of the environment on epigenetics comes again from the "Dutch Famine" study, which showed that 60 years after the event, exposure to the famine that occurred early during pregnancy resulted in the hypomethylation of the insulin growth factor 2 gene (*IGF2*), a key developmental gene. On the contrary, exposure to famine late in gestation did not change *IGF2* methylation compared with that of age-matched controls (Heijmans et al., 2008). Similarly, birth weight in perigestationally exposed infants was significantly higher than that of newborns exposed to famine late in pregnancy, but was not associated with *IGF2* methylation (Heijmans et al., 2008). This study represented one of the first examples of the effect of environmental conditions on the fetus; however, it must be pointed out that many other stressing conditions during pregnancy and later in life were not taken into consideration in the statistical model. Moreover, other relevant phenotypic consequences besides weight at birth could not be assessed.

Strong support of the hypothesis that environmental factors could influence the epigenome comes from studies in animal models that show that certain dietary supplements can alter in utero the methylation status of genes involved in the obesity, coat color, and phenotype of newborn mice (Dolinoy et al., 2007; Waterland and Jirtle, 2003).

Maternal stress and nurturing can alter the epigenome of offspring (Weaver et al., 2005, 2004); this effect is an example of "behavioral epigenetics" and is considered to depend on a mother's behavior early after birth and can be reversed by the intraventricular administration of methyl donors or HDAC inhibitors (Weaver et al., 2005). It is difficult to predict how these findings will have a real clinical impact; indeed, epigenetic traits were modulated by central administration of bioactive molecules, a route that will be difficult to utilize in daily medical practice. Therefore, it will be necessary to determine whether epigenetic effects can also be modulated by more conventional delivery routes. or ideally, through diet.

In mammals, parental genomes are extensively demethylated after fertilization, although complex methods to maintain methylation imprinting and preserve histone variants are developed by the cell (Brykczynska et al., 2010; Messerschmidt, 2012). Epigenetic reprogramming clears up the acquired epigenetic signatures so that development can follow the information contained in the DNA sequence.

Studies in animal models have reported that the phenotypic effects induced by several chemicals such as diethylstilbestrol (DES) (Newbold et al., 2006) and vinclozolin (Anway et al., 2005) may be transgenerational; i.e., they can be passed through F0 to F2 and F3. The inheritance of acquired characters was first proposed by botanists (Lamarck, *Philosophie Zoologique*, 1809) and demonstrated for plants and lower organisms. In mammals, epigenetic inheritance is limited to intergenerational effects as in the case of intrauterine exposure, and it is uncertain whether epigenetic changes seen in somatic cells have a germline origin. Recent surveys of the literature raised several questions on transgenerational epigenetic inheritance, and in this respect, it is worth noting that the transmission of traits to next generations through purely epigenetic mechanisms in the absence of DNA sequence variations has never been formally documented (Grossniklaus et al., 2013; Heard and Martienssen, 2014).

EPIGENETIC EFFECTS OF INDUSTRY ENVIRONMENTAL POLLUTANTS

The health effects of toxic chemicals as a result of work-related or leisure exposures in utero and after birth are well established, and every year new molecules are tested by governmental and independent agencies for their effects in vitro and in vivo (http://publications.iarc.fr/Book-And-Report-Series/Iarc-Monographs-On-The-Evaluation-Of-Carcinogenic-Risks-To-Humans).

The risk of exposure to genotoxic chemicals is often monitored through identification of the damage to the DNA sequence that has occurred in cells. Other chemicals that promote carcinogenesis have mainly epigenetic effects that can be monitored by also looking at the genome function (Baccarelli and Bollati, 2009).

Metals are common environmental contaminants that induce generalized alterations in the DNA methylation pattern of DNA. Metals induce oxidative stress and DNA damage, and damaged DNA is less efficiently targeted by DNMTs. The effects of metals on the epigenome have been tested primarily in vitro, but most results have been validated in vivo either in animal models or in humans who have been professionally exposed to these pollutants.

Arsenic is detoxified by methylation utilizing methyl donor S-adenosyl-methionine (SAM), which is also required by DNMTs for the methylation of DNA. Reduced SAM results in reduced DNMT activity and global hypomethylation (Zhao et al., 1997). Several pathologies have been associated with prenatal and postnatal exposure to arsenic, including lung and bladder cancers (Smith et al., 2006; Marsit et al., 2006). Cadmium is a nonmutational carcinogen that alters DNA methylation, inducing oxidative stress, and possibly through direct interaction with DNMTs induces hypomethylation and reactivation of proto-oncogenes (Filipic and Hei, 2004; Takiguchi et al., 2003). Nickel and chromium have multiple epigenetic effects, altering not only DNA methylation but also histone modifications that deacetylate H3 and H4 and alter the methylation and phosphorylation of H3 (Lee et al., 1995; Ke et al., 2006; Chen et al., 2006; Govindarajan et al., 2002; Kondo et al., 2006).

Trichloroethylene and di- and trichloroacetic acids are other industrial contaminants that are carcinogenic for the liver. These chemicals decrease the methylation of *c-Jun* and *c-Myc* in animal models and increase the expression of these oncogenes, which can be prevented by methionine administration as a methyl donor (Tao et al., 1999). Benzene is a common low-level occupational contaminant (e.g., for gas station and road construction workers) that at high concentrations increases the risk of acute leukemia. Exposure to benzene is associated with global hypomethylation and hypermethylation of the p15 tumor-suppressor gene (Bollati et al., 2007).

We live "immersed" in the atmosphere, so it is not surprising that air is one of the most important vehicles for potentially harmful chemicals associated with increased morbidity for cardiovascular diseases and lung cancer (Vineis and Husgafvel-Pursiainen, 2005; Brook et al., 2004). Many different potentially harmful substances are included in the broad category of air pollutants, such as particulate matter (PM), polycyclic aromatic hydrocarbons (PAHs), benzo[a]pyrene (B[a]P), nitrogen dioxide, and black carbon (BC).

B[a]P and PAH are Group 1 carcinogens derived from the partial combustion of fossil fuels. Occupational studies have demonstrated that workers at oil refineries have altered methylation levels in repetitive DNA sequences (Alu and LINE-1) as well as in specific genes such as the DNA repair gene MGMT (Duan et al., 2013; Pavanello et al., 2010). Hypomethylation of LINE-1 is observed in cancer cells as part of the generalized hypomethylation seen in cancer and is involved in chromatin repackaging. Moreover, unmethylated Alu and LINE-1 act as transposable elements contributing to chromosomal instability (Peluso et al., 2012; Madrigano et al., 2011). Similarly, PM and BC were found to have demethylating effects on Alu and LINE-1 sequences that seem to be the preferential targets for many air contaminants (Tarantini et al., 2009).

Intrauterine exposure to PAH induces epigenetic alterations and DNA adducts; global hypomethylation was found in animal models and in nucleated blood cells from the umbilical cord depending on the mother's level of exposure to PAH (Herbstman et al., 2012; Janssen et al., 2013).

B[a]P is the principal carcinogen in tobacco smoke, although it is also an industrial pollutant; B[a]P hypermethylates tumor-suppressor genes such as *RAR-β2* and *HIC1* that are inactivated in many tumors including lung cancer (Peluso et al., 2012; Ye and Xu, 2010).

Limited information exists on histone modifications after exposure to air pollutants. In an animal study, deacetylation of H3K14 after exposure to B[a]P was observed in a testosterone regulatory gene, *StAR*, and resulted in reduced sperm count in the exposed animals (Liang et al., 2012).

Information is also limited about the effect of air contaminants on miRNAs. Microarray analysis was utilized to study the effect of diesel exhaust on bronchial epithelial cells and showed that approximately 200 miRNAs had changed their level of expression (Jardim et al., 2009). Interestingly, the 12 most altered miRNAs are involved in inflammation and tumorigenesis. Changes in miRNA expression could potentially have major effects because they act on multiple pathways as described previously.

EFFECTS ON THE EPIGENOME OF ENDOCRINE-DISRUPTING CHEMICALS

Endocrine-disrupting chemicals (EDCs) are substances that perform hormonelike activities that can alter endocrine function by intervening in either endogenous hormone production or endocrine physiologic activity (Schug et al., 2011). Although many EDCs are industrial pollutants, they also are found normally in the human environment as widespread components utilized daily such as phthalates (substances that render plastic more flexible), dioxins and dioxin derivatives (persistent by-products of industrial activities and waste incineration), pesticides, and chlorinated compounds. By altering the endocrine status of the exposed individual, EDCs interfere with many aspects of normal development including reproduction and psychosocial behavior (De Coster and van Larebeke, 2012; Zoeller et al., 2012). Exposure to EDC is of more concern when it occurs in utero, particularly at certain stages of development, or in newborns, because it can have a significant impact on long-term health outcomes.

Hormone signaling is required for normal development, and steroid hormones induce epigenetic modifications that persist beyond the perinatal age. It has been observed that EDCs can alter regular hormone programming and that epigenetic mechanisms are involved in the reprogramming exerted by these agents (Fleisch et al., 2012).

Bisphenol A (BPA) is likely one of the most common EDCs found in the environment and is used in the manufacture of polycarbonate and epoxy resins that are widely employed by households and with health and dental tools. BPA has estrogenic activities and is consumed through canned food and water contained in plastic bottles (Husain and Qayyum, 2013). It has been estimated that its intake through food is approximately 7 μg/person/day (Diamanti-Kandarakis et al., 2009). Animal models showed that BPA crosses the placenta and induces several adverse effects in endocrine, reproductive, and immune systems (Doherty et al., 2010; Veiga-Lopez et al., 2013) and increases the risk of prostate cancer in rats (Ho et al., 2006). Human studies suggest that exposure to BPA during development may increase the risk of breast cancer (Weng et al., 2010). In humans, exposure in utero results in the accumulation of BPA in the amniotic fluid (Ikezuki et al., 2002), and several studies indicate that this molecule can alter the expression of many genes and miRNAs that are part of important pathways; these changes can persist in adult life (reviewed in Vaiserman, 2014).

BPA has multiple epigenetic effects and alters the methylation patterns of many signaling genes in different pathways and of homeobox genes; these changes can be reverted if the diet is supplemented with methyl donors such as folic acid or vitamin B12 (Bromer et al., 2010; Dolinoy et al., 2007; Smith and Taylor, 2007). A recent epigenome-wide methylation analysis on human fetuses that were exposed to BPA revealed a complex array of methylation alteration across the genome as well as sequence-specific alterations (Faulk et al., 2015).

DES, another example of an EDC, is an estrogen antagonist that was utilized from 1938 to 1971 to prevent miscarriage. In 1977, it was demonstrated that intrauterine exposure to DES is associated with a high risk of vaginal cancer (Robboy et al., 1977). DES causes epigenetic reprogramming of genes of the estrogen pathway and of the histone methyltransferase *EZH2* that is linked to the development of several types of cancer, including breast cancer (Li et al., 2003, 1997; Skinner, 2011; Doherty et al., 2010). Moreover, DES treatment in mice resulted in the increased expression of *DNMT1* and *DNMT3* and in hypermethylation of *HOXA10* (Bromer et al., 2009), a gene involved in stemness and hypermethylated in cancer (Di Vinci et al., 2012, 2011; Murat et al., 2008).

Chlorine insecticides, such as DDT, methoxychlor (MXC), vinclozolin, and permethrin, are widespread environmental contaminants and well-established EDCs. In experimental models, gestational exposure to MXC results in several abnormalities such as reduced ovulatory rate, alteration of sexual development, and testis involution (Stouder and Paoloni-Giacobino, 2011). Some of these effects are likely determined by methylation and downregulation of the estrogen receptor-β gene and by overexpression of DNMT3b (Armenti et al., 2008; Zama and Uzumcu, 2009). Permethrin is a common insecticide that is also used to treat scabies by direct application on the skin. Although it is considered a safe agent, permethrin is classified by the EPA as a suspect human carcinogen based on rodent studies that also induced heart damage in rats (Vadhana et al., 2011). Treatment of gestating rats with clinically relevant dosages of a mixture of permethrin and an insect repellant resulted in sperm abnormalities and alterations of DNA methylation evenly distributed without the clear-cut involvement of a specific biological pathway (Manikkam et al., 2012). Vinclozolin has antiandrogenic effects and increases the risk of gonadal abnormalities and cancer in many organs such as the kidney and prostate, and also to the immune system (Anway et al., 2005, 2006a,b). An epigenome-wide analysis conducted on mice identified differently methylated regions in 66 promoters distributed throughout the entire genome (Guerrero-Bosagna et al., 2012) and in small ncRNAs involved in a variety of different pathways (Schuster et al., 2016). Although these studies reported a transgenerational effect for EDCs up to F4 in rodents, other reports failed to confirm the transgenerational effects of these compounds (Schneider et al., 2008, 2013; Inawaka et al., 2009).

Also, dioxins, which are industrial contaminants mentioned previously, are EDCs with anti-estrogenic effects that can induce alterations in the reproductive system (Somm et al., 2013). Exposure to dioxins early in pregnancy results in reduced fetal growth, increased methyltransferase activity, and hypermethylation of the imprinted genes H19 and IGF2 (Wu et al., 2004). The effects of dioxins are very heterogeneous, and many genes and pathways are involved depending on gestational age at the time of exposure in utero.

EFFECTS OF LIFESTYLE ON THE EPIGENOME

"Lifestyle" can be defined as the habits, standards, and attitudes that collectively constitute the mode of living and interactions for an individual or group. A given lifestyle thus includes a large number of factors, avoidable and unavoidable, that interact with the genotype and can change genome function. Food and food habits along with smoking, alcohol consumption, UV light, and oxidative stress can all interfere with epigenetic marks. Tobacco smoke (other than B[a]P, whose epigenetic effects were described in an earlier section of this chapter (Herbstman et al., 2012; Liang et al., 2012; Ye and Xu, 2010)) contains a complex mix of carcinogens. Smoke condensate alters the epigenetic asset and the expression of pro-inflammatory and oxidative stress genes and induces epigenetic alterations commonly found in cancer cells in the lungs of exposed rodents (Liu et al., 2010; Marwick et al., 2004). Drinking alcohol is another recreational activity that has been linked to cancer development, although the carcinogenicity of alcohol has never been demonstrated (Boyle et al., 2003). Moreover, conflicting reports on the association between methylation of tumor-suppressor genes in colorectal cancer developed in patients with low folate and high alcohol intakes were published (de Vogel et al., 2008; van Engeland et al., 2003). It must be pointed out that some studies were flawed by the utilization of qualitative methylation analysis technologies that may be inadequate when dealing with the small methylation differences that likely result from environmental and dietary factors. The combination of alcohol and tobacco-smoking abuse has been proven to increase the risk of developing different cancers by about 50-fold, and many different genetic and epigenetic mechanisms seem to be responsible for this increased risk (Tong et al., 2014). The oxidative metabolites generated by alcohol catabolism increase oxidative stress and hyperacetylate H3 in hepatocytes (Chaudhary et al., 2013). This in turn results in the overexpression of inflammatory cytokines (Kendrick et al., 2010), and chronic inflammation leads to cell apoptosis, regeneration, and eventually cancer.

Besides alcohol abuse, dietary habits have been extensively studied in relation to their implications in health; a proper diet, along with the control of alcohol and tobacco consumption, has become the standard hallmark of a "good lifestyle." It is generally believed that food components can modulate epigenetic marks through complex and redundant pathways that change methyl donor availability and therefore DNMT activity and DNA methylation as outlined in Fig. 18.4. The nutrients most implicated in the control of DNA methylation are primarily the cofactors involved in one-carbon metabolism (such as folate, vitamins B6 and B12, choline, and methionine); in particular, folic acid administration to colon cancer patients and healthy individuals resulted in increased methylation in colon mucosa and lymphocytes (Cravo et al., 1994; Kim et al., 2001; Pufulete et al., 2005). Other dietary components can affect DNA methylation—selenium and zinc can impact one-carbon metabolism, and genistein and polyphenols are considered DNMT inhibitors (Davis et al., 2000; Day et al., 2002; Fang et al., 2003, 2005, 2007; Taylor et al., 2009). Dietary factors also can influence histone marks that are intrinsically more complex than DNA methylation; in this respect, the enzymes involved in histone modifications and the availability of the enzyme substrates can be affected. Sodium butyrate produced by colon bacterial flora from carbohydrates, and polyphenols from garlic and soy, are inhibitors of histone deacetylases (Demary et al., 2001; Rada-Iglesias et al., 2007), whereas polyphenols from green tea, curcumin, and copper inhibit histone acetyltransferases (Kang et al., 2005; Lin et al., 2005;

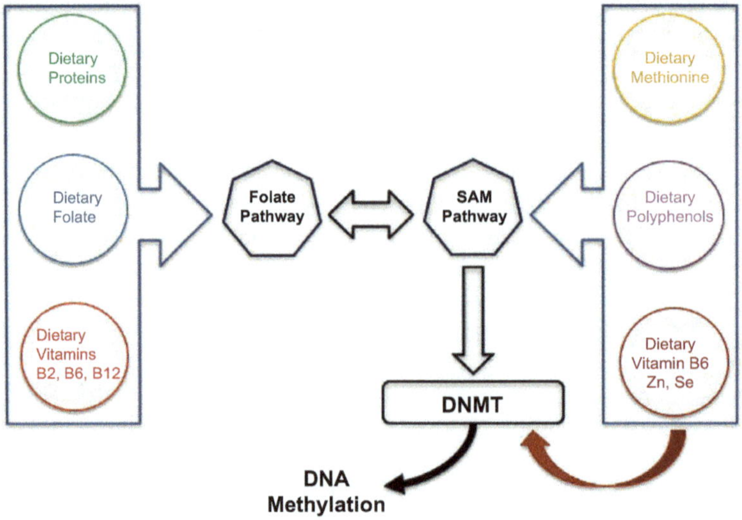

FIGURE 18.4 Schematic representation of the interaction between dietary compounds, folate and SAM pathways, and DNA methylation.

Choi et al., 2009), and epigallocatechin-3-gallate from green tea inhibits histone methyltransferase (Balasubramanian et al., 2010).

As we discussed earlier, exposure to epigenetic modifiers during gestation can affect the epigenome of the fetus; however, several questions need to be answered. Most of these studies were carried out in vitro or in animal models; it is likely that the duration of treatment, its scheduling, and dosage, as well as the time window during gestation, are variables that have impacts that should be carefully evaluated for the effect being measured. Therefore, the major question that needs to be answered is if "epigenetic" nutrients taken at physiological dosages with the diet have a biological consequence. Overall, systematic and unequivocal studies on this subject in humans are still limited, and at least for some molecules such as genistein and other soy isoflavones with estrogen receptor-binding properties, these studies have been focused on toxicology, reproductive effects, and carcinogenic properties (reviewed in Taylor et al., 2009).

Dietary intervention trials with genistein on a small cohort of postmenopausal women showed the absence of estrogenic effects (Marini et al., 2008), and other studies showed a modest antitumor effect for isoflavones in breast cancer patients, although the number of patients was limited and the study lacked stringent controls (Sartippour et al., 2004).

As of June 2017, the Clinical Trial Database of the USA (https://clinicaltrials.gov/) listed 62 phase 1, 2, and 3 trials that utilized genistein, mostly in breast and prostate cancer but also in other tumors as well as nontumor diseases. Incomplete results are available for a minority of the 37 completed studies and do not show substantial advantages in the arms that included genistein alone, or in combination, over placebo.

A major limit of studies on diet, lifestyle, and epigenetics is that the analyses were focused either on the level of global methylation or on targets already known to undergo epigenetic modifications in cancer and aging. Thus, the results obtained so far might be biased toward common end-effects while overlooking more biologically relevant targets. Whole-genome studies based on microarray or next generation sequencing to detect DNA methylation and histone modification variations in response to diet or other environmental and lifestyle events are ideal tools for investigating the plasticity of the epigenome in medium-density CpG regions outside of CGIs (Irizarry et al., 2009; Kim et al., 2011; Zhang et al., 2015). The potential of these technologies has been outlined in several studies. For instance, intrauterine growth retardation in rodents was indeed characterized by methylation changes in approximately 1400 loci (Thompson et al., 2010), and caloric restriction in adult obese women was associated with methylation changes in 35 loci, some of them linked to weight control (Bouchard et al., 2010). In the same study, transcriptomic analysis showed changes of expression for approximately 600 genes involved in several pathways, mainly in angiogenesis and cerebellar long-term depression (Bouchard et al., 2010). These unbiased approaches will undoubtedly reveal new insights into the epigenomic and transcriptomic effects of diet and dietary factors.

Epigenetics, Lifestyle, and Chemoprevention

The effect of lifestyle on health is directly linked to the issue of chemoprevention, which is defined as the use of natural or synthetic substances to prevent the development of cancer (or other diseases) in the general population or in the population at risk. Chemoprevention is not a substitute for therapy, but can also be useful for cancer patients in reducing the risk of tumor relapse. Intuitively, abolishing tobacco smoke, having a balanced diet, and physical exercise are essential aspects of a "good lifestyle," but the general idea behind chemoprevention is that these positive effects and the genetic background of the individual can be further improved by the appropriate use of molecules that, for example, will influence the transition from healthy to cancerous cells. Two well-known examples of chemoprevention are the use of antiestrogens such as tamoxifen to reduce the risk of breast cancer recurrence, raloxifene to lower the risk of breast cancer in postmenopausal women, and nonsteroidal anti-inflammatory drugs such as aspirin or of metformin to reduce the average risk of cancers (Drew et al., 2016; Gadducci et al., 2016; Heckman-Stoddard et al., 2016).

The utilization of molecules acting on the epigenome with the purpose of preventing the development of a disease has been pioneered in cancer because of the beneficial effect of modulating DNA methylation that is hypothesized for tumor cells (Issa, 2007). The rationale for epigenetic chemoprevention of cancer is that by erasing methyl marks on DNA and repressive postsynthetic modifications on histones, dormant tumor-suppressor genes can be reactivated to counteract the effects of active oncogenes. This simplistic view does not consider the relative nonspecificity of epigenetic therapies, because, for example, the removal of the DNA methylation brake could in fact have detrimental effects by activating silenced oncogenes. Nevertheless, inhibitors of DNA methylation have therapeutic potential and are utilized in clinical practice because cancer cells, more than normal cells, largely depend on DNA methylation to maintain their phenotype (Issa, 2007).

Epidemiological studies were taken as a support for the concept that dietary factors could exert protective functions against cancer. It is known that the incidence of breast cancer in Asian women is significantly lower than that of Western women, and although the genetic background likely contributes to this difference, it was shown that Asian women who migrated to Western countries had a breast cancer incidence comparable to that of Western women (Parkin, 1989; Ziegler et al., 1993). The lower incidence of breast cancer in Asian women was attributed at least in part to the diet, and it was proposed that the higher intake of fibers and soy that contain high levels of isoflavones, as well as a lower intake of fat, exerted a protective effect not only in breast cancer, but also in other types of cancer including "big killers" such as prostate and colon cancer (Rose et al., 1986).

Case–control studies where the effects of phytoestrogens (which include isoflavonoids and lignans) on breast cancer incidence were carried out in pre- or postmenopausal women, but yielded conflicting results and did not clarify the issue (Mangtani and Silva Idos, 1998; Lee et al., 1991; De Stefani et al., 1997; Hirose et al., 1995; Chen et al., 2014). More recently, a metaanalysis that considered 23 case–control and 5 cohort studies concluded that a limited protective effect of soy food was evident in Asian women only in the case–control studies (Zhong and Zhang, 2012). Two other metaanalyses concluded that the effect of isoflavones on breast cancer showed a limited reduction of risk, but overall the results were confounding because of poor patient stratification and misclassification, lack of dose response, and in general because the studies were poorly designed (Delaune et al., 2009; Trock et al., 2006).

It has been proposed that phytoestrogens may have protective properties against prostate cancer (Adlercreutz and Mazur, 1997). As with breast cancer, prostate cancer incidence is lower in Asian men than it is for men in Western countries, but Japanese men who emigrate to the USA have a prostate cancer incidence comparable to that of US natives. Japanese men exhibit lower activity of 5α-reductase, which is considered to have a role in prostate cancer (Ross et al., 1992), and this might be a consequence of the high intake of soy isoflavones (Evans et al., 1995). It must be noted that the inhibitory effect in vitro of isoflavones over 5α-reductase occurs at a concentration far outside the physiological range, so additional factors besides isoflavones must be involved in the reduction of prostate cancer risk in the Asian population.

Most early chemopreventive studies appear to be inconclusive also because of poor study design without rigorous randomization. As of June 2017, the USA Clinical Trial Database listed 72 trials (ongoing, terminated, or withdrawn) that were classified as "chemoprevention" studies where dietary factors and other molecules were utilized as chemopreventive agents on the basis of in vitro and preclinical studies. Partial or complete results are available for only seven of the completed and closed studies.

Study NCT00585637 was aimed at determining, in 328 individuals, the dosage of vitamin D necessary to reach a plasma concentration predicted to reduce the risk of developing prostate or colon cancer in the black American population. In this study, the protective effect of vitamin D was not directly assessed (Chandler et al., 2014).

Study NCT00363805 was aimed at determining whether polyphenol or green tea could protect former smokers with chronic obstructive disease from developing lung cancer. The 178 patients enrolled were divided into three arms. Also, in this case the endpoints were essentially biochemical and related to changes in the expression of oxidative stress markers. The results showed reduction in one marker (8-F2-isoprostane). No statistical analysis was provided, and the results of the study, terminated in 2013, have not been published.

Study NCT00335504 was aimed at determining whether the cholesterol-reducing agent atorvastatin could be used to protect 85 patients enrolled in the study, in four arms, from colorectal cancer. The end points utilized were the change of aberrant cryptic foci, and the effect on cell proliferation (Ki67) and apoptosis. The statistical analysis of the results did not show significant differences among the study groups, and the results were not published. The aim of trial NCT00253643 was to determine whether fish oil and/or green tea supplement could protect patients with preneoplastic lesions from cancer, with 85 patients enrolled. The end points of the study were biochemical (fatty acid synthase expression and cell proliferation by Ki76 staining), no significant differences were observed with the treatment, and the results of the study were not published.

Study NCT00303823 was aimed at determining the effect of green tea extract in preventing cervical cancer in patients with human papillomavirus and low-grade cervical intraepithelial neoplasia, with 98 eligible patients randomized into two arms. The end points of the study were the clearance of oncogenic HPV and the histologic and cytological signs of disease. No difference was found between treated and control patients in terms of complete or partial response or progression of the disease and clearance from HPV infection (Garcia et al., 2014).

Study NCT00073073 examined the protective effect of the aromatase inhibitor exemestane in 42 postmenopausal women at high risk of breast cancer. Several end points were considered and included changes in mammographic density and bone density at 1 and 2 years in addition to serum hormone and protein changes. Small but significant modification of breast density was observed between treated and control patients as well as the reduction of estrogens, testosterone, and the estrogen-responsive gene TFF1. These results have led to the hypothesis that breast density and TFF1 are potential biomarkers of breast cancer–preventive effects, although the effect of this treatment on the development of cancer was not examined (Gatti-Mays et al., 2016).

The last study with complete results, NCT00153816, was a randomized double-blind trial that examined the effect of vitamin D/calcium on the prevention of colorectal cancer in 2813 patients enrolled between 2004 and 2008 in 11 centers. The study, terminated in 2015, failed to demonstrate a protective effect from vitamin D/calcium (Baron et al., 2015).

CONCLUSION

The role of epigenetics in the control of gene expression in physiologic and pathologic conditions is well established (Choi and Lee, 2013; Gosden and Feinberg, 2007), and because of their reversibility, epigenetic conditions are considered ideal therapeutic targets. Indeed, "epigenetic drugs" are a rapidly expanding field of molecular pharmacology and have entered the standard clinical practice for a few diseases (Campbell and Tummino, 2014; Rius and Lyko, 2012). In 2014 the Clinical Trials Database listed 195 trials involving "epigenetic drugs," and this number had increased to 375 by June 2017, a demonstration of the steep increase in the interest for this type of therapeutic approach. Much in vitro experimental evidence and fewer studies in experimental animals indicate that epigenetic marks can be modulated by environmental and dietary factors and in general by lifestyle. These findings have rapidly become the bridge between well-established chemopreventive measures such as antiestrogen therapy in breast cancer patients after surgery and standard radio- and chemotherapy, as well as the heterogeneous and somehow confused field of "natural" medicine. In this respect, active molecules are present in many natural products and herbs and are believed to have great potential for protection from diseases by acting on the epigenome (for a recent review, see Huang et al., 2016).

Several experimental strategies have been devised to test this hypothesis, and as discussed in this chapter, and epidemiological and much laboratory data have in principle confirmed the biological effect of several molecules, either purified or administered through the diet. These findings have been taken as a mechanistic explanation for the different cancer incidences observed in different populations. The results of some case–control studies support the epidemiological and preclinical findings; however, these studies have been criticized because of poor design that did not fully account for important confounding factors and therefore were deemed not clinically applicable (Trock et al., 2006; Delaune et al., 2009).

The issue of the role of dietary components in disease prevention was recently more rigorously addressed in randomized trials. Only a limited number of these studies were completed, and partial results were made available to the scientific community. Unfortunately, the end points of most of these studies were indirect because they considered only the changes of expression of biomarkers that may be associated to a "preventive" effect. In some instances, the difference between treated individuals and those that received placebo was significant, but the design of the studies did not allow for demonstration, as an end effect, of the chemopreventive potential of the molecules because the number of individuals enrolled was small and the observation time was too short. Only studies NCT00335504, NCT00303823, and NCT00153816 considered either the protective effect of the molecule or changes related to disease progression such as the presence of atypical cells, the clearance of oncogenic viruses, or the proliferation index. These studies failed to demonstrate the chemopreventive activity of the molecules.

Does the evidence accumulated so far warrant further studies, or are we entitled to dismiss the hypothesis of the relation between the epigenetic effect of dietary components and chemoprevention? A very simple explanation is that the positive association between certain dietary habits and the reduced risk of developing certain diseases is the logical consequence of a "healthier" lifestyle, and the addition of bioactive compounds provides only a marginal and hardly measurable further reduction of risk. Hopefully, the completion of ongoing studies will provide the necessary framework for designing new, rigorous studies that will unequivocally demonstrate whether dietary natural compounds have a chemopreventive role in humans.

References

Adlercreutz, H., Mazur, W., 1997. Phyto-oestrogens and Western diseases. Ann. Med. 29, 95–120.

Allfrey, V.G., Faulkner, R., Mirsky, A.E., 1964. Acetylation and methylation of histones and their possible role in the regulation of RNA synthesis. Proc. Natl. Acad. Sci. U.S.A. 51, 786–794.

Allfrey, V.G., Mirsky, A.E., 1964. Structural modifications of histones and their possible role in the regulation of RNA synthesis. Science 144, 559.

Alvarez-Errico, D., Vento-Tormo, R., Ballestar, E., 2017. Genetic and epigenetic determinants in autoinflammatory diseases. Front. Immunol. 8, 318.

Anway, M.D., Cupp, A.S., Uzumcu, M., Skinner, M.K., 2005. Epigenetic transgenerational actions of endocrine disruptors and male fertility. Science 308, 1466–1469.

Anway, M.D., Leathers, C., Skinner, M.K., 2006a. Endocrine disruptor vinclozolin induced epigenetic transgenerational adult-onset disease. Endocrinology 147, 5515–5523.

Anway, M.D., Memon, M.A., Uzumcu, M., Skinner, M.K., 2006b. Transgenerational effect of the endocrine disruptor vinclozolin on male spermatogenesis. J. Androl. 27, 868–879.

Aranda, S., Mas, G., Di Croce, L., 2015. Regulation of gene transcription by Polycomb proteins. Sci. Adv. 1, e1500737.

Armenti, A.E., Zama, A.M., Passantino, L., Uzumcu, M., 2008. Developmental methoxychlor exposure affects multiple reproductive parameters and ovarian folliculogenesis and gene expression in adult rats. Toxicol. Appl. Pharmacol. 233, 286–296.

Armitage, J.A., Khan, I.Y., Taylor, P.D., Nathanielsz, P.W., Poston, L., 2004. Developmental programming of the metabolic syndrome by maternal nutritional imbalance: how strong is the evidence from experimental models in mammals? J. Physiol. 561, 355–377.

Baccarelli, A., Bollati, V., 2009. Epigenetics and environmental chemicals. Curr. Opin. Pediatr. 21, 243–251.

Balasubramanian, S., Adhikary, G., Eckert, R.L., 2010. The Bmi-1 polycomb protein antagonizes the (-)-epigallocatechin-3-gallate-dependent suppression of skin cancer cell survival. Carcinogenesis 31, 496–503.

Banelli, B., Carra, E., Barbieri, F., Wurth, R., Parodi, F., Pattarozzi, A., Carosio, R., Forlani, A., Allemanni, G., Marubbi, D., Florio, T., Daga, A., Romani, M., 2015. The histone demethylase KDM5A is a key factor for the resistance to temozolomide in glioblastoma. Cell Cycle 14, 3418–3429.

Banelli, B., Daga, A., Forlani, A., Allemanni, G., Marubbi, D., Pistillo, M.P., Profumo, A., Romani, M., 2017. Small molecules targeting histone demethylase genes (KDMs) inhibit growth of temozolomide-resistant glioblastoma cells. Oncotarget 8, 34896–34910.

Banik, A., Kandilya, D., Ramya, S., Stunkel, W., Chong, Y.S., Dheen, S.T., 2017. Maternal factors that induce epigenetic changes contribute to neurological disorders in offspring. Genes (Basel) 8.

Bansal, A., Pinney, S.E., 2017. DNA methylation and its role in the pathogenesis of diabetes. Pediatr. Diabetes 18, 167–177.

Barker, D.J., 1990. The fetal and infant origins of adult disease. BMJ 301, 1111.

Baron, J.A., Barry, E.L., Mott, L.A., Rees, J.R., Sandler, R.S., Snover, D.C., Bostick, R.M., Ivanova, A., Cole, B.F., Ahnen, D.J., Beck, G.J., Bresalier, R.S., Burke, C.A., Church, T.R., Cruz-Correa, M., Figueiredo, J.C., Goodman, M., Kim, A.S., Robertson, D.J., Rothstein, R., Shaukat, A., Seabrook, M.E., Summers, R.W., 2015. A trial of calcium and vitamin D for the prevention of colorectal adenomas. N. Engl. J. Med. 373, 1519–1530.

Bartel, D.P., 2004. MicroRNAs: genomics, biogenesis, mechanism, and function. Cell 116, 281–297.

Bateson, P., Gluckman, P., Hanson, M., 2014. The biology of developmental plasticity and the Predictive Adaptive Response hypothesis. J. Physiol. 592, 2357–2368.

Bauer, A.J., Martin, K.A., 2017. Coordinating regulation of gene expression in cardiovascular disease: interactions between chromatin modifiers and transcription factors. Front. Cardiovasc. Med. 4, 19.

Bollati, V., Baccarelli, A., Hou, L., Bonzini, M., Fustinoni, S., Cavallo, D., Byun, H.M., Jiang, J., Marinelli, B., Pesatori, A.C., Bertazzi, P.A., Yang, A.S., 2007. Changes in DNA methylation patterns in subjects exposed to low-dose benzene. Cancer Res. 67, 876–880.

Bouchard, L., Rabasa-Lhoret, R., Faraj, M., Lavoie, M.E., Mill, J., Perusse, L., Vohl, M.C., 2010. Differential epigenomic and transcriptomic responses in subcutaneous adipose tissue between low and high responders to caloric restriction. Am. J. Clin. Nutr. 91, 309–320.

Boyle, P., Autier, P., Bartelink, H., Baselga, J., Boffetta, P., Burn, J., Burns, H.J., Christensen, L., Denis, L., Dicato, M., Diehl, V., Doll, R., Franceschi, S., Gillis, C.R., Gray, N., Griciute, L., Hackshaw, A., Kasler, M., Kogevinas, M., Kvinnsland, S., La Vecchia, C., Levi, F., Mcvie, J.G., Maisonneuve, P., Martin-Moreno, J.M., Bishop, J.N., Oleari, F., Perrin, P., Quinn, M., Richards, M., Ringborg, U., Scully, C., Siracka, E., Storm, H., Tubiana, M., Tursz, T., Veronesi, U., Wald, N., Weber, W., Zaridze, D.G., Zatonski, W., Zur Hausen, H., 2003. European Code against Cancer and scientific justification: third version (2003). Ann. Oncol. 14, 973–1005.

Bradley, P., 1991. Fetal and infant origins of adult disease. BMJ 302, 113.

Bromer, J.G., Wu, J., Zhou, Y., Taylor, H.S., 2009. Hypermethylation of homeobox A10 by in utero diethylstilbestrol exposure: an epigenetic mechanism for altered developmental programming. Endocrinology 150, 3376–3382.

Bromer, J.G., Zhou, Y., Taylor, M.B., Doherty, L., Taylor, H.S., 2010. Bisphenol-A exposure in utero leads to epigenetic alterations in the developmental programming of uterine estrogen response. FASEB J. 24, 2273–2280.

Brook, R.D., Franklin, B., Cascio, W., Hong, Y., Howard, G., Lipsett, M., Luepker, R., Mittleman, M., Samet, J., Smith Jr., S.C., Tager, I., Expert Panel on Population and Prevention Science of the American Heart Association, 2004. Air pollution and cardiovascular disease: a statement for healthcare professionals from the expert panel on population and prevention science of the American heart association. Circulation 109, 2655–2671.

Bruce, K.D., Cagampang, F.R., Argenton, M., Zhang, J., Ethirajan, P.L., Burdge, G.C., Bateman, A.C., Clough, G.F., Poston, L., Hanson, M.A., Mcconnell, J.M., Byrne, C.D., 2009. Maternal high-fat feeding primes steatohepatitis in adult mice offspring, involving mitochondrial dysfunction and altered lipogenesis gene expression. Hepatology 50, 1796–1808.

Brykczynska, U., Hisano, M., Erkek, S., Ramos, L., Oakeley, E.J., Roloff, T.C., Beisel, C., Schubeler, D., Stadler, M.B., Peters, A.H., 2010. Repressive and active histone methylation mark distinct promoters in human and mouse spermatozoa. Nat. Struct. Mol. Biol. 17, 679–687.

Campbell, R.M., Tummino, P.J., 2014. Cancer epigenetics drug discovery and development: the challenge of hitting the mark. J. Clin. Investig. 124, 64–69.

Chandler, P.D., Scott, J.B., Drake, B.F., Ng, K., Manson, J.E., Rifai, N., Chan, A.T., Bennett, G.G., Hollis, B.W., Giovannucci, E.L., Emmons, K.M., Fuchs, C.S., 2014. Impact of vitamin D supplementation on inflammatory markers in African Americans: results of a four-arm, randomized, placebo-controlled trial. Cancer Prev. Res. (Phila.) 7, 218–225.

Chaudhary, P., Sharma, R., Sahu, M., Vishwanatha, J.K., Awasthi, S., Awasthi, Y.C., 2013. 4-Hydroxynonenal induces G2/M phase cell cycle arrest by activation of the ataxia telangiectasia mutated and Rad3-related protein (ATR)/checkpoint kinase 1 (Chk1) signaling pathway. J. Biol. Chem. 288, 20532–20546.

Chen, H., Ke, Q., Kluz, T., Yan, Y., Costa, M., 2006. Nickel ions increase histone H3 lysine 9 dimethylation and induce transgene silencing. Mol. Cell Biol. 26, 3728–3737.

Chen, M., Rao, Y., Zheng, Y., Wei, S., Li, Y., Guo, T., Yin, P., 2014. Association between soy isoflavone intake and breast cancer risk for pre- and post-menopausal women: a meta-analysis of epidemiological studies. PLoS One 9, e89288.

Chittock, E.C., Latwiel, S., Miller, T.C., Muller, C.W., 2017. Molecular architecture of polycomb repressive complexes. Biochem. Soc. Trans. 45, 193–205.

Choi, J.D., Lee, J.S., 2013. Interplay between epigenetics and genetics in cancer. Genomics Inform. 11, 164–173.

Choi, K.C., Jung, M.G., Lee, Y.H., Yoon, J.C., Kwon, S.H., Kang, H.B., Kim, M.J., Cha, J.H., Kim, Y.J., Jun, W.J., Lee, J.M., Yoon, H.G., 2009. Epigallocatechin-3-gallate, a histone acetyltransferase inhibitor, inhibits EBV-induced B lymphocyte transformation via suppression of RelA acetylation. Cancer Res. 69, 583–592.

Colvin, H., Mori, M., 2017. Getting to the heart of the matter in cancer: novel approaches to targeting cancer stem cells. Proc. Jpn. Acad. Ser. B Phys. Biol. Sci. 93, 146–154.

Cravo, M., Fidalgo, P., Pereira, A.D., Gouveia-Oliveira, A., Chaves, P., Selhub, J., Mason, J.B., Mira, F.C., Leitao, C.N., 1994. DNA methylation as an intermediate biomarker in colorectal cancer: modulation by folic acid supplementation. Eur. J. Cancer Prev. 3, 473–479.

Davis, C.D., Uthus, E.O., Finley, J.W., 2000. Dietary selenium and arsenic affect DNA methylation in vitro in Caco-2 cells and in vivo in rat liver and colon. J. Nutr. 130, 2903–2909.

Day, J.K., Bauer, A.M., Desbordes, C., Zhuang, Y., Kim, B.E., Newton, L.G., Nehra, V., Forsee, K.M., Macdonald, R.S., Besch-Williford, C., Huang, T.H., Lubahn, D.B., 2002. Genistein alters methylation patterns in mice. J. Nutr. 132, 2419S–2423S.

De Coster, S., Van Larebeke, N., 2012. Endocrine-disrupting chemicals: associated disorders and mechanisms of action. J. Environ. Public Health 2012, 713696.

De Stefani, E., Correa, P., Ronco, A., Mendilaharsu, M., Guidobono, M., Deneo-Pellegrini, H., 1997. Dietary fiber and risk of breast cancer: a case-control study in Uruguay. Nutr. Cancer 28, 14–19.

De Vogel, S., Bongaerts, B.W., Wouters, K.A., Kester, A.D., Schouten, L.J., De Goeij, A.F., De Bruine, A.P., Goldbohm, R.A., Van Den Brandt, P.A., Van Engeland, M., Weijenberg, M.P., 2008. Associations of dietary methyl donor intake with MLH1 promoter hypermethylation and related molecular phenotypes in sporadic colorectal cancer. Carcinogenesis 29, 1765–1773.

Deaton, A.M., Bird, A., 2011. CpG islands and the regulation of transcription. Genes Dev. 25, 1010–1022.

Dechassa, M.L., Sabri, A., Pondugula, S., Kassabov, S.R., Chatterjee, N., Kladde, M.P., Bartholomew, B., 2010. SWI/SNF has intrinsic nucleosome disassembly activity that is dependent on adjacent nucleosomes. Mol. Cell 38, 590–602.

Delaune, A., Landreneau, K., Hire, K., Dillard, B., Cox, M., Payne, K., Powell, C., 2009. Credibility of a meta-analysis: evidence-based practice concerning soy intake and breast cancer risk in premenopausal women. Worldviews Evid. Based Nurs. 6, 160–166.

Demary, K., Wong, L., Spanjaard, R.A., 2001. Effects of retinoic acid and sodium butyrate on gene expression, histone acetylation and inhibition of proliferation of melanoma cells. Cancer Lett. 163, 103–107.

Di Vinci, A., Brigati, C., Casciano, I., Banelli, B., Borzi, L., Forlani, A., Ravetti, G.L., Allemanni, G., Melloni, I., Zona, G., Spaziante, R., Merlo, D.F., Romani, M., 2012. HOXA7, 9, and 10 are methylation targets associated with aggressive behavior in meningiomas. Transl. Res. 160, 355–362.

Di Vinci, A., Casciano, I., Marasco, E., Banelli, B., Ravetti, G.L., Borzi, L., Brigati, C., Forlani, A., Dorcaratto, A., Allemanni, G., Zona, G., Spaziante, R., Gohlke, H., Gardin, G., Merlo, D.F., Mantovani, V., Romani, M., 2011. Quantitative methylation analysis of HOXA3, 7, 9, and 10 genes in glioma: association with tumor WHO grade and clinical outcome. J. Cancer Res. Clin. Oncol. 138, 35–47.

Diamanti-Kandarakis, E., Bourguignon, J.P., Giudice, L.C., Hauser, R., Prins, G.S., Soto, A.M., Zoeller, R.T., Gore, A.C., 2009. Endocrine-disrupting chemicals: an Endocrine Society scientific statement. Endocr. Rev. 30, 293–342.

Doherty, L.F., Bromer, J.G., Zhou, Y., Aldad, T.S., Taylor, H.S., 2010. In utero exposure to diethylstilbestrol (DES) or bisphenol-A (BPA) increases EZH2 expression in the mammary gland: an epigenetic mechanism linking endocrine disruptors to breast cancer. Horm. Cancer 1, 146–155.

Dolinoy, D.C., Huang, D., Jirtle, R.L., 2007. Maternal nutrient supplementation counteracts bisphenol A-induced DNA hypomethylation in early development. Proc. Natl. Acad. Sci. U.S.A. 104, 13056–13061.

Drew, D.A., Cao, Y., Chan, A.T., 2016. Aspirin and colorectal cancer: the promise of precision chemoprevention. Nat. Rev. Cancer 16, 173–186.

Du, M., Tong, J., Zhao, J., Underwood, K.R., Zhu, M., Ford, S.P., Nathanielsz, P.W., 2010. Fetal programming of skeletal muscle development in ruminant animals. J. Anim. Sci. 88, E51–E60.

Du, M., Zhao, J.X., Yan, X., Huang, Y., Nicodemus, L.V., Yue, W., Mccormick, R.J., Zhu, M.J., 2011. Fetal muscle development, mesenchymal multipotent cell differentiation, and associated signaling pathways. J. Anim. Sci. 89, 583–590.

Duan, H., He, Z., Ma, J., Zhang, B., Sheng, Z., Bin, P., Cheng, J., Niu, Y., Dong, H., Lin, H., Dai, Y., Zhu, B., Chen, W., Xiao, Y., Zheng, Y., 2013. Global and MGMT promoter hypomethylation independently associated with genomic instability of lymphocytes in subjects exposed to high-dose polycyclic aromatic hydrocarbon. Arch. Toxicol. 87, 2013–2022.

Dunford, A.R., Sangster, J.M., 2017. Maternal and paternal periconceptional nutrition as an indicator of offspring metabolic syndrome risk in later life through epigenetic imprinting: a systematic review. Diabetes Metab. Syndr. 11 (2), S655–S662.

Eriksson, J.G., Forsen, T., Tuomilehto, J., Winter, P.D., Osmond, C., Barker, D.J., 1999. Catch-up growth in childhood and death from coronary heart disease: longitudinal study. BMJ 318, 427–431.

Evans, B.A., Griffiths, K., Morton, M.S., 1995. Inhibition of 5 alpha-reductase in genital skin fibroblasts and prostate tissue by dietary lignans and isoflavonoids. J. Endocrinol. 147, 295–302.

Fabbri, M., Garzon, R., Cimmino, A., Liu, Z., Zanesi, N., Callegari, E., Liu, S., Alder, H., Costinean, S., Fernandez-Cymering, C., Volinia, S., Guler, G., Morrison, C.D., Chan, K.K., Marcucci, G., Calin, G.A., Huebner, K., Croce, C.M., 2007. MicroRNA-29 family reverts aberrant methylation in lung cancer by targeting DNA methyltransferases 3A and 3B. Proc. Natl. Acad. Sci. U.S.A. 104, 15805–15810.

Fang, M., Chen, D., Yang, C.S., 2007. Dietary polyphenols may affect DNA methylation. J. Nutr. 137, 223S–228S.

Fang, M.Z., Chen, D., Sun, Y., Jin, Z., Christman, J.K., Yang, C.S., 2005. Reversal of hypermethylation and reactivation of p16INK4a, RARbeta, and MGMT genes by genistein and other isoflavones from soy. Clin. Cancer Res. 11, 7033–7041.

Fang, M.Z., Wang, Y., Ai, N., Hou, Z., Sun, Y., Lu, H., Welsh, W., Yang, C.S., 2003. Tea polyphenol (-)-epigallocatechin-3-gallate inhibits DNA methyltransferase and reactivates methylation-silenced genes in cancer cell lines. Cancer Res. 63, 7563–7570.

Faulk, C., Kim, J.H., Jones, T.R., Mceachin, R.C., Nahar, M.S., Dolinoy, D.C., Sartor, M.A., 2015. Bisphenol A-associated alterations in genome-wide DNA methylation and gene expression patterns reveal sequence-dependent and non-monotonic effects in human fetal liver. Environ. Epigenet. 1.

Filipic, M., Hei, T.K., 2004. Mutagenicity of cadmium in mammalian cells: implication of oxidative DNA damage. Mutat. Res. 546, 81–91.

Fleisch, A.F., Wright, R.O., Baccarelli, A.A., 2012. Environmental epigenetics: a role in endocrine disease? J. Mol. Endocrinol. 49, R61–R67.

Forhead, A.J., Fowden, A.L., 2009. The hungry fetus? Role of leptin as a nutritional signal before birth. J. Physiol. 587, 1145–1152.

Gadducci, A., Biglia, N., Tana, R., Cosio, S., Gallo, M., 2016. Metformin use and gynecological cancers: a novel treatment option emerging from drug repositioning. Crit. Rev. Oncol. Hematol. 105, 73–83.

Garcia, F.A., Cornelison, T., Nuno, T., Greenspan, D.L., Byron, J.W., Hsu, C.H., Alberts, D.S., Chow, H.H., 2014. Results of a phase II randomized, double-blind, placebo-controlled trial of Polyphenon E in women with persistent high-risk HPV infection and low-grade cervical intraepithelial neoplasia. Gynecol. Oncol. 132, 377–382.

Gatti-Mays, M.E., Venzon, D., Galbo, C.E., Singer, A., Reynolds, J., Makariou, E., Kallakury, B., Heckman-Stoddard, B.M., Korde, L., Isaacs, C., Warren, R., Gallagher, A., Eng-Wong, J., 2016. Exemestane use in postmenopausal women at high risk for invasive breast cancer: evaluating biomarkers of efficacy and safety. Cancer Prev. Res. (Phila.) 9, 225–233.

Generali, E., Ceribelli, A., Stazi, M.A., Selmi, C., 2017. Lessons learned from twins in autoimmune and chronic inflammatory diseases. J. Autoimmun. 83, 51–61.

Gosden, R.G., Feinberg, A.P., 2007. Genetics and epigenetics—nature's pen-and-pencil set. N. Engl. J. Med. 356, 731–733.

Govindarajan, B., Klafter, R., Miller, M.S., Mansur, C., Mizesko, M., Bai, X., Lamontagne Jr., K., Arbiser, J.L., 2002. Reactive oxygen-induced carcinogenesis causes hypermethylation of p16(Ink4a) and activation of MAP kinase. Mol. Med. 8, 1–8.

Grossniklaus, U., Kelly, W.G., Ferguson-Smith, A.C., Pembrey, M., Lindquist, S., 2013. Transgenerational epigenetic inheritance: how important is it? Nat. Rev. Genet. 14, 228–235.

Guerrero-Bosagna, C., Covert, T.R., Haque, M.M., Settles, M., Nilsson, E.E., Anway, M.D., Skinner, M.K., 2012. Epigenetic transgenerational inheritance of vinclozolin induced mouse adult onset disease and associated sperm epigenome biomarkers. Reprod. Toxicol. 34, 694–707.

Havas, K., Flaus, A., Phelan, M., Kingston, R., Wade, P.A., Lilley, D.M., Owen-Hughes, T., 2000. Generation of superhelical torsion by ATP-dependent chromatin remodeling activities. Cell 103, 1133–1142.

He, L., Hannon, G.J., 2004. MicroRNAs: small RNAs with a big role in gene regulation. Nat. Rev. Genet. 5, 522–531.

Heard, E., Martienssen, R.A., 2014. Transgenerational epigenetic inheritance: myths and mechanisms. Cell 157, 95–109.

Heckman-Stoddard, B.M., Gandini, S., Puntoni, M., Dunn, B.K., Decensi, A., Szabo, E., 2016. Repurposing old drugs to chemoprevention: the case of metformin. Semin. Oncol. 43, 123–133.

Heerboth, S., Lapinska, K., Snyder, N., Leary, M., Rollinson, S., Sarkar, S., 2014. Use of epigenetic drugs in disease: an overview. Genet. Epigenet. 6, 9–19.

Heijmans, B.T., Tobi, E.W., Stein, A.D., Putter, H., Blauw, G.J., Susser, E.S., Slagboom, P.E., Lumey, L.H., 2008. Persistent epigenetic differences associated with prenatal exposure to famine in humans. Proc. Natl. Acad. Sci. U.S.A. 105, 17046–17049.

Herbstman, J.B., Tang, D., Zhu, D., Qu, L., Sjodin, A., Li, Z., Camann, D., Perera, F.P., 2012. Prenatal exposure to polycyclic aromatic hydrocarbons, benzo[a]pyrene-DNA adducts, and genomic DNA methylation in cord blood. Environ. Health Perspect. 120, 733–738.

Herman, J.G., Baylin, S.B., 2003. Gene silencing in cancer in association with promoter hypermethylation. N. Engl. J. Med. 349, 2042–2054.

Hiort, O., 2013. The differential role of androgens in early human sex development. BMC Med. 11, 152.

Hirose, K., Tajima, K., Hamajima, N., Inoue, M., Takezaki, T., Kuroishi, T., Yoshida, M., Tokudome, S., 1995. A large-scale, hospital-based case-control study of risk factors of breast cancer according to menopausal status. Jpn. J. Cancer Res. 86, 146–154.

Ho, S.M., Tang, W.Y., Belmonte De Frausto, J., Prins, G.S., 2006. Developmental exposure to estradiol and bisphenol A increases susceptibility to prostate carcinogenesis and epigenetically regulates phosphodiesterase type 4 variant 4. Cancer Res. 66, 5624–5632.

Huang, Z., Huang, Q., Ji, L., Wang, Y., Qi, X., Liu, L., Liu, Z., Lu, L., 2016. Epigenetic regulation of active Chinese herbal components for cancer prevention and treatment: a follow-up review. Pharmacol. Res. 114, 1–12.

Husain, Q., Qayyum, S., 2013. Biological and enzymatic treatment of bisphenol A and other endocrine disrupting compounds: a review. Crit. Rev. Biotechnol. 33, 260–292.

Huxley, R., Neil, A., Collins, R., 2002. Unravelling the fetal origins hypothesis: is there really an inverse association between birthweight and subsequent blood pressure? Lancet 360, 659–665.

Hwang, J.Y., Aromolaran, K.A., Zukin, R.S., 2017. The emerging field of epigenetics in neurodegeneration and neuroprotection. Nat. Rev. Neurosci. 18, 347–361.

Ikezuki, Y., Tsutsumi, O., Takai, Y., Kamei, Y., Taketani, Y., 2002. Determination of bisphenol A concentrations in human biological fluids reveals significant early prenatal exposure. Hum. Reprod. 17, 2839–2841.

Inawaka, K., Kawabe, M., Takahashi, S., Doi, Y., Tomigahara, Y., Tarui, H., Abe, J., Kawamura, S., Shirai, T., 2009. Maternal exposure to anti-androgenic compounds, vinclozolin, flutamide and procymidone, has no effects on spermatogenesis and DNA methylation in male rats of subsequent generations. Toxicol. Appl. Pharmacol. 237, 178–187.

Irizarry, R.A., Ladd-Acosta, C., Wen, B., Wu, Z., Montano, C., Onyango, P., Cui, H., Gabo, K., Rongione, M., Webster, M., Ji, H., Potash, J.B., Sabunciyan, S., Feinberg, A.P., 2009. The human colon cancer methylome shows similar hypo- and hypermethylation at conserved tissue-specific CpG island shores. Nat. Genet. 41, 178–186.

Issa, J.P., 2007. DNA methylation as a therapeutic target in cancer. Clin. Cancer Res. 13, 1634–1637.

Jambhekar, A., Anastas, J.N., Shi, Y., 2017. Histone lysine demethylase inhibitors. Cold Spring Harb. Perspect. Med. 7.

Janssen, B.G., Godderis, L., Pieters, N., Poels, K., Kicinski, M., Cuypers, A., Fierens, F., Penders, J., Plusquin, M., Gyselaers, W., Nawrot, T.S., 2013. Placental DNA hypomethylation in association with particulate air pollution in early life. Part. Fibre Toxicol. 10, 22.

Jardim, M.J., Fry, R.C., Jaspers, I., Dailey, L., Diaz-Sanchez, D., 2009. Disruption of microRNA expression in human airway cells by diesel exhaust particles is linked to tumorigenesis-associated pathways. Environ. Health Perspect. 117, 1745–1751.

Kadoch, C., Hargreaves, D.C., Hodges, C., Elias, L., Ho, L., Ranish, J., Crabtree, G.R., 2013. Proteomic and bioinformatic analysis of mammalian SWI/SNF complexes identifies extensive roles in human malignancy. Nat. Genet. 45, 592–601.

Kang, J., Chen, J., Shi, Y., Jia, J., Zhang, Y., 2005. Curcumin-induced histone hypoacetylation: the role of reactive oxygen species. Biochem. Pharmacol. 69, 1205–1213.

Kanherkar, R.R., Bhatia-Dey, N., Csoka, A.B., 2014. Epigenetics across the human lifespan. Front. Cell Dev. Biol. 2, 49.

Ke, Q., Davidson, T., Chen, H., Kluz, T., Costa, M., 2006. Alterations of histone modifications and transgene silencing by nickel chloride. Carcinogenesis 27, 1481–1488.

Kendrick, S.F., O'boyle, G., Mann, J., Zeybel, M., Palmer, J., Jones, D.E., Day, C.P., 2010. Acetate, the key modulator of inflammatory responses in acute alcoholic hepatitis. Hepatology 51, 1988–1997.

Kim, J.H., Dhanasekaran, S.M., Prensner, J.R., Cao, X., Robinson, D., Kalyana-Sundaram, S., Huang, C., Shankar, S., Jing, X., Iyer, M., Hu, M., Sam, L., Grasso, C., Maher, C.A., Palanisamy, N., Mehra, R., Kominsky, H.D., Siddiqui, J., Yu, J., Qin, Z.S., Chinnaiyan, A.M., 2011. Deep sequencing reveals distinct patterns of DNA methylation in prostate cancer. Genome Res. 21, 1028–1041.

Kim, Y.I., Baik, H.W., Fawaz, K., Knox, T., Lee, Y.M., Norton, R., Libby, E., Mason, J.B., 2001. Effects of folate supplementation on two provisional molecular markers of colon cancer: a prospective, randomized trial. Am. J. Gastroenterol. 96, 184–195.

Kondo, K., Takahashi, Y., Hirose, Y., Nagao, T., Tsuyuguchi, M., Hashimoto, M., Ochiai, A., Monden, Y., Tangoku, A., 2006. The reduced expression and aberrant methylation of p16(INK4a) in chromate workers with lung cancer. Lung Cancer 53, 295–302.

Kraus, T.F., Guibourt, V., Kretzschmar, H.A., 2015. 5-Hydroxymethylcytosine, the "Sixth Base", during brain development and ageing. J. Neural. Transm. 122, 1035–1043.

Lee, H.P., Gourley, L., Duffy, S.W., Esteve, J., Lee, J., Day, N.E., 1991. Dietary effects on breast-cancer risk in Singapore. Lancet 337, 1197–1200.

Lee, H.S., Herceg, Z., 2014. The epigenome and cancer prevention: a complex story of dietary supplementation. Cancer Lett. 342, 275–284.

Lee, Y.W., Klein, C.B., Kargacin, B., Salnikow, K., Kitahara, J., Dowjat, K., Zhitkovich, A., Christie, N.T., Costa, M., 1995. Carcinogenic nickel silences gene expression by chromatin condensation and DNA methylation: a new model for epigenetic carcinogens. Mol. Cell Biol. 15, 2547–2557.

Levitt, N.S., Steyn, K., De Wet, T., Morrell, C., Edwards, R., Ellison, G.T., Cameron, N., 1999. An inverse relation between blood pressure and birth weight among 5 year old children from Soweto, South Africa. J. Epidemiol. Community Health 53, 264–268.

Levy, E., Spahis, S., Bigras, J.L., Delvin, E., Borys, J.M., 2017. The epigenetic machinery in vascular dysfunction and hypertension. Curr. Hypertens. Rep. 19, 52.

Li, S., Hursting, S.D., Davis, B.J., Mclachlan, J.A., Barrett, J.C., 2003. Environmental exposure, DNA methylation, and gene regulation: lessons from diethylstilbesterol-induced cancers. Ann. N.Y. Acad. Sci. 983, 161–169.

Li, S., Washburn, K.A., Moore, R., Uno, T., Teng, C., Newbold, R.R., Mclachlan, J.A., Negishi, M., 1997. Developmental exposure to diethylstilbestrol elicits demethylation of estrogen-responsive lactoferrin gene in mouse uterus. Cancer Res. 57, 4356–4359.

Liang, J., Zhu, H., Li, C., Ding, Y., Zhou, Z., Wu, Q., 2012. Neonatal exposure to benzo[a]pyrene decreases the levels of serum testosterone and histone H3K14 acetylation of the StAR promoter in the testes of SD rats. Toxicology 302, 285–291.

Lin, C., Kang, J., Zheng, R., 2005. Oxidative stress is involved in inhibition of copper on histone acetylation in cells. Chem. Biol. Interact. 151, 167–176.

Liu, F., Killian, J.K., Yang, M., Walker, R.L., Hong, J.A., Zhang, M., Davis, S., Zhang, Y., Hussain, M., Xi, S., Rao, M., Meltzer, P.A., Schrump, D.S., 2010. Epigenomic alterations and gene expression profiles in respiratory epithelia exposed to cigarette smoke condensate. Oncogene 29, 3650–3664.

Lumey, L.H., Stein, A.D., Kahn, H.S., Van Der Pal-De Bruin, K.M., Blauw, G.J., Zybert, P.A., Susser, E.S., 2007. Cohort profile: the Dutch Hunger Winter families study. Int. J. Epidemiol. 36, 1196–1204.

Madrigano, J., Baccarelli, A., Mittleman, M.A., Wright, R.O., Sparrow, D., Vokonas, P.S., Tarantini, L., Schwartz, J., 2011. Prolonged exposure to particulate pollution, genes associated with glutathione pathways, and DNA methylation in a cohort of older men. Environ. Health Perspect. 119, 977–982.

Mangtani, P., Silva Idos, S., 1998. Phyto-oestrogens and breast cancer. Lancet 351, 138–139, 137; author reply.

Manikkam, M., Tracey, R., Guerrero-Bosagna, C., Skinner, M.K., 2012. Pesticide and insect repellent mixture (permethrin and DEET) induces epigenetic transgenerational inheritance of disease and sperm epimutations. Reprod. Toxicol. 34, 708–719.

Marini, H., Bitto, A., Altavilla, D., Burnett, B.P., Polito, F., Di Stefano, V., Minutoli, L., Atteritano, M., Levy, R.M., D'anna, R., Frisina, N., Mazzaferro, S., Cancellieri, F., Cannata, M.L., Corrado, F., Frisina, A., Adamo, V., Lubrano, C., Sansotta, C., Marini, R., Adamo, E.B., Squadrito, F., 2008. Breast safety and efficacy of genistein aglycone for postmenopausal bone loss: a follow-up study. J. Clin. Endocrinol. Metab. 93, 4787–4796.

Marsit, C.J., Karagas, M.R., Schned, A., Kelsey, K.T., 2006. Carcinogen exposure and epigenetic silencing in bladder cancer. Ann. N.Y. Acad. Sci. 1076, 810–821.

Marwick, J.A., Kirkham, P.A., Stevenson, C.S., Danahay, H., Giddings, J., Butler, K., Donaldson, K., Macnee, W., Rahman, I., 2004. Cigarette smoke alters chromatin remodeling and induces proinflammatory genes in rat lungs. Am. J. Respir. Cell Mol. Biol. 31, 633–642.

Messerschmidt, D.M., 2012. Should I stay or should I go: protection and maintenance of DNA methylation at imprinted genes. Epigenetics 7, 969–975.

Modabbernia, A., Velthorst, E., Reichenberg, A., 2017. Environmental risk factors for autism: an evidence-based review of systematic reviews and meta-analyses. Mol. Autism 8, 13.

Murat, A., Migliavacca, E., Gorlia, T., Lambiv, W.L., Shay, T., Hamou, M.F., De Tribolet, N., Regli, L., Wick, W., Kouwenhoven, M.C., Hainfellner, J.A., Heppner, F.L., Dietrich, P.Y., Zimmer, Y., Cairncross, J.G., Janzer, R.C., Domany, E., Delorenzi, M., Stupp, R., Hegi, M.E., 2008. Stem cell-related "self-renewal" signature and high epidermal growth factor receptor expression associated with resistance to concomitant chemoradiotherapy in glioblastoma. J. Clin. Oncol. 26, 3015–3024.

Natoli, G., Testa, G., De Santa, F., 2009. The future therapeutic potential of histone demethylases: a critical analysis. Curr. Opin. Drug Discov. Dev. 12, 607–615.

Newbold, R.R., Padilla-Banks, E., Jefferson, W.N., 2006. Adverse effects of the model environmental estrogen diethylstilbestrol are transmitted to subsequent generations. Endocrinology 147, S11–S17.

Nile, C.J., Read, R.C., Akil, M., Duff, G.W., Wilson, A.G., 2008. Methylation status of a single CpG site in the IL6 promoter is related to IL6 messenger RNA levels and rheumatoid arthritis. Arthritis Rheum. 58, 2686–2693.

Olins, A.L., Olins, D.E., 1974. Spheroid chromatin units (v bodies). Science 183, 330–332.

Parkin, D.M., 1989. Cancers of the breast, endometrium and ovary: geographic correlations. Eur. J. Cancer Clin. Oncol. 25, 1917–1925.

Parodi, F., Carosio, R., Ragusa, M., Di Pietro, C., Maugeri, M., Barbagallo, D., Sallustio, F., Allemanni, G., Pistillo, M.P., Casciano, I., Forlani, A., Schena, F.P., Purrello, M., Romani, M., Banelli, B., 2016. Epigenetic dysregulation in neuroblastoma: a tale of miRNAs and DNA methylation. Biochim. Biophys. Acta 1859, 1502–1514.

Pavanello, S., Pesatori, A.C., Dioni, L., Hoxha, M., Bollati, V., Siwinska, E., Mielzynska, D., Bolognesi, C., Bertazzi, P.A., Baccarelli, A., 2010. Shorter telomere length in peripheral blood lymphocytes of workers exposed to polycyclic aromatic hydrocarbons. Carcinogenesis 31, 216–221.

Peluso, M., Bollati, V., Munnia, A., Srivatanakul, P., Jedpiyawongse, A., Sangrajrang, S., Piro, S., Ceppi, M., Bertazzi, P.A., Boffetta, P., Baccarelli, A.A., 2012. DNA methylation differences in exposed workers and nearby residents of the Ma Ta Phut industrial estate, Rayong, Thailand, 1753-60; discussion Int. J. Epidemiol. 41, 1761–1763.

Poddar, S., Kesharwani, D., Datta, M., 2017. Interplay between the miRNome and the epigenetic machinery: implications in health and disease. J. Cell. Physiol. 232, 2938–2945.

Pogribny, I.P., Pogribna, M., Christman, J.K., James, S.J., 2000. Single-site methylation within the p53 promoter region reduces gene expression in a reporter gene construct: possible in vivo relevance during tumorigenesis. Cancer Res. 60, 588–594.

Pufulete, M., Al-Ghnaniem, R., Khushal, A., Appleby, P., Harris, N., Gout, S., Emery, P.W., Sanders, T.A., 2005. Effect of folic acid supplementation on genomic DNA methylation in patients with colorectal adenoma. Gut 54, 648–653.

Rada-Iglesias, A., Enroth, S., Ameur, A., Koch, C.M., Clelland, G.K., Respuela-Alonso, P., Wilcox, S., Dovey, O.M., Ellis, P.D., Langford, C.F., Dunham, I., Komorowski, J., Wadelius, C., 2007. Butyrate mediates decrease of histone acetylation centered on transcription start sites and down-regulation of associated genes. Genome Res. 17, 708–719.

Ravelli, A.C., Van Der Meulen, J.H., Michels, R.P., Osmond, C., Barker, D.J., Hales, C.N., Bleker, O.P., 1998. Glucose tolerance in adults after prenatal exposure to famine. Lancet 351, 173–177.

Ravelli, G.P., Stein, Z.A., Susser, M.W., 1976. Obesity in young men after famine exposure in utero and early infancy. N. Engl. J. Med. 295, 349–353.

Reynolds, L.P., Borowicz, P.P., Caton, J.S., Vonnahme, K.A., Luther, J.S., Hammer, C.J., Maddock Carlin, K.R., Grazul-Bilska, A.T., Redmer, D.A., 2010. Developmental programming: the concept, large animal models, and the key role of uteroplacental vascular development. J. Anim. Sci. 88, E61–E72.

Reynolds, L.P., Caton, J.S., 2012. Role of the pre- and postnatal environment in developmental programming of health and productivity. Mol. Cell. Endocrinol. 354, 54–59.

Reynolds, L.P., Caton, J.S., Redmer, D.A., Grazul-Bilska, A.T., Vonnahme, K.A., Borowicz, P.P., Luther, J.S., Wallace, J.M., Wu, G., Spencer, T.E., 2006. Evidence for altered placental blood flow and vascularity in compromised pregnancies. J. Physiol. 572, 51–58.

Riggs, A.D., Jones, P.A., 1983. 5-methylcytosine, gene regulation, and cancer. Adv. Cancer Res. 40, 1–30.

Rius, M., Lyko, F., 2012. Epigenetic cancer therapy: rationales, targets and drugs. Oncogene 31, 4257–4265.

Robboy, S.J., Scully, R.E., Welch, W.R., Herbst, A.L., 1977. Intrauterine diethylstilbestrol exposure and its consequences: pathologic characteristics of vaginal adenosis, clear cell adenocarcinoma, and related lesions. Arch. Pathol. Lab Med. 101, 1–5.

Rose, D.P., Boyar, A.P., Wynder, E.L., 1986. International comparisons of mortality rates for cancer of the breast, ovary, prostate, and colon, and per capita food consumption. Cancer 58, 2363–2371.

Ross, R.K., Bernstein, L., Lobo, R.A., Shimizu, H., Stanczyk, F.Z., Pike, M.C., Henderson, B.E., 1992. 5-alpha-reductase activity and risk of prostate cancer among Japanese and US white and black males. Lancet 339, 887–889.

Sartippour, M.R., Rao, J.Y., Apple, S., Wu, D., Henning, S., Wang, H., Elashoff, R., Rubio, R., Heber, D., Brooks, M.N., 2004. A pilot clinical study of short-term isoflavone supplements in breast cancer patients. Nutr. Cancer 49, 59–65.

Sato, F., Tsuchiya, S., Meltzer, S.J., Shimizu, K., 2011. MicroRNAs and epigenetics. FEBS J. 278, 1598–1609.

Schneider, S., Kaufmann, W., Buesen, R., Van Ravenzwaay, B., 2008. Vinclozolin—the lack of a transgenerational effect after oral maternal exposure during organogenesis. Reprod. Toxicol. 25, 352–360.

Schneider, S., Marxfeld, H., Groters, S., Buesen, R., Van Ravenzwaay, B., 2013. Vinclozolin—no transgenerational inheritance of anti-androgenic effects after maternal exposure during organogenesis via the intraperitoneal route. Reprod. Toxicol. 37, 6–14.

Schug, T.T., Janesick, A., Blumberg, B., Heindel, J.J., 2011. Endocrine disrupting chemicals and disease susceptibility. J. Steroid Biochem. Mol. Biol. 127, 204–215.

Schuster, A., Skinner, M.K., Yan, W., 2016. Ancestral vinclozolin exposure alters the epigenetic transgenerational inheritance of sperm small noncoding RNAs. Environ. Epigenet. 2.

Shen, H., Xu, W., Lan, F., 2017. Histone lysine demethylases in mammalian embryonic development. Exp. Mol. Med. 49, e325.

Shi, Y., Lan, F., Matson, C., Mulligan, P., Whetstine, J.R., Cole, P.A., Casero, R.A., 2004. Histone demethylation mediated by the nuclear amine oxidase homolog LSD1. Cell 119, 941–953.

Skinner, M.K., 2011. Environmental epigenetic transgenerational inheritance and somatic epigenetic mitotic stability. Epigenetics 6, 838–842.

Smith, A.H., Marshall, G., Yuan, Y., Ferreccio, C., Liaw, J., Von Ehrenstein, O., Steinmaus, C., Bates, M.N., Selvin, S., 2006. Increased mortality from lung cancer and bronchiectasis in young adults after exposure to arsenic in utero and in early childhood. Environ. Health Perspect. 114, 1293–1296.

Smith, C.C., Taylor, H.S., 2007. Xenoestrogen exposure imprints expression of genes (Hoxa10) required for normal uterine development. FASEB J. 21, 239–246.

Somm, E., Stouder, C., Paoloni-Giacobino, A., 2013. Effect of developmental dioxin exposure on methylation and expression of specific imprinted genes in mice. Reprod. Toxicol. 35, 150–155.

Stanner, S.A., Bulmer, K., Andres, C., Lantseva, O.E., Borodina, V., Poteen, V.V., Yudkin, J.S., 1997. Does malnutrition in utero determine diabetes and coronary heart disease in adulthood? Results from the Leningrad siege study, a cross sectional study. BMJ 315, 1342–1348.

Stouder, C., Paoloni-Giacobino, A., 2011. Specific transgenerational imprinting effects of the endocrine disruptor methoxychlor on male gametes. Reproduction 141, 207–216.

Takiguchi, M., Achanzar, W.E., Qu, W., Li, G., Waalkes, M.P., 2003. Effects of cadmium on DNA-(Cytosine-5) methyltransferase activity and DNA methylation status during cadmium-induced cellular transformation. Exp. Cell Res. 286, 355–365.

Tao, L., Ge, R., Xie, M., Kramer, P.M., Pereira, M.A., 1999. Effect of trichloroethylene on DNA methylation and expression of early-intermediate protooncogenes in the liver of B6C3F1 mice. J. Biochem. Mol. Toxicol. 13, 231–237.

Tarantini, L., Bonzini, M., Apostoli, P., Pegoraro, V., Bollati, V., Marinelli, B., Cantone, L., Rizzo, G., Hou, L., Schwartz, J., Bertazzi, P.A., Baccarelli, A., 2009. Effects of particulate matter on genomic DNA methylation content and iNOS promoter methylation. Environ. Health Perspect. 117, 217–222.

Taylor, C.K., Levy, R.M., Elliott, J.C., Burnett, B.P., 2009. The effect of genistein aglycone on cancer and cancer risk: a review of in vitro, preclinical, and clinical studies. Nutr. Rev. 67, 398–415.

Thame, M., Osmond, C., Wilks, R.J., Bennett, F.I., Mcfarlane-Anderson, N., Forrester, T.E., 2000. Blood pressure is related to placental volume and birth weight. Hypertension 35, 662–667.

Thomas, J.O., Kornberg, R.D., 1975. An octamer of histones in chromatin and free in solution. Proc. Natl. Acad. Sci. U.S.A. 72, 2626–2630.

Thompson, R.F., Fazzari, M.J., Niu, H., Barzilai, N., Simmons, R.A., Greally, J.M., 2010. Experimental intrauterine growth restriction induces alterations in DNA methylation and gene expression in pancreatic islets of rats. J. Biol. Chem. 285, 15111–15118.

Tong, G.X., Liang, H., Chai, J., Cheng, J., Feng, R., Chen, P.L., Geng, Q.Q., Shen, X.R., Wang, D.B., 2014. Association of risk of gastric cancer and consumption of tobacco, alcohol and tea in the Chinese population. Asian Pac. J. Cancer Prev. 15, 8765–8774.

Trock, B.J., Hilakivi-Clarke, L., Clarke, R., 2006. Meta-analysis of soy intake and breast cancer risk. J. Natl. Cancer Inst. 98, 459–471.

Tyagi, M., Imam, N., Verma, K., Patel, A.K., 2016. Chromatin remodelers: we are the drivers!! Nucleus 7, 388–404.

Vadhana, M.S., Carloni, M., Nasuti, C., Fedeli, D., Gabbianelli, R., 2011. Early life permethrin insecticide treatment leads to heart damage in adult rats. Exp. Gerontol. 46, 731–738.

Vaiserman, A., 2014. Early-life exposure to endocrine disrupting chemicals and later-life health outcomes: an epigenetic bridge? Aging Dis. 5, 419–429.

Valko, M., Leibfritz, D., Moncol, J., Cronin, M.T., Mazur, M., Telser, J., 2007. Free radicals and antioxidants in normal physiological functions and human disease. Int. J. Biochem. Cell Biol. 39, 44–84.

Van Engeland, M., Weijenberg, M.P., Roemen, G.M., Brink, M., De Bruine, A.P., Goldbohm, R.A., Van Den Brandt, P.A., Baylin, S.B., De Goeij, A.F., Herman, J.G., 2003. Effects of dietary folate and alcohol intake on promoter methylation in sporadic colorectal cancer: The Netherlands cohort study on diet and cancer. Cancer Res. 63, 3133–3137.

Veiga-Lopez, A., Luense, L.J., Christenson, L.K., Padmanabhan, V., 2013. Developmental programming: gestational bisphenol-A treatment alters trajectory of fetal ovarian gene expression. Endocrinology 154, 1873–1884.

Vidali, G., Gershey, E.L., Allfrey, V.G., 1968. Chemical studies of histone acetylation. The distribution of epsilon-N-acetyllysine in calf thymus histones. J. Biol. Chem. 243, 6361–6366.

Vineis, P., Husgafvel-Pursiainen, K., 2005. Air pollution and cancer: biomarker studies in human populations. Carcinogenesis 26, 1846–1855.

Wang, J., Tang, J., Lai, M., Zhang, H., 2014. 5-Hydroxymethylcytosine and disease. Mutat. Res. Rev. Mutat. Res. 762C, 167–175.

Waterland, R.A., Jirtle, R.L., 2003. Transposable elements: targets for early nutritional effects on epigenetic gene regulation. Mol. Cell Biol. 23, 5293–5300.

Weaver, I.C., Cervoni, N., Champagne, F.A., D'alessio, A.C., Sharma, S., Seckl, J.R., Dymov, S., Szyf, M., Meaney, M.J., 2004. Epigenetic programming by maternal behavior. Nat. Neurosci. 7, 847–854.

Weaver, I.C., Champagne, F.A., Brown, S.E., Dymov, S., Sharma, S., Meaney, M.J., Szyf, M., 2005. Reversal of maternal programming of stress responses in adult offspring through methyl supplementation: altering epigenetic marking later in life. J. Neurosci. 25, 11045–11054.

Weng, Y.I., Hsu, P.Y., Liyanarachchi, S., Liu, J., Deatherage, D.E., Huang, Y.W., Zuo, T., Rodriguez, B., Lin, C.H., Cheng, A.L., Huang, T.H., 2010. Epigenetic influences of low-dose bisphenol A in primary human breast epithelial cells. Toxicol. Appl. Pharmacol. 248, 111–121.

Wood, C.E., 2014. Estrogen in the fetus. Adv. Exp. Med. Biol. 814, 217–228.

Wu, G., Bazer, F.W., Wallace, J.M., Spencer, T.E., 2006. Board-invited review: intrauterine growth retardation: implications for the animal sciences. J. Anim. Sci. 84, 2316–2337.

Wu, Q., Ohsako, S., Ishimura, R., Suzuki, J.S., Tohyama, C., 2004. Exposure of mouse preimplantation embryos to 2,3,7,8-tetrachlorodibenzo-p-dioxin (TCDD) alters the methylation status of imprinted genes H19 and Igf2. Biol. Reprod. 70, 1790–1797.

Xiang, Z., Yang, Y., Chang, C., Lu, Q., 2017. The epigenetic mechanism for discordance of autoimmunity in monozygotic twins. J. Autoimmun. 83, 43–50.

Xu, F., Mao, C., Ding, Y., Rui, C., Wu, L., Shi, A., Zhang, H., Zhang, L., Xu, Z., 2010. Molecular and enzymatic profiles of mammalian DNA methyltransferases: structures and targets for drugs. Curr. Med. Chem. 17, 4052–4071.

Ye, F., Xu, X.C., 2010. Benzo[a]pyrene diol epoxide suppresses retinoic acid receptor-beta2 expression by recruiting DNA (cytosine-5-)-methyltransferase 3A. Mol. Cancer 9, 93.

Zama, A.M., Uzumcu, M., 2009. Fetal and neonatal exposure to the endocrine disruptor methoxychlor causes epigenetic alterations in adult ovarian genes. Endocrinology 150, 4681–4691.

Zhang, W., Spector, T.D., Deloukas, P., Bell, J.T., Engelhardt, B.E., 2015. Predicting genome-wide DNA methylation using methylation marks, genomic position, and DNA regulatory elements. Genome Biol. 16, 14.

Zhao, C.Q., Young, M.R., Diwan, B.A., Coogan, T.P., Waalkes, M.P., 1997. Association of arsenic-induced malignant transformation with DNA hypomethylation and aberrant gene expression. Proc. Natl. Acad. Sci. U.S.A. 94, 10907–10912.

Zhao, J., Sun, B.K., Erwin, J.A., Song, J.J., Lee, J.T., 2008. Polycomb proteins targeted by a short repeat RNA to the mouse X chromosome. Science 322, 750–756.

Zhong, X., Zhang, C., 2012. Soy food intake and breast cancer risk: a meta-analysis. Wei Sheng Yan Jiu 41, 670–676.

Ziegler, R.G., Hoover, R.N., Pike, M.C., Hildesheim, A., Nomura, A.M., West, D.W., Wu-Williams, A.H., Kolonel, L.N., Horn-Ross, P.L., Rosenthal, J.F., Hyer, M.B., 1993. Migration patterns and breast cancer risk in Asian-American women. J. Natl. Cancer Inst. 85, 1819–1827.

Zoeller, R.T., Brown, T.R., Doan, L.L., Gore, A.C., Skakkebaek, N.E., Soto, A.M., Woodruff, T.J., Vom Saal, F.S., 2012. Endocrine-disrupting chemicals and public health protection: a statement of principles from the Endocrine Society. Endocrinology 153, 4097–4110.

Zullo, A., Sommese, L., Nicoletti, G., Donatelli, F., Mancini, F.P., Napoli, C., 2017. Epigenetics and type 1 diabetes: mechanisms and translational applications. Transl. Res. 185, 85–93.

Index

'*Note*: Page numbers followed by "f" indicate figures and "t" indicate tables.'

I

Kolibri 360 GmbH
Ferdinand-Jühlke-Straße 7
99095 Erfurt, Deutschland
produktsicherheit@kolibri360.de